Geometric Gradient

Geometric Series Present Worth:

To Find P $(P/A, g, i, n)$
Given A_1, g When $i = g$ $P = A_1[n(1+i)^{-1}]$

To Find P $(P/A, g, i, n)$
Given A_1, g When $i \neq g$ $P = A_1\left[\dfrac{1-(1+g)^n(1+i)^{-n}}{i-g}\right]$

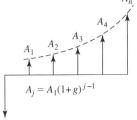

$A_j = A_1(1+g)^{j-1}$

Compound Interest

i = Interest rate per interest period*.

n = Number of interest periods.

P = A present sum of money.

F = A future sum of money. The future sum F is an amount, n interest periods from the present, that is equivalent to P with interest rate i.

A = An end-of-period cash receipt or disbursement in a uniform series continuing for n periods, the entire series equivalent to P or F at interest rate i.

G = Uniform period-by-period increase or decrease in cash receipts or disbursements; the arithmetic gradient.

g = Uniform *rate* of cash flow increase or decrease from period to period; the geometric gradient.

r = Nominal interest rate per interest period*.

m = Number of compounding subperiods per period*.

---

*Normally the interest period is one year, but it could be something else.

To Find	Excel
P	$-\text{PV}(i, n, A, F, \text{Type})$
A	$-\text{PMT}(i, n, P, F, \text{Type})$
F	$-\text{FV}(i, n, A, P, \text{Type})$
n	$\text{NPER}(i, A, P, F, \text{Type})$
i	$\text{RATE}(n, A, P, F, \text{Type}, \text{guess})$
P	$\text{NPV}(i, \text{CF}_1 : \text{CF}_n)$
i	$\text{IRR}(\text{CF}_0 : \text{CF}_n)$

ENGINEERING ECONOMIC ANALYSIS

ENGINEERING ECONOMIC ANALYSIS

THIRTEENTH EDITION

Donald G. Newnan
San Jose State University

Ted G. Eschenbach
University of Alaska Anchorage

Jerome P. Lavelle
North Carolina State University

Neal A. Lewis
University of New Haven

New York   Oxford
OXFORD UNIVERSITY PRESS

Oxford University Press is a department of the University of Oxford. It furthers
the University's objective of excellence in research, scholarship, and education
by publishing worldwide. Oxford is a registered trade mark of Oxford University
Press in the UK and certain other countries.

Published in the United States of America by Oxford University Press
198 Madison Avenue, New York, NY 10016, United States of America.

CIP Data is on file at the Library of Congress

ISBN 978–0–19–029690–2

9 8 7 6 5 4 3 2 1
Printed by Edwards Brothers Malloy, United States of America

Brian Newnan, my chemical engineering nephew,
who helped guide this book forward
DN

Richard Corey Eschenbach, for his lifelong example
of engineering leadership and working well with others
TE

To my lovely wife Christine
and sweet daughters Gabrielle, Veronica,
Miriam, Regina, and Magdalen,
who all inspire me daily to be my best!
JL

My wife Joan,
for her continued support
NL

CONTENTS IN BRIEF

CONTENTS

O ur goal has been, and still is, to provide an easy-to-understand and up-to-date presentation of engineering economic analysis for today's students. That means the book's writing style must promote the reader's understanding. We humbly note that our approach has been well received by engineering professors—and more importantly, by engineering students through multiple editions.

Hallmarks of this Book

Since it was first published, this text has become the market-leading book for the engineering economy course. It has always been characterized by

- **A focus on practical applications.** One way to encourage students to read the book, and to remember and apply what they have learned in this course, is to make the book interesting. And there is no better way to do that than to infuse the book with real-world examples, problems, and vignettes.
- **Accessibility.** We meet students where they are. Most don't have any expertise in accounting or finance. We take the time to explain concepts carefully while helping students apply them to engineering situations.
- **Superior instructor and student support packages.** To make this course easier to understand, learn, and teach, Oxford University Press has produced the best support package available. We offer more for students and instructors than any competing text.

Strengths of the 13th Edition

- New to this edition is chapter appendix material on investing, diversification, and personal finance. This builds on the loans, savings, and other personal finance examples that have long been used to motivate students and engage them with engineering economy concepts. Our first goal is force the realization that engineering economy *really does matter.* Second, personal financial success contributes to success as a student, as an engineer, and most importantly in life.
- New to this edition is coverage of multiple objectives using simple additive models starting in Chapter 1. With this simple model students can incorporate the important nonmonetary factors found in many engineering economy applications.

- Green engineering and ethics have more coverage with *new problem parts and icons* in every chapter. Ethics questions continue to be part of the Questions to Consider in the vignettes.
- Factor notation has long provided a clear way to describe and understand engineering economic calculations supported by tables—which is continued in this edition. Spreadsheet annuity functions with the same assumptions as tabulated factors are presented in a visual 5-BUTTON format. The two approaches are mutually reinforcing for faster and deeper student understanding. As detailed in Appendix B, students can also use a financial calculator or an HP 33s or 35s—which can be used on the FE exam.
- Over 500 problems are new or revised of the 1457 in the text. To minimize errors and typos, solutions in the *Instructor's Manual* were completed and text corrections were made *before* the book was finalized for printing.

  - Over 10% of these problems have been contributed by adopters with class-tested fresh examples and new question formats.
  - Problem headings link with chapter headings to assist faculty in choosing problems and students in studying similar problems. Problems are ordered from easy to hard within each heading.
  - There is an answer icon next to most even-numbered problems with answers in Appendix E.

    - Instructors can easily pick a preferred mix of problems with and without answers.
    - Students can work on assigned problems without knowing the answer until they are done (or stuck).
    - Students can do extra problems and check their own answers.

- Each chapter opens with a list of *keywords,* which are **boldfaced** when first explained and indexed for later reference.
- Spreadsheet coverage has been designed to support student learning and engineering practice. Faculty can choose from no coverage to heavy reliance. In either case the text has been designed to support student self-directed learning.

  - Appendix A supports spreadsheet novices and demonstrates the use of data blocks and relative/absolute addresses.
  - Spreadsheet annuity functions are introduced beginning with Example 3–5; spreadsheet block functions are covered late in Chapter 4 after factor approaches for arithmetic and geometric gradients where annuity functions cannot be used. This structure allows a mutually reinforcing presentation of tabulated factors and spreadsheet annuity functions.
  - Other spreadsheet functions including XNPV, XIRR, SUMPRODUCT, and GOAL SEEK are presented when they will allow or speed solution in applications of engineering economy concepts.
  - Problems in Chapters 12, 13, and 14 on taxes, replacement analysis, and inflation tend to involve more calculations than other chapters so spreadsheets are particularly useful.

- Renewed in this edition is a focus on selecting and revisiting examples with different economic measures, different tools, and increasing complexity.
- To support classes where some students rent or buy a used text, the 54 cases in *Cases in Engineering Economy* 2nd are now posted on the student website, their solutions on the instructor website, and each chapter concludes with a list of recommended cases.
- The chapter opening vignettes are replaced, revised, and updated to ensure they are up to date and relevant.

Changes in conceptual coverage are detailed by chapter.

- Chapter 1 (*Making Economic Decisions*) now introduces multiple objectives. The electric vehicle vignette is new to emphasize vehicle performance.
- Chapter 2 (*Estimating Engineering Costs and Benefits*) now includes coverage of internal and external costs with an emphasis on green engineering.
- Chapter 3 (*Interest and Equivalence*) reduces continuous compounding to coverage of the effective interest rate and drops coverage of the almost never used continuous compounding factors.
- Chapter 4 (*Equivalence for Repeated Cash Flows*) has reordered the factor and spreadsheet annuity function coverage to be mutually reinforcing. It drops coverage of the almost never used continuous compounding factors.
- Chapter 5 (*Present Worth Analysis*) now includes XNPV for cash flows that occur on specific dates rather than evenly spaced periods.
- Chapter 7 (*Rate of Return Analysis*) now includes XIRR for cash flows that occur on specific dates rather than evenly spaced periods.
- Chapter 9 (*Other Analysis Techniques*) now includes discounted payback period and Appendix 9A (*Investing for Retirement and Other Future Needs*) has been added to prepare students for likely retirement saving environments.
- Chapter 10 (*Uncertainty in Future Events*) now includes balancing risk and return using multiple objectives. Appendix 10A (*Diversification Reduces Risk*) has been added so that students can apply it when investing.
- Chapter 12 (*Income Taxes for Corporations*) now includes bonus depreciation and demonstrates how multiple depreciation methods can be used together. Appendix 12A (*Taxes and Personal Financial Decision Making*) has added coverage of insurance and personal budgeting. As in every edition, this material has been updated for changes in the tax code.
- Chapter 13 (*Replacement Analysis*) now matches the replacement map to real-world applications of replacement analysis. Formal equations for replacement analysis are now included.
- Chapter 14 (*Inflation and Price Changes*) has a new vignette on price trends in solar technologies.
- Chapter 15 (*Selection of a Minimum Attractive Rate of Return*) adds a new section on the MARR for individuals.

- Chapter 16 (*Economic Analysis in the Public Sector*) adds design to cost and compares it with minimizing life cycle costs.
- "Trust Me: You'll Use This" (between chapters 2 and 3) focuses on applications to the lives of students and graduates.

The Superior Newnan Support Package

The supplement package for this text has expanded with each edition, and it has been updated and expanded even further for this edition. No competing text has a more extensive support package. The package features the following:

- A Study Guide by Ed Wheeler of the University of Tennessee, Martin (ISBN 978-0-19-029702-2) is available for individual purchase and is also available free of charge when packaged with the textbook.
- The book is accompanied by an extensive set of materials that are presented, free of charge, at www.oup.com/us/newnan. These include:
  - A set of tutorials on engineering economy applications of Excel by Julie L. Fortune of the University of Alabama in Huntsville.
  - A set of 54 cases provides realistic, complex problems for further study. These cases, written by William Peterson and Ted Eschenbach, also include three chapters on case analysis and an example case solution.
  - Spreadsheet problem modules, written by Thomas Lacksonen of the University of Wisconsin Stout.
  - Interactive multiple-choice problems, written by Paul Schnitzler of the University of South Florida and William Smyer of Mississippi State University.
  - Additional practice FE Exam problems, authored by Karen Thorsett, University of Phoenix.

Instructors will find an updated and expanded set of resources available on Oxford's Ancillary Resource Center. Please contact your Oxford University Press representative for access.

- An exam file written and edited by Meenakshi Sundaram of Tennessee Technological University.
- PowerPoint lecture notes for all chapters by Neal Lewis of the University of New Haven.
- An Instructor's Manual by John M. Usher of Mississippi State University and Lawrence Samuelson of Tri-State University and the authors, with complete solutions to all end-of-chapter problems.
- The compound interest tables from the textbook are available for adopting professors who prefer to give closed-book exams. The tables are on the website as PDF files that can be printed in part or in total.

Acknowledgments

Many people have directly or indirectly contributed to the content of the book. We have been influenced by our educations and our university colleagues, and by students who have provided invaluable feedback on content and form. We are particularly grateful to the following professors and students for their contributions in the form of insights, reviews, contributed problems, and contributed vignettes to this and previous editions: Kate D. Abel, Stevens Institute of Technology; Francisco Aguíñiga, Texas A&M University–Kingsville; Magdy Akladios, University of Houston–Clear Lake; Yasser Alhenawi, University of Evansville; Benjamin Armbruster, Northwestern University; Baabak Ashuri, Georgia Institute of Technology; M. Affan Badar, Indiana State University; Kailash Bafna, Western Michigan University; Biswanath Bandyopadhyay, University of North Dakota; Robert Baston, University of Alabama–Tuscaloosa; Richard H. Bernhard, North Carolina State University; Marsha Berry, U.S. Army Research, Development & Engineering Command; Edgar Blevins, Southern University; Rebeca Book, Pittsburg State University; William Brown, West Virginia University at Parkersburg; Patrick Brunese, Purdue University; Mark Budnik, Valparaiso University; Karen M. Bursic, University of Pittsburgh; Peter A. Cerenzio, Cerenzio & Panaro Consulting Engineers; Linda Chattin, Arizona State University; Lijian Chen, University of Louisville; Steven Chiesa, Santa Clara University; Tracy Christofero, Marshall University; Paul Componation, University of Texas at Arlington; Jennifer Cross, Texas Tech University; Gene Dixon, East Carolina University; Emmanuel Donkor, George Washington University; Colin Drummond, Case Western Reserve University; Julie Drzymalski, Western New England University; John Easley, Louisiana Tech University–Ruston; David Elizandro, Tennessee Technological University; Alberto Garcia, University of Tennessee; Mostafa Ghandehari, University of Texas at Arlington; Dolores K. Gooding, University of South Florida; Johnny R. Graham, University of North Carolina at Charlotte; Tarun Gupta, Western Michigan University; Safwat H. Shakir Hanna, Prairie View A&M University; Craig Harvey, Louisiana State University; Oliver Hedgepeth, American Public University; Morgan E. Henrie, University of Alaska Anchorage; Joseph R. Herkert, North Carolina State University; Hamed Kashani, Georgia Institute of Technology; Paul Kauffmann, East Carolina University; Khoiat Kengskool, Florida International University; Adeel Khalid, Southern Polytechnic State University; David Kieser, Indiana University-Purdue University at Indianapolis; Changhyun Kwon, SUNY Buffalo; Marcial Lapp, University of Michigan; Tony Lima, CSU East Bay; Barry Liner, George Mason University; Daniel P. Loucks, Cornell University; Muslim Majeed, Carleton University; Louis Manz, University of Texas, San Antonio; Jessica Matson, Tennessee Technological University; Brooke Mayer, Arizona State University; Paul R. McCright, University of South Florida; Dale McDonald, Midwestern State University; Nina Miville, University of Miami; Gary Moynihan, University of Alabama; David W. Naylor, UNC Charlotte; Kim LaScola Needy, University of Arkansas; Gillian M. Nicholls, Southeast Missouri State University; Charles Nippert, Widener University; Benedict N. Nwokolo, Grambling State University; John O'Haver, University of Mississippi; Darren Olson, Central Washington University; Jani Macari Pallis, University of Bridgeport & Cislunar Aerospace, Inc.; Renee Petersen, Washington State University; William Peterson, Minnesota State University, Mankato; Letitia M. Pohl, University of Arkansas; Md. Mamunur Rashid, University of Massachusetts–Lowell; Kevin A. Rider,

West Virginia University; Saeid Sadri, JSC Nazarbayev University; Thobias Sando, University of North Florida; Scott Schultz, Mercer University; Dhananjai B. Shah, Cleveland State University; Deepak Sharma, California State University–Fullerton; Michael Shenoda, University of North Texas; James Simonton, University of Tennessee Space Institute; Don Smith, Texas A&M University; William N. Smyer, Mississippi State University; Hansuk Sohn, New Mexico State University; Musaka E. Ssemakula, Wayne State University; John Stratton, Rochester Institute of Technology; Meenakshi R. Sundaram, Tennessee Technological University; Robert Swan, Drexel University; Walter Towner, Worcester Polytechnic Institute; William R. Truran, Stevens Institute of Technology; Francis M. Vanek, Cornell University; Ed Wheeler, University of Tennessee at Martin; John Whittaker, University of Alberta; Gregory Wiles, Southern Polytechnic State University; Nong Ye, Arizona State University; Xiaoyan Zhu, University of Tennessee–Knoxville.

The following individuals are new additions to our list of contributors of vignettes, text or solution reviews, problems, and insight: Eva Andrijcic, Rose-Hulman Institute of Technology; Biswanath Bandyopadhyay, University of North Dakota; Robert Batson, University of Alabama–Tuscaloosa; Edgar R. Blevins, Southern University; Emmanuel A. Donkor, George Washington University; Ona Egbue, University of Minnesota Duluth; Tarun Gupta, Western Michigan University; Safwat H. Shakir Hanna, Prairie View A&M University; Craig Harvey, Louisiana State University; Muslim A. Majeed, Carleton University; Hector Medina, Liberty University; Nina D Miville, University of Miami; Gana Natarajan, Oregon State University; Ean Ng, Oregon State University; Md. Mamunur Rashid, University of Massachusetts–Lowell; Michael Shenoda, University of North Texas; Hansuk Sohn, New Mexico State University; John Stratton, Rochester Institute of Technology; Robert H. Swan, Jr., Drexel University; Xiaoyan Zhu, University of Tennessee–Knoxville.

John Whittaker of the University of Alberta wrote a new Chapter 8 for the first Canadian edition and many improvements he made in the second and third Canadian editions have been incorporated. Neal Lewis of the University of New Haven has contributed significantly to the last two editions, but for this edition we were fortunate to add him to the team of coauthors. Thank you, Neal, from Ted and Jerome. Finally, we were helped by the professors who participated in the market survey for this book, and whose collective advice helped us shape this new edition.

Finally, our largest thanks go to the professors (and their students) who have developed the products that support this text: Julie L. Fortune, University of Alabama in Huntsville; Thomas Lacksonen, University of Wisconsin–Stout; Shih Ming Lee, Florida International University; David Mandeville, Oklahoma State University; William Peterson, Minnesota State University, Mankato; Lawrence Samuelson, Tri-State University; Paul Schnitzler, University of South Florida; William Smyer, Mississippi State University; Meenakshi Sundaram, Tennessee Technological University; Karen Thorsett, University of Phoenix; John M. Usher, Mississippi State University; Ed Wheeler, University of Tennessee at Martin.

Textbooks are produced through the efforts of many people. Dan Kaveney, Christine Mahon, Patrick Lynch, and Megan Carlson have worked to make this a timely and improved edition. Keith Faivre managed the text's design and production. Dorothy Bauhoff copyedited the manuscript, and Mary Anne Shahidi proofread the page proofs.

The sales force at OUP has maintained the text's leading position and ensured the flow of adopter feedback to the authors.

This book remains the best text on the market in large part because of feedback from users. We would appreciate hearing comments about the book, including being informed of any errors that have snuck in despite our best attempts to eradicate them. We also look forward to adding problems and vignettes in the next edition that adopters have found effective for their students. Please write us c/o the Engineering Editor at Oxford University Press, 198 Madison Avenue, New York, NY 10016, or email us directly. Thanks for using the Newnan book!

Don Newnan
Ted Eschenbach
tgeschenbach@alaska.edu
Jerome Lavelle
jerome_lavelle@ncsu.edu
Neal Lewis
nlewis@newhaven.edu

ENGINEERING ECONOMIC ANALYSIS

MAKING ECONOMIC DECISIONS

Electric Vehicles

I n 2008 President Obama called for one million electric cars to be owned by Americans. However, by the end of 2015 only 300,000 were on the road. Why? The reasons included cost, battery range, and performance. Have the barriers changed?

In 2016, GM produced the Chevy Bolt EV, the first electric car under $30,000. Given that the average price of a new vehicle in the U.S. is $34,000, this class is large and ripe for competition. At introduction the Chevy Bolt EV had a 200-mile-per-charge electric battery, with an expected improvement of 30% by 2022. As the battery efficiency improves, sales of electric cars will increase. In turn, higher sales will support lower manufacturing costs and reward technological improvements.

Can electric vehicles perform as well as gas-powered cars? Currently, fast automobiles can go from zero to 60 mph in under 6 seconds, while performance cars can do it in 3 to 4 seconds. The Tesla, with 2 electric motors, can go from zero to 60 mph in 2.8 seconds. The two quickest production cars are hybrids. Being the fastest might require electric motors.

These examples show that the electric car market seems ready for explosive growth. Andy Palmer, the CEO of Aston Martin, was quoted in *Auto News* as saying, "it's inevitable that the entire industry will shift over to electricity, if only because it's the most plausible way to deliver the power drivers expect."

The widespread adoption of electric vehicles would be a disruption to the current industry. Engineers who design electric vehicles will have different constraints. Electric motors and batteries will replace combustion engines, emission controls, fuel storage, and drive trains. This should reduce the weight and increase the available space in the vehicle,

making the electric car faster and more efficient. Design decisions will determine manufacturing costs and influence the sales price. In addition, the supply, production, and service chain will be affected by a conversion to electric vehicles.

The outlook for electric vehicles is positive for many reasons. As manufacturing experience increases, electric vehicles can be designed to surpass the performance of conventional cars, raising customer expectations. As economies of scale bring the price of electric vehicles down, more consumers can afford them and more manufacturers can profit from making them. All of this should be good news for the environment. ■ ■ ■

Contributed by Kate D. Abel, Stevens Institute of Technology

QUESTIONS TO CONSIDER

1. What marketplace dynamics drive or suppress developing electric vehicles? What role should government play?
2. Develop a list of concerns and questions that consumers might have regarding the conversion to electric vehicles. Which are economic and which are non-economic factors?
3. From a manufacturer's viewpoint, what are the major concerns, potential problems, and overriding goals of producing an electric vehicle? How do these affect the price charged for the vehicle?
4. Are there ethical aspects in the shift from gasoline-powered vehicles to electric vehicles? List these and determine how they could be or should be resolved and by whom.

After Completing This Chapter...

The student should be able to:

- Distinguish between simple and complex problems.
- Discuss the role and purpose of engineering economic analysis.
- Describe and give examples of the nine steps in the *economic decision-making process.*
- Select appropriate economic criteria for use with different types of problems.
- Describe common ethical issues in engineering economic decision making.
- Solve engineering problems with current costs.
- Solve problems that have multiple objectives.

Key Words

absolute address	fixed input	resolving consequences
benefit	fixed output	shadow price
brainstorming	green engineering	societal costs
cost	maximizing profit	value engineering
criteria	model building	what-if analysis
data block	multiple objectives	
decision making	overhead	

This book is about making decisions. **Decision making** is a broad topic, for it is a major aspect of everyday human existence. This book develops the tools to properly analyze and solve the economic problems that are commonly faced by engineers. Even very complex situations can be broken down into components from which sensible solutions are produced. If one understands the decision-making process and has tools for obtaining realistic comparisons between alternatives, one can expect to make better decisions.

Our focus is on solving problems that confront firms in the marketplace, but many examples are problems faced in daily life. Let us start by looking at some of these problems.

A SEA OF PROBLEMS

A careful look at the world around us clearly demonstrates that we are surrounded by a sea of problems. There does not seem to be any exact way of classifying them, simply because they are so diverse in complexity and "personality." One approach arranges problems by their *difficulty*.

Simple Problems

Many problems are pretty simple, and good solutions do not require much time or effort.

- Should I pay cash or use my credit card?
- Do I buy a semester parking pass or use the parking meters?
- Shall we replace a burned-out motor?
- If we use three crates of an item a week, how many crates should we buy at a time?

Intermediate Problems

At a higher level of complexity we find problems that are primarily economic.

- Shall I buy or lease my next car?
- Which equipment should be selected for a new assembly line?
- Which materials should be used as roofing, siding, and structural support for a new building?
- Shall I buy a 1- or 2-semester parking pass?
- What size of transformer or air conditioner is most economical?

Some numeric examples of operational economics follow the section on ethics later in this chapter.

Complex Problems

Complex problems are a mixture of *economic, political,* and *humanistic* elements.

- Honda Motors in North America illustrates complex problems. In Alliston, Ontario, they employ 4000 workers and manufacture the Acura MDX, ZDX, CSX, and Civic. In Lincoln, Alabama, they employ 4000 workers and manufacture the Odyssey, Pilot, Ridgeline, and Acura MDX. Any decision allocating production must consider, along with economic aspects: reactions of the American, Canadian, Japanese, and Mexican governments; the North American Free Trade Agreement; labor unions in three countries; and the 2014 opening of a second Mexican plant in Celaya.
- The selection of a dating partner (who may later become a permanent partner) is obviously complex. Economic analysis can be of little or no help.
- A firm's annual budget allocates resources and all projects are economically evaluated. The budget process is also heavily influenced by noneconomic forces such as power struggles, geographical balancing, and impact on individuals, programs, and profits. For multinational corporations there are even national interests to be considered.

The chapter's final section presents one approach to more complex problems.

THE ROLE OF ENGINEERING ECONOMIC ANALYSIS

Engineering economic analysis is most suitable for intermediate problems and the economic aspects of complex problems. They have these qualities:

1. The problem is *important enough* to justify our giving it serious thought and effort.
2. The problem can't be worked in one's head—that is, a careful analysis *requires that we organize* the problem and all the various consequences.
3. The problem has *economic aspects* important in reaching a decision.

When problems meet these three criteria, engineering economic analysis is useful in seeking a solution. Since vast numbers of problems in the business world (and in one's personal life) meet these criteria, engineering economic analysis is often required.

Examples of Engineering Economic Analysis

Engineering economic analysis focuses on costs, revenues, and benefits that occur at different times. For example, when a civil engineer designs a road, a dam, or a building, the construction costs occur in the near future; but the benefits to users begin only when construction is finished and then continue for a long time.

In fact nearly everything that engineers design calls for spending money in the design and building stages, and only after completion do revenues or benefits occur—usually for years. Thus the economic analysis of costs, benefits, and revenues occurring over time is called *engineering* economic analysis.

Engineering economic analysis is used by firms and government agencies to answer many different questions.

- *Which engineering projects are worthwhile?* Has the mining or petroleum engineer shown that the mineral or oil deposit is worth developing?
- *Which engineering projects should have a higher priority?* Has the industrial engineer shown which factory improvement projects should be funded with the available dollars?
- *How should the engineering project be designed?* Has the mechanical or electrical engineer chosen the most economical motor size? Has the civil or mechanical engineer chosen the best thickness for insulation? Has the aeronautical engineer made the best trade-offs between (1) lighter materials that are expensive to buy but cheaper to fly and (2) heavier materials that are cheap to buy and more expensive to fly?

Engineering economic analysis can also be used to answer questions that are personally important.

- *How to achieve long-term financial goals:* How much should you save each month to buy a house, retire, or fund a trip around the world? Is going to graduate school a good investment—will your additional earnings in later years balance the cost of attending and your lost income while in graduate school?
- *How to compare different ways to finance purchases:* Is it better to finance your car purchase by using the dealer's low interest rate loan or by taking an available rebate and borrowing money from your bank or credit union?
- *How to make short- and long-term investment decisions:* Should you buy a 1- or 2-semester parking pass? Is a higher salary better than stock options?

THE DECISION-MAKING PROCESS

Decision making may take place by default; that is, a person may not consciously recognize that an opportunity for decision making exists. This fact leads to our first element in a definition of decision making—there must be at least two alternatives available. If only one course of action is available, there is nothing to decide. The only alternative is to proceed with the single available course of action. (It is rather unusual to find that there are no alternative courses of action. More frequently, alternatives simply are not recognized.)

At this point we might conclude that the decision-making process consists of choosing from among alternative courses of action. But this is an inadequate definition. Consider the following situation.

At a race track, a bettor was uncertain about which horse to bet on in the next race. He closed his eyes and pointed his finger at the list of horses printed in the racing program. Upon opening his eyes, he saw that he was pointing to horse number 4. He hurried off to bet on that horse.

Does this racehorse selection represent decision making? Yes, (assuming the bettor had already ruled out the "do-nothing" alternative of placing no bet). But the particular method of deciding seems inadequate and irrational. We want to deal with rational decision making.

Rational Decision Making

Rational decision making is a complex process that contains nine essential elements, which are shown in Figure 1–1. Although these nine steps are shown sequentially, it is common for a decision maker to repeat steps, take them out of order, and do steps simultaneously. For example, when a new alternative is identified more data will be required. Or when the outcomes are summarized, it may become clear that the problem needs to be redefined or new goals established.

The value of this sequential diagram is to show all the steps that are usually required, and to show them in a logical order. Occasionally we will skip a step entirely. For example, a new alternative may be so clearly superior that it is immediately adopted at Step 4 without further analysis. The following sections describe the elements listed in Figure 1–1.

1. Recognize the Problem

The starting point in rational decision making is recognizing that a problem exists.

Some years ago, for example, it was discovered that several species of ocean fish contained substantial concentrations of mercury. The decision-making process began with this recognition of a problem, and the rush was on to determine what should be done. Research revealed that fish taken from the ocean decades before and preserved in laboratories also contained similar concentrations of mercury. Thus, the problem had existed for a long time but had not been recognized.

FIGURE 1–1 One possible flowchart of the decision process.

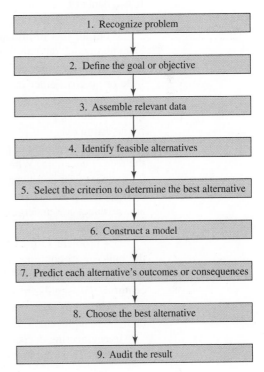

In typical situations, recognition is obvious and immediate. An auto accident, an overdrawn check, a burned-out motor, an exhausted supply of parts all produce the recognition of a problem. Once we are aware of the problem, we can solve it as best we can. Many firms establish programs for total quality management (TQM) or continuous process improvement (CPI) that are designed to identify problems so that they can be solved.

2. Define the Goal or Objective

The goal or objective can be an overall goal of a person or a firm. For example, a personal goal could be to lead a pleasant and meaningful life, and a firm's goal is usually to operate profitably. The presence of multiple, conflicting goals is often the foundation of complex problems.

But an objective need not be an overall goal of a business or an individual. It may be quite narrow and specific: "I want to pay off the loan on my car by May," or "The plant must produce 300 golf carts in the next 2 weeks," are more limited objectives. Thus, defining the objective is the act of exactly describing the task or goal.

3. Assemble Relevant Data

To make a good decision, one must first assemble good information. In addition to all the published information, there is a vast quantity of information that is not written down anywhere but is stored as individuals' knowledge and experience. There is also information that remains ungathered. A question like "How many people in your town would be interested in buying a pair of left-handed scissors?" cannot be answered by examining published data or by asking any one person. Market research or other data gathering would be required to obtain the desired information.

From all this information, what is relevant in a specific decision-making process? Deciding which data are important and which are not may be a complex task. The availability of data further complicates this task. Published data are available immediately at little or no cost; other data are available from specific knowledgeable people; still other data require surveys or research to assemble the information. Some data will be of high quality—that is, precise and accurate, while other data may rely on individual judgment for an estimate.

If there is a published price or a contract, the data may be known exactly. In most cases, the data is uncertain. What will it cost to build the dam? How many vehicles will use the bridge next year and twenty years from now? How fast will a competing firm introduce a competing product? How will demand depend on growth in the economy? Future costs and revenues are uncertain, and the range of likely values should be part of assembling relevant data.

The problem's time horizon is part of the data that must be assembled. How long will the building or equipment last? How long will it be needed? Will it be scrapped, sold, or shifted to another use? In some cases, such as for a road or a tunnel, the life may be centuries with regular maintenance and occasional rebuilding. A shorter time period, such as 50 years, may be chosen as the problem's time horizon, so that decisions can be based on more reliable data.

In engineering decision making, an important source of data is a firm's own accounting system. These data must be examined quite carefully. Accounting data focuses on past information, and engineering judgment must often be applied to estimate current and future values. For example, accounting records can show the past cost of buying computers, but engineering judgment is required to estimate the future cost of buying computers.

Financial and cost accounting are designed to show accounting values and the flow of money—specifically **costs** and **benefits**—in a company's operations. When costs are directly related to specific operations, there is no difficulty; but there are other costs that are not related to specific operations. These indirect costs, or **overhead,** are usually allocated to a company's operations and products by some arbitrary method. The results are generally satisfactory for cost-accounting purposes but may be unreliable for use in economic analysis.

To create a meaningful economic analysis, we must determine the *true* differences between alternatives, which might require some adjustment of cost-accounting data. The following example illustrates this situation.

EXAMPLE 1–1

A firm's printing department charges the other departments for its services to recover its monthly costs. For example, the charge to run 30,000 copies for the shipping department is:

Direct labor	$228
Materials and supplies	294
Overhead costs	271
Cost to shipping department	$793

The shipping department checks with a commercial printer, which would print the same 30,000 copies for $688. The shipping department foreman wants to have the work done externally. The in-house printing department objects to this. The general manager has asked you to recommend what should be done.

SOLUTION

Some of the printing department's output reveals the firm's costs, prices, and other financial information. Thus, the printing department is necessary to prevent disclosing such information to people outside the firm. The firm cannot switch to an outside printer for all needs.

A review of the cost-accounting charges reveals nothing unusual. The charges made by the printing department cover direct labor, materials and supplies, and overhead. The allocation of indirect costs is a customary procedure in cost-accounting systems (see Chapter 17 for more). It can be misleading for decision making, as the following discussion indicates.

The shipping department would reduce its cost by $105 (= $793 – $688) by using the outside printer. In that case, how much would the printing department's costs decline, and which solution is better for the firm?

1. *Direct Labor.* If the printing department had been working overtime, then the overtime could be reduced or eliminated. But, assuming no overtime, how much would the saving be? It seems unlikely that an employee could be fired or even put on less than a 40-hour

work week. Thus, although there might be a $228 saving, it is much more likely that there will be no reduction in direct labor.

2. *Materials and Supplies.* There would be a $294 saving in materials and supplies.

3. *Allocated Overhead Costs.* There will be no reduction in the printing department's monthly overhead, and in fact the firm will incur $50 of additional expenses in purchasing and accounting for processing the purchase order, invoice, and payment.

The firm will save $294 in materials and supplies, will spend $50 in purchasing and accounting, and may or may not save $228 in direct labor if the printing department no longer does the shipping department work. The maximum saving would be $294 + 228 − 50 = $472. Either value of $294 or $472 is less than the $688 the firm would pay the outside printer. The shipping department should not be allowed to send its printing to the outside printer.

Gathering cost data presents other difficulties. One way to look at the financial consequences—costs and benefits—of various alternatives is as follows.

- *Market Consequences.* These consequences have an established price in the marketplace. We can quickly determine raw material prices, machinery costs, labor costs, and so forth.

- *Extra-Market Consequences.* There are other items that are not directly priced in the marketplace. But by indirect means, a price may be assigned to these items. (Economists call these prices **shadow prices.**) Examples might be the cost of an employee injury or the value to employees of going from a 5-day to a 4-day, 40-hour week.

- *Intangible Consequences.* Numerical economic analysis probably never fully describes the real differences between alternatives. The tendency to leave out consequences that do not have a significant impact on the analysis itself, or on the conversion of the final decision into actual money, is difficult to resolve or eliminate. How does one evaluate the potential loss of workers' jobs due to automation? What is the value of landscaping around a factory? These and a variety of other consequences may be left out of the numerical calculations, but they must be considered in reaching a decision.

4. Identify Feasible Alternatives

One must keep in mind that unless the best alternative is considered, the result will always be suboptimal.[1] Two types of alternatives are sometimes ignored. First, in many situations a do-nothing alternative is feasible. This may be the "Let's keep doing what we are now doing," or the "Let's not spend any money on that problem" alternative. Second, there are often feasible (but unglamorous) alternatives, such as "Patch it up and keep it running for another year before replacing it."

---

[1]A group of techniques called value analysis or **value engineering** is used to examine past decisions and current trade-offs in designing alternatives.

There is no way to ensure that the best alternative is among the alternatives being considered. One should try to be certain that all conventional alternatives have been listed and then make a serious effort to suggest innovative solutions. Sometimes a group of people considering alternatives in an innovative atmosphere—**brainstorming**—can be helpful. Even impractical alternatives may lead to a better possibility. The payoff from a new, innovative alternative can far exceed the value of carefully selecting between the existing alternatives.

Any good listing of alternatives will produce both practical and impractical alternatives. It would be of little use, however, to seriously consider an alternative that cannot be adopted. An alternative may be infeasible for a variety of reasons. For example, it might violate fundamental laws of science, require resources or materials that cannot be obtained, violate ethics standards, or conflict with the firm's strategy. Only the feasible alternatives are retained for further analysis.

5. Select the Criterion to Determine the Best Alternative

The central task of decision making is choosing from among alternatives. How is the choice made? Logically, to choose the best alternative, we must define what we mean by *best*. There must be a **criterion,** or set of **criteria,** to judge which alternative is best. Now, we recognize that *best* is on one end of the following relative subjective judgment:

Worst	Bad	Fair	Good	Better	Best

relative subjective judgment spectrum

Since we are dealing in *relative terms,* rather than *absolute values,* the choice will be the alternative that is relatively the most desirable. Consider a driver found guilty of speeding and given the alternatives of a $475 fine or 3 days in jail. In absolute terms, neither alternative is good. But on a relative basis, one simply makes the best of a bad situation.

There may be an unlimited number of ways that one might judge the various alternatives. Several possible criteria are:

- Create the least disturbance to the environment.
- Improve the distribution of wealth among people.
- Minimize the expenditure of money.
- Ensure that the benefits to those who gain from the decision are greater than the losses of those who are harmed by the decision.[2]
- Minimize the time to accomplish the goal or objective.
- Minimize unemployment.
- Maximize profit.

---

[2]This is the Kaldor criterion.

Selecting the criterion for choosing the best alternative will not be easy if different groups support different criteria and desire different alternatives. The criteria may conflict. For example, minimizing unemployment may require increasing the expenditure of money. Or minimizing environmental disturbance may conflict with minimizing time to complete the project. The disagreement between management and labor in collective bargaining (concerning wages and conditions of employment) reflects a disagreement over the objective and the criterion for selecting the best alternative.

The last criterion—maximize profit—is the one normally selected in engineering decision making. When this criterion is used, all problems fall into one of three categories: neither input nor output fixed, fixed input, or fixed output.

Neither input nor output fixed. The first category is the general and most common situation, in which the amount of money or other inputs is not fixed, nor is the amount of benefits or other outputs. For example:

- A consulting engineering firm has more work available than it can handle. It is considering paying the staff for working evenings to increase the amount of design work it can perform.
- One might wish to invest in the stock market, but the total cost of the investment is not fixed, and neither are the benefits.
- A car battery is needed. Batteries are available at different prices, and although each will provide the energy to start the vehicle, the useful lives of the various products are different.

What should be the criterion in this category? Obviously, to be as economically efficient as possible, we must maximize the difference between the return from the investment (benefits) and the cost of the investment. Since the difference between the benefits and the costs is simply profit, a businessperson would define this criterion as **maximizing profit.**

Fixed input. The amount of money or other input resources (like labor, materials, or equipment) is fixed. The objective is to effectively utilize them. For economic efficiency, the appropriate criterion is to maximize the benefits or other outputs. For example:

- A project engineer has a budget of $350,000 to overhaul a portion of a petroleum refinery.
- You have $300 to buy clothes for the start of school.

Fixed output. There is a fixed task (or other output objectives or results) to be accomplished. The economically efficient criterion for a situation of fixed output is to minimize the costs or other inputs. For example:

- A civil engineering firm has been given the job of surveying a tract of land and preparing a "record of survey" map.
- You must choose the most cost-effective design for a roof, an engine, a circuit, or other component.

For the three categories, the proper economic criteria are:

Category	Economic Criterion
Neither input nor output fixed	Maximize profit = value of outputs − cost of inputs.
Fixed input	Maximize the benefits or other outputs.
Fixed output	Minimize the costs or other inputs.

6. Constructing the Model

At some point in the decision-making process, the various elements must be brought together. The *objective, relevant data, feasible alternatives,* and *selection criterion* must be merged. For example, if one were considering borrowing money to pay for a car, there is a mathematical relationship between the loan's variables: amount, interest rate, duration, and monthly payment.

Constructing the interrelationships between the decision-making elements is frequently called **model building** or **constructing the model.** To an engineer, modeling may be a scaled *physical representation* of the real thing or system or a *mathematical equation,* or set of equations, describing the desired interrelationships. In economic decision making, the model is usually mathematical.

In modeling, it is helpful to represent only that part of the real system that is important to the problem at hand. Thus, the mathematical model of the student capacity of a classroom might be

$$\text{Capacity} = \frac{lw}{k}$$

where l = length of classroom, in meters
$\quad\quad w$ = width of classroom, in meters
$\quad\quad k$ = classroom arrangement factor

The equation for student capacity of a classroom is a very simple model; yet it may be adequate for the problem being solved.

7. Predicting the Outcomes for Each Alternative

A model and the data are used to predict the outcomes for each feasible alternative. As was suggested earlier, each alternative might produce a variety of outcomes. Selecting a motorcycle, rather than a bicycle, for example, may make the fuel supplier happy, the neighbors unhappy, the environment more polluted, and one's savings account smaller. But, to avoid unnecessary complications, we assume that decision making is based on a single criterion for measuring the relative attractiveness of the various alternatives. As will be shown in Example 1–5, one can devise a single composite criterion that is the weighted average of several different choice criteria.

To choose the best alternative, the outcomes for each alternative must be stated in a *comparable* way. Usually the consequences of each alternative are stated in terms of money, that is, in the form of costs and benefits. **Resolving the consequences** is done with all monetary and nonmonetary consequences. The consequences can also be categorized as follows:

Market consequences—where there are established market prices available

Extra-market consequences—no direct market prices, so priced indirectly

Intangible consequences—valued by judgment, not monetary prices.

In the initial problems we will examine, the costs and benefits occur over a short time period and can be considered as occurring at the same time. In other situations the various costs and benefits take place in a longer time period. The result may be costs at one point in time followed by periodic benefits. We will resolve these in the next chapter into a *cash flow diagram* to show the timing of the various costs and benefits.

For these longer-term problems, the most common error is to assume that the current situation will be unchanged for the do-nothing alternative. In reality if a firm does nothing new then current profits will shrink or vanish as a result of the actions of competitors and the expectations of customers. As another example, traffic congestion normally increases over the years as the number of vehicles increases—doing nothing does not imply that the situation will not change.

8. Choosing the Best Alternative

Earlier we said that choosing the best alternative may be simply a matter of determining which alternative best meets the selection criterion. But the solutions to most problems in economics have market consequences, extra-market consequences, and intangible consequences. Since the intangible consequences of possible alternatives are left out of the numerical calculations, they should be introduced into the decision-making process at this point. The alternative to be chosen is the one that best meets the choice criterion after considering both the numerical consequences and the consequences not included in the monetary analysis.

During the decision-making process certain feasible alternatives are eliminated because they are dominated by other, better alternatives. For example, shopping for a computer on-line may allow you to buy a custom-configured computer for less money than a stock computer in a local store. Buying at the local store is feasible, but dominated. While eliminating dominated alternatives makes the decision-making process more efficient, there are dangers.

Having examined the structure of the decision-making process, we can ask, When is a decision made, and who makes it? If one person performs *all* the steps in decision making, then she is the decision maker. *When* she makes the decision is less clear. The selection of the feasible alternatives may be the key item, with the rest of the analysis a methodical process leading to the inevitable decision. We can see that the decision may be drastically affected, or even predetermined, by the way in which the decision-making process is carried out. This is illustrated by the following example.

Liz, a young engineer, was assigned to develop an analysis of additional equipment needed for the machine shop. The single criterion for selection was that the equipment should be the most economical, considering both initial costs and future operating costs. A little investigation by Liz revealed three practical alternatives:

1. A new specialized lathe
2. A new general-purpose lathe
3. A rebuilt lathe available from a used-equipment dealer

A preliminary analysis indicated that the rebuilt lathe would be the most economical. Liz did not like the idea of buying a rebuilt lathe, so she decided to discard that alternative. She prepared a two-alternative analysis that showed that the general-purpose lathe was more economical than the specialized lathe. She presented this completed analysis to her manager. The manager assumed that the two alternatives presented were the best of all feasible alternatives, and he approved Liz's recommendation.

At this point we should ask: Who was the decision maker, Liz or her manager? Although the manager signed his name at the bottom of the economic analysis worksheets to authorize purchasing the general-purpose lathe, he was merely authorizing what already had been made inevitable, and thus he was not the decision maker. Rather Liz had made the key decision when she decided to discard the most economical alternative from further consideration. The result was a decision to buy the better of the two *less economically desirable* alternatives.

9. Audit the Results

An audit of the results is a comparison of what happened against the predictions. Do the results of a decision analysis reasonably agree with its projections? If a new machine tool was purchased to save labor and improve quality, did it? If so, the economic analysis seems to be accurate. If the savings are not being obtained, what was overlooked? The audit may help ensure that projected operating advantages are ultimately obtained. On the other hand, the economic analysis projections may have been unduly optimistic. We want to know this, too, so that the mistakes that led to the inaccurate projection are not repeated. Finally, an effective way to promote *realistic* economic analysis calculations is for all people involved to know that there *will* be an audit of the results!

ETHICS

You must be mindful of the ethical dimensions of engineering economic analysis and of your engineering and personal decisions. This text can only introduce the topic, and we hope that you will explore this subject in greater depth.

Ethics can be described variously; however, a common thread is the concept of distinguishing between right and wrong in decision making. Ethics includes establishing systems of beliefs and moral obligations, defining values and fairness, and determining duty and guidelines for conduct. Ethics and ethical behavior are important because when people behave in ethical ways, individuals and society benefit. Usually the ethical choice is reasonably clear, but there are ethical dilemmas with conflicting moral imperatives. Consider an overloaded and sinking lifeboat. If one or more passengers are thrown into the shark-infested waters, the entire lifeboat can be saved. How is the decision made, how is it implemented, and who if anyone goes into the water? Ethical dilemmas also exist in engineering and business contexts. Ethical decision making requires the understanding of problem context, choices, and associated outcomes.

Ethical Dimensions in Engineering Decision Making

Ethical issues can arise at every stage of the integrated process for engineering decision making described in Figure 1–1. Ethics is such an important part of professional and business decision making that ethical codes or standards of conduct exist for professional engineering societies, small and large organizations, and every individual. Written professional codes are common in the engineering profession, serving as a reference basis for new engineers and a basis for legal action against engineers who violate the code.

One such example is the Code of Ethics of the National Society of Professional Engineers (NSPE). Here is NSPE's fundamental canon of ethical behavior for engineering: Engineers, in the fulfillment of their professional duties, shall:

- Hold paramount the safety, health and welfare of the public.
- Perform services only in areas of their competence.
- Issue public statements only in an objective and truthful manner.
- Act for each employer or client as faithful agents or trustees.
- Avoid deceptive acts.
- Conduct themselves honorably, responsibly, ethically and lawfully so as to enhance the honor, reputation, and usefulness of the profession.

In addition, NSPE has Rules of Practice and Professional Obligations for its members. Most engineering organizations have similar written standards. For all engineers difficulties arise when they act contrary to these written or internal codes, and opportunities for ethical dilemmas are found throughout the engineering decision-making process. Table 1–1 provides examples of ethical lapses that can occur at each step of the decision-making process.

TABLE 1–1 Example Ethical Lapses by Decision Process Step

Decision Process Step	Example Ethical Lapses
1. Recognize the problem	• "Looking the other way," that is, not to recognize the problem—due to bribes or perhaps fear of retribution for being a "whistle-blower"
2. Define the goal or objective	• Favoring one group of stakeholders by focusing on their objective for a project
3. Assemble relevant data	• Using faulty or inaccurate data
4. Identify feasible alternatives	• Leaving legitimate alternatives out of consideration
5. Select the criterion to determine the best alternative	• Considering only monetary consequences when there are other significant consequences
6. Construct a model	• Using a short horizon that favors one alternative over another
7. Predict each alternative's outcomes or consequences	• Using optimistic estimates for one alternative and pessimistic ones for the other alternatives
8. Choose the best alternative	• Choosing an inferior alternative, one that is unsafe, adds unnecessary cost for the end user, harms the environment, etc.
9. Audit the result	• Hiding past mistakes

Ethical dilemmas for engineers may arise in connection with engineering economic analysis in many situations. Following are examples of a few of these.

Gaining Knowledge and Building Trust Versus Favors for Influence

Consider these three situations:

- The salesman for a supplier of HVAC (heating, ventilating, and air conditioning) equipment invites a mechanical engineer and spouse to come along on the company jet for a users' conference at a vacation resort.
- Same salesman and same engineer, but the invitation is for a day of golfing at an exclusive club.
- Same salesman invites the same engineer to lunch.

In each case the salesman is trying to "get the order," and there is likely to be some mix of business—discussing specifications—and pleasure. The first case, which brings up the largest ethical questions, also has the largest business justification. This is the opportunity to meet other users of the products and see displays of the product line. Often, firms and government agencies have strict guidelines that dictate behavior in these situations.

Cost, Quality, and Functionality

One of the most common conflicts in the conceptual and design phase involves the trade-offs between cost, quality, and required functionality. Most modern products entail many thousands of decisions by designers that ultimately affect the cost and quality for the end user.

- A designer in an engineering consulting firm knows that a "gold-plated" solution would be very profitable for his firm (and for his bonus). This solution may also provide excellent reliability and require little maintenance cost.
- Engineers in the consumer durables division of a multinational company know that by using lower-quality connectors, fasteners, and subcomponents they can lower costs and improve the firm's market position. In addition, they know that these design elements have only a limited usable life, and the firm's most profitable business is repairs and extended warranties.

The Environment We Live In

Projects for transportation and power generation typically must consider environmental impacts in their design and in deciding whether the project should be done in any form. Who incurs the costs for the project, and who receives the benefits? Many other engineering products are designed to promote recycling, reduce energy usage, and reduce pollution. Ethical issues can be particularly difficult because there are often stakeholders with opposing viewpoints, and some of the data may be uncertain and hard to quantify.

Green engineering design includes the effects of environmental impacts and gives consideration to life-cycle sustainability issues. In this context, **societal costs** are the negative impacts of a project or product. Reducing these societal costs is the goal of environmental fees and regulation. For the opening vignette on electric vehicles, examples of the social costs of combustion-engine automobiles include tailpipe emissions and the negative environmental impact of mining, refining, and distributing gasoline/diesel fuels. Other examples of difficult choices include:

- Protecting the habitat of an endangered species versus flood control projects that protect people, animals, and structures.
- Meeting the needs for electrical power when all choices have some negative environmental impacts:
  - Hydroelectric—reservoir covers land and habitat
  - Coal—underground mining can be dangerous, open-pit mining damages habitat, and burning the coal can cause air pollution
  - Nuclear—disposal of radioactive waste
  - Fuel oil—air pollution and economic dependence
  - Wind—visual pollution of wind farms; birds killed by whirling blades
- Determining standards for pollutants: Is 1 part per million OK, or is 1 part per billion needed?

Safety and Cost

Some of the most common and most difficult ethical dilemmas involve trade-offs between safety and cost. If a product is "too safe," then it will be too expensive, and it will not be used. Also sometimes the cost is incurred by one party and the risk by another.

- Should the oil platform be designed for the 100-year, 500-year, or 1000-year hurricane?
- Should the auto manufacturer add run-flat tires, stability control, side-cushion airbags, and rear-seat airbags to every car?
- Should a given product design go through another month of testing?
- Are stainless steel valves required, or is it economically better to use less corrosion-resistant valves and replace them more frequently?

Emerging Issues and "Solutions"

Breaches of the law by corporate leaders of Enron, Tyco, and other firms have led to attempts to prevent, limit, and expose financial wrongdoing within corporations. One part of the solution has been the Sarbanes–Oxley Act of 2002, which imposed requirements on firm executives and auditing accounting firms, as well as penalties for violations.

Globalization is another area of increasing importance for ethical considerations. One reason is that different ethical expectations prevail in the world's various countries and regions. A second reason is that jobs may be moved to another country based on differences in cost, productivity, environmental standards, and so on. What may be viewed as a sweatshop from a U.S. perspective may be viewed as a wonderful opportunity to support many families from the perspective of a less developed nation.

Importance of Ethics in Engineering and Engineering Economy

Many times engineers and firms try to act ethically, but mistakes are made—the data were wrong, the design was changed, or the operating environment was different than expected. In other cases, a choice was made between expediency (profit) and ethics. For example, some engineers and managers within VW chose to manipulate diesel vehicle performance during emission testing. As of 2016, estimates of international costs to VW exceeded $38B. The firm and management are driven by the need to make a profit, and they expect the engineer to identify when safety will be compromised.

Ethics in engineering economic analysis focuses on how well and how honestly the decision-making process is conducted—the data, method of analysis, recommendations, and follow-up. The first step in avoiding problems is to recognize that ethical issues exist and to make them an explicit part of your decision-making process.

As a student, you've no doubt heard discussions about cheating on exams, plagiarism on written reports, violating university drinking and drug use policies, accepting one job while continuing to interview for others, and selling student sports tickets to nonstudents. You've made your own decisions about your behavior, and you've established patterns of behavior.

You should know that your professors care deeply about the ethical decisions you make at school. Your ethical habits there form a foundation for the character of your work and personal behavior after graduation.

Often recent engineering graduates are asked, "What is the most important thing you want from your supervisor?" The most common response is mentoring and opportunities to learn and progress. When employees with 5, 15, 25, or more years of experience are asked the same question, the most common response at all experience levels is *integrity*. This is what your subordinates, peers, and superiors will expect and value the most from you. Integrity is the foundation for long-term career success.

ENGINEERING DECISION MAKING FOR CURRENT COSTS

Some of the easiest forms of engineering decision making deal with problems of alternative *designs, methods,* or *materials*. If results of the decision occur in a very short period of time, one can quickly add up the costs and benefits for each alternative. Then, using the suitable economic criterion, the best alternative can be identified. Three example problems illustrate these situations.

EXAMPLE 1–2

A concrete aggregate mix must contain at least 31% sand by volume for proper batching. One source of material, which has 25% sand and 75% coarse aggregate, sells for $3 per cubic meter (m^3). Another source, which has 40% sand and 60% coarse aggregate, sells for $4.40/m^3. Determine the least cost per cubic meter of blended aggregates.

SOLUTION

The least cost of blended aggregates results from using just enough higher-cost material to meet the minimum 31% proportion of sand.

$$\text{Let } x = \text{Portion of blended aggregates from \$3.00/m}^3 \text{ source}$$
$$1 - x = \text{Portion of blended aggregates from \$4.40/m}^3 \text{ source}$$

Sand Balance

$$x(0.25) + (1 - x)(0.40) = 0.31$$
$$-.15x = -.09 \Rightarrow x = 0.60$$

The 60%/40% blended aggregate will cost

$$0.60(\$3.00) + 0.40(\$4.40) = 1.80 + 1.76 = \$3.56/m^3$$

EXAMPLE 1–3

A machine part is manufactured at a unit cost of 40¢ for material and 15¢ for direct labor. An investment of \$500,000 in tooling is required. The order calls for 3 million pieces. Halfway through the order, managers learn that a new method of manufacture can be put into effect that will reduce the unit costs to 34¢ for material and 10¢ for direct labor—but it will require \$100,000 for additional tooling. This tooling will not be useful for future orders. Other costs are allocated at 2.5 times the direct labor cost. What, if anything, should be done?

SOLUTION

Since there is only one way to handle the first 1.5 million pieces, our problem concerns only the second half of the order. While the arithmetic can easily be done on a calculator, in the real world problems like these are usually done using a spreadsheet. This allows easy substitution of "better" numbers for the initial estimates and supports **what-if analysis**. The first spreadsheet shows the data entry stage of the problem. These values form the problem's **data block** (see Appendix A). Note that we want a clear, compact table, so columns of these values are alternated with calculation columns for our two alternatives.

	A	B	C	D	E
1	1,500,000	Number of pieces			
2	2.5	Other cost $/direct labor $			
3		A: Present Method		B: New method	
4	Costs	unit	total	unit	total
5	Material	0.4		0.34	
6	Direct labor	0.15		0.1	
7	Other				
8	Added tooling			$100,000	

The second spreadsheet includes column F to show the formulas for the cells in column E. Note that the formulas in cells E5, E6, and C6 are all copied from cell C5. Because the C5 formula was originally written as =B5*$A$1, the **absolute address** of $A$1 does not change when copied. Note: Appendix A discusses how to efficiently do this and other examples of addressing alternatives that maximize the flexibility of copying formulas.

The most efficient way to create the formulas is to:

- Write the formula for C5 as = "click on B5" * "click on A1" "F4 or Apple T to toggle to $A$1"
- Copy it to C6
- Write the formula for C7 (including an absolute address)
- Copy C5:C7 to E5:E7

Select E5:E9 and click on the "sum" formula button. This can be copied to C9.

	A	B	C	D	E	F
1	1,500,000	Number of pieces				
2	2.5	Other cost $/direct labor $				
3		A: Present Method		B: New method		
4	Costs	unit	total	unit	total	
5	Material	0.4	$600,000	0.34	$510,000	=D5*$A$1
6	Direct labor	0.15	$225,000	0.1	$150,000	=D6*$A$1
7	Other		$562,500		$375,000	=E6*$A$2
8	Added tooling			$100,000	$100,000	
9	Total		$1,387,500		$1,135,000	
10				Possible savings	$252,500	

Looking at the results, we can see that much of the total $252,500 in savings comes from the reduced value of other costs. Thus, before making a final decision, one should closely examine the *other costs* to see whether they do, in fact, vary as the *direct labor cost* varies. Assuming they do, the decision would be to change the manufacturing method.

EXAMPLE 1–4

In the design of a cold-storage warehouse, the specifications call for a maximum heat transfer through the warehouse walls of 30,000 joules per hour per square meter of wall when there is a 30°C temperature difference between the inside surface and the outside surface of the insulation. The two insulation materials being considered are as follows:

Insulation Material	Cost per Cubic Meter	Conductivity $(J\text{-}m/m^2\text{-}°C\text{-}hr)$
Rock wool	$12.50	140
Foamed insulation	14.00	110

The basic equation for heat conduction through a wall is

$$Q = \frac{K(\Delta T)}{L}$$

where Q = heat transfer, in J/hr/m^2 of wall
  K = conductivity, in J-m/m^2-°C-hr
  ΔT = difference in temperature between the two surfaces, in °C
  L = thickness of insulating material, in meters

Which insulation material should be selected?

SOLUTION

Two steps are needed to solve the problem. First, the required thickness of each of the available materials must be calculated. Then, since the problem is one of providing a fixed output (heat transfer through the wall limited to a fixed maximum amount), the criterion is to minimize the input (cost).

Required Insulation Thickness

Rock wool	$30,000 = \dfrac{140(30)}{L}$,	$L = 0.14$ m
Foamed insulation	$30,000 = \dfrac{110(30)}{L}$,	$L = 0.11$ m

Cost of Insulation per Square Meter of Wall

Unit cost = Cost/m^3 × Insulation thickness, in meters

Rock wool   Unit cost = $12.50 × 0.14 m = $1.75/m^2

Foamed insulation   Unit cost = $14.00 × 0.11 m = $1.54/m^2

The foamed insulation is the lesser-cost alternative. However, there is a constraint that must be considered. How thick is the available wall space?

This constraint suggests a better problem definition—how should the available wall thickness be used? A better decision criterion that looks at total costs (not just insulation costs) requires engineering economy and the time value of money to decide what the maximum heat transfer should be. What is the cost of more insulation versus the cost of cooling the warehouse over its life?

From a broader perspective energy use and emissions are both environmental issues. Thus any government regulations on insulation standards should consider costs and benefits from a societal perspective, and not just from the firm's.

WHEN MORE THAN ECONOMICS IS INVOLVED

Consider the moderately complex problem of which job offer to accept. Example 1–5 shows a simple way to address this **multiple-objective** problem. These models should:

- Include all important objectives.
- Weight the relative importance of the objectives.
- Select an objective and rate all alternatives. Then repeat for all objectives.
- Disqualify alternatives that do not meet the minimum performance requirements of one or more objectives.

This example uses simple 0 to 10 rating scales. Since the weights are stated as percentages (or their decimal equivalents), the totals show how close to a perfect 10 each alternative is.

Multi-objective models do much more than calculate a measure of each alternative's attractiveness. Constructing the model enforces a level of clarity about the importance of each objective and how each alternative performs. The model also communicates those assumptions and estimates to others, who may suggest changes. Since there may be multiple iterations in arriving at the final model, spreadsheets are particularly effective here.

Examples in later chapters will show how to convert numeric values to a 0 to 10 point scale. For those who want to search the web for additional examples, this is an *additive* model, because the scores are added together. This is also a *compensatory* model, because strength on one objective can compensate for a weakness on another objective.

Example 1–5 is linked to an individual's financial and life decision making. But this situation can also be viewed from the firm's or government agency's perspective. Which applicant(s) should receive offer(s) of employment? In that case, evaluations from multiple individuals might be combined for the overall total.

EXAMPLE 1–5

A senior undergraduate has received four job offers, but the salary on one is unacceptably low. The other three offers have been rated on three criteria or objectives, with a scale of 0 = barely acceptable and 10 = outstanding! *Job* considers the salary relative to the local cost of housing and the job itself. The latter was hard to estimate because it considered the initial job, growth prospects, the firm, and the industry. *Family* is important to this senior, but the senior wanted to live the right distance away—neither too close nor too far. *Livability* covers the senior's desires on community size, climate, commuting time, and overall political balance. The senior weighted the importance of the three criteria at 50%, 30%, and 20% respectively. Given the following table of ratings, which job offer should the senior accept?

Offer	Job	Family	Livability
A	4	9	5
B	8	5	4
C	6	3	8

SOLUTION

None of the job offers is ideal in any respect, and each has some aspect that is less attractive than the other offers. Comparing the total values, offer B is the most attractive. This table is the result of many hours of thinking, and more model iterations would not be useful. Thus offer B should be accepted.

There are many ways to write the formula, but the easiest uses the function SUMPRODUCT. As shown, the function uses a fixed address for the weights, so the formula for offer A can be copied for the other offers.

	A	B	C	D	E
1		Job	Family	Livability	
2	Weight	50%	30%	20%	
3	Offer				Total
4	A	4	9	5	5.7
5	B	8	5	4	6.3
6	C	6	3	8	5.5
7					
8			=SUMPRODUCT($B$2:$D$2,B4:D4)		

SUMMARY

Classifying Problems

Many problems are simple and thus easy to solve. Others are of intermediate difficulty and need considerable thought and/or calculation to properly evaluate. These intermediate problems tend to have a substantial economic component and to require economic analysis. Complex problems, on the other hand, often contain people elements, along with political and economic components. Economic analysis is still very important, but the best alternative must be selected by considering all criteria—not just economics.

The Decision-Making Process

Rational decision making uses a logical method of analysis to select the best alternative from among the feasible alternatives. The following nine steps can be followed sequentially, but decision makers often repeat some steps, undertake some simultaneously, and skip others altogether.

1. Recognize the problem.
2. Define the goal or objective: What is the task?
3. Assemble relevant data: What are the facts? Is more data needed, and is it worth more than the cost to obtain it?
4. Identify feasible alternatives.

5. Select the criterion for choosing the best alternative: possible criteria include political, economic, environmental, and social. The single criterion may be a composite of several different criteria.

6. *Mathematically model* the various interrelationships.

7. Predict the outcomes for each alternative.

8. Choose the best alternative.

9. Audit the results.

Engineering decision making refers to solving substantial engineering problems in which economic aspects dominate and economic efficiency is the criterion for choosing from among possible alternatives. It is a particular case of the general decision-making process. Some of the unusual aspects of engineering decision making are as follows:

1. Cost-accounting systems, while an important source of cost data, contain allocations of indirect costs that may be inappropriate for use in economic analysis.

2. The various consequences—costs and benefits—of an alternative may be of three types:

   (a) Market consequences—there are established market prices.

   (b) Extra-market consequences—there are no direct market prices, but prices can be assigned by indirect means.

   (c) Intangible consequences—valued by judgment, not by monetary prices.

3. The economic criteria for judging alternatives can be reduced to three cases:

   (a) When neither input nor output is fixed: maximize profit, which equals the difference between benefits and costs.

   (b) For fixed input: maximize benefits or other outputs.

   (c) For fixed output: minimize costs or other inputs.

   The first case states the general rule from which both the second and third cases may be derived.

4. To choose among the alternatives, the market consequences and extra-market consequences are organized into a cash flow diagram. We will see in Chapter 3 that engineering economic calculations can be used to compare differing cash flows. These outcomes are compared against the selection criterion. From this comparison *plus* the consequences not included in the monetary analysis, the best alternative is selected.

5. An essential part of engineering decision making is the postaudit of results. This step helps to ensure that projected benefits are obtained and to encourage realistic estimates in analyses.

Importance of Ethics in Engineering and Engineering Economy

One of the gravest responsibilities of an engineer is protecting the safety of the public, clients, and/or employees. In addition, the engineer can be responsible for the economic performance of projects and products on which bonuses and jobs depend. Not surprisingly, in this environment one of the most valued personal characteristics is integrity.

Decision Making with Current Costs

When all costs and benefits occur within a brief period of time, the time value of money is not a consideration. We still must use the criteria of maximizing profit, minimizing cost, or maximizing benefits.

PROBLEMS

Key to icons: **A** = Answer in Appendix E; **G** = Green, which may include environmental ethics; **E** = Ethics other than green.

Many end-of-chapter problems are primarily numerical, but others require more discussion—especially the case studies and questions linked to ethics. Section C in Chapter 2 of *Cases in Engineering Economy* 2nd on the student website may be helpful for the more discussion-oriented questions.

Decision Making

1-1   Think back over your past academic year and decisions that you made. List a few decisions that you would classify as simple, intermediate, and complex. What did you learn about your decision making by the way you approached these decisions?

1-2   Some of the following problems would be suitable for solution by engineering economic analysis. Which ones are they?

(*a*) Would it be better to buy a hybrid car?

(*b*) Should an automatic machine be purchased to replace three workers now doing a task by hand?

(*c*) Would it be wise to enroll for an early morning class to avoid traveling during the morning traffic rush hours and thus improve fuel efficiency?

(*d*) Would you be better off if you changed your major?

(*e*) Should you work more and borrow less even if it delays your graduation?

(*f*) Should a corporate farm build waste mitigation ponds or continue using a contracted service?

1-3   Which one of the following problems is *most* suitable for analysis by engineering economic analysis?

(*a*) One of your two favorite sandwich shops offers a 10-punch loyalty card and the other does not. Where should you stop today?

(*b*) A woman has $150,000 in a bank checking account that pays no interest. She can either invest it immediately at a desirable interest rate

or wait a week and know that she will be able to obtain an interest rate that is 0.15% higher.

(*c*) Joe backed his car into a tree, damaging the fender. He has car insurance that will pay for the fender repair. But if he files a claim for payment, they may charge him more for car insurance in the future.

1-4   If you have $1000 and could make the right decisions, how long would it take you to become a millionaire? Explain briefly what you would do.

1-5   One can find books on "How I Made My Millions" in any bookstore. In some cases the authors seem to plan to make millions by selling that book. Do you think this is ethical? How would you lay out the factors to analyze this question?

1-6   The owner of a small machine shop has just lost one of his larger customers. The solution to his problem, he says, is to fire three machinists to balance his workforce with his current level of business. The owner says it is a simple problem with a simple solution.

(*a*) The three machinists disagree. Why?

(*b*) What are the ethical factors from the perspective of the owner and the workers?

1-7   Designing a chair for use in a classroom seems like a simple task. Make an argument for how this can be considered a complex decision and include environmental and ethical factors in your argument.

1-8   Toward the end of the twentieth century, the U.S. government wanted to save money by closing a small portion of its domestic military installations. While many people agreed that saving money was a desirable goal, people in areas potentially affected by a closing soon reacted negatively. Congress finally selected a panel whose task was to develop a list of installations to close, with the legislation specifying that Congress could not alter the list. Since the goal was to save money, why was this problem so hard to solve?

1-9 The college bookstore has put pads of engineering computation paper on sale at half price. What is the minimum and maximum number of pads you might buy during the sale? Explain.

1-10 Consider the five situations described. Which one situation seems most suitable for solution by economic analysis?

(a) John has met two college students that interest him. Beth is a music major who is lots of fun to be with. Alice is a fellow engineering student, but she does not like to party. John wonders what to do.

(b) You drive periodically to the post office to send or pick up packages. The parking meters cost $1 for 15 minutes—about the time required for medium length lines. If parking fines cost $20, do you put money in the meter or not?

(c) The cost of car insurance varies widely from company to company. Should you check with several insurance companies when your policy comes up for renewal?

(d) There is a special local sales tax ("sin tax") on a variety of things that the town council would like to remove from local distribution. As a result, a store has opened up just outside the town and offers an abundance of these specific items at prices about 30% less than is charged in town. Should you shop there?

(e) One of your professors mentioned that you have a poor attendance record in her class. You wonder whether to drop the course now or wait to see how you do on the first midterm exam. Unfortunately, the course is required for graduation.

1-11 A car manufacturer is considering locating an assembly plant in your region.
G

(a) List two simple, two intermediate, and two complex problems associated with this proposal.

(b) What is NIMBY? Does this come into play for this complex decision?

1-12 Consider the following situations. Which ones appear to represent rational decision making? Explain.

(a) Joe's best friend has decided to become a civil engineer, so Joe has decided that he will also become a civil engineer.

(b) Jill needs to get to the university from her home. She bought a car and now drives to the university each day. When Jim asks her why she didn't buy a bicycle instead, she replies, "Gee, I never thought of that."

(c) Don needed a wrench to replace the spark plugs in his car. He went to the local automobile supply store and bought the cheapest one they had. It broke before he had finished replacing all the spark plugs in his car.

1-13 Identify possible objectives for NASA. For your favorite of these, how should alternative plans to achieve the objective be evaluated?

1-14 Suppose you have just 2 hours to determine how many students would be interested in a highway trash pickup event. Give a step-by-step outline of how you would proceed.
E

1-15 A college student determines he will have only half of the cost for university housing available for the coming year. List five feasible alternatives.

1-16 Think about the issue of implementing renewable energies in the U.S. Research/find an instance where a decision was made to implement without adequately looking at other potential alternative solutions.
E

1-17 If there are only two alternatives available and both are unpleasant and undesirable, what should you do?

1-18 The three economic criteria for choosing the best alternative are maximize the difference between output and input, minimize input, and maximize output. For each of the following situations, what is the correct economic criterion?
A

(a) A manufacturer can sell up to two full shifts of production at a fixed price. As production is increased, unit costs increase as a result of overtime pay and so forth. The manufacturer's criterion should be _____.

(b) An architectural and engineering firm has been awarded the contract to design a wharf with fixed performance specifications for a petroleum company. The engineering firm's criterion for its client should be _____.

(c) An off-campus bookstore is choosing its target used/new split for next year. Its criterion should be _____

(d) At an auction of antiques, a bidder for a particular porcelain statue would be trying to _____.

1-19 As in Problem 1-18, state the correct economic criterion for each of the following situations.

(a) The engineering student club raffled off a donated car; tickets sold for $5 each or three for $10. When the students were selling tickets, they noted that many people had trouble deciding whether to buy one or three tickets. This indicates the buyers' criterion was _____.

(b) A student organization bought a soft-drink machine and then had to decide whether to charge 75¢, $1, or $1.25 per drink. The organization recognized that the number of soft drinks sold would depend on the price charged. Eventually the decision was made to charge $1. The criterion was _____.

(c) In many cities, grocery stores find that their sales are much greater on days when they advertise special bargains. However, the advertised special prices do not appear to increase the total physical volume of groceries sold by a store. This leads us to conclude that many shoppers' criterion is _____.

(d) A recently graduated engineer has decided to return to school in the evenings to obtain a master's degree. He feels it should be accomplished in a manner that will allow him the maximum amount of time for his regular day job plus time for recreation. In working for the degree, he will _____.

1-20 Seven criteria are given in the chapter for judging which is the best alternative. After reviewing the list, devise three additional criteria that might be used.

1-21 Suppose you are assigned the task of determining the route of a new highway through an older section of town. The highway will require that many older homes be either relocated or torn down. Two possible criteria that might be used in deciding exactly where to locate the highway are:

(a) Ensure that there are benefits to those who gain from the decision and that no one is harmed by the decision.

(b) Ensure that the benefits to those who gain from the decision are greater than the losses of those who are harmed by the decision.

Which criterion will you select to use in determining the route of the highway? Explain.

1-22 For the project in Problem 1-21, identify the major costs and benefits. Which are market consequences, which are extra-market consequences, and which are intangible consequences?

1-23 You must fly to another city for a Friday meeting. If you stay until Sunday morning your ticket will be $250, rather than $800. Hotel costs are $200 per night. Compare the economics with reasonable assumptions for meal expenses. What intangible consequences may dominate the decision?

1-24 In the fall, Jay Thompson decided to live in a university dormitory. He signed a dorm contract under which he was obligated to pay the room rent for the full college year. One clause stated that if he moved out during the year, he could sell his dorm contract to another student who would move into the dormitory as his replacement. The dorm cost was $5000 for the two semesters, which Jay had already paid.

A month after he moved into the dorm, he decided he would prefer to live in an apartment. That week, after some searching for a replacement to fulfill his dorm contract, Jay had two offers. One student offered to move in immediately and to pay Jay $300 per month for the eight remaining months of the school year. A second student offered to move in the second semester and pay $2500 to Jay.

Jay estimates his food cost per month is $500 if he lives in the dorm and $450 if he lives in an apartment with three other students. His share of the apartment rent and utilities will be $400 per month. Assume each semester is $4\frac{1}{2}$ months long. Disregard the small differences in the timing of the disbursements or receipts.

(a) What are the three alternatives available to Jay?

(b) Evaluate the cost for each of the alternatives.

(c) What do you recommend that Jay do?

1-25 An electric motor on a conveyor burned out. The foreman told the plant manager that the motor had to be replaced. The foreman said that there were no alternatives and asked for authorization to order the replacement. In this situation, is any decision making taking place? If so, who is making the decision(s)?

1-26 Ⓖ A farmer must decide what combination of seed, water, fertilizer, and pest control will be most profitable and environmentally conscious for the coming year. The local agricultural college did a study of this farmer's situation and prepared the following table.

Plan	Direct Cost/Acre	Extra-market Cost/Acre	Income/Acre
A	$750	$150	$1200
B	800	450	1400
C	1000	250	1500
D	1300	200	1650

The last page of the college's study was torn off, and hence the farmer is not sure which plan the agricultural college recommends. Which plan should the farmer adopt considering:

(a) only the direct costs,
(b) both the direct and extra-market costs?

1-27 Identify the alternatives, outcomes, criteria, and process for the selection of your college major. Did you make the best choice for you?

1-28 Describe a major problem you must address in the next two years. Use the techniques of this chapter to structure the problem and recommend a decision.

1-29 Apply the steps of the decision-making process from this chapter and develop plans to achieve one each of your 5-year, 10-year, and 25-year goals.

1-30 One strategy for solving a complex problem is to break it into a group of less complex problems and then find solutions to the smaller problems. The result is the solution of the complex problem. Give an example in which this strategy will work. Then give another example in which this strategy will not work.

Ethics

1-31 **E** When you make professional decisions involving investments in engineering projects, what criteria will you use?
Contributed by D. P. Loucks, Cornell University

1-32 **E** What are ethics?
Contributed by D. P. Loucks, Cornell University

1-33 **E** A student accepts a full-time job in November, but a better job comes before graduation in May. What are the ethical dimensions of the student's decision? Would you take the better job? Why or why not?

1-34 **E** Suppose you are an engineer working in a private engineering firm and you are asked to sign

documents verifying information that you believe is not true. You like your work and your colleagues in the firm, and your family depends on your income. What criteria can you use to guide your decision regarding this issue?
Contributed by D. P. Loucks, Cornell University

1-35 **E** Find the ethics code for the professional society of your major.

(a) Summarize its key points.
(b) What are its similarities and differences in comparison to NSPE's ethics code?

1-36 **E** Use a personal example or a published source to analyze what went wrong or right with respect to ethics at the assigned stage(s) of the decision-making process.

(a) Recognize problem.
(b) Define the goal or objective.
(c) Assemble relevant data.
(d) Identify feasible alternatives.
(e) Select the criterion for determining the best alternative.
(f) Construct a model.
(g) Predict each alternative's outcomes or consequences.
(h) Choose the best alternative.
(i) Audit the result.

For problems 1–37 to 1–49:

(a) What ethical issues can arise—personal, business, and/or environmental?
(b) Use local, state, national, or international news sources to identify an example situation.
(c) Summarize and analyze the ethical issues, including relevant laws, regulations, codes, and processes.

1-37 **G** Municipal assemblies, school boards, transit boards, and municipal utility boards are responsible for public infrastructure, such as roads and schools. Especially for this responsibility, engineers bring skills, knowledge, and perspectives that can improve public decision making. Often the public role is a part-time one; engineers that fulfill it will also have full-time jobs as employees or owners of engineering firms.

1-38 **G** Increasing population and congestion often are addressed through road improvement projects. These may pit the interests of homeowners and business owners in the project area against the interests of people traveling through the improvement project and environmental activists.

1-39 Stadiums for professional sports teams often involve
E some level of municipal support. Some businesses and home owners benefit, while others do not; some pay more in taxes, while others pay less.

1-40 Economic development and redevelopment often
E require significant acreage that is assembled by acquiring smaller parcels. Sometimes this is done through simple purchase, but the property of an "unwilling seller" can be acquired through the process of eminent domain.

1-41 State governments use a variety of advisory and
E regulatory bodies. Example responsibilities include oversight of professional engineering licensing and the pricing and operation of regulated utilities. Often the public role is a part-time one, and engineers that fulfill it will also have full-time jobs as employees or owners of engineering firms.

1-42 Many engineers work in state governments, and
E some are in high-profile roles as legislators, department commissioners, and so on. Many of these individuals move between working in the private and public sectors.

1-43 In the U.S., regulation of payment for overtime hours
E is done at the state and federal levels. Because most engineering work is accomplished through projects, it is common for engineers to be asked or required to work overtime as projects near deadlines. Sometimes the overtime is paid at time and a half, sometimes as straight time, and sometimes the engineer's salary is treated as a constant even when overtime occurs. In a particular firm, engineering interns, engineers, and partners may be treated the same or differently.

1-44 At the federal government level, the economic con-
E sequences of decisions can be very large. Firms hire lobbyists, legislators may focus on their constituents, and advocacy organizations promote their own agendas. In addition, sometimes some of the players are willing to be unethical.

1-45 At both state and federal levels, legislators can
E be involved in "pork barrel" funding of capital projects. These projects may even bypass the economic evaluation using engineering economy that normal projects are subject to.

1-46 At the international level, a common ethics issue
G important to engineering and project justification is that of environmental regulation. Often different nations have different environmental standards, and a project or product might be built in either location.

1-47 At the international level, a common ethics issue
E important to engineering and project justification is that of worker health and safety. Often different nations have different standards, and a project or product could be built in either location.

1-48 At the international level, engineering decisions are
G critical in matters of "sustainable development," a common ethics issue.

1-49 At the international level, questions arise about
E whether the U.S. ban on bribery is practical or appropriate. In some countries government workers are very poorly paid, and they can support their families only by accepting money to "grease" a process.

1-50 In the 1970s the Ford Motor Company sold its subcompact Pinto model with known design defects. In particular, the gas tank's design and location led to rupture, leaks, and explosion in low-speed, rear-impact collisions. Fifty-nine people burned to death in Pinto accidents. In a cost–benefit analysis weighing the cost of fixing the defects ($11 per vehicle) versus the firm's potential liability for lawsuits on behalf of accident victims, Ford had placed the value of a human life at $200,000. Ford eventually recalled 1.4 million Pintos to fix the gas tank problem for a cost of $30 million to $40 million. In addition the automaker ultimately paid out millions more in liability settlements and incurred substantial damage to its reputation.

(*a*) Critique Ford's actions from the perspective of the NSPE Code of Ethics.

(*b*) One well-known ethical theory, utilitarianism, suggests that an act is ethically justified if it results in the "greatest good for the greatest number" when all relevant stakeholders are considered. Did Ford's cost–benefit analysis validly apply this theory?

(*c*) What should engineers do when the product they are designing has a known safety defect with an inexpensive remedy?

Contributed by Joseph R. Herkert, North Carolina State University

1-51 The decision-making process used to launch the Challenger shuttle has been extensively analyzed. Briefly summarize the key institutional groups, how the decision was made, and the ethical principles that may have been compromised.

1-52 One of the elements in the flooding of New Orleans during Hurricane Katrina was the failure of some of the levees that protected the city. Outline the role that ethical failures by engineers may have played in this situation. How could society structure decision making to minimize such failures?

1-53
G Hurricane Sandy's flooding of New York City high-lighted the vulnerability of coastal cities to extreme weather events, which are becoming more common. Strengthening and protecting infrastructure and the environment before the fact can be very expensive—and perhaps never needed. The possible availability of after-the-fact disaster aid can distort economic perspectives. Why is minimizing economic, environmental, and human costs related to extreme weather such a difficult problem for public infrastructure?

Current Costs

1-54
A A manufacturing firm has received a contract to assemble 1000 units of test equipment in the next year. The firm must decide how to organize its assembly operation. Skilled workers, at $33 per hour each, can individually assemble the test equipment in 2.6 hours per unit. Alternatively, teams of four less skilled workers (at $19 per hour each) can assemble a unit in one hour. Which approach is more economical?

1-55
G Two manufacturing firm, located in cities 90 miles apart, both send their trucks four times a week to the other city full of cargo and return empty. Each driver costs $275 per day with benefits (the round trip takes all day) and each firm has truck operating costs of $1.20 a mile.

(*a*) How much could each firm save weekly if each sent its truck twice a week and hauled the other firm's cargo on the return trip?

(*b*) What would the savings be if there was a $0.20 per mile emissions tax on all business truck travel?

1-56
A
G An oil company is considering adding a more envi-ronmentally friendly grade of fuel at its service stations. To do this, an additional 3000-gallon tank must be buried at each station. Discussions with tank fabricators indicate that the least expensive tank would be cylindrical with minimum surface area. What size tank should be ordered?

1-57 Cathy Gwynn for a class project is analyzing a "Quick Shop" grocery store. The store emphasizes quick service, a limited assortment of grocery items, and higher prices. Cathy wants to see if the store hours (currently 0600 to 0100) can be changed to make the store more profitable.

Time Period	Daily Sales in the Time Period
0600–0700	$ 40
0700–0800	80
0800–0900	120
0900–1200	400
1200–1500	500
1500–1800	600
1800–2100	800
2100–2200	200
2200–2300	60
2300–2400	120
2400–0100	40

The cost of the groceries sold averages 70% of sales. The incremental cost to keep the store open, including the clerk's wage and other operat-ing costs, is $22 per hour. To maximize profit, when should the store be opened, and when should it be closed?

1-58
A Willie Lohmann travels from city to city for busi-ness. Every other year he buys a used car for about $15,000. The dealer allows about $8000 as a trade-in allowance, so Willie spends $7000 every other year for a car. Willie keeps accurate records of his expenses, which total 32.3¢ per mile. Willie's employer has two plans to reimburse car expenses:

A. Actual expenses: Willie will receive all his oper-ating expenses, and $3500 each year for the car's decline in value.

B. Standard mileage rate: Willie will receive 56.5¢ per mile but no operating expenses and no depre-ciation allowance.

If Willie travels 18,000 miles per year, which method gives him the larger reimbursement? At what annual mileage do the two methods give the same reimbursement?

1-59 If you rent a car, you can (1) return it with a full gas tank, (2) return it without filling it and pay $5.45/gallon, or (3) accept a fixed price of $50 for gas. The local price is $3.95/gallon for gasoline, and you expect this car to get 28 miles per gallon. The car has a 16-gallon tank. What choice should you make if you expect to drive:

(*a*) 150 miles?

(*b*) 300 miles?

(*c*) 400 miles?

(*d*) How do your answers change if stopping at the filling station takes 15 minutes and your time is worth $12/hr?

1-60 Your car gets 24 miles per gallon (mpg) at 60 miles
Ⓐ per hour (mph) and 20 mpg at 70 mph. At what speed should you make a 500-mile trip:

(*a*) If gas costs $3 per gallon and your time is worth $18/hr?

(*b*) If gas costs $4 per gallon and your time is worth $12/hr?

(*c*) If gas costs $5 per gallon and your time is worth $9/hr?

(*d*) Build a spreadsheet (see Appendix A) to calculate the total trip cost for gas costs of $2, $3, $4, and $5 and values of time of $6, $9, $12, $15, and $18 per hour. Do two tables: one at 60 mph and one at 70 mph.

1-61 A city needs to choose area rubbish disposal
Ⓖ areas.

Area A: A gravel pit has a capacity of 16 million cubic meters. Owing to the possibility of high groundwater the Regional Water Pollution Control Board has restricted the lower 2 million cubic meters of fill to inert material only (earth, concrete, asphalt, paving, brick, etc.). This must be purchased and hauled to this area for the bottom fill.

Area B: Capacity is 14 million cubic meters. For 20% of the city, the haul is the same distance as for Area A. The round-trip haul is 5 miles longer for 60% of the city, and 2 miles shorter for 20% of the city.

Assume the following conditions:

- Cost of inert material placed in Area A will be $9.40/m^3.
- Average speed of trucks from last pickup to disposal site is 25 miles per hour.
- The rubbish truck and a two-man crew will cost $210 per hour.
- Truck capacity of 4$^1/_2$ tons per load or 20 m^3.
- Sufficient cover material is available at all areas.

Which of the sites do you recommend?

1-62 A firm is planning to manufacture a new product. As
Ⓐ the selling price is increased, the quantity that can be sold decreases. Numerically the sales department estimates:

$$P = \$350 - 0.2Q$$

where $P =$ selling price per unit

$Q =$ quantity sold per year

On the other hand, management estimates that the average unit cost of manufacturing and selling the product will decrease as the quantity sold increases. They estimate

$$C = \$40Q + \$20,000$$

where $C =$ cost to produce and sell Q per year

The firm's management wishes to maximize profit. What quantity should the decision makers plan to produce and sell each year and what profit will be earned?

1-63 The vegetable buyer for a group of grocery stores has decided to sell packages of sprouted grain in the vegetable section of the stores. The product is perishable, and any remaining unsold after one week in the store is discarded. The supplier will deliver the packages to the stores, arrange them in the display space, and remove and dispose of any old packages. The price the supplier will charge the stores depends on the size of the total weekly order for all the stores.

Weekly Order	Price per Package
Less than 1000 packages	70¢
1000–1499	56
1500–1999	50
2000 or more	46

The vegetable buyer estimates the quantity that can be sold per week, at various selling prices, as follows:

Selling Price	Packages Sold per Week
$1.20	300
.90	600
.80	1200
.66	1800
.52	2300

The sprouted grain will be sold at the same price in all the grocery stores.

(a) How many packages should be purchased per week, and at which of the five prices listed above should they be sold?

(b) Build a spreadsheet (see Appendix A) to calculate the profit for every combination of selling price and weekly order size.

1-64 Jim Jones, a motel owner, noticed that just down the street the "Motel 46" advertises a $46-per-night room rental rate on its sign. As a result, this competitor has rented all 80 rooms every day by late afternoon. Jim, on the other hand, does not advertise his rate, which is $64 per night, and he averages only a 68% occupancy of his 50 rooms.

There are a lot of other motels nearby, but only Motel 46 advertises its rate on its sign. (Rates at the other motels vary from $48 to $99 per night.) Jim estimates that his actual incremental cost per night for each room rented, rather than remaining vacant, is $12. This $12 pays for all the cleaning, laundering, maintenance, utilities, and so on. Jim believes his eight alternatives are:

Alternative		Resulting Occupancy Rate
	Advertise and Charge	
1	$45 per night	100%
2	52 per night	94
3	58 per night	80
4	64 per night	66
	Do Not Advertise and Charge	
5	$58 per night	70%
6	64 per night	68
7	72 per night	66
8	78 per night	56

What should Jim do? Show how you reached your conclusion.

1-65 A grower estimates that if he picks his apple crop now, he will obtain 1000 boxes of apples, which he can sell at $30 per box. However, he thinks his crop will increase by 120 boxes of apples for each week he delays picking, but that the price will drop at a rate of $1.50 per box per week; in addition, he estimates that approximately 20 boxes per week will spoil for each week he delays picking.

(a) When should he pick his crop to obtain the largest total cash return? How much will he receive for his crop at that time?

(b) Build a spreadsheet (see Appendix A) to calculate the profit for 0, 1, 2, ..., 6 weeks.

1-66 On her first engineering job, Joy Hayes was given the responsibility of determining the production rate for a new product. She has assembled the data presented in the graphs. Note that costs are in $1000s.

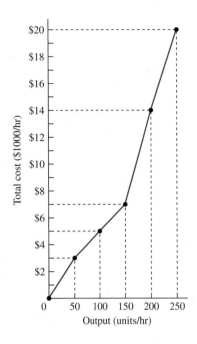

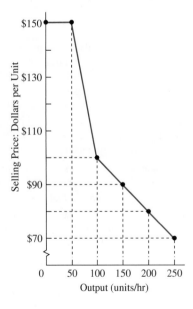

(a) Select an appropriate economic criterion and estimate the production rate based upon it.

(b) Joy's boss told Joy: "I want you to maximize output with minimum input." Joy wonders if it is possible to meet her boss's criterion. What would you tell her?

Multiple Objectives

1-67 Use the data in Example 1–5.

(a) What is the total score for each offer if the three objectives have the same weight?

(b) Holding livability's weight constant, how important does family have to be for offer A to be the best choice? Remember that the weights must sum to 1.

(c) Holding family's weight constant, how important does livability have to be for offer C to be the best choice?

1-68 A graduating senior has been accepted by three universities for an M.S. in engineering. Two criteria have been identified. The first is the program and university's academic ranking. The second is the cost. A third criteria of location was initially considered, but then the student recognized that it is only for about a year, and applications were only made to acceptable schools. The student is currently enrolled in the first university, which is rated as a 5 for academic rank and a 10 for cost. The second is a larger out-of-state public university, which is rated as an 8 for academic rank and a 6 for cost. The third is a prestigious private school, which is rated as a 10 academically and a 3 for its higher cost.

(a) What is the total score for each school if the two objectives have the same weight?

(b) If academic rank is has a weight of 75%, what is the total score for each school?

Minicases

1-69 Pick a decision involving multiple objectives that you must make. Estimate each objective's weighted importance, rate the alternative choices on each objective, and develop the totals for a model like Example 1–5.

1-70 Green engineering is a design construct that explic-
ⓖ itly considers environmental and sustainability factors within the design process. It seeks to promote
responsible use of limited resources, and to produce environmentally ethical and safe engineering
products, goods, and services. Using the web, find
two lists of principles that have been suggested by
different groups.

(a) Write a short paragraph on each list of principles
that describes the who, when, and why of their
formation.

(b) Compare and contrast the two lists. Which do
you think is best and why? Is there anything that
you see is missing from both lists?

CASES

The following cases from *Cases in Engineering Economy* **(www.oup.com/us/newnan) are suggested as matched with this chapter.**

CASE 1 **New Office Equipment**
Student develops requirements for new
copier machines with little given data.
Organizational thinking required.

CASE 2 **Budgeting Issues**
Strategies for a group's operating budget
request. One focus is ethics.

ESTIMATING ENGINEERING COSTS AND BENEFITS

LightTUBe

The Tullahoma Utilities Board (TUB) installed a $3.1 million Automated Metering Information System in its full service area in Tennessee. The process uses special meters, each equipped with a radio transmitter. They collect information on water and electric usage at residences and businesses, and then forward the information to collectors mounted on nearby poles. The usage information is then relayed using LightTUBe, the utility's fiber optics network, eliminating the need for employees to physically read utility meters. It provides not only real-time monitoring capability for residential and commercial users but also the ability to monitor the entire system's health and loads.

The change to the automated system allows TUB customers to better understand and actively manage their consumption patterns. Customers are able to access a portal to examine their use habits and reduce their bills by doing household tasks that require more electricity during off-peak hours. The new system also reduces personnel costs, provides better information on leak detection, outage management, and any theft of service.

After the system was economically justified and built for automated metering, TUB found that there was significant unused bandwidth on the fiber optic network. It now operates the LightTUBe network throughout the City of Tullahoma to provide Fiber to the Premise (FTTP), producing high quality video, high speed Internet, and telephone services. Light-TUBe provides voice, video, and data services over 250 miles of Fiber Optic cable. TUB serves nearly 10,000 residential, commercial, and industrial customers.

This "hybrid" business model crosses traditional boundaries associated with public utilities, phone companies, and Internet/cable TV providers. Traditionally, public utilities

operate with a set profit margin, which allows them to minimize the cost to the customer while maintaining operational and future growth requirements. Most cable TV, Internet, and phone companies operate on a private sector business model in which profits are maximized. ■ ■ ■

Contributed by James Simonton, University of Tennessee Space Institute

QUESTIONS TO CONSIDER

1. How do you estimate the cost—and ultimately the benefit—of such a "hybrid" operation with the traditional utility segment of the business (water, electricity, and sewage) working to minimize cost, while the LighTUBe segment works to maximize profit?

2. Discuss the ethical issues related to using public funds for an operation that could be seen as a for profit venture that is in competition with private sector companies.

3. In general, private sector project cost estimating may seem the same as public sector estimating. The real difference in this case would be the ability of TUB to consider both what it collects and consumer surplus value. How would a project that blurs the line between the two be handled?

4. When performing an economic analysis of "hybrid" projects, what tools would be appropriate to establish the benefit for justification purposes?

After Completing This Chapter...

The student should be able to:

- Define and use costs and benefits of many types including: average, external, fixed, incremental, internal, marginal, opportunity, sunk, and variable costs and benefits.
- Provide specific examples of how and why these concepts are important.
- Define engineering estimating for costs and benefits.
- Explain the three types of engineering estimate, as well as common difficulties encountered in making engineering estimates.
- Use several common mathematical estimating models in estimating costs and benefits.
- Discuss the impact of the *learning curve* on estimates.
- State the relationship between cost estimating and estimating project benefits.
- Draw *cash flow diagrams* to show project costs and benefits.

Key Words

average cost	incremental cost	recurring cost
book cost	internal cost	rough estimate
breakeven	learning curve	segmenting model
cash cost	life-cycle cost	semidetailed estimate
cash flow diagram	marginal cost	sunk cost
cost and price index	nonrecurring cost	triangulation
detailed estimate	opportunity cost	variable cost
estimation by analogy	per-unit model	work breakdown structure
external cost	power-sizing model	
fixed cost	profit–loss breakeven chart	

Estimating the engineering costs and benefits of proposed decision choices is "where the numbers come from." In this chapter we describe cost and benefit concepts and methods. These include fixed and variable costs, marginal and average costs, sunk and opportunity costs, recurring and nonrecurring benefits and costs, incremental cash costs, book costs, and life-cycle costs. We then describe the various types of estimates and difficulties sometimes encountered. The models that are described include unit factor, segmenting, cost indexes, power sizing, triangulation, and learning curves. The chapter discusses estimating benefits, developing cash flow diagrams, and drawing these diagrams with spreadsheets.

FIXED, VARIABLE, MARGINAL, AND AVERAGE COSTS

Fixed costs are constant or unchanging regardless of the level of output or activity. In contrast, **variable** costs depend on the level of output or activity. A **marginal** cost is the variable cost for one more unit, while the **average** cost is the total cost divided by the number of units.

In a production environment, for example, fixed costs, such as those for factory floor space and equipment, remain the same even though production quantity, number of employees, and level of work-in-process may vary. Labor costs are classified as a *variable* cost because they depend on the number of employees and the number of hours they work. Thus *fixed* costs are level or constant regardless of output or activity, and *variable* costs are changing and related to the level of output or activity.

As another example, many universities charge full-time students a fixed cost for 12 to 18 hours and a cost per credit hour for each credit hour over 18. Thus for full-time students who are taking an overload (>18 hours), there is a variable cost that depends on the level of activity, but for most full-time students tuition is a fixed cost.

This example can also be used to distinguish between *marginal* and *average* costs. A marginal cost is the cost of one more unit. This will depend on how many credit hours the student is taking. If currently enrolled for 12 to 17 hours, adding one more is free. The marginal cost of an additional credit hour is $0. However, for a student taking 18 or more hours, the marginal cost equals the variable cost of one more hour.

To illustrate average costs, suppose the cost of 12 to 18 hours is $3600 per term and overload credits are $240/hour. If a student takes 12 hours, the *average* cost is $3600/12 = $300 per credit hour. If the student were to take 18 hours, the *average* cost would decrease to $3600/18 = $200 per credit hour. If the student takes 21 hours, the *average* cost is \$205.71 per credit hour [\$3600 + (3 × \$240)]/21. Average cost is thus calculated by dividing the total cost for all units by the total number of units. Decision makers use **average** cost to attain an overall cost picture of the investment on a per unit basis.

   Marginal cost is used to decide whether an additional unit should be made, purchased, or enrolled in. For our example full-time student, the marginal cost of another credit is \$0 or \$240 depending on how many credits the student has already signed up for.

EXAMPLE 2–1

The Federation of Student Societies of Engineering (FeSSE) wants to offer a one-day training course to help students in job hunting and to raise funds. The organizing committee is sure that they can find alumni, local business people, and faculty to provide the training at no charge. Thus the main costs will be for space, meals, handouts, and advertising.

   The organizers have classified the costs for room rental, room setup, and advertising as fixed costs. They also have included the meals for the speakers as a fixed cost. Their total of \$225 is pegged to a room that will hold 40 people. So if demand is higher, the fixed costs will also increase.

   The variable costs for food and bound handouts will be \$20 per student. The organizing committee believes that \$35 is about the right price to match value to students with their budgets. Since FeSSE has not offered training courses before, they are unsure how many students will reserve seats.

   Develop equations for FeSSE's total cost and total revenue, and determine the number of registrations that would be needed for revenue to equal cost.

SOLUTION

Let x equal the number of students who sign up. Then,

$$\text{Total cost} = \$225 + \$20x$$

$$\text{Total revenue} = \$35x$$

   To find the number of student registrants for revenue to equal cost, we set the equations equal to each other and solve.

$$\text{Total cost} = \text{Total revenue}$$

$$\$225 + \$20x = \$35x$$

$$\$225 = (\$35 - \$20)x$$

$$x = 225/15 = 15 \text{ students}$$

   While this example has been defined for a student engineering society, we could just as easily have described this as a training course to be sponsored by a local chapter of a professional technical society. The fixed cost and the revenue would increase by a factor of about 10, while the variable cost would probably double or triple.

   If a firm were considering an in-house short course, the cost of the in-house course would be compared with the cost per employee (a variable cost) for enrolling employees in external training.

From Example 2–1 we see how it is possible to calculate total fixed and total variable costs. Furthermore, these values can be combined into a single **total** cost equation as follows:

$$\text{Total cost} = \text{Total fixed cost} + \text{Total variable cost} \qquad (2\text{-}1)$$

Example 2–1 developed *total cost* and *total revenue* equations to describe a training course proposal. These equations can be used to create what is called a **profit–loss breakeven chart** (see Figure 2–1). A plot of revenues against costs for various levels of output (activity) allows one to illustrate a *breakeven point* (in terms of costs and revenue) and regions of *profit* and *loss*. These terms can be defined as follows.

Breakeven point: The activity level at which total costs are *equal to* the revenue (or savings) generated. This is the level at which one "just breaks even."

Profit region: The variable x is greater than the breakeven point and total revenue is *greater* than total costs.

Loss region: The variable x is less than the breakeven point and total revenue is *less* than total costs.

Notice in Figure 2–1 that the *breakeven* point for the number of persons in the training course is 15 people. For more than 15 people, FeSSE will make a profit. If fewer than 15 sign up, there will be a net loss.

The fixed costs of our simple model are in reality *fixed* over a range of values for x. In Example 2–1, that range was 1 to 40 students. If *zero* students signed up, then the course could be canceled and many of the fixed costs would not be incurred. Some costs such as advertising might already have been spent, and there might be cancellation fees. If more than 40 students signed up, then greater costs for larger rooms or multiple sessions would be incurred. The model is valid only within the range named.

When modeling a specific situation, we often use *linear* variable costs and revenues. However, sometimes the relationship may be nonlinear. For example, employees are often paid at 150% of their hourly rate for overtime hours, so that production levels requiring overtime have higher variable costs. Total cost in Figure 2–2 is a fixed cost of $3000 plus

FIGURE 2–1 Profit–loss breakeven chart for Example 2–1.

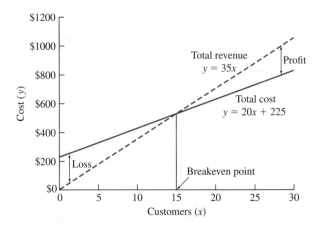

FIGURE 2-2 Nonlinear variable costs.

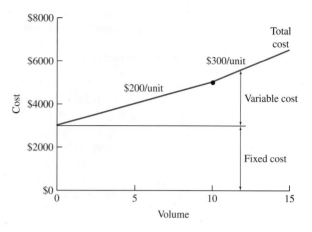

a variable cost of $200 per unit for straight-time production of up to 10 units and $300 per unit for overtime production of up to 5 more units.

Figure 2–2 can also be used to illustrate marginal and average costs. At a volume of 5 units the marginal cost is $200 per unit, while at a volume of 12 units the marginal cost is $300 per unit. At 5 units the average cost is $800 per unit, or $(3000 + 200 \times 5)/5$. At 12 units the average cost is $467 per unit, or $(3000 + 200 \times 10 + 300 \times 2)/12$.

Sunk Costs

A **sunk cost** is money already spent as a result of a *past* decision. If only 5 students signed up for the training course in Example 2–1, the advertising costs would be a *sunk cost*.

Sunk costs must be ignored in engineering economic analysis because current decisions cannot change the past. For example, dollars spent last year to purchase new production machinery is money that is *sunk:* the money has already been spent—there is nothing that can be done now to change that action. As engineering economists we deal with present and future opportunities.

Many times it is difficult not to be influenced by sunk costs. Consider 100 shares of stock in XYZ, Inc., purchased for $15 per share last year. The share price has steadily declined over the past 12 months to a price of $10 per share today. Current decisions must focus on the $10 per share that could be attained today (as well as future price potential), not the $15 per share that was paid last year. The $15 per share paid last year is a *sunk cost* and has no influence on present opportunities.

As another example, when Regina was a sophomore, she purchased a newest-generation laptop from the college bookstore for $2000. By the time she graduated, the most anyone would pay her for the computer was $400 because the newest models were faster and cheaper and had more capabilities. For Regina, the original purchase price was a *sunk cost* that has no influence on her present opportunity to sell the laptop at its current market value of $400.

When we get to Chapters 11 and 12 on depreciation and income taxes, we will find an exception to the rule of *ignore sunk costs*. When an asset is sold or disposed of, then the sunk cost of what was paid for it is important in figuring out how much is owed in taxes. This exception applies only to the after-tax analysis of capital assets.

Opportunity Costs

An **opportunity cost** is associated with using a resource in one activity instead of another. Every day firms use resources to accomplish various tasks—forklifts transport materials, engineers design products and processes, assembly lines make a product, and parking lots provide parking for employees' vehicles. There are costs for these intended purposes. These are also forgone opportunity costs. For instance, the assembly line could produce a different product, and the parking lot could be rented out, used as a building site, or converted into a small airstrip. Each alternative use would provide some benefit to the firm. These opportunity costs can be included, or they can be addressed by considering that alternative use as another decision-making choice.

As an example, suppose a college student may travel through Europe over the summer break. The student should estimate all the *out-of-pocket* cash costs for air travel, lodging, meals, entertainment, and train passes. Suppose this amounts to $3000 for a 10-week period—which the student can afford. However, the *true* cost includes not only *out-of-pocket* cash costs but also the *opportunity cost*. By taking the trip, the student is giving up the *opportunity* to earn $5000 as a summer intern. The student's total cost is thus $8000.

Remember that opportunity costs are really foregone benefits. When those benefits are not chosen they become costs. The key is to make a choice whereby the *actual* benefits realized outweigh the foregone benefits not chosen. Example 2–2 shows how opportunity costs are part of decisions about idle or under-used assets. What benefit is foregone by keeping the pumps?

EXAMPLE 2–2

A distributor of electric pumps must decide what to do with a "lot" of old electric pumps purchased 3 years ago. Soon after the distributor purchased the lot, technology advances made the old pumps less desirable to customers. The pumps are becoming obsolescent as they sit in inventory. The pricing manager has the following information.

Distributor's purchase price 3 years ago	$ 7,000
Distributor's storage costs to date	1,000
Distributor's list price 3 years ago	9,500
Current list price of the same number of new pumps	12,000
Amount offered for the old pumps from a buyer 2 years ago	5,000
Current price the lot of old pumps would bring	3,000

Looking at the data, the pricing manager has concluded that the price should be set at $8000. This is the money that the firm has "tied up" in the lot of old pumps ($7000 purchase and $1000 storage), and it was reasoned that the company should at least recover this cost. Furthermore, the pricing manager has argued that an $8000 price would be $1500 less than the list price from 3 years ago, and it would be $4000 less than what a lot of new pumps would cost ($12,000 − $8000). What would be your advice on price?

SOLUTION

Let's look more closely at each of the data items.

Distributor's purchase price 3 years ago: This is a sunk cost that should not be considered in setting the price today.

Distributor's storage costs to date: The storage costs for keeping the pumps in inventory are sunk costs; that is, they have been paid. Hence they should not influence the pricing decision.

Distributor's list price 3 years ago: If there have been no willing buyers in the past 3 years at this price, it is unlikely that a buyer will emerge in the future. This past list price should have no influence on the current pricing decision.

Current list price of newer pumps: Newer pumps now include technology and features that have made the older pumps less valuable. Directly comparing the older pumps to those with new technology is misleading. However, the price of the new pumps and the value of the new features help determine the market value of the old pumps.

Amount offered by a buyer 2 years ago: This once was an opportunity. At the time of the offer, the company chose to keep the lot and thus the $5000 offered became an opportunity cost for keeping the pumps. This amount should not influence the current pricing.

Current price the lot could bring: The price a willing buyer in the marketplace offers is called the asset's *market value*. This $3000 is the relevant opportunity cost for decision making.

From this analysis, it is easy to see the flaw in the pricing manager's reasoning. In an engineering economist analysis we deal only with *today's* and prospective *future* opportunities. It is impossible to go back in time and change decisions that have been made. Thus, the pricing manager should recommend to the distributor that the price be set at the current value that a buyer assigns to the item: $3000.

Recurring and Nonrecurring Costs

Recurring costs refer to any expense that is known and anticipated and that occurs at regular intervals. **Nonrecurring costs** are one-of-a-kind expenses that occur at irregular intervals and thus are sometimes difficult to plan for or anticipate from a budgeting perspective.

Examples of recurring costs include those for resurfacing a highway and reshingling a roof. Annual expenses for maintenance and operation are also recurring expenses. Examples of nonrecurring costs include the cost of installing a new machine (including any facility modifications required), the cost of augmenting equipment based on older technology to restore its usefulness, emergency maintenance expenses, and the disposal or close-down costs associated with ending operations.

In engineering economic analyses *recurring costs* are modeled as cash flows that occur at regular intervals (such as every year or every 5 years). Their magnitude can be estimated, and they can be included in the overall analysis. *Nonrecurring costs* can be handled easily in our analysis if we are able to anticipate their timing and size. However, this is not always so easy to do.

There are also recurring and nonrecurring benefits. A nonrecurring benefit is a single cash inflow that is anticipated today or in the future, such as the proceeds from selling a house, business, vehicle, or any other asset. Personal examples include a graduation gift, retirement plan lump sum distribution, lottery winnings, or inheritance. Engineering projects are often intended to produce recurring benefits that continue for months, years, or decades. Examples include sales of a new product, faster travel on a safer bridge or highway, attending events at a sports arena or theater, and the services of schools, hospitals, and libraries.

Incremental Costs

One of the fundamental principles in engineering economic analysis is that in choosing between competing alternatives, the focus is on the *differences* between those alternatives. This is the concept of **incremental costs.** For instance, one may be interested in comparing two options to lease a vehicle for personal use. The two lease options may have several specifics for which costs are the same. However, there may be incremental costs associated with one option but not with the other. In comparing the two leases, the focus should be on the differences between the alternatives, not on the costs that are the same.

The principle described above for costs also holds true for the incremental benefits of competing alternatives. Consider the case of lease options for the vehicle. The benefits associated with each option were assumed to be the same, and thus we were only interested in the incremental cost differences. However, what if the benefits of the two options were different? In this case your focus would be on the differences of the costs *and the benefits* associated with each option.

EXAMPLE 2–3

Philip is choosing between model *A* (a budget model) and model *B* (with more features and a higher purchase price). What *incremental costs* would Philip incur if he chose model *B* instead of the less expensive model *A*?

Cost Items	Model A	Model B
Purchase price	$10,000	$17,500
Installation costs	3,500	5,000
Annual maintenance costs	2,500	750
Annual utility expenses	1,200	2,000
Disposal costs after useful life	700	500

SOLUTION

We are interested in the incremental or *extra* costs that are associated with choosing model *B* instead of model *A*. To obtain these we subtract model *A* costs from model *B* costs for each category (cost item) with the following results.

Cost Items	(Model *B* Cost – *A* Cost)	Incremental Cost of *B*
Purchase price	17,500 – 10,000	$7500
Installation costs	5,000 – 3,500	1500
Annual maintenance costs	750 – 2,500	–1750/yr
Annual utility expenses	2,000 – 1,200	800/yr
Disposal costs after useful life	500 – 700	–200

Notice that for the cost categories given, the incremental costs of model *B* are both positive and negative. Positive incremental costs mean that model *B* costs more than model *A*, and negative incremental costs mean that there would be a *savings* (reduction in cost) if model *B* were chosen instead.

As described in the problem statement, because model *B* has more features, the decision must also include incremental benefits offered by those features rather than focussing only on costs.

Cash Costs Versus Book Costs

A *cash cost* requires the cash transaction of dollars "out of one person's pocket" into "the pocket of someone else." When you buy dinner for your friends or make your monthly car payment you are incurring a **cash cost** or **cash flow.** Cash costs and cash flows are the basis for engineering economic analysis.

Book costs do not require the transaction of dollars "from one pocket to another." Rather, **book costs** are cost effects from past decisions that are recorded "in the books" (accounting books) of a firm. In one common book cost, asset depreciation (which we discuss in Chapter 11), the expense paid for a particular business asset is "written off" on a company's accounting system over a number of periods. Book costs do not ordinarily represent cash flows and thus are not included in engineering economic analysis. One exception to this is the impact of asset depreciation on tax payments—which are cash flows and are included in after-tax analyses.

CONSIDERING ALL COSTS

Life-Cycle Costs

The products, goods, and services designed by engineers all progress through a **life cycle** very much like the human life cycle. People are conceived, go through a growth phase, reach their peak during maturity, and then gradually decline and expire. The same general pattern holds for products, goods, and services. As with humans, the duration of the different phases, the height of the peak at maturity, and the time of the onset of decline and termination all vary depending on the individual product, good, or service. Figure 2–3 illustrates the typical phases that a product, good, or service progresses through over its life cycle.

Beginning ————————————————— Time ————————————————→ End

Needs Assessment and Justification Phase	Conceptual or Preliminary Design Phase	Detailed Design Phase	Production or Construction Phase	Operational Use Phase	Decline and Retirement Phase
Requirements	Impact Analysis	Allocation of Resources	Product, Goods, & Services Built	Operational Use	Declining Use
Overall Feasibility	Proof of Concept	Detailed Specifications	All Supporting Facilities Built	Use by Ultimate Customer	Phase Out
Conceptual Design Planning	Prototype/ Breadboard	Component and Supplier Selection	Operational Use Planning	Maintenance and Support	Retirement
	Development and Testing	Production or Construction Phase		Processes, Materials, and Methods Used	Responsible Disposal
	Detailed Design Planning			Decline and Retirement Planning	

FIGURE 2–3 Typical life cycle for products, goods, and services.

Life-cycle costing refers to the concept of designing products, goods, and services with a full and explicit recognition of the associated costs over the various phases of their life cycles. Since *all* costs over the life cycle are considered, this is the correct approach to economic decision making. This contrasts with decisions by firms, agencies, or individuals that only consider acquisition costs.

Design Changes and Cost Impacts

Two key concepts in life-cycle costing are that the later design changes are made, the higher the costs, and that decisions made early in the life cycle tend to "lock in" costs and benefits that will be incurred later. Figure 2–4 illustrates how costs are committed early in the product life cycle—nearly 70–90% of all costs are set during the design phases. At the same time, as the figure shows, only 10–30% of cumulative life-cycle costs have been spent. In addition, notice that as life-cycle costs are committed, ultimate life-cycle benefits are set. This highlights the fact that early decisions not only commit resources, but also bound the benefits that will be realized by the product, good, or service.

Figure 2–5 reinforces these concepts by illustrating that later product changes are more costly and that earlier changes are easier (and less costly) to make. When planners try to save money at an early design stage, the result is often a poor design that results in change orders during construction and prototype development. These changes, in turn, are more costly than working out a better design would have been.

From Figures 2–4 and 2–5 we see that the time to consider all life-cycle effects and make design changes is during the needs and conceptual/preliminary design phases— before a lot of dollars are committed. Some of the life-cycle effects that engineers should consider at design time include product costs for liability, production, material, testing and quality assurance, and maintenance and warranty.

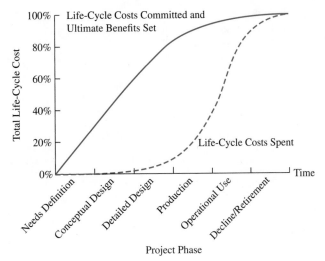

FIGURE 2–4 Cumulative life-cycle costs committed and dollars spent.

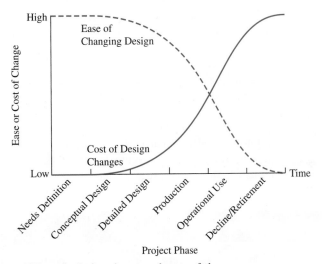

FIGURE 2–5 Ease of life-cycle design change and costs of change.

Internal and External Costs

An important extension of life-cycle costing by green or sustainable engineering is to consider both internal and external costs in a design, project, or product. **Internal costs** are incurred and paid by the firm and are used to calculate the production cost. This cost is part of decisions on pricing and ultimately profitability. In contrast, **external costs** are outside the firm's normal cost accounting and thus do not directly affect the price of goods/services. Green engineering focuses on the environmental impacts of inputs, such as electric power, or outputs, such as discarded packaging. Examples include costs of disposal, decommissioning, or landfilling; effects on animals and habitats; degradation of air

FIGURE 2–6 Green design goal of reducing *all* costs and increasing profit.

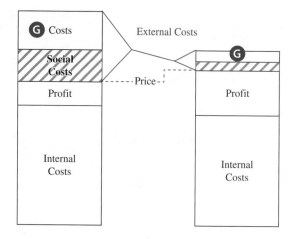

and/or water quality and quality of life; and managing wastes and pollutants. Often the focus is designing for repair, reuse, or recycling, rather than disposal as landfill.

Figure 2–6 also includes social costs, such as for full-time workers who need Medicaid or food stamps—costs that are paid through taxes rather than by employers. Other examples include poor working conditions and communities and individuals impacted by layoffs, plant closures, and so on.

While trade-offs between external and internal costs are common, the goal of green engineering is shown in Figure 2–6: lower external costs by thoughtful attention; lower internal costs by innovation; and increase profits as buyers respond to a better corporate reputation and *green* choices for consumers.

Internal and external costs are particularly important in public-sector applications of engineering economy. Figure 16-1's smokestack and plant boundary emphasizes the internal/external costs discussed here.

EXAMPLE 2–4

JPL Enterprises Co. includes potential environmental costs in their design process. From the following list, identify the internal and external costs associated with a new mountaintop ski resort project.

- Site costs
- Land costs
- Water quality costs
- Legal costs
- Community costs
- Design costs
- Habitat costs
- Viewshed costs
- Recreational costs
- Administrative costs
- Roadway costs
- Labor costs
- Materials costs
- Equipment costs
- Overhead costs
- Construction costs

SOLUTION

Internal costs include the site, land, legal, design, administrative, labor, materials, equipment, overhead, and construction categories.

External (societal) costs include items such as:

Water quality costs: impacts on downstream water sources and aquifers

Community costs: prospective loss of culture, quaintness, community values

Habitat costs: effect on natural habitat for native/migratory animals

Viewshed costs: negative impact on the visual sight lines

Recreational costs: impact on hikers, fishermen, hunters, birdwatchers, and others

Roadway costs: cost to community to build roads and bridges (developers sometimes share these costs with local, state and federal entities)

ESTIMATING BENEFITS

Along with estimating the costs of proposed projects, engineering economists must often also quantify the anticipated benefits. Example benefits include sales from products, revenues from bridge tolls and electric power sales, cost reductions from reduced material or labor costs, less time spent in traffic jams, and reduced risk of flooding. Many engineering projects are undertaken precisely to secure the benefits.

Uncertainty associated with benefit estimates is asymmetric, with broader limits for negative outcomes. Compared to costs, benefits are more likely to be overestimated, thus an example set of limits might be (-50%, $+20\%$). Another important difference between cost and benefit estimation is that many costs of engineering projects occur in the near future (design and construction), but benefits are further into the future—thus more uncertainty is typical.

The estimation of economic benefits is an important step that should not be overlooked. Most of the models, concepts, and issues that apply in estimating costs also apply to estimating economic benefits.

THE ESTIMATING PROCESS

Engineering economic analysis focuses on the future consequences of current decisions. Because these consequences are in the future, usually they must be estimated and cannot be known with certainty. Estimates that may be needed in engineering economic analysis include purchase costs, annual revenue, yearly maintenance, interest rates for investments, annual labor and insurance costs, equipment salvage values, and tax rates.

Estimating is the foundation of economic analysis. As in any analysis procedure, the outcome is only as good as the numbers used to reach the decision. For example, a person who wants to estimate her federal income taxes for a given year could do a very detailed analysis, including social security deductions, retirement savings deductions, itemized personal deductions, exemption calculations, and estimates of likely changes to the tax code. However, this very technical and detailed analysis will be grossly inaccurate if poor data are used to predict the next year's income. Thus, to ensure that an analysis is a reasonable evaluation of future events, it is very important to make careful estimates.

Types of Cost Estimates

We can define three general types of estimate whose purposes, accuracies, and underlying methods are quite different.

Rough estimates: Order-of-magnitude estimates used for high-level planning, for determining project feasibility, and in a project's initial planning and evaluation phases. Rough estimates tend to involve back-of-the-envelope numbers with limited detail or accuracy. The intent is to quantify and consider the order of magnitude of the numbers involved. These estimates require minimum resources to develop, and their accuracy is generally −30 to +60%.

Notice the nonsymmetry in the estimating error. This is because decision makers tend to underestimate the magnitude of costs (negative economic effects). Also, as Murphy's law predicts, there seem to be more ways for results to be worse than expected than there are for the results to be better than expected.

Semidetailed estimates: Used for budgeting purposes at a project's conceptual or preliminary design stages. These estimates are more detailed, and thus require additional time and resources. Greater sophistication is used in developing semidetailed estimates than the rough-order type, and their accuracy is generally −15 to +20%.

Detailed estimates: Used during a project's detailed design and contract bidding phases. These estimates are made from detailed quantitative models, blueprints, product specification sheets, and vendor quotes. Detailed estimates involve the most time and resources to develop and thus are much more accurate than rough or semidetailed estimates. The accuracy of these estimates is generally −3 to +5%.

The upper limits of +60% for rough order, +20% for semidetailed, and +5% for detailed estimates are based on construction data for plants and infrastructure. Final costs for software, research and development, and new military weapons often have much higher corresponding percentages.

Accuracy of Estimate

In considering the three types of estimates it is important to recognize that each has its unique purpose, place, and function in a project's life cycle. Rough estimates are used for general feasibility activities and ranking possible projects; semidetailed estimates support budgeting and preliminary design decisions; and detailed estimates are used for establishing design details and contracts. As one moves from rough to detailed design estimates, one also moves from less to more accuracy. As a result "significant digits" become more important with detailed estimates as opposed to rough estimates. For example, at the feasibility phase of a large construction project one might estimate costs to the nearest million dollars when looking at several design decisions. However, when contracts are signed after detailed design they will be to the dollar.

When both costs and benefits are estimated for a decision situation one should balance the order of accuracy of each. One should not estimate costs to the nearest $100 while estimating benefits to the nearest $1000. Such an imbalance may skew a comparison of the true difference in costs and benefits for a proposed action.

Continuity in perspective in estimating costs and benefits is another important accuracy issue. We had previously mentioned that most people tend to underestimate costs and overestimate benefits in isolation. Care must be taken to balance one's perspective to ensure a consistent approach to quantifying both costs and benefits.

Differences in degree of accuracy and cost-benefit perspective may result in an inaccurate analysis and thus favor one decision choice over another. Care must be taken from the outset to mitigate or eliminate these effects.

Cost Versus Accuracy Trade-off

Increasing the accuracy of an estimate is not a free thing—it requires added time and resources. Figure 2–7 illustrates the trade-off between accuracy and cost. In general, in engineering economic analysis, the resources spent must be justified by the need for detail in the estimate. From the figure we see that low accuracy estimates should have low costs, and high accuracy estimates will have higher costs. This relationship applies for cost estimates as well as estimating the benefits associated with a prospective choice. As a rule, we should not spend resources developing accuracy that is not warranted by the use and purpose of that estimate. For example, during the project feasibility phase we would not want to use our people, time, and money to develop detailed estimates for infeasible alternatives that will be quickly eliminated from further consideration.

FIGURE 2–7 Accuracy versus cost trade-off in estimating costs and benefits.

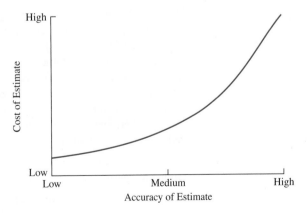

Difficulties in Estimation

Estimating is difficult because the future is unknown. With few exceptions (such as with legal contracts) it is difficult to foresee future economic consequences exactly. In this section we discuss several aspects of estimating that make it a difficult task.

One-of-a-Kind Estimates

Estimated parameters can be for one-of-a-kind or first-run projects. The first time something is done, it is difficult to estimate both the costs required to design, produce, and maintain a product over its life cycle as well as the anticipated benefits. Consider the projected cost estimates that were developed for the first NASA missions. The U.S. space program initially had no experience with human flight in outer space; thus the development of the cost estimates for design, production, launch, and recovery of the astronauts, flight hardware, and payloads was a "first-time experience." In addition, estimating the benefits of space exploration, such as advances in aircraft design, database management, surveying, water preservation, telemetry, and forest management (to name a few) were initially difficult to envision. The same is true for any endeavor lacking local or global historical cost data. New products or processes that are unique and fundamentally different make estimating costs difficult.

The good news is that there are very few one-of-a-kind estimates to be made in engineering design and analysis. Nearly all new technologies, products, and processes have "close cousins" that have led to their development. The concept of **estimation by analogy** allows one to use knowledge about well-understood activities to anticipate costs and benefits for new activities. In the 1950s, at the start of the military missile program, aircraft companies drew on their in-depth knowledge of designing and producing aircraft when they bid on missile contracts. As another example, consider the problem of estimating the production labor requirements for a brand new product, X. A company may use its labor knowledge about Product Y, a similar type of product, to build up the estimate for X. Thus, although "first-run" estimates are difficult to make, estimation by analogy can be an effective tool.

Time and Effort Available

Our ability to develop engineering estimates is constrained by time and person-power availability. In an ideal world, it would *cost nothing* to use *unlimited resources* over an *extended period of time*. However, reality requires the use of limited resources in fixed intervals of time. Thus for a rough estimate only limited effort is used.

Constraints on time and person-power can make the overall estimating task more difficult. If the estimate does not require as much detail (such as when a rough estimate is the goal), then time and personnel constraints may not be a factor. When detail is necessary and critical (such as in legal contracts), however, requirements must be anticipated and resource use planned.

Estimator Expertise

Consider two common expressions: *The past is our greatest teacher*, and *Knowledge is power*. These simple axioms hold true for much of what we encounter during life, and they are true in engineering estimating as well. The more experienced and knowledgeable the engineering estimator is, the less difficult the estimating process will be, the more accurate the estimate will be, the less likely it is that a major error will occur, and the more likely it is that the estimate will be of high quality.

How is experience acquired in industry? One approach is to assign inexperienced engineers relatively small jobs, to create expertise and build familiarity with products and processes. Another strategy used is to pair inexperienced engineers with mentors who have vast technical experience. Technical boards and review meetings conducted to "justify the numbers" also are used to build knowledge and experience. Finally, many firms maintain databases of their past estimates and the costs that were actually incurred.

ESTIMATING MODELS

This section develops several estimating models that can be used at the rough, semide-tailed, or detailed design levels. For rough estimates the models are used with rough data, likewise for detailed design estimates they are used with detailed data.

Per-Unit Model

The **per-unit model** uses a "per unit" factor, such as cost per square foot, to develop the estimate desired. This is a very simplistic yet useful technique, especially for develop-ing estimates of the rough or order-of-magnitude type. The per-unit model is commonly used in the construction industry. As an example, you may be interested in a new home that is constructed with a certain type of material and has a specific construction style. Based on this information a contractor may quote a cost of $65 per square foot for your home. If you are interested in a 2000-square-foot floor plan, your cost would thus be: 2000 × 65 = $130,000. Other examples where per unit factors are used for both costs and benefits include

- Service cost per customer
- Safety cost per employee
- Gasoline cost per mile
- Cost of defects per batch
- Maintenance cost per window
- Mileage cost per vehicle
- Utility cost per square foot of floor space
- Housing cost per student
- Sales per customer region
- Revenue per acre
- Fee per transaction
- Royalty per book
- Revenue per mile
- Quality improvement per training hour
- Rent per square footage
- Sales increase per representative

It is important to note that the per-unit model does not make allowances for economies of scale (the fact that higher quantities usually cost less on a per-unit basis). In most cases, however, the model can be effective at getting the decision maker "in the ballpark" of likely costs and benefits, and it can be very accurate if accurate data are used.

EXAMPLE 2–5

Gaber Land Corp. is evaluating a 4-acre waterfront property for development into rental condominiums. The front 2-acre lot is more expensive to purchase than the rear 2-acre lot, and condo leases closer to the waterfront can be more expensive than those units in the rear. Gaber is considering a design that includes a 32-unit building on each lot. Data includes the following:

Initial Costs

Lot purchase prices: $400,000/acre front lot, $100,000/acre back lot

Legal fees, applications, permits, etc.: $80,000

Site clearing and preparation: $3000/acre

Paving roadways, parking, curbs, and sidewalks: 25% of total lot at $40,000/acre

Construction costs: $3,000,000 per building

Recurring Costs

Taxes and insurance: $5000/month per building

Landscaping: 25% of lot at $1000/acre/month

Security: $1000/building + $1500/month

Other costs: $2000/month

Revenue (assume 90% annual occupancy)

Front lot units: $2500/unit/month

Rear lot units: $1750/unit/month

Other revenue: $5000/month

Answer the following: (1) Use the concept of the per-unit model to estimate the total initial cost, annual cost, and annual revenue of this prospective project, and (2) If you made the simplifying assumption of no changes to costs and revenues for 10 years, estimate the profitability of this prospective investment ignoring the effects of money's value over time.

SOLUTION

(1) Using the per unit model:

Total Initial Cost		
Purchase price: $(400,000 \times 2) + (100,000 \times 2)$	=	$1,000,000
Legal costs	=	80,000
Site clearing & preparation: 3000×4	=	12,000
Roadways, etc.: $(.25 \times 4) \times 40,000$	=	40,000
Construction: $3,000,000 \times 2$	=	6,000,000
		$7,132,000

Annual Cost

Taxes and insurance: $20,000 \times 12$	=	$240,000
Landscaping: $(.25 \times 4) \times 1000 \times 12$	=	12,000
Security: $(1000 \times 2) + (1500 \times 12)$	=	20,000
Other costs: 2000×12	=	24,000
		$296,000

Annual Revenue

Front lot leases: $(32 \times 2500 \times 12) \times .90$	=	$864,000
Rear lot leases: $(32 \times 1750 \times 12) \times .90$	=	604,800
Other revenue: 5000×12	=	60,000
		$1,528,800

(2) Using the fundamental relationship that Net Profit = Revenue – Costs:

$$\text{Net Profit} = [1{,}528{,}800 \times 10] - [7{,}132{,}000 + (296{,}000 \times 10)]$$
$$= \$5{,}196{,}000$$

Segmenting Model

The **segmenting model** can be described as "divide and conquer." An estimate is decomposed into its individual components, estimates are made at those lower levels, and then the estimates are aggregated (added) back together. It is much easier to estimate at the lower levels because they are more readily understood. This approach is common in engineering estimating in many applications and for any level of accuracy needed. In estimating costs for the condominiums in Example 2–5, the estimate was segmented into the initial and monthly costs, and monthly revenues. The example illustrated the segmenting model (division of the overall estimate into various categories and activities, such as costs and benefits) together with the unit factor model to make the subestimates for each category. Example 2–6 provides another example of the segmenting approach.

EXAMPLE 2–6

Clean Lawn Corp., a manufacturer of yard equipment, is planning to introduce a new high-end industrial-use lawn mower called the Grass Grabber. The Grass Grabber is designed as a walk-behind, self-propelled mower. Clean Lawn engineers have been asked by the accounting department to estimate the material costs for the new mower. The material cost estimate will be used, along with estimates for labor and overhead to evaluate the potential of this new model.

The engineers decide to decompose the design specifications for the Grass Grabber into its components, estimate the material costs for each of the components, and then sum these costs up to obtain the overall estimate. The engineers are using a segmenting approach to build up their estimate. After careful consideration, the engineers have divided the mower into the following major subsystems: chassis, drive train, controls, and cutting/collection system. Each of these was further divided as appropriate, and unit material costs were estimated at this lowest of levels as follows:

Cost Item	Unit Material Cost Estimate	Cost Item	Unit Material Cost Estimate
A. Chassis		**C. Controls**	
A.1 Deck	$ 7.40	C.1 Handle assembly	$ 3.85
A.2 Wheels	10.20	C.2 Engine linkage	8.55
A.3 Axles	4.85	C.3 Blade linkage	4.70
	$22.45	C.4 Speed control linkage	21.50
B. Drive train		C.5 Drive control assembly	6.70
B.1 Engine	$38.50	C.6 Cutting height adjuster	7.40
B.2 Starter assembly	5.90		$52.70
B.3 Transmission	5.45	**D. Cutting/Collection system**	
B.4 Drive disc assembly	10.00	D.1 Blade assembly	$10.80
B.5 Clutch linkage	5.15	D.2 Side chute	7.05
B.6 Belt assemblies	7.70	D.3 Grass bag and adapter	7.75
	$72.70		$25.60

The total material cost estimate of $173.45 was calculated by summing up the estimates for each of the four major subsystem levels (chassis, drive train, controls, and cutting/collection system). It should be noted that this cost represents only the material portion of the overall cost to produce the mowers. Other costs would include labor and overhead items.

In Example 2–6 the engineers at Clean Lawn Corp. decomposed the cost estimation problem into logical elements. The scheme they used of decomposing cost items and numbering the material components (A.1, A.1, A.2, etc.) is known as a **work breakdown structure.** This technique is commonly used in engineering cost estimating and project management of large products, processes, or projects. A work breakdown structure decomposes a large "work package" into its constituent parts, which can then be estimated or managed individually. In Example 2–6 the work breakdown structure of the Grass Grabber has three levels. At the top level is the product itself, at the second level are the four major subsystems, and at the third level are the individual cost items. Imagine what the product work breakdown structure for a modern commercial airliner looks like. Then imagine trying to manage an aircraft's design, engineering, construction, and costing without a tool like the work breakdown structure.

Cost and Price Indexes

Indexes are dimensionless numerical values that reflect historical change in engineering (and other) costs and prices. They can reflect relative cost/price change in either individual items (labor, materials, utilities) or groups of costs (consumer costs, producers' costs). Single item indexes are called *commodity specific indexes* and those that group costs or prices are called composite indexes. Equation 2–2 gives the formulation for how indexes can be used to update historical values—where the ratio of the cost or price index numbers at two points in time (A and B) is equivalent to the dollar cost ratio of the item at the same time.

$$\frac{\text{Cost or price at time } A}{\text{Cost or price at time } B} = \frac{\text{Index value at time } A}{\text{Index value at time } B} \tag{2-2}$$

EXAMPLE 2–7

Miriam is interested in estimating the annual labor and material costs for a new production facility. She was able to obtain the following cost data:

Labor Costs

- Labor cost index value was at 124 ten years ago and is 188 today.
- Annual labor costs for a similar facility were $575,500 ten years ago.

Material Costs

- Material cost index value was at 544 three years ago and is 715 today.
- Annual material costs for a similar facility were $2,455,000 three years ago.

SOLUTION

Miriam will use Equation 2–2 to develop her cost estimates for annual labor and material costs.

Labor

$$\frac{\text{Annual cost today}}{\text{Annual cost 10 years ago}} = \frac{\text{Index value today}}{\text{Index value 10 years ago}}$$

$$\text{Annual cost today} = \frac{188}{124} \times \$575,500 = \$872,500$$

Materials

$$\frac{\text{Annual cost today}}{\text{Annual cost 3 years ago}} = \frac{\text{Index value today}}{\text{Index value 3 years ago}}$$

$$\text{Annual cost today} = \frac{715}{544} \times \$2,455,000 = \$3,227,000$$

Cost and price index data are collected and published by several private and public sources in the U.S. (and world). The U.S. government publishes data through the Bureau of Labor Statistics of the Department of Commerce. The *Statistical Abstract of the United States* publishes cost indexes for labor, construction, and materials. Another useful source for engineering cost price index data is the *Engineering News Record*.

Power-Sizing Model

The **power-sizing model** is used to estimate the costs of industrial plants and equipment. The model "scales up" or "scales down" known costs, thereby accounting for economies of scale that are common in industrial plant and equipment costs. Consider the cost to build a refinery. Would it cost twice as much to build the same facility with double the capacity? It is unlikely. The *power-sizing model* uses the exponent (x), called the *power-sizing exponent,* to reflect economies of scale in the size or capacity:

$$\frac{\text{Cost of equipment } A}{\text{Cost of equipment } B} = \left(\frac{\text{Size (capacity) of equipment } A}{\text{Size (capacity) of } B}\right)^x \qquad (2\text{-}3)$$

where x is the power-sizing exponent, costs of A and B are at the same point in time (same dollar basis), and size or capacity is in the same physical units for both A and B.

The power-sizing exponent (x) can be 1.0 (indicating a linear cost-versus-size/capacity relationship) or greater than 1.0 (indicating *dis*economies of scale), but it is usually less than 1.0 (indicating economies of scale). Generally the size ratio should be less than 2, and it should never exceed 5. This model works best in a "middle" range—not very small or very large.

Exponent values for plants and equipment of many types may be found in several sources, including industry reference books, research reports, and technical journals. Such exponent values may be found in *Perry's Chemical Engineers' Handbook, Plant Design and Economics for Chemical Engineers,* and *Preliminary Plant Design in Chemical Engineering.* Table 2–1 gives power-sizing exponent values for several types of

TABLE 2–1 Example Power-Sizing Exponent Values

Equipment or Facility	Size Range	Power-Sizing Exponent
Blower, centrifugal	10,000–100,000 ft^3/min	0.59
Compressor	200–2100 hp	0.32
Crystallizer, vacuum batch	500–7000 ft^2	0.37
Dryer, drum, single atmospheric	10–100 ft^2	0.40
Fan, centrifugal	20,000–70,000 ft^3/min	1.17
Filter, vacuum rotary drum	10–1500 ft^2	0.48
Lagoon, aerated	0.05–20 million gal/day	1.13
Motor	5–20 hp	0.69
Reactor, 300 psi	100–1000 gal	0.56
Tank, atmospheric, horizontal	100–40,000 gal	0.57

industrial facilities and equipment. The exponent given applies only to equipment within the size range specified.

In Equation 2–3 equipment costs for both *A* and *B* occur at the same point in time. This equation is useful for scaling equipment costs but *not* for updating those costs. When the time of the desired cost estimate is different from the time in which the scaling occurs (per Equation 2–3) cost indexes accomplish the time updating. Thus, in cases like Example 2–8 involving both scaling and updating, we use the power-sizing model together with cost indexes.

EXAMPLE 2–8

Based on her work in Example 2–7, Miriam has been asked to estimate the cost today of a 2500-ft^2 heat exchange system for the new plant being analyzed. She has the following data.

- Her company paid $50,000 for a 1000-ft^2 heat exchanger 5 years ago.
- Heat exchangers within this range of capacity have a power-sizing exponent (*x*) of 0.55.
- Five years ago the Heat Exchanger Cost Index (HECI) was 1306; it is 1487 today.

SOLUTION

Miriam will first use Equation 2–3 to scale up the cost of the 1000-ft^2 exchanger to one that is 2500 ft^2 using the 0.55 power-sizing exponent.

$$\frac{\text{Cost of 2500-ft}^2 \text{ equipment}}{\text{Cost of 1000-ft}^2 \text{ equipment}} = \left(\frac{2500\text{-ft}^2 \text{ equipment}}{1000\text{-ft}^2 \text{ equipment}}\right)^{0.55}$$

$$\text{Cost of 2500-ft}^2 \text{ equipment} = \left(\frac{2500}{1000}\right)^{0.55} \times 50,000 = \$82,800$$

Miriam knows that the $82,800 reflects only the scaling up of the cost of the 1000-ft^2 model to a 2500-ft^2 model. Now she will use Equation 2–2 and the HECI data to estimate the cost of a 2500-ft^2 exchanger today. Miriam's cost estimate would be

$$\frac{\text{Equipment cost today}}{\text{Equipment cost 5 years ago}} = \frac{\text{Index value today}}{\text{Index value 5 years ago}}$$

$$\text{Equipment cost today} = \frac{1487}{1306} \times \$82,800 = \$94,300$$

Triangulation

Triangulation is used in engineering surveying. A geographical area is divided into triangles from which the surveyor is able to map points within that region by using three fixed points and horizontal angular distances to locate fixed points of interest (e.g., property line reference points). Since any point can be located with two lines, the third line represents an extra perspective and check. We will not use trigonometry to arrive at our cost and benefit estimates, but we can use the concept of triangulation. We should approach our economic estimate from different perspectives because such varied perspectives add richness, confidence, and quality to the estimate. **Triangulation** in estimating costs and benefits might involve using multiple sources of data or multiple quantitative models. As decision makers, we should always seek out varied perspectives.

Improvement and the Learning Curve

One common phenomenon observed, regardless of the task being performed, is that as the number of repetitions increases, performance becomes faster and more accurate. This is the concept of learning and improvement in the activities that people perform. From our own experience we all know that our fiftieth repetition is done in much less time than we needed to do the task the first time.

The **learning curve** captures the relationship between task performance and task repetition. In general, as output *doubles*, the unit production time will be reduced to some fixed percentage, the **learning-curve percentage** or **learning-curve rate.** For example, it may take 300 minutes to produce the third unit in a production run involving a task with a 95% learning time curve. In this case the sixth unit doubles the output, so it will take $300(0.95) = 285$ minutes to produce. Sometimes the learning curve is also known as the progress curve, improvement curve, experience curve, or manufacturing progress function.

Equation 2–4 gives an expression that can be used for time estimating for repetitive tasks.

$$T_N = T_{initial} \times N^b \qquad (2\text{-}4)$$

where T_N = time required for the Nth unit of production

$T_{initial}$ = time required for the first (initial) unit of production

N = number of completed units (cumulative production)

b = learning-curve exponent (slope of the learning curve on a log–log plot)

As just given, a learning curve is often referred to by its percentage learning slope. Thus, a curve with $b = -0.074$ is a 95% learning curve because $2^{-0.074} = 0.95$. This equation uses 2 because the learning-curve percentage applies for doubling cumulative production. The learning-curve exponent is calculated by using Equation 2–5.

$$b = \frac{\log(\text{learning curve expressed as a decimal})}{\log 2} \qquad (2\text{-}5)$$

EXAMPLE 2–9

Calculate the time required to produce the hundredth unit of a production run if the first unit took 32.0 minutes to produce and the learning-curve rate for production is 80%.

SOLUTION

Using Equation 2–4, we write

$$T_{100} = T_1 \times 100^b$$

$$T_{100} = T_1 \times 100^{\log 0.80/\log 2.0}$$

$$T_{100} = 32.0 \times 100^{-0.3219}$$

$$T_{100} = 7.27 \text{ minutes}$$

It is particularly important to account for the learning-curve effect if the production run involves a small number of units instead of a large number. When thousands or even millions of units are being produced, early inefficiencies tend to be "averaged out" because of the larger batch sizes. However, in the short run, inefficiencies of the same magnitude can lead to rather poor estimates of production time requirements, and thus production cost estimates may be understated. Consider Example 2–10 and the results that might be observed if the learning-curve effect is ignored. Notice in this example that a "steady-state" time is given. Steady state is the time at which the physical constraints of performing the task prevent the achievement of any more learning or improvement.

EXAMPLE 2–10

Green Energy Inc., a clean energy equipment manufacturer, is responding to a *request for bids* to produce 20 2.0 MW wind turbines for a new wind farm planned in a coastal municipality. From previous projects, they assembled the following data. Estimate the manufacturing labor cost to include in the bid.

- The learning curve rate for labor is 85%.
- Steady-state manufacturing will be reached with the 16th unit.
- The steady-state production rate per unit is 100 hours, with 15 workers per unit.
- The average labor rate with benefits is $25 per hour.

SOLUTION

From the learning-curve rate we can calculate the value of the learning-curve exponent, b. Then from the time required to produce the 16th unit, we can calculate the time required to produce the first unit.

$$b = \log 0.85 / \log 2 = -0.2345$$

$$T_{16} = T_1 \times 16^{-0.2345}$$

$$100 = T_1 \times 16^{-0.2345}$$

$$T_1 = 100 \times 16^{0.2345} = 191.6 \text{ hours}$$

Now that we know the two parameters for Equation 2.4, it is easier to use a spreadsheet for the rest of our calculations and to plot the data.

	A	B	C	D	E
1	–0.2345	learning curve rate	Unit	Time to	Cum. Time
2	191.6	time for first unit	(N)	Produce Nth	1 to N
3			1	191.6	191.6
4	100	steady state time	2	162.8	354.4
5	16	steady state unit	3	148.1	502.5
6			4	138.4	640.9
7	20	# units	5	131.4	772.3
8	15	workers per unit	6	125.9	898.1
9	25	labor rate ($/hr)	7	121.4	1019.5
10			8	117.6	1137.2
11	=$A$2*C11^$A$1		9	114.4	1251.6
12			10	111.7	1363.3
13			11	109.2	1472.5
14	Cost if steady state assumed		12	107.0	1579.4
15	$ 750,000	=A4*A7*A8*A9	13	105.0	1684.4
16			14	103.2	1787.6
17			15	101.5	1889.1
18			16	100.0	1989.1
19			17	100	2089.1
20			18	100	2189.1
21	Cost of 20 units		19	100	2289.1
22	$ 895,927	=A9*A8*E22	20	100	2389.1
23					
24	19.5%	increase due to learning curve			

From the spreadsheet the total labor cost estimate would have been underestimated by 19.5% had Green Energy not included learning-curve effects in the estimate. The underestimate would have lowered their bid and increased the chance that they would win the project. Had they won the bid this would have affected the project's profitability and with a 20% error, they would probably have lost money on the project.

Figures 2–8 and 2–9 illustrate learning curves using the Green Energy data in columns C and D of the spreadsheet. When plotted on a linear scale the time per unit decreases as a declining rate. When plotted on a *log-log* scale, the relationship is a straight line through the 16th unit, when a steady state is reached. The straight line is because the 2nd, 4th, 8th, and 16th units have a production time that is 85% of the 1st, 2nd, 4th, and 8th units respectively.

FIGURE 2-8 Learning curve of time vs. number of units.

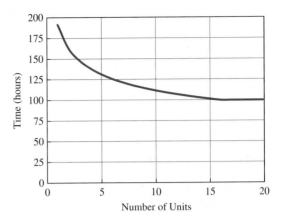

FIGURE 2-9 Learning curve on log–log scale.

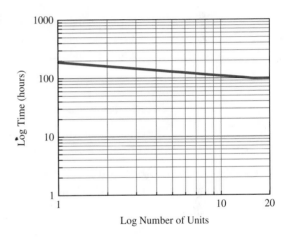

CASH FLOW DIAGRAMS

The costs and benefits of engineering projects occur over time and are summarized on a cash flow diagram (CFD). Specifically, a CFD illustrates the size, sign, and timing of individual cash flows. In this way the CFD is the basis for engineering economic analysis.

A **cash flow diagram** is created by first drawing a segmented time-based horizontal line, divided into time units. The time units on the CFD can be years, months, quarters, or any other consistent time unit. Then at each time at which a cash flow will occur, a vertical arrow is added—pointing down for costs and up for revenues or benefits. These cash flows are drawn to scale.

FIGURE 2–10 An example cash flow diagram (CFD).

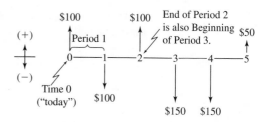

Unless otherwise stated, cash flows are **assumed** to occur at time 0 or at the **end** of each period. Consider Figure 2–10, the CFD for a specific investment opportunity whose cash flows are described as follows:

Timing of Cash Flow	Size of Cash Flow
At time zero (now or today)	A positive cash flow of $100
1 time period from today	A negative cash flow of $100
2 time periods from today	A positive cash flow of $100
3 time periods from today	A negative cash flow of $150
4 time periods from today	A negative cash flow of $150
5 time periods from today	A positive cash flow of $50

Categories of Cash Flows

The expenses and receipts due to engineering projects usually fall into one of the following categories:

First cost ≡ expense to build or to buy and install

Operating and maintenance (O&M) ≡ annual expense, such as electricity, labor, and minor repairs

Salvage value ≡ receipt at project termination for sale or transfer of the equipment (can be a salvage cost)

Revenues ≡ annual receipts due to sale of products or services

Overhaul ≡ major capital expenditure that occurs during the asset's life

Individual projects will often have specific costs, revenues, or user benefits. For example, annual operating and maintenance (O&M) expenses on an assembly line might be divided into direct labor, power, and other. Similarly, a public-sector dam project might have its annual benefits divided into flood control, agricultural irrigation, and recreation.

Drawing a Cash Flow Diagram

The cash flow diagram shows when all cash flows occur. Look at Figure 2–10 and the $100 positive cash flow at the end of period 2. From the time line one can see that this cash flow can also be described as occurring at the *beginning* of period 3. Thus, in a CFD the end of *period t* is the same time as the beginning of *period t* + 1. Beginning-of-period cash flows (such as rent, lease, and insurance payments) are thus easy to handle: just draw your CFD and put them in where they occur. Thus O&M, salvages, revenues, and overhauls are assumed to be end-of-period cash flows.

The choice of time 0 is arbitrary. For example, it can be when a project is analyzed, when funding is approved, or when construction begins. When construction periods are assumed to be short, first costs are assumed to occur at time 0, and the first annual revenues and costs start at the end of the first period. When construction periods are long, time 0 is usually the date of commissioning—when the facility comes on stream.

Perspective is also important when one is drawing a CFD. Consider the simple transaction of paying $5000 for some equipment. To the firm buying the equipment, the cash flow is a cost and hence negative in sign. To the firm selling the equipment, the cash flow is a revenue and positive in sign. This simple example shows that a consistent perspective is required when one is using a CFD to model the cash flows of a problem. One person's cash outflow is another person's inflow.

Often two or more cash flows occur in the same year, such as an overhaul and an O&M expense or the salvage value and the last year's O&M expense. Combining these into one total cash flow per year would simplify the cash flow diagram. However, it is

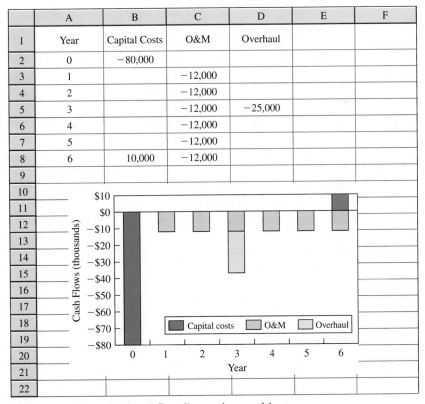

	A	B	C	D	E	F
1	Year	Capital Costs	O&M	Overhaul		
2	0	−80,000				
3	1		−12,000			
4	2		−12,000			
5	3		−12,000	−25,000		
6	4		−12,000			
7	5		−12,000			
8	6	10,000	−12,000			
9						

FIGURE 2–11 Example of cash flow diagram in spreadsheets.

better to show each individually, to ensure a clear connection from the problem statement to each cash flow in the diagram.

Drawing Cash Flow Diagrams with a Spreadsheet

One simple way to draw cash flow diagrams with "arrows" proportional to the size of the cash flows is to use a spreadsheet to draw a stacked bar chart. The data for the cash flows are entered, as shown in the table part of Figure 2–11. To make a quick graph, select cells A1 to D8, which are the years and the three columns of the cash flows. Then select insert and chart, choose column chart, and select the stack option. Except for choosing the scale for the y axis, choosing the y-axis value for the x-axis, and adding titles, the cash flow diagram is done. Refer to the appendix for a review of basic spreadsheet use. (*Note*: A bar chart labels periods rather than using an x axis with arrows at times 0, 1, 2, . . .)

SUMMARY

This chapter has introduced several concepts and definitions for estimating the costs and benefits of proposed engineering projects. Covered were several cost concepts: fixed and variable, marginal and average, sunk, opportunity, recurring and nonrecurring, incremental, cash and book, and life-cycle. **Fixed costs** are constant and unchanging as volumes change, while **variable costs** change as output changes. Fixed and variable costs are used to find a breakeven value between costs and revenues, as well as the regions of net profit and loss. A **marginal cost** is for one more unit, while the **average cost** is the total cost divided by the number of units.

Sunk costs result from past decisions; they should not influence our attitude toward current and future opportunities. Remember, "sunk costs are sunk." **Opportunity costs** involve the benefit that is forgone when we choose to use a resource in one activity instead of another. **Recurring costs** can be planned and anticipated expenses; **nonrecurring costs** are one-of-a-kind costs that are often more difficult to anticipate.

Incremental costs are economic consequences associated with the differences between two choices of action. **Cash costs** are also known as **out-of-pocket costs** that represent actual cash flows. **Book costs** do not result in the exchange of money, but rather are costs listed in a firm's accounting books. **Life-cycle costs** are all costs that are incurred over the life of a product, process, or service. Thus engineering designers must consider life-cycle costs when choosing materials and components, tolerances, processes, testing, safety, service and warranty, and disposal. Historically, firms often focused on internal costs, but green engineering and its consideration of external costs have become increasingly influential.

Cost estimating is the process of "developing the numbers" for engineering economic analysis. Unlike a textbook, the real world does not present its challenges with neat problem statements that provide all the data. **Rough estimates** give us order-of-magnitude numbers and are useful for high-level and initial planning as well as judging the feasibility of alternatives. **Semidetailed estimates** are more accurate than rough-order estimates, thus requiring more resources (people, time, and money) to develop. These estimates are used in preliminary design and budgeting. **Detailed estimates** generally have an accuracy of −3% to 5%. They are used during the detailed design and contract bidding phases of a project.

Difficulties are common in developing estimates. **One-of-a-kind estimates** will have no basis in earlier work, but this disadvantage can be addressed through **estimation by analogy.** Lack of time is best addressed by planning and by matching the estimate's detail to the purpose—one should not spend money developing a detailed estimate when only a rough estimate is needed. **Estimator expertise** must be developed through work experiences and mentors.

Several general models and techniques for developing cost and benefit estimates were discussed. The **per-unit** and **segmenting models** use different levels of detail and costs and benefits per square foot or other unit. **Cost index data** are useful for updating historical costs to formulate current estimates. The **power-sizing model** is useful for scaling up or down a known cost quantity to account for economies of scale, with different power-sizing exponents for industrial plants and equipment of different types. **Triangulation** suggests that one should seek varying perspectives when developing estimates of a project's costs and benefits. Different information sources, databases, and analytical models can all be used to create unique perspectives. As the number of task repetitions increases, efficiency improves because of learning or improvement. This is summarized in the **learning-curve percentage,** where doubling the cumulative production reduces the time to complete the task, which equals the learning-curve percentage times the current production time.

Cash flow estimation must include project benefits. These include labor cost savings, avoided quality costs, direct revenue from sales, reduced catastrophic risks, improved traffic flow, and cheaper power supplies. Benefits are frequently overestimated resulting in an "optimist's bias." **Cash flow diagrams** are used to model the positive and negative cash flows of potential investment opportunities. These diagrams provide a consistent view of the problem (and the alternatives) to support economic analysis.

PROBLEMS

Key to icons: **A** = Answer in Appendix E; **G** = Green, which may include environmental ethics; **E** = Ethics other than green.

Fixed, Variable, Average, and Marginal Costs

2-1 **G** A New York renewable energy company pays $0.18 per kilowatt-hour (kWh) for the first 10,000 units of electricity each month and $0.15/kWh for all remaining units. If a firm uses 25,000 kWh/month, what is its average and marginal cost?

2-2 **A** One of your firm's suppliers discounts prices for larger quantities. The first 1000 parts are $13 each. The next 2000 are $12 each. All parts in excess of 3000 cost $11 each. What are the average cost and marginal cost per part for the following quantities?

(a) 500
(b) 1500
(c) 2500
(d) 3500

2-3 A new machine comes with 200 free service hours over the first year. Additional time costs $180 per hour. What are the average and marginal costs per hour for the following quantities?

(a) 125
(b) 225
(c) 325

2-4 **A** Venus Robotics can produce 23,000 robots a year on its daytime shift. The fixed manufacturing costs per year are $2 million and the total labor cost is $9,109,000. To increase its production to 46,000 robots per year, Venus is considering adding a second shift. The unit labor cost for the second shift would be 25% higher than the day shift, but the total fixed manufacturing costs would increase only to $2.4 million from $2 million.

(a) Compute the unit manufacturing cost for the daytime shift.

(b) Would adding a second shift increase or decrease the unit manufacturing cost at the plant?

2-5 A labor-intensive process has a fixed cost of $338,000 and a variable cost of $143 per unit. A capital-intensive (automated) process for the same product has a fixed cost of $1,244,000 and a variable cost of $92.50 per unit. How many units must be produced and sold at $197 each for the automated process to be preferred to the labor-intensive process?

Contributed by Paul R. McCright, University of South Florida

2-6 CleanTech manufactures equipment to mitigate the environmental effects of waste.

(*a*) If Product *A* has fixed expenses of $15,000 per year and each unit of product has a $0.20 variable cost, and Product *B* has fixed expenses of $5000 per year and a $0.50 variable cost, at what number of units of annual production will *A* have the same overall cost as *B*?

(*b*) As a manager at CleanTech what other data would you need to evaluate these two products?

2-7 Heinrich is a manufacturing engineer with the Miller Company. He has determined the costs of producing a new product to be as follows:

Equipment cost: $288,000/year
Equipment salvage value at EOY5 = $41,000
Variable cost per unit of production: $14.55
Overhead cost per year: $48,300

If the Miller Company uses a 5-year planning horizon and the product can be sold for a unit price of $39.75, how many units must be produced and sold each year to break even? *Contributed by Paul R. McCright, University of South Florida*

2-8 An assembly line can produce 90 units per hour. The line's hourly cost is $4500 on straight time (the first 8 hours). Workers are guaranteed a minimum of 6 hours. There is a 50% premium for overtime, however, productivity for overtime drops by 5%. What are the average and marginal costs per unit for the following daily quantities?

(*a*) 450
(*b*) 600
(*c*) 720
(*d*) 900

2-9 Christine Lynn travels from her home to a remote island. Her trip involves: car travel of 250 miles, air travel of 400 miles, and a boat ride of 75 miles. She is interested in calculating the average fuel cost per mile (per person) of her trip. Assume the fuel efficiency of car, air, and boat travel is 20, 0.20, and 2 miles per gallon, respectively, and that fuel cost per gallon is equal for all and is $3.00/gallon. She was alone in the car and among 180 people on the plane

and 15 on the boat. What is the average fuel cost per mile, per person?

2-10 This month your vendor invoiced $50,000 in testing charges for your production run. The unit cost for testing is twice as much for each of the first 500 units per month as compared to each unit over 500. If we shipped 750 units to the vendor this month, find:

(*a*) Average cost per unit;
(*b*) Cost per unit below the price break point;
(*c*) Marginal cost for the 600th unit.

2-11 The Country Fields Retirement Community charges $6000/month for a single senior citizen to reside in an efficiency apartment with assisted living care. The facility has operating expenses of $600,000 per month. Staffing levels are dependent on the number of residents. Each senior who enters the community requires additional food, personal care, and support staff time. The estimated cost for each person is $4000 per person per month.

(*a*) How many senior citizen residents does the facility need to have in order to reach the breakeven point?
(*b*) What is the company's annual profit or loss if they maintain an average residency level of 350 senior citizens?

Contributed by Gillian Nicholls, Southeast Missouri State University

2-12 The Ozzie Chocolate Company is preparing to offer a new product in its candy offerings, the Minty Dark Chocolate Bite bar. Material costs per new candy bar are $0.20 for chocolate, $0.01 for sugar, and $0.02 for mint flavoring. Labor costs of the new product are approximately $0.12 per bar. Adding a production line devoted to the new candy will cost $250,000 per year.

(*a*) If the sales price is $1.25 per candy bar, how many must the company make per year in order to break even? Assume that each bar made is sold at full price.
(*b*) What is the company's profit or loss if they make and sell 300,000 candy bars at the $1.25 price in the first year?
(*c*) About 20% of the food consumed in the U.S. is imported. Production in many industries has been offshored. What ethical issues do companies face when presented with the decision to move operations?

Contributed by Gillian Nicholls, Southeast Missouri State University

2-13 A small machine shop, with 30 hp of connected load, purchases electricity at the following monthly rates (assume any demand charge is included in this schedule) per hp of connected load:

First 50 kWh at 12.6¢ per kWh

Next 50 kWh at 10.6¢ per kWh

Next 150 kWh at 9.0¢ per kWh

Over 250 kWh at 7.7¢ per kWh

The shop uses 2800 kWh per month.

(a) Calculate the monthly bill for this shop. What are the marginal and average costs per kilowatt-hour?

(b) A contract for additional business would require more operating hours per day. This will use an extra 1200 kWh per month. What is the "cost" of this additional energy? What is this per kilowatt-hour?

(c) New machines would reduce the labor time required on certain operations. These will increase the connected load by 10 hp, but since they will operate only on certain special jobs, will add only 100 kWh per month. In a study to determine the economy of installing these new machines, what should be considered as the "cost" of this energy? What is this per kilowatt-hour?

2-14 Two automatic systems for dispensing maps are being compared by the state highway department. The accompanying breakeven chart of the comparison of these systems (System I vs. System II) shows total yearly costs for the number of maps dispensed per year for both alternatives. Answer the following questions.

(a) What is the fixed cost for System I?

(b) What is the fixed cost for System II?

(c) What is the variable cost per map dispensed for System I?

(d) What is the variable cost per map dispensed for System II?

(e) What is the breakeven point in terms of maps dispensed at which the two systems have equal annual costs?

(f) For what range of annual number of maps dispensed is System I recommended?

(g) For what range of annual number of maps dispensed is System II recommended?

(h) At 3000 maps per year, what are the marginal and average map costs for each system?

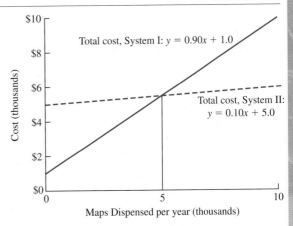

2-15 A privately owned summer camp for youngsters has the following data for a 12-week session:

Charge per camper	$480 per week
Fixed costs	$192,000 per session
Variable cost per camper	$320 per week
Capacity	200 campers

(a) Develop the mathematical relationships for total cost and total revenue.

(b) What is the total number of campers that will allow the camp to *just break even*?

(c) What is the profit or loss for the 12-week session if the camp operates at 80% capacity?

(d) What are marginal and average costs per camper at 80% capacity?

(e) Would it be ethical to charge campers different rates depending on their family's socioeconomic status? Identify and describe two points pro and two points con for such a policy.

2-16 Two new rides are being compared by a local amusement park in terms of their annual operating costs. The two rides would generate the same level of revenue (thus the focus on costs). The Tummy Tugger has fixed costs of $10,000 per year and variable costs of $2.50 per visitor. The Head Buzzer has fixed costs of $4000 per year and variable costs of $4 per visitor.

(a) Mathematically find the number of visitors per year for the two rides to have equal annual costs.

(b) Develop a breakeven graph to show:

- Accurate total cost lines for the two alternatives (show line, slopes, and equations).

- The breakeven point for the two rides in terms of number of visitors.

- The ranges of visitors per year where each alternative is preferred.

2-17 Consider the accompanying breakeven graph for an investment, and answer the following questions.

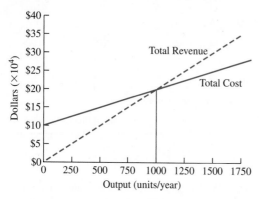

Output (units/year)

(a) Give the equation for total revenue for x units per year.

(b) Give the equation for total costs for x units per year.

(c) What is the "breakeven" level of x?

(d) If you sell 1500 units this year, will you have a profit or loss? How much?

(e) At 1500 units, what are your marginal and average costs?

2-18 Quatro Hermanas, Inc. is investigating implementing some new production machinery as part of its operations. Three alternatives have been identified, and they have the following fixed and variable costs:

Alternative	Annual Fixed Costs	Annual Variable Costs per Unit
A	$100,000	$20.00
B	200,000	5.00
C	150,000	7.50

Determine the ranges of production (units produced per year) over which each alternative would be recommended up to 30,000 units per year.

2-19 Three alternative designs have been created by engineers for a new machine that inspects solar power photovoltaic cells for home use. The costs for the three designs (where x is the annual production rate) follow:

Design	Fixed Cost	Variable Cost ($/x)
A	$100,000	20.5x
B	350,000	10.5x
C	600,000	8.0x

(a) Management is interested in the production interval of 0–150,000 cells per year. Mathematically determine the production volume over which each design (A or B or C) would be chosen.

(b) Depict your solution from part (a) graphically, clearly labeling your axes and including a *title* for the graph, so that management can easily see the following:

 i. Accurate total cost lines for each alternative (show line, slopes, and line equations).

 ii. Any relevant breakeven or crossover points.

 iii. Ranges of annual production where each alternative is preferred.

(c) Our decision rule is to minimize total cost across the range of output. How would your analysis approach change if some alternative produced an ethical dilemma? Describe your thinking.

2-20 A painting operation is performed by a production worker at a labor cost of $1.40 per unit. A robot spray-painting machine, costing $15,000, would reduce the labor cost to $0.20 per unit. If the device would be valueless at the end of 3 years, what is the minimum number of units that would have to be painted each year to justify the purchase of the robot machine?

2-21 Mr. Sam Spade, the president of Ajax, recently read in a report that a competitor named Bendix has the following relationship between cost and production quantity:

$$C = \$3,000,000 - \$18,000Q + \$75Q^2$$

where C = total manufacturing cost per year and Q = number of units produced per year.

A newly hired employee, who previously worked for Bendix, tells Mr. Spade that Bendix is now producing 110 units per year. If the selling price remains unchanged, Sam wonders if Bendix is likely to increase the number of units produced per year in the near future. He asks you to look at the information and tell him what you are able to deduce from it.

2-22 A small company manufactures a certain product. Variable costs are $20 per unit and fixed costs are $10,875. The price–demand relationship for this product is $P = -0.25D + 250$, where P is the unit sales price of the product and D is the annual demand.

● Total cost = Fixed cost + Variable cost

● Revenue = Demand × Price

● Profit = Revenue − Total cost

Set up your graph with dollars on the *y* axis (between 0 and $70,000) and, on the *x* axis, demand *D*: (units produced or sold), between 0 and 1000 units.

(*a*) Develop the equations for total cost and total revenue.

(*b*) Find the breakeven quantity.

(*c*) What profit is earned if total revenue is maximized?

(*d*) What is the company's maximum possible profit?

(*e*) Graph the solutions to each part.

2-23 A firm believes a product's sales volume (*S*) depends on its unit selling price (*P*) as $S = \$100 - P$. The production cost (*C*) is $\$1000 + 10S$.

(*a*) Graph the sales volume (*S*) from 0 to 100 on the *x* axis, total cost and total income from $0 to $2500 on the *y* axis, $C = \$1000 + 10S$, and plot the curve of total income. Mark the breakeven points on the graph.

(*b*) Determine the breakeven point (lowest sales volume at which total sales income just equals total production cost).

(*c*) Determine the sales volume (*S*) at which the firm's profit is a maximum.

Sunk and Other Costs

2-24 Define each of the costs below as either sunk, opportunity, cash, or book:

(*a*) Amount you could have sold a piece of equipment for last month.

(*b*) Value claimed for tax purposes on a depreciated soda filling machine.

(*c*) Price paid to a consultant for a feasibility study on a prospective project.

(*d*) Monthly utility expense.

(*e*) Salary you could have earned while you were on non-paid leave.

(*f*) Value of floor space in a warehouse facility left empty the past 10 quarters.

(*g*) Low price of a stock that you chose not to invest in.

2-25 Define the difference between a "cash cost" and a "book cost." Is engineering economic analysis concerned with both types of cost? Give an example of each, and provide the context in which it is important.

2-26 In your own words, develop a statement of what the authors mean by "life-cycle costs." It is important for a firm to be aware of life-cycle costs. Can you explain why?

2-27 Most engineering students own a computer. What costs have you incurred at each stage of your computer's life cycle? Estimate the total cost of ownership. Estimate the benefits of ownership. Has it been worth it?

2-28 In looking at Figures 2–4 and 2–5, restate in your own words what the authors are trying to get across with these graphs. Do you agree that this is an important effect for companies? Explain.

2-29 In the context of green engineering, what is the difference between internal and external costs? Which type is easier to estimate and quantify, and why?

2-30 List and classify your costs in this academic year as recurring or nonrecurring.

2-31 Last year to help with your New Year's resolutions you purchased a $500 piece of fitness equipment. However, you use it only once a week on average. It is December, and you can sell the equipment for $200 (to someone with a New Year's resolution) and rely on the university gym until you graduate in May. If you don't sell until May, you will get only $100. If you keep the heavy piece, you'll have to pay $25 to move it to the city of your new job (where you interned last summer). There is no convenient gym at the new location. What costs and intangible consequences are relevant to your decision? What should you do?

2-32 You are reevaluating the industrial heat pump choice
Ⓖ that was made last year by your boss. The expected energy savings have not occurred because the pump is too small. Choice *A*, at $90,000, is to replace the pump with one that is the right size and sell the old one. Choice *B*, at $100,000, is to buy a small pump to use in tandem with the existing pump. The two-pump solution has slightly higher maintenance costs, but it is more flexible and reliable. What criteria should you use? Which choice would you recommend, and why?

2-33 Consider the situation of owning rental properties that local university students rent from you for the academic year.

(*a*) Develop a set of costs that you could classify as recurring and others that could be classified as nonrecurring.

(*b*) Research and list the ethical issues that student housing landlords face. What recourse do renters have when they encounter unethical landlords?

2-34 A pump has failed in a facility that will be com-
Ⓐ pletely replaced in 3 years. A brass pump costing $6000 installed will last 3 years. However, a used

stainless steel pump that should last 3 more years has been sitting in the maintenance shop for a year. The pump cost $13,000 new. The accountants say the pump is worth $7000 now. The maintenance supervisor says that it will cost an extra $500 to reconfigure the pump for the new use and that he could sell it used (as is) for $4000.

(a) What is the book cost of the stainless steel pump?
(b) What is the opportunity cost of the stainless steel pump?
(c) How much cheaper or more expensive would it be to use the stainless steel pump rather than a new brass pump?

2-35 Owning, operating, and maintaining an automobile carries with it private internal costs to the driver, as well as public external societal costs. List five internal and external costs.

2-36 Bob Johnson decided to buy a new home. After looking at tracts of new homes, he decided that a custom-built home was preferable. He hired an architect to prepare the plans for a fee of $7000. While a building contractor was working on a bid to construct the home on a lot Bob already owned, Bob found a standard house plans on the Internet for $200 that he and his wife liked better. Bob then asked the contractor to provide a bid to construct this "stock plan" home. The building contractor submitted the following bids:

Custom-designed home	$258,000
Stock plan home	261,000

Bob was willing to pay the extra $3000 for the stock-plan home. Bob's wife, however, felt they should go ahead with the custom-designed home, for, as she put it, "We can't afford to throw away a set of plans that cost $7000." Bob agreed, but he disliked the thought of building a home that is less desirable than the stock plan home. Which house would you advise him to build? Explain.

2-37 Identify the following costs as either internal to the firm or external to society:

Surveying	Lost animal habitat	Materials
Land	Public health	Reduced fisheries
Reduced water quality	Recreational loss	Administrative
	Roadway congestion	Air quality
Legal	Community	Noise pollution
Overhead	Construction	Virus tolerances
Roadway	Species extinction	Health benefits
Design	Waste storage	
Equipment	Labor	

2-38 A paper manufacturer creates industrial waste that is treated before being released into a local waterway. The EPA strictly monitors the level of this harmful waste. To maintain low levels of parts-per-million (PPM) of harmful waste requires expense on the part of the firm, and as a result they try to stay just under the allowable level. There is currently a 25% chance at each semi-annual inspection that the firm will receive a $400,000 fine.

(a) How much should the firm be willing to pay on an annual basis to lease a new process that will decrease the risk of non-compliance to 5% per inspection?
(b) If the EPA fine represents the external cost, what is the average cost to society for each scenario?

2-39 LED light bulbs are more energy efficient, last longer, cost more, and have a smaller environmental impact from manufacture to recycling or disposal. The question is which costs less: incandescent, compact fluorescent (CFL), or LED lights?

	Incandescent	CFL	LED
Cost/bulb	$0.50	$2	$7
Life in hours	1250	8333	25,000
Watts used	60	14	7

(a) Ignore the time value of money. What is the average cost per hour for 25,000 hours of lighting in a community with $0.15/kWh electricity for each choice?
(b) If the light is on 5%, 30%, or 60% of the time, how long until does it take for the 25,000 hours?
(c) Which type of light should a homeowner buy?

Estimating Process

2-40 In the text we describe three effects that complicate the process of making estimates to be used in engineering economy analyses. List these three effects and comment on which might be the most influential.

2-41 Develop an estimate for each of the following situations.

(a) The cost of a 500-mile automobile trip, if gasoline is $4 per gallon, vehicle wear and tear is $0.65 per mile, and our vehicle gets 25 miles per gallon.
(b) The total number of hours in the average human life, if the average life is 75 years.
(c) The number of days it takes to travel around the equator using a hot air balloon, if the balloon

averages 100 miles per day, the diameter of the earth is ~4000 miles.

(*d*) The total area in square miles of the United States of America, if Kansas is an average-sized state. Kansas has an area of 390 miles × 200 miles.

2-42 A local minor league baseball team plays home games in a stadium that has four sections that each hold 2000 fans. On an average night, attendance in each higher section is filled to 70% the capacity of one below it. Estimate attendance on an average night if the bottom section is completely filled: What percent of the stadium is full?

2-43 Your dining room (11′ × 13′) has a hardwood floor discolored by uneven fading from sunlight exposure, and your living room (12′ × 12′) has wall-to-wall carpet ruined by your aging dog. You have obtained some cost figures (all per square foot) from a local flooring store for refinished and new hardwood floors. The cost to purchase and install red oak flooring is $3.55 (includes tax), and labor is $1.50. Refinishing costs the same $1.50. For new or old floors the labor/material/tax cost to apply a new light stain is $1.00 and two coats of polyurethane cost another $1.00. The store's sales manager estimates that you should include a 5% wastage factor on the wood purchase. Assume you will be removing and disposing of the carpet yourself without cost. Estimate the cash cost of this project. *Contributed by Gillian Nicholls, Southeast Missouri State University*

Segmenting Models

2-44 Northern Tundra Telephone (NTT) has received a
(A) contract to install emergency phones along a new 100-mile section of the Snow-Moose Turnpike. Fifty emergency phone systems will be installed about 2 miles apart. The material cost of a unit is $125. NTT will need to run underground communication lines that cost NTT $7500 per mile (including labor) to install. There will also be a one-time cost of $10,000 to network these phones into NTT's current communication system.

(*a*) Develop a cost estimate of the project from NTT's perspective.

(*b*) If NTT adds a profit margin of 35% to its costs, how much will it cost the state to fund the project?

2-45 You and your spouse are planning a second honeymoon to the Cayman Islands this summer and would like to have your house painted while you are

away. Estimate the total cost of the paint job from the information given below, where:

$$\text{Cost}_{\text{total}} = \text{Cost}_{\text{paint}} + \text{Cost}_{\text{labor}} + \text{Cost}_{\text{fixed}}$$

Paint information: Your house has a surface area of 6000 ft^2. One can of paint can cover 300 ft^2. You are estimating the cost to put on *two coats* of paint for the entire house.

Number of Cans Purchased	Cost per Can
First 10 cans purchased	$15.00
Second 15 cans purchased	$12.50
Up to next 50 cans purchased	$12

Labor information: You plan to hire five painters who will paint for 8 hours per day each. You estimate that the job will require 4.5 days of their painting time. The painter's rate is $18 per hour.

Fixed cost information: The painting company charges $1200 per job to cover travel expenses, clothing, cloths, thinner, administration, and so on.

2-46 You want a mountain cabin built for weekend trips,
(A) vacations, to host family, and perhaps eventually to retire in. After discussing the project with a local contractor, you receive an estimate that the total construction cost of your 2000-ft^2 lodge will be $250,000. The percentage of costs is broken down as follows:

Cost Items	Percentage of Total Costs
Construction permits, legal and title fees	8
Roadway, site clearing, preparation	15
Foundation, concrete, masonry	13
Wallboard, flooring, carpentry	12
Heating, ventilation, air conditioning (HVAC)	13
Electric, plumbing, communications	10
Roofing, flooring	12
Painting, finishing	17
	100

(*a*) What is the cost per square foot of the 2000-ft^2 lodge?

(b) If you are also considering a 4000-ft^2 layout option, estimate your construction costs if:

    i. All cost items (in the table) change proportionately to the size increase.

    ii. The first two cost items do not change at all; HVAC changes by 50%; and all others are proportionate.

2-47 SungSam, Inc. is designing a new digital camcorder that is projected to have the following per-unit costs to manufacture:

Cost Categories	Unit Costs
Materials costs	$ 63
Labor costs	24
Overhead costs	110
Total unit cost	$197

SungSam adds 30% to its manufacturing cost for corporate profit.

(a) What unit profit would SungSam realize on each camcorder?

(b) What is the overall cost to produce a batch of 10,000 camcorders?

(c) What would SungSam's profit be on the batch of 10,000 if historical data show that 1% of product will be scrapped in manufacturing, 3% of finished product will go unsold, and 2% of *sold* product will be returned for refund?

(d) How much can SungSam afford to pay for a contract that would lock in a 50% reduction in the unit material cost previously given? If SungSam does sign the contract, the sales price will not change.

Indexes and Sizing Models

2-48 Estimate the cost of expanding a planned new clinic by 20,000 ft^2. The appropriate capacity exponent is 0.66, and the budget estimate for 200,000 ft^2 was $15 million.

2-49 A new aerated sewage lagoon is required in a small town. Earlier this year one was built on a similar site in an adjacent city for $2.3 million. The new lagoon will be 65% larger.

(a) Use the data in Table 2–1 to estimate the cost of the new lagoon,

(b) What are lagoon systems? Are they thought to be environmentally friendly? What are the pros and cons of this technology?

2-50 Fifty years ago, Grandma Bell purchased a set of gold-plated dinnerware for $55, and last year you inherited it. Unfortunately a fire at your home destroyed the set. Your insurance company is at a loss to define the replacement cost and has asked your help. You do some research and find that the Aurum Flatware Cost Index (AFCI) for gold-plated dinnerware, which was 112 when Grandma Bell bought her set, is at 2050 today. Use the AFCI to update the cost of Bell's set to today's cost to show to the insurance company.

2-51 Your boss is the director of reporting for the Athens County Construction Agency (ACCA). It has been his job to track the cost of construction in Athens County. Twenty-five years ago he created the ACCA Cost Index to track these costs. Costs during the first year of the index were $12 per square foot of constructed space (the index value was set at 100 for that first year). This past year a survey of contractors revealed that costs were $90 per square foot. What index number will your boss publish in his report for this year? If the index value was 600 last year, what was the cost per square foot last year?

2-52 A refinisher of antiques named Constance has been so successful with her small business that she is planning to expand her shop. She is going to start enlarging her shop by purchasing the following equipment.

Equipment	Original Size	Cost of Original Equipment	Power-Sizing Exponent	New Equipment Size
Varnish bath	50 gal	$3500	0.80	75 gal
Power scraper	3/4 hp	$250	0.22	1.5 hp
Paint booth	3 ft^3	$3000	0.6	12 ft^3

(a) What would be the *net* cost to Constance to obtain this equipment? Assume that she can trade the old equipment in for 15% of its original cost. Assume there has been no inflation in equipment prices.

(b) Suggest a green engineering approach to constance for disposing of the solvents and lacquers used in her business.

2-53 Refer to Problem 2-52 and now assume the prices for the equipment that Constance wants to replace have not been constant. Use the cost index data for each piece of equipment to update the costs to the price that would be paid today. Develop the overall cost for Constance, again assuming the 15% trade-in allowance for the old equipment. Trade-in value is based on original cost.

Original Equipment	Cost Index When Originally Purchased	Cost Index Today
Varnish bath	154	171
Power scraper	780	900
Paint booth	49	76

2-54 Five years ago, when the relevant cost index was
A 120, a nuclear centrifuge cost $40,000. The centrifuge had a capacity of separating 1500 gallons of ionized solution per hour. Today, it is desired to build a centrifuge with capacity of 4500 gallons per hour, but the cost index now is 300. Assuming a power-sizing exponent to reflect economies of scale, x, of 0.75, use the power-sizing model to determine the approximate cost (expressed in today's dollars) of the new reactor.

2-55 Padre works for a trade magazine that publishes lists of power-sizing exponents (PSE) that reflect economies of scale for developing engineering estimates of various types of equipment. Padre has been unable to find any published data on the VMIC machine and wants to list its *PSE* value in his next issue. Given the following data calculate the *PSE* value that Padre should publish. (*Note:* The VMIC-100 can handle twice the volume of a VMIC-50.)

> Cost of VMIC-100 today $250,000
> Cost of VMIC-50 5 years ago $95,000
> VMIC equipment index today = 350
> VMIC equipment index 5 years ago = 235

2-56 Remerowski Corporation Inc. asks you to estimate
A the cost to purchase a new piece of production equipment. The company purchased this same type of equipment in the past for $10,000. The original equipment had a capacity of 2000 units, while the new equipment has a capacity of 1000 units. The power-sizing exponent for this type of equipment is 0.28. In addition, the cost index for this type of equipment was 126 when the original unit was purchased and is now 160. If Remerowski uses an 8% annual interest rate to evaluate equipment purchases, estimate the cost to purchase the new piece.

Contributed by Gillian Nicholls, Southeast Missouri State University

2-57 Sage Lorimer owns a boutique that sells jewelry made by her husband and other local artists. She is considering expanding the business by opening a second store. The current store has 1200 square feet of retail space and 400 square feet of space for administrative offices. For the new store she would like to double the retail space and triple the administrative space of the current store. At the time the current shop was purchased, retail space sold for $75 per square foot and general office space cost $30 per square foot. The cost indexes at the time were 153 for retail space and 120 for office space. The costs indexes are now 140 for retail space and 132 for office space. Estimate the cost for Sage to purchase a second store at this time. *Contributed by Gillian Nicholls, Southeast Missouri State University*

Learning-Curve Models

2-58 If 200 labor hours were required to produce the
A 1st unit in a production run and 60 labor hours were required to produce the 7th unit, what was the *learning-curve rate* during production?

2-59 Ima New is a recent employee who initially requires 20 minutes to complete a job task. If she experiences an 80% learning rate, how much time will it take her to complete the sixth task?

2-60 Rose is a project manager at the civil engineering
A consulting firm of Sands, Gravel, Concrete, and Waters, Inc. She has been collecting data on a project in which concrete pillars were being constructed; however not all the data are available. She has been able to find out that the 10th pillar required 260 person-hours to construct and that a 75% learning curve applied. She is interested in calculating the time required to construct the 1st and 20th pillars. Compute the values for her.

2-61 Home Building Inc. (HBI) seeks to schedule manual labor for 18 new homes being constructed. Historical data leads HBI to apply a 92% learning curve rate to the manual labor portions of the project. If the first home requires 3500 manual labor hours to build, estimate the time required to build (*a*) the fifth house, (*b*) the tenth house, (*c*) all 18 houses. What would the manual labor estimate be for all 18 of the HBI houses in the problem above if the learning curve rate is (*d*) 70%, (*e*) 75%, (*f*) 80%?

2-62 Ima Neworker requires 30 minutes to produce her first unit of output. If her learning curve rate is 65%, how many units will be produced before the output rate exceeds 12 units per hour?

2-63 An industrial engineering consulting firm usually observes a 90% learning curve rate in the installation of enterprise level software with its clients. If the first installation required 75 hours, estimate the time required for (a) the fifth, (b) the tenth, and (c) the twentieth installations. (d) Research the AMCF Code of Ethics. How are these similar to and different from engineering society ethics statements from your discipline?

2-64 Sally Statistics is implementing a system of statistical process control (SPC) charts in her factory in an effort to reduce the overall cost of scrapped product. The current cost of scrap is X per month. If an 80% learning curve is expected in the use of the SPC charts to reduce the cost of scrap, what would the *percentage reduction* in monthly scrap cost be after the charts have been used for 12 months? (*Hint:* Model each month as a unit of production.)

2-65 A new product made from recycled bio-plastics needs 20 labor hours to complete the build for the first unit. If production operates at a 85% learning curve rate, calculate the average labor hours required per unit to produce five units.

2-66 New technicians in an oncology department process patients at rates shown below. Steady state occurs at the eighth unit.

Patient number:	1	2	3	4	5	6	7	8
Process time (min):	75	61	52	50	45	43	40	39

(a) Calculate the learning curve rate for units 1–8.
(b) What is the total time needed to process 12 patients?

Benefit Estimation

2-67 Estimating benefits is often more difficult than cost estimation. Use the example of car ownership to describe the complicating factors in estimating the costs and benefits.

2-68 The authors mention in a few places in the chapter that benefits are overestimated while costs tend to be underestimated—they called this the "optimist's bias." What do they mean by this, and what effect can it have on the fair evaluation of projects?

2-69 Create a 2 × 2 table labeling the columns "costs" and "benefits," and the rows "recurring" and "non-recurring costs." Think about your life and list two to three items/examples in each cell.

2-70 Projects A and B have the same cost to implement, maintain, and dispose of over their life cycles. What impact will any incremental benefits play in the comparison of the alternatives?

2-71 Many types of employees are on performance-based contracts—which combine fixed and variable salary amounts depending on the performance (productivity) of the worker. Give examples of employees who are on these types of contracts.

2-72 Develop a statement that expresses the extent to which cost-estimating topics also apply to estimating benefits. Provide examples to illustrate.

2-73 Large projects, such as a new tunnel under the Hudson, the Big Dig in Boston, the Denver airport, a new military jet, and a natural gas pipeline from Alaska to the Midwest, often take 5 to 15 years from concept to completion.

(a) Should benefit and cost estimates be adjusted for the greater influences and impacts of inflation, government regulatory changes, and changing local economic environments? Why or why not?
(b) How does the public budget-making process interact with the goal of accurate benefit and cost estimating for these large projects?

Contributed by Morgan Henrie, University of Alaska Anchorage

2-74 A home run king's team pays him a base salary of $15.0 million per season plus $100,000 per home run over 50. If the king hits 70 home runs next season, what will his salary be for the year?

Cash Flow Diagrams

2-75 On December 1, Al Smith purchased a car for $25,000. He paid $7500 immediately and agreed to pay three additional payments of $8000 each (which includes principal and interest) at the end of 1, 2, and 3 years. Maintenance for the car is projected at $1000 at the end of the first year and $2000 at the end of each subsequent year. Al expects to sell the car at the end of the fourth year (after paying for the maintenance work) for $12,000. Using these facts, prepare a table of cash flows.

2-76 Bonka Toys is considering a robot that will cost $75,000 to buy. After 5 years its salvage value will be $18,000. An overhaul costing $10,000 will be needed in Year 3. O&M costs will be $2000 per year. Draw the cash flow diagram.

2-77 Pine Village needs some additional recreation fields. Construction will cost $300,000, and annual O&M expenses are $85,000. The city council estimates that the value of added youth leagues is about $200,000 annually. In year 6 another $75,000 will be needed to refurbish the fields. The salvage value is estimated to be $50,000 after 10 years. Draw the cash flow diagram.

2-78 Identify your major cash flows for the current school term as first costs, O&M expenses, salvage values, revenues, overhauls, and so on. Using a week as the time period, draw the cash flow diagram.

2-79 HiTech Inc. is investing in a new production line to manufacture their newest high volume consumer product. Design costs for the past three years, leading up to production, have been $0.75 million per year. Today production machinery totaling $3.65 million is being purchased to equip the new line. Operating and maintenance expenses will total $50,000 per year, and materials and overhead costs will be $1.75 million annually. The new line will operate for 7 years, at which time equipment will be sold at 10% of purchase price. Revenue from the new product is expected to be $2.0 million the first year, increasing by $0.50 million per year for the next 4 years, then decreasing by $0.75 million in years 6 and 7. HiTech has asked you to create a cash flow table and diagram.

Multiple Objectives A

2-80 A firm is bidding on a contract for a government
G building, which will be awarded using an augmented lowest bidder model. Most points will be awarded based on the firm's bid, but 3 extra points will be awarded if (1) the firm is minority, female, or veteran owned, or (2) the design will satisfy the platinum level LEED (Leadership in Energy and Environmental Design) criteria. The government estimate is $32M. The firm does not satisfy the ownership category, but it is deciding whether or not to develop its bid to meet the LEED standard. The firm's rough estimate (including profit) is $28M without meeting LEED. The estimated cost for meeting the LEED criteria is $650,000. The government evaluation function is

$$\text{Score} = (\text{Gov't. Estimate / Bid}) \times 100$$
$$+ \{0 \text{ or } 3\} \text{ bonus points.}$$

Should the firm's bid be developed with or without a platinum LEED design?

2-81 A large international firm has decided to include
G external costs in its decision making about its products. The firm's leaders believe this approach is better than focusing on meeting a variety of changing national regulations. While the firm is developing its knowledge base on estimating social and environmental costs, it plans to add 5% of the external costs to its internal costs. In the long run, its goal is use a 20% value for external costs. One process now costs $25M per year in internal costs, with an estimated $8M in external costs. Rough estimates indicate that spending an extra $1.2M internally would reduce the external costs to $3M. What total *cost* values should be used to evaluate this change?

(*a*) Now
(*b*) Later with a 20% weight

CASES

The following cases from *Cases in Engineering Economy* (www.oup.com/us/newnan) are suggested as matched with this chapter.

CASE 3 **Wildcat Oil in Kasakstan**
Rough order-of-magnitude estimation of total facility cost and annual revenue. Option for NPV and different size facility.

CASE 25 **Raster Blaster**
Questions to guide students. Includes breakeven analysis.

TRUST ME: YOU'LL USE THIS

Contributed by Kate Abel, Stevens Institute of Technology

- Engineering in the real world is about money.
  - Projects cost money now. The benefits may come in for decades.
- Engineering economy also applies to your personal life.
  - This "trust me" introduces some of the ways.

Most of us borrow money using credit cards or loans for cars, college, houses, etc.

- How much do you trust the person offering the loan? Are there hidden fees? Do payments match the verbal terms?
- With these tools you can calculate the payments. Know the rules of the game so you cannot be taken advantage of.

Your credit score is between 350 (very bad) and 850 (outstanding).

- What does it depend on?
  - Do you have a credit history? It helps if you do.
  - What is your debt/income ratio? Not too high is best.
  - Do you make payments on time? Very important.
- Why does it matter?
  - Follows you for the rest of your life.
  - Lower credit scores $\longrightarrow$ loan not available or at a higher interest rate.
  - Some employers use credit scores to screen resumes.

Why save now, when you can borrow when needed? See also Chapters 4–7.

- If young parents invested $200 a month at 7% for 10 years; then the $24,000 in savings would earn $11,000 in interest. A college fund of $35,000 results.
- Suppose instead a $35,000 loan is used. If paid back over the same 10 years and at the same 7% interest rate, then an additional $13,000 in interest or a total of $48,000 would be paid.

Which debt to pay off first? Goal is to minimize interest paid. See also Chapters 6 and 7.

- The best approach is to pay off the highest interest debt as fast as possible, and just pay the minimum on the other debts.
- Payday loans have the highest rate.
- Credit card debts are usually next.

- If you've missed a payment or paid late, some credit cards may charge you 25–40% interest. So making at least the minimum payment on time each month is very important.
- Pay credit card balances in full each month if you can, then most cards charge *no* interest for the next month.
- Sometimes the best strategy is to pay off a credit card with a lower interest rate and a lower balance, and then use only that card for new charges which are paid in full each month, while you pay off any other credit card balances.

When should one save money vs. pay off debt? See Appendix 12A.

- Having an emergency fund is important! Cars, computers, phones, . . . may be stolen or need repair—now. A job can be lost or sickness/injury may mean you can't work for a while. An emergency fund comes first.
- Saving is paying yourself and it should become a habit as early in life as possible.
- Often investment returns are higher than low cost debt such as a mortgage.

Why save and invest when I have so much college debt? See Chapters 4 and 7 and Appendix 9A.

- The best investment tool is the *power of time.*
- If you want $20,000 in 15 years for a down payment on a house, how much must you save assuming a 6% return?

  - Start today and $69/month is enough.
  - Start after 5 years and $123/month is needed.
  - Start after 10 years and $286/month is needed.
  - That $69 a month may be a challenge right out of college, but it will feel like nothing later on.

- You may feel you can't afford to save. It seems easier to save later when your income is higher, but your expenses will also be higher. See Appendix 12A.

Other important questions you will learn how to answer in Chapters 4–7 and Appendices 9A, 10A, and 12A.

- What's the real interest rate on the car loan at 0% the dealer is advertising?
- How long will it take to pay off this debt if I pay the minimum, make an extra payment, or pay $100 extra each month?
- What are reasonable expectations for different types of investment?
- How is insurance included in financial planning?

3

INTEREST AND EQUIVALENCE

A Prescription for Success

A pharmaceutical company manufactured a prescription drug that contained a tablet inside another tablet. This inner tablet, called a "core," needed to be in the exact center of the larger tablet, and tolerances were measured in tenths of millimeters. The process was not robust, and the placement of the inner tablet sometimes drifted, requiring the scrapping of off-quality product and the adjustment of the tablet press. This resulted in significant scrap and tablet press downtime.

A process change was invented in Japan to correct the problem, using a new process to place the inner tablet in the die of the press that made the outer tablet. There were three of these tablet presses in use in the U.S., but modifications were made to one tablet press as a test. The modification to the first press cost $27,000. During the first batch the modified press ran the entire batch without a quality problem and without quality losses. The batch finished compressing in 16 hours, which was considerably faster than the typical time of 24 hours (however, core centering problems could cause a delay of several days).

Additional test batches were run, all with excellent results. A detailed quality examination proved that the modification performed as desired, reducing variation and nearly eliminating product quality scrap. The other two tablet presses were later modified. The total cost for all modifications, including spare parts, was $90,000.

Over the next year, the results of the change were analyzed. Product yield increased from 92.4% to 96.5%. Because less of the expensive active drug was scrapped and instead became good product, each 1% increase in yield was valued at $2.4 million per year. Operating efficiency improved, resulting in higher output because of less scrap and less downtime due to quality problems. Production plans called for 240 batches to be processed over the year after the tablet press modification was made. This product was produced daily, but production was reduced from three shifts to two because of the improved efficiency. Production planning could now plan effectively; they knew that a batch could be processed in two shifts, not one to five days.

Year-end accounting showed $10 million saved in the first year. Because the product's patent was about to expire, production was expected to be greatly reduced beyond this time. ■ ■ ■

QUESTIONS TO CONSIDER

1. One year of production had a value of $240 million. What is the value of one batch of product?

2. How many batches needed to be produced to break even on the initial $27,000 investment? (Assume all batches improved the yield by 4.2%. Do not consider the time value of money.)

3. If the first-year savings is considered to be a single end-of-year cash flow, and the entire $90,000 investment is considered to occur at time 0, what is the present value of the project? Assume an interest rate of 15%.

4. If one batch is produced per day, how often are the savings actually compounded?

5. Does a company face any ethical considerations when it improves process efficiency resulting in lost labor hours for employees? Discuss and explain.

After Completing This Chapter...

The student should be able to:

- Define and provide examples of the *time value of money*.
- Distinguish between *simple* and *compound interest,* and use compound interest in engineering economic analysis.
- Explain *equivalence* of cash flows.
- Solve problems using the single payment compound interest formulas.
- Distinguish and apply *nominal* and *effective interest rates*.

Key Words

annual percentage rate (APR)	effective interest rate	simple interest
cash flows	equivalence	technique of equivalence
compound interest	interest	time value of money
continuous compounding	nominal interest rate	
disbursement	receipt	

In the first chapter, we discussed situations where the economic consequences of an alternative were immediate or took place in a very short period of time, as in Example 1–2 (design of a concrete aggregate mix) or Example 1–3 (change of manufacturing method). We totaled the various positive and negative aspects, compared our results, and could quickly reach a decision. But can we do the same if the economic consequences occur over a considerable period of time?

No we cannot, because *money has value over time.* Would you rather (1) receive $1000 today or (2) receive $1000 ten years from today? Obviously, the $1000 today has more value. Money's value over time is expressed by an interest rate. This chapter describes two introductory concepts involving the *time value of money:* interest and cash flow equivalence. Later in the chapter, nominal and effective interest are discussed. Finally, equations are derived for situations where interest is compounded continuously.

Most examples in this chapter are loans and savings accounts for individuals. This keeps the examples simple and emphasizes that engineering economy is part of everyday life, and not just used for evaluation of large engineering projects. Values in the hundreds or thousands are also easier to read. All of the extra zeros in millions or hundreds of millions can get in the way of understanding.

COMPUTING CASH FLOWS

Installing expensive machinery in a plant obviously has economic consequences that occur over an extended period of time. If the machinery was bought on credit, the simple process of paying for it could take several years. What about the usefulness of the machinery? Certainly it was purchased because it would be a beneficial addition to the plant. These favorable consequences may last as long as the equipment performs its useful function. In these circumstances, we describe each alternative as cash **receipts** or **disbursements** at different points in **time.** Since earlier cash flows are more valuable than later cash flows, we cannot just add them together. Instead, each alternative is resolved into a set of **cash flows,** as in Examples 3–1 and 3–2.

Example 3–1 is the iconic engineering economy problem for both teaching and practice. There is a cash flow at the beginning, the annual cash flows are estimated to be uniform, and there is a cash flow at the end. Until the cash flow diagram is instantly visualized as you read the problem, it should be sketched. (Just like a free-body diagram in statics.) Figure 2–11 shows how a spreadsheet can be used to create a cash flow diagram.

EXAMPLE 3–1

A machine will cost $30,000 to purchase. Annual operating and maintenance costs (O&M) will be $2000. The machine will save $10,000 per year in labor costs. The salvage value of the machine after 5 years will be $7000. Draw the cash flow diagram.

SOLUTION

Even though some cash flows occur at the same time, they should not be combined. Not only does showing each cash flow clearly link the text and the diagram, we will find that it also makes it easier to calculate equivalent values.

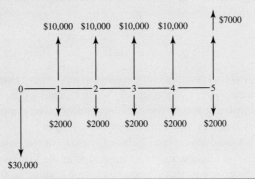

EXAMPLE 3–2

A man borrowed $1000 from a bank at 8% interest. He agreed to repay the loan in two end-of-year payments. At the end of each year, he will repay half of the $1000 principal amount plus the interest that is due. Compute the borrower's cash flow.

SOLUTION

In engineering economic analysis, we normally refer to the beginning of the first year as "time 0." At this point the man receives $1000 from the bank. (A positive sign represents a receipt of money and a negative sign, a disbursement.) The time 0 cash flow is +$1000.

At the end of the first year, the man pays 8% interest for the use of $1000 for that year. The interest is $0.08 \times \$1000 = \$80$. In addition, he repays half the $1000 loan, or $500. Therefore, the end-of-year-1 cash flow is −$580.

At the end of the second year, the payment is 8% for the use of the balance of the principal ($500) for the one-year period, or $0.08 \times 500 = \$40$. The $500 principal is also repaid for a total end-of-year-2 cash flow of −$540. The cash flow is:

End of Year	Cash Flow
0 (now)	+$1000
1	−580
2	−540

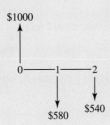

Techniques for comparing the value of money at different dates are the foundation of engineering economic analysis. We must be able to compare, for example, a low-cost pump with a higher-cost pump. If there were no other consequences, we would obviously prefer the low-cost one. But what if the higher-cost pump is more durable? Then we must consider whether to spend more money now to postpone the future cost of a replacement pump. This chapter will provide the methods for comparing the alternatives to determine which pump is preferred.

TIME VALUE OF MONEY

We often find that the monetary consequences of any alternative occur over a substantial period of time—say, a year or more. When monetary consequences occur in a short period of time, we simply add up the various sums of money and obtain a net result. But can we treat money this way when the time span is greater?

Which would you prefer, $100 cash today or the assurance of receiving $100 a year from now? You might decide you would prefer the $100 now because that is one way to be certain of receiving it. But suppose you were convinced that you would receive the $100 one year hence. Now what would be your answer? A little thought should convince you that it *still* would be more desirable to receive the $100 now. If you had the money now, rather than a year hence, you would have the use of it for an extra year. And if you had no current use for $100, you could let someone else pay you to use it.

Money is quite a valuable asset—so valuable that people are willing to pay to have money available for their use. Money can be rented in roughly the same way one rents an apartment; only with money, the charge is called **interest** instead of rent. The importance of interest is demonstrated by banks and savings institutions continuously offering to pay for the use of people's money, to pay interest.

If the current interest rate is 9% per year and you put $100 into the bank for one year, how much will you receive back at the end of the year? You will receive your original $100 together with $9 interest, for a total of $109. This example demonstrates the time preference for money: we would rather have $100 today than the assured promise of $100 one year hence; but we might well consider leaving the $100 in a bank if we knew it would be worth $109 one year hence. This is because there is a **time value of money** in the form of the willingness of banks, businesses, and people to pay interest for the use of various sums.

Simple Interest

While simple interest is rarely used, it is our starting point for a better understanding of compound interest. **Simple interest** is interest that is computed only on the original sum, not on accrued interest. Thus if you were to loan a present sum of money P to someone at a simple annual interest rate i (stated as a decimal) for a period of n years, the amount of interest you would receive from the loan would be

$$\text{Total interest earned} = P \times i \times n = Pin \tag{3-1}$$

At the end of n years the amount of money due you, F, would equal the amount of the loan P plus the total interest earned. That is, the amount of money due at the end of the loan would be

$$F = P + Pin \tag{3-2}$$

or $F = P(1 + in)$.

EXAMPLE 3–3

You have agreed to loan a friend $5000 for 5 years at a simple interest rate of 8% per year. How much interest will you receive from the loan? How much will your friend pay you at the end of 5 years?

SOLUTION

$$\text{Total interest earned} = Pin = \$5000 \times 0.08 \times 5 \text{ yr} = \$2000$$

$$\text{Amount due at end of loan} = P + Pin = 5000 + 2000 = \$7000$$

In Example 3–3 the interest earned at the end of the first year is $(5000)(0.08)(1) = \$400$, but this money is not paid to the lender until the end of the fifth year. As a result, the borrower has the use of the $400 for 4 years without paying any interest on it. This is how simple interest works, and it is easy to see why lenders seldom agree to make simple interest loans.

Compound Interest

With simple interest, the amount earned (for invested money) or due (for borrowed money) in one period does not affect the principal for interest calculations in later periods. However, this is not how interest is normally calculated. In practice, interest is computed by the **compound interest** method. For a loan, any interest owed but not paid at the end of the year is added to the balance due. Then the next year's interest is calculated on the unpaid balance due, which includes the unpaid interest from the preceding period. In this way, compound interest can be thought of as *interest on top of interest*. This distinguishes compound interest from simple interest. In this section, the remainder of the book, and in practice you should **assume that the rate is a compound interest rate**. The very few exceptions, such as some end-of-chapter problems, will clearly state "simple interest."

EXAMPLE 3–4

To highlight the difference between simple and compound interest, rework Example 3–3 using an interest rate of 8% per year compound interest. How will this change affect the amount that your friend pays you at the end of 5 years?

Original loan amount (original principal) $= \$5000$

Loan term $= 5$ years

Interest rate charged $= 8\%$ per year compound interest

SOLUTION

In the following table we calculate on a year-to-year basis the total dollar amount due at the end of each year. Notice that this amount becomes the principal upon which interest is calculated in the next year (this is the compounding effect).

Year	Total Principal (P) on Which Interest Is Calculated in Year n	Interest (I) Owed at the End of Year n from Year n's Unpaid Total Principal	Total Amount Due at the End of Year n, New Total Principal for Year $n+1$
1	$5000	$5000 \times 0.08 = 400$	$5000 + 400 = 5400$
2	5400	$5400 \times 0.08 = 432$	$5400 + 432 = 5832$
3	5832	$5832 \times 0.08 = 467$	$5832 + 467 = 6299$
4	6299	$6299 \times 0.08 = 504$	$6299 + 504 = 6803$
5	6803	$6803 \times 0.08 = 544$	$6803 + 544 = 7347$

The total amount due at the end of the fifth year is $7347. This is $347 more than you received for loaning the same amount, for the same period, at simple interest. This is because of the effect of interest being earned on top of interest.

Repaying a Debt

To better understand the mechanics of interest, let us say that $5000 is owed and is to be repaid in 5 years, together with 8% annual interest. There are a great many ways in which debts are repaid; for simplicity, we have selected four specific ways for our example. Table 3–1 tabulates the four plans.

Plan 1 (Constant Principal), like Example 3–2, repays $1/n^{\text{th}}$ of the principal each year. So in Plan 1, $1000 will be paid at the end of each year plus the interest due at the end of the year for the use of money to that point. Thus, at the end of Year 1, we will have had the use of $5000. The interest owed is $8\% \times \$5000 = \$400$. The end-of-year payment is $1000 principal *plus* $400 interest, for a total payment of $1400. At the end of Year 2, another $1000 principal plus interest will be repaid on the money owed during the year. This time the amount owed has declined from $5000 to $4000 because of the Year 1 $1000 principal payment. The interest payment is $8\% \times \$4000 = \$320$, making the end-of-year payment a total of $1320. As indicated in Table 3–1, the series of payments continues each year until the loan is fully repaid at the end of the Year 5.

In *Plan 2 (Interest Only)* only the interest due is paid each year, with no principal payment. Instead, the $5000 owed is repaid in a lump sum at the end of the fifth year. The end-of-year payment in each of the first 4 years of Plan 2 is $8\% \times \$5000 = \$400$. The fifth year, the payment is $400 interest *plus* the $5000 principal, for a total of $5400.

Plan 3 (Constant Payment) calls for five equal end-of-year payments of $1252 each. In Example 4–3, we will show how the figure of $1252 is computed. By following the computations in Table 3–1, we see that a series of five payments of $1252 repays a $5000 debt in 5 years with interest at 8%.

Plan 4 (All at Maturity) repays the $5000 debt like Example 3–4. In this plan, no payment is made until the end of Year 5, when the loan is completely repaid. Note what happens at the end of Year 1: the interest due for the first year—$8\% \times \$5000 = \$400$—is not paid; instead, it is added to the debt. At the second year then, the debt has increased to $5400. The Year 2 interest is thus $8\% \times \$5400 = \$432$. This amount, again unpaid, is added to the debt, increasing it further to $5832. At the end of Year 5, the total sum due has grown to $7347 and is paid at that time.

TABLE 3–1 Four Plans for Repayment of $5000 in 5 Years with Interest at 8%

(a) Year	(b) Amount Owed at Beginning of Year	(c) Interest Owed for That Year, 8% × (b)	(d) Total Owed at End of Year, (b) + (c)	(e) Principal Payment	(f) Total End-of-Year Payment
Plan 1: Constant principal payment *plus* interest due.					
1	$5000	$ 400	$5400	$1000	$1400
2	4000	320	4320	1000	1320
3	3000	240	3240	1000	1240
4	2000	160	2160	1000	1160
5	1000	80 / $1200	1080	1000 / $5000	108C / $6200
Plan 2: Annual interest payment and principal payment at end of 5 years.					
1	$5000	$ 400	$5400	$ 0	$ 400
2	5000	400	5400	0	400
3	5000	400	5400	0	400
4	5000	400	5400	0	400
5	5000	400 / $2000	5400	5000 / $5000	5400 / $7000
Plan 3: Constant annual payments.					
1	$5000	$ 400	$5400	$ 852	$1252*
2	4148	331	4479	921	1252
3	3227	258	3485	994	1252
4	2233	178	2411	1074	1252
5	1159	93 / $1260	1252	1159 / $5000	1252 / $6260
Plan 4: All payment at end of 5 years.					
1	$5000	$ 400	$5400	$ 0	$ 0
2	5400	432	5832	0	0
3	5832	467	6299	0	0
4	6299	504	6803	0	0
5	6803	544 / $2347	7347	5000 / $5000	7347 / $7347

*The exact value is $1252.28, which has been rounded to an even dollar amount.

Note that when the $400 interest was not paid at the end of Year 1, it was added to the debt and, in Year 2 there was interest charged on this unpaid interest. That is, the $400 of unpaid interest resulted in 8% × $400 = $32 of additional interest charge in Year 2. That $32, together with 8% × $5000 = $400 interest on the $5000 original debt, brought the total interest charge at the end of the Year 2 to $432. Charging interest on unpaid interest is called **compound interest.**

With Table 3–1 we have illustrated four different ways of accomplishing the same task, that is, to repay a debt of $5000 in 5 years with interest at 8%. Having described the alternatives, we will now use them to present the important concept of *equivalence*.

EQUIVALENCE

When we are indifferent as to whether we have a quantity of money now or the assurance of some other sum of money in the future, or series of future sums of money, we say that the present sum of money is **equivalent** to the future sum or series of future sums at a specified interest rate.

If an industrial firm believed 8% was a reasonable interest rate, it would have no particular preference about whether it received $5000 now or the payments from any plan in Table 3–1. In fact, *all four repayment plans are equivalent to each other and to $5000 now at 8% interest.*

Equivalence is an essential concept in engineering economic analysis. In Chapter 2, we saw how an alternative could be represented by a cash flow table. For example, consider the cash flows from Table 3–1:

Year	Plan 1	Plan 2	Plan 3	Plan 4
1	−$1400	−$400	−$1252	$0
2	−1320	−400	−1252	0
3	−1240	−400	−1252	0
4	−1160	−400	−1252	0
5	−1080	−5400	−1252	−7347
Total	−$6200	−$7000	−$6260	−$7347
Interest Paid	$1200	$2000	$1260	$2347

If you were given your choice between the alternatives, which one would you choose? Obviously the plans have cash flows that are different, and you cannot compare the totals, the cash flows, or the interest paid. To make a decision, we must use the **technique of equivalence**.

We can determine an **equivalent value** at some point in time for each plan, based on a selected interest rate. Then we can judge the relative attractiveness of the alternatives, not from their cash flows, but from comparable equivalent values. Since each plan repays a *present* sum of $5000 with interest at 8%, all plans are equivalent to $5000 *now* at an interest rate of 8%. This cannot be deduced from the given cash flows alone. It is necessary to learn this by determining the equivalent values for each alternative at some point in time, which in this case is "the present."

Difference in Repayment Plans

The four plans computed in Table 3–1 are equivalent in nature but different in structure. Table 3–2 repeats the end-of-year payment schedules from Table 3–1 and also graphs each plan to show the debt still owed at any point in time. Since $5000 was borrowed at the beginning of the first year, all the graphs begin at that point. We see, however, that the

TABLE 3–2 End-of-Year Payment Schedules and Their Graphs

Plan 1 (Constant Principal): At end of each year pay $1000 principal *plus* interest due.

Year	End-of-Year Payment
1	$1400
2	1320
3	1240
4	1160
5	1080
	$6200

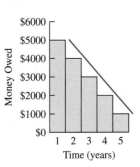

Plan 2 (Interest Only): Pay interest due at end of each year and principal at end of 5 years.

Year	End-of-Year Payment
1	$ 400
2	400
3	400
4	400
5	5400
	$7000

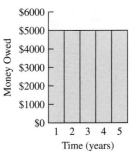

Plan 3 (Constant Payment): Pay in five equal end-of-year payments.

Year	End-of-Year Payment
1	$1252
2	1252
3	1252
4	1252
5	1252
	$6260

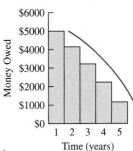

Plan 4 (All at Maturity): Pay principal and interest in one payment at end of 5 years.

Year	End-of-Year Payment
1	$ 0
2	0
3	0
4	0
5	7347
	$7347

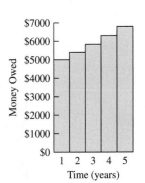

four plans result in quite different amounts of money owed at any other point in time. In Plans 1 and 3, the money owed declines as time passes. With Plan 2 the debt remains constant, while Plan 4 increases the debt until the end of the fifth year. These graphs show an important difference among the repayment plans—the areas under the curves differ greatly. Since the axes are *Money Owed* and *Time,* the area is their product: Money owed × Time, in dollar-years.

The dollar-years for the four plans would be as follows:

	Dollar-Years			
	Plan 1	**Plan 2**	**Plan 3**	**Plan 4**
Money owed in Year 1 × 1 year	$ 5,000	$ 5,000	$ 5,000	$ 5,000
Money owed in Year 2 × 1 year	4,000	5,000	4,148*	5,400
Money owed in Year 3 × 1 year	3,000	5,000	3,227*	5,832
Money owed in Year 4 × 1 year	2,000	5,000	2,233*	6,299
Money owed in Year 5 × 1 year	1,000	5,000	1,159*	6,803
Total dollar-years	$15,000	$25,000	$15,767	$29,334

* Chapter 4 details the calculations of the amount owed each year with a constant total $ payment.

With the area under each curve computed in dollar-years, the ratio of total interest paid to area under the curve may be obtained:

Plan	Total Interest Paid	Area Under Curve (dollar-years)	Ratio of Total Interest Paid to Area Under Curve
1	$1200	15,000	0.08
2	2000	25,000	0.08
3	1260	15,767	0.08
4	2347	29,334	0.08

We see that the ratio of total interest paid to the area under the curve is constant and equal to 8%. Stated another way, the total interest paid equals the interest rate *times* the area under the curve.

From our calculations, we more easily see why the repayment plans require the payment of different total sums of money, yet are actually equivalent. The key factor is that the four repayment plans provide the borrower with different quantities of dollar-years. Since dollar-years times interest rate equals the interest charge, the four plans result in different total interest charges.

Equivalence Is Dependent on Interest Rate

In the example of Plans 1 to 4, all calculations were made at an 8% interest rate. At this interest rate, it has been shown that all four plans are equivalent to a present sum of $5000. But what would happen if we were to change the interest rate?

If the interest rate were increased to 9%, we know that the interest payment for each plan would increase, and the current calculated repayment schedules (Table 3–1, column **f**) would repay a sum *less* than the principal of $5000. By some calculations (to be explained later in this chapter and in Chapter 4), the equivalent present sum that each plan will repay at 9% interest is:

Plan	Repay a Present Sum of
1	$4877
2	4806
3	4870
4	4775

At the higher 9% interest, each of the repayment plans of Table 3–1 repays a present sum less than $5000. And, they do not repay the *same* present sum. Plan 1 would repay $4877 with 9% interest, while Plan 4 would repay only $4775. Thus, with interest at 9%, none of the plans is economically equivalent, for each will repay a different present sum. The series of payments were equivalent at 8%, but not at 9%. This leads to the conclusion **that equivalence is dependent on the interest rate.** Changing the interest rate destroys the equivalence between the series of payments.

Could we create revised repayment schemes that would be equivalent to $5000 now with interest at 9%? Yes, of course we could—but the revised plans would not be equivalent at 8%.

Thus far we have discussed computing equivalent present sums for a cash flow. But the technique of equivalence is not limited to a present computation. Instead, we could compute the equivalent sum for a cash flow at any point in time. We could compare alternatives in "Equivalent Year 10" dollars rather than "now" (Year 0) dollars. Furthermore, the equivalence need not be a single sum; it could be a series of payments or receipts. In Plan 3 of Table 3–1, the series of equal payments was equivalent to $5000 now. But the equivalency works both ways. Suppose we ask, What is the equivalent equal annual payment continuing for 5 years, given a present sum of $5000 and interest at 8%? The answer is $1252.

Differences in Economically Equivalent Plans

While Plans 1 to 4 are economically equivalent at 8% interest, they represent different approaches to paying back $5000. For example, most consumers when buying a car or a home are offered a repayment schedule similar to Plan 3 with a uniform payment. With a uniform repayment schedule the borrower is paying *all* of the interest due from that period along with at least a small payment toward reducing the balance due. Lenders prefer to reduce the risk of repayment problems by offering either Plan 1 or Plan 3. Borrowers prefer Plan 3 to Plan 1 because it has a lower initial payment for the same principal.

Plan 2 has only an interest payment every period, and the principal is repaid at the end. As will be described in Chapter 5, most long-term borrowing by firms and governments is done by issuing bonds that fit this pattern. Short-term borrowing by firms and governments often follows Plan 4.

Thus economically equivalent cash flows may have differences, such as the risk of non-payment, that are important for decision making.

SINGLE PAYMENT COMPOUND INTEREST FORMULAS

To facilitate equivalence computations, a series of **interest formulas** will be derived. We use the following notation:

i = *interest rate per interest period;* in the equations the interest rate is stated as a decimal (that is, 9% interest is 0.09)

n = *number of interest periods*

P = *a present sum of money*

F = *a future sum of money* at the end of the nth interest period, which is equivalent to P with interest rate i

Suppose a present sum of money P is invested for one year[1] at interest rate i. At the end of the year, we should receive back our initial investment P, together with interest equal to iP, or a total amount $P + iP$. Factoring P, the sum at the end of one year is $P(1 + i)$.

Let us assume that, instead of removing our investment at the end of one year, we agree to let it remain for another year. How much would our investment be worth at the end of Year 2? The end-of-first-year sum $P(1 + i)$ will draw interest in the second year of $iP(1 + i)$. This means that at the end of Year 2 the total investment will be

$$P(1 + i) + i[P(1 + i)]$$

This may be rearranged by factoring out $P(1 + i)$, which gives

$$P(1 + i)(1 + i) \quad \text{or} \quad P(1 + i)^2$$

If the process is continued for Year 3, the end-of-the-third-year total amount will be $P(1 + i)^3$; at the end of n years, we will have $P(1 + i)^n$. The progression looks like:

Year	Amount at Beginning of Interest Period	+ Interest for Period	= Amount at End of Interest Period
1	P	$+ iP$	$= P(1 + i)$
2	$P(1 + i)$	$+ iP(1 + i)$	$= P(1 + i)^2$
3	$P(1 + i)^2$	$+ iP(1 + i)^2$	$= P(1 + i)^3$
n	$P(1 + i)^{n-1}$	$+ iP(1 + i)^{n-1}$	$= P(1 + i)^n$

In other words, a present sum P increases in n periods to $P(1 + i)^n$. We therefore have a relationship between a present sum P and its equivalent future sum, F.

$$\text{Future sum} = (\text{Present sum}) (1 + i)^n$$

$$F = P(1 + i)^n \tag{3-3}$$

This is the **single payment compound amount formula** and is written in factor or functional notation as

$$F = P(F/P, i, n) \tag{3-4}$$

---

[1]A more general statement is to specify "one interest period" rather than "one year." Since it is easier to visualize one year, the derivation uses one year as the interest period.

The notation in parentheses $(F/P, i, n)$ can be read as follows:

To find a future sum F, given a present sum, P, at an interest rate i per interest period, and n interest periods hence.

or

Find F, given P, at i, over n.

Functional notation is designed so that the compound interest factors may be written in an equation in an algebraically correct form. In Equation 3–4, for example, the functional notation is interpreted as

$$F = P\left(\frac{F}{P}\right)$$

which is dimensionally correct. Without proceeding further, we can see that when we derive a compound interest factor to find a present sum P, given a future sum F, the factor will be $(P/F, i, n)$; so, the resulting equation would be

$$P = F(P/F, i, n)$$

which is dimensionally correct.

EXAMPLE 3–5

If \$500 were deposited in a bank savings account, how much would be in the account 3 years from now if the bank paid 6% interest compounded annually?

FORMULA SOLUTION

From the viewpoint of the person depositing the \$500, the cash flows are:

The present sum P is \$500. The interest rate per interest period is 6%, and in 3 years there are three interest periods. The future sum F is found by using Equation 3–3, where $P = \$500$, $i = 0.06$, $n = 3$, and F is unknown:

$$F = P(1 + i)^n = 500(1 + 0.06)^3 = \$595.50$$

Thus if we deposit \$500 now at 6% interest, there will be \$595.50 in the account in three years.

TABLE SOLUTION

The equation $F = P(1 + i)^n$ need not be solved. Instead, *the single payment compound amount factor, $(F/P, i, n)$, is readily found in the tables given in Appendix C.*[2] In this case the factor is

$$(F/P, 6\%, 3)$$

Knowing $n = 3$, locate the proper row in the 6% table of Appendix C. To find F given P, look in the first column, which is headed "Single Payment, Compound Amount Factor": or F/P for $n = 3$, we find 1.191.

Thus,

$$F = 500(F/P, 6\%, 3) = 500(1.191) = \$595.50$$

5-BUTTON SOLUTION

Appendix B describes spreadsheet annuity functions, financial calculators, and programmable scientific calculators that can be used instead of or in addition to the tables. Programmable scientific calculators can be used on the FE exam.

These spreadsheet functions and calculators make the same assumptions as the tabulated factors. But fewer numbers have to be entered, and many calculations can be completed much more quickly. Many problems can be solved by entering four values chosen from $i, n, P, A,$ and F, and solving for the fifth value. Thus, these solutions are shown as 5-BUTTON SOLUTIONS.

As detailed in Appendix B, this solution corresponds to the more powerful spreadsheet and financial calculator functions that can also be programmed into HP calculators that can be used on the FE exam. Note that the A or PMT variable is the focus of Chapter 4, and it is 0 here.

	A	B	C	D	E	F	G	H	I	
1	Problem	i	n	PMT	PV	FV	Solve for	Answer	Formula	
2	Exp. 3-5	6%	3	0	-500		FV	$595.51	=FV(B2,C2,D2,E2)	
3									FV(rate, nper, pmt, [pv],[type])	
4										

BANK'S POINT OF VIEW

Before leaving this problem, let's draw another diagram of it, this time from the bank's point of view.

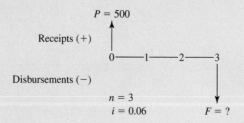

Footnote

This indicates that the bank receives $500 now and must make a disbursement of F at the end of 3 years. The computation, from the bank's point of view, is

$$F = 500(F/P, 6\%, 3) = 500(1.191) = \$595.50$$

This is exactly the same as what was computed from the depositor's viewpoint, since this is just the other side of the same transaction. The bank's future disbursement equals the depositor's future receipt.

If we take $F = P(1 + i)^n$ and solve for P, then

$$P = F\frac{1}{(1 + i)^n} = F(1 + i)^{-n}$$

This is the **single payment present worth formula.** The equation

$$P = F(1 + i)^{-n} \tag{3-5}$$

in our notation becomes

$$P = F(P/F, i, n) \tag{3-6}$$

EXAMPLE 3–6

If you wish to have $800 in a savings account at the end of 4 years, and 5% interest will be paid annually, how much should you put into the savings account now?

FORMULA SOLUTION

$$F = \$800, \qquad i = 0.05, \qquad n = 4, \qquad P = \text{unknown}$$
$$P = F(1 + i)^{-n} = 800(1 + 0.05)^{-4} = 800(0.8227) = \$658.16$$

Thus to have $800 in the savings account at the end of 4 years, we must deposit $658.16 now.

TABLE SOLUTION

$$P = F(P/F, i, n) = \$800(P/F, 5\%, 4)$$

From the compound interest tables,

$$(P/F, 5\%, 4) = 0.8227$$

$$P = \$800(0.8227) = \$658.16$$

5-BUTTON SOLUTION

As detailed in Appendix B, this solution corresponds to the spreadsheet and financial calculator functions that can also be programmed into HP calculators that can be used on the FE exam. Note that the *A* or *PMT* variable is the focus of Chapter 4, and it is 0 here.

Note that a positive future value results in a negative present value.

	A	B	C	D	E	F	G	H	I	
1	Problem	*i*	*n*	*PMT*	*PV*	*FV*	Solve for	Answer	Formula	
2	Exp. 3-6	5%	4	0		800	PV	-$658.16	=PV(B2,C2,D2,F2)	
3									PV(rate, nper, pmt, **[fv]**,[type])	
4										

Here the problem has an exact answer. In many situations, however, the answer is rounded off, since it can be only as accurate as the input information on which it is based.

It is useful to examine how compound interest *adds up*, by examining the future values from Eq. 3-3 over time for different interest rates. Example 3–7 builds a spreadsheet to tabulate and graph a set of $(F/P, i, n)$ values.

EXAMPLE 3–7

Tabulate the future value factor for interest rates of 5%, 10%, and 15% for *n*'s from 0 to 20 (in 5's).

SOLUTION

This example has been built in Excel and some of the described short-cuts may not work in other spreadsheet packages—and they may work differently for you even in Excel as menus and capabilities are continually evolving.

It is probably most efficient to simply enter the three interest rates in row 1. However, it is probably more efficient to enter the first two years of 0 and 5, select them, and then extend the selection to autofill the values of 10, 15, and 20.

Then the formula $(1 + i)^n$ is entered into cell B2 as =(1+B$1)^$A2. Note that the interest rates are all in row 1, so that the B$1 cell reference fixes the row as an absolute address. Since the years are all in column A, the $A2 cell reference fixes the column as an absolute address (see Appendix A for a more complete explanation of relative and absolute references). The formula for cell D6 is included in the table to show how the relative and absolute addresses change as the formula is copied. In the example the formatting of cell B2 was changed to 3 decimals before doing the copying.

This formula can be copied into the rest of column B by simply double-clicking on the cell's lower right corner. Then select cells B2:B6, and drag the lower right corner to D6. The table is complete.

	A		B	C	D	E	F
1	*n*	*i*	5%	10%	15%		
2	0		1.000	1.000	1.000		
3	5		1.276	1.611	2.011		
4	10		1.629	2.594	4.046		
5	15		2.079	4.177	8.137		
6	20		2.653	6.727	16.367	=(1+D$1)^$A6	

To graph the values, it is easiest to select cells A2:D6 and INSERT a scatter or *xy* plot. In some versions of Excel, the figure may look strange until the data are selected and the row and column are switched (A2:A6 needed to be the *x* variable). The 3 series are then edited one-by-one so that the interest rates cells were entered as NAMES for the series. The data series was reordered (up/down arrows under SELECT DATA) so that the 15% legend label would be on top and the 5% label on the bottom (to match the graph).

So that the curves would "fill" the figure, each axis was selected so it could be formatted for maximum values and for the *y*-axis the number of decimals. Finally under CHART LAYOUT titles were added to the chart and each axis. Since this was designed for print, the data series also had to be edited to change the line-type rather than relying on color to distinguish the series.

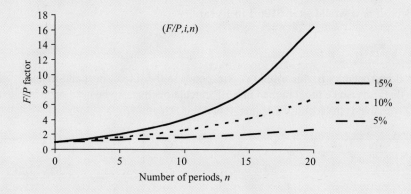

NOMINAL AND EFFECTIVE INTEREST

EXAMPLE 3–8

Suppose the bank changed the interest policy in Example 3–5 to "6% interest, compounded quarterly." For this situation, how much money would be in the account at the end of 1 year, assuming a $100 deposit now?

TABLE SOLUTION

First, we must be certain to understand the meaning of *6% interest, compounded quarterly*. There are two elements:

6% interest: Unless otherwise described, it is customary to assume that the stated interest is for a one-year period. *If the stated interest is for any other period, the time frame must be clearly stated.*

Compounded quarterly: This indicates there are four interest periods per year; that is, an interest period is 3 months long.

We know that the 6% interest is an annual rate because if it were for a different period, it would have been stated. Since we are dealing with four interest periods per year, it follows that the interest rate per interest period is $1\frac{1}{2}\%$ $(= 6\%/4)$. For the total 1-year duration, there are 4 interest periods. Thus

$$P = \$100, \qquad i = 0.06/4 = 0.015, \qquad n = 1 \times 4 = 4, \qquad F = \text{unknown}$$

$$F = P(1 + i)^n = P(F/P, i, n)$$

$$= \$100(1 + 0.015)^4 = \$100(F/P, 1^1/_2\%, 4)$$

$$= \$100(1.061) = \$106.1$$

A \$100 deposit now would yield \$106.1 in 1 year.

5-BUTTON SOLUTION

As detailed in Appendix B, this solution corresponds to the spreadsheet and financial calculator functions that can also be programmed into HP calculators that can be used on the FE exam. The *A* or *PMT* variable is the focus of Chapter 4, and it is 0 here.

	A	B	C	D	E	F	G	H	I
1	Problem	i	n	*PMT*	*PV*	*FV*	Solve for	Answer	Formula
2	Exp. 3-8	1.5%	4	0	-100		FV	-\$106.14	=FV(B2,C2,D2,E2)

Nominal interest rate per year, r, is the annual interest rate without considering the effect of any compounding.

In Example 3–8, the bank pays $1^1/_2\%$ interest every 3 months. The nominal interest rate per year, r, therefore, is $4 \times 1^1/_2\% = 6\%$. The federal government mandates that lenders provide the **annual percentage rate (APR)** for any loan. For credit cards and many consumer loans this is the nominal interest rate. Calculating the APR for a mortgage with origination fees and points is more complex.

Effective interest rate per year, i_a, is the *annual* interest rate taking into account the effect of any compounding during the year.

In Example 3–8, we saw that $100 left in the savings account for one year increased to $106.14, so the interest paid was $6.14. The effective interest rate per year, i_a, is $6.14/$100.00 = 0.0614 = 6.14\%$. The effective annual interest rate earned on savings accounts, certificates of deposit (CD), bonds, and so on is often called the annual percentage yield (**APY**). To calculate the effective annual interest rate i_a, we use the following variables:

$$r = \text{nominal interest rate per interest period (usually one year)}$$

$$i = \text{effective interest rate per interest period}$$

$$m = \text{number of compounding subperiods per time period}$$

Using the method presented in Example 3–8, we can derive the equation for the effective interest rate. If a $1 deposit were made to an account that compounded interest m times per year and paid a nominal interest rate per year, r, the *interest rate per compounding subperiod* would be r/m, and the total in the account at the end of one year would be

$$\$1\left(1 + \frac{r}{m}\right)^m \qquad \text{or simply} \qquad \left(1 + \frac{r}{m}\right)^m$$

If we deduct the $1 principal sum, the expression would be

$$\left(1 + \frac{r}{m}\right)^m - 1$$

Therefore,

Effective annual interest rate $\qquad i_a = \left(1 + \frac{r}{m}\right)^m - 1$ $\qquad$ (3-7)

where $r = $ nominal interest rate per year

$\qquad m = $ number of compounding subperiods per year

Or, substituting the effective interest rate per compounding subperiod, $i = (r/m)$,

Effective annual interest rate $\qquad i_a = (1 + i)^m - 1$ $\qquad$ (3-8)

where $i = $ effective interest rate per compounding subperiod

$\qquad m = $ number of compounding subperiods per year

Either Equation 3–7 or 3–8 may be used to compute an effective interest rate per year. This and many other texts use i_a for the effective annual interest rate. The Fundamentals of Engineering (FE) exam uses i_e for the effective annual interest rate in its supplied material.

EXAMPLE 3–9

If a credit card charges $1^1/_2\%$ interest every month, what are the nominal and effective interest rates per year?

SOLUTION

$$\text{Nominal interest rate per year} \quad r = 12 \times 1^1/_2\% = 18\%$$

$$\text{Effective annual interest rate} \quad i_a = \left(1 + \frac{r}{m}\right)^m - 1$$

$$= \left(1 + \frac{0.18}{12}\right)^{12} - 1 = 0.1956$$

$$= 19.56\%$$

Alternately,

$$\text{Effective annual interest rate} \quad i_a = (1 + i)^m - 1$$

$$= (1 + 0.015)^{12} - 1 = 0.1956 = 19.56\%$$

5-BUTTON SOLUTION

This solution corresponds to the spreadsheet and financial calculator functions. See Appendix B for details.

If $PV = -1$, then $FV = 1+$ effective rate (for the given number of compounding periods and interest rate per compounding period). The complete formula for the interest rate is *effective rate* $= (FV - PV)/PV$ since the total period is 1 year.

	A	B	C	D	E	F	G	H	I
1	Problem	i	n	PMT	PV	FV	Solve for	Answer	Formula
2	Exp. 3-9	1.50%	12	0	-1		FV	1.1956	=FV(B2,C2,D2,E2)

So subtracting 1 from the FV value and restating as an interest rate, the effective rate is 19.56%.

SPREADSHEET SOLUTION

This problem can also be solved with the function =EFFECT(nominal_rate, npery).

	A	B	C	D	E
1	Nominal annual rate, r	Periods per year, m	Effective rate, i_a	Answer	Spreadsheet Function
2	18.00%	12		19.56%	=EFFECT(A2,B2)

There is a corresponding =NOMINAL(effect_rate, npery) function to convert effective rates to nominal.

EXAMPLE 3–10

A payday lender lends money on the following terms: "If I give you $100 today, you will write me a check for $120, which you will redeem or I will cash on your next payday." Noting that calculated rates would be even higher for closer paydays, assume that the payday is two weeks away.

    (a) What nominal interest rate per year (r) is the lender charging?

    (b) What effective interest rate per year (i_a) is the lender charging?

    (c) If the lender started with $100 and was able to keep it, as well as all the money received, loaned out at all times, how much money does the lender have at the end of one year?

SOLUTION TO PART a

$$F = P(F/P, i, n)$$

$$120 = 100(F/P, i, 1)$$

$$(F/P, i, 1) = 1.2$$

Therefore, $i = 20\%$ per 2-week period

$$\text{Nominal interest rate per year} = 26 \times 0.20 = 5.20 = 520\%$$

SOLUTION TO PART b

Effective annual interest rate $i_a = \left(1 + \dfrac{r}{m}\right)^m - 1$

$$= \left(1 + \frac{5.20}{26}\right)^{26} - 1 = 114.48 - 1 = 113.48 = 11{,}348\%$$

Or

    Effective annual interest rate $i_a = (1 + i)^m - 1$

$$= (1 + 0.20)^{26} - 1 = 113.48 = 11{,}348\%$$

SOLUTION TO PART c

$$F = P(1 + i)^n = 100(1 + 0.20)^{26} = \$11{,}448$$

$$= \$100 \text{ principal} + \$11{,}348 \text{ interest}$$

5-BUTTON/ SPREADSHEET SOLUTION

This solution corresponds to the spreadsheet and financial calculator functions.

If $PV = -1$, then $FV = 1+$ effective rate (for the given number of compounding periods and interest rate per compounding period). The complete formula for the interest rate is *effective rate* $= (FV - PV)/PV$ since the total period is 1 year.

	A	B	C	D	E	F	G	H	I
1	Problem	i	n	PMT	PV	FV	Solve for	Answer	Formula
2	Exp. 3-10		1	0	-100	120	i	20%	=RATE(C2,D2,E2,F2)
3									
4	Nominal	20%	26				r	520%	=B3*C3
5									
6	Effective	520%	26				i_a	11348%	=EFFECT(B6,C6)
7	or	20%	26	0	-1		FV	114.48	=FV(B9,C9,D9,E9)
8							i_a	113.48	=FV-1
9									
10	FV	20%	26	0	-100		FV	$11,448	=FV(B8,C8,D8,E8)

One should note that i was described earlier simply as the interest rate per interest period. We were describing the effective interest rate without making any fuss about it. A more precise definition, we now know, is that i is the *effective* interest rate per interest period. Although it seems more complicated, we are describing the same exact situation, but with more care.

The nominal interest rate r is often given for a one-year period (but it could be given for either a shorter or a longer time period). In the special case of a nominal interest rate that is given per compounding subperiod, the effective interest rate per compounding subperiod, i_m, equals the nominal interest rate per subperiod, r.

In the typical effective interest computation, there are multiple compounding subperiods ($m > 1$). The resulting effective interest rate is either the solution to the problem or an intermediate solution, which allows us to use standard compound interest factors to proceed to solve the problem.

For **continuous compounding** the effective annual interest rate is,

$$i_a = e^r - 1 \tag{3-9}$$

Table 3–3 tabulates the effective interest rate for a range of compounding frequencies and nominal interest rates. It should be noted that when a nominal interest rate is compounded annually, the nominal interest rate equals the effective interest rate. Also, it should be noted that increasing the frequency of compounding to semiannually or monthly matters more than increasing the frequency to daily or continuously.

Continuous compounding and cash flows that occur throughout a year would be a *better theoretical* model for most engineering economy problems. However, such models are extremely rare and are becoming more so because they are not supported by Excel functions. Thus, continuous compounding is most interesting as the limit for an effective interest rate, as shown in Example 3−11.

TABLE 3–3 Nominal and Effective Interest

Nominal Interest Rate per Year	Effective Annual Interest Rate, i_a When Nominal Rate Is Compounded				
r (%)	Yearly	Semiannually	Monthly	Daily	Continuously
1	1%	1.0025%	1.0046%	1.0050%	1.0050%
2	2	2.0100	2.0184	2.0201	2.0201
3	3	3.0225	3.0416	3.0453	3.0455
4	4	4.0400	4.0742	4.0809	4.0811
5	5	5.0625	5.1162	5.1268	5.1271
6	6	6.0900	6.1678	6.1831	6.1837
8	8	8.1600	8.3000	8.3278	8.3287
10	10	10.2500	10.4713	10.5156	10.5171
15	15	15.5625	16.0755	16.1798	16.1834
25	25	26.5625	28.0732	28.3916	28.4025

EXAMPLE 3–11   (Example 3–10 Revisited)

If the savings bank in Example 3–10 changes its interest policy to 6% interest, compounded continuously, what are the nominal and the effective interest rates?

SOLUTION

The nominal interest rate remains at 6% per year.

$$\text{Effective interest rate} = e^r - 1$$
$$= e^{0.06} - 1 = 0.0618$$
$$= 6.18\%$$

SUMMARY

This chapter describes cash flow tables, the time value of money, and equivalence. The single payment compound interest formulas were derived. These concepts and these interest formulas are the foundation of the rest of this book and the practice of engineering economy.

Time value of money: The continuing offer of banks to pay interest for the temporary use of other people's money is ample proof that there is a time value of money. Thus, we would always choose to receive $100 today rather than the promise of $100 to be paid at a future date.

Equivalence: What sum would a person be willing to accept a year hence instead of $100 today? If a 9% interest rate is appropriate, $109 would be required a year hence. If $100 today and $109 a year from today are considered equally desirable, we say the two sums of money are equivalent. But, if we decided that a 12% interest rate is applicable, then $109 a year from today would no longer be equivalent to $100 today. Equivalence depends on the interest rate.

This chapter also defined simple interest, where interest does not carry over and become part of the principal in subsequent periods. Unless otherwise specified, all interest rates in this text are compound rates.

Single Payment Formulas

These formulas are for compound interest, which is used in engineering economy.

$$\textbf{Compound amount} \qquad F = P(1 + i)^n \ \ = P(F/P, i, n)$$
$$\textbf{Present worth} \qquad P = F(1 + i)^{-n} = F(P/F, i, n)$$

where i = interest rate per interest period (stated as a decimal)

n = number of interest periods

P = a present sum of money

F = a future sum of money at the end of the nth interest period that is equivalent

   to P with interest rate i

Nominal Annual Interest Rate, *r*

The annual interest rate without considering the effect of any compounding. Also called the annual percentage rate (APR).

Effective Annual Interest Rate, i_a

The annual interest rate taking into account the effect of any compounding during the year.
   Effective annual interest rate (periodic compounding):

$$i_a = \left(1 + \frac{r}{m}\right)^m - 1$$

or

$$i_a = (1 + i)^m - 1$$

Effective annual interest rate (continuous compounding):

$$i_a = e^r - 1$$

PROBLEMS

Key to icons: **A** = Answer in Appendix E;
G = Green, which may include environmental
ethics; **E** = Ethics other than green.

Simple and Compound Interest

3-1 How would you describe the difference between simple and compound interest to a class of 3rd graders? If you are not sure if the interest that you are paying is simple or compound, how will you find out?

3-2 **A** **E** A student borrowed $5000 from her parents and agreed to repay it at the end of 4 years, together with 6% simple interest.

(a) How much is repaid?
(b) Do you ever borrow or lend money with family and friends? Given the time value of money, is it ethical not to pay interest? How does the amount or duration of the loan matter?

3-3 A $5000 loan was to be repaid with 5% simple annual interest. A total of $6000 was paid. How long had the loan been outstanding?

Contributed by Hamed Kashani, Saeid Sadri, and Baabak Ashuri, Georgia Institute of Technology

3-4 **A** At an interest rate of 10% per year, $100,000 today is equivalent to how much a year from now?

Contributed by Hamed Kashani, Saeid Sadri, and Baabak Ashuri, Georgia Institute of Technology

3-5 Calculate the interest and total amount due at the end of the loan for both simple and compound interest.

	Loan	Years	Rate
(a)	$ 1000	2	5%
(b)	$ 1500	5	6%
(c)	$10,000	10	10%
(d)	$25,000	15	15%
(e)	$47,750	20	20%

3-6 **A** How long will it take for an investment to double at a 6% per year _____?

(a) simple interest rate
(b) compound interest rate

Contributed by Hamed Kashani, Saeid Sadri, and Baabak Ashuri, Georgia Institute of Technology

3-7 **G** Carolina Land Trust purchases private land for clean waterway conservation. A benefactor invested $1,000,000 for 15 years and gave the principal and accrued interest of $1,200,000 to the trust. (a) What simple interest rate did the investment earn? (b) Had the interest been compounded annually at that rate, what would have been the value of the gift?

Equivalence

3-8 Which is more valuable, $20,000 received now or $5000 per year for 4 years? Why? Explain the term "time value of money" in your own words.

3-9 Magdalen, Miriam, and Mary June were asked to consider two different cash flows: $500 that they could receive today and $1000 that would be received 3 years from today. Magdalen wanted the $500 dollars today, Miriam chose to collect $1000 in 3 years, and Mary June was indifferent between these two options. Can you offer an explanation of the choice made by each woman?

Contributed by Hamed Kashani, Saeid Sadri, and Baabak Ashuri, Georgia Institute of Technology

3-10 (a) If $100 at Time "0" will be worth $110 a year later and was $90 a year ago, compute the interest rate for the past year and the interest rate next year.

(b) Assume that $90 invested a year ago will return $110 a year from now. What is the annual interest rate in this situation?

3-11 A firm has borrowed $5,000,000 for 5 years at 10% per year compound interest. The firm will make no payments until the loan is due, when it will pay off the interest and principal in one lump sum. What is the total payment?

Contributed by Hamed Kashani, Saeid Sadri, and Baabak Ashuri, Georgia Institute of Technology

3-12 **A** What sum of money now is equivalent to $8250 two years later, if interest is 4% per 6-month period?

3-13 The following cash flows are equivalent in value if the interest rate is i. Which one is more valuable if the interest rate is $2i$?

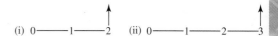

Single Payment Factors

3-14 **A** Solve the diagram for the unknown Q assuming an 8% interest rate.

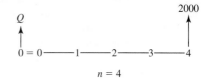

3-15 A student has inherited $50,000. If it is placed in a savings account that earns 3% interest, how much is in the account in 30 years?

3-16 We know that a certain piece of equipment will cost
Ⓐ $150,000 in 5 years. How much must be deposited today using 6% interest to pay for it?

3-17 You are planning to withdraw $100 in Year 1, $150 in Year 3, and $200 in Year 5. At a 5% interest rate, what is the present worth of these withdrawals?
Contributed by Gana Natarajan, Oregon State University

3-18 Suppose that $4000 is deposited in an account that
Ⓐ earns 5% interest. How much is in the account

(*a*) after 5 years?

(*b*) after 10 years?

(*c*) after 20 years?

(*d*) after 50 years?

(*e*) after 100 years?

3-19 An inheritance will be $25,000. The interest rate for the time value of money is 6%. How much is the inheritance worth now, if it will be received

(*a*) in 10 years?

(*b*) in 20 years?

(*c*) in 35 years?

(*d*) in 50 years?

3-20 Rita borrows $5000 from her parents. She repays
Ⓐ them $6000. What is the interest rate if she pays the $6000 at the end of

(*a*) Year 2?

(*b*) Year 3?

(*c*) Year 5?

(*d*) Year 10?

3-21 A savings account earns 2.5% interest. If $3000 is invested, how many years is it until each of the following amounts is on deposit?

(*a*) $3394

(*b*) $3655

(*c*) $4035

(*d*) $5165

3-22 to 3-24 *Contributed by Paul R. McCright, University of South Florida*

3-22 Alvin's Uncle Arnold gave him $16,000 from selling
Ⓐ the old family farm. Alvin wants to start college and have $12,000 available to buy a used car when he graduates in 4 years. Alvin earns 2% in his savings account. How much can he spend on a motorcycle

now and still have enough to grow to the $12,000 he needs when he graduates?

3-23 Ace Manufacturing is building a new Platinum Level
Ⓖ LEED certified facility that will cost $44M. Ace will borrow $40M from First National Bank and pay the remainder immediately as a down payment. Ace will pay 7% interest but will make no payments for 4 years, at which time the entire amount will be due.

(*a*) How large will Ace's payment be?

(*b*) What is LEED, by whom was it devised, and why?

3-24 Maheera can get a certificate of deposit (CD) at his
Ⓐ bank that will pay 2.4% annually for 10 years. If he places $5530 in this CD, how much will it be worth when it matures?

(*a*) Use the formula.

(*b*) Use the interest tables and interpolation.

(*c*) Use a calculator or spreadsheet for a 5-button solution.

3-25 How much must you invest now at 4.2% interest to accumulate $175,000 in 46 years?

3-26 In 1995 an anonymous private collector purchased a
Ⓐ painting by Picasso entitled *Angel Fernandez de Soto* for $29,152,000. The picture depicts Picasso's friend de Soto seated in a Barcelona cafe drinking absinthe. The painting was done in 1903 and was valued then at $600. If the painting was owned by the same family until its sale in 1995, what rate of return did they receive on the $600 investment?

3-27 In 1990 Mrs. John Hay Whitney sold her painting by Renoir, *Au Moulin de la Galette,* depicting an open-air Parisian dance hall, for $71 million. The buyer also had to pay the auction house commission of 10%, or a total of $78.1 million. The Whitney family had purchased the painting in 1929 for $165,000.

(*a*) What rate of return did Mrs. Whitney receive on the investment?

(*b*) Was the rate of return really as high as you computed in (*a*)? Explain.

3-28 The local bank pays 4% interest on savings deposits.
Ⓐ In a nearby town, the bank pays 1% per quarter. A man who has $3000 to deposit wonders whether the higher interest paid in the nearby town justifies driving there. If all money is left in the account for 2 years, how much interest would he obtain from the out-of-town bank?

3-29 One thousand dollars is borrowed for one year at an interest rate of $1/2\%$ per month. If the same sum of money could be borrowed for the same period at an interest rate of 6% per year, how much could be saved in interest charges?

3-30 A sum of money invested at 2% per 6-month period
Ⓐ (semiannually) will double in amount in approximately how many years?

3-31 The tabulated factors for $i = 2.5\%$ to 12% stop at $n = 100$. How can they be used to calculate $(P/F, i, 150)$? $(P/F, i, 200)$?

3-32 If lottery winnings of Q are invested now at an inter-
Ⓐ est rate of 9%, how much is available to help fund an early retirement in Year 25?

3-33 A firm paid $160,000 for a building site two years ago. It is now worth $200,000, and the firm's plans have changed so that no building is planned. The firm estimates that the land will be worth $240,000 in four years. If the firm's interest rate is 5%, what should it do?

3-34 What is the present worth of a two-part legal settle-
Ⓐ ment if the interest rate is 4%? $100,000 is received at the end of Year 1 and $400,000 at the end of Year 5.

3-35 An R&D lab will receive $250,000 when a proposed contract is signed, a $200,000 progress payment at the end of Year 1, and $400,000 when the work is completed at the end of Year 2. What is the present worth of the contract at 15%?

3-36 An engineer invested $5000 in the stock market. For
Ⓐ the first 6 years the average return was 9% annually, and then it averaged 3% for 4 years. How much is in the account after 10 years?

3-37 Camila Vega made an investment of $10,000 in a savings account 10 years ago. This account paid interest of 4% for the first 4 years and 6% interest for the remaining 6 years. How much is this investment worth now?

Nominal and Effective Interest Rates

3-38 A thousand dollars is invested in Green Bonds for 7
Ⓐ months at an interest rate of 1% per month. What is
Ⓖ the nominal interest rate?

(a) What is the effective interest rate?

(b) What are Green Bonds? What type of projects do the bonds encourage?

3-39 A firm charges its credit customers 12/3% interest per month. What is the effective interest rate?

3-40 If the nominal annual interest rate is 9% com-
Ⓐ pounded quarterly, what is the effective annual interest rate?

3-41 A local store charges 11/2% each month on the unpaid balance for its charge account. What nominal annual interest rate is being charged? What is the effective interest rate?

3-42 What interest rate, compounded quarterly, is
Ⓐ equivalent to a 9.31% effective interest rate?

3-43 A bank advertises it pays 4% annual interest, compounded daily, on savings accounts, provided the money is left in the account for 5 years. What is the effective annual interest rate?

3-44 At the Central Furniture Company, customers who
Ⓐ buy on credit pay an effective annual interest rate of
Ⓔ 16.1%, based on monthly compounding.

(a) What is the nominal annual interest rate that they pay?

(b) Research the effective annual interest rates charged on a credit card that you or a friend has. How does the rate change if there is a late or skipped payment? What drives these rates? Are the rates ethical? Why or why not?

3-45 A student bought a $75 used guitar and agreed to pay for it with a single $80 payment at the end of 3 months. What is the nominal annual interest rate? What is the effective interest rate?

3-46 A bank is offering to sell 6-month certificates of
Ⓐ deposit for $9800. At the end of 6 months, the bank will pay $10,000 to the certificate owner. Compute the nominal annual interest rate and the effective annual interest rate.

3-47 A firm spent $2 million for new equipment that
Ⓖ reduces greenhouse gases in their operations. The process improvement saved them $14,000 in the first month.

(a) What is the first month's rate of return on this investment? Express that value on an annual effective basis.

(b) If this investment was judged not to make economic sense, should some entity pay to reduce such emissions? Who and why?

3-48 Steelgrave Financing offers payday loans. The firm
Ⓐ charges a $10 interest fee for a two-week period on a $300 loan. What are the nominal and effective annual interest rates on this loan? *Contributed by Gillian Nicholls, Southeast Missouri State University*

3-49 The treasurer of a firm noted that many invoices were received with the following terms of payment: "2%—10 days, net 30 days." Thus, if the bill is paid within 10 days of its date, he could deduct 2%. Or the full amount would be due 30 days from the invoice date. Assuming a 20-day compounding period, the 2% deduction for prompt payment is equivalent to what effective annual interest rate?

3-50 First Bank is sending university alumni an invitation to obtain a credit card, with the name of their university written on it, for a nominal 9.9% interest per year after 6 months of 0% interest. Interest is compounded monthly. If you fail to make the minimum payment in any month, your interest rate could increase (without notice) to a nominal 19.99% per year. Calculate the effective annual interest rates the credit company is charging in both cases.
Contributed by D. P. Loucks, Cornell University

3-51 Mona Persian is considering a new investment fund with a semiannual interest rate of 2.5%. Any money she invests would have to be left in it for at least five years if she wanted to withdraw it without a penalty.

(a) What is the nominal interest rate?
(b) What is the annual effective interest rate?
(c) If Mona deposits $10,000 in the fund now, how much should it be worth in five years?

Contributed by Gillian Nicholls, Southeast Missouri State University

3-52 A department store charges 13/4% interest per month, compounded continuously, on its customers' charge accounts. What is the nominal annual interest rate? What is the effective interest rate?

3-53 A bank pays 4% nominal annual interest on special three-year certificates. What is the effective annual interest rate if interest is compounded

(a) Every three months?
(b) Daily?
(c) Continuously?

3-54 A friend was left $50,000 by his uncle. He has decided to put it into a savings account for the next year or so. He finds there are varying interest rates at savings institutions: 23/8% compounded annually, 21/4% compounded quarterly, and 21/8% compounded continuously. He wishes to select the savings institution that will give him the highest return on his money. What interest rate should he select?

3-55 Jill deposited $8000 into a bank for 6 months. At the end of that time, she withdrew the money and received $8250. If the bank paid interest based on continuous compounding:

(a) What was the effective annual interest rate?
(b) What was the nominal annual interest rate?

Minicases

3-56 The local garbage company charges $6 a month for garbage collection. It had been their practice to send out bills to their 100,000 customers at the end of each 2-month period. Thus, at the end of February it would send a bill to each customer for $12 for garbage collection during January and February.

Recently the firm changed its billing date: it now sends out the 2-month bills after one month's service has been performed. Bills for January and February, for example, are sent out at the end of January. The local newspaper points out that the firm is receiving half its money before the garbage collection. This unearned money, the newspaper says, could be temporarily invested for one month at 1% per month interest by the garbage company to earn extra income.

Compute how much extra income the garbage company could earn each year if it invests the money as described by the newspaper.

3-57 The Apex Company sold a water softener to Marty Smith. The price of the unit was $350. Marty asked for a deferred payment plan, and a contract was written. Under the contract, the buyer could delay paying for the water softener if he purchased the coarse salt for recharging the softener from Apex. At the end of 2 years, the buyer was to pay for the unit in a lump sum, with interest at a quarterly rate of 1.5%. According to the contract, if the customer ceased buying salt from Apex at any time prior to 2 years, the full payment due at the end of 2 years would automatically become due.

Six months later, Marty decided to buy salt elsewhere and stopped buying from Apex, whereupon Apex asked for the full payment that was to have been due 18 months hence. Marty was unhappy about this, so Apex offered as an alternative to accept the $350 with interest at 10% per semiannual period for the 6 months. Which alternative should Marty accept? Explain.

3-58 The U.S. recently purchased $1 billion of 30-year zero-coupon bonds from a struggling foreign nation. The bonds yield 41/2% per year interest. Zero coupon means the bonds pay no annual interest payments. Instead, all interest is at the end of 30 years.

A U.S. senator objected, claiming that the correct interest rate for bonds like this is 7 1/4%. The result, he said, was a multimillion dollar gift to the foreign country without the approval of Congress. Assuming the senator's rate is correct, how much will the foreign country have saved in interest?

EQUIVALENCE FOR REPEATED CASH FLOWS

Student Solar Power

Indiana State University (ISU) mechanical and manufacturing engineering technology students undertook a project to design a photovoltaic (PV) system to make use of solar energy in the College of Technology Building. The students considered different types of PV solar tracking systems and selected a two-axis tracking system (north/south for summer/winter and east/west for morning/evening). This PV system was designed to provide emergency lighting for the building as well as a learning opportunity for students concerning renewable energy and its functioning.

The system included four PV panels of 123 watts each with a life of 25 years as per the manufacturer's specifications. Panel tilt was fixed at 45 degrees. A two-axis system was able to track altitude and azimuth for Terre Haute, where ISU is located. Most of the electrical parts, such as the converter, PLC controller, and wiring, were provided for free by the college CIM Lab. More recently, engineering economy students have analyzed the system's economic viability, based on the city's available solar potential, as a case study. ■ ■ ■

Contributed by M. Affan Badar, Indiana State University

QUESTIONS TO CONSIDER

1. The panels were purchased by ISU about five years ago. Should the purchase cost be considered as sunk cost?

2. How much difference does the longitude of a city make if the same panel is installed in a different city (east or west coast of the U.S.)? Can latitude make a difference in the value of the summer/winter shift?

3. How important are latitude and yearly days of sunshine in system economics?

4. What costs must be considered, and how can they be estimated over time? In particular, battery performance and costs are getting better every year. Given this trend, how do you determine the system's equivalent present cost?

5. Electricity generated from the system is a saving, as this amount of electricity is not going to be purchased from the utility company (or, if it's a grid-connected system, electricity can be sold back to the utility company). How do you compute the annual dollar amount of the saving? Do the panels decline in efficiency each year?

After Completing This Chapter...

The student should be able to:

- Solve problems with the uniform series compound interest factors and annuity functions.

- Use arithmetic and geometric gradients to solve appropriately modeled problems.

- Understand when and why the assumed uniformity of *A*, *G*, and *g* is a good engineering economic model.

- Use spreadsheets and financial functions to model and solve engineering economic analysis problems.

Key Words

annuity	geometric gradient series	spreadsheet block functions
arithmetic gradient series	ordinary annuity	uniform payment series
data block	scenario	uniform rate
feasibility analysis	spreadsheet annuity functions	what-if analysis

Chapter 3 presented the fundamental components of engineering economic analysis, including formulas to compute equivalent single sums of money at different points in time. Most problems we will encounter are much more complex. This chapter develops formulas for cash flows that are a uniform series or are increasing on an arithmetic or geometric gradient.

UNIFORM SERIES COMPOUND INTEREST FORMULAS

Many times we will find uniform series of receipts or disbursements. Automobile loans, house payments, and many other loans are based on a **uniform payment series.** Future costs and benefits are often estimated to be the *same* or *uniform* every year. This is the simplest assumption; it is often sufficiently accurate; and often there is no data to support what might be a more accurate model.

Since most engineering economy problems define "a period" as one year, the uniform cash flow is an *annual* cash flow denoted by A for all period lengths. More formally, A is defined as

$A =$ an end-of-period cash receipt or disbursement in a uniform series, continuing for n periods

Engineering economy practice and textbooks (including this one) generally assume that cash flows after Time 0 are end-of-period cash flows. Thus the equations and tabulated values of A and F assume end-of-period timing. Tables for beginning- or middle-of-period assumptions have been built, but they are rarely used. When the uniform cash flow is assumed to be on end-of-period cash flow, it is an **annuity** or **ordinary annuity**.

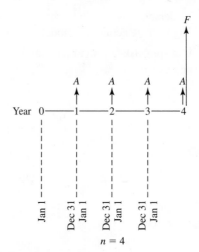

FIGURE 4–1 The general relationship between A and F.

The horizontal line in Figure 4–1 is a representation of time with four interest periods illustrated. Uniform payments A have been placed at the end of each interest period, and there are as many A's as there are interest periods n. (Both these conditions are specified in the definition of A.) Figure 4–1 uses January 1 and December 31, but other 1-year or other length periods could be used.

In Chapter 3's section on single payment formulas, we saw that a sum P at one point in time would increase to a sum F in n periods, according to the equation

$$F = P(1 + i)^n$$

We will use this relationship in our uniform series derivation.

Looking at Figure 4–1, we see that if an amount A is invested at the end of each year for 4 years, the total amount F at the end of 4 years will be the sum of the compound amounts of the individual investments.

In the general case for n years,

$$F = A(1 + i)^{n-1} + \cdots + A(1 + i)^3 + A(1 + i)^2 + A(1 + i) + A \qquad (4\text{-}1)$$

Multiplying Equation 4-1 by $(1 + i)$, we have

$$(1 + i)F = A(1 + i)^n + \cdots + A(1 + i)^4$$
$$+ A(1 + i)^3 + A(1 + i)^2 + A(1 + i) \qquad (4\text{-}2)$$

Factoring out A and subtracting Equation 4-1 gives

$$(1 + i)F = A[(1 + i)^n + \cdots + (1 + i)^4 + (1 + i)^3 + (1 + i)^2 + (1 + i)]$$
$$- \quad F = A[(1 + i)^{n-1} + \cdots + (1 + i)^3 + (1 + i)^2 + (1 + i) + 1]$$
$$\overline{\quad iF = A[(1 + i)^n - 1] \quad} \qquad (4\text{-}3)$$

TABLE 4–1 Excel Functions for Use with Annuities

To find the equivalent P	$-$PV $(i, n, A, F, \text{Type})$
To find the equivalent A	$-$PMT$(i, n, P, F, \text{Type})$
To find the equivalent F	$-$FV $(i, n, A, P, \text{Type})$
To find n	NPER $(i, A, P, F, \text{Type})$
To find i	Rate $(n, A, P, F, \text{Type, guess})$

Solving Equation 4-3 for F gives the **uniform series compound amount factor**

$$F = A \left[\frac{(1 + i)^n - 1}{i} \right] = A(F/A, i\%, n) \tag{4-4}$$

These annuity functions, which are listed in Table 4-1 and detailed in Appendix B, are the spreadsheet version of the 5-BUTTON SOLUTION calculations. The spreadsheet annuity functions list four variables chosen from n, A, P, F, and i, and solve for the fifth variable. The *Type* variable is optional. If it is omitted or 0, then the A value is assumed to be the end-of-period cash flow. If the A value represents the beginning-of-period cash flow, then a value of 1 can be entered for the Type variable.

When solving for PV, PMT, or FV, entering positive values produces a negative result. This sign convention, which seems odd to some students, is explained in Appendix B. The sign convention affects spreadsheets and calculators in the same way, and must be managed. The difficulty becomes minor with a little practice.

We suggest here and in Appendix B that you build a spreadsheet calculator with one row for each of the functions listed in Table 4−1. We suggest that it look like a 5-BUTTON SOLUTION with all 5 functions. Then when you're solving homework problems you can simply create new rows in your spreadsheet by copying the rows that you need. The cash flow diagram and your understanding of the factors and functions determine what the values are. The spreadsheet just does the arithmetic.

Appendix B also describes financial and programmable scientific calculators that can be used instead of or in addition to the tables. Two programmable scientific calculators can be used on the FE exam. These spreadsheet functions and calculators make the same assumptions as the tabulated factors.

EXAMPLE 4–1

You deposit $500 in a credit union at the end of each year for 5 years. The credit union pays 5% interest, compounded annually. Immediately after the fifth deposit, how much can you withdraw from your account?

TABLE SOLUTION

Both diagrams of the five deposits and the desired computation of the future sum F duplicates the situation for the uniform series compound amount formula

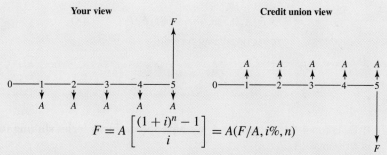

Your view   **Credit union view**

$$F = A \left[\frac{(1 + i)^n - 1}{i} \right] = A(F/A, i\%, n)$$

where $A = \$500, n = 5, i = 0.05, F =$ unknown. Filling in the known variables gives

$$F = \$500(F/A, 5\%, 5) = \$500(5.526) = \$2763$$

There will be $2763 in the account following the fifth deposit.

5-BUTTON SOLUTION

As detailed in Appendix B, this solution corresponds to the more powerful spreadsheet and financial calculator functions that can also be programmed into HP calculators that can be used on the FE exam.

	A	B	C	D	E	F	G	H	I
1	Problem	i	n	PMT	PV	FV	Solve for	Answer	Formula
2	Exp. 4-1	5%	5	-500	0		FV	$2,762.82	=FV(B2,C2,D2,E2)

Because the spreadsheet function is not rounded off like the interest factor, the spreadsheet answer is slightly more precise.

EXAMPLE 4–2

A new engineer wants to save money for down payment on a house. The initial deposit is $685, and $375 is deposited at the end of each month. The savings account earns interest at an annual nominal rate of 6% with monthly compounding. How much is on deposit after 48 months?

5-BUTTON SOLUTION

Because deposits are made monthly, the nominal annual interest rate of 6% must be converted to $1/2\%$ per month for the 48 months. Note both the initial and periodic deposits are negative cash flows from the engineer to the savings account.

	A	B	C	D	E	F	G	H	I
1	Problem	i	n	PMT	PV	FV	Solve for	Answer	Formula
2	Exp. 4-2	0.5%	48	-375	-685		FV	$21,156.97	=FV(B2,C2,D2,E2)

TABLE SOLUTION

Because there are multiple types of cash flows, the uniform series compound amount factor (Eq. 4-4) must be used for the monthly deposit and the compound amount factor (Eq. 3-4) for the initial deposit.

$$F = 375(F/A, 0.5\%, 48) + 685(F/P, 0.5\%, 48)$$
$$= 375(54.098) + 685(1.270)$$
$$= \$21,156.7$$

If Equation 4-4 is solved for A, we have the **uniform series sinking fund**[1] factor, which is written as $(A/F, i, n)$.

$$A = F\left[\frac{i}{(1 + i)^n - 1}\right]$$
$$= F(A/F, i\%, n) \tag{4-5}$$

EXAMPLE 4–3

Jim Hayes wants to buy some electronic equipment for $1000. Jim has decided to save a uniform amount at the end of each month so that he will have the required $1000 at the end of one year. The local credit union pays 6% interest, compounded monthly. How much does Jim have to deposit each month?

TABLE SOLUTION

In this example,

$$F = \$1000, \quad n = 12, \quad i = 1/2\%, \quad A = \text{unknown}$$
$$A = 1000(A/F, 1/2\%, 12) = 1000(0.0811) = \$81.10$$

Jim would have to deposit $81.10 each month.

5-BUTTON SOLUTION

As detailed in Appendix B, this solution corresponds to the more powerful spreadsheet and financial calculator functions that can also be programmed into HP calculators that can be used on the FE exam.

	A	B	C	D	E	F	G	H	I
1	Problem	i	n	PMT	PV	FV	Solve for	Answer	Formula
2	Exp. 4-3	5%	12		0	1000	PMT	-$81.07	=PMT(B2,C2,E2,F2)

If we use the sinking fund formula (Equation 4-5) and substitute for F the single payment compound amount formula (Equation 3-3), we obtain the **uniform series capital recovery**

---

[1]A *sinking fund* is a separate fund into which one makes a uniform series of money deposits (A) to accumulate a desired future sum (F) by the end of period n.

factor, which has the notation $(A/P, i, n)$.

$$A = F \left[\frac{i}{(1+i)^n - 1} \right] = P(1+i)^n \left[\frac{i}{(1+i)^n - 1} \right]$$

$$A = P \left[\frac{i(1+i)^n}{(1+i)^n - 1} \right] = P(A/P, i\%, n) \tag{4-6}$$

The name *capital recovery factor* comes from asking the question, How large does the annual return, A, have to be to "recover" the capital, P, that is invested at Time 0? In other words, find A given P. This is illustrated in Example 4–4.

EXAMPLE 4–4

An energy-efficient machine costs \$5000 and has a life of 5 years. If the interest rate is 8%, how much must be saved every year to recover the cost of the capital invested in it?

TABLE SOLUTION

$$P = \$5000, \qquad n = 5, \qquad i = 8\%, \qquad A = \text{unknown}$$

$$A = P(A/P, 8\%, 5) = 5000(0.2505) = \$1252$$

The required annual savings to recover the capital investment is \$1252.

5-BUTTON SOLUTION

	A	B	C	D	E	F	G	H	I
1	Problem	i	n	*PMT*	*PV*	*FV*	Solve for	Answer	Formula
2	Exp. 4-4	8.0%	5		-5000	0	PMT	\$1,252.28	=PMT(B2,C2,E2,F2)

In Example 4–4, with interest at 8%, a present sum of \$5000 is equivalent to five equal end-of-period disbursements of \$1252. This is another way of stating Plan 3 of Table 3–1. The method for determining the annual payment that would repay \$5000 in 5 years with 8% interest has now been explained. The calculation is simply

$$A = 5000(A/P, 8\%, 5) = 5000(0.2505) = \$1252$$

If the capital recovery formula (Equation 4-6) is solved for the present sum P, we obtain the uniform series present worth formula or **uniform series present worth factor.**

$$P = A \left[\frac{(1+i)^n - 1}{i(1+i)^n} \right] = A(P/A, i\%, n) \tag{4-7}$$

EXAMPLE 4–5   (Example 3–1 Revisited)

A machine will cost $30,000 to purchase. Annual operating and maintenance costs (O&M) will be $2000. The machine will save $10,000 per year in labor costs. The salvage value of the machine after 5 years will be $7000. Calculate the machine's present worth for an interest rate of 10%.

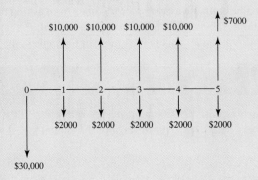

TABLE SOLUTION

Because there are multiple types of cash flows, the uniform series present worth factor (Eq. 4–7) must be used for the annual savings and O&M costs in conjunction with the present worth factor (Eq. 3–5) for the salvage value. No factor is need for the first cost since it is already a present worth.

$$P = -30,000 + (10,000 - 2000)(P/A, 10\%, 5) + 7000(P/F, 10\%, 5)$$

$$= -30,000 + 8000(3.791) + 7000(0.6209)$$

$$= \$4672$$

5-BUTTON SOLUTION

The n is 5 for both the annual series of $8000 (= $10,000 - $2000) and the salvage value of $7000, so the present worth of both can be found in one step.

	A	B	C	D	E	F	G	H	I
1	Problem	i	n	PMT	PV	FV	Solve for	Answer	Formula
2	Exp. 4-5	10%	5	8000		7000	PV	-$34,672.74	=PV(B2,C2,D2,F2)
3							change sign	$34,672.74	=-H2
4							first cost	-$30,000.00	-$30,000.00
5							PW	$4,672.74	=H3+H4

EXAMPLE 4–6   (Example 3–1 and 4–5 Revisited)

What is the rate of return on the machine in Example 4–5?

5-BUTTON SOLUTION

This is one type of problem where the annuity functions of spreadsheets and calculators save a lot of arithmetic. This is a one-step solution with the rate function.

	A	B	C	D	E	F	G	H	I
1	Problem	*i*	*n*	*PMT*	*PV*	*FV*	Solve for	Answer	Formula
2	Exp. 4-6		5	8000	-30000	7000	RATE	15.38%	=RATE(C2,D2,E2,F2)

TABLE SOLUTION

The starting point for finding the rate of return on the machine is the same equation that was developed in Example 4–5. However, instead of specifying an interest rate of 10%, that is now our unknown. We are solving for the interest rate that makes the total present worth equal to 0. We are finding the interest rate that makes the $30,000 first cost equivalent to the other cash flows.

$$0 = -30,000 + (10,000 - 2000)(P/A, i, 5) + 7000(P/F, i, 5)$$

Adding up the cash flows gives a total of $17,000 ($= -30,000 + 8000 \times 5 + 7000$) at an interest rate of 0%. Example 4–4 tells us that the present worth was $4672 at 10%. So the rate of return is higher than 10%. There is a bit of trial and error in solving this problem with the tabulated factors, so this solution saves some time by using the answer from the 5-BUTTON SOLUTION to choose a 15% interest rate for our next calculation. It is common to use a subscript on the P to indicate the interest rate.

$$P_{15} = -30,000 + (10,000 - 2000)(P/A, 15\%, 5) + 7000(P/F, 15\%, 5)$$

$$= -\$30,000 + 8000(3.352) + 7000(0.4972)$$

$$= \$296.4$$

A higher interest rate is needed to discount the future receipts to below $30,000. The next higher tabulated rate is 18%.

$$P_{18} = -30,000 + (10,000 - 2000)(P/A, 18\%, 5) + 7000(P/F, 18\%, 5)$$

$$= -30,000 + 8000(3.127) + 7000(0.4371)$$

$$= -\$1924.3$$

We know that the interest rate is between 15% and 18%—and it is much closer to 15%. Doing an approximation before we start the arithmetic of interpolation helps catch arithmetic and data entry errors. The interpolation equation can be built using the simple tabular graphic that follows and knowing that $a/b = c/d$ so $a = b(c/d)$.

Interest Rate   Present Worth

$$b\begin{bmatrix} a\begin{bmatrix} 15\% \\ i \\ 18\% \end{bmatrix} \end{bmatrix} \qquad \begin{matrix} \$296.4 \\ 0 \\ -\$1924.3 \end{matrix} \begin{bmatrix} c \\ \end{bmatrix} d$$

$$\text{Rate of return } i = 15\% + a$$

$$= 15\% + b(c/d)$$

$$= 15\% + 3\%(296.4 - 0)/(296.4 - (-1924.3))$$

$$= 15\% + 3\% \times 296.4/2220.7$$

$$= 15\% + 0.40\% = 15.4\%$$

EXAMPLE 4–7

A new engineer buys a car with 0% down financing from the dealer. The cost with all taxes, registration, and license fees is $15,732. If each of the 48 monthly payments is $398, what is the monthly interest rate? What is the effective annual interest rate?

5-BUTTON SOLUTION

The engineer is receiving the car valued at $15,732 at time 0. So that is the PV. The payment is a negative cash flow. The unknown is the monthly interest rate.

	A	B	C	D	E	F	G	H	I
1	Problem	i	n	PMT	PV	FV	Solve for	Answer	Formula
2	Exp. 4-7		48	-$398	$15,732	0	i	0.822%	=RATE(C2,D2,E2,F2)
3		monthly						annual	
4	Effective	0.822%	12	0	-1		FV	1.1033	=FV(B4,C4,D4,E4)
5	or						i_a	10.33%	=FV-1
6	Nominal	0.822%	12				r	9.87%	=B6*C6
7	Effective	9.87%	12				i_a	10.33%	=EFFECT(B7,C7)

CASH FLOWS THAT DO NOT MATCH BASIC PATTERNS

EXAMPLE 4–8

A student is borrowing $1000 per year for 3 years. The loan will be repaid 2 years later at a 15% interest rate. Compute how much will be repaid. This is F in the following cash flow table and diagram.

Year	Cash Flow
1	+1000
2	+1000
3	+1000
4	0
5	$-F$

SOLUTION

We see that the cash flow diagram does not match the sinking fund factor diagram: F occurs two periods later, rather than at the same time as the last A. Since the diagrams do not match, the problem is more difficult than those we've discussed so far. The approach to use in this situation is to convert the cash flow from its present form into standard forms, for which we have compound interest factors and compound interest tables.

One way to solve this problem is to consider the cash flow as a series of single payments P and then to compute their sum F. In other words, the cash flow is broken into three parts, each one of which we can solve.

$$F = F_1 + F_2 + F_3 = 1000(F/P, 15\%, 4) + 1000(F/P, 15\%, 3) + 1000(F/P, 15\%, 2)$$

$$= 1000(1.749) + 1000(1.521) + 1000(1.322)$$

$$= \$4592$$

ALTERNATE SOLUTION

A second approach is to calculate an equivalent F_3 at the end of Period 3.

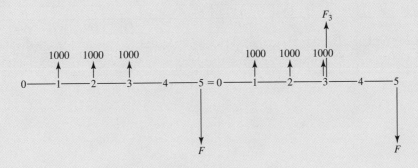

Looked at this way, we first solve for F_3.

$$F_3 = 1000(F/A, 15\%, 3) = 1000(3.472) = \$3472$$

Now F_3 can be considered a present sum P in the diagram

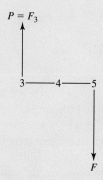

and so

$$F = F_3(F/P, 15\%, 2)$$
$$= 3472(1.322)$$
$$= \$4590$$

The \$2 difference in calculated values is due to rounding in the interest tables. This two-step solution can be combined into a single equation:

$$F = 1000(F/A, 15\%, 3)(F/P, 15\%, 2)$$
$$= 1000(3.472)(1.322)$$
$$= \$4590$$

5-BUTTON SOLUTION

First find F_3 as in the previous solution; then that becomes the P for finding F.

	A	B	C	D	E	F	G	H
1	Problem	i	n	PMT	PV	FV	Solve for	Answer
2	Exp. 4-8	15%	3	1000	0		F_3	-\$3,472.50
3		15%	2	0	3472.50		F	-\$4,592.38
4					=-H2			

EXAMPLE 4–9

Consider the following situation, where P is deposited into a savings account and three withdrawals are made. Find P.

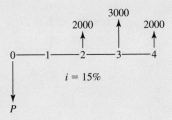

The diagram is not in a standard form, indicating that there will be a multiple-step solution. The solutions that follow are the most common approaches that are used. Understanding why all of them are correct requires understanding *economic equivalence*—which is the foundation of engineering economy. Not only does visualizing and understanding the different ways help develop overall understanding. Each way is sometimes the easiest way to solve a particular problem.

SOLUTION 1

$$P = P_1 + P_2 + P_3$$
$$= 2000(P/F, 15\%, 2) + 3000(P/F, 15\%, 3) + 2000(P/F, 15\%, 4)$$
$$= 2000(0.7561) + 3000(0.6575) + 2000(0.5718)$$
$$= \$4628$$

SOLUTION 2

The second approach converts each withdrawal into an equivalent value at the end of Period 4.

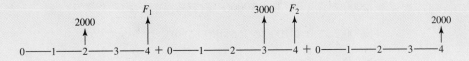

$$F = F_1 + F_2 + 2000$$

$$= 2000(F/P, 15\%, 2) + 3000(F/P, 15\%, 1) + 2000$$

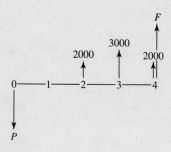

The relationship between P and F in the diagram is

$$P = F(P/F, 15\%, 4)$$

Combining the two equations, we have

$$P = [F_1 + F_2 + 2000](P/F, 15\%, 4)$$

$$= [2000(F/P, 15\%, 2) + 3000(F/P, 15\%, 1) + 2000](P/F, 15\%, 4)$$

$$= [2000(1.322) + 3000(1.150) + 2000](0.5718)$$

$$= \$4628$$

SOLUTION 3

The third approach finds how much would have to be deposited (P_1) at $t = 1$ and then converts that into an equivalent value (P) at $t = 0$.

$$P = P_1(P/F, 15\%, 1) \qquad\qquad P_1 = 2000(P/A, 15\%, 3) + 1000(P/F, 15\%, 2)$$

Combining, we have

$$P = [2000(P/A, 15\%, 3) + 1000(P/F, 15\%, 2)] \times (P/F, 15\%, 1)$$

$$= [2000(2.283) + 1000(0.7561)](0.8696)$$

$$= \$4628$$

5-BUTTON SOLUTION

Solution 1 is probably the easiest approach, with changes in *n* and *F* as needed.

	A	B	C	D	E	F	G	H
1	Problem	*i*	*n*	*PMT*	*PV*	*FV*	Solve for	Answer
2	Exp. 4-9	15%	2	0		2000	P_1	-$1,512.29
3		15%	3	0		3000	P_2	-$1,972.55
4		15%	4	0		2000	P_3	-$1,143.51
5							P	-$4,628.34

ECONOMIC EQUIVALENCE VIEWED AS A MOMENT DIAGRAM

The similarity between cash flow and free body diagrams allows an analogy that helps some students better understand economic equivalence.[2] Think of the cash flows as forces that are always perpendicular to the axis. Then the time periods become the distances along the axis.

When we are solving for unknown forces in a free body diagram, we know that a moment equation about any point will be in equilibrium. Typically moments are calculated by using the right-hand rule so that counterclockwise moments are positive, but it is also possible to define moments so that a clockwise rotation is positive. We are assuming that **clockwise rotations are positive**. This allows us to make the normal assumptions that positive forces point up and positive distances from the force to the pivot point are measured from left to right. Thus, negative forces point down, and negative distances are measured from the force on the right to the pivot point on the left. These assumptions are summarized in the following diagram.

With these assumptions we can write force moment equations at equilibrium for the following diagram.

---

[2]We thank David Elizandro and Jessica Matson of Tennessee Tech for developing, testing, and describing this approach (*The Engineering Economist*, 2007, Vol. 52, No. 2, "Taking a Moment to Teach Engineering Economy," pp. 97–116).

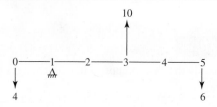

For example, in the force moment equation about Point 1, the force at 0 is –4, and the distance from the force to the pivot point is 1. Similarly the force at 5 is –6, and the distance from the force on the right to the pivot point on the left is –4. The equilibrium equation for force moments about Point 1 is

$$0 = -4 \times 1 + 10 \times (-2) + -6 \times (-4)$$

To write the cash flow moment equation for the cash flow diagram we need:

1. A sign convention for cash flows, such that positive values point up.
2. A way to measure the moment arm for each cash flow. This moment arm must be measured as $(1 + i)^T$, where T is the number of periods measured *from* the cash flow *to* the pivot point or axis of rotation.
   a. Thus the sign of the distance is moved to the exponent.
   b. For cash flows at the pivot point, $T = 0$, and $(1 + i)^0 = 1$.

We can redraw the simple example with an unknown present cash flow, P.

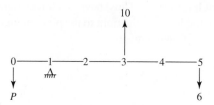

Since P is drawn as a negative cash flow, we put a minus sign in front of it when we write the cash flow moment equation. To rewrite the cash flow moment equation at Year 1, we use the distances from the diagram as the exponents for $(1 + i)^t$:

$$0 = -P \times (1 + i)^1 + 10 \times (1 + i)^{-2} + -6 \times (1 + i)^{-4}$$

If $i = 5\%$, the value of P can be calculated.

$$0 = -P \times 1.05^1 + 10 \times 1.05^{-2} + -6 \times 1.05^{-4}$$
$$P = 10 \times 1.05^{-3} + -6 \times 1.05^{-5} = 8.64 - 4.70 = 3.94$$

EXAMPLE 4–10

For the cash flow diagram in Example 4–8 (repeated here), write the cash flow moment equations at Years 0, 3, and 5. Solve for F when $i = 15\%$.

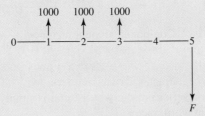

SOLUTION

With Year 0 as the pivot point, the cash flow moment equation is

$$0 = 1000 \times (1 + i)^{-1} + 1000 \times (1 + i)^{-2} + 1000 \times (1 + i)^{-3} - F \times (1 + i)^{-5}$$

With Year 3 as the pivot point, the cash flow moment equation is

$$0 = 1000 \times (1 + i)^{2} + 1000 \times (1 + i)^{1} + 1000 \times (1 + i)^{0} - F \times (1 + i)^{-2}$$

With Year 5 as the pivot point, the cash flow moment equation is

$$0 = 1000 \times (1 + i)^{4} + 1000 \times (1 + i)^{3} + 1000 \times (1 + i)^{2} - F \times (1 + i)^{0}$$

In each case, the cash flow moment equation simplifies to

$$F = 1000 \times (1 + i)^{4} + 1000 \times (1 + i)^{3} + 1000 \times (1 + i)^{2}$$

If $i = 15\%$, then

$$F = 1000 \times (1.749 + 1.521 + 1.323) = \$4592.3$$

RELATIONSHIPS BETWEEN COMPOUND INTEREST FACTORS

From the derivations, we see there are several simple relationships between the compound interest factors. They are summarized here.

Single Payment

$$\text{Compound amount factor} = \frac{1}{\text{Present worth factor}}$$

$$(F/P, i, n) = \frac{1}{(P/F, i, n)} \tag{4-8}$$

Uniform Series

$$\text{Capital recovery factor} = \frac{1}{\text{Present worth factor}}$$

$$(A/P, i, n) = \frac{1}{(P/A, i, n)} \tag{4-9}$$

$$\text{Compound amount factor} = \frac{1}{\text{Sinking fund factor}}$$

$$(F/A, i, n) = \frac{1}{(A/F, i, n)} \tag{4-10}$$

The uniform series present worth factor is simply the sum of the n terms of the single payment present worth factor

$$(P/A, i, n) = \sum_{t=1}^{n} (P/F, i, t) \tag{4-11}$$

For example:

$$(P/A, 5\%, 4) = (P/F, 5\%, 1) + (P/F, 5\%, 2) + (P/F, 5\%, 3) + (P/F, 5\%, 4)$$
$$3.546 = 0.9524 + 0.9070 + 0.8638 + 0.8227$$

The uniform series compound amount factor equals 1 *plus* the sum of $(n - 1)$ terms of the single payment compound amount factor

$$(F/A, i, n) = 1 + \sum_{t=1}^{n-1} (F/P, i, t) \tag{4-12}$$

For example,

$$(F/A, 5\%, 4) = 1 + (F/P, 5\%, 1) + (F/P, 5\%, 2) + (F/P, 5\%, 3)$$
$$4.310 = 1 + 1.050 + 1.102 + 1.158$$

The uniform series capital recovery factor equals the uniform series sinking fund factor *plus* i:

$$(A/P, i, n) = (A/F, i, n) + i \tag{4-13}$$

For example,

$$(A/P, 5\%, 4) = (A/F, 5\%, 4) + 0.05$$
$$0.2820 = 0.2320 + 0.05$$

This may be proved as follows:

$$(A/P, i, n) = (A/F, i, n) + i$$

$$\left[\frac{i(1 + i)^n}{(1 + i)^n - 1} \right] = \left[\frac{1}{(1 + i)^n - 1} \right] + i$$

Multiply by $(1 + i)^n - 1$ to get

$$i(1 + i)^n = i + i(1 + i)^n - i = i(1 + i)^n$$

ARITHMETIC GRADIENT

It frequently happens that the cash flow series is not of constant amount A. Instead, there is a uniformly increasing series as shown:

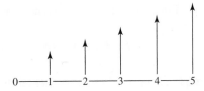

Cash flows of this form may be resolved into two components:

The *first* cash flow in the **arithmetic gradient series** is zero. Thus, G is the *change* from period to period. The gradient (G) series normally is used along with a uniform series (A).

$$P = A(P/A, i, n) + G(P/G, i, n)$$

Pause here and look at the tables for $n = 1$. What is the value of $(P/G, i, 1)$ for any i?

Derivation of Arithmetic Gradient Factors

The arithmetic gradient is a series of increasing cash flows as follows:

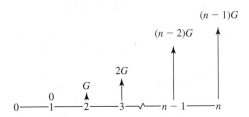

The arithmetic gradient series may be thought of as a series of individual cash flows that can individually be converted to equivalent final cash flows at the end of period n.

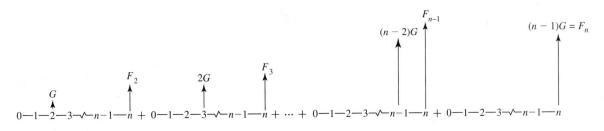

The value of F for the sum of the cash flows $= F_2 + F_3 + \cdots + F_{n-1} + F_n$, or

$$F = G(1 + i)^{n-2} + 2G(1 + i)^{n-3} + \cdots + (n - 2)(G)(1 + i)^1 + (n - 1)G \quad (4\text{-}14)$$

Multiply Equation 4–14 by $(1 + i)$ and factor out G, or

$$(1 + i)F = G[(1 + i)^{n-1} + 2(1 + i)^{n-2} + \cdots + (n - 2)(1 + i)^2 + (n - 1)(1 + i)^1] \quad (4\text{-}15)$$

Rewrite Equation 4–14 to show other terms in the series,

$$F = G[(1 + i)^{n-2} + \cdots + (n - 3)(1 + i)^2 + (n - 2)(1 + i)^1 + n - 1] \quad (4\text{-}16)$$

Subtracting Equation 4—6 from Equation 4–15, we obtain

$$F + iF - F = G[(1 + i)^{n-1} + (1 + i)^{n-2} + \cdots + (1 + i)^2 + (1 + i)^1 + 1] - nG \quad (4\text{-}17)$$

In the derivation of Equation 4–4, the terms inside the brackets of Equation 4–17 were shown to equal the series compound amount factor:

$$[(1 + i)^{n-1} + (1 + i)^{n-2} + \cdots + (1 + i)^2 + (1 + i)^1 + 1] = \frac{(1 + i)^n - 1}{i}$$

Thus, Equation 4–17 becomes

$$iF = G\left[\frac{(1 + i)^n - 1}{i}\right] - nG$$

Rearranging and solving for F, we write

$$F = \frac{G}{i}\left[\frac{(1 + i)^n - 1}{i} - n\right] \quad (4\text{-}18)$$

Multiplying Equation 4–18 by the single payment present worth factor gives

$$P = \frac{G}{i}\left[\frac{(1 + i)^n - 1}{i} - n\right]\left[\frac{1}{(1 + i)^n}\right]$$

$$= G\left[\frac{(1 + i)^n - in - 1}{i^2(1 + i)^n}\right]$$

$$(P/G, i, n) = \left[\frac{(1 + i)^n - in - 1}{i^2(1 + i)^n}\right] \quad (4\text{-}19)$$

Equation 4–19 is the **arithmetic gradient present worth factor.** Multiplying Equation 4-18 by the sinking fund factor, we have

$$A = \frac{G}{i}\left[\frac{(1+i)^n - 1}{i} - n\right]\left[\frac{i}{(1+i)^n - 1}\right] = G\left[\frac{(1+i)^n - in - 1}{i(1+i)^n - i}\right]$$

$$(A/G,i,n) = \left[\frac{(1+i)^n - in - 1}{i(1+i)^n - i}\right] = \left[\frac{1}{i} - \frac{n}{(1+i)^n - 1}\right] \tag{4-20}$$

Equation 4–20 is the **arithmetic gradient uniform series factor.**

EXAMPLE 4–11

Andrew has purchased a new car. He wishes to set aside enough money in a bank account to pay the maintenance for the first 5 years. It has been estimated that the maintenance cost of a car is as follows:

Year	Maintenance Cost
1	$120
2	150
3	180
4	210
5	240

Assume the maintenance costs occur at the end of each year and that the bank pays 5% interest. How much should Andrew deposit in the bank now?

SOLUTION

The cash flow diagram may be broken into its two components:

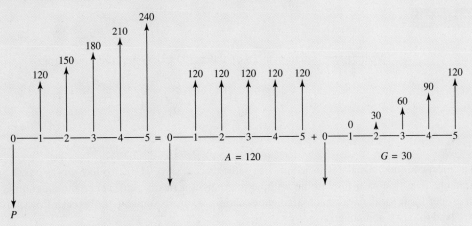

Both components represent cash flows for which compound interest factors have been derived. The first is a uniform series present worth, and the second is an arithmetic gradient series present worth:

$$P = A(P/A, 5\%, 5) + G(P/G, 5\%, 5)$$

Note that the value of n in the gradient factor is 5, not 4. In deriving the gradient factor, the cash flow in the first period is zero followed by $(n-1)$ terms containing G. Here there are four terms containing G, and it is a 5-period gradient.

$$P = 120(P/A, 5\%, 5) + 30(P/G, 5\%, 5)$$
$$= 120(4.329) + 30(8.237)$$
$$= 519 + 247$$
$$= \$766$$

Andrew should deposit $766 in the bank now.

EXAMPLE 4–12

On a certain piece of machinery, it is estimated that the maintenance expense will be as follows:

Year	Maintenance
1	$100
2	200
3	300
4	400

What is the equivalent uniform annual maintenance cost for the machinery if 6% interest is used?

SOLUTION

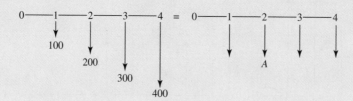

The most common mistake with arithmetic gradients is to solve problems like this as only a $100 gradient. The first cash flow in the arithmetic gradient series is zero, hence the diagram is *not* in proper form for the arithmetic gradient equation. As in Example 4–9, the cash flow must be resolved into two components:

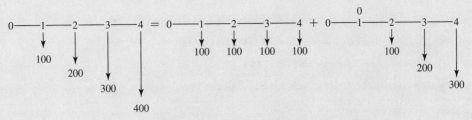

$$A = 100 + 100(A/G, 6\%, 4) = 100 + 100(1.427) = \$242.70$$

The equivalent uniform annual maintenance cost is $242.70.

EXAMPLE 4–13

Demand for a new product will decline as competitors enter the market. If interest is 10%, what is an equivalent uniform value?

Year	Revenue
1	$24,000
2	18,000
3	12,000
4	6,000

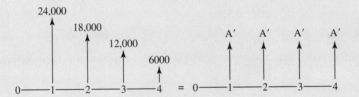

SOLUTION

The projected cash flow is still a cash flow in Period 1 ($24,000) that defines the uniform series. However, now the gradient or change each year is −$6000.

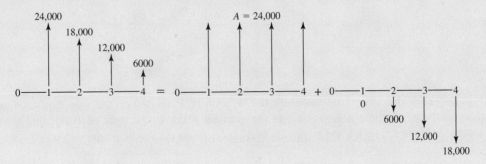

$$A' = 24{,}000 - 6000(A/G, 10\%, 4)$$
$$= 24{,}000 - 6000(1.381)$$
$$= \$15{,}714$$

The projected equivalent uniform value is $15,714 per year.

EXAMPLE 4–14

A car's warranty is 3 years. Upon expiration, annual maintenance starts at $150 and then climbs $25 per year until the car is sold at the end of Year 7. Use a 10% interest rate and find the present worth of these expenses.

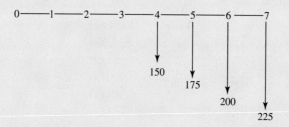

SOLUTION

With the uniform series and arithmetic gradient series present worth factors, we can compute a present sum P'.

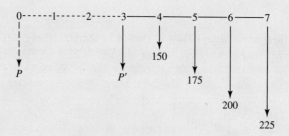

It is important that you closely examine the location of P'. Based on the way the factor was derived, there will be one zero value in the gradient series to the right of P'. (If this seems strange or incorrect, review the beginning of this section on arithmetic gradients.)

$$P' = A(P/A, i, n) + G(P/G, i, n)$$
$$= 150(P/A, 10\%, 4) + 25(P/G, 10\%, 4)$$
$$= 150(3.170) + 25(4.378) = 475.50 + 109.45 = 584.95$$

Then

$$P = P'(P/F, 10\%, 3) = 584.95(0.7513) = \$439.47$$

Reality and the Assumed Uniformity of A, G, and g

The reality of engineering projects is that the annual revenues from selling a new product or annual benefits from using a new highway change each year as demand and traffic levels change. Most annual cash flows are not really uniform.

Thus, why do we define and start with an A that is a uniform annual cost, a G that is a uniform annual gradient, and a g (next section) that is a uniform annual rate of increase?

1. It is easier to start with simpler models. We use cash flow tables and spreadsheets when needed for more complex models.
2. These model cash flows are the basis of the formulas and tabulated factors that are often used in engineering economic analysis.
3. Often in the real world, engineering economy is applied in a **feasibility** or preliminary **analysis**. At this stage, annual cash flows for costs and revenues are typically estimated using A, G, and/or g. Not enough is known about the problem for more detailed estimates.

GEOMETRIC GRADIENT

Earlier, we saw that the arithmetic gradient is applicable where the period-by-period change in a cash receipt or payment is a *constant* amount. There are other situations where the period-by-period change is a **uniform rate,** g. Often **geometric gradient series** can be traced to population levels or other levels of activity where changes over time are best modeled as a percentage of the previous year. The percentage or *rate* is constant over time, rather than the *amount* of the change as in the arithmetic gradient.

For example, if the maintenance costs for a car are $100 the first year and they increase at a uniform rate, g, of 10% per year, the cash flow for the first 5 years would be as follows:

Year			Cash Flow
1	100.00	=	$100.00
2	$100.00 + 10\%(100.00) = 100(1 + 0.10)^1$	=	110.00
3	$110.00 + 10\%(110.00) = 100(1 + 0.10)^2$	=	121.00
4	$121.00 + 10\%(121.00) = 100(1 + 0.10)^3$	=	133.10
5	$133.10 + 10\%(133.10) = 100(1 + 0.10)^4$	=	146.41

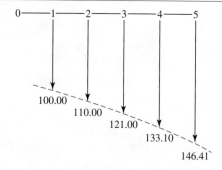

From the table, we can see that the maintenance cost in any year is

$$\$100(1 + g)^{t-1}$$

Stated in a more general form,

$$A_t = A_1(1 + g)^{t-1} \tag{4-21}$$

where g = uniform *rate* of cash flow increase/decrease from period
to period, that is, the geometric gradient
A_1 = value of cash flow at Year 1 ($100 in the example)
A_t = value of cash flow at any year t

Since the present worth P_t of any cash flow A_t at interest rate i is

$$P_t = A_t(1 + i)^{-t} \tag{4-22}$$

we can substitute Equation 4-21 into Equation 4-22 to get

$$P_t = A_1(1 + g)^{t-1}(1 + i)^{-t}$$

This may be rewritten as

$$P_t = A_1(1 + i)^{-1}\left(\frac{1+g}{1+i}\right)^{t-1} \tag{4-23}$$

The present worth of the entire gradient series of cash flows may be obtained by expanding Equation 4-23:

$$P = A_1(1 + i)^{-1}\sum_{t=1}^{n}\left(\frac{1+g}{1+i}\right)^{t-1} \tag{4-24}$$

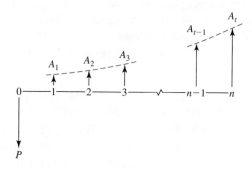

In the general case, where $i \neq g$, Equation 4-23 may be written out as follows:

$$P = A_1(1+i)^{-1} + A_1(1+i)^{-1}\left(\frac{1+g}{1+i}\right) + A_1(1+i)^{-1}\left(\frac{1+g}{1+i}\right)^2$$

$$+ \cdots + A_1(1+i)^{-1}\left(\frac{1+g}{1+i}\right)^{n-1} \qquad (4\text{-}25)$$

Let $a = A_1(1+i)^{-1}$ and $b = (1+g)/(1+i)$. Equation 4-25 becomes

$$P = a + ab + ab^2 + \cdots + ab^{n-1} \qquad (4\text{-}26)$$

Multiply Equation 4-26 by b:

$$bP = ab + ab^2 + ab^3 + \cdots + ab^{n-1} + ab^n \qquad (4\text{-}27)$$

Subtract Equation 4-27 from Equation 4-26:

$$P - bP = a - ab^n$$
$$P(1-b) = a(1-b^n)$$
$$P = \frac{a(1-b^n)}{1-b}$$

Replacing the original values for a and b, we obtain

$$P = A_1(1+i)^{-1}\left[\frac{1-\left(\dfrac{1+g}{1+i}\right)^n}{1-\left(\dfrac{1+g}{1+i}\right)}\right] = A_1\left[\frac{1-\left(\dfrac{1+g}{1+i}\right)^n}{(1+i)-\left(\dfrac{1+g}{1+i}\right)(1+i)}\right]$$

$$= A_1\left[\frac{1-(1+g)^n(1+i)^{-n}}{1+i-1-g}\right]$$

$$P = A_1\left[\frac{1-(1+g)^n(1+i)^{-n}}{i-g}\right] \qquad (4\text{-}28)$$

where $i \neq g$.

The expression in the brackets of Equation 4-28 is the **geometric series present worth factor**

$$(P/A,g,i,n) = \left[\frac{1-(1+g)^n(1+i)^{-n}}{i-g}\right], \qquad \text{where } i \neq g \qquad (4\text{-}29)$$

In the special case of $i = g$, Equation 4-28 becomes

$$P = A_1 n(1+i)^{-1}$$

$$(P/A, g, i, n) = [n(1+i)^{-1}], \qquad \text{where } i = g \qquad (4\text{-}30)$$

EXAMPLE 4–15

The first-year maintenance cost for a new car is estimated to be $100, and it increases at a uniform rate of 10% per year. Using an 8% interest rate, calculate the present worth (PW) of the cost of the first 5 years of maintenance.

STEP-BY-STEP SOLUTION

Year n			Maintenance Cost		$(P/F, 8\%, n)$		PW of Maintenance
1	100.00	=	100.00	×	0.9259	=	$ 92.59
2	100.00 + 10%(100.00)	=	110.00	×	0.8573	=	94.30
3	110.00 + 10%(110.00)	=	121.00	×	0.7938	=	96.05
4	121.00 + 10%(121.00)	=	133.10	×	0.7350	=	97.83
5	133.10 + 10%(133.10)	=	146.41	×	0.6806	=	99.65
							$480.42

SOLUTION USING GEOMETRIC SERIES PRESENT WORTH FACTOR

$$P = A_1 \left[\frac{1 - (1 + g)^n (1 + i)^{-n}}{i - g} \right], \qquad \text{where } i \neq g$$

$$= 100.00 \left[\frac{1 - (1.10)^5 (1.08)^{-5}}{-0.02} \right] = \$480.42$$

The present worth of cost of maintenance for the first 5 years is $480.42.

SPREADSHEETS FOR ECONOMIC ANALYSIS

Spreadsheets: *The* Tool for Engineering Practice

Spreadsheets are used to model annual revenues from selling a new product or annual benefits from using a new highway that change each year. Often this follows a geometric gradient, which spreadsheets can easily include; but more complex patterns are common. For example, cumulative sales and market penetration of a new product often follow a technology growth curve or *S-curve*, where the volume in each year must be individually estimated.

Similarly, if a monthly model of energy use is built, then fluctuations in air-conditioning and heating costs over the year can be accurately modeled in a spreadsheet. As a final example, consider the construction cost for a large project such as new tunnels under the Hudson River, the new airport that Denver built, a new power plant, or a new professional sports stadium. The construction period may be 3 to 10 years long, and the costs will be spread in a project-unique way over that time span. Spreadsheets are a powerful tool to model and analyze complex projects.

Spreadsheets are used in most real-world applications of engineering economy. Common tasks include the following:

1. Constructing tables of cash flows.
2. Using annuity functions to calculate a P, F, A, n, or i.
3. Using a block function to find the present worth or internal rate of return for a table of cash flows.
4. Making graphs for analysis and convincing presentations (virtually all graphs in text).
5. Conducting what-if analysis for different assumed values of problem variables (see scenarios in Example 4–16 and Chapter 9).

Constructing tables of cash flows relies mainly on spreadsheet basics that are covered in Appendix A. These basics include using and naming spreadsheet variables, using a **data block** to enter the variable values, understanding the difference between absolute and relative addresses when copying a formula, and formatting a cell.

Spreadsheet Block Functions for Cash Flow Tables

Cash flows can be specified period by period as a block of values. These cash flows are analyzed by **block functions** that identify the row or column entries for which a present worth or an internal rate of return should be calculated. In Excel the two functions are NPV(i,values) and IRR(values,guess).

Economic Criterion	Excel Function	Values for Periods
Net present value	NPV(i,values)	1 to n
Internal rate of return	IRR(values,guess); guess argument is optional	0 to n

Excel's IRR function can be used to find the interest rate for a loan with irregular payments (other applications are covered in Chapter 7).

These block functions make different assumptions about the range of years included. The NPV(i,values) function assumes that Year 0 is **not** included, while IRR(values,guess) assumes that Year 0 is included. These functions require that a cash flow be identified for each period. You cannot leave cells blank even if the cash flow is $0. The cash flows for 1 to n are assumed to be end-of-period flows. All periods are assumed to be the same length of time.

Also, the NPV function returns the present worth equivalent to the cash flows, unlike the PV annuity function, which returns the negative of the equivalent value.

For cash flows involving only constant values of P, F, and A this block approach seems to be inferior to the annuity functions. However, this is a conceptually easy approach for more complicated cash flows, such as arithmetic or geometric gradients. Suppose the years (row 1) and the cash flows (row 2) are specified in columns B through E.

	A	B	C	D	E	F
1	Year	0	1	2	3	4
2	Cash flow	−25,000	6000	8000	10,000	12,000

NPV(.08,C2:F2)

IRR(B2:F2)

If an interest rate of 8% is assumed, then the present worth of the cash flows can be calculated as $=$B2+NPV(.08,C2:F2), which equals \$4172.95. This is the present worth equivalent to the five cash flows, rather than the negative of the present worth equivalent returned by the PV annuity function. The internal rate of return calculated using IRR(B2:F2) is 14.5%. Notice how the NPV function does not include the Year 0 cash flow in B2 but the IRR function does.

- PW $=$ B2+NPV(.08,C2:F2)   NPV range without Year 0
- IRR $=$ IRR(B2:F2)   IRR range with Year 0

The following example illustrates the ease with which cash flow tables can model and find the present worth and rate of return for arithmetic and geometric gradients. It also illustrates how spreadsheets can be used to model other patterns of cash flows and to answer "what-if" questions by examining multiple **scenarios**.

EXAMPLE 4–16

A new product will cost \$750,000 to design, test prototypes, and set up for production. Net revenue the first year is projected to be \$225,000. Marketing is unsure whether future year revenues will (a) increase by \$25,000 per year as the product's advantages become more widely known or (b) decrease by 10% per year due to competition. A third pattern of increasing by \$25,000 for one year and then decreasing by 10% per year has been suggested as being more realistic. The firm evaluates projects with a 12% interest rate, and it believes that this product will have a 5-year life. Calculate the present worth and rate of return for each scenario.

SOLUTION

The first cost, interest rate, and year 1 revenues are the same for all three scenarios so they are entered into the **data block** in A1:B3. Since the gradients are different for each scenario, the type of gradient is described in rows 5 and 6 and the value of each gradient is stated in row 7. The present worth formulas in row 16 must each include the year 0 cash flows in row 9.

	A	B	C	D	E
1	$750,000	First cost			
2	12%	Interest rate			
3	$225,000	Year 1 net revenue			
4	Scenario	a	b	c	
5	Gradient	Arithmetic	Geometric		
6		G	g	both	
7	Value	$ 25,000	-10%		
8	Year				
9	0	-$750,000	-$750,000	-$750,000	
10	1	225,000	225,000	225,000	=$A$3
11	2	250,000	202,500	250,000	=B10+B7
12	3	275,000	182,250	225,000	=C10*(1+$C$7)
13	4	300,000	164,025	202,500	
14	5	325,000	147,623	182,250	=D13*(1+$C$7) .
15					
16	PW	$221,000	-$69,948	$42,448	=D9+NPV($A$2,D10:D14)
17	Rate of return	22.6%	7.9%	14.4%	=IRR(D9:D14)

Under scenarios (a) and (c), the new product is profitable. However, under scenario (b) the new product has a negative present worth and a rate of return below the 12% interest rate the firm uses. If scenario (c) truly is more realistic, then the project should go ahead.

COMPOUNDING PERIOD AND PAYMENT PERIOD DIFFER

When the various time periods in a problem match, we generally can solve the problem by simple calculations. Thus in Example 4–4, where we had $5000 in an account paying 8% interest, compounded annually, the five equal end-of-year withdrawals are simply computed as follows:

$$A = P(A/P, 8\%, 5) = 5000(0.2505) = \$1252$$

Consider how this simple problem becomes more difficult if the compounding period is changed so that it no longer matches the annual withdrawals.

EXAMPLE 4–17

On January 1, a woman deposits $5000 in a credit union that pays 8% nominal annual interest, compounded quarterly. She wishes to withdraw all the money in five equal yearly sums, beginning December 31 of the first year. How much should she withdraw each year?

SOLUTION

Since the 8% nominal annual interest rate r is compounded quarterly, we know that the effective interest rate per interest period, i, is 2%; and there are a total of $4 \times 5 = 20$ interest periods in 5 years. For the equation $A = P(A/P, i, n)$ to be used, there must be as many periodic withdrawals as there are interest periods, n. In this example we have 5 withdrawals and 20 interest periods.

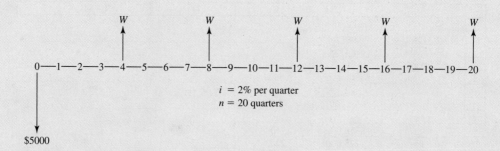

To solve the problem, we must adjust it so that it is in one of the standard forms for which we have compound interest factors. This means we must first compute either an equivalent A for each 3-month interest period or an effective i for each time period between withdrawals. Let's solve the problem both ways.

SOLUTION 1

Compute an equivalent A for each 3-month time period.
If we had been required to compute the amount that could be withdrawn quarterly, the diagram would have been as follows:

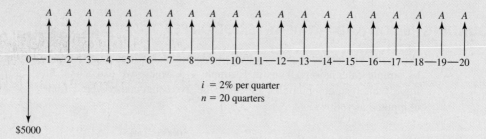

$$A = P(A/P, i, n) = 5000(A/P, 2\%, 20) = 5000(0.0612) = \$306$$

Looking at each one-year period,

$$W = A(F/A, i, n) = 306(F/A, 2\%, 4) = 306(4.122)$$

$$= \$1260$$

SOLUTION 2

Compute an effective i for the time period between withdrawals.

Between withdrawals, W, there are four interest periods, hence $m = 4$ compounding subperiods per year. The nominal interest rate per year, r, is 8%, so the effective annual interest rate is:

$$i_a = \left(1 + \frac{r}{m}\right)^m - 1 = \left(1 + \frac{0.08}{4}\right)^4 - 1$$

$$= 0.0824 = 8.24\% \text{ per year}$$

Now the problem may be redrawn as follows:

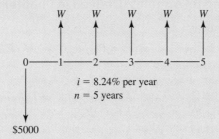

$$i = 8.24\% \text{ per year}$$
$$n = 5 \text{ years}$$

The annual withdrawal W is found using the capital recovery factor:

$$W = P(A/P, i, n) = 5000(A/P, 8.24\%, 5)$$

$$= P\left[\frac{i(1+i)^n}{(1+i)^n - 1}\right] = 5000\left[\frac{0.0824(1 + 0.0824)^5}{(1 + 0.0824)^5 - 1}\right]$$

$$= 5000(0.2520) = \$1260$$

The depositor should withdraw $1260 per year.

SUMMARY

The compound interest formulas described in this chapter, along with those in Chapter 3, will be referred to throughout the rest of the book. It is very important that the reader understand the concepts presented and how these formulas are used. The following notation is used consistently:

$i =$ effective interest rate per interest period[3] (stated as a decimal)

$n =$ number of interest periods

$P =$ a present sum of money

---

[3]Normally the interest period is one year, but it could be some other period (e.g., quarter, month, half-year).

F = a future sum of money at the end of the nth interest period, which is equivalent to P with interest rate i

A = an end-of-period cash receipt or disbursement in a uniform series continuing for n periods; the entire series equivalent to P or F at interest rate i

G = uniform period-by-period increase or decrease in cash receipts or disbursements; the arithmetic gradient

g = uniform rate of cash flow increase or decrease from period to period; the geometric gradient

r = nominal interest rate per interest period (see footnote 3)

i_a = effective annual interest rate

m = number of compounding subperiods per period (see footnote 3)

Single Payment Formulas (Derived in Chapter 3)

Compound amount:

$$F = P(1 + i)^n = P(F/P, i, n)$$

Present worth:

$$P = F(1 + i)^{-n} = F(P/F, i, n)$$

Uniform Series Formulas

Compound amount:

$$F = A\left[\frac{(1 + i)^n - 1}{i}\right] = A(F/A, i, n)$$

Sinking fund:

$$A = F\left[\frac{i}{(1 + i)^n - 1}\right] = F(A/F, i, n)$$

Capital recovery:

$$A = P\left[\frac{i(1 + i)^n}{(1 + i)^n - 1}\right] = P(A/P, i, n)$$

Present worth:

$$P = A\left[\frac{(1 + i)^n - 1}{i(1 + i)^n}\right] = A(P/A, i, n)$$

Arithmetic Gradient Formulas

Arithmetic gradient present worth:

$$P = G\left[\frac{(1 + i)^n - in - 1}{i^2(1 + i)^n}\right] = G(P/G, i, n)$$

Arithmetic gradient uniform series:

$$A = G\left[\frac{(1 + i)^n - in - 1}{i(1 + i)^n - i}\right] = G\left[\frac{1}{i} - \frac{n}{(1 + i)^n - 1}\right] = G(A/G, i, n)$$

Geometric Gradient Formulas

Geometric series present worth, where $i \neq g$:

$$P = A_1 \left[\frac{1 - (1 + g)^n (1 + i)^{-n}}{i - g} \right] = A_1(P/A, g, i, n)$$

Geometric series present worth, where $i = g$:

$$P = A_1 [n(1 + i)^{-1}] = A_1(P/A, g, i, n) = A_1(P/A, i, i, n)$$

Spreadsheet Functions

Annuity functions
=PV(i, n, pmt, [fv], [type])
=FV(i, n, pmt, [pv], [type])
=PMT(i, n, pv, [fv], [type])
=NPER(i, pmt, pv, [fv], [type])
=RATE(n, pmt, pv, [fv], [type], [guess])

Block functions for cash flow tables
=NPV(rate, CF_1:CF_n)
=IRR(CF_0:CF_n, [guess])

Nominal and effective interest rate functions
=EFFECT(nominal_rate, npery)
=NOMINAL(effective_rate, npery)

PROBLEMS

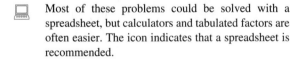

Most of these problems could be solved with a spreadsheet, but calculators and tabulated factors are often easier. The icon indicates that a spreadsheet is recommended.

Uniform Annual Cash Flows

4-1 Rose recently graduated in engineering. Her employer will give her a raise of $6500 per year if she passes the FE exam (Fundamentals of Engineering).

    (*a*) Over a career of 45 years, what is the present worth of the raise if the interest rate is 4%?

    (*b*) What is the future worth at Year 45?

    (*c*) Incentive pay systems can create ethical dilemmas in the workplace. Describe one each from the perspective of the employer and the employee.

4-2 Jose graduated in engineering 5 years ago. His employer will give him a raise of $10,000 per year if he passes the PE exam (Professional Engineer).

    (*a*) Over a career of 35 years, what is the present worth of the raise if the interest rate is 8%?

    (*b*) What is the future worth at year 35?

4-3 Brad will graduate next year. When he begins working, he plans to deposit $6000 at the end of each year into a retirement account that pays 6% interest. How much will be in his account after 40 deposits? *Contributed by Paul R. McCright, University of South Florida*

4-4 If the university's College of Engineering can earn 4% on its investments, how much should be in its savings account to fund one $7500 scholarship each year for 10 years? *Contributed by Paul R. McCright, University of South Florida*

4-5 How much must be invested now at 8% interest to produce $5000 at the end of every year for 10 years?

4-6 A man buys a car for $18,000 with no money down. He pays for the car in 30 equal monthly payments with interest at 12% per annum, compounded monthly. What is his monthly loan payment?

4-7   A car may be purchased with a $3500 down payment now and 72 monthly payments of $480. If the interest rate is 9% compounded monthly, what is the price of the car?

4-8   (A) A manufacturing firm spends $500,000 annually for a required safety inspection program. A new monitoring technology would eliminate the need for such inspection. If the interest rate is 10% per year, how much can the firm afford to spend on this new technology? The firm wants to recover its investment in 15 years.
Contributed by Hamed Kashani, Saeid Sadri, Baabak Ashuri, Georgia Institute of Technology

4-9   A city engineer knows that she will need $25 million in 3 years to implement new automated toll booths on a toll road in the city. Traffic on the road is estimated to be 3 million vehicles per year. How much per vehicle should the toll be to cover the cost of the toll booth replacement project in 3 years? Interest is 8%. (Simplify your analysis by assuming that the toll receipts are received at year end as a lump sum.)

4-10   (A) For what value of *n*, based on a 3 1/2 interest rate, do these cash flows have a present value of 0?

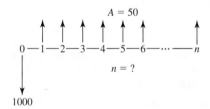

4-11   The cash flows have a present value of 0. Compute the value of *n*, assuming a 10% interest rate.

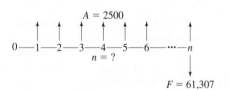

4-12   (A) How many months will it take to pay off a $525 debt, with monthly payments of $15 at the end of each month, if the interest rate is 18%, compounded monthly?

4-13   Assume that you save $1 a day for 50 years, that you deposit it in the bank at the end of each month,

and that there are 30.5 days per month (you save $30.50 each month). How much do you have after 50 years, if:

(*a*) The bank does not pay any interest.

(*b*) The bank pays 1/2% per month interest.

Contributed by Hamed Kashani, Saeid Sadri, and Baabak Ashuri, Georgia Institute of Technology

4-14   (A) A local finance company will loan $10,000 to a homeowner. It is to be repaid in 24 monthly payments of $499 each. The first payment is due 30 days after the $10,000 is received. What interest rate per month are they charging?

4-15   The cash flows have a present value of 0. Compute the value of *J*, assuming a 6% interest rate.

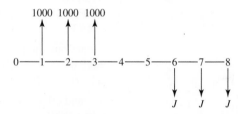

4-16   (A) For diagrams (*a*) to (*c*), compute the unknown values *B*, *C*, *V*, using the minimum number of compound interest factors.

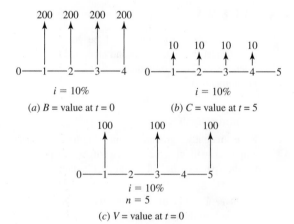

4-17   (G) The automated toll booth system from Problem 4-9 improves traffic flow and thus reduces greenhouse emissions in the city. Rather than wait three years, the city can do it now with a 2% loan. If the toll is set at $0.75 per vehicle, and the state subsidizes the project at a rate of $0.25 per vehicle, how many years will it take for the city to pay off the loan?

4-18 A company deposits $2000 in a bank at the end of every year for 10 years. The company makes no deposits during the subsequent 5 years. If the bank pays 8% interest, how much would be in the account at the end of 15 years?

4-19 Kelsey Construction has purchased a crane that comes with a 5-year warranty. Repair costs are expected to average $5000 per year beginning in Year 6 when the warranty expires. Determine the present worth of the crane's repair costs over its 15-year life. The interest rate is 10%.

4-20 A company buys a machine for $12,000, which it agrees to pay for in five equal annual payments, beginning one year after the date of purchase, at an annual interest rate of 4%. Immediately after the second payment, the terms of the agreement are changed to allow the balance due to be paid off in a single payment the next year. What is the final single payment?

4-21 Using linear interpolation, determine the value of (*a*) (*F/A*, 11%, 15) and (*b*) (*F/P*, 16%, 25) from the compound interest tables. Compute this same value using the equation or a 5-BUTTON SOLUTION. Why do the values differ?

4-22 A student is buying a new car. The car's price is $16,500, the sales tax is 8%, and the title, license, and registration fee is $450 to be paid in cash. The dealer offers to finance 90% of the car's price for 48 months at a nominal interest rate of 9% per year, compounded monthly.

(*a*) How much cash is paid when the car is purchased?

(*b*) How much is the monthly payment?

Contributed by Hamed Kashani, Saeid Sadri, and Baabak Ashuri, Georgia Institute of Technology

4-23 Jennifer Creek is saving up for a new car. She wants to finance no more than $10,000 of the $26,000 estimated price in two years. She deposits $5000 into a savings account now and will make monthly deposits for the next two years. If the savings account pays a nominal interest rate of 3% per year with monthly compounding, how much must she deposit each month? *Contributed by Gillian Nicholls, Southeast Missouri State University*

4-24 Tori is planning to buy a car. The maximum payment she can make is $3400 per year, and she can get a car loan at her credit union for 7.3% interest. Assume her payments will be made at the end of each year 1–4. If Tori's old car can be traded in for $3325, which

is her down payment, what is the most expensive car she can purchase? *Contributed by Paul R. McCright, University of South Florida*

4-25 Determine the breakeven resale price 15 years from now of an apartment house that can be bought today for $250,000. Its annual net income is $22,000. The owner wants a 10% annual return on her investment. *Contributed by D. P. Loucks, Cornell University*

4-26 Martin pays rent of $500 per month for the 9-month academic year. He is going to travel the world this summer and won't be working. How much must he set aside in his savings account for the 3-month summer to cover his rent for next year? The savings account earns 3% with monthly compounding.

4-27 Tameshia deposits $5500 in her retirement account every year. If her account pays an average of 6% interest and she makes 38 deposits before she retires, how much money can she withdraw in 20 equal payments beginning one year after her last deposit? *Contributed by Paul R. McCright, University of South Florida*

4-28 A young engineer wishes to become a millionaire by the time she is 60 years old. She believes that by careful investment she can obtain a 15% rate of return. She plans to add a uniform sum of money to her investment program each year, beginning on her 20th birthday and continuing through her 59th birthday. How much money must the engineer set aside in this project each year?

4-29 What amount will be required to purchase, on an engineers's 40th birthday, an annuity to provide him with 25 equal semiannual payments of $10,000 each, the first to be received on his 60th birthday, if nominal interest is 4% compounded semiannually?

4-30 A man wants to help provide a college education for his young daughter. He can afford to invest $600/yr for the next 4 years, beginning on the girl's 4th birthday. He wishes to give his daughter $4000 on her 18th, 19th, 20th, and 21st birthdays, for a total of $16,000. Assuming 5% interest, what uniform annual investment will he have to make on the girl's 8th through 17th birthdays?

4-31 To provide for a college education for her son, a woman opened an escrow account in which equal deposits were made. The first deposit was made on January 1, 1998, and the last deposit was made on January 1, 2015. The yearly college expenses including tuition were estimated to be $9000, for each of the 4 years. Assuming the interest rate to be 4.5%, how much did the mother have to deposit each year in the escrow account for the son to draw $9000 per year for 4 years beginning January 1, 2013?

4-32 Jennifer is saving up for the closing costs ($3500) and down payment on a home. For a better interest rate and savings on mortgage insurance, she must have a down payment of 20%. She can afford a monthly payment of $600 based on her current earnings and expenses. The amount available for the mortgage is reduced by an estimated $100 per month to cover home insurance and real estate taxes. The current nominal annual interest rate is 3% for a 30-year fixed-rate mortgage loan. How much of a loan can she afford? What is the corresponding house price? How much must she save?
Contributed by Gillian Nicholls, Southeast Missouri State University

4-33 Abby and Jason are building a new house. They obtained a construction loan of $100,000, which will be rolled over into a conventional 20-year mortgage when the house is completed in 14 months. Simple interest of 1/2% per month will be charged on the construction loan. The 20-year mortgage will carry a 6% interest rate with monthly payments. What is the monthly payment that Abby and Jason will make? If they make each payment as scheduled for the life of the 20-year mortgage, how much total interest will they pay on the house? *Contributed by Ed Wheeler, University of Tennessee at Martin*

4-34 Jerry bought a house for $400,000 and made an $80,000 down payment. He obtained a 30-year loan for the remaining amount. Payments were made monthly. The nominal annual interest rate was 6%. After 10 years (120 payments) he sold the house and paid off the loan's remaining balance.

(*a*) What was his monthly loan payment?

(*b*) What must he have paid (in addition to his regular 120th monthly payment) to pay off the loan?

4-35 If $i = 5\%$, for what value of B is the present value = 0.

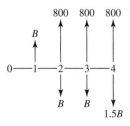

4-36 An engineer graduates at age 22, and she gets a job that pays $60,000 per year. She wants to invest enough to fund her own retirement without relying on an employer pension program or Social Security. Her goal is to have $1 million saved for retirement at age 67. She is relatively confident that her investments will earn an average interest rate of at least 4% per year.

(*a*) Assume that she makes equal annual deposits starting on her 23rd birthday and continuing through her 67th birthday. How much must she invest each year to meet her goal?

(*b*) Suppose she invests the same amount from part (a) every year starting on her 33rd birthday. How much money will she have in the account on her 67th birthday under this scenario?

Contributed by Gillian Nicholls, Southeast Missouri State University

4-37 On January 1, Frank bought a used car for $7200 and agreed to pay for it as follows: 1/4 down payment; the balance to be paid in 36 equal monthly payments; the first payment due February 1; an annual interest rate of 9%, compounded monthly.

(*a*) What is the amount of Frank's monthly payment?

(*b*) During the summer, Frank made enough money to pay off the entire balance due on the car as of October 1. How much did Frank owe on October 1?

4-38 An engineering student bought a car at a local used car lot. Including tax and insurance, the total price was $15,000. He is to pay for the car in 13 equal monthly payments, beginning with the first payment immediately (the first payment is the down payment). Nominal interest on the loan is 12%, compounded monthly. After six payments he decides to sell the car. A buyer agrees to pay off the loan in full and to pay the engineering student $2000. If there are no penalty charges for this early payment of the loan, how much will the car cost the new buyer?

4-39 Liam dreams of starting his own business to import consumer electronic products to his home country. He estimates he can earn 5% on his investments and that he will need to have $300,000 at the end of year 10 if he wants to give his business a good solid foundation. He now has $28,850 in his account, and he believes he can save $12,000 each year from his income, beginning now. He plans to marry at about the end of Year 6 and will skip the investment contribution that year. How far below or above his $300,000 goal will he be? *Contributed by Paul R. McCright, University of South Florida*

4-40 Table 3–1 presented four plans for the repayment of $5000 in 5 years with interest at 8%. Still another way to repay the $5000 would be to make four annual end-of-year payments of $750 each, followed by a final payment at the end of the fifth year. How much would the final payment be?

4-41 An engineer borrowed $3000 from the bank, payable in six equal end-of-year payments at 8%. The bank agreed to reduce the interest on the loan if interest rates declined before the loan was fully repaid. After the third payment, the bank agreed to reduce the interest rate on the remaining debt to $6^1/4\%$. What was the amount of the equal annual end-of-year payments for each of the first 3 years? What was the amount of the equal annual end-of-year payments for each of the last 3 years?

4-42 A woman made 5 annual end-of-year purchases of $2000 worth of common stock. The stock paid no dividends. Then after holding the stock for 10 years, she sold all the stock for $25,000. What interest rate did her investment earn?

4-43 A bank recently announced an "instant cash" plan for holders of its bank credit cards. A cardholder may receive cash from the bank up to a preset limit (about $500). There is a special charge of the minimum of $15 or 4% made at the time the "instant cash" is sent to the cardholders. Each month the bank charges $1\frac{1}{2}\%$ on the unpaid balance. The monthly payment, including interest, may be as little as $10. Assume the cardholder makes the minimum monthly payment of $10. How many months are required to repay $150 of instant cash? How much interest is paid?

Moment Diagram Equations

4-44 Write the cash flow equivalence equation as a moment equation for Problem 4-15.

(a) About Year 4
(b) About Year 5

4-45 Write the cash flow equivalence equation as a moment equation about Year 4 for Problem 4-16.

(a) Include B
(b) Include C
(c) Include V

4-46 Write the cash flow equivalence equation as a moment equation about Year 3 for Problem 4-35.

Relationships Between Factors

4-47 How can the tables be used to compute $(P/A, 5\%, 150)$? $(P/A, 7\%, 200)$?

4-48 For some interest rate i and some number of interest periods n, the uniform series capital recovery factor is 0.1408 and the sinking fund factor is 0.0408. What is the interest rate? What is n?

4-49 For some interest rate i and some number of interest periods n, the uniform series capital recovery factor is 0.1468 and the sinking fund factor is 0.0268. What is the interest rate? What is n?

4-50 Derive an equation to find the end-of-year future sum F that is equivalent to a series of n beginning-of-year payments B at interest rate i. Then use the equation to determine the future sum F equivalent to six B payments of $100 at 8% interest.

4-51 Prove the following relationships algebraically

(a) $(A/F, i, n) = (A/P, i, n) - i$
(b) $(P/F, i, n) = (P/A, i, n) - (P/A, i, n - 1)$
(c) $(P/A, i, n) = (P/F, i, 1) + (P/F, i, 2) + \cdots + (P/F, i, n)$
(d) $(F/A, i, n) = [(F/P, i, n) - 1]/i$

Arithmetic Gradients

4-52 Assume a 10% interest rate and find S, T, and X.

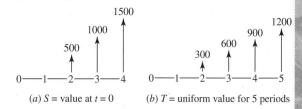

(a) S = value at $t = 0$ (b) T = uniform value for 5 periods

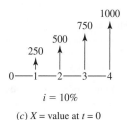

$i = 10\%$

(c) X = value at $t = 0$

4-53 For diagrams (*a*) to (*d*), compute the present values of the cash flows.

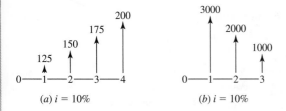

(*a*) *i* = 10%

(*b*) *i* = 10%

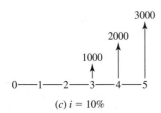

(*c*) *i* = 10%

4-54 It is estimated that the maintenance cost on a new car will be $500 the first year. Each subsequent year, this cost is expected to increase by $150. How much would you need to set aside when you bought a new car to pay all future maintenance costs if you planned to keep the vehicle for 10 years? Assume interest is 5% per year.

4-55 A firm expects to install smog control equipment on the exhaust of a gasoline engine. The local smog control district has agreed to pay to the firm a lump sum of money to provide for the first cost of the equipment and maintenance during its 10-year useful life. At the end of 10 years the equipment, which initially cost $10,000, is valueless. The firm and the smog control district have agreed that the following are reasonable estimates of the end-of-year maintenance costs:

Year 1	$75	Year 6	$200
2	100	7	225
3	125	8	250
4	150	9	275
5	175	10	300

Assuming interest at 6% per year, how much should the smog control district pay to the firm now to provide for the first cost of the equipment and its maintenance for 10 years?

4-56 The council members of a small town have decided that the earth levee that protects the town from flooding should be rebuilt and strengthened. The town engineer estimates that the cost of the work at the end of the first year will be $85,000. He estimates that in subsequent years the annual repair costs will decline by $10,000, making the second-year cost $75,000; the third-year $65,000, and so forth. The council members want to know what the equivalent present cost is for the first 5 years of repair work if interest is 4%.

4-57 A construction firm can achieve a $15,000 cost savings in Year 1 and increasing by $2000 each year for the next 5 years by converting their diesel engines for biodiesel fuel.

   (*a*) At an interest rate of 15%, what is the equivalent annual worth of the savings?

   (*b*) What is biodiesel fuel and why is it considered green energy?

Contributed by Hamed Kashani, Saeid Sadri, and Baabak Ashuri, Georgia Institute of Technology

4-58 Helen can earn 3% interest in her savings account. Her daughter Roberta is 11 years old today. Suppose Helen deposits $4000 today, and one year from today she deposits $1000. Each year she increases her deposit by $500 until she makes her last deposit on Roberta's 18th birthday. How much is on deposit after the 18th birthday, and what is the annual equivalent of her deposits? *Contributed by Paul R. McCright, University of South Florida*

4-59 Perry is a freshman. He estimates that the cost of tuition, books, room and board, transportation, and other incidentals will be $21,000 this year. He expects these costs to rise about $1500 each year while he is in college. If it will take him 5 years to earn his BS, what is the present cost of his degree at an interest rate of 4%? If he earns an extra $15,000 annually for 40 years, what is the present worth of his degree? *Contributed by Paul R. McCright, University of South Florida*

4-60 A sports star can sign a 6-year contract that starts at $12M with increases of $3M each year for his expected playing career of 6 years. It is also possible to sign a contract that starts at $8M for the first year and then increases at $2M each year for 10 years (note some income is deferred until after he retires). If his interest rate for the time value of money is 8%, what is the value of each choice?

Contributed by Hamed Kashani, Saeid Sadri, and Baabak Ashuri, Georgia Institute of Technology

4-61 Use a 7% interest rate to compute the present value of the cash flows.

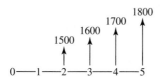

4-62 Compute the present value of the cash flows.

Ⓐ

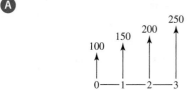

4-63 The cash flows have a present value equal 0. Compute the value of D in the diagram.

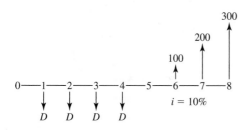

4-64 These cash flow transactions are said to be equivalent in terms of economic desirability at an interest rate of 12% compounded annually. Determine the unknown value A.

Ⓐ

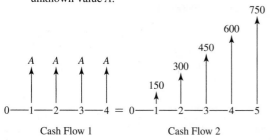

4-65 For what value of P in the cash flow diagram does the present value equal 0?

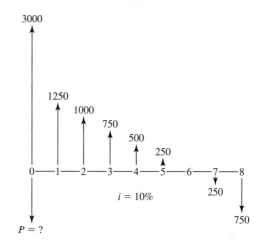

4-66 A debt of $5000 can be repaid, with interest at 8%, by the following payments. How much is X?

Ⓐ

Year	Payment
1	$ 500
2	1000
3	1500
4	2000
5	X

4-67 A man is buying a new eco-friendly electric riding mower. There will be no maintenance cost during the first 2 years because the mower is sold with 2 years free maintenance. For the third year, the maintenance is estimated at $100. In subsequent years the maintenance cost will increase by $50 per year.

Ⓖ

(a) How much would need to be set aside now at 8% interest to pay the maintenance costs on the tractor for the first 6 years of ownership?

(b) What are the advantages and disadvantages of electric powered riding mowers?

4-68 A college student is buying a new sub-compact car, which costs $16,500 plus 8% sales tax. The title, license, and registration fees are $900. The dealer offers her a financing program that starts with a monthly payment of $300, and each successive payment will increase by a constant dollar amount x. The dealer offers to finance 80% of the car's price for 48 months at a nominal interest rate of 6% per year, compounded monthly.

(a) How much is the constant amount x?
(b) How much is the 48$^\text{th}$ payment?

Contributed by Hamed Kashani, Saeid Sadri, and Baabak Ashuri, Georgia Institute of Technology

4-69 The following cash flows are equivalent in value if the interest rate is i. Which one is more valuable if the interest rate is $2i$?

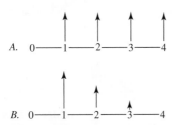

4-70 The following cash flows are equivalent in value if the interest rate is i. Which one is more valuable if the interest rate is $2i$?

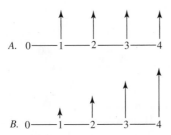

4-71 Using a 10% interest rate, for what value of B does the present value equal 0?

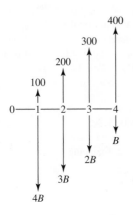

4-72 Consider the following cash flow:

Year	Cash Flow
0	−$100
1	50
2	60
3	70
4	80
5	140

If the present value is zero, which one of the following is correct?

A. $100 = 50 + 10(A/G, i, 5) + 50(P/F, i, 5)$

B. $1 = \dfrac{50(P/A, i, 5) + 10(P/G, i, 5) + 50(P/F, i, 5)}{100}$

C. $100(A/P, i, 5) = 50 + 10(A/G, i, 5)$

D. None of the equations are correct.

4-73 Consider the following cash flow:

Year	Cash Flow
1	$1000
2	$850
3	$700
4	$550
5	$400
6	$400
7	$400
8	$400

Which of the equations below is correct to compute the present value of the cash flows at 8% interest?

A. $P = 1000(P/A, i, 8) - 150(P/G, i, 8) + 150(P/G, i, 4)(P/F, i, 4)$

B. $P = 400(P/A, i, 8) + 600(P/A, i, 5) - 150(P/G, i, 4)$

C. $P = 150(P/G, i, 4) + 850(P/A, i, 4) + 400(P/A, i, 4)(P/F, i, 4)$

Geometric Gradients

4-74 The market for a product is expected to increase at an annual rate of 8%. First-year sales are estimated at $60,000, the horizon is 15 years, and the interest rate is 10%. What is the present value?

4-75 Fred is evaluating whether a more efficient motor
Ⓖ with a life of 5 years should be installed on an
assembly line. If the interest rate is 10%, what is
the present value of the energy savings?

(*a*) Energy savings are estimated at $4000 for the
first year, then increasing by 7% annually.

(*b*) What if the energy savings are increasing by
12% annually?

4-76 A set of cash flows begins at $20,000 the first year,
Ⓐ with an increase each year until $n = 10$ years. If the
interest rate is 8%, what is the present value when

(*a*) the annual increase is $2000?

(*b*) the annual increase is 10%?

4-77 A set of cash flows begins at $80,000 the first year,
with an increase each year until $n = 10$ years. If the
interest rate is 5%, what is the present value when

(*a*) the annual increase is $10,000?

(*b*) the annual increase is 15%?

4-78 A set of cash flows begins at $20,000 the first year,
Ⓐ with an increase each year until $n = 10$ years. If the
interest rate is 10%, what is the present value when

(*a*) the annual increase is $2000?

(*b*) the annual increase is 10%?

4-79 A set of cash flows begins at $20,000 the first year,
with a decrease each year until $n = 10$ years. If the
interest rate is 7%, what is the present value when

(*a*) the annual decrease is $2000?

(*b*) the annual decrease is 10%?

4-80 Suzanne is a recent chemical engineering graduate
Ⓐ who has been offered a 5-year contract at a remote
location. She has been offered two choices. The first
is a fixed salary of $75,000 per year. The second has
a starting salary of $65,000 with annual raises of 5%
starting in Year 2. (For simplicity, assume that her
salary is paid at the end of the year, just before her
annual vacation.) If her interest rate is 9%, which
should she take?

4-81 The football coach at a midwestern university
was given a 5-year employment contract that paid
$800,000 the first year, and increased at an 8% uni-
form rate in each subsequent year. At the end of the
first year's football season, the alumni demanded
that the coach be fired. The alumni agreed to buy
his remaining years on the contract by paying him
the equivalent present sum, computed using a 6%
interest rate. How much will the coach receive?

4-82 A contractor estimates maintenance costs for a
new backhoe to be $500 for the first month with
a monthly increase of 0.75%. The contractor can buy
a 4-year maintenance contract for $20,000 at any
point. If the contract is purchased at the same time
as the backhoe is purchased, the dealer has offered a
10% discount. Use $i = 1\%$ per month. What should
the contractor do?
*Contributed by Hamed Kashani, Saeid Sadri, and
Baabak Ashuri, Georgia Institute of Technology*

4-83 Eddie is a production engineer for a major supplier
of component parts for cars. He has determined that
a robot can be installed on the production line to
replace one employee. The employee earns $20 per
hour and benefits worth $8 per hour for a total
annual cost of $58,240 this year. Eddie estimates
this cost will increase 6% each year. The robot will
cost $16,500 to operate for the first year with costs
increasing by $1500 each year. The firm uses an
interest rate of 15% and a 10-year planning horizon.
The robot costs $75,000 installed and will have a
salvage value of $5000 after 10 years. Should Eddie
recommend that purchase of the robot?
*Contributed by Paul R. McCright, University of
South Florida*

4-84 An engineer will deposit 15% of her salary each year
into a retirement fund. If her current annual salary is
$80,000 and she expects that it will increase by 5%
each year, what will be the present worth of the fund
after 35 years if it earns 5% per year?
*Contributed by Hamed Kashani, Saeid Sadri, and
Baabak Ashuri, Georgia Institute of Technology*

4-85 Mark Johnson invests a fixed percentage of his salary
at the end of each year. This year he invested $1500.
For the next 5 years, he expects his salary to increase
8% annually, and he plans to increase his savings
at the same rate. How much will the investments be
worth at the end of 6 years if the average increase in
the stock market is

(*a*) 8%?

(*b*) 5%?

(*c*) 3%?

4-86 Zachary has opened a retirement account that will
pay 5% interest each year. He plans to deposit 10%
of his annual salary into the account for 39 years
before he retires. His first year's salary is $52,000,
and he expects the salary to grow 4% each year.
How much will be in his account after he makes the
last deposit? What uniform amount can he withdraw

from the account for 25 years beginning one year after his last deposit?

Contributed by Paul R. McCright, University of South Florida

4-87 A 25-year-old engineer is opening an individual retirement account (IRA) at a bank. Her goal is to accumulate $1 million in the account by the time she retires from work in 40 years. The bank manager estimates she may expect to receive 6% nominal annual interest, compounded quarterly, throughout the 40 years. The engineer believes her income will increase at a 5% annual rate during her career. She wishes to start her IRA with as low a deposit as possible and increase it at a 5% rate each year. Assuming end-of-year deposits, how much should she deposit the first year?

Nominal and Effective Interest Rates

4-88 Pete borrows $10,000 to purchase a used car. He
Ⓐ   must repay the loan in 48 equal end-of-period monthly payments. Interest is calculated at $1\frac{1}{2}\%$ per month. Determine the following:

(*a*) The amount of the monthly payment
(*b*) The nominal annual interest rate
(*c*) The effective annual interest rate

4-89 Picabo borrows $1000. To repay the amount, she makes 12 equal monthly payments of $93.12. Determine the following:

(*a*) The monthly interest rate
(*b*) The nominal annual interest rate
(*c*) The effective annual interest rate

4-90 In the 1500s King Henry VIII borrowed money from
Ⓐ   his bankers on the condition that he pay 5% of the loan at each fair (there were four fairs per year) until he had made 40 payments. At that time the loan would be considered repaid. What effective annual interest did King Henry pay?

4-91 Quentin has been using his credit card too much. His plan is to use only cash until the balance of $8574 is paid off. The credit card company charges 18% interest, compounded monthly. What is the effective interest rate? How much interest will he owe in the first month's payment? If he makes monthly payments of $225, how long until it is paid off?

Contributed by Paul R. McCright, University of South Florida

4-92 One of the largest car dealers in the city advertises a
Ⓐ   3-year-old car for sale as follows:
Ⓔ
Cash price $13,750, or a down payment of $1375 with 45 monthly payments of $361.23.

Susan DeVaux bought the car and made a down payment of $2000. The dealer charged her the same interest rate used in his advertised offer.

(*a*) What is the monthly interest rate? How much will Susan pay each month for 45 months? What effective interest rate is being charged?
(*b*) Find the Ethics Guide of the National Automobile Autodealers Association (NADA). If you were selling an auto on-line would you practice these guidelines?

4-93 You are taking a $5000 loan. You will pay it back in four equal amounts, paid every 6 months starting 5 years from now. The interest rate is 12% compounded semiannually. Calculate:

(*a*) The effective interest rate
(*b*) The amount of each semiannual payment
(*c*) The total interest paid

4-94 The *Bawl Street Journal* costs $580, payable now, for a 2-year subscription. The newspaper is published 252 days per year (5 days per week, except holidays). If a 10% nominal annual interest rate, compounded quarterly, is used:

(*a*) What is the effective annual interest rate in this problem?
(*b*) Compute the equivalent interest rate per $1/252$ of a year.
(*c*) What is a subscriber's cost per copy of the newspaper, taking interest into account?

4-95 A bank is offering a loan of $10,000 with a nominal interest rate of 6% compounded monthly, payable in 72 months. There is a loan origination fee of 3% that is taken out from the loan amount.

(*a*) What is the monthly payment?
(*b*) What is the effective interest rate?

4-96 A local car dealer offers a customer a 2-year car loan of $10,000 using "add-on" interest.

Money to pay for car	$10,000
Two years' interest at 7%: $2 \times 0.07 \times 10,000$	1400
	$11,400

$$24 \text{ monthly payments} = \frac{11,400}{24} = \$475.00$$

The first payment must be made in 30 days. What are the nominal and effective annual interest rates?

4-97 A local lending institution advertises the "51–50 Club." A person may borrow $2000 and repay $51 for the next 50 months, beginning 30 days after receiving the money. Compute the nominal annual interest rate for this loan. What is the effective interest rate?

4-98 The **Rule of 78s** is a commonly used method of computing the amount of interest when the balance of a 1-year loan is repaid in advance. Adding the numbers from month 1 to month 12 equals 78.

Now the first month's interest is 12/78 of the year's interest. The second month's interest is 11/78 of the year's interest. Thus after 11 months the total interest charged would be 77/78 of the total year's interest.

Helen borrowed $10,000 at 15% annual interest, compounded monthly. The loan was to be repaid in 12 equal end-of-period payments. After making the first two payments she decided to pay off the balance along with the third payment. Calculate the amount of this additional sum

(*a*) based on the rule of 78s.

(*b*) based on exact economic analysis methods.

(*c*) How close is the approximation?

Spreadsheets for Loans

4-99 Develop a complete amortization table for a loan of $20,000, to be paid back in 18 uniform monthly installments, based on an interest rate of 6%. The amortization table should include the following column headings:

> Payment Number, Principal Owed, Interest Owed, Principal Paid, and Balance Due Monthly Payment

You must also show the equations used to calculate each column of the table.

4-100 Five annual payments at an interest rate of 9% are made to repay a loan of $6000. Build the table that shows the balance due, principal payment, and interest payment for each payment. What is the annual payment? What interest is paid in the last year?

4-101 Using the loan and payment plan developed in Problem 4-99, determine the month that the final payment is due, and the amount of the final payment, if $1500 is paid for Payment 5 and $2500 is paid for Payment 10.

4-102 A newly graduated engineer bought furniture for $900 from a local store. Monthly payments for 1 year will be made. Interest is computed at a nominal rate of 6%. Build the table that shows the balance due, principal payment, and interest payment for each payment. What is the monthly payment? What interest is paid in the last month?

4-103 Calculate and print out the balance due, principal payment, and interest payment for each period of a used-car loan. The nominal interest is 6% per year, compounded monthly. Payments are made monthly for 3 years. The original loan is for $18,000.

4-104 Calculate and print out the balance due, principal payment, and interest payment for each period of a new car loan. The nominal interest is 8% per year, compounded monthly. Payments are made monthly for 5 years. The original loan is for $27,000.

4-105 For the used car loan of Problem 4-103, graph the monthly payment.

(*a*) As a function of the interest rate (5–15%).

(*b*) As a function of the number of payments (24–48).

4-106 For the new car loan of Problem 4-104, graph the monthly payment.

(*a*) As a function of the interest rate (4–14%).

(*b*) As a function of the number of payments (36–84).

4-107 Develop a general-purpose spreadsheet to calculate the balance due, principal payment, and interest payment for each period of a loan. The user inputs to the spreadsheet will be the loan amount, the number of payments per year, the number of years payments are made, and the nominal interest rate. Submit printouts of your analysis of a loan in the amount of $15,000 at 8.9% nominal rate for 36 months and for 60 months of payments.

4-108 Use the spreadsheet developed for Problem 4-107 to analyze 180-month and 360-month house loan payments for a $100,000 mortgage loan at a nominal interest rate of 4.5%. Submit a graph of the interest and principal paid over time. What is the total interest paid for each number of payments?

Spreadsheets for Gradients

4-109 What is the present worth of cash flows that begin at $10,000 and increase at 5% per year for 10 years? The interest rate is 12%.

4-110 What is the present worth of cash flows that begin
(A) at $20,000 and increase at 7% per year for 10 years?
    The interest rate is 9%.

4-111 What is the present worth of cash flows that begin at
    $30,000 and decrease at 10% per year for 5 years?
    The interest rate is 15%.

4-112 What is the present worth of cash flows that begin at
(A) $50,000 and decrease at 12% per year for 10 years?
    The interest rate is 8%.

4-113 Net revenues at an older manufacturing plant will
    be $2 million for this year. The net revenue will
    decrease 15% per year for 5 years, when the assem-
    bly plant will be closed (at the end of Year 6). If the
    firm's interest rate is 10%, calculate the PW of the
    revenue stream.

4-114 Your beginning salary is $70,000. You deposit 12%
    at the end of each year in a savings account that
    earns 3% interest. Your salary increases by 2% per
    year. What value does your savings book show after
    40 years?

4-115 The market volume for widgets is increasing by 15%
    per year from current profits of $300,000. Investing
    in a design change will allow the profit per widget to
    stay steady; otherwise profits will drop 3% per year.
    What is the present worth of the design change over
    the next 5 years? Ten years? The interest rate is 9%.

4-116 In an effort to be more environmentally conscious,
(G) a homeowner may upgrade a furnace that runs on
    fuel oil to a natural gas unit. The investment will be
    $2500 installed. The cost of the natural gas will aver-
    age $60 per month over the year, instead of the $145
    per month that the fuel oil costs. Assume energy
    costs increase 3% per year. If the interest rate is 9%
    per year, how long will it take to recover the initial
    investment?

Compounding and Payment Periods Differ

4-117 Upon the birth of his first child, Dick Jones decided
    to establish a savings account to partly pay for his
    son's education. He plans to deposit $200 per month
    in the account, beginning when the boy is 13 months
    old. The savings and loan association has a cur-
    rent interest policy of 3% per annum, compounded
    monthly, paid quarterly. Assuming no change in the
    interest rate, how much will be in the savings account
    when Dick's son becomes 16 years old?

4-118 Ann deposits $1000 at the end of each month into
(A) her bank savings account. The bank paid 6% nominal
    interest, compounded and paid quarterly. No interest

was paid on money not in the account for the full
3-month period. How much was in Ann's account at
the end of 3 years?

4-119 What is the present worth of a series of equal quar-
    terly payments of $5000 that extends over a period
    of 8 years if the interest rate is 12% compounded
    monthly?

4-120 What single amount on April 1, 2015, is equivalent
    to a series of equal, semiannual cash flows of $1000
    that starts with a cash flow on January 1, 2013,
    and ends with a cash flow on January 1, 2022? The
    interest rate is 12% and compounding is quarterly.

4-121 A contractor wishes to set up a special fund by
    making uniform semiannual end-of-period deposits
    for 20 years. The fund is to provide $10,000 at the
    end of each of the last 5 years of the 20-year period.
    If interest is 8%, compounded semiannually, what is
    the required semiannual deposit?

4-122 The State University is considering funding options
    for a new engineering building on campus. The
    money has been raised for the construction costs and
    now the focus is raising funds for the annual upkeep
    and maintenance (U&M) expenses. For this build-
    ing, contractors will be hired with a series of 3-year
    agreements over the 30 years. Under each contract
    the university will pay $125,000 at the beginning of
    each 3-year agreement to cover all U&M building
    expenses over that 3-year period. The first 3-year
    agreement begins when the building opens.
        A wealthy alumnus has agreed to donate enough
    at the building's opening to cover the U&M expenses
    over the 30-year term. If money invested by the
    school's engineering foundation earns 6% interest
    compounded quarterly, how much must be donated?

4-123 A series of monthly cash flows is deposited into
    an account that earns 12% nominal interest com-
    pounded monthly. Each monthly deposit is equal to
    $2100. The first monthly deposit occurred on June 1,
    2012 and the last monthly deposit will be on January
    1, 2019. The account (the series of monthly deposits,
    12% nominal interest, and monthly compounding)
    also has equivalent quarterly withdrawals from it.
    The first quarterly withdrawal is equal to $5000 and
    occurred on October 1, 2012. The last $5000 with-
    drawal will occur on January 1, 2019. How much
    remains in the account after the last withdrawal?

4-124 Jing, a recent engineering graduate, never took engi-
    neering economics. When she graduated, she was
    hired by a prominent engineering firm. The earnings

from this job allowed her to deposit $1000 each quarter into a savings account. There were two banks that offered savings accounts in her town (a small town!). The first bank was offering 4.5% interest compounded continuously. The second bank offered 4.6% compounded monthly. Jing decided to deposit in the first bank because it offered continuous compounding. Did she make the right decision?

Minicases

4-125 Assume that you plan to retire 40 years from now and that you expect to need $2M to support the lifestyle that you want.

    (*a*) If the interest rate is 10%, is the following statement approximately true? "Waiting 5 years to start saving doubles what you must deposit each year."

    (*b*) If the interest rate is 12%, is the required multiplier higher or lower than for the 10% rate in (*a*)?

    (*c*) At what interest rate is the following statement exactly true? "Waiting 5 years to start saving doubles what you must deposit each year."

4-126 For winners of the California SuperLotto Plus, the choice is between a lump sum and annual payments that increase from 2.5% for the first year to 2.7% for the second year and then increase by 0.1% per year to 5.1% for the 26th payment. The lump sum is equal to the net proceeds of bonds purchased to fund the 26 payments. This is estimated at 45% to 55% of the lump sum amount. At what interest rate is the present worth of the two payment plans equivalent if the lump sum is 45%? If it is 55%?

CASES

The following case from *Cases in Engineering Economy* (www.oup.com/us/newnan) is suggested as matched with this chapter.

CASE **27** **Can Crusher**

    Basic time value of money, costs, and breakeven analysis.

PRESENT WORTH ANALYSIS

The Present Value of 30 Years of Benefits

The Columbia River, from its headwaters west of Banff in the Canadian Rockies (Columbia Lake, elevation 820 m) flows 2000 km through British Columbia and the state of Washington before entering the Pacific Ocean at Astoria, Oregon. Measured by the volume of its flow, the Columbia is the largest river flowing into the Pacific from North America. Measured by elevation drop (0.41 m/km compared to the Mississippi's 0.12 m/km), the Columbia alone possesses one-third of the hydroelectric potential of the U.S. Because of the steep mountain trenches and high snowfall in its catchment area, the Columbia's water levels used to fluctuate wildly, and vulnerable areas along its banks were subject to seasonal flooding.

Anxious to exploit that hydroelectric power and control flooding, the U.S. and Canadian governments negotiated the Columbia River Treaty. It was signed in 1961 and ratified by the U.S. in 1961 and by Canada in 1964.

> The Treaty requires Canada to store 15.5 Million Acre Feet [the volume of water that would cover an acre of land one foot deep], for flood control in perpetuity. This storage was accomplished with the construction of the Duncan, Hugh Keenleyside, and Mica Dams in Canada. In return for constructing the dams and regulating the water levels, the Province of British Columbia is entitled to half of the electrical downstream power benefits that the water generates on the dams located in the U.S. Canada was also entitled to half of the estimated value of the future flood control benefits in the U.S. in a one-time payment of $69.9 million. (*Virtual Museum of Canada*)

It took Canada several years to ratify the treaty because British Columbia in the 1950s was sparsely populated and not very industrialized, and the prospects of turning a large part of

the interior of the province into a reservoir to provide benefits that the residents could not use was not appealing to the premier, W. A. C. Bennett. The impasse was resolved when the U.S. agreed to pay in cash to British Columbia the discounted present value of the first 30 years of the province's benefit entitlement. The treaty said:

> The purchase price of the entitlement shall be $254,400,000, in United States funds as of October 1, 1964, subject to adjustment, in the event of an earlier payment of all or part thereof, to the then present worth, at a discount rate of 4.5% per annum.

The 30 years was up in 1994, and British Columbia then started receiving annual benefits. Today, Canada's 50% share in the downstream benefits is worth annually about $250 to 350 million, which is paid to the government of British Columbia. ■ ■ ■

Contributed by John Whittaker, University of Alberta

QUESTIONS TO CONSIDER

1. Canadian Premier Bennett used the $254 million to finance the Portage Mountain Dam (later renamed the W. A. C. Bennett Dam) and an associated 2730-megawatt power station in northern British Columbia. Was this a good use for the money from the Canadian perspective?

2. Negotiating downstream rights can be a curious tangle. The people at the bottom of the hill can argue that the water is going to flow there anyway and they are free to use it as they want. The people upstream can argue that while the water is on their land, they should be free to use or divert it as they please. Who is right- and wrong-minded in this situation? What are the ethical considerations from both perspectives?

3. Was 4.5%, the discount rate used in 1964, a reasonable one for the governments to choose? Explain and justify your conclusion.

After Completing This Chapter...

The student should be able to:

- Define and apply the *present worth criteria*.
- Compare two competing alternative choices using present worth (PW).
- Apply the PW model in cases with equal, unequal, and infinite project lives.
- Compare multiple alternatives using the PW criteria.
- Develop and use spreadsheets to make *present worth* calculations.
- Compute bond prices and yields.

Key Words

analysis period	investment	par (face) value
capitalized cost	least common multiple	planning horizon
coupon interest rate	minimum attractive	present worth analysis
economic efficiency	rate of return (MARR)	project life
financing	net present worth criterion	XNPV

In Chapters 3 and 4 we accomplished two important tasks. First, we presented the concept of equivalence. We can compare cash flows only if we can resolve them into equivalent values. Second, we derived a series of compound interest factors to find those equivalent values.

ASSUMPTIONS IN SOLVING ECONOMIC ANALYSIS PROBLEMS

One of the difficulties of problem solving is that most problems tend to be very complicated. It becomes apparent that *some* simplifying assumptions are needed to make complex problems manageable. The trick, of course, is to solve the simplified problem and still be satisfied that the solution is applicable to the *real* problem! In the subsections that follow, we will consider six different items and explain the customary assumptions that are made. These assumptions apply to all problems and examples, unless other assumptions are given.

End-of-Year Convention

As we indicated in Chapter 4, economic analysis textbooks and practice follow the end-of-period convention. This makes "*A*" a series of end-of-period receipts or disbursements.

A cash flow diagram of *P*, *A*, and *F* for the end-of-period convention is as follows:

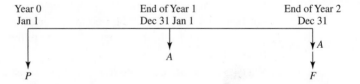

If one were to adopt a middle-of-period convention, the diagram would be:

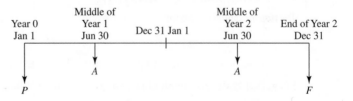

As the diagrams illustrate, only *A* shifts; *P* remains at the beginning of the period and *F* at the end of the period, regardless of the convention. The compound interest tables in Appendix C are based on the end-of-period convention.

Viewpoint of Economic Analysis Studies

When we make economic analysis calculations, we must proceed from a point of reference. Generally, we will want to take the point of view of a total firm when doing industrial economic analyses. Example 1–1 vividly illustrated the problem: a firm's shipping department decided it could save money by outsourcing its printing work rather than by using the in-house printing department. An analysis from the viewpoint of the shipping department supported this, as it could get for $688 the same printing it was paying $793 for in-house. Further analysis showed, however, that its printing department costs would decline *less* than using the commercial printer would save. From the viewpoint of the firm, the net result would be an increase in total cost.

From Example 1–1 we see it *is* important that the **viewpoint of the study** be carefully considered. Selecting a narrow viewpoint, like that of the shipping department, may result in a suboptimal decision from the firm's viewpoint. It is the total firm's viewpoint that is used in industrial economic analyses. For public-sector problems, the combined viewpoint of the government and the citizenry is chosen, since for many public projects the benefits of faster commuting, newer schools, and so on are received by individuals and the costs are paid by the government.

Sunk Costs

We know that it is the *differences between alternatives* that are relevant in economic analysis. Events that have occurred in the past really have no bearing on what we should do in the future. When the judge says, "$200 fine or 3 days in jail," the events that led to these unhappy alternatives really are unimportant. It is the current and future differences between the alternatives that *are* important. Past costs, like past events, have no bearing on deciding between alternatives unless the past costs somehow affect the present or future costs. In general, past costs do not affect the present or the future, so we refer to them as *sunk costs* and disregard them.

Borrowed Money Viewpoint

In most economic analyses, the proposed alternatives inevitably require money to be spent, and so it is natural to ask the source of that money. Thus, each problem has two monetary aspects: one is the **financing**—the obtaining of money; the other is the **investment**—the spending of money. Experience has shown that these two concerns should be distinguished. When separated, the problems of obtaining money and of spending it are both logical and straightforward. Failure to separate financing and investment sometimes produces confusing results and poor decision making.

The conventional assumption in economic analysis is that the money required to finance alternatives is considered to be *obtained from the bank or the firm at interest rate i.*

Effect of Inflation and Deflation

For the present we will assume that prices are stable or stated in constant-value dollars. This means that a machine that costs $5000 today can be expected to cost the same amount several years hence. Inflation and deflation may need to be considered for after-tax analysis and for cost and revenues whose inflation rates differ from the economy's inflation rates (see Chapter 14), but we assume stable or constant dollar prices for now.

Income Taxes

Income taxes, like inflation and deflation, must be considered to find the real payoff of a project. However, taxes will often affect alternatives similarly, allowing us to compare choices without considering income taxes. We will introduce income taxes into economic analyses in Chapter 12.

ECONOMIC CRITERIA

We have shown how to manipulate cash flows in a variety of ways, and we can now solve many kinds of compound interest problems. But engineering economic analysis is more than simply solving interest problems. The decision process (see Figure 1–1) requires that the outcomes of feasible alternatives be arranged so that they may be judged for **economic efficiency** in terms of the selection criterion. The economic criteria were previously stated in general terms, and they are restated in Table 5–1.

We will now examine ways to resolve engineering problems, so that criteria for economic efficiency can be applied.

Equivalence provides the logic by which we may adjust the cash flow for a given alternative into some equivalent sum or series. We must still choose which comparable units to use. In this chapter we will learn how analysis can resolve alternatives into *equivalent present consequences*, referred to simply as **present worth analysis.** Chapter 6 will show how alternatives are converted into an *equivalent uniform annual cash flow*, and Chapter 7 solves for the interest rate at which favorable consequences—that is, *benefits*—are equivalent to unfavorable consequences—or *costs*.

As a general rule, any economic analysis problem may be solved by any of these three methods. This is true because *present worth*, *annual cash flow*, and *rate of return* are exact

TABLE 5-1 Present Worth Analysis

Input/Output	Situation	Criterion
Neither input nor output is fixed	Typical, general case	Maximize (present worth of benefits *minus* present worth of costs), that is, maximize net present worth
Fixed input	Amount of money or other input resources are fixed	Maximize present worth of benefits or other outputs
Fixed output	There is a fixed task, benefit, or other output to be accomplished	Minimize present worth of costs or other inputs

methods that will always yield the same recommendation for selecting the best alternative from among a set of mutually exclusive alternatives. Remember that "mutually exclusive" means that selecting one alternative precludes selecting any other alternative. For example, constructing a gas station and constructing a drive-in restaurant on a particular piece of vacant land are mutually exclusive alternatives.

Some problems, however, may be more easily solved by one method. Present worth analysis is most frequently used to determine the present value of future money receipts and disbursements. It would help us, for example, to determine the present worth of an income-producing property, like an oil well or an apartment house. If the future income and costs are known, then we can use a suitable interest rate to calculate the property's present worth. This should provide a good estimate of the price at which the property could be bought or sold. Another application is valuing stocks or bonds based on the anticipated future benefits of ownership.

TIME PERIOD FOR ANALYSIS

In present worth analysis, careful consideration must be given to the time period covered by the analysis. Usually the task to be accomplished has a time period associated with it. The consequences of each alternative must be considered for this period of time, which is usually called the **analysis period, planning horizon,** or **project life.**

The analysis period for an economy study should be determined from the situation. In some industries with rapidly changing technologies, a rather short analysis period or planning horizon might be in order. Industries with more stable technologies (like steelmaking) might use a longer period (say, 10–20 years), while government agencies frequently use analysis periods extending to 50 years or more.

Three different analysis-period situations are encountered in economic analysis problems with multiple alternatives:

1. The useful life of each alternative equals the analysis period.
2. The alternatives have useful lives different from the analysis period.
3. There is an infinite analysis period, $n = \infty$.

Useful Lives Equal the Analysis Period

Since different lives and an infinite analysis period present some complications, we will begin with four examples in which the useful life of each alternative equals the analysis period.

EXAMPLE 5–1

A firm will install one of two mechanical devices to reduce costs. Both devices have useful lives of 5 years and no salvage value. Device A costs $10,000 and can be expected to result in $3000 savings annually. Device B costs $13,500 and will provide cost savings of $3000 the first year but savings will increase $500 annually, making the second-year savings $3500, the third-year savings $4000, and so forth. With interest at 7%, which device should the firm purchase?

SOLUTION

The analysis period can conveniently be selected as the useful life of the devices, or 5 years. The appropriate decision criterion is to choose the alternative that maximizes the net present worth of benefits minus costs.

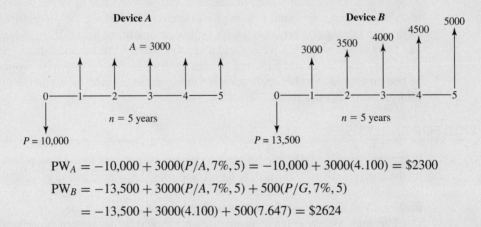

$$PW_A = -10,000 + 3000(P/A, 7\%, 5) = -10,000 + 3000(4.100) = \$2300$$

$$PW_B = -13,500 + 3000(P/A, 7\%, 5) + 500(P/G, 7\%, 5)$$

$$= -13,500 + 3000(4.100) + 500(7.647) = \$2624$$

Device *B* has the larger present worth and is the preferred alternative.

EXAMPLE 5–2

Wayne County will build an aqueduct to bring water in from the upper part of the state. It can be built at a reduced size now for $300 million and be enlarged 25 years hence for an additional $350 million. An alternative is to construct the full-sized aqueduct now for $400 million.

Both alternatives would provide the needed capacity for the 50-year analysis period. Maintenance costs are small and may be ignored. At 6% interest, which alternative should be selected?

TABLE SOLUTION

This problem illustrates staged construction. The aqueduct may be built in a single stage, or in a smaller first stage followed many years later by a second stage to provide the additional capacity when needed.

For the Two-Stage Construction

$$PW \text{ of cost} = \$300 \text{ million} + 350 \text{ million}(P/F, 6\%, 25)$$

$$= \$300 \text{ million} + 81.6 \text{ million}$$

$$= \$381.6 \text{ million}$$

For the Single-Stage Construction

$$\text{PW of cost} = \$400 \text{ million}$$

The two-stage construction has a smaller present worth of cost and is the preferred construction plan.

5-BUTTON SOLUTION

	A	B	C	D	E	F	G	H
1	Problem	*i*	*n*	*PMT*	*PV*	*FV*	Solve for	Answer
2	Two-Stage	6%	25	0		-350	PV	$81.55
3							year 0	$300.00
4							total	$381.55

This is less than the $400M for the one-stage construction, so the two-stage is preferred.

EXAMPLE 5–3

A firm is trying to decide which of two weighing scales it should install to check a package-filling operation in the plant. The ideal scale would allow better control of the filling operation, hence less overfilling. If both scales have lives equal to the 6-year analysis period, which one should be selected? Assume an 8% interest rate.

Alternatives	Cost	Uniform Annual Benefit	End-of-Useful-Life Salvage Value
Atlas scale	$2000	$450	$200
Tom Thumb scale	3000	600	700

TABLE SOLUTION

Atlas Scale

$$
\begin{aligned}
\text{PW of benefits} - \text{PW of cost} &= 450(P/A, 8\%, 6) + 200(P/F, 8\%, 6) - 2000 \\
&= 450(4.623) + 200(0.6302) - 2000 \\
&= 2080 + 126 - 2000 = \$206
\end{aligned}
$$

Tom Thumb Scale

$$
\begin{aligned}
\text{PW of benefits} - \text{PW of cost} &= 600(P/A, 8\%, 6) + 700(P/F, 8\%, 6) - 3000 \\
&= 600(4.623) + 700(0.6302) - 3000 \\
&= 2774 + 441 - 3000 = \$215
\end{aligned}
$$

The salvage value of each scale, it should be noted, is simply treated as another positive cash flow. Because the PWs for the two alternatives are nearly identical, it is likely that there are other tangible or intangible differences that should determine the decision. If there are no such differences, buy the Tom Thumb equipment.

5-BUTTON SOLUTION

	A	B	C	D	E	F	G	H
1	Alternative	*i*	*n*	*PMT*	*PV*	*FV*	Solve for	Answer
2	Atlas	8%	6	450		200	PV	-$2,206
3					-2000		total = -H2+E3	$206
4								
5	Tom Thumb	8%	6	600		700	PV	-$3,215
6					-3000		total = -H8+E9	$215

In Examples 5–1 and 5–3, we compared two alternatives and selected the one in which present worth of benefits *minus* present worth of cost was a maximum. The criterion is called the **net present worth criterion** and written simply as **NPW:**

Net present worth = Present worth of benefits − Present worth of cost

$$\text{NPW} = \text{PW of benefits} - \text{PW of cost} \qquad (5\text{-}1)$$

The field of engineering economy and this text often use present worth (PW), present value (PV), net present worth (NPW), and net present value (NPV) as synonyms. Sometimes, as in the foregoing definition, *net* is included to emphasize that both costs and benefits have been considered.

Useful Lives Different from the Analysis Period

In present worth analysis, there always must be an identified analysis period. It follows, then, that each alternative must be considered for the entire period. In Examples 5–1 to 5–3, the useful life of each alternative was equal to the analysis period. While often this is true, in many situations at least one alternative will have a useful life different from the analysis period. This section describes one way to evaluate alternatives with lives different from the study period.

For present worth calculations, it is important that we select an analysis period and judge the consequences of each of the alternatives during that period. As such, in Example 5–4 it is not a fair comparison to compare the NPW of Pump *A* over its 12-year life against the NPW of Pump *B* over its 6-year life.

The firm, its economic environment, and the specific situation are important in selecting an analysis period. If Pump *A* (Example 5–4) has a useful life of 12 years, and Pump *B* will last 6 years, one method is to select an analysis period that is the **least common multiple** of their useful lives. Thus we would compare the 12-year life of Pump *A* against an initial purchase of Pump *B* *plus* its replacement with new Pump *B* in 6 years. The result is to judge the alternatives on the basis of a 12-year requirement.

EXAMPLE 5–4

Two pumps are being considered for purchase. If interest is 7%, which pump should be bought? Their maintenance costs are the same.

	Pump A	Pump B
Initial cost	$7000	$5000
End-of-useful-life salvage value	1200	1000
Useful life, in years	12	6

TABLE SOLUTION

Since the maintenance costs are the same, they can be omitted from the comparison. The present worth of Pump A over 12 years is

$$PW_A = -7000 + 1200(P/F, 7\%, 12)$$
$$= -7000 + 1200(0.4440)$$
$$= -\$6467$$

For a common analysis period of 12 years, we need to replace Pump B at the end of its 6-year useful life. If we assume that another pump B' can be obtained, having the same $5000 initial cost, $1000 salvage value, and 6-year life, the cash flow will be as follows:

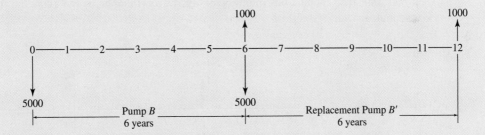

For the 12-year analysis period, the present worth for Pump B is

$$PW_B = -5000 + 1000(P/F, 7\%, 6) - 5000(P/F, 7\%, 6)$$
$$\quad + 1000(P/F, 7\%, 12)$$
$$= -5000 + (1000 - 5000)(0.6663) + 1000(0.4440)$$
$$= -\$7221$$

By assuming that the shorter-life equipment is replaced by equipment with identical economic consequences, we have avoided a lot of calculations. Select Pump A.

5-BUTTON SOLUTION

	A	B	C	D	E	F	G	H	I
		i	*n*	*PMT*	*PV*	*FV*	Solve for	Answer	Formula
1	Exp. 5-4								
2	Pump A	7%	12	0		1200	PV	-$533	=PV(B2,C2,D2,F2)
3								$533	change sign
4								-$7,000	Initial cost
5								-$6,467	PW
6	Pump B yr 6	7%	6	0		-4000	PV	$2,665	=PV(B6,C6,D6,F6)
7	yr 12	7%	12	0		1000	PV	-$444	=PV(B7,C7,D7,F7)
8					-7000			-$2,221	sum & change sign
9	yr 0							-$5,000	Initial cost
10								-$7,221	PW

We have seen that setting the analysis period equal to the least common multiple of the lives of the two alternatives seems reasonable in Example 5–4. However, what if the alternatives had useful lives of 7 and 13 years? Here the least common multiple of lives is 91 years. An analysis period of 91 years hardly seems realistic. Instead, a suitable analysis period should be based on how long the equipment is likely to be needed. This may require that terminal values be estimated for the alternatives at some point prior to the end of their useful lives.

As Figure 5–1 shows, it is not necessary for the analysis period to equal the useful life of an alternative or some multiple of the useful life. To properly reflect the situation at the end of the analysis period, an estimate is required of the market value of the equipment at that time. The calculations might be easier if everything came out even, but this is not essential.

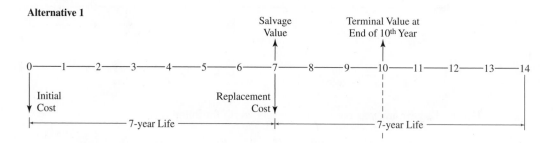

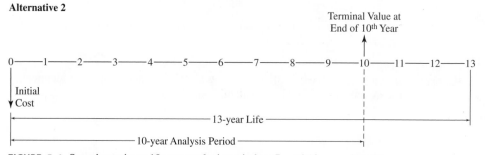

FIGURE 5–1 Superimposing a 10-year analysis period on 7- and 13-year alternatives.

EXAMPLE 5–5

A diesel manufacturer is considering the two alternative production machines graphically depicted in Figure 5–1. Specific data are as follows:

	Alt. 1	Alt. 2
Initial cost	$50,000	$75,000
Estimated salvage value at end of useful life	$10,000	$12,000
Useful life of equipment, in years	7	13

The manufacturer uses an interest rate of 8% and wants to use the PW method to compare these alternatives over an analysis period of 10 years. To do so, the market values at year 10 must be estimated. Alt. 1 will be 3 years into its "second" life and Alt. 2 will be nearing the end of its "first" life.

	Alt. 1	Alt. 2
Estimated market value, end of 10-year analysis period	$20,000	$15,000

SOLUTION

In this case, the decision maker is setting the analysis period at 10 years rather than accepting a common multiple of the lives of the alternatives, or assuming that the period of needed service is infinite (to be discussed in the next section). This is a legitimate approach—perhaps the diesel manufacturer will be phasing out this model at the end of the 10-year period. In any event, we need to compare the alternatives over 10 years.

As illustrated in Figure 5–1, we may assume that Alternative 1 will be replaced by an identical machine after its 7-year useful life. Alternative 2 has a 13-year useful life. The diesel manufacturer has provided an estimated market value of the equipment at the time of the analysis period. We can compare the two choices over 10 years as follows:

$$PW \text{ (Alt. 1)} = -50,000 + (10,000 - 50,000)(P/F, 8\%, 7) + 20,000(P/F, 8\%, 10)$$
$$= -50,000 - 40,000(0.5835) + 20,000(0.4632)$$
$$= -\$64,076$$

$$PW \text{ (Alt. 2)} = -75,000 + 15,000(P/F, 8\%, 10)$$
$$= -75,000 + 15,000(0.4632)$$
$$= -\$68,052$$

To minimize PW of costs the diesel manufacturer should select Alt. 1.

Infinite Analysis Period: Capitalized Cost

Some present worth analyses use an infinite analysis period ($n = \infty$). In governmental analyses, a service or condition sometimes must be maintained for an infinite period. The need for roads, dams, pipelines, and so on, is sometimes considered to be permanent. In these situations a present worth of cost analysis would have an infinite analysis period. We call this particular analysis **capitalized cost.**

Infinite lives are rare in the private sector, but a similar assumption of "indefinitely long" horizons is sometimes made. This assumes that the facility will need electric motors, mechanical HVAC equipment, and forklifts as long as it operates and that the facility will last far longer than any individual unit of equipment. So the equipment can be analyzed as though the problem horizon is *infinite* or indefinitely long.

Capitalized cost is the present sum of money that would need to be set aside now, at some interest rate, to yield the funds required to provide the service (or whatever) indefinitely. To accomplish this, the money set aside for future expenditures must not decline. The interest received on the money set aside can be spent, but not the principal. When one stops to think about an infinite analysis period (as opposed to something relatively short, like a hundred years), we see that an unchanged principal sum is essential.

In Chapter 3 we saw that

$$\text{Principal sum} + \text{Interest for the period} = \text{Amount at end of period, or}$$

$$P \quad + \quad iP \quad = \quad P + iP$$

If we spend iP, then in the next interest period the principal sum P will again increase to $P + iP$. Thus, we can again spend iP.

This concept may be illustrated by a numerical example. Suppose you deposited $200 in a bank that paid 4% interest annually. How much money could be withdrawn each year without reducing the balance in the account below the initial $200? At the end of the first year, the $200 would have earned 4%($200) = $8 interest. If this interest were withdrawn, the $200 would remain in the account. At the end of the second year, the $200 balance would again earn 4%($200) = $8. This $8 could also be withdrawn and the account would still have $200. This procedure could be continued indefinitely and the bank account would always contain $200. If more or less than $8 is withdrawn, the account will either increase to ∞ or decrease to 0.

The year-by-year situation would be depicted like this:

Year 1: $200 initial $P \rightarrow 200 + 8 = 208$

$$\text{Withdrawal } iP = -\ 8$$

$$\overline{}$$

Year 2: $200 $\rightarrow 200 + 8 = 208$

$$\text{Withdrawal } iP = \underline{-\ 8}$$
$$\$200$$

and so on

Thus, for any initial present sum P, there can be an end-of-period withdrawal of A equal to iP each period, and these withdrawals can continue forever without diminishing the initial sum P. This gives us the basic relationship:

$$\text{For} \quad n = \infty, \quad A = iP$$

This relationship is the key to capitalized cost calculations. Earlier we defined capitalized cost as the present sum of money that would need to be set aside at some interest rate to yield the funds to provide the desired task or service forever. Capitalized cost is therefore the P in the equation $A = iP$. It follows that:

$$\textbf{Capitalized cost} \quad P = \frac{A}{i} \tag{5-2}$$

If we can resolve the desired task or service into an equivalent A, the capitalized cost can be computed. The following examples illustrate such computations.

EXAMPLE 5–6

How much should one set aside to pay $5000 per year for maintenance on a small park if interest is assumed to be 4%? For perpetual maintenance, the principal sum must remain undiminished after the annual disbursement is made.

SOLUTION

$$\text{Capitalized cost } P = \frac{\text{Annual disbursement } A}{\text{Interest rate } i}$$

$$P = \frac{5000}{0.04} = \$125,000$$

One should set aside $125,000.

EXAMPLE 5–7

A city plans a pipeline to transport water from a distant watershed area to the city. The pipeline will cost $8 million and will have an expected life of 70 years. The city expects to keep the water line in service indefinitely. Compute the capitalized cost, assuming 7% interest.

SOLUTION

The capitalized cost equation

$$P = \frac{A}{i}$$

is simple to apply when there are end-of-period disbursements A. Here we have renewals of the pipeline every 70 years. To compute the capitalized cost, it is necessary to first compute an end-of-period disbursement A that is equivalent to $8 million every 70 years.

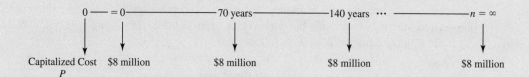

The $8 million disbursement at the end of each 70-year period may be resolved into an equivalent A.

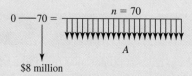

$$A = F(A/F, i, n) = \$8 \text{ million}(A/F, 7\%, 70)$$

$$= \$8 \text{ million}(0.00062)$$

$$= \$4960$$

Each 70-year period is identical to this one, and the infinite series is shown in Figure 5–2.

$$\text{Capitalized cost } P = 8 \text{ million} + \frac{A}{i} = 8 \text{ million} + \frac{4960}{0.07}$$

$$= \$8,071,000$$

FIGURE 5–2 Using the sinking fund factor to compute an infinite series.

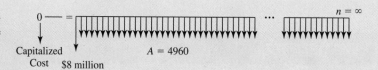

ALTERNATE SOLUTION 1

Instead of solving for an equivalent end-of-period payment A based on a *future* $8 million disbursement, we could find A, given a *present* $8 million disbursement.

$$A = P(A/P, i, n) = 8 \text{ million}(A/P, 7\%, 70)$$

$$= 8 \text{ million}(0.0706) = \$565,000$$

On this basis, the infinite series is shown in Figure 5–3. Carefully note the difference between this and Figure 5–2. Now:

$$\text{Capitalized cost } P = \frac{A}{i} = \frac{565{,}000}{0.07} = \$8{,}071{,}000$$

FIGURE 5–3 Using the capital recovery factor to compute an infinite series.

$n = \infty$

$A = 565{,}000$

ALTERNATE SOLUTION 2

Another way of solving the problem is to assume the interest period is 70 years long and compute an equivalent interest rate for the 70-year period. Then the capitalized cost may be computed by using Equation 3–8 for $m = 70$

$$i_{70\,\text{yr}} = (1 + i_{1\,\text{yr}})^{70} - 1 = (1 + 0.07)^{70} - 1 = 112.989$$

$$\text{Capitalized cost} = 8 \text{ million} + \frac{8 \text{ million}}{112.989} = \$8{,}070{,}803$$

MULTIPLE ALTERNATIVES

So far the discussion has been based on examples with only two alternatives. But multiple-alternative problems may be solved by exactly the same methods. (The only reason for avoiding multiple alternatives was to simplify the examples.) Examples 5–8 and 5–9 have multiple alternatives.

EXAMPLE 5–8

A contractor has been awarded the contract to construct a 6-mile-long tunnel in the mountains. During the 5-year construction period, the contractor will need water from a nearby stream. She will construct a pipeline to carry the water to the main construction yard. An analysis of costs for various pipe sizes is as follows:

	Pipe Sizes (in.)			
	2	**3**	**4**	**6**
Installed cost of pipeline and pump	$22,000	$23,000	$25,000	$30,000
Cost per hour for pumping	$1.20	$0.65	$0.50	$0.40

At the end of 5 years, the pipe and pump will have a salvage value equal to the cost of removing them. The pump will operate 2000 hours per year. The lowest interest rate at which the contractor is willing to invest money is 7%. (The minimum required interest rate for invested money is called the **minimum attractive rate of return,** or MARR.) Select the alternative with the least present worth of cost.

SOLUTION

We can compute the present worth of cost for each alternative. For each pipe size, the present worth of cost is equal to the installed cost of the pipeline and pump plus the present worth of 5 years of pumping costs.

	Pipe Size (in.)			
	2	**3**	**4**	**6**
Installed cost of pipeline and pump	$22,000	$23,000	$25,000	$30,000
1.20×2000 hr $\times (P/A, 7\%, 5)$	9,840			
0.65×2000 hr $\times 4.100$		5,330		
0.50×2000 hr $\times 4.100$			4,100	
0.40×2000 hr $\times 4.100$				3,280
Present worth of cost	$31,840	$28,330	$29,100	$33,280

Select the 3-in. pipe, since the lowest present worth of cost.

EXAMPLE 5–9

An investor paid $8000 to a consulting firm to analyze possible uses for a small parcel of land on the edge of town that can be bought for $30,000. In their report, the consultants suggested four alternatives:

	Alternatives	Total Investment Including Land*	Uniform Net Annual Benefit	Terminal Value at End of 20 yr
A	Do nothing	$ 0	$ 0	$ 0
B	Vegetable market	50,000	5,100	30,000
C	Gas station	95,000	10,500	30,000
D	Small motel	350,000	36,000	150,000

*Includes the land and structures but does not include the $8000 fee to the consulting firm.

Assuming 10% is the minimum attractive rate of return, what should the investor do?

SOLUTION

Alternative *A* is the "do-nothing" alternative. Generally, one feasible alternative is to remain in the present status and do nothing. The investor could decide that the most attractive alternative is not to purchase the property and develop it.

We note that if he does nothing, then the total venture would not be a very satisfactory one. This is because the investor spent $8000 for professional advice on the possible uses of the property. But because the $8000 is a past cost, it is a **sunk cost.** The only relevant costs in an economic analysis are *present* and *future* costs. Sunk costs are gone and cannot be allowed to affect future planning. (Past costs may be relevant in computing depreciation charges and income taxes, but nowhere else.) The past should not deter the investor from making the best decision now.

This problem is one of neither fixed input nor fixed output, so our criterion will be to maximize the present worth of benefits *minus* the present worth of cost; that is, to maximize net present worth.

Alternative *A*, Do Nothing

$$NPW = 0$$

Alternative *B*, Vegetable Market

$$NPW = -50{,}000 + 5100(P/A, 10\%, 20) + 30{,}000(P/F, 10\%, 20)$$

$$= -50{,}000 + 5100(8.514) + 30{,}000(0.1486)$$

$$= -50{,}000 + 43{,}420 + 4460$$

$$= -\$2120$$

Alternative *C*, Gas Station

$$NPW = -95{,}000 + 10{,}500(P/A, 10\%, 20) + 30{,}000(P/F, 10\%, 20)$$

$$= -95{,}000 + 89{,}400 + 4460 = -\$1140$$

Alternative *D*, Small Motel

$$NPW = -350{,}000 + 36{,}000(P/A, 10\%, 20) + 150{,}000(P/F, 10\%, 20)$$

$$= -350{,}000 + 306{,}500 + 22{,}290 = -\$21{,}210$$

The criterion is to maximize net present worth. In this situation, one alternative has NPW equal to zero, and three alternatives have negative values for NPW. We will select the best of the four alternatives, namely, the do-nothing Alt. *A*, with NPW equal to zero.

5-BUTTON SOLUTION

	A	B	C	D	E	F	G	H
		i	*n*	*PMT*	*PV*	*FV*	Solve for	Answer
1	Alternative							
2	Vegetable market	10%	20	5100		30,000	PV	-$47,878
3					-50,000		total = -H2+E3	-$2,122
4								
5	Gas station	10%	20	10,500		30,000	PV	-$93,852
6					-95,000		total = -H5+E6	-$1,148
7								
8	Small Motel	10%	20	36,000		150,000	PV	-$328,785
9					-350,000		total = -H8+E9	-$21,215

Since the total present worth of each action alternative is negative, the best choice is to do nothing. Note that in this case the PV of the positive annual revenue and positive salvage value is negative. To compute the present worth of each alternative, we must subtract the PV values in H2, H5, and H8.

APPLICATIONS AND COMPLICATIONS

EXAMPLE 5–10

Two pieces of construction equipment are being analyzed:

Year	Alt. *A*	Alt. *B*
0	−$2000	−$1500
1	1000	700
2	850	300
3	700	300
4	550	300
5	400	300
6	400	400
7	400	500
8	400	600

At an 8% interest rate, which alternative should be selected?

SPREADSHEET SOLUTION

In this case the cash flows are so irregular that it is easiest and clearer to treat them as individual data block entries. Time can be saved by dragging or double clicking on the cell corners to copy the cashflows and gradients.

	A	B	C
1	8%	interest rate	
2	Year	Alt. *A*	Alt. *B*
3	0	-2000	-1500
4	1	1000	700
5	2	850	300
6	3	700	300
7	4	550	300
8	5	400	300
9	6	400	400
10	7	400	500
11	8	400	600
12		$ 1,588.42	$ 936.16
13			
14	=B3+NPV($A$1,B4:B11)		
15		=C3+NPV($A$1,C4:C11)	

TABLE SOLUTION

Alternative *A*

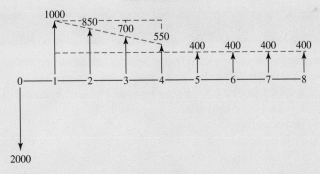

$$\text{PW of benefits} = 400(P/A, 8\%, 8) + (1000 - 400)(P/A, 8\%, 4) - 150(P/G, 8\%, 4)$$

$$= 400(5.747) + 600(3.312) - 150(4.650) = 3588.50$$

$$\text{PW of cost} = 2000$$

$$\text{Net present worth} = 3588.50 - 2000 = +\$1588.50$$

Alternative B

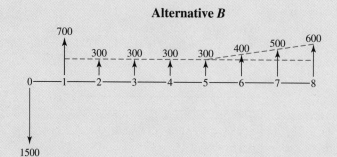

$$\text{PW of benefits} = 300(P/A, 8\%, 8) + (700 - 300)(P/F, 8\%, 1)$$
$$+ \, 100(P/G, 8\%, 4)(P/F, 8\%, 4)$$
$$= 300(5.747) + 400(0.9259) + 100(4.650)(0.7350)$$
$$= 2436.24$$
$$\text{PW of cost} = 1500$$
$$\text{Net present worth} = 2436.24 - 1500$$
$$= +\$936.24$$

To maximize NPW, choose Alt. A.

EXAMPLE 5–11

A piece of land may be purchased for $610,000 to be strip-mined for the underlying coal. Annual net income will be $200,000 for 10 years. At the end of the 10 years, the surface of the land will be restored as required by a federal law on strip mining. The reclamation will cost $1.5 million more than the resale value of the land after it is restored. Using a 10% interest rate, determine whether the project is desirable.

SOLUTION

The investment opportunity may be described by the following cash flow:

Year	Cash Flow (thousands)
0	−$610
1–10	+200 (per year)
10	−1500

$$NPW = -610 + 200(P/A, 10\%, 10) - 1500(P/F, 10\%, 10)$$

$$= -610 + 200(6.145) - 1500(0.3855)$$

$$= -610 + 1229 - 578$$

$$= +\$41$$

Since NPW is positive, the project seems desirable. However, at interest rates below 4.07% or above 18.29%, the NPW is negative, indicating an undesirable project. This indicates that the project is undesirable at 4% and desirable at 10%. This does not make sense. The results warn us that NPW may not always be a proper criterion for judging whether or not an investment should be undertaken. In this example the disbursements ($610,000 + $1,500,000) exceed the benefits (10 × $200,000), which certainly does not portray a desirable investment. Thus Example 5–11 shows that NPW calculations in certain infrequent conditions can lead to unreliable results. Very large environmental cleanup costs can cause this, as in this example. Appendix 7A addresses this in more detail.

Spreadsheets make it easy to build more accurate models with shorter time periods. When one is using factors, it is common to assume that costs and revenues are uniform for n years. With spreadsheets it is easy to use 120 months instead of 10 years, and the cash flows can be estimated for each month. For example, energy costs for air conditioning peak in the summer, and in many areas there is little construction during the winter.

Example 4–14 illustrated how spreadsheets can be used for arithmetic and geometric gradients. The latter are important, because cash flows that depend on population often increase at $x\%$ per year, such as for electric power and transportation costs. Example 5–12 is another illustration of how spreadsheets support models that more closely match reality.

EXAMPLE 5–12

Regina Industries has a new product whose sales are expected to be 1.2, 3.5, 7, 5, and 3 million units per year over the next 5 years. Production, distribution, and overhead costs decline 10% per year from $140 per unit in the first year. The price will be $200 per unit for the first 2 years and then $180, $160, and $140 for the next 3 years. The remaining R&D and production costs are $300 million. If i is 15%, what is the present worth of the new product?

SOLUTION

All of the variable values are entered in the spreadsheet's data block, except for the yearly volume. Since each year is different, these values are simply entered into column B. Note that the gradient for revenue per unit first makes a difference in Year 3.

	A	B	C	D	E	F	G
1	$300	First cost ($M)					
2	$140	Initial unit cost					
3	-10%	Geometric gradient for cost					
4	$200	Initial unit revenue					
5	-$20	Arithmetic change in unit revenue for years 3 to 5					
6	15%	Interest rate					
7	Year	Volume (M)	Revenue unit	Cost unit	Net revenue unit	Cash flow	
8	0					-$300.0	=-$A$1
9	1	1.2	$200	$140	$60	72.0	
10	2	3.5	200	126	74	259.0	=B10*E10
11	3	7.0	180	113	67	466.2	
12	4	5.0	160	102	58	289.7	
13	5	3.0	140	92	48	144.4	
14							
15			=D12*(1+$A$3)		PW ($M)	$502	
16		=C12+$A$5				=F8+NPV($A$6,F9:F13)	

With a present worth of over $500M, this is a desirable project.

Example 5–13 uses the **XNPV** spreadsheet function, which finds the present worth of a series of cash flows that occur on specific dates. The NPV function assumes that time 0 is one period before the first cash flow, so time 0 must *not* be included in the function's range. XNPV is different. Time 0 must be specified just like every other date—even if there is no cash flow at time 0. The time 0 date specifies the *when* for the net present value or present worth. The interest rate for the XNPV function is an effective annual rate, and XNPV is using calendar dates, so leap year has an extra day and the daily interest rate is slightly smaller in leap years.

EXAMPLE 5–13

A construction firm has just won a contract for $2.25 million. If work is completed on schedule, it will receive a series of progress payments. The first $100,000 will be received 15 days after the contract is signed. Further payments of $450,000 each will be received at 45 and 180 days. The remaining $1.25 million will be received 390 days after the contract is signed. If the firm's interest rate is 9.5%, what is the present worth of the progress payments on the contract signing date of September 26, 2016?

Notice that the XNPV function specifies the range of cash flows and then the range of dates.

	A	B	C
1	9.50%	interest rate	
2			
3	Days	Dates	Cash flows
4	0	9/26/2016	0
5	15	10/11/2016	$100,000
6	45	11/10/2016	450,000
7	180	3/25/2017	450,000
8	390	10/21/2017	1,250,000
9			
10		=$B$4+A8	
11	$2,109,403	=XNPV(A1,C4:C8, **B4:B8**)	

The two key differences between the NPV and XNPV functions can be seen in Example 5–13. XNPV includes:

- The time 0 cash flow, as in cell C4.
- A final argument of the range of dates for the cash flows, as in cells B4 to B8.

Bond Pricing

The calculation in Example 5–14 is done routinely when bonds are bought and sold during their life. Bonds are issued at a **face or par value** (usually $1000), which is received when the bonds mature. There is a **coupon interest rate**, which is set when the bond is originally issued or sold. The term *coupon interest* dates from the time when bonds were paper rather than electronic, and a paper coupon was detached from the bond to be redeemed in cash. The interest paid equals the coupon rate times the face value. This interest is usually paid in two semiannual payments, although some bonds have other compounding periods. A cash flow diagram for the remaining interest payments and the final face value is used with a current interest rate to calculate a present worth. This is the bond's price.

Bond pricing calculations are important because selling bonds is the principal source of new external funds for existing firms. Bonds are also how governments fund many projects for bridges, schools, airports, dams, and so on. Bonds are also considered a safer way for individuals, pension funds, and other entities to invest. A bond is a contract to deliver interest payments when due and the face value of the bond at maturity. The cash flows are on known dates and of known amounts. When interest rates rise above the coupon rate, those future cash flows are discounted more and the bond sells at a discount. When interest rates fall below the coupon rate, future cash flows are discounted less and the bond sells at a premium. Bonds of the federal government are safer than bonds of a firm or many cities and nations. The U.S. government is less likely to *default* or not pay interest or face value when due.

EXAMPLE 5–14

A 15-year municipal bond was issued 5 years ago. Its coupon interest rate is 8%, interest payments are made semiannually, and its face value is $1000. If the current market interest rate is 12.36%, what should be the bond's price? *Note*: The issuer of the bond (city, state, company) makes interest payments to the bondholder (at the coupon rate), as well as a final value payment.

TABLE SOLUTION

The first 5 years are past, and there are 20 more semiannual payments. The coupon interest rate is the nominal annual rate or APR. Half that, or 4%, of $1000 or $40 is paid at the end of each 6-month period.

The bond's price is the PW of the cash flows that will be received if the bond is purchased. The cash flows are $40 at the end of each of the 20 semiannual periods and the face value of $1000 at the end of period 20.

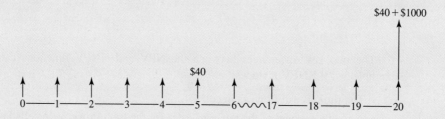

Since the $40 in interest is received semiannually, the market or effective annual interest rate (i_a) must be converted to a semiannual rate. Using Equation 3-8, we obtain

$$(1+i)^2 = 1 + i_a = 1.1236$$

$$(1+i) = 1.06$$

$$i = 6\% \text{ effective semiannual interest rate}$$

$$PW = 40(P/A, 6\%, 20) + 1000(P/F, 6\%, 20)$$

$$= 40(11.470) + 1000(0.3118) = \$770.6$$

The $770.6 is the discounted price; that is, the PW at 12.36% of the cash flows from the $1000 bond. The $229.4 discount raises the investment's rate of return from a nominal 8% for the face value to 12.36% on an investment of $770.6.

5-BUTTON SOLUTION

	A	B	C	D	E	F	G	H
1	Problem	i	n	*PMT*	*PV*	*FV*	Solve for	Answer
2	Exp. 5-14	6%	20	40		1000	PV	-$770.60

This example also illustrates why it is better to separately state cash flows. At the end of period 20, there are two cash flows, $40 and $1000. The $40 is part of the 20-period uniform series, and the $1000 is a single cash flow. All of these numbers come directly from the problem statement. If the two final cash flows are combined into $1040, then the $40 uniform series has only 19 periods—and it is easy to err and forget that change.

SUMMARY

Present worth analysis is suitable for almost any economic analysis problem. But it is particularly desirable when we wish to know the present worth of future costs and benefits. We frequently want to know the value today of such things as income-producing assets, stocks, and bonds.

For present worth analysis, the proper economic criteria are:

Neither input nor output is fixed	Maximize (PW of benefits − PW of costs) or, more simply stated: Maximize NPW
Fixed input	Maximize the PW of benefits
Fixed output	Minimize the PW of costs

To make valid comparisons, we need to analyze each alternative in a problem over the same **analysis period** or **planning horizon.** If the alternatives do not have equal lives, some technique must be used to achieve a common analysis period. One method is to select an analysis period equal to the least common multiple of the alternative lives. Another method is to select an analysis period and estimate end-of-analysis-period salvage values for the alternatives.

Capitalized cost is the present worth of cost for an infinite analysis period ($n = \infty$). When $n = \infty$, the fundamental relationship is $A = iP$.

The numerous assumptions routinely made in solving economic analysis problems include the following.

1. Present sums (P) are beginning-of-period, and all series receipts or disbursements (A) and future sums (F) occur at the end of the interest period. The compound interest tables were derived on this basis.

2. In industrial economic analyses, the point of view for computing the consequences of alternatives is that of the total firm. Narrower views can result in suboptimal solutions.

3. Only the differences between the alternatives are relevant. Past costs are sunk costs and generally do not affect present or future costs. For this reason they are ignored.

4. The investment problem is isolated from the financing problem. We generally assume that all required money is borrowed at interest rate i.

5. For now, stable prices are assumed. The inflation–deflation problem is deferred to Chapter 14. Similarly, our discussion of income taxes is deferred to Chapter 12.

6. Often uniform cash flows or arithmetic gradients are reasonable assumptions. However, spreadsheets simplify the finding of PW in more complicated problems.

PROBLEMS

Present Value of One Alternative

5-1 A project has a net present worth of −$20,000 as of January 1, 2022. If a 4% interest rate is used, what is the project NPW as of December 31, 2017?

5-2 The annual income from a rented house is $24,000. The annual expenses are $6000. If the house can be sold for $245,000 at the end of 10 years, how much could you afford to pay for it now, if you considered 9% to be a suitable interest rate?
(A)

5-3 A machine costs $250,000 to purchase and will provide $60,000 a year in benefits. The company plans to use the machine for 12 years and then will sell the machine for scrap, receiving $15,000. The company interest rate is 10%. Should the machine be purchased?

5-4 IBP Inc. is considering establishing a new machine to automate a meatpacking process. The machine will save $50,000 in labor annually. The machine can be purchased for $200,000 today and will be used for 10 years. It has a salvage value of $10,000 at the end of its useful life. The new machine will require an annual maintenance cost of $9000. The corporation has a minimum rate of return of 10%. Do you recommend automating the process?
(A)

5-5 A student has a job that leaves her with $300 per month in disposable income. She decides that she will use the money to buy a car. Before looking for a car, she arranges a 100% loan whose terms are $300 per month for 48 months at 9% nominal annual interest. What is the maximum car purchase price that she can afford with her loan?

5-6 The student in Problem 5-5 finds a car she likes and the dealer offers to arrange financing. His terms are 6% interest for 72 months and no down payment. The car's sticker price is $24,000. How expensive a car can she afford on the dealer's terms?
(A)

5-7 A road building contractor has received a major highway construction contract that will require 50,000 m³ of crushed stone each year for 5 years. The stone can be obtained from a quarry for $7.80/m³. As an alternative, the contractor has decided to try to buy the quarry. He believes that if he owned the quarry, the stone would cost him only $6.30/m³. He thinks he could resell the quarry at the end of 5 years for $200,000.
(A)
(G)

(a) If the contractor uses a 12% interest rate, how much would he be willing to pay for the quarry?

(b) The extraction of needed quarry stone and other aggregates via mining can pose serious social consequences to various stakeholders. Provide a list of concerns and mitigating potential actions that the road-building contractor may want to consider if he purchases the quarry.

5-8 Compute the present value, P, for the following cash flows.
(A)

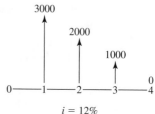

$i = 12\%$

5-9 You supervise an aging production line that constantly needs maintenance and new parts. Last month you spent $25,000 replacing a failed controller. Should the following plan be accepted if the interest rate is 15%? The net installed cost of the new line to replace the old line is $600,000 with a useful life of six years. In the first year its operating cost will be $100,000, and it will generate annual revenues of $300,000. Each year the operating cost will increase by $5000 and the revenues will fall by $15,000. After six years the equipment will have a value of $100,000 in the next re-building of the line.
Contributed by Yasser Alhenawi, University of Evansville

5-10 (a) How much would the owner of a building be justified in paying for a sprinkler system that will save $1000 a year in insurance premiums if the system has to be replaced every 15 years and has a salvage value equal to 10% of its initial cost? Assume money is worth 5%.
(A)
(E)

(b) Research the International Code Council (ICC). Find and read their code of ethics. Why are organizations such as ICC important from a building safety perspective?

5-11 A wholesale company has signed a contract with a supplier to purchase goods for $2,000,000 annually. The first purchase will be made now to be followed by 10 more. Determine the contract's present worth at a 7% interest rate.
Contributed by Hamed Kashani, Saeid Sadri, and Baabak Ashuri, Georgia Institute of Technology

5-12 Compute the present value, P, for the following cash
A flows.

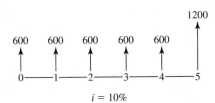

$$i = 10\%$$

5-13 By installing some elaborate inspection equipment
on its assembly line, the Robot Corp. can avoid hir-
ing an extra worker who would have earned $36,000
a year in wages and an additional $9500 a year in
employee benefits. The inspection equipment has a
6-year useful life and no salvage value. Use a nominal 18% interest rate in your calculations. How much
can Robot afford to pay for the equipment if the
wages and worker benefits were to have been paid

(*a*) At the end of each year
(*b*) Monthly

Explain why the answer in (*b*) is larger.

5-14 On February 1, the Miro Company needs to purchase
A some office equipment. The company is short of cash
and expects to be short for several months. The treasurer has said that he could pay for the equipment as
follows:

Date	Payment
April 1	$150
June 1	300
Aug. 1	450
Oct. 1	600
Dec. 1	750

A local office supply firm will agree to sell
the equipment to Miro now and accept payment
according to the treasurer's schedule. If interest
will be charged at 3% every 2 months, with compounding once every 2 months, how much office
equipment can the Miro Company buy now? What
is the effective interest rate?

5-15 A stonecutter, assigned to carve the headstone for a
well-known engineering economist, began with the
following design.

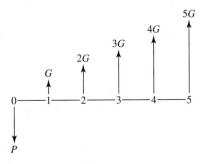

He started the equation

$$P = G(P/G, i, 6)$$

He realized he had made a mistake, but he does not
want to discard the stone and start over. What one
compound interest factor can be added to make the
equation correct?

$$P = G(P/G, i, 6)(\quad, i, \quad)$$

5-16 A young industrial engineer analyzed some equip-
A ment to replace one production worker. The present
E worth of employing one less production worker just
equaled the present worth of the equipment costs,
assuming a 10-year useful life for the equipment. It
was decided not to buy the equipment.

A short time later, the production workers won
a new 3-year union contract that granted them an
immediate 40¢-per-hour wage increase, plus an additional 25¢-per-hour wage increase in years 2 and 3.
Assume that in future years, a 25¢-per-hour wage
increase will be granted.

(*a*) By how much does the present worth of replacing one production employee increase? Assume
an interest rate of 8%, a single 8-hour shift, and
250 days per year.
(*b*) What are the ethical issues of replacing workers
with advanced technologies from the firm's, the
workers', and society's perspective?

5-17 Use a geometric gradient formula to compute the
present value, P, for the following cash flows.

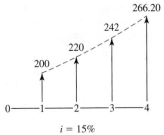

$$i = 15\%$$

5-18 A firm has installed a manufacturing line for packaging materials. The firm plans to produce 50 tons of packing peanuts at $5000 per ton annually for 4 years, and then 80 tons of packing peanuts per year at $5500 per ton for the next 6 years. What is the present worth of the expected income? The firm's interest rate is 18% per year.
Contributed by Hamed Kashani, Saeid Sadri, and Baabak Ashuri, Georgia Institute of Technology

5-19 Annual maintenance costs for a particular section of highway pavement are $2500. The placement of a new surface would reduce the annual maintenance cost to $500 per year for the first 3 years and to $1000 per year for the next 7 years. After 10 years the annual maintenance would again be $2500. If maintenance costs are the only saving, what investment can be justified for the new surface? Assume interest at 5%.

5-20 Luis is responsible for buying some specialized manufacturing equipment that has a purchase price of $10,000 and annual operating costs of $1000. The vendor is offering a special buyer incentive that provides free maintenance for the first four years. After that time, the maintenance is $500 per year over the 10-year life, and there is an overhaul expense in year 5 of $2000. The equipment has a salvage value of $1000. If the interest rate is 8%, what is the present value?
Contributed by Gillian Nicholls, Southeast Missouri State University

5-21 A new office building was constructed 5 years ago by a consulting engineering firm. At that time the firm obtained a bank loan for $600,000 with a 12% annual interest rate, compounded quarterly. The loan terms call for equal quarterly payments for 10 years. The loan also allows for its prepayment at any time without penalty.

The firm proposes to refinance the loan through an insurance company. The new loan would be for a 20-year term with an interest rate of 8% per year, compounded quarterly. The insurance company requires the payment of a 5% loan initiation charge (often described as a "5-point loan fee"), which will be added to the starting balance.

(*a*) What is the balance due on the original mortgage if 16 payments have been made?

(*b*) What is the difference between the old and new quarterly payments?

5-22 Argentina is considering constructing a bridge across the Rio de la Plata to connect its northern coast to the southern coast of Uruguay. If this bridge is constructed, it will reduce the travel time from Buenos Aires, Argentina, to São Paulo, Brazil, by over 10 hours, and there is the potential to significantly improve the flow of manufactured goods between the two countries. The cost of the new bridge, which will be the longest bridge in the world, spanning over 50 miles, will be $700 million. The bridge will require an annual maintenance of $10 million for repairs and upgrades and is estimated to last 80 years. It is estimated that 550,000 vehicles will use the bridge during the first year of operation and an additional 50,000 vehicles per year until the tenth year. The annual traffic for the remainder of the life of the bridge will be 1,000,000 vehicles per year. These data are based on a toll charge of $90 per vehicle. The Argentine government requires a minimum rate of return of 9% to proceed with the project.

(*a*) Does this project provide sufficient revenues to offset its costs?

(*b*) What considerations are there besides economic factors in deciding whether to construct the bridge?

5-23 Compute the present value, P, for the following cash flows.

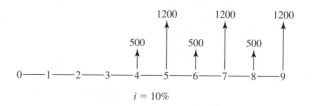

$i = 10\%$

Lives Match

5-24 Walt Wallace Construction Enterprises may buy a new dump truck with a 10-year life. Interest is 5%. The cash flows for two likely models are as follows:

Model	First Cost	Annual Operating Cost	Annual Income	Salvage Value
A	$50,000	$12,000	$19,000	$10,000
B	80,000	11,000	22,000	30,000

(*a*) Using present worth analysis, which truck should the firm buy, and why?

(*b*) Before the construction company can close the deal, the dealer sells out of Model *B* and cannot get any more. What should the firm do now, and why?

5-25 A new tennis court complex is planned. Each of two alternatives will last 18 years, and the interest rate is 7%. Use present worth analysis to determine which should be selected.
Contributed by D. P. Loucks, Cornell University

	Construction Cost	Annual O&M
A	$500,000	$25,000
B	640,000	10,000

5-26 Two alternative courses of action have the following
Ⓐ schedules of disbursements:

Year	A	B
0	−$13,000	
1	0	−$1000
2	0	−2000
3	0	−3000
4	0	−4000
5	0	−5000
	−$13,000	−$15,000

Based on a 10% interest rate, which alternative should be selected?

5-27 If produced by Method *A*, a product's initial capital cost will be $100,000, its annual operating cost will be $20,000, and its salvage value after 3 years will be $20,000. With Method *B* there is a first cost of $150,000, an annual operating cost of $10,000, and a $50,000 salvage value after its 3-year life. Based on a present worth analysis at a 15% interest rate, which method should be used?
Contributed by Hamed Kashani, Saeid Sadri, and Baabak Ashuri, Georgia Institute of Technology

5-28 For a new punch press. Company *A* charges
Ⓐ $250,000 to deliver and install it. Company *A* has estimated that the machine will have operating and maintenance (O & M) costs of $4000 a year. You

estimate an annual benefit of $89,000. Company *B* charges $205,000 to deliver and install the device. Company *B* has estimated O & M of the press at $4300 a year. You estimate an annual benefit of $86,000. Both machines will last 5 years and can be sold for $15,000. Use an interest rate of 12%. Which machine should your company buy?

5-29 Quinton's refrigerator has just died. He can get a
Ⓖ basic refrigerator or a more efficient and environmentally friendly refrigerator with an Energy Star designation. Quinton earns 4% compounded annually on his investments, he wants to consider a 10-year planning horizon, and he will use present worth analysis to determine the best alternative. What is your recommendation? *Contributed by Paul R. McCright, University of South Florida*

	Basic Unit	Energy Star Unit
Initial cost	700	800
Delivery and installation	60	60
Professional servicing (year 5)	100	100
Annual energy costs	120	55
Salvage value (year 10)	150	175

5-30 A battery manufacturing plant has been ordered to
Ⓐ cease discharging acidic waste liquids containing
Ⓖ mercury into the city sewer system. As a result, the firm must now adjust the pH and remove the mercury from its waste liquids. Quotations from three firms are included in the following table of costs.

Bidder	Installed Cost	Annual Operating Cost	Annual Income from Mercury Recovery	Salvage Value
Foxhill Instrument	$ 35,000	$8000	$2000	$20,000
Quicksilver	40,000	7000	2200	0
Almaden	100,000	2000	3500	0

If the installation will last 20 years and money is worth 10%, which equipment should be purchased?

5-31 Teléfono Mexico is expanding its facilities to serve a new manufacturing plant. The new plant will require 2000 telephone lines this year, and another 2000 lines after expansion in 10 years. The plant will operate for 30 years.

Option 1 Provide one cable now with capacity to serve 4000 lines. The cable will cost $20,000 and annual maintenance costs will be $1500.

Option 2 Provide a cable with capacity to serve 2000 lines now and a second cable to serve the other 2000 lines in 10 years. Each cable will cost $15,000 and will have an annual maintenance of $1000.

The telephone cables will last at least 30 years, and the cost of removing the cables is offset by their salvage value.

(a) Which alternative should be selected, assuming a 10% interest rate?

(b) Will your answer change if the demand for additional lines occurs in 5 years?

5-32 A consulting engineer has been hired to advise a town how best to proceed with the construction of a 200,000-m^3 water supply reservoir. Since only 120,000 m^3 of storage will be required for the next 25 years, an alternative to building the full capacity now is to build the reservoir in two stages. Initially, the reservoir could be built with 120,000 m^3 of capacity and then, 25 years hence, the additional 80,000 m^3 of capacity could be added by increasing the height of the reservoir. If interest is computed at 4%, which construction plan is preferred?

	Construction Cost	Annual Maintenance Cost
Build in two stages		
First stage	$14,200,000	$75,000
Second stage	12,600,000	add $25,000
Build full capacity now	22,400,000	100,000

5-33 In order to improve evacuation routes out of New Orleans in the event of another major disaster such as Hurricane Katrina, the Louisiana Department of Transportation (L-DoT) is planning to construct an additional bridge across the Mississippi River. The department uses an interest rate of 8% and plans a 50-year life for either bridge. Which design has the better present worth? *Contributed by Paul R. McCright, University of South Florida*

All Costs in $M	Suspension Bridge	Cantilever Bridge
Initial construction	$585	$470
Land acquisition	120	95
Annual O&M	2	3
Annual increase	4%	0.3
Major maintenance (Year 25)	185	210
Salvage cost	30	27

5-34 Javier is an IE at Lobos Manufacturing. He has been studying process line G to determine if an automated system would be preferred to the existing labor-intensive system. If Lobos wants to earn at least 25% and uses a 15-year planning horizon, which alternative is preferred? *Contributed by Paul R. McCright, University of South Florida*

	Labor Intensive	Automated
Initial cost	$ 0	$ 55,000
Installation cost	0	15,500
First-year O&M	1,500	4,800
Annual increase	200	600
First-year labor costs	116,000	41,000
Annual increase	4%	4%
Salvage value (EOY15)	5,000	19,000

5-35 In an analysis one alternative has a net present worth of $4200, based on a 6-year analysis period that matches its useful life. A 9% interest rate was used. The replacement will be an identical item with the same cost, benefits, and useful life. Using a 9% interest rate, compute the net present worth for a 12-year analysis period.

Lives Differ

5-36 Use an 8-year analysis period and a 10% interest rate to determine which alternative should be selected:

	A	B
First cost	$5300	$10,700
Uniform annual benefit	$1800	$2100
Useful life, in years	4	8

5-37 The Larkspur Furniture Company needs a new grinder. Compute the present worth for these mutually exclusive alternatives and identify which you would recommend given $i = 6\%$ per year. Larkspur uses a 10-year planning horizon.

Alternative	A	B
Initial cost	$4500	$5500
Annual costs	$300	$400
Salvage value	$500	$0
Life	5 years	10 years

Contributed by Gillian Nicholls, Southeast Missouri State University

5-38 Which process line should be built for a new chemical? The expected market for the chemical is 20 years. A 20% rate is used to evaluate new process facilities, which are compared with present worth. How much does the better choice save?

	First Cost	O&M Cost/year	Salvage	Life
A	$18M	$5M	$4M	10 years
B	25M	3M	6M	20 years

5-39 Which equipment is preferred if the firm's interest rate is 15%? In PW terms how great is the difference?

Alternative	A	B
First cost	$45,000	$35,000
Annual O&M	3,900	4,200
Salvage value	0	5,000
Overhaul (Year 6)	10,000	Not required
Life, in years	10	5

5-40 A weekly business magazine offers a 1-year subscription for $48 and a 3-year subscription for $116. If you thought you would read the magazine for at least the next 3 years, and consider 20% as a minimum rate of return, which way would you purchase the magazine: With three 1-year subscriptions or a single 3-year subscription?

5-41 A man had to have the muffler replaced on his 2-year-old car. The repairman offered two alternatives. For $250 he would install a muffler guaranteed for 2 years. But for $400 he would install a muffler guaranteed "for as long as you own the car." Assuming the present owner expects to keep the car for about 3 more years, which muffler would you advise him to have installed if you thought 10% was a suitable interest rate and the less expensive muffler would only last 2 years?

5-42 North City must choose between two new snow-removal machines. The SuperBlower has a $70,000 first cost, a 20-year life, and an $8000 salvage value. At the end of 9 years, it will need a major overhaul costing $19,000. Annual maintenance and operating costs are $9000. The Sno-Mover will cost $50,000, has an expected life of 10 years, and has no salvage value. The annual maintenance and operating costs are expected to be $12,000. Using a 12% interest rate, which machine should be chosen?

5-43 A new alloy can be produced by Process A, which costs $200,000 to implement. The operating cost will be $10,000 per quarter with a salvage value of $25,000 after its 2-year life. Process B will have a first cost of $250,000, an operating cost of $15,000 per quarter, and a $40,000 salvage value after its 4-year life. The interest rate is 8% per year compounded quarterly. What is the present value difference between A and B?
Contributed by Hamed Kashani, Saeid Sadri, and Baabak Ashuri, Georgia Institute of Technology

5-44 An elevator company has redesigned their product to be 50% more energy efficient than hydraulic designs. Two designs are being considered for implementation in a new building.

(a) Given an interest rate of 20% which bid should be accepted?

Alternatives	A	B
Installed cost	$45,000	$54,000
Annual cost	2700	2850
Salvage value	3000	4500
Life, in years	10	15

(b) Research and list a few attributes that the company might be using in the elevators' major systems: drive, cab, hoist, and control mechanisms that make them more energy efficient.

5-45 The Crockett Land Winery must replace its present grape-pressing equipment. The two alternatives are the Quik-Skwish and the Stomp-Master. The annual operating costs increase by 12% each year as the machines age. If the interest rate is 9%, which press should be chosen?

	Quik-Skwish	Stomp-Master
First cost	$350,000	$500,000
Annual operating costs	28,000	22,500
Salvage value	35,000	50,000
Useful life, in years	5	10

5-46 A railroad branch line to a landfill site is to be constructed. It is expected that the railroad line will be used for 15 years, after which the landfill site will be closed and the land turned back to agricultural use. The railroad track and ties will be removed at that time.

In building the railroad line, either treated or untreated wood ties may be used. Treated ties have an installed cost of $6 and a 10-year life; untreated ties are $4.50 with a 6-year life. If at the end of 15 years the ties then in place have a remaining useful life of 4 years or more, they will be used by the railroad elsewhere and have an estimated salvage value of $3 each. Any ties that are removed at the end of their service life, or too close to the end of their service life to be used elsewhere, can be sold for 50¢ each.

(a) Determine the most economical plan for the initial railroad ties and their replacement for the 15-year period. Make a present worth analysis assuming 8% interest.
(b) Plastics in landfills do not degrade quickly and the land may not be useful for agriculture. Research and write a summary of what is being done with landfills now to make them more earth friendly and reduce negative impacts.

Perpetual Life

5-47 Dr. Fog E. Professor is retiring and wants to endow a chair of engineering economics at his university. It is expected that he will need to cover an annual cost of $250,000 forever. What lump sum must he donate to the university today if the endowment will earn 5% interest?

5-48 An elderly lady decided to distribute most of her considerable wealth to charity and to keep for herself only enough money to provide for her living. She feels that $3000 a month will amply provide for her needs. She will establish a trust fund at a bank that pays 3% interest, compounded monthly. Upon her death, the balance is to be paid to her niece, Susan. If she deposits enough money to last forever, how much will Susan receive when her aunt dies?

5-49 A local symphony association offers memberships as follows:

Continuing membership, per year	$ 300
Patron lifetime membership	5000

The patron membership has been based on the symphony association's belief that it can obtain a 3% rate of return. If you agree that 3% is appropriate, would you be willing to purchase the patron membership? Explain why or why not.

5-50 A depositor puts $25,000 in a savings account that pays 5% interest, compounded semiannually. Equal annual withdrawals are to be made from the account, beginning one year from now and continuing forever. What is the maximum annual withdrawal?

5-51 The local botanical society wants to ensure that the gardens in the town park are properly cared for. The group recently spent $100,000 to plant the gardens. The members want to set up a perpetual fund to provide $100,000 for future replantings every 10 years.

(a) If interest is 5%, how much money is needed for the fund?
(b) If the last replanting is in year 100, how much is needed for the fund?

5-52 What amount of money deposited 50 years ago at 4% interest would provide a perpetual payment of $25,000 a year beginning this year?

5-53 A subdivision developer must construct a sewage treatment plant and deposit sufficient money in a perpetual trust fund to pay the $5000 per year operating cost and to replace the treatment plant every 40 years. The plant and future replacement plants will cost $150,000 each. If the trust fund earns 8% interest, what is the developer's capitalized cost?

5-54 Use capitalized cost to determine which type of road surface is preferred on a particular section of highway. Use a 12% interest rate.

	A	**B**
Initial cost	$500,000	$700,000
Annual maintenance	35,000	25,000
Periodic resurfacing	350,000	450,000
Resurfacing interval	10 years	15 years

5-55 A new bridge project is being evaluated at $i = 5\%$. Recommend an alternative based on the capitalized cost for each.

	Construction Cost	Annual O&M	Life (years)
Concrete	$50 million	$250,000	70
Steel	40 million	500,000	50

5-56 A new stadium is being evaluated at $i = 5\%$. Recommend an alternative for the main structural material based on the capitalized cost for each.

	Construction Cost	Annual O&M	Life (years)
Concrete	$27 million	$1.2 million	60
Steel	21 million	1.6 million	50

5-57 **(G)** An open-pit mine must fund an account now to pay for maintenance of a tailing pond in perpetuity (after the mine shuts down in 30 years). The costs until shutdown are part of the mine's operating costs. The maintenance costs begin in 31st year at $300,000 annually.

(*a*) How much must be deposited now if the fund will earn 5% interest? How much does this change if the interest rate is 4%?

(*b*) What is a tailing pond? Is the mine building this because they are a good corporate citizen or because they are required to do so? By whom? Could it be both?

5-58 **(A)** A firm wants to sponsor a new engineering lab at a local university. This requires $2.5M to construct the lab, $1.2M to equip it, and $600,000 every 5 years for new equipment. What is the required endowment if the university will earn 6% interest on the funds?

5-59 A city has developed a plan to provide for future municipal water needs. The plan proposes an aqueduct that passes through 500 feet of tunnel in a nearby mountain. The first alternative is to build a full-capacity tunnel now for $556,000. The second

alternative is to build a half-capacity tunnel now (cost = $402,000), which should be adequate for 20 years, and then to build a second parallel half-capacity tunnel. The maintenance cost of the tunnel linings every 10 years for the full-capacity tunnel is $40,000 and for each half-capacity tunnel it is $32,000.

The friction losses in the half-capacity tunnels will be greater than in the full-capacity tunnel. The estimated additional pumping costs in each half-capacity tunnel will be $2000 per year. Based on capitalized cost and a 7% interest rate, which alternative should be selected?

5-60 **(A)** A rather wealthy man decides to arrange for his descendants to be well educated. He wants each child to have $60,000 for his or her education. He plans to set up a perpetual trust fund so that six children will receive this assistance in each generation. He estimates that generations will be spaced 25 years apart. He expects the trust to be able to obtain a 4% rate of return and the first recipients to receive the money 10 years hence. How much money should he now set aside in the trust?

5-61 Kansas Public Service Company wishes to determine the capitalized worth of a new windmill at an interest rate of 9% and following costs.

Purchase	$ 725,000	Installation	$ 143,000
Annual O&M	12,000	Overhaul (Year 25)	260,000
Expected life	40 years	Salvage value	32,000

Contributed by Paul R. McCright, University of South Florida

5-62 Compute the present value, P, for the following cash flows (assume series repeats forever).

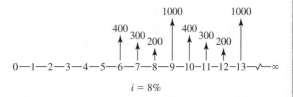

$i = 8\%$

Multiple Alternatives

5-63 **(G)** A steam boiler is needed as part of the design of a new plant. The boiler can be fired by natural gas, fuel oil, or coal. A cost analysis shows that natural gas would be the cheapest at $30,000; for fuel oil it would be $55,000; and for coal it would be

$180,000. If natural gas is used rather than fuel oil, the annual fuel cost will decrease by $7500. If coal is used rather than fuel oil, the annual fuel cost will be $15,000 per year less.

(a) Assuming 8% interest, a 20-year analysis period, and no salvage value, which is the most economical installation?

(b) Find resources and build a *green effects* table to compare the major advantages and disadvantages of using natural gas, fuel oil, or coal to heat the steam boiler. Use *Advantages* and *Disadvantages* as the column headers and then each choice as a row.

5-64 Austin General Hospital is evaluating new lab equipment. The interest rate is 15% and in each case the equipment's useful life is 4 years. Use NPW analysis to pick which company you should purchase from.

Ⓐ

Company	A	B	C
First cost	$15,000	$25,000	$20,000
O&M costs	1,600	400	900
Annual benefit	8,000	13,000	11,000
Salvage value	3,000	6,000	4,500

5-65 A building contractor obtained bids for some asphalt paving, based on a specification. Three paving subcontractors quoted the following prices and terms of payment:

Paving Co.	Price	Payment Schedule
Quick	$85,000	50% payable immediately
		25% payable in 6 months
		25% payable at year end
Tartan	82,000	Payable immediately
Faultless	84,000	25% payable immediately
		75% payable in 6 months

The building contractor uses a 12% nominal interest rate, compounded monthly. Which paving subcontractor should be awarded the paving job?

5-66 Six mutually exclusive alternatives, *A–F*, are being examined. For an 8% interest rate, which alternative should be selected? What is the highest present worth? Each alternative has a 6-year useful life.

Ⓐ

	Initial Cost	Uniform Annual Benefit
A	$ 20	$ 6.00
B	35	9.25
C	55	13.38
D	60	13.78
E	80	24.32
F	100	24.32

5-67 The management of an electronics manufacturing firm believes it is desirable to automate its production facility. The automated equipment would have no salvage value at the end of a 10-year life. The plant engineering department has suggested eight mutually exclusive alternatives. If the firm expects a 12% rate of return, which plan, if any, should it adopt?

Plan	Initial Cost (thousands)	Net Annual Benefit (thousands)
1	$265	$51
2	220	39
3	180	26
4	100	15
5	300	57
6	130	23
7	245	47
8	165	33

5-68 Consider *A–E*, five mutually exclusive alternatives:

Ⓐ

	A	B	C	D	E
Initial cost	$600	$600	$600	$600	$600
Uniform annual benefits					
For first 5 years	100	100	100	150	150
For last 5 years	50	100	110	0	50

The interest rate is 10%. If all the alternatives have a 10-year useful life and no salvage value, which alternative should be selected?

5-69 A firm is considering three mutually exclusive alternatives as part of a production improvement program. The alternatives are:

	A	B	C
Installed cost	$10,000	$15,000	$20,000
Annual benefit	1,625	1,530	1,890
Useful life (yrs)	10	20	20

The salvage value of each alternative is zero. At the end of 10 years, Alternative A could be replaced with another A with identical cost and benefits.

(a) Which alternative should be selected if interest is 6%?
(b) 3%,
(c) If there is a difference between parts (a) and (b) can you explain it?

5-70 The following costs are associated with three tomato-peeling machines being considered for use in a canning plant. If the canning company uses an interest rate of 12%, which is the best alternative? Use NPW to make your decision. (*Note:* Consider the least common multiple as the study period.)

Machine	A	B	C
First cost	$52,000	$63,000	$67,000
O&M costs	15,000	9,000	12,000
Annual benefit	38,000	31,000	37,000
Salvage value	13,000	19,000	22,000
Useful life (yrs)	4	6	12

5-71 Consider the following three alternatives. There is also a "do nothing" alternative.

	A	B	C
Cost	$50	$30	$40
Net annual benefit	12	4.5	6
Useful life, in years	5	10	10

At the end of the 5-year useful life of A, a replacement is made. If a 10-year analysis period and a 8% interest rate are selected, which is the preferred alternative?

5-72 A cost analysis is to be made to determine what, if anything, should be done in a situation offering three "do-something" and one "do-nothing" alternatives. Estimates of the cost and benefits are as follows.

Alternatives	Cost	Uniform Annual Benefit	End-of-Useful-Life Salvage Value	Useful Life (years)
1	$500	$135	$ 0	5
2	600	100	250	5
3	700	100	180	10
4	0	0	0	0

Use a 10-year analysis period. At the end of 5 years, Alternatives 1 and 2 may be replaced with identical alternatives. Which alternative should be selected?

(a) If an 8% interest rate is used?
(b) If a 12% interest rate is used?

5-73 Given the following data, use present worth analysis to find the best alternative, A, B, or C.

	A	B	C
Initial cost	$10,000	15,000	$12,000
Annual benefit	6,000	10,000	5,000
Salvage value	1,000	−2,000	3,000
Useful life	2 years	3 years	4 years

Use an analysis period of 12 years and 15% interest.

5-74 An investor has carefully studied a number of companies and their common stock. From his analysis, he has decided that the stocks of six firms are the best of the many he has examined. They represent about the same amount of risk, and so he would like to determine one single stock in which to invest. He plans to keep the stock for 4 years and requires a 10% minimum attractive rate of return.

Which stock if any, should the investor consider buying?

Common Stock	Price per Share	Annual End-of-Year Dividend per Share	Estimated Price at End of 4 Years
Western House	$23³/₄	$1.25	$32
Fine Foods	45	4.50	45
Mobile Motors	30⁵/₈	0	42
Spartan Products	12	0	20
U.S. Tire	33³/₈	2.00	40
Wine Products	52¹/₂	3.00	60

Nominal and Effective Interest

5-75 Assume monthly car payments of $500 per month for 4 years and an interest rate of 0.75% per month. What initial principal or PW will this repay?

5-76 Assume annual car payments of $6000 for 4 years
Ⓐ and an interest rate of 9% per year. What initial principal or PW will this repay?

5-77 Assume annual car payments of $6000 for 4 years and an interest rate of 9.381% per year. What initial principal or PW will this repay?

5-78 Why do the values in Problems 5-75, 5-76, and 5-77 differ?

5-79 Assume mortgage payments of $1000 per month for 30 years and an interest rate of 0.5% per month. What initial principal or PW will this repay?

5-80 Assume annual mortgage payments of $12,000 for
Ⓐ 30 years and an interest rate of 6% per year. What initial principal or PW will this repay?

5-81 Assume annual mortgage payments of $12,000 for 30 years and an interest rate of 6.168% per year. What initial principal or PW will this repay?

5-82 Why do the values in Problems 5-79, 5-80, and 5-81 differ?

🖥 **Applications and Complications**

5-83 A construction project has the following end-of-month costs. Calculate the PW at a nominal interest rate of 18%.

January	$ 30,000	May	$620,000
February	50,000	June	460,000
March	210,000	July	275,000
April	530,000	August	95,000

5-84 A factory has averaged the following monthly heat-
Ⓐ ing and cooling costs over the last 5 years. Calculate the PW at a nominal interest rate of 12%.

January	$25,000	July	$29,000
February	19,000	August	33,000
March	15,000	September	19,000
April	9,000	October	8,000
May	12,000	November	16,000
June	18,000	December	28,000

5-85 Ding Bell Imports requires a return of 15% on all projects. If Ding is planning an overseas development project with these cash flows, what is the project's net present value?

Year	1	2	3	4	5	6	7
Net Cash ($)	−60,000	−110,000	20,000	40,000	80,000	100,000	60,000

5-86 Maverick Enterprises is planning a new product.
Ⓐ Annual sales, unit costs, and unit revenues are as tabulated; the first cost of R&D and setting up the assembly line is $42,000. If i is 10%, what is the PW?

Year	Annual Sales	Cost/unit	Price/unit
1	$ 5,000	$3.50	$6
2	6,000	3.25	5.75
3	9,000	3.00	5.50
4	10,000	2.75	5.25
5	8,000	2.5	4.5
6	4,000	2.25	3

5-87 Northern Engineering is analyzing a mining project. Annual production, unit costs, and unit revenues are in the table. The first cost of the mine setup is $6 million. If i is 15%, what is the PW?

Year	Annual Production (tons)	Cost per Ton	Price per Ton
1	90,000	$25	$35
2	120,000	20	36
3	120,000	22	37
4	100,000	24	38
5	80,000	26	39
6	60,000	28	40
7	40,000	30	41

Date	Cash Flow
4/16/2018	−$5000
12/10/2018	1500
4/1/2019	1800
7/22/2019	2000
11/11/2019	1900
2/4/2020	1500
6/2/2020	1200

XNPV

5-88 **A** An investment has the following cash flows. Use the XNPV function to find the present worth as of December 11, 2017. The interest rate is 8%.

Date	Cash Flow
12/11/2017	−$10,000
1/8/2018	50
7/2/2018	300
1/7/2019	400
7/1/2019	500
1/6/2020	10,700

5-89 An engineering firm is doing design work on a client's project. It has $40,000 in expenses at the beginning of each month starting in February 2018 through December 2018. The client has agreed to a payment schedule, if the firm meets milestone delivery dates. Use the XNPV function to find the present worth as of December 11, 2017. The interest rate is 12%.

Date	Income
1/1/2018	$150,000
5/14/2018	150,000
8/6/2018	150,000
10/22/2018	150,000
1/1/2019	200,000

5-90 **A** Use the XNPV function to find the present worth on January 1, 2018. The interest rate is 10%.

5-91 A cold remedy's cash flows for one season's cycle are shown below. Use XNPV to find the present worth as of June 1, 2018 if the MARR is 15%.

Date	Cash Flow (in $M)
6/1/2018	−$30
9/1/2018	12
11/1/2018	2
12/17/2018	6
1/7/2019	8
2/25/2019	5
4/1/2019	2

Bonds

5-92 **A** A corporate bond has a face value of $1000 with maturity date 20 years from today. The bond pays interest semiannually at a rate of 4% nominal per year based on the face value. The interest rate paid on similar corporate bonds has decreased to a current rate of 2%. Determine the market value of the bond.

5-93 You bought a $1000 corporate bond for $900 three years ago. It is paying $30 in interest at the end of every 6 months, and it matures in 5 more years.

(a) Compute its coupon rate.

(b) Compute its current value, assuming the market interest rate for such investments is 4% per year, compounded semiannually.

Contributed by D. P. Loucks, Cornell University

5-94 An investor is considering buying a 20-year
Ⓐ corporate bond. The bond has a face value of $1000
and pays 6% interest per year in two semiannual payments. To receive 8% interest, compounded semiannually, how much should be paid for the bond?

*Problems 5-95 to 5-98 contributed by Meenakshi
Sundaram, Tennessee Tech University*

5-95 A 4% coupon rate bond has a face value of $1000,
pays interest semiannually, and will mature in 10
years. If the current market rate is 2% interest
compounded semiannually, what is the bond's price?

5-96 A Treasury bond with a face value of $5000 and a
Ⓐ coupon rate of 6% payable semiannually was bought
when the market's nominal rate was 8%. The bond
matures 20 years from now. What was paid for the
bond?

5-97 A zero-coupon bond (coupon rate = 0%) has a face
value of $10,000 and a maturity date in 5 years. The
current market interest rate is a nominal 6%, compounded quarterly. How much should be paid for the
bond?

5-98 A city government wants to raise $3 million by issu-
ing bonds. By ballot proposition, the bond's coupon
interest rate was set at 8% per year with semiannual
payments. However, market interest rates have risen
to a nominal 9% interest rate. If the bonds mature in
20 years, how much will the city raise from issuing
$3M in bonds.

Minicases

5-99 Bayview's growth is constrained by mountains on
one side and the bay on the other. A bridge across
the bay is planned, but which plan is best? It can be
built with a single deck to meet the needs of the next
20 years, or it can be built with two decks to meet
the needs of the next 50 years. The piers can also be
built to support two decks, but with only one deck
being built now.

    Building it all now will cost $160M, and leav-
ing the top deck for later will save $40M. Building
that top deck later will cost $70M including the cost
of traffic disruption. A single-deck bridge will cost
$105M now and $120M in 20 years. Deck maintenance is $1.4M per year per deck. Pier maintenance
is $1.2M per year per bridge. If the interest rate is
5%, which design should be built?

If the two-deck bridge is built immediately, then
dedicated lanes for buses, carpools, and bicycles can
be added. To economically evaluate this use, estimate the cost of the underutilized capacity for the
bridge.

5-100 Florida Power and Light has committed to building
a solar power plant. JoAnne, an IE working for FPL,
has been tasked with evaluating the three current
designs. FPL uses an interest rate of 10% and a
20-year horizon.

Design 1: Flat Solar Panels
A field of "flat" solar panels angled to best catch the
sun will yield 2.6 MW of power and will cost $87
million initially with first-year operating costs at $2
million, growing $250,000 annually. It will produce
electricity worth $6.9 million the first year and will
increase by 8% each year thereafter.

Design 2: Mechanized Solar Panels
A field of mechanized solar panels rotates from side
to side so that they are always positioned parallel to
the sun's rays, maximizing the production of electricity. This design will yield 3.1 MW of power and
will cost $101 million initially with first-year operat-
ing costs at $2.3 million, growing $300,000 annually.
It will produce electricity worth $8.8 million the first
year and will increase 8% each year thereafter.

Design 3: Solar Collector Field
This design uses a field of mirrors to focus the sun's
rays onto a boiler mounted in a tower. The boiler
then produces steam and generates electricity the
same way a coal-fired plant operates. This system
will yield 3.3 MW of power and will cost $91 mil-
lion initially with first-year operating costs at $3
million, growing $350,000 annually. It will produce
electricity worth $9.7 million the first year and will
increase 8% each year thereafter.

5-101 Your grandparents are asking you for advice on
when they should start collecting social security
payments. If they wait until age 66, they will collect $2000 per month; but if they start collecting at
age 62, they will collect $1500 per month. Assume
they live to be 85, and simplify by assuming annual
payments.

(*a*) When do the higher payments catch up in total
dollars received with the lower payment that
starts earlier?

(*b*) If their interest rate is 6%, which plan has a
higher PW?

CASES

The following cases from *Cases in Engineering Economy* (www.oup.com/us/newnan) are suggested as matched with this chapter.

CASE 7 **The Board Looks to You**

Bond valuation with realistic business details including early-call premium.

Medium difficulty. Good example of employer taking a small fact and making big assumptions about individual talents which implies why engineering students must keep learning.

CASE 8 **Picking a Price**

Simplified real estate analysis. Demonstrates analysis to screen before acquiring more data and further decision making.

CASE 9 **Recycling?**

Financial analysis of recycling cardboard and selling unusable pallets to recycler.

ANNUAL CASH FLOW ANALYSIS

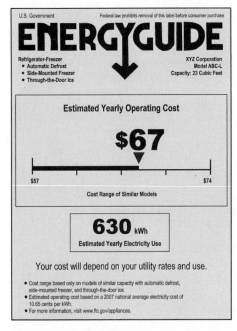

U.S. Government   Federal law prohibits removal of this label before consumer purchase.

ENERGYGUIDE

Refrigerator-Freezer
• Automatic Defrost
• Side-Mounted Freezer
• Through-the-Door Ice

XYZ Corporation
Model ABC-L
Capacity: 23 Cubic Feet

Estimated Yearly Operating Cost

$67

$57 $74

Cost Range of Similar Models

630 kWh
Estimated Yearly Electricity Use

Your cost will depend on your utility rates and use.

• Cost range based only on models of similar capacity with automatic defrost, side-mounted freezer, and through-the-door ice.
• Estimated operating cost based on a 2007 national average electricity cost of 10.65 cents per kWh.
• For more information, visit www.ftc.gov/appliances.

Typical FTC required EnergyGuide label for a clothes washer. According to the California Energy Commission, the typical American household does 400 loads of laundry per year using 40 gallons of water per load with a typical, non–ENERGY STAR clothes washer. An ENERGY STAR appliance would reduce water and energy consumption by 40%. For more information see: http://www.energystar.gov/.

Are More Efficient Appliances Cost Effective?

A typical residence is used for 75 years, and its major appliances last from 10 to 30 years. The long-term trend in energy prices (1–3% per year) and reduced operating costs are not usually enough to induce home owners to replace existing appliances—except when replacing failed appliances.

One driver in the slow adoption of energy efficient appliances is that many investment decisions are made by home builders, landlords, and property managers, rather than those who pay the monthly energy bills. When consumers do decide to replace appliances, EnergyGuide labels show the typical energy cost so that different models can be compared by all consumers, not just by those who have studied engineering economy.

From the EnergyGuide label for a clothes washer, the range of energy usage for comparable ENERGY STAR clothes washers goes from 113 kWh/year to 680 kWh/year. The energy consumption estimate is based on ratings of electricity and water usage factors. If electricity costs $0.08/kWh, the estimated electricity costs for comparable models ranges from $9.04/year to $54.40/year.

When selecting a clothes washer, the consumer should consider initial cost, the annual energy worth (benefit or cost?), and select a model that suits the need. Choices are based on needs; making choices requires tools that are part of the skill set provided by engineering economics. ■ ■ ■

Contributed by Gene Dixon, East Carolina University

QUESTIONS TO CONSIDER

1. Should average use be based on family size or lifestyle? How can lifestyle be properly analyzed? For example, a family of two in which one or more work outdoors in construction may have larger laundry demands than a family of four in which the parents work in offices. What about the impact of geography and climate; for example, does wearing shorts year-round (versus jeans) create a significant difference in laundry demands?

2. How would a home builder, landlord, or property manager justify the use of ENERGY STAR appliances in pricing properties and competing with other home builders, landlords, or property managers who may use less efficient but still ENERGY STAR-rated appliances?

3. Based on type, major appliances have life spans of 7 to 15 years, with refrigerators having the longest life. If a family stays in a residence on average for eight years, what considerations should a family use in deciding whether or not to keep their major appliances when moving?

4. Which appliance replacement decisions would be more sensitive to changes in the cost of electricity? How would you explain your answer to someone not skilled in engineering economics?

5. EnergyGuide labels are not required for TVs, ranges, ovens, and clothes dryers. Why do you think this is true?

After Completing This Chapter...

The student should be able to:

- Define *equivalent uniform annual cost (EUAC)* and *equivalent uniform annual benefits (EUAB)*.

- Resolve an engineering economic analysis problem into its annual cash flow equivalent.

- Conduct an *equivalent uniform annual worth (EUAW) analysis* for a single investment.

- Use EUAW, EUAC, and EUAB to compare alternatives with equal, common multiple, or continuous lives, or over some fixed study period.

- Develop and use spreadsheets to analyze loans for purposes of building an amortization table, calculating interest versus principal, finding the balance due, and determining whether to pay off a loan early.

- Use annuity due for beginning of period cash flows such as leases, insurance, and tuition payments.

Key Words

amortization schedule	infinite analysis period	salvage value
annuity due		

This chapter is devoted to annual cash flow analysis—the second of the three major analysis techniques. With present worth analysis, we resolved an alternative into an equivalent net present worth, a present worth of cost, or a present worth of benefit. Here we compare alternatives based on their equivalent annual cash flows: the equivalent uniform annual cost (EUAC), the equivalent uniform annual benefit (EUAB), or their difference, the equivalent uniform annual worth: (EUAW) = (EUAB − EUAC).

To prepare for a discussion of annual cash flow analysis, we will review some annual cash flow calculations, then examine annual cash flow criteria.

ANNUAL CASH FLOW CALCULATIONS

Resolving Cash Flows to an Annual Equivalent

In annual cash flow analysis, the goal is to convert money to an equivalent uniform annual cost or benefit. The simplest case is to convert a present sum P to a series of equivalent uniform end-of-period cash flows. This is illustrated in Example 6–1.

EXAMPLE 6–1   **(Example 3–1 and 4–5 Revisited)**

A machine will cost $30,000 to purchase. Annual operating and maintenance costs (O&M) will be $2000. The machine will save $10,000 per year in labor costs. The salvage value of the machine after 5 years will be $7000. Calculate the machine's equivalent uniform annual worth (EUAW) for an interest rate of 10%.

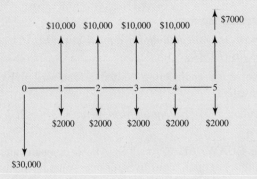

TABLE SOLUTION

Because there are multiple types of cash flows, the capital recovery factor (Eq. 4–6) must be used for the initial cost and the sinking fund factor (Eq. 4–5) for the salvage value. No factor is needed for the annual savings and O&M costs since they are already annual cash flows.

$$EUAW = -30{,}000(A/P, 10\%, 5) + (10{,}000 - 2000) + 7000(A/F, 10\%, 5)$$

$$= -30{,}000(0.2638) + 8000 + 7000(0.1638)$$

$$= \$1232$$

5-BUTTON SOLUTION

The n is 5 for both the initial cash flow of $-\$30{,}000$ and the salvage value of $\$7000$, so the annual equivalent of both can be found in one step.

	A	B	C	D	E	F	G	H	I
1	Problem	i	n	PMT	PV	PV	Solve for	Answer	Formula
2	Exp. 6-1	10%	5		-30,000	7000	PMT	$6,767	=PMT(B2,C2,E2,F2)
3							change sign	-6,767	=-H2
4							annual revenue	10,000	
5							annual O&M	-2,000	
6							EUAW	$1,233	=SUM(H3:H5)

Capital Recovery Costs

When there is an initial disbursement P followed by a salvage value S, the annual cost or *capital recovery cost* of a project may be computed in three different ways.

$$EUAC = P(A/P, i, n) - S(A/F, i, n) \qquad (6\text{-}1)$$

$$EUAC = (P - S)(A/F, i, n) + Pi \qquad (6\text{-}2)$$

$$EUAC = (P - S)(A/P, i, n) + Si \qquad (6\text{-}3)$$

Each of the three calculations gives the same results. In practice, the first method is the most commonly used. This approach simply treats the cash flows as occurring at the beginning and the end. The second approach can be described as considering the drop in value $(P - F)$ at the end and including the interest on the capital tied up from the beginning. The third approach can be described as considering the drop in value $(P - F)$ at the beginning and including the interest on the capital tied up until the end. As shown in solutions 2 and 3 of Example 6–2, Equation 4–13 can also be used to link these equations.

EXAMPLE 6–2

Consider only the capital costs from Example 6–1. A machine will cost $30,000 to purchase. The salvage value of the machine after 5 years will be $7000. Calculate the machine's capital recovery cost (EUAC) for an interest rate of 10%.

SOLUTIONS

For this situation, the problem may be solved by means of three different formulas using tabulated factors. The first approach can also be solved directly using a 5-button calculator.

SOLUTION 1

$$EUAC = P(A/P, i, n) - S(A/F, i, n) \tag{6-1}$$

$$= 30,000(A/P, 10\%, 5) - 7000(A/F, 10\%, 5)$$

$$= 30,000(0.2638) - 7000(0.1638)$$

$$= \$6767$$

This method reflects the annual cost of the cash disbursement minus the annual benefit of the future resale value.

5-BUTTON SOLUTION

The n is 5 for both the initial cash flow of $-\$30,000$ and the salvage value of 7000, so the annual equivalent of both can be found in one step.

	A	B	C	D	E	F	G	H	I
1	Problem	i	n	PMT	PV	PV	Solve for	Answer	Formula
2	Exp. 6-2	10%	5		-30,000	7000	PMT	$6,767	=PMT(B2,C2,E2,F2)

SOLUTION 2

Equation 6–1 describes a relationship that may be modified by an identity presented in Chapter 4:

$$(A/P, i, n) = (A/F, i, n) + i \tag{4-13}$$

Substituting this into Equation 6-1 gives

$$EUAC = P(A/F, i, n) + Pi - S(A/F, i, n)$$

$$= (P - S)(A/F, i, n) + Pi \tag{6-2}$$

$$= (30,000 - 7000)(A/F, 10\%, 5) + 30,000(0.10)$$

$$= 23,000(0.1638) + 3000$$

$$= \$6767$$

This method computes the equivalent annual cost due to the unrecovered $28,000 when the machine is sold, and it adds annual interest on the $30,000 investment.

SOLUTION 3

If the value for $(A/F, i, n)$ from Equation 4–13 is substituted into Equation 6–1, we obtain

$$EUAC = P(A/P, i, n) - S(A/P, i, n) + Si$$

$$= (P - S)(A/P, i, n) + Si \tag{6-3}$$

$$= (30,000 - 7000)(A/P, 10\%, 5) + 7000(0.10)$$

$$= 23,000(0.2638) + 700$$

$$= \$6767$$

This method computes the annual cost of the $23,000 decline in value during the 10 years, plus interest on the $7000 tied up in the machine as the salvage value.

EXAMPLE 6–3

The maintenance costs for a generator have been recorded over its 5-year life. Compute the generator's equivalent uniform annual cost (EUAC) assuming 7% interest and end-of-year cash flows.

Year	Maintenance Cost
1	$545
2	590
3	635
4	680
5	725

TABLE SOLUTION

This is an arithmetic gradient series plus a uniform annual cost, as follows:

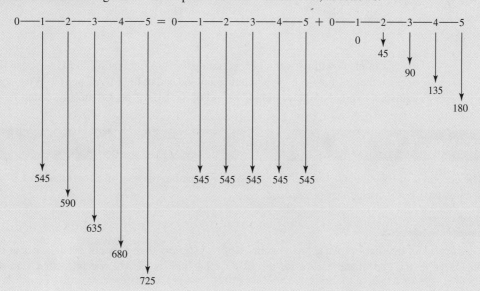

$$EUAC = 545 + 45(A/G, 7\%, 5)$$

$$= 545 + 45(1.865)$$

$$= \$629$$

Rather than using a gradient, this example could have been solved by using 5 (P/F) factors. However, this not only takes longer, but it multiplies the chance of an error in table look-up, keying in the numbers, and so on. Using more than a few individual (P/F) factors should be avoided whenever possible.

SPREADSHEET SOLUTION

This example has a gradient, so the first step is to find the present cost of the cash flows using the NPV function. Then a PMT annuity function is used to find the EUAC. Since the table is of costs not cash flows, the PMT function must include a minus sign. Since this example has only costs, we know that the EUAC must be a positive number.

	A	B	C
1	7%	interest rate	
2	5	horizon	
3	545	first year maintenance	
4	45	maintenance gradient	
5	Year	Maintenance	
6	0	0	
7	1	545	
8	2	590	
9	3	635	
10	4	680	
11	5	725	
12	Present cost	$2,578.71	=B4+NPV($A$1,B5:B9)
13	EUAC	$628.92	=PMT(A1,A2,-B10)

Example 6–4 illustrates a very common issue in calculating an EUAC: cash flows may occur sometime between the beginning and the end of the EUAC period. When this occurs, equivalent values at the beginning or end of the EUAC period must be found first.

EXAMPLE 6–4

The generator in Example 6–3 needed a repair and overhaul at the end of Year 3 that cost $750. What is the EUAC, including this expense?

TABLE SOLUTION

The repair and overhaul expense is an example of a cash flow that does not occur either at the beginning or the end of the period for the EUAC. To find its EUAC, we must first find the equivalent value at the beginning (PW) or at the end (FW) of the EUAC period. Then a second factor is used to find the annual equivalent. This can be done in one equation as follows:

$$EUAC = 545 + 45(A/G, 7\%5) + 750(P/F, 7\%, 3)(A/P, 7\%5)$$

$$= 545 + 45(1.865) + 750(0.8163)(0.2439) = \$778.24$$

SPREADSHEET SOLUTION

When adding the repair and overhaul expense to the spreadsheet for Example 6–3, it is best to use a new column. This ensures that the calculations for the gradient remain simple.

	A	B	C	D	E
1	7%	interest rate			
2	5	horizon			
3	545	first year maintenance			
4	45	maintenance gradient			
5	750	repair in yr 3			
6	Year	Maintenance	Repair	Total cost	
7	0	0		0	
8	1	545		545	
9	2	590		590	
10	3	635	750	1385	
11	4	680		680	
12	5	725		725	
13			Present cost	$3,190.93	=D7+NPV($A$1,D8:D12)
13			EUAC	$778.24	=PMT($A$1,A2,-D13)

The examples have shown four essential points concerning cash flow calculations:

1. There is a direct relationship between the present worth of cost and the equivalent uniform annual cost. It is

$$EUAC = (PW \text{ of cost})(A/P, i, n)$$

2. In a problem, expending money increases the EUAC, while receiving money—for example, from an item's salvage value—decreases the EUAC.

3. When there are irregular cash disbursements over the analysis period, a convenient method of solution is to first determine the PW of cost; then use the equation in Item 1 to calculate the EUAC.

4. Where there is an arithmetic gradient, EUAC may be rapidly computed by using the arithmetic gradient uniform series factor, $(A/G, i, n)$.

ANNUAL CASH FLOW ANALYSIS

The criteria for economic efficiency are presented in Table 6–1. One notices immediately that the table is quite similar to Table 5–1. It is apparent that, if you are maximizing the

TABLE 6–1 Annual Cash Flow Analysis

Input/Output	Situation	Criterion
Neither input nor output is fixed	Typical, general situation	Maximize equivalent uniform annual worth (EUAW = EUAB − EUAC)
Fixed input	Amount of money or other input resources is fixed	Maximize equivalent uniform annual benefit (maximize EUAB)
Fixed output	There is a fixed task, benefit, or other output to be accomplished	Minimize equivalent uniform annual cost (minimize EUAC)

present worth of benefits, simultaneously you must be maximizing the equivalent uniform annual worth. This is illustrated in Example 6–5.

EXAMPLE 6–5   (Example 5–1 Revisited)

A firm is considering which of two devices to install to reduce costs. Both devices have useful lives of 5 years with no salvage value. Device A costs $10,000 and can be expected to result in $3000 savings annually. Device B costs $13,500 and will provide cost savings of $3000 the first year; however, savings will increase $500 annually, making the second-year savings $3500, the third-year savings $4000, and so forth. With interest at 7%, which device should the firm purchase?

SOLUTION

Device A

$$EUAW = -10,000(A/P, 7\%, 5) + 3000$$
$$= -10,000(0.2439) + 3000 = \$561$$

Device B

$$EUAW = -13,500(A/P, 7\%, 5) + 3000 + 500(A/G, 7\%, 5)$$
$$= -13,500(0.2439) + 3000 + 500(1.865) = \$640$$

To maximize EUAW, select Device B.

Example 6–5 was presented earlier, as Example 5–1, where we found:

$$PW_A = -10{,}000 + 3000(P/A, 7\%, 5)$$
$$= -10{,}000 + 3000(4.100) = \$2300$$

This is converted to EUAW by multiplying by the capital recovery factor:

$$EUAW_A = 2300(A/P, 7\%, 5) = 2300(0.2439) = \$561$$

Similarly, for machine B

$$PW_B = -13{,}500 + 3000(P/A, 7\%, 5) + 500(P/G, 7\%, 5)$$
$$= -13{,}500 + 4000(4.100) + 500(7.647) = \$2624$$

and, hence,

$$EUAW_B = 2624(A/P, 7\%, 5) = 2624(0.2439)$$
$$= \$640$$

We see, therefore, that it is easy to convert the present worth analysis results into the annual cash flow analysis results. We could go from annual cash flow to present worth just as easily, by using the series present worth factor. And, of course, both methods show that Device B is the preferred alternative.

EXAMPLE 6–6

Three alternatives are being considered to improve an assembly line along with the "do-nothing" alternative. Each of Plans A, B, and C has a 10-year life and a salvage value equal to 10% of its original cost.

	Plan A	Plan B	Plan C
Installed cost of equipment	$15,000	$25,000	$33,000
Material and labor savings per year	14,000	9,000	14,000
Annual operating expenses	8,000	6,000	6,000
End-of-useful life salvage value	1,500	2,500	3,300

If interest is 8%, which plan, if any, should be adopted?

TABLE SOLUTION

Since neither installed cost nor output benefits are fixed, the economic criterion is to maximize EUAW = EUAB − EUAC.

	Plan *A*	Plan *B*	Plan *C*
Equivalent uniform annual benefit (EUAB)			
Material and labor per year	$14,000	$9,000	$14,000
Salvage value $(A/F, 8\%, 10)$	104	172	228
EUAB =	$14,104	$9,172	$14,228
Equivalent uniform annual cost (EUAC)			
Installed cost $(A/P, 8\%, 10)$	$ 2,235	$3,725	$ 4,917
Annual operating expenses	8,000	6,000	6,000
EUAC =	$10,235	$9,725	$10,917
EUAW = EUAB − EUAC =	$ 3,869	−$ 553	$ 3,311

Based on our criterion of maximizing EUAW, Plan *A* is the best of the four alternatives. Since the do-nothing alternative has EUAW = 0, it is a more desirable alternative than Plan *B*.

5-BUTTON SOLUTION

	A	B	C	D	E	F	G	H
1	Problem	*i*	*n*	*PMT*	*PV*	*FV*	Solve for	Answer
2	Plan A	8%	10		-15,000	1500	PMT	$2,132
3	Plan B	8%	10		-25,000	2500	PMT	3,553
4	Plan C	8%	10		-33,000	3300	PMT	4,690
5					PMT sign change	+M & L savings	−Operating	EUAW
6			Plan A	-$2,132	$14,000	-$8,000	$3,868
7			Plan B	-3,553	9,000	-6,000	-553
8			Plan C	-4,690	14,000	-6,000	3,310

ANALYSIS PERIOD

In Chapter 5, we saw that the analysis period is an important consideration in computing present worth comparisons. In such problems, a common analysis period must be used for all alternatives. In annual cash flow comparisons, we again have the analysis period question. Example 6–7 will help in examining the problem.

EXAMPLE 6–7   (Example 5–4 Revisited)

Two pumps are being considered for purchase. If interest is 7%, which pump should be bought?

	Pump A	Pump B
Initial cost	$7000	$5000
End-of-useful-life salvage value	1200	1000
Useful life, in years	12	6

TABLE SOLUTION

The annual cost for 12 years of Pump A can be found by using Equation 6-4:

$$EUAC = (P - S)(A/P, i, n) + Si$$
$$= (7000 - 1200)(A/P, 7\%, 12) + 1200(0.07)$$
$$= 5800(0.1259) + 84 = \$814$$

Now compute the annual cost for 6 years of Pump B:

$$EUAC = (5000 - 1000)(A/P, 7\%, 6) + 1000(0.07)$$
$$= 4000(0.2098) + 70 = \$909$$

For a common analysis period of 12 years, we need to replace Pump B at the end of its 6-year useful life. If we assume that another pump B′ can be obtained, having the same $5000 initial cost, $1000 salvage value, and 6-year life, it will have the same $909 EUAC over its 6-year life. Analyzing the cash flows over 12 years will confirm this as follows:

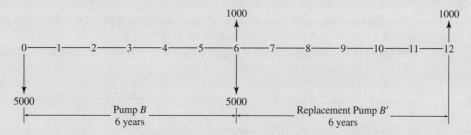

For the 12-year analysis period, the annual cost for Pump B is

$$EUAC = [5000 - 1000(P/F, 7\%, 6) + 5000(P/F, 7\%, 6)$$
$$- 1000(P/F, 7\%, 12)] \times (A/P, 7\%, 12)$$
$$= [5000 - 1000(0.6663) + 5000(0.6663) - 1000(0.4440)] \times (0.1259)$$
$$= (5000 - 666 + 3331 - 444)(0.1259)$$
$$= (7211)(0.1259) = \$909$$

The annual cost of B for the 6-year analysis period is the same as the annual cost for the 12-year analysis period. Thus the lengthy calculation of EUAC for 12 years of Pump B and B' was not needed. Assuming that the shorter-life equipment is replaced by equipment with identical economic consequences avoids a lot of calculations. Select Pump A.

5-BUTTON SOLUTION

The EUAC of Pump A over 12 years is a 1-step calculation. However, finding the EUAC of Pump B either requires breaking the problem into several steps or recognizing that both 6-year periods have the same EUAC.

	A	B	C	D	E	F	G	H
1	Problem	*i*	*n*	*PMT*	*PV*	*FV*	Solve for	Answer
2	Pump A	7%	12		-7000	1500	PMT	$797
3	Pump B	7%	6		-5000	1000	PMT	909

Note that the sign assumption of the PMT function produces results that match the previous EUAC calculation.

Analysis Period Equal to Alternative Lives

If the analysis period for an economy study coincides with the useful life for each alternative, then the economy study is based on this analysis period.

Analysis Period a Common Multiple of Alternative Lives

When the analysis period is a common multiple of the alternative lives (for example, in Example 6–7, the analysis period was 12 years with 6- and 12-year alternative lives), a "replacement with an identical item with the same costs, performance, and so forth" is frequently assumed. This means that when an alternative has reached the end of its useful life, we assume that it will be replaced with an identical item. As shown in Example 6–7, the result is that the EUAC for Pump B with a 6-year useful life is equal to the EUAC for the entire analysis period based on Pump B *plus* the replacement unit, Pump B'.

Under these circumstances of identical replacement, we can compare the annual cash flows computed for alternatives based on their own service lives. In Example 6–7, the annual cost for Pump A, based on its 12-year service life, was compared with the annual cost for Pump B, based on its 6-year service life.

Analysis Period for a Continuing Requirement

Many times an economic analysis is undertaken to determine how to provide for a more or less continuing requirement. One might need to pump water from a well as a continuing requirement. There is no distinct analysis period. In this situation, the analysis period is assumed to be long but undefined.

If, for example, we had a continuing requirement to pump water and alternative Pumps A and B had useful lives of 7 and 11 years, respectively, what should we do? The customary assumption is that Pump A's annual cash flow (based on a 7-year life) may be compared to Pump B's annual cash flow (based on an 11-year life). This is done without much concern that the least common multiple of the 7- and 11-year lives is 77 years. This comparison of "different-life" alternatives assumes identical replacement (with identical costs, performance, etc.) when an alternative reaches the end of its useful life.

This continuing requirement, which can also be described as an *indefinitely long horizon*, is illustrated in Example 6–8. Since this is longer than the lives of the alternatives, we can make the best decision possible given current information by minimizing EUAC or maximizing EUAW or EUAB. At a later time, we will make another replacement and there will be more information on costs *at that time*.

EXAMPLE 6–8

Pump *B* in Example 6–7 is now believed to have a 9-year useful life. Assuming the same initial cost and salvage value, compare it with Pump *A* using the same 7% interest rate.

TABLE SOLUTION

If we assume that the need for *A* or *B* will exist for some continuing period, the comparison of costs per year for the unequal lives is an acceptable technique. For 12 years of Pump *A*:

$$\text{EUAC} = (7000 - 1500)(A/P, 7\%, 12) + 1500(0.07) = \$797$$

For 9 years of Pump *B*:

$$\text{EUAC} = (5000 - 1000)(A/P, 7\%, 9) + 1000(0.07) = \$684$$

For minimum EUAC, select Pump *B*.

5-BUTTON SOLUTION

	A	B	C	D	E	F	G	H
1	Problem	*i*	*n*	*PMT*	*PV*	*FV*	Solve for	Answer
2	Pump A	7%	12		-7000	1500	PMT	$797
3	Pump B	7%	9		-5000	1000	PMT	684

With the longer assumed life, Pump *B* now has the lowest EUAC.

Infinite Analysis Period

At times we have an alternative with a limited (finite) useful life in an **infinite analysis period** situation. The equivalent uniform annual cost may be computed for the limited life. The assumption of identical replacement (replacements have identical costs, performance, etc.) is often appropriate. Based on this assumption, the same EUAC occurs for each replacement of the limited-life alternative. The EUAC for the infinite analysis period is therefore equal to the EUAC computed for the limited life. With identical replacement,

$$\text{EUAC}_{\text{infinite analysis period}} = \text{EUAC}_{\text{for limited life } n}$$

A somewhat different situation occurs when there is an alternative with an infinite life in a problem with an infinite analysis period:

$$\text{EUAC}_{\text{infinite analysis period}} = P(A/P, i, \infty) + \text{Any other annual costs}$$

When $n = \infty$, we have $A = Pi$ and, hence, $(A/P, i, \infty)$ equals i.

$$\text{EUAC}_{\text{infinite analysis period}} = Pi + \text{Any other annual costs}$$

EXAMPLE 6–9

In the construction of an aqueduct to expand the water supply of a city, there are two alternatives for a particular portion of the aqueduct. Either a tunnel can be constructed through a mountain, or a pipeline can be laid to go around the mountain. If there is a permanent need for the aqueduct, should the tunnel or the pipeline be selected for this particular portion of the aqueduct? Assume a 6% interest rate.

SOLUTION

	Tunnel Through Mountain	Pipeline Around Mountain
Initial cost	$5.5 million	$5 million
Maintenance	0	0
Useful life	Permanent	50 years
Salvage value	0	0

Tunnel

For the tunnel, with its permanent life, we want $(A/P, 6\%, \infty)$. For an infinite life, the capital recovery is simply interest on the invested capital. So $(A/P, 6\%, \infty) = i$, and we write

$$\text{EUAC} = Pi = \$5.5 \text{ million}(0.06)$$

$$= \$330,000$$

Pipeline

$$\text{EUAC} = \$5 \text{ million}(A/P, 6\%, 50)$$

$$= \$5 \text{ million}(0.0634) = \$317,000$$

For fixed output, minimize EUAC. Select the pipeline.

The difference in annual cost between a long life and an infinite life is small unless an unusually low interest rate is used. In Example 6–9 the tunnel is assumed to be permanent. For comparison, compute the annual cost if an 85-year life is assumed for the tunnel.

$$\text{EUAC} = \$5.5 \text{ million}(A/P, 6\%, 85)$$
$$= \$5.5 \text{ million}(0.0604)$$
$$= \$332,000$$

The difference in time between 85 years and infinity is great indeed; yet the difference in annual costs equivalent is only $2000.

Some Other Analysis Period

The analysis period in a particular problem may be something other than one of the four we have so far described. It may be equal to the life of the shorter-life alternative, the longer-life alternative, or something entirely different. One must carefully examine the consequences of each alternative throughout the analysis period and, in addition, see what differences there might be in salvage values, and so forth, at the end of the analysis period.

EXAMPLE 6–10

Suppose that Alternative 1 has a 7-year life and a salvage value at the end of that time. The replacement cost at the end of 7 years may be more or less than the original cost. If the replacement is retired prior to 7 years, it will have a terminal value that exceeds the end-of-life salvage value. Alternative 2 has a 13-year life and a terminal value whenever it is retired. If the situation indicates that 10 years is the proper analysis period, set up the equations to compute the EUAC for each alternative. Use results from Example 5–5 to compute the results.

SOLUTION

Alternative 1

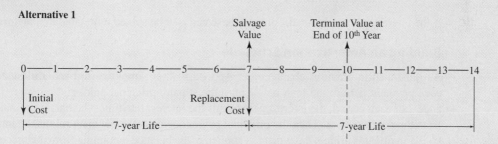

Alternative 2

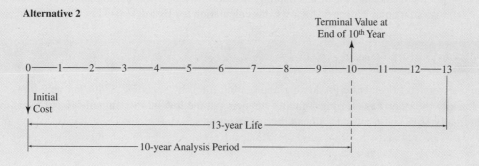

Alternative 1

$$EUAC_1 = [\text{Initial cost} + (\text{Replacement cost} - \text{Salvage value})(P/F, \ i, 7)$$
$$- (\text{Terminal value})(P/F, \ i, 10)](A/P, \ i, 10)$$
$$= 64{,}076(A/P, 8\%, 10) \quad \text{using results from Example 5-5}$$
$$= 64{,}076(0.1490) = \$9547$$

Alternative 2

$$EUAC_2 = [\text{Initial cost} - (\text{Terminal value})(P/F, \ i, 10)](A/P, \ i, 10)$$
$$= 69{,}442(A/P, 8\%, 10) \quad \text{using results from Example 5-5}$$
$$= 69{,}442(0.1490) = \$10{,}347$$

Select Alternative 1.

ANALYZING LOANS

Loan and bond payments are made by firms, agencies, and individuals. Usually, the payments in each period are constant. Spreadsheets make it easy to:

- Calculate the loan's amortization schedule
- Demonstrate how a payment is split between principal and interest
- Find the balance due on a loan
- Calculate the number of payments remaining on a loan.

If the interest rate is tabulated, some of these can also be solved with tabulated factors.

Building an Amortization Schedule

As illustrated in previous chapters and Appendix A, an **amortization schedule** lists for each payment period: the loan payment, interest paid, principal paid, and remaining balance. For each period, the interest paid equals the interest rate times the balance remaining from the period before. Then the principal payment equals the payment minus the interest paid. Finally, this principal payment is applied to the balance remaining from the preceding period to calculate the new remaining balance. As a basis for comparison with spreadsheet loan functions, Figure 6-1 shows this calculation for Example 6-11.

EXAMPLE 6-11

An engineer wanted to celebrate graduating and getting a job by spending $2400 on new furniture. Luckily the store was offering 6-month financing at the low interest rate of 6% per year nominal (really $1/2\%$ per month). Calculate the amortization schedule.

SOLUTION

The first step is to calculate the monthly payment:

$$A = 2400(A/P, \tfrac{1}{2}\%, 6) = 2400(0.1696) = \$407.04$$

or

	A	B	C	D	E	F	G	H
1	Problem	i	n	PMT	PV	FV	Solve for	Answer
2	Exp. 6-11	0.5%	6		2400	0	PMT	-$407.03

With this information the engineer can use a spreadsheet like Figure 6–1 to obtain the amortization schedule.

	A	B	C	D	E
1	2400	Initial balance			
2	0.50%	i			
3	6	N			
4	$407.03	Payment	= −PMT(A2,A3,A1)		
5					
6			Principal	Ending	
7	Month	Interest	Payment	Balance	
8	0			2400.00	=A1
9	1	12.00	395.03	2004.97	=D8−C9
10	2	10.02	397.00	1607.97	
11	3	8.04	398.99	1208.98	
12	4	6.04	400.98	807.99	
13	5	4.04	402.99	405.00	
14	6	2.03	405.00	0.00	
15				=$A$4−B14	
16				=Payment−Interest	
17			=$A$2*D13		
18			=rate*previous balance		

FIGURE 6–1 Amortization schedule for furniture loan.

Finding the Balance Due on a Loan

An amortization schedule is one used to calculate the balance due on a loan. A second, easier way is to remember that the balance due equals the present worth of the remaining payments. Interest is paid in full after each payment, so later payments are simply based on the balance due.

EXAMPLE 6–12

A car is purchased with a 48-month, 9% nominal loan with an initial balance of $15,000. What is the balance due halfway through the 4 years?

TABLE SOLUTION

The first step is to calculate the monthly payment, at a monthly interest rate of $3/4\%$. This equals

$$\text{Payment} = 15{,}000(A/P, 0.75\%, 48) = (15{,}000)(0.0249) = \$373.5$$

Note that there are only three significant digits in the tabulated factor. After 24 payments and with 24 left, the remaining balance equals $(P/A, i, N_{\text{remaining}})$ payment

$$\text{Balance} = (P/A, 0.75\%, 24)\$373.5 = (21.889)(373.28) = \$8176$$

Thus halfway through the repayment schedule, 54.5% (= \$8176/\$15,000) of the original balance is still owed.

5-BUTTON SOLUTION

	A	B	C	D	E	F	G	H
1	Problem	i	n	*PMT*	*PV*	*FV*	Solve for	Answer
2	Monthly payment	0.75%	48		15,000	0	PMT	-$373.28
3	Balance due 1/2 way	0.75%	24	-$373.28		0	PV	$8,170.68

How Much to Interest? How Much to Principal?

For a loan with constant payments, we can answer these questions for any period without the full amortization schedule. For a loan with constant payments, the functions IPMT and PPMT directly answer these questions. For simple problems, both functions have four arguments $(i, t, n, -P)$, where t is the time period being calculated. Both functions have optional arguments that permit adding a balloon payment (an F) and changing from end-of-period payments to beginning-of-period payments.

For example, consider Period 4 of Example 6–11. The spreadsheet formulas give the same answer as shown in Figure 6–1.

$$\text{Interest period 4} = \text{IPMT}(0.5\%, 4, 6, -2400) = \$6.04$$
$$\text{Principal payment period 4} = \text{PPMT}(0.5\%, 4, 6, -2400) = \$400.98$$

This can also be solved with factors by first finding the balance due after 3 periods.

$$\text{Interest period 4} = 407.04(P/A, 0.5\%, 3) \times \text{interest rate}$$
$$= 407.04(2.970)(0.005) = \$6.04$$
$$\text{Principal payment} = 407.04 - 6.04 = \$401.00$$

Pay Off Debt Sooner by Increasing Payments

Paying off debt can be a good investment because the investment earns the rate of interest on the loan. For example, this could be 8% for a mortgage, 10% for a car loan, or 19% for a credit card. When one is making extra payments on a loan, the common question is: How much sooner will the debt be paid off? Until the debt is paid off, the savings from any early payments are essentially locked up, since the same payment amount is owed each month. An early payment reduces the principal owed—which reduces future interest which increases future principal payments. This compounding cycle further speeds paying off the loan.

The first reason that spreadsheets and TVM calculators are convenient relates to fractional interest rates. For example, an auto loan might be at a nominal rate of 13% with monthly compounding or 1.08333% per month. The second reason is that the function NPER or the n key calculates the number of periods remaining on a loan. Thus we can calculate how much difference is made by one extra payment or by increasing all payments by x%. Extra payments are applied entirely to principal, so the interest rate, remaining balance, and payment amounts are all known. $N_{remaining}$ equals NPER(i, payment, remaining balance) with optional arguments for beginning-of-period cash flows and balloon payments. The signs of the payment and the remaining balance must be different.

EXAMPLE 6–13

Maria has a 7.5% mortgage with monthly payments for 30 years. Her original balance was $100,000, and she just made her twelfth payment. Each month she also pays into a reserve account, which the bank uses to pay her fire and liability insurance ($900 annually) and property taxes ($1500 annually). By how much does she shorten the loan if she makes an extra *loan* payment today? If she makes an extra *total* payment? If she increases each total payment to 110% of her current total payment?

5-BUTTON SOLUTION

	A	B	C	D	E	F	G	H
1	Problem	*i*	*n*	*PMT*	*PV*	*FV*	Solve for	Answer
2	Monthly payment	0.625%	360		100,000	0	PMT	-$699.21
3	After 12 payments	0.625%	348	-$699.21		0	PV	$99,078.17
4								
5	Extra loan payment now	0.625%	348	-$699.21	$98,378.95	0	*n*	339.5
6	Extra total payment now	0.625%	348	-$699.21	$98,178.95	0	*n*	337.1
7	110% payment to end	0.625%	348	-$789.14	$99,078.17	0	*n*	246.5

Formulas can be shown for all cells by clicking on icon within the Formula Auditing panel.

	A	B	C	D	E	F	G	H
1	Problem	*i*	*n*	*PMT*	*PV*	*FV*	Solve for	Answer
2	Monthly payment	=0.075/12	360		100000	0	PMT	=PMT(B2,C2,E2,F2)
3	After 12 payments	=0.075/12	=$C$2-12	=$H$2		0	PV	=PV(B3,C3,D3,F3)
4								
5	Extra loan payment now	=0.075/12	=$C$2-12	=$H$2	=$H$3+D5	0	*n*	=NPER(B5,D5,E5,F5)
6	Extra total payment now	=0.075/12	=$C$2-12	=$H$2	=$H$3+D6-200	0	*n*	=NPER(B6,D6,E6,F6)
7	110% payment to end	=0.075/12	=$C$2-12	=($H$2-200)*1.1+200	=$H$3	0	*n*	=NPER(B7,D7,E7,F7)

ANNUITY DUE

Equivalent uniform annual worths and costs have been defined as end-of-period values or ordinary annuities. However, there are some uniform cash flows that are beginning-of-period cash flows. These cash flows are referred to as an **annuity due**. Lease, rent, insurance, and tuition payments are normally beginning-of-the-period cash flows. Some things must be paid for in advance. Example 6–14 illustrates 3 different ways to convert these beginning-of-period cash flows to present worth or EUAC values.

EXAMPLE 6–14

Find the present worth and equivalent uniform monthly cost of lease payments of $1200 per month for a year. The monthly interest rate is 1%.

SOLUTION

The easiest way to find the present worth of the lease payments is to use the *Type* argument in the PV function. When the *Type* argument is non-zero (a value of 1 is normally used) then beginning-of-period cash flows are assumed for *only* the uniform cash flow. For comparison purposes, the present worth for end-of-period payments is also shown.

	A	B	C	D	E	F	G	H	I
1	Problem	i	n	PMT	PV	FV	Solve for	Answer	Formula
2	Exp. 6-14	1%	12	1200			PV	-$13,641	=PV(B2,C2,D2,F2,1)
3	ordinary								
4	annuity	1%	12	1200			PV	-$13,506	=PV(B2,C2,D2,F2)

The second method for calculating the present worth is to recognize that the first payment is already at time 0 and that there are payments at the end of the next 11 months. The beginnings of months 2 to 12 are the same as the endings of months 1 to 11.

$$PW = -1200 - 1200(P/A, 1\%, 11)$$

$$= -1200 - 1200(10.368) = -\$13,642$$

To calculate an equivalent uniform monthly cost or ordinary annuity, the easiest approach is to calculate an end-of-period equivalent for each beginning-of-period value by multiplying by $(1 + i)$. This could also be calculated from the present worth value with an (A/P) factor or PMT function.

$$\text{EUmonthlyC} = 1200(1 + 0.01) = \$1212$$

SUMMARY

Annual cash flow analysis is the second of the three major methods of resolving alternatives into comparable values. When an alternative has an initial cost P and salvage value S, there are three ways of computing the equivalent uniform annual cost:

- $\text{EUAC} = P(A/P, i, n) - S(A/F, i, n)$ (6-1)
- $\text{EUAC} = (P - S)(A/F, i, n) + Pi$ (6-2)
- $\text{EUAC} = (P - S)(A/P, i, n) + Si$ (6-3)

All three equations give the same answer. This quantity is also known as the *capital recovery cost* of the project.

The relationship between the present worth of cost and the equivalent uniform annual cost is

- $\text{EUAC} = (\text{PW of cost})(A/P, i, n)$

The three annual cash flow criteria are:

Neither input nor output fixed	Maximize EUAW = EUAB − EUAC
For fixed input	Maximize EUAB
For fixed output	Minimize EUAC

In present worth analysis there must be a common analysis period. Annual cash flow analysis, however, allows some flexibility provided the necessary assumptions are suitable in the situation being studied. The analysis period may be different from the lives of the alternatives, and provided the following criteria are met, a valid cash flow analysis may be made.

1. When an alternative has reached the end of its useful life, it is assumed to be replaced by an identical replacement (with the same costs, performance, etc.).
2. The analysis period is a common multiple of the useful lives of the alternatives, or there is a continuing or perpetual requirement for the selected alternative.

If neither condition applies, it is necessary to make a detailed study of the consequences of the various alternatives over the entire analysis period with particular attention to the difference between the alternatives at the end of the analysis period.

There is very little numerical difference between a long-life alternative and a perpetual alternative. As the value of n increases, the capital recovery factor approaches i. At the limit, $(A/P, i, \infty) = i$.

One of the most common uniform payment series is the repayment of loans. Spreadsheets and TVM calculators are useful in analyzing loans (balance due, interest paid, etc.) for several reasons: they have specialized functions, different loan lives are easy, and any interest rate can be used.

PROBLEMS

Annual Calculations

6-1   Compute the EUAB for these cash flows.

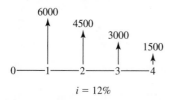

$i = 12\%$

6-2 Compute the EUAB for these cash flows based on a
Ⓐ   10% interest rate.

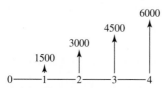

6-3 A production machine costs $40,000 and has a 7-year
Ⓖ   useful life. At the end of 7 years, it can be sold for
$10,000.

(a) If interest is 10% compounded quarterly, what is
the equivalent annual cost of ownership of this
machine over its useful life?

(b) The company that buys the 7-year old machine
uses it for an additional 5 years and then wants
to responsibly dispose of it. What *green* options
does the company have? Research and report
to the firm on what happens to most industrial
machinery after its useful life.

6-4 An electronics firm invested $60,000 in a precision
Ⓐ   inspection device. It cost $4000 to operate and main-
tain in the first year and $3000 in each later year.
At the end of 4 years, the firm changed their inspection procedure, eliminating the need for the device.
The purchasing agent was very fortunate to sell the
inspection device for $60,000, the original price.
Compute the equivalent uniform annual cost during
the 4 years the device was used. Assume interest at
10% per year.

6-5 A firm is buying an adjacent 10-acre parcel for a
future plant expansion. The price has been set at
$30,000 per acre. The payment plan is 25% down,

and the balance 2 years from now. If the transaction interest rate is 9% per year, what are the two
payments?
*Contributed by Hamed Kashani, Saeid Sadri, and
Baabak Ashuri, Georgia Institute of Technology*

6-6 When he started work on his twenty-second birthday,
Ⓐ   D. B. Cooper decided to invest money each month
with the objective of becoming a millionaire by the
time he reaches age 65. If he expects his investments
to yield 18% per annum, compounded monthly, how
much should he invest each month?

6-7 The average age of engineering students at
graduation is a little over 23 years. This means
that the working career of most engineers is almost
exactly 500 months. How much would an engineer
need to save each month to accrue $5 million by the
end of her working career? Assume a 9% interest
rate, compounded monthly.

6-8 An engineer wishes to have $3 million by the time
he retires in 40 years. Assuming 9% nominal interest,
compounded continuously, what annual sum must he
set aside?

6-9 To reduce her personal carbon footprint, Zooey is
Ⓖ   buying a new hybrid. She has negotiated a price of
$21,900 and will trade in her old car for $2350.
She will put another $850 with it and borrow the
remainder at 6% interest compounded monthly for
4 years. How large will her monthly payments be?
*Contributed by Paul R. McCright, University of
South Florida*

6-10 Zwango Plus Manufacturing expects that fixed costs
Ⓐ   of keeping its Zephyr Hills Plant operating will
be $1.4M this year. If the fixed costs increase by
$100,000 each year, what is the EUAC for a 10-year
period? Assume the interest rate is 12%.
*Contributed by Paul R. McCright, University of
South Florida*

6-11 A firm purchased some equipment at a very favorable price of $30,000. The equipment reduced costs
by $1000 per year during 8 years of use. After 8
years, the equipment was sold for $40,000. Assum-
ing interest at 9%, did the equipment purchase prove
to be desirable?
*Contributed by Hamed Kashani, Saeid Sadri, and
Baabak Ashuri, Georgia Institute of Technology*

6-12 If $i = 6\%$, compute the EUAB over 6 years that is
A equivalent to the two receipts shown.

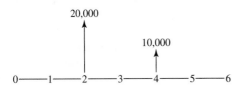

6-13 Compute the EUAC for these cash flows.

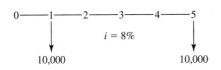

6-14 For the diagram, compute the value of D that results
A in a net equivalent uniform annual worth (EUAW)
of 0.

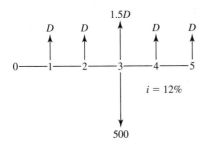

6-15 Amanda and Blake have found a house, which
owing to a depressed real estate market costs only
$201,500. They will put $22,000 down and finance
the remainder with a 30-year mortgage loan at 3%
interest (compounded monthly).

(a) How much is their monthly loan payment?
(b) How much interest will they pay in the second
payment?
(c) They will also have the following expenses:
property taxes of $2100, homeowners' insurance
of $1625, and $290 mortgage insurance (in case
one of them dies before the loan is repaid, a
requirement of the bank). These annual amounts
are paid in 12 installments and added to the loan
payment. What is their full monthly cost?
(d) If they can afford $1200 per month, can Amanda
and Blake afford this house?

*Contributed by Paul R. McCright, University of
South Florida*

6-16 How much should a new graduate pay in 10 equal
A annual payments, starting 2 years from now, in order
to repay a $30,000 loan he has received today? The
interest rate is 6% per year.
*Contributed by Hamed Kashani, Saeid Sadri, and
Baabak Ashuri, Georgia Institute of Technology*

6-17 A firm manufactures and sells high quality business printers and ink toners. Each printer sells for
$650 and each toner for $100. The average user
keeps the printer for 5 years and consumes 4 toners
every year. In response to a recent significant drop
in printer sales (which will reduce future toner sales
as well) the firm wants to lower the printer price to
$500. Assume that income from toner sales occurs at
year-end and the firm's cost of capital is 10%. How
much of an increase is needed in the toner price to
cover the loss in printer price? *Contributed by Yasser
Alhenawi, University of Evansville*

6-18 A motorcycle is for sale for $26,000. The dealer is
A willing to sell it on the following terms:

> No down payment; pay $440 at the end of
> each of the first 4 months; pay $840 at the
> end of each month after that until the loan
> has been paid in full.

At a 12% annual interest rate compounded monthly,
how many $840 payments will be required?

6-19 A couple is saving for their newborn daughter's college education. She will need $25,000 per year for
a four-year college program, which she will start
when she is 18. What uniform deposits starting 2
years from now and continuing through year 17 are
needed, if the account earns interest at 4%?
*Contributed by Hamed Kashani, Saeid Sadri, and
Baabak Ashuri, Georgia Institute of Technology*

6-20 There is an annual receipt of money that varies from
A $100 to $300 in a fixed pattern that repeats forever.
If interest is 10%, compute the EUAB, also continuing forever, that is equivalent to the fluctuating
disbursements.

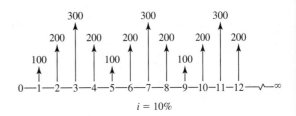

6-21 An engineer has a fluctuating future budget for the maintenance of a particular machine. During each of the first 5 years, $10,000 per year will be budgeted. During the second 5 years, the annual budget will be $15,000 per year. In addition, $5000 will be budgeted for an overhaul of the machine at the end of the fourth year, and again at the end of the eighth year. What uniform annual expenditure would be equivalent, if interest is 8% per year?

6-22 The maintenance foreman of a plant in reviewing his records found that maintenance costs on a large press had increased with sales of a product that will decline in the future.

5 years ago	$ 600
4 years ago	700
3 years ago	800
2 years ago	900
Last year	1000

He believes that maintenance will be $900 this year and will decrease by $100 yearly for 4 years. What will be the equivalent uniform annual maintenance cost for the 10-year period? Assume interest at 8%.

6-23 If interest is 12%, what is the EUAB?

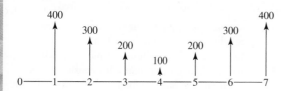

6-24 A machine has a first cost of $150,000, an annual operation and maintenance cost of $2500, a life of 10 years, and a salvage value of $30,000. At the end of Years 4 and 8, it requires a major service, which costs $20,000 and $10,000, respectively. At the end of Year 5, it will need to be overhauled at a cost of $45,000. What is the equivalent uniform annual cost of owning and operating this particular machine if interest is 5%?

6-25 LaQuesha Jackson has made a considerable fortune. She wishes to start a perpetual scholarship for engineering students at her school. The scholarship will provide a student with an annual stipend of $10,000 for each of 4 years (freshman through senior), plus an additional $5000 during the senior year to cover job search expenses. Assume that students graduate in 4

years, a new award is given every 4 years, and the money is paid at the beginning of each year with the first award at the beginning of Year 1. The interest rate is 10%.

(a) Determine the equivalent uniform annual cost (EUAC) of providing the scholarship.
(b) How much money must LaQuesha donate?

6-26 Linda O'Shay deposited $30,000 in a savings account as a perpetual trust. She believes the account will earn 7% annual interest during the first 10 years and 5% interest thereafter. The trust is to provide a uniform end-of-year scholarship at the university. What uniform amount could be used for the student scholarship each year, beginning at the end of the first year and continuing forever?

6-27 Curtis Lowe must pay his property taxes in two equal installments on December 1 and April 1. The two payments are for taxes for the fiscal year that begins on July 1 and ends the following June 30. Curtis purchased a home on September 1. Assuming the annual property taxes remain at $3400 per year for the next several years, Curtis plans to open a savings account and to make uniform monthly deposits the first of each month. The account is to be used to pay the taxes when they are due.

To open the account, Curtis deposits a lump sum equivalent to the monthly payments that will not have been made for the first year's taxes. The savings account pays 4% interest, compounded quarterly (March 31, June 30, September 30, and December 31). How much money should Curtis put into the account when he opens it on September 1? What uniform monthly deposit should he make from that time on?

6-28 Your company must make a $500,000 balloon payment on a lease 2 years and 9 months from today. You have been directed to deposit an amount of money quarterly, beginning today, to provide for the $500,000 payment. The account pays 4% per year, compounded quarterly. What is the required quarterly deposit? *Note:* Lease payments are due at the beginning of the quarter.

6-29 If the owner earns 5% interest on her investments, determine the equivalent annual cost of owning a car with the following costs (EOY = end of year).

Initial down payment = $2200
Annual payments = $5500, EOY1–EOY4
Prepaid insurance = $1500, growing 8% annually

Gas & oil & minor maintenance = $2000, growing 10% annually

Replacement tires = $650 at EOY4 & $800 at EOY8

Major maintenance = $2400 at EOY5

Salvage value = $3750 at EOY9

Contributed by Paul R. McCright, University of South Florida

Annual Comparisons

6-30
Ⓐ The Johnson Company pays $2000 a month to a trucker to haul wastepaper and cardboard to the city dump. The material could be recycled if the company were to buy a $60,000 hydraulic press baler and spend $30,000 a year for labor to operate the baler. The baler has an estimated useful life of 30 years and no salvage value. Strapping material would cost $2000 per year for the estimated 500 bales a year that would be produced. A wastepaper company will pick up the bales at the plant and pay Johnson $23 per bale for them. Use an annual cash flow analysis and an interest rate of 8% to recommend whether it is economical to install and operate the baler.

6-31
Ⓖ A municipal power plant uses natural gas from an existing pipeline at an annual cost of $40,000 per year. A new pipeline would initially cost $100,000, but it would reduce the annual cost to $10,000 per year.

(*a*) Assume an analysis period of 25 years and no salvage value for either pipeline. The interest rate is 6%. Using the equivalent uniform annual cost (EUAC), should the new pipeline be built?

(*b*) The power plant uses natural gas. What are some of the non-economic benefits to the municipality of this energy source over others? Develop three primary advantages and disadvantages.

6-32
Ⓐ A construction firm needs a new small loader. It can be leased from the dealer for 3 years for $5500 per year including all maintenance, or it can be purchased for $20,000. The firm expects the loader to have a salvage value of $7000 after 7 years. The maintenance will be $500 the first year and then it will increase by $300 each year. The firm's interest rate is 12% per year. Compare the EUACs for leasing and buying the loader.
Contributed by Hamed Kashani, Saeid Sadri, and Baabak Ashuri, Georgia Institute of Technology

6-33 Hinson's Homegrown Farms needs a new irrigation system. System one will cost $145,000, have annual maintenance costs of $10,000, and need an overhaul at the end of year six costing $30,000. System two will have first-year maintenance costs of $5000 with increases of $500 each year thereafter. System two would not require an overhaul. Both systems will have no salvage value after 12 years. If Hinson's cost of capital is 4%, using annual worth analysis determine the maximum Hinson's should be willing to pay for system two. *Contributed by Ed Wheeler, University of Tennessee at Martin*

6-34
Ⓐ When he financed his firm's building, Al Silva borrowed $280,000 at 10% interest to be repaid in 25 equal annual end-of-year payments. After making 10 payments, Al found he could refinance the loan at 9% interest for the remaining 15 years.

To refinance the loan, Al must pay the balance due plus a prepayment penalty charge of 2% of the balance due, and he must pay a $1000 service charge on the new loan. All payments are financed by the new loan. Should Al refinance the loan, assuming that he will keep the firm's building for the next 15 years? Use an annual cash flow analysis.

6-35 Claude James, a salesman, needs a new car for business use. He expects to be promoted to a supervisory job after 3 years, and he will no longer be "on the road." The company reimburses salesmen each month at the rate of 55¢ per mile driven. Claude believes he should use a 12% interest rate. If the car could be sold for $7500 at the end of 3 years, which method should he use to obtain it?

A. Pay cash: the price is $26,000.

B. Lease the car: the monthly charge is $700 on a 36-month lease, payable at the end of each month; at the end of the 3-year period, the car is returned to the leasing company.

C. Lease the car with an option to buy at the end of the lease: pay $720 a month for 36 months; at the end of that time, Claude could buy the car, if he chooses, for $7000.

6-36
Ⓐ An RV manufacturer estimates that annual profits will increase if a mobile model is built and taken to trade shows to market their product line. A finance and engineering team has looked at the issue and has developed two options:

1. A large model can be developed at a cost of $70,000, and it should increase annual profits by $30,000 per year.

2. A small model can be developed for $55,000, but it will only increase annual profits by $12,000 per year.

The salvage value for the large model is $5000 more than the small model after their common useful life of 5 years, and it costs $1000 more a year to transport to the trade shows. The manufacturer uses an interest rate of 20%. Use an annual worth comparison to make a recommendation on which, if either, option should be chosen.

6-37 **E** The town of Dry Gulch needs more water from Pine Creek. The town engineer has selected two plans for comparison: a *gravity plan* (divert water at a point 10 miles up Pine Creek and pipe it by gravity to the town) and a *pumping plan* (divert water at a point closer to town). The pumping plant would be built in two stages, with half-capacity installed initially and the other half installed 10 years later.

(a) Use an annual cash flow analysis to find which plan is more economical. The analysis will assume a 40-year life, 10% interest, and no salvage value.

	Gravity	Pumping
Initial investment	$2,800,000	$1,400,000
Investment in 10th year	None	200,000
Operation and maintenance	10,000/yr	25,000/yr
Power cost		
Average first 10 years	None	50,000/yr
Average next 30 years	None	100,000/yr

(b) The situation described in this problem is a common one for cities that are upstream and downstream of a water supply. Search using the term "water ethics" and write a short paragraph that describes what you find from the perspectives of Dry Gulch and Pine Creek.

6-38 **A** A manufacturer is considering replacing a production machine tool. The new machine, costing $37,000, would have a life of 4 years and no salvage value, but would save the firm $5000 per year in direct labor costs and $2000 per year in indirect labor costs. The existing machine tool was purchased 4 years ago at a cost of $40,000. It will last 4 more years and will have no salvage value. It could be sold now for $10,000 cash. Assume that money is worth 8% and that differences in taxes, insurance, and so forth are negligible. Use an annual cash flow analysis to determine whether the new machine should be purchased.

6-39 **G** Two possible routes for a power line are under study. In both cases the power line will last 15 years, have

no salvage value, have annual property taxes of 2% of first cost, and have a yearly power loss of $500/km.

	Around the Lake	Under the Lake
Length	15 km	5 km
First cost	$9000/km	$25,000/km
Maintenance	$200/km/yr	$300/km/yr

(a) If 6% interest is used, should the power line be routed around the lake or under the lake?
(b) Search using the term "environmental impact of power lines" and write a short paragraph on what you learn about this topic. What aspects of the environment seem to be most affected, and what policies and technologies are being used to eliminate or mitigate concerns?

6-40 **A** **G** An oil refinery must now begin sending its waste liquids through a costly treatment process before discharging them. The engineering department estimates costs at $300,000 for the first year. It is estimated that if process and plant alterations are made, the waste treatment cost will decline $30,000 each year. As an alternate, a specialized firm, Hydro-Clean, has offered a contract to process the waste liquids for 10 years for $150,000 per year. Either way, there should be no need for waste treatment after 10 years. Use an 8% interest rate and annual cash flow analysis to determine whether the Hydro-Clean offer should be accepted.

6-41 Bill Anderson buys a car every 2 years as follows: initially he makes a down payment of $12,000 on a $30,000 car. The balance is paid in 24 equal monthly payments with annual interest at 12%. When he has made the last payment on the loan, he trades in the 2-year-old car for $12,000 on a new $30,000 car, and the cycle begins over again.

Doug Jones decided on a different purchase plan. He thought he would be better off if he paid $30,000 cash for a new car. Then he would make a monthly deposit in a savings account so that, at the end of 2 years, he would have $18,000 in the account. The $18,000 plus the $12,000 trade-in value of the car will allow Doug to replace his 2-year-old car by paying $30,000 for a new one. The bank pays 3% interest, compounded monthly.

(a) What is Bill's monthly loan payment?
(b) What is Doug's monthly savings account deposit?

(c) Why is Doug's monthly savings account deposit smaller than Bill's payment?

6-42 One of two mutually exclusive alternatives must be
Ⓐ selected. Alternative *A* costs $30,000 now for an annual benefit of $8450. Alternative *B* costs $50,000 now for an annual benefit of $14,000. Using a 15% nominal interest rate, compounded continuously, which do you recommend? Solve by annual cash flow analysis.

6-43 North Plains Biofuels (NPB) has negotiated a con-
Ⓖ tract with an oil firm to sell 150,000 barrels of ethanol per year, for 10 years beginning in end of year (EOY) 4. The oil firm will pay NPB $10M annually beginning from EOY0 to EOY3 and then $110 per barrel.

(a) If NPB uses an interest rate of 15%, which method should be used to produce the biofuels? Annual O&M costs increase 2% per year and raw materials costs 3% per year.

	Corn	Algae
Purchase of land (EOY0)	$1,900,000	$ 3,800,000
Facility construction (EOY1)	$5,300,000	$7,100,000
Annual O&M (EOY4)	$2,450,000	$2,800,000
Raw materials (EOY4)	$1,500,000	$250,000
Salvage value (EOY13)	$3,000,000	$3,600,000

(b) What are biofuels? How are they used in the automotive industry, and what are the pros and cons of this use? What are some of the important non-auto industry uses of biofuels?

Contributed by Paul R. McCright, University of South Florida

6-44 Which car has a lower EUAC if the owner can earn
Ⓐ 5% in his best investment? *Contributed by Paul R. McCright, University of South Florida*

	Midsize	Hybrid
Initial cost	$19,200	$25,500
Annual maintenance	1,000	1,500
Annual gas & oil	2,500	1,200
(increasing 15% yearly)		
Salvage value (Year 8)	8,000	10,000

Different Lives

6-45 A pump is needed for 10 years at a remote location. The pump can be driven by an electric motor if a power line is extended to the site. Otherwise, a gasoline engine will be used. Use an annual cash flow analysis and an 6% interest rate. How should the pump be powered?

	Gasoline	Electric
First cost	$2400	$9000
Annual operating cost	1200	250
Annual maintenance	300	50
Salvage value	300	600
Life, in years	5	10

6-46 A firm is choosing between machines that perform
Ⓐ the same task in the same time. Assume the minimum attractive return is 8%. Which machine would you choose?

	Machine *X*	Machine *Y*
First cost	$5000	$8000
Estimated life, in years	5	12
Salvage value	0	$2000
Annual maintenance cost	0	150

6-47 Alternative *B* may be replaced with an identical item every 20 years at the same cost and annual benefit. Using a 9% interest rate and an annual cash flow analysis, which alternative should be selected?

	A	*B*
Cost	$100,000	$150,000
Uniform annual benefit	30,000	45,000
Useful life, in years	∞	20

6-48 Road Runner LLC (RRL) is considering three alternate routes in the desert. RRL uses a MARR of 4%. Using equivalent annual worth over the least common multiple horizon, which choice is best?

	Route 105	Route 205	Route 305
First cost	$500,000	$450,000	$400,000
Savings/year	100,000	125,000	90,000
Life	6	4	5

Contributed by Ean Ng and Gana Natarajan, Oregon State University

6-49 A suburban taxi company is considering buying taxis with diesel engines instead of gasoline engines. The cars average 80,000 km a year. Use an annual cash flow analysis to determine the more economical choice if interest is 6%.

	Diesel	Gasoline
Vehicle cost	$24,000	$19,000
Useful life, in years	5	4
Fuel cost per liter	88¢	92¢
Mileage, in km/liter	16	11
Annual repairs	$ 900	$ 700
Annual insurance premium	1,000	1,000
End-of-useful-life resale value	4,000	6,000

6-50 Assuming that Alternatives *B* and *C* are replaced with identical units at the end of their useful lives, and an 8% interest rate, which alternative should be selected? Use an annual cash flow analysis.

	A	B	C
Cost	$10,000	$15,000	$20,000
Annual benefit	1,000	1,762	5,548
Useful life (yrs)	∞	20	5

6-51 The manager in a canned food processing plant is trying to decide between two labeling machines.

	Machine A	Machine B
First cost	$15,000	$25,000
Maintenance and operating costs	1,600	400
Annual benefit	8,000	13,000
Salvage value	3,000	6,000
Useful life, in years	6	10

Assume an interest rate of 6%. Use annual cash flow analysis to determine which machine should be chosen.

6-52 A college student has been looking for new tires. The student feels that the warranty period is a good estimate of the tire life and that a 10% interest rate is appropriate. Using an annual cash flow analysis, which tire should be purchased?

Tire Warranty (months)	Price per Tire
12	$39.95
24	59.95
36	69.95
48	90.00

6-53 Carp, Inc. wants to evaluate two methods of packaging their products. Use an interest rate of 15% and annual cash flow analysis to decide which is the most desirable alternative.

	A	B
First cost	$700,000	$1,700,000
O&M costs (yr 1)	18,000	29,000
+ Cost gradient	+900/yr	+750/yr
Annual benefit	154,000	303,000
Salvage value	142,000	210,000
Useful life, in years	10	20

6-54 The analysis period is 10 years, but there will be no replacement for Alternative *B* after 5 years. Based on a 15% interest rate, which alternative should be selected? Use an annual cash flow analysis.

	A	B
Cost	$5000	$18,000
Uniform annual benefit	1500	6,000
Useful life, in years	10	5

6-55 A small manufacturing company is evaluating trucks for delivering their products. Truck *A* has a first cost of $32,000, its operating cost will be $5500 per year, and its salvage after 3 years will be $7000. Truck *B* has a first cost of $37,000, an operating cost of $5200, and a resale value of $12,000 after 4 years. At an interest rate of 12% which model should be chosen? *Contributed by Hamed Kashani, Saeid Sadri, and Baabak Ashuri, Georgia Institute of Technology*

6-56 Dick Dickerson Construction, Inc. has asked you to help them select a new backhoe. You have a choice between a wheel-mounted version, which costs $60,000 and has an expected life of 5 years and a salvage value of $2000, and a track-mounted one, which costs $80,000, with a 7-year life and an expected salvage value of $10,000. Both machines will achieve the same productivity. Interest is 8%. Which one will you recommend? Use an annual worth analysis.

6-57 A job can be done with Machine *A* that costs $12,500 and has annual end-of-year maintenance costs of $5000; its salvage value after 3 years is $2000. Or the job can be done with Machine *B*, which costs $15,000 and has end-of-year maintenance costs of $4000 and a salvage value of $1500 at the end of 4 years. These investments can be repeated in the future, and your work is expected to continue indefinitely. Use present worth, annual worth, and capitalized cost to compare the machines. The interest rate is 5%/year. *Contributed by D. P. Loucks, Cornell University*

6-58 Hospitality Enterprises is planning to build a new 112 room inn in Martin. The initial cost of land leases and construction is anticipated to be $3.4 million. The annual operating and maintenance costs are expected to average $25,000 for the 20-year life of the inn. Every 4 years the interior of the inn must be painted at a cost of $15,000.

The exterior must be painted and refurbished every 5 years at a cost of $60,000. The carpet and furniture must be replaced every 6 years at a cost of $100,000. Every 8 years $80,000 will be spent on paving and striping the parking areas. The inn will have a net demolition cost of $100,000 at the end of its life. If the MARR for Hospitality is 5%, determine the EUAC for the inn. *Contributed by Ed Wheeler, University of Tennessee at Martin*

🖥 Spreadsheets and Loans

6-59 A new car is purchased for $22,000 with a 10% down, 6% loan. The loan is for 4 years. After making 30 monthly payments, the owner wants to pay off the loan's remaining balance. How much is owed?

6-60 A year after buying her car, Anita has been offered a job in Europe. Her car loan is for $15,000 at a 9% nominal interest rate for 60 months. If she can sell the car for $12,000, how much does she get to keep after paying off the loan?

6-61 (*a*) You are paying off a debt at a nominal 8% per year by paying $400 at the end of each quarter for the next year. Find the interest paid in the last $400 payment.

(*b*) If this debt were to be paid off in two equal payments of $1650 at the end of this year and next year, find the interest paid in the first $1650 payment. Again the loan rate is a nominal 8% per year compounded quarterly.

Contributed by D. P. Loucks, Cornell University

6-62 A student loan totals $30,000 at graduation. The interest rate is 6%, and there will be 60 payments beginning 1 month after graduation. What is the monthly payment? What is owed after the first 2 years of payments?

6-63 The student in Problem 6-62 received $2500 as a graduation present. If an extra $2500 is paid at Month 1, when is the final payment made? How much is it?

6-64 Sam can afford to spend $500 per month on a car. He figures he needs half of it for gas, parking, and insurance. He has been to the bank, and they will loan him 100% of the car's purchase price. (*Note*: If he had a down payment saved, then he could borrow at a lower rate.)

(*a*) If his loan is at a nominal 12% annual rate over 36 months, what is the most expensive car he can purchase?

(b) The car he likes costs $14,000 and the dealer will finance it over 60 months at 12%. Can he afford it? If not, for how many months will he need to save his $500 per month?

(c) What is the highest interest rate he can pay over 60 months and stay within his budget if he buys the $14,000 car now?

6-65 EnergyMax Engineering constructed a small office building for their firm 5 years ago. They financed it with a bank loan for $450,000 over 15 years at 6% interest with quarterly payments and compounding. The loan can be repaid at any time without penalty. The loan can be refinanced through an insurance firm for 4% over 20 years—still with quarterly compounding and payments. The new loan has a 5% loan initiation fee, which will be added to the new loan.

(a) What is the balance due on the original mortgage (20 payments have been made in the last 5 years)?

(b) How much will EnergyMax's payments drop with the new loan?

(c) How much longer will the proposed loan run?

6-66 Suppose you graduate with a debt of $42,000 that you or someone must repay. One option is to pay off the debt in constant amounts at the beginning of each month over the next 10 years at a nominal annual interest rate of 10%.

(a) What is the constant beginning-of-month payment?

(b) Of the first payment, what is the interest and the principal paid?

(c) Of the last payment, what is the interest and the principal paid?

(d) How are student loans treated in bankruptcy? What are the practical and ethical reasons for and against treating student loans differently from other loans?

Contributed by D. P. Loucks, Cornell University

6-67 An $11,000 mortgage has a 30-year term and a 6% nominal interest rate.

(a) What is the monthly payment?

(b) After the first year of payments, what is the outstanding balance?

(c) How much interest is paid in Month 13? How much principal?

6-68 A 30-year mortgage for $95,000 is issued at a 9% nominal interest rate.

(a) What is the monthly payment?

(b) How long does it take to pay off the mortgage, if $1000 per month is paid?

(c) How long does it take to pay off the mortgage, if double payments are made?

6-69 Solve Problem 6-39(a) for the breakeven first cost per kilometer of going under the lake.

6-70 Redo Problem 6-49 to calculate the EUAW of the alternatives as a function of miles driven per year to see if there is a crossover point in the decision process. Graph your results.

6-71 Develop a spreadsheet to solve Problem 6-37(a). What is the breakeven cost of the additional pumping investment in Year 10?

Annuity Due

6-72 You are leasing some furniture for your apartment. The monthly lease is for $395/month, payable at the beginning of the month. If your personal cost of capital is 3%, what is the present value of a one-year lease?

6-73 It is the first of October, and you are developing cost estimates for creating an engineering consulting business with a small group of friends. Liability insurance beginning January 1 will cost $475 per month, payable at the beginning of each month. What is the PW of this insurance for the first year as of today's date? The firm's MARR is 12%.

6-74 You are interested in leasing a new car for 36 months.

- The value of the car is $22,555.
- You must pay $3025 at signing, which does not include the first month's lease payment.
- The monthly lease cost for the car is $154 for 36 months.
- At the end of the lease, you will need to pay a lease termination fee of $2000.
- The interest rate for this type of new car is 1.90% APR.

Calculate the present worth of leasing the car.

6-75 Your grandmother set up an annuity, in which she will receive a monthly payment from the bank of

$1200 per month, payable at the beginning of each month. She paid $200,000 for this annuity. The bank says they are paying an interest rate of 4%. How many months does the bank plan to pay the annuity?

Minicases

6-76 An office building should last 60 years, but this owner will sell it at 20 years for 40% of its construction cost. For the first 20 years it can be leased as Class A space, which is all this owner operates. When the building is sold, the land's cost will be recovered in full.

$2.2M	Land
$4.1M	Building
$640,000	Annual operating and maintenance
4%	Annual property taxes and insurance
	(% of initial investment)

(a) If the owner wants a 12% rate of return, what is the required monthly leasing cost?

(b) Assuming that the building is vacant 5% of the time, what is the required monthly lease?

(c) What is an example monthly cost per square foot for Class A space in your community?

6-77 A 30-unit apartment building should last 35 years, when it will need to be either replaced or undergo major renovation. Assume the building's value at 35 years will be 10% of its construction cost. Assume it

will be sold and that the land's cost will be recovered in full.

$3.2M	Land
$4.8M	Building
$850,000	Annual operating and maintenance
6%	Annual property taxes and insurance
	(% of initial investment)
12%	Vacancy rate

(a) If the owner wants a 15% rate of return, what is the required monthly leasing cost for each unit?

(b) If turning 2 units into an exercise facility would decrease the vacancy rate by 5%, would that be a good decision?

CASES

The following cases from *Cases in Engineering Economy* (www.oup.com/us/newnan) are suggested as matched with this chapter.

CASE 6   Lease a Lot

Compares leasing and ownership. Both financing and investing decisions are needed.

CASE 10   The Cutting Edge

Make versus buy and machine selection.

CASE 27   Harbor Delivery Service

Focus is treatment of sunk costs. More complicated than most. Some discoveries in the data gathering process. Solution uses equation rather than cash flow table.

RATE OF RETURN ANALYSIS

ISBN 978-0-19-533541-5

Bar Codes Give a Number; RFID Codes Tell a Story and May Become the Spies You Buy

The shift from manual price tags, shipping labels, and baggage tags to bar codes revolutionized operations by reducing costs and errors. The shift to Radio Frequency Identification (RFID) is revolutionizing operations again. A bar code speaks five or six times in its lifetime; RFID codes can speak many times each second. RFID is in your car keys, the EZ Pass on your windshield, a rail pass in China, your new Smart Passport, a subway pass in Washington, DC, some employee IDs, and garment labels. RFID chips are used to tightly control inventory from razor blades to computers and to monitor temperatures of Alaska Wild Seafood from packaging to restaurant delivery. Today's wholesalers, retailers, manufacturers, and military decision makers trust RFID technology to create real-time inventory tracking.

In 2003, RFID tags cost about 30¢ each, which made them too expensive to embed in low-cost products like soap and shaving cream. RFID tags with batteries for tracking truckloads or containers cost more than $1000. Today, a major manufacturer of RFID tags quotes a price of 5¢ each for batches of at least a billion tags. Thus manufacturers of products from jeans to shampoo are using RFID tags. The growth rate for sales to the retail sector is estimated to exceed 30% annually through 2020.

Part of the story that RFID tags tell may be about the person who is wearing them. RFID tags emit data that can track and trace goods from the time they leave the manufacturing plant until you purchase them at your favorite retail store—and take them home. While the cost of an RFID tag for a garment may be down to pennies, what is the real cost and benefit to a company that may invest several million dollars in technology to track items—and possibly the person wearing that item? ■ ■ ■

Contributed by Oliver Hedgepeth, American Public University, author of RFID Metrics, 2nd, CRC Press (2017)

QUESTIONS TO CONSIDER

1. A clothing manufacturer is considering using RFID tags to track all the new season's clothing. What cost and benefit factors should be considered? Is there a cost beyond the initial capital investment?

2. How would cost and benefit considerations differ for those who have already invested in RFID versus those who are only considering it?

3. Consider that the military is moving 1500 tactical vehicles and 1000 containers per month from the Middle East to the United States to recycle, refurbish, or send to the trash pile. What are some of the military cost and benefit factors that are different from commercial or for-profit companies doing similar retrograde activities?

4. Some consumers have already expressed concern about RFID tags being in their children's clothing, calling the tags an invasion of privacy. What is the ethical and social cost associated with such RFID tags for a retail company?

5. In response to consumer worries, RFID tag manufacturers are developing "kill" technologies to allow consumers to disable RFID tags after goods are purchased. How might this affect consumer attitudes and company costs?

After Completing This Chapter...

The student should be able to:

- Evaluate project cash flows with the *internal rate of return* (IRR) measure.
- Plot a project's present worth (PW) against the interest rate.
- Use an *incremental* rate of return analysis to evaluate competing alternatives.
- Develop and use spreadsheets to make IRR and incremental rate of return calculations.
- Identify when multiple roots exist and select the correct alternative. (Appendix 7A)

Key Words

balloon payment	incremental rate of return	NPW plot
increment of borrowing	internal rate of return	rate of return
increment of investment	MARR	XIRR

In this chapter we will examine four aspects of rate of return, the third major analysis method. First, the meaning of "rate of return" is explained; second, calculating the rate of return is illustrated; third, rate of return analysis problems are presented; and fourth, incremental analysis is presented. In an appendix to the chapter, we describe difficulties sometimes encountered when computing an interest rate for cash flow series with multiple sign changes.

Rate of return is the most frequently used measure in industry. Problems in computing the rate of return sometimes occur, but its major advantage is that it is a single figure of merit that is readily understood.

Consider these statements:

- The net present worth on a project is $32,000.
- The equivalent uniform annual net benefit is $2800.
- The project will produce a 23% rate of return.

While none of these statements tells the complete story, the third one measures the project's desirability in terms that are widely and easily understood. Thus, this measure is accepted by engineers and business leaders alike.

There is another advantage to rate of return analysis. In both present worth and annual cash flow calculations, one must select an interest rate before calculations start—and the exact value may be a difficult and controversial item. In rate of return analysis, no interest rate is introduced into the calculations (except as described in Appendix 7A). Instead, we compute the rate of return (more accurately called *internal rate of return*) from the cash flows. To decide how to proceed, the calculated rate of return is compared with a preselected **minimum attractive rate of return,** or simply **MARR**. This is the same value of *i* used for present worth and annual cash flow analysis, but the comparison is after calculations are complete. Often the project's IRR is well above or below the MARR so an exact value for the MARR is not needed.

INTERNAL RATE OF RETURN

Internal rate of return is the interest rate at which the present worth and equivalent uniform annual worth are equal to 0.

This definition is easy to remember, and it also tells us how to solve for the rate of return. In earlier chapters we did this when we solved for the interest rate on a loan or investment.

Other definitions based on the unpaid balance of a loan or the unrecovered investment can help clarify why this rate of return is also called the *internal rate of return* or IRR. In Chapter 3 we examined four plans to repay $5000 in 5 years with interest at 8% (Table 3–1). In each case the amount loaned ($5000) and the loan duration (5 years) were the same. Yet the total interest paid to the lender varied from $1200 to $2347. In each case the lender received 8% interest each year on the amount of money actually owed. And, at the end of 5 years, the principal and interest payments exactly repaid the $5000 debt with interest at 8%. We say the lender received an "8% rate of return."

> **Internal rate of return** can also be defined as the interest rate paid on the unpaid balance of a *loan* such that the payment schedule makes the unpaid loan balance equal to zero when the final payment is made.

Instead of lending money, we might invest $5000 in a machine tool with a 5-year useful life and an equivalent uniform annual benefit of $1252. The question becomes, What rate of return would we receive on this investment?

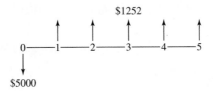

We recognize these cash flows as reversed from Plan 3 of Table 3–1. We know that five payments of $1252 are equivalent to a present sum of $5000 when interest is 8%. Therefore, the rate of return on this investment is 8%.

> **Internal rate of return** can also be defined as the interest rate earned on the unrecovered *investment* such that the payment schedule makes the unrecovered investment equal to zero at the end of the investment's life.

It must be understood that the 8% rate of return does not mean an annual return of 8% on the $5000 investment, or $400 in each of the 5 years with $5000 returned at the end of Year 5. Instead, each $1252 payment represents an 8% return on the unrecovered investment *plus* a partial return of the investment. This may be tabulated as follows:

Year	Cash Flow	Unrecovered Investment at Beginning of Year	8% Return on Unrecovered Investment	Investment Repayment at End of Year	Unrecovered Investment at End of Year
0	−$5000				
1	1252	$5000	$ 400	$ 852	$4148
2	1252	4148	331	921	3227
3	1252	3227	258	994	2233
4	1252	2233	178	1074	1159
5	1252	1159	93	1159	0
			$1260	$5000	

This cash flow represents a $5000 investment with benefits that produce an 8% rate of return on the unrecovered investment. Except for the language shift to return on unrecovered balance from interest on balance due, the table above is like the amortization schedules covered in Chapter 6.

Although the definitions of internal rate of return are stated differently for a loan and for an investment, there is only one fundamental concept. It is that **the internal rate**

of return is the interest rate at which the benefits are equivalent to the costs, or the present worth (PW) is 0. Since we are describing the funds that remain within the investment throughout its life, the resulting rate of return is described as the internal rate of return, i.

CALCULATING RATE OF RETURN

To calculate a rate of return on an investment, we must convert the various consequences of the investment into a cash flow series. Then we solve the cash flow series for the unknown value of the internal rate of return (IRR). Five forms of the cash flow equation are as follows:

$$PW \text{ of benefits} - PW \text{ of costs} = 0 \qquad (7\text{-}1)$$

$$\frac{PW \text{ of benefits}}{PW \text{ of costs}} = 1 \qquad (7\text{-}2)$$

$$Present \text{ worth} = Net \text{ present worth}^1 = 0 \qquad (7\text{-}3)$$

$$EUAW = EUAB - EUAC = 0 \qquad (7\text{-}4)$$

$$PW \text{ of costs} = PW \text{ of benefits} \qquad (7\text{-}5)$$

The five equations represent the same concept in different forms. They relate costs and benefits with the IRR as the only unknown. The calculation of rate of return is illustrated by the following examples.

EXAMPLE 7–1

An engineer invests $5000 at the end of every year for a 40-year career. If the engineer wants $1 million in savings at retirement, what interest rate must the investment earn?

While this example has been defined in terms of an engineer's personal finances, we could just as easily have said, "a mining firm makes annual deposits of $50,000 into a reclamation fund for 40 years. If the firm must have $10 million when the mine is closed, what interest rate must the investment earn?" Since the annual deposit and required final amount are 10 times larger the answer is obviously the same.

TABLE SOLUTION

Using the net future worth version of Equation 7–3, we write

$$Net \text{ FW} = 0 = -\$5000(F/A, i, 40) + \$1,000,000$$

---

[1]Remember that present value (PV), present worth (PW), net present value (NPV), and net present worth (NPW) are synonyms in practice and in this text.

Rewriting, we see that

$$(F/A, i, 40) = \$1{,}000{,}000/\$5000 = 200$$

We then look at the compound interest tables for the value of i where $(F/A, i, 40) = 200$. If no tabulated value of i gives this value, then we will interpolate, using the two closest values or solve exactly with a calculator or spreadsheet. In this case $(F/A, 0.07, 40) = 199.636$, which to three significant digits equals 200. Thus the required rate of return for the investment is 7%.

5-BUTTON SOLUTION

	A	B	C	D	E	F	G	H
1	Problem	i	n	PMT	PV	FV	Solve for	Answer
2	Exp. 7-1		40	-5000	0	1,000,000	i	7.007%

EXAMPLE 7–2

An investment resulted in the following cash flow. Compute the rate of return.

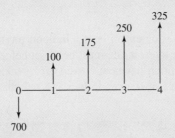

TABLE SOLUTION

$$\text{EUAW} = \text{EUAB} - \text{EUAC} = 0 = 100 + 75(A/G, i, 4) - 700(A/P, i, 4)$$

Here, we have two different interest factors in the equation, and we will solve the equation by trial and error. The EUAW value is a function of i. Try $i = 5\%$ first:

$$\text{EUAW}_{5\%} = 100 + 75(A/G, 5\%, 4) - 700(A/P, 5\%, 4)$$
$$= 100 + 75(1.439) - 700(0.2820)$$
$$= \text{EUAB} - \text{EUAC} = 207.9 - 197.4 = +10.5$$

The EUAW is too high. If the interest rate is increased, EUAW will decrease. Try $i = 8\%$:

$$\text{EUAW}_{8\%} = 100 + 75(A/G, 8\%, 4) - 700(A/P, 8\%, 4)$$
$$= 100 + 75(1.404) - 700(0.3019)$$
$$= \text{EUAB} - \text{EUAC} = 205.3 - 211.3 = -6.0$$

This time the EUAW is too low. We see that the true rate of return is between 5% and 8%. Try $i = 7\%$:

$$\text{EUAW}_{7\%} = 100 + 75(A/G, 7\%, 4) - 700(A/P, 7\%, 4)$$
$$= 100 + 75(1.416) - 700(0.2952)$$
$$= \text{EUAB} - \text{EUAC} = 206.2 - 206.6 = -0.4$$

The IRR is about 7%. Interpolating between the $\text{EUAW}_{6\%}$ and $\text{EUAW}_{7\%}$ values would yield a more precise answer.

SPREADSHEET SOLUTION

Since there is a gradient a 5-BUTTON SOLUTION is not possible. Best practice is to build or re-use a spreadsheet with a data block. However, it is also possible to build a small spreadsheet with hard coded values for one-time use to find that the rate of return is 6.91%.

	A	B	C	D	E	F	G	H
1	Year	0	1	2	3	4	Solve for	Answer
2	Cash flow	-700	100	175	250	325	IRR	6.91%
3				=C2+75				=IRR(B2:F2)

Example 7–3 is an example of a **balloon payment** loan. A lender may offer a lower interest rate, if the interest rate must be re-set before the loan is completely paid off. Typically the final balloon payment is paid off by taking out a new loan. The borrower may prefer the lower rate now, even if there is a risk that the rate may be higher later.

EXAMPLE 7–3

A firm borrows $300,000 to be repaid with 5 annual payments of $45,000 and a final balloon payment of $170,000. What interest rate is the firm paying on this loan?

5-BUTTON SOLUTION

The first step is to draw the cash flow diagram.

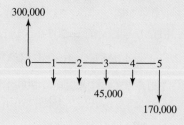

This can be solved in one step with a spreadsheet or calculator annuity function.

	A	B	C	D	E	F	G	H	I
1	Problem	i	n	*PMT*	*PV*	*FV*	Solve for	Answer	Formula
2	Exp. 7-3		5	-45,000	300,000	-170,000	RATE	7.546%	=RATE(C2,D2,E2,F2)

TABLE SOLUTION

This can also be solved with tabulated factors. A good starting point is the present worth equation.

$$PW_i = 300,000 - 45,000(P/A, i, 5) - 170,000(P/F, i, 5)$$

$PW_{7\%}$ is $-\$5717$ and $PW_{8\%}$ is $\$4629$. Thus the interpolated interest rate is:

$$i = 0.07 + 0.01(5717)/(5717 + 4629) = 7.553\%$$

The interest rates are shown with an extra decimal point to illustrate the difference between the exact and the interpolated values here.

EXAMPLE 7–4

A local firm sponsors a student loan program for the children of employees. No interest is charged until graduation, and then the interest rate is 5%. Maria borrows $9000 per year, and she graduates after 4 years. Since tuition must be paid ahead of time, assume that she borrows the money at the start of each year.

   If Maria makes five equal annual payments, what is each payment? Use the cash flow from when she started borrowing the money to when it is all paid back, and then calculate the internal rate of return for Maria's loan. Is this arrangement attractive to Maria?

TABLE SOLUTION

Maria owes $36,000 at graduation. The first step is to calculate the five equal annual payments to repay this loan at 5%.

$$\text{Loan payment} = \$36,000(A/P, 5\%, 5) = 36,000(0.2310) = \$8316$$

   Maria receives $9000 by borrowing at the start of each year. She graduates at the end of Year 4. At the end of Year 4, which is also the beginning of Year 5, interest starts to accrue. She makes her first payment at the end of Year 5, which is one year after graduation.

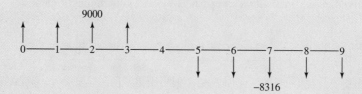

The next step is to write the present worth equation in factor form, so that we can apply Equation 7-3 and set it equal to 0. This equation has three factors, so we will have to solve the problem by picking interest rates and substituting values. The present worth value is a function of i.

$$PW_i = 9000[1 + (P/A, i, 3)] - 8316(P/A, i, 5)(P/F, i, 4)$$

The first two interest rates used are 0% (because it is easy) and 3% because the subsidized rate will be below the 5% that is charged after graduation.

At 0% any P/A factor equals n, and any P/F factor equals 1.

$$PW_{0\%} = 9000(4) - 8316(5) = -5180$$

$$PW_{3\%} = 9000(1 + 2.829) - 8316(4.580)(0.8885) = 620.5$$

Since PW_i has opposite signs for 0 and 3, there is a value of i between 0 and 3% which is the IRR. Because the value for 3% is closer to 0, the IRR will be closer to 3%. Try 2% next.

$$PW_{2\%} = 9000(1 + 2.884) - 8316(4.713)(0.9238) = -1251$$

As shown in Figure 7–1, interpolating between 2 and 3% leads to

$$IRR = 2\% + (3\% - 2\%)[1251/(1251 + 620.5)] = 2.67\%.$$

This rate is quite low, and it makes the loan look like a good choice.

FIGURE 7–1 Plot of PW versus interest rate i.

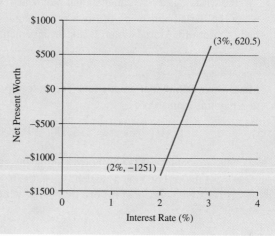

SPREADSHEET SOLUTION

Note that cell F6 cannot be blank—a zero must be entered. If it is blank, then the cell is ignored by the IRR function and the IRR is calculated as though the cash flows occur through year 8.

	A	B	C	D	E	F	G
1	$9,000	Annual tuition			Year	Cash flow	
2	4	# years in college			0	$9,000	
3	5%	Interest rate after graduation			1	9,000	
4	5	# payments			2	9,000	
5					3	9,000	
6	$36,000	Amount owed at graduation			4	0	Cannot be blank!!
7	-$8,315.09	Annual payment			5	-8,315	
8					6	-8,315	
9		=PMT(A3,A4,A6,0)			7	-8,315	
10					8	-8,315	
11					9	-8,315	
12					IRR	2.66%	=IRR(F2:F11)

We can prove that the rate of return in Example 7–4 is very close to 2.66% by tabulating how much Maria owes each year, the interest that accrues, and her borrowings and payments. This is called amortization table.

Year	Amount Owed at Start of Year	Interest at 2.66%	Cash Flow at End of Year	Amount Owed at End of Year
1	$ 9,000	$ 239	$ 9000	$ 18,239
2	18,239	485	9000	27,725
3	27,725	737	9000	37,462
4	37,462	996	0	38,459
5	38,459	1023	−8315	31,166
6	31,166	829	−8315	23,680
7	23,680	630	−8315	15,995
8	15,995	425	−8315	8,106
9	8,106	216	−8315	6*

*This small amount owed indicates that the interest rate is slightly less than 2.66%.

If in Figure 7–1 net present worth (NPW) had been computed for a broader range of values of i, Figure 7–2 would have been obtained. From this figure it is apparent that the error resulting from linear interpolation increases as the interpolation width increases.

Plot of NPW Versus Interest Rate i

The plot of NPW versus interest rate i is an important source of information. For a cash flow where borrowed money is repaid, the NPW plot would appear as in Figure 7–3. The borrowed money is received early in the time period with a later repayment of an equal

FIGURE 7-2 Replot of NPW versus interest rate *i* over a larger range of values.

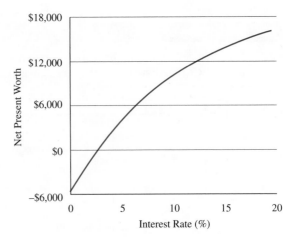

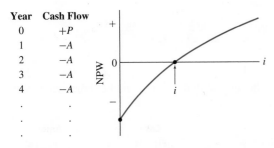

Year	Cash Flow
0	+P
1	−A
2	−A
3	−A
4	−A
.	.
.	.
.	.

FIGURE 7.3 Typical NPW plot for borrowed money.

Year	Cash Flow
0	−P
1	+Benefit A
2	+A
3	+A
4	+A
.	.
.	.
.	.

FIGURE 7.4 NPW plot for a typical investment.

sum, plus payment of interest on the borrowed money. In all cases in which interest is charged and the amount borrowed is fully repaid, the NPW at 0% will be negative.

For a cash flow representing an investment followed by benefits from the investment, the plot of NPW versus *i* (we will call it an **NPW plot** for convenience) would have the form of Figure 7–4. As the interest rate increases, future benefits are discounted more heavily and the NPW decreases.

Thus, interest is a charge for the use of someone else's money or a receipt for letting others use our money. The interest rate is almost always positive, but negative interest rates do occur. A loan with a forgiveness provision (not all principal is repaid) can have a negative rate. Some investments perform poorly and have negative rates.

EXAMPLE 7–5

A new corporate bond with a coupon rate of 8% was initially sold by a stockbroker to an investor for $1000. The issuing corporation promised to pay the bondholder $40 interest on the $1000 face value of the bond every 6 months, and to repay the $1000 at the end of 10 years. After one year the bond was sold by the original buyer for $950.

(a) What rate of return did the original buyer receive on his investment?

(b) What rate of return can the new buyer (paying $950) expect to receive if he keeps the bond for its remaining 9-year life?

TABLE SOLUTION TO PART a

The original bondholder sold the bond for less ($950) than the purchase price ($1000). So the semiannual rate of return is less than the 4% semiannual interest rate on the bond.

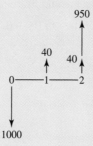

Since $40 is received each 6 months, we will solve the problem using a 6-month interest period.

$$PW = -1000 + 40(P/A, i, 2) + 950(P/F, i, 2)$$

$$PW_{1\frac{1}{2}\%} = -1000 + 40(1.956) + 950(0.9707) = \$0.41$$

The interest rate per 6 months, $IRR_{6\,mon}$, is very close to $1\frac{1}{2}\%$. This means the nominal (annual) interest rate is $2 \times 1.5\% = 3\%$. The effective (annual) interest rate or IRR is $(1 + 0.015)^2 - 1 = 3.02\%$.

TABLE SOLUTION TO PART b

The new buyer will redeem the bond for more ($1000) than the purchase price ($950). So the semiannual rate of return is more than the 4% semiannual interest rate on the bond.

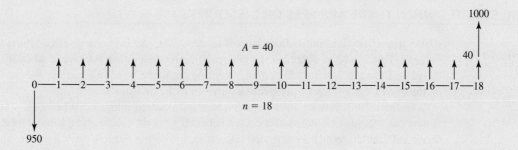

Given the same $40 semiannual interest payments, for 6-month interest periods we write

$$PW = -950 + 40(P/A, i, 18) + 1000(P/F, i, 18)$$

$$PW_{5\%} = -950 + 40(11.690) + 1000(0.4155) = -66.90$$

Try a lower interest rate, say, $i = 4\%$:

$$PW_{4\%} = -950 + 40(12.659) + 1000(0.4936) = 49.96$$

The value of the 6-month rate i, is between 4 and 5%. By interpolation,

$$i = 4\% + (1\%) \left(\frac{49.96 - 0}{49.96 - (-66.90)} \right) = 4.43\%$$

The nominal interest rate is $2 \times 4.43\% = 8.86\%$. The effective interest rate or IRR is $(1 + 0.0443)^2 - 1 = 9.05\%$.

SPREADSHEET SOLUTION

Entering the information from the cash flow diagrams into the spreadsheet produces more precise answers.

	A	B	C	D	E	F	G	H
1	Problem	i	n	PMT	PV	FV	Solve for	Answer
2	Exp. 7-5a		2	40	-1000	950	$i_{6\,mon}$	1.519%
3		3.04%	2				effective	3.061%
4	Exp. 7-5b		18	40	-950	1000	$i_{6\,mon}$	4.408%
5		8.82%	2				effective	9.011%
6		=C5*H4						=EFFECT(B5,C5)

In Example 7–5a the $50 change lowers the interest rate by 1% for 1 year and in 7–5b it raises it by 0.4% for 9 years.

INTEREST RATES WHEN THERE ARE FEES OR DISCOUNTS

Often when firms and individuals borrow money, there are fees charged in addition to the interest. This can be as simple as the underwriting fee that a firm is charged when it sells a bond. In Example 7–6 we add that underwriting fee to the bond in Example 7–5, and look at the bond from the firm's perspective rather than the investor's.

Example 7–7 shows how a cash discount is "unstated interest." When one is buying a new car, this may be stated as a choice between a "cash" rebate and a low-interest loan. Fees and cash discounts are unstated interest that raise the interest rate on a loan, and the true cost of the loan is its internal rate of return. When there is a down payment, as in

Example 7–7, the fee or discount is larger relative to the amount borrowed and the interest rate being paid increases.

Example 7–7 also uses incremental analysis to compare two different ways to borrow money. Later in this chapter, we will show that this is a requirement in applying rate of return analysis. Chapter 8 will show different ways to apply this for multiple alternatives.

EXAMPLE 7–6   (Example 7–5 Revisited)

The corporate bond in Example 7–5 was part of a much larger offering that the firm arranged with the underwriter. Each of the bonds had a face value of $1000 and a life of 10 years. Since $40 or 4% of the face value was paid in interest every 6 months, the bond had a nominal or coupon interest rate of 8% per year. If the firm paid the underwriter a 1% fee to sell the bond, what is the effective annual interest rate that the firm is paying on the bond?

TABLE SOLUTION

From the firm's perspective, it receives $1000 minus the fee at time 0, then it pays interest every 6 months for 10 years, and then it pays $1000 to redeem the bond. The 1% fee reduces what the firm receives when the bond is sold to $990. The interest payments are $40 every 6 months. This is easiest to model using twenty 6-month periods.

$$PW_i = 990 - 40(P/A, i, 20) - 1000(P/F, i, 20)$$

Since the nominal interest rate is 4% every 6 months, we know that the fee will raise this some. So let us use the next higher table of 4.5%.

$$PW_{4.5\%} = 990 - 40(P/A, 4.5\%, 20) - 1000(P/F, 4.5\%, 20)$$
$$= 990 - 40(13.008) - 1000(0.4146) = \$55.08$$

We know that the PW of the interest and final bond payoff is $1000 at 4%.

$$PW_{4\%} = 990 - 1000 = -10$$

Now we interpolate to find the interest rate for each 6-month period.

$$i = 4\% + (4.5\% - 4\%) \times 10/(10 + 55.08) = 4.077\%$$

The effective annual rate is

$$i_a = 1.04077^2 - 1 = 0.0832 = 8.32\%$$

SPREADSHEET SOLUTION

As discussed in the table solution, the firm receives 1% less than the bond's face value at time 0, and 4% of the face value is paid as interest every 6 months. Thus, the cash flows begin with a positive $990, and the interest payments and final payment of the face value are negative cash flows.

	A	B	C	D	E	F	G	H
1	Problem	i	n	PMT	PV	FV	Solve for	Answer
2	Exp. 7-6		20	-40	990	-1000	$i_{6\,mon}$	4.074%
3		8.15%	2				effective	8.314%
4		=C3*H2						=EFFECT(B3,C3)

EXAMPLE 7–7

A manufacturing firm may decide to buy an adjacent property so that it can expand its ware-house. If financed through the seller, the property's price is $300,000, with 20% down and the balance due in five annual payments at 12%. The seller will accept 10% less if cash is paid. The firm does not have $270,000 in cash, but it can borrow this amount from a bank. What is the rate of return or IRR for the loan offered by the seller?

TABLE SOLUTION

One choice is to pay $270,000 in cash. The other choice is to pay $60,000 down and five annual payments computed at 12% on a principal of $240,000. The annual payments equal

$$240,000(A/P, 12\%, 5) = 240,000(0.2774) = \$66,576$$

The two ways of borrowing the money and the *incremental* difference between them can be summarized in a cash flow table. The incremental difference is the result of subtracting the second set of cash flows from the first.

Year	Pay Cash	Borrow from Property Owner	Incremental Difference
0	−$270,000	−$60,000	−$210,000
1		−66,576	66,576
2		−66,576	66,576
3		−66,576	66,576
4		−66,576	66,576
5		−66,576	66,576

This choice between (1) −$270,000 now and (2) −$60,000 now and −$66,576 annually for 5 years can be stated by setting them equal in PW terms, as follows:

$$-270,000 = -60,000 - 66,576(P/A, \text{IRR}, 5)$$

The true amount borrowed is $210,000, which is the difference between the $60,000 down payment and the $270,000 cash price. This is the incremental amount paid or invested by the firm instead of borrowing from the seller. Collecting the terms, the PW equation is the same as the PW equation for the incremental difference.

$$0 = -210{,}000 + 66{,}576(P/A, \text{IRR}, 5)$$

In either case, the final equation for finding the IRR is

$$\text{PW} = 0 = -210{,}000 + 66{,}576(P/A, \text{IRR}, 5)$$

$$(P/A, IRR, 5) = 3.154$$

Looking in the tables we see that the IRR is between 15% and 18%. Interpolating gives us the following:

$$\text{IRR} = 15\% + (18\% - 15\%)[(3.352 - 3.154)/(3.352 - 3.127)] = 17.6\%$$

on a borrowed amount of $210,000. This is a high rate of interest, so that borrowing from a bank and paying cash to the property owner is better.

5-BUTTON SOLUTION

With a 20% down payment, the loan payment is on a remaining principal of $240,000. As discussed in the table solution, the firm is really borrowing the $210,000 difference between the down payment and the available cash price. The cash discount is unstated interest rate that raises the true rate on the loan to 17.62%.

	A	B	C	D	E	F	G	H
1	Problem	i	n	PMT	PV	FV	Solve for	Answer
2	Payment	12%	5		240,000	0	PMT	-$66,578
3	Rate		5	-66,578	210,000	0	i	17.62%

LOANS AND INVESTMENTS ARE EVERYWHERE

Examples 7–3 to 7–7 were about borrowing money through bonds and loans, but many applications for the rate of return are stated in other ways. Example 7–8 is a common problem on university campuses—buying parking permits for an academic year or a term at a time.

Buying a year's parking permit is investing more money now to avoid paying for another shorter permit later. Choosing to buy a shorter permit is a loan, where the money saved by not buying the annual permit is borrowed to be repaid with the cost of the second semester permit.

EXAMPLE 7–8

An engineering student is deciding whether to buy two 1-semester parking permits or an annual permit. The annual parking permit costs $180 due August 15th; the semester permits are $130 due August 15th and January 15th. What is the rate of return for buying the annual permit?

TABLE SOLUTION

Before we solve this mathematically, let us describe it in words. We are equating the $50 cost difference now between the two permits with the $130 cost to buy another semester permit in 5 months. Since the $130 is 2.6 times the $50, it is clear that we will get a high interest rate.

This is most easily solved by using monthly periods, and the payment for the second semester is 5 months later. The cash flow table adds a column for the incremental difference to the information in the cash flow diagram.

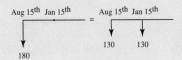

Time	Annual Pass	Two Semester Passes	Incremental Difference
Aug. 15th	−$180	−$ 130	−$ 50
Jan. 15th		−130	130

Setting the two PWs equal to each other we have

$$-\$180 = -\$130[1 + (P/F, i_{mon}, 5)]$$

$$(P/F, i_{mon}, 5) = -\$50/-\$130 = 5/13 = 0.3846$$

Rather than interpolating, we can use the formula for the P/F factor.

$$1/(1 + i_{mon})^5 = 0.3846$$

$$(1 + i_{mon})^5 = 2.600$$

$$1 + i_{mon} = 1.2106$$

$$i_{mon} = 21.06\%, \text{ which is an extremely high rate per month}$$

On an annual basis, the effective interest rate is $(1.2106^{12} - 1) = 891\%$. Unless the student is planning to graduate in January, it is clearly better to buy the permit a year at a time.

5-BUTTON SOLUTION

The first step as detailed in the table solution is to identify the incremental cash flows between the two choices. Paying an extra $50 in August saves $130 in January. The resulting interest rate of 21.06% per month is extremely high!

Compounding the monthly interest rate for 12 periods with a PV $= -1$ leads to an FV that equals 1 + effective interest rate. Subtracting 1 from the FV value of 9.907 shows the annual rate is 890.7%. Unless the student is planning to graduate in January or sell the car, at this university permits should only be purchased by the year.

	A	B	C	D	E	F	G	H
1	Problem	i	n	PMT	PV	FV	Solve for	Answer
2	Exp. 7-8		5	0	-50	130	i_{mon}	21.06%
3		21.06%	12	0	-1		FV	9.907
4							effective annual rate	890.7%

Examples 7–9 and 7–10 reference two common situations faced by firms and by individuals—buying insurance with a choice of payment plans and deciding whether to buy or lease a vehicle or equipment. Like the parking permit case, these situations can be described as investing more now to save money later, or borrowing money now to be paid later. Lease, insurance, and tuition payments are annuity cash flows that have beginning-of-period payments (see also Example 6–14).

EXAMPLE 7–9

An engineering firm can pay for its liability insurance on an annual or quarterly basis. If paid quarterly, the insurance costs $10,000. If paid annually, the insurance costs $35,000. What is the rate of return for paying on annual basis?

TABLE SOLUTION

This is most easily solved by using quarterly periods. As shown in the cash flow diagram and table, insurance must be paid at the start of each quarter. The cash flow table adds a column for the incremental difference to the information in the cash flow diagram.

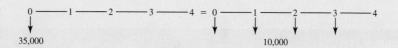

Quarter	Annual Payment	Quarterly Payments	Incremental Difference
0	−$35,000	−$10,000	−$25,000
1		−10,000	10,000
2		−10,000	10,000
3		−10,000	10,000

Setting the two PWs equal to each other we have

$$-\$35{,}000 = -\$10{,}000(1 + (P/A, i_{qtr}, 3))$$

$$(P/A, i_{qtr}, 3) = -\$25{,}000/ - \$10{,}000 = 2.5$$

$$(P/A, 9\%, 3) = 2.531$$

$$(P/A, 10\%, 3) = 2.487$$

$$i_{qtr} = 0.09 + (0.10 - 0.09)(2.531 - 2.5)/(2.531 - 2.487)$$

$$= 0.09 + 0.01(0.7045) = 9.705\%$$

On an annual basis the effective interest rate is $(1.09705^4 - 1) = 44.8\%$.

Unless the firm is planning to go out of business, it is clearly better to buy the insurance a year at a time.

5-BUTTON SOLUTION

The incremental cash flows were identified as −$25,000 at time 0 and $10,000 at the end of the first three quarters. The resulting interest rate is shown in the first calculation of the 5-BUTTON SOLUTION. It is 9.70% for 3 months, which is a high rate.

The second calculation shows four quarterly payments and an initial investment of −$35,000. These annuity due payments are identified as beginning-of-period cash flows by using BEGIN on a financial calculator or Type=1 in a spreadsheet.

Compounding the quarterly interest rate for 4 periods with a PV = −1 leads to a FV that equals 1 + effective interest rate. Subtracting 1 from the FV value of 1.448 shows the annual rate is 44.8%, which means the firm should buy its insurance a year at a time.

	A	B	C	D	E	F	G	H
1	Problem	i	n	*PMT*	*PV*	*FV*	Solve for	Answer
2	Exp. 7-9		3	10,000	-25,000	0	i_{qtr}	9.70%
3	using Type=1 or Begin		4	10,000	-35,000	0	i_{qtr}	9.70%
4		9.70%	4	0	-1		FV	1.448
5							effective annual rate	44.8%

EXAMPLE 7–10

Mountain Environmental Consulting may buy some field equipment for $40,000 or lease it for $2500 per month. In either case, the equipment will be replaced in 2 years. The salvage value of the equipment after 2 years would be $6000. What is the IRR or cost of the lease?

TABLE SOLUTION

The first step is to summarize the cash flows in a table. We must use monthly periods and remember that lease fees are paid at the start of the period.

Month	Lease	Buy	Buy – Lease
0	−$2500	−$40,000	−$37,500
1–23	−2500		2,500
24	0	6,000	6,000

The easiest way to analyze this example is to simply set the present worths of buying and leasing equal.

$$-\$2500 - \$2500(P/A, i_{mon}, 23) = -\$40,000 + \$6000(P/F, i_{mon}, 24)$$
$$0 = \$37,500 - \$2500(P/A, i_{mon}, 23) - \$6000(P/F, i_{mon}, 24)$$

The cash flow table was arranged so that the alternative with the lower cost now (leasing) was listed first. This ensures that the column of incremental cash flows (buy − lease) is an investment—the negative cash flows come first. An additional $37,500 must be paid or invested at time 0 to buy rather than lease.

The order of the two alternatives can be reversed. If Mountain Environmental leases the equipment, it avoids the expenditure of $37,500 at time 0. It incurs a cost of $2500 at the ends of Months 1 through 23, and it gives up the salvage value of $6000 at the end of Month 24. This cash flow pattern (positive at time 0 and negative in later years) occurs because leasing is a way to borrow money.

The column of incremental difference describes the decision to buy rather than lease, which invests $37,500 more now to avoid the later lease payments.

This equation is easier to solve if $n = 24$ for both factors. There is a total cash flow of −$6000 at the end of Month 24. This cash flow can be split into −$2500 and −$3500. Then both the uniform and single payment periods continue to the end of Month 24.

$$0 = \$37,500 - \$2500(P/A, i_{mon}, 24) - \$3500(P/F, i_{mon}, 24)$$
$$PW_{0\%} = \$37,500 - \$2500(24) - \$3500(1) = -\$26,000$$
$$PW_{5\%} = \$37,500 - \$2500(13.799) - \$3500(0.3101) = \$1917$$

The answer is clearly between 0% and 5%, and closer to 5%. Try 4% next.

$$PW_{4\%} = \$37,500 - \$2500(15.247) - \$3500(0.3901) = -\$1983$$
$$i_{mon} = 4\% + (5\% - 4\%)(1983)/(1983 + 1917) = 4.51\%$$

The effective annual rate is

$$i_a = 1.0451^{12} - 1 = 0.6975 = 69.8\%$$

5-BUTTON SOLUTION

As noted above, using $n = 24$ makes things easier. The resulting interest rate is shown in the first calculation of the 5-BUTTON SOLUTION. It is 4.49% per month, which is a very high rate.

The second calculation identifies the payments as beginning-of-period cash flows by using BEGIN on a financial calculator or Type=1 in a spreadsheet. Then the initial investment is −$40,000 and the final salvage value is $6000.

Compounding the monthly interest rate for 12 periods with a PV = −1 leads to a FV that equals 1 + effective interest rate. Subtracting 1 from the FV value of 1.694 shows the annual rate is 69.4%, which means the firm should buy (not lease) the equipment.

	A	B	C	D	E	F	G	H
1	Problem	i	n	PMT	PV	FV	Solve for	Answer
2	Exp. 7-10		24	2,500	-37,500	3500	i_{qtr}	4.49%
3	using Type=1 or Begin		24	2,500	-40,000	6000	i_{qtr}	4.49%
4		4.49%	12	0	-1		FV	1.694
5							effective annual rate	69.4%

In this case leasing is an extremely expensive way to obtain the equipment.

INCREMENTAL ANALYSIS

When there are two alternatives, rate of return analysis is performed by computing the **incremental rate of return**—ΔIRR—on the difference between the alternatives. In Examples 7–7 through 7–10 this was done by setting the present worths of the two alternatives equal to each other. Chapter 8 will address incremental analyses with more than 2 alternatives.

Now we will calculate the increment more formally. Since we want to look at increments of investment, the cash flow for the difference between the alternatives is computed by taking the higher initial-cost alternative *minus* the lower initial-cost alternative (as in Example 7–10). Since the increment is an investment, a higher rate of return is better.

Two-Alternative Situation	Decision	Want Higher Rate
ΔIRR ≥ MARR	Choose the higher-cost alternative	Invest increment at ΔIRR
ΔIRR < MARR	Choose the lower-cost alternative	Invest increment elsewhere at MARR

Example 7–11 illustrates a *very* important point that was not part of our earlier examples of parking permits, insurance, and equipment to be bought or leased. In those cases, the decision to purchase the insurance, permit, or equipment had already been made. We only needed to decide the best way to obtain it. In Example 7–11, there is a *do-nothing* alternative that must be considered.

Because Example 7–11 includes a do-nothing alternative, the rates of return of each alternative have meaning. When rates are high, a common mistake is to select the alternative with the highest rate. Example 7–11 shows why this is incorrect. If both mutually exclusive alternatives are acceptable, it is the incremental rate that determines which one should be chosen. This will be applied to multiple alternatives in Chapter 8.

EXAMPLE 7–11

You may select one of two mutually exclusive alternatives. Doing nothing is allowed. (*Note*: Engineering economists often use the term "mutually exclusive alternatives" to emphasize that selecting one alternative precludes selecting any other.) The alternatives are as follows:

Year	Alt. 1	Alt. 2
0	−$1000	−$2000
1	+1500	+2800

Any money not invested here may be invested elsewhere at the MARR of 6%. If you can choose at most one alternative one time, using the internal rate of return (IRR) analysis method, which one would you select?

SOLUTION

For Alternative 1, if $1000 increases to $1500 in a year, then the rate of return must be 50% (= 500/1000). For Alternative 2, if $2000 increases to $2800 in a year, then the rate of return must be 40% (= 800/2000). Both rates are *very* attractive, so both alternatives are clearly better than doing nothing and investing elsewhere at 6%. To decide which is better—incremental analysis is needed.

Year	Alt. 1	Alt. 2	Alt. 2 − Alt. 1
0	−$1000	−$2000	−$1000
1	1500	2800	1300

One can see that if $1000 increases to $1300 in one year, the interest rate must be 30%, which is far higher than the 6% MARR. The additional $1000 investment to obtain Alt. 2 is superior to investing the $1000 elsewhere at 6%. The interest rate can also be easily calculated using a future worth equation.

$$FW = 0 = -1000(1 + i) + 1300$$

$$1 + i = 1300/1000$$

$$i = 30\%$$

To understand an intuitively appealing—but *incorrect*—approach to Example 7–11, consider the rate of return for each alternative. The interest rates for Alternatives 1 and 2

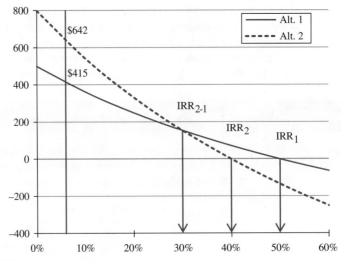

FIGURE 7–6 PW vs. i for two alternatives.

were 50% and 40%. The higher rate of return of Alt. 1 is attractive, but Alt. 1 is not the correct solution. This analysis only compares each alternative with doing nothing rather than with the other alternative.

Figure 7–6 shows why comparing the alternatives with nothing is not enough. Those rates of 50% and 40% are very attractive, but it is not correct to choose the alternative with the highest rate. It is also not correct to choose the largest project with a good rate of return. The alternatives must be directly compared with incremental analysis.

On the graph we can see that Alternative 1 has a rate of return of 50%, Alternative 2 has a rate of 40%, and the increment Alternative 2-1 has a rate of 30%. All of these rates are high compared with an MARR of 6%. The incremental rate determines that Alternative 2 is better.

These results are consistent with a present worth analysis. At 6% Alternative 2 clearly has a higher PW. The exact values are $415 for Alternative 1 and $642 for Alternative 2.

EXAMPLE 7–12

Consider the alternatives in Example 7–11 again, but this time compute the interest rate on the increment (Alt. 1 − Alt. 2) instead of (Alt. 2 − Alt. 1). How do you interpret the results?

SOLUTION

The signs of the incremental cash flows are reversed.

Year	Alt. 1 – Alt. 2
0	$1000
1	−1300

The present worth equation is:

$$PW = 0 = +1000 - 1300/(1 + i)$$

$$1 + i = \frac{1300}{1000} \implies i = 30\%$$

Once again the interest rate is found to be 30%. The critical question is, What does the 30% represent? The incremental cash flows do *not* represent an investment, but rather a loan. It is as if we borrowed $1000 in Year 0 ($1000 represents a receipt of money) and repaid it in Year 1 (−$1300 represents a disbursement). The 30% interest rate means this is the amount *we would pay* for the loan.

This is not a desirable borrowing scenario. The MARR on investments is 6%; it is reasonable to assume our maximum interest rate on borrowing would be less. Here the interest rate is 30%, which means the borrowing is undesirable. Alternative 1 is preferred to Alternative 2—the same conclusion reached in Example 7–11.

Example 7–12 illustrated that one can analyze either **increments of investment** or **increments of borrowing.** When looking at increments of investment, we accept the increment when the incremental rate of return equals or exceeds the minimum attractive rate of return ($\Delta IRR \geq MARR$). When looking at increments of borrowing, we accept the increment when the incremental interest rate is less than or equal to the *minimum* attractive rate of return ($\Delta IRR \leq MARR$). One way to avoid much of the possible confusion is to organize the solution to any problem so that one is examining increments of investment. This is illustrated in the next example.

EXAMPLE 7–13   (Examples 5–1 and 6–5 Revisited)

A firm is considering which of two devices to install to reduce costs. Both devices have useful lives of 5 years and no salvage value. Device *A* costs $10,000 and can be expected to result in $3000 savings annually. Device *B* costs $13,500 and will provide cost savings of $3000 the first year but will increase $500 annually, making the second-year savings $3500, the third-year savings $4000, and so forth. For a 7% MARR, which device should the firm purchase?

This problem has been solved by present worth analysis (Example 5–1) and annual cost analysis (Example 6–5). This time we will use rate of return analysis, which must be done on the incremental investment.

Year	Device A	Device B	Difference Between Alternatives: Device B − Device A
0	−$10,000	−$13,500	−$3500
1	3000	3000	0
2	3000	3500	500
3	3000	4000	1000
4	3000	4500	1500
5	3000	5000	2000

For the difference between the alternatives, write a single equation with i as the only unknown.

$$PW(i) = 0 = -3500 + 500(P/G, i, 5)$$

$$(P/G, i, 5) = 7; \text{ so } i \text{ is between } 9\%, (P/G, 9\%, 5) = 7.111$$

$$\text{and } 10\%, (P/G, 10\%, 5) = 6.862$$

$$i = 9\% + (10\% - 9\%)(7.111 - 7)/(7.111 - 6.862) = 9.45\%$$

The 9.45% IRR is greater than the 7% MARR; therefore, the increment is desirable. Reject Device A and choose Device B.

	A	B	C	D	E	F	G
1	$3,500	Incremental initial cost for B vs. A			Year	Cash flow	
2	0	Initial incremental savings for B vs. A			0	-$3,500	
3	$500	Gradient in incremental savings for B vs. A			1	0	=A2
4	5	Horizon			2	500	=F3+$A$3
5					3	1,000	
6					4	1,500	
7					5	2,000	
8					IRR	9.44%	=IRR(F2:F7)

Example 7–15 will show how GOAL SEEK can be used to solve for the interest rate that makes two values, such as NPV_A and NPV_B, equal.

ANALYSIS PERIOD

In discussing present worth analysis and annual cash flow analysis, an important consideration is the analysis period. This is also true in rate of return analysis. The solution method for two alternatives is to examine the differences between the alternatives. Clearly, the examination must cover the selected analysis period. For now, we can only suggest that the assumptions made should reflect one's perception of the future as accurately as possible.

In Examples 7–14 and 7–15 the analysis period is a common multiple of the alternative service lives and identical replacement is assumed. Example 7–14 explicitly includes a second Machine X that is identical to the first. Example 7–15 implicitly assumes identical repetitions until the lives match. The two approaches are shown to demonstrate their consistency.

EXAMPLE 7–14

Two machines are being considered for purchase. If the MARR is 10%, which machine should be bought? Use an IRR analysis comparison.

All $ values in 1000s	Machine X	Machine Y
Initial cost	$200	$700
Uniform annual benefit	95	120
End-of-useful-life salvage value	50	150
Useful life, in years	6	12

SOLUTION

The solution is based on a 12-year analysis period and a replacement machine X that is identical to the present Machine X. The cash flow for the differences between the alternatives is as follows:

Year	Machine X	Machine Y	Difference Between Alternatives (Alts): Machine Y − Machine X
0	−$200	−$700	−$500
1	95	120	25
2	95	120	25
3	95	120	25
4	95	120	25
5	95	120	25
6	95 50 −200	120	25 150
7	95	120	25
8	95	120	25
9	95	120	25
10	95	120	25
11	95	120	25
12	95 50	120 150	25 100

$$PW = -500 + 25(P/A, i, 12) + 150(P/F, i, 6) + 100(P/F, i, 12)$$

The cash flow sum over years 1 to 12 is $550, which is only a little greater than the $500 additional cost. This indicates that the rate of return is quite low. Try $i = 1\%$.

$$PW_{1\%} = -500 + 25(11.255) + 150(0.942) + 100(0.887) = 11$$

The interest rate is too low. Try $i = 1\frac{1}{2}\%$:

$$PW_{1.5\%} = -500 + 25(10.908) + 150(0.914) + 100(0.836) = -6$$

The internal rate of return on the $Y - X$ increment, IRR_{Y-X}, is about 1.3%, far below the 10% minimum attractive rate of return. The additional investment to obtain Machine Y yields an unsatisfactory rate of return, therefore X is the preferred alternative.

Many problems compare mutually exclusive alternatives for a continuing requirement where the horizon is longer than the lives of the current alternatives. In Example 6–8 pumps with lives of 9 and 12 years were compared, and the pump with the lower EUAC was recommended. The discussion before the example noted that this common approach assumes identical cost repetition for a horizon equal to the least common multiple of the lives.

To analyze that problem for an incremental rate of return, a horizon of 36 years could be used with repetitions like Machine X in Example 7–14. Example 7–15 in the next section illustrates another approach.

GOAL SEEK

Examples throughout this chapter have shown that the spreadsheet functions are particularly useful in calculating internal rates of returns (IRRs). If a cash flow diagram can be reduced to at most one P, one A, and/or one F, then the RATE *annuity function* can be used. Otherwise the IRR *block function* is used with a cash flow in each period.

Example 7–15 explains the spreadsheet feature, GOAL SEEK. In general, this is used to vary one cell in a spreadsheet in order to achieve a goal for another cell. The goal may be to maximize or minimize the cell's value or to make it equal to a specified value. In this case, the goal will be to make the difference between two EUACs equal zero.

Example 7–15 shows how to calculate incremental rates of return when alternatives have different lives, and it is appropriate to assume that alternatives repeat with the same costs within the indefinite continuing requirement.

EXAMPLE 7–15   (Example 6–7 Revisited)

Compare two pumps that are being used for a continuing indefinite requirement. What is the incremental rate of return for buying Pump A, which is longer lasting and more expensive?

	Pump A	Pump B
Initial cost	$7000	$5000
End-of-useful-life salvage value	1500	1000
Useful life, in years	12	6

SPREADSHEET SOLUTION

The incremental rate of return is the interest rate that will make the EUACs for each pump equal. In Example 6–7 an interest rate of 7% was used, and Pump A had the lower EUAC at $797 per year (Pump B was $909). That incremental rate is what is earned on cash flows that start with an extra $2000 at time 0.

This rate of return cannot be solved for using the RATE function, since the alternatives have different lives. Instead an interest rate is specified for one alternative and then also used for the other alternative. Annual equivalent payments are calculated for each alternative, and then GOAL SEEK is used to vary the interest rate. This is much easier than building a cash flow table for the 12 years it takes for Pumps A and B to match when they end.

GOAL SEEK is found under the DATA tab in Excel by selecting WHAT-IF ANALYSIS. The 5 button calculations shown in rows 2 and 3 are the EUAC calculations for Example 6–7. Rows 5 and 6 show the result when GOAL SEEK is applied to make the two PMT values the same—the incremental rate of return is 13.32%. The difference of the EUACs is calculated in cell H7. Then GOAL SEEK changes the interest rate (B5) until the difference is 0.

Figure 7–7 graphs the EUAC for each pump for interest rates from 0% to +20%. It verifies this result. The graph in Figure 7–7 is a powerful tool for understanding the EUACs at different interest rates. The cross-over interest rate is the incremental rate of return. Graphs like this are a key tool in Chapter 8 to examine which alternatives and/or projects are preferred at different interest rates.

	A	B	C	D	E	F	G	H
1	Problem	*i*	*n*	*PMT*	*PV*	*FV*	Solve for	Answer
2	Pump A	7%	12		-7000	1500	PMT	$797
3	Pump B	7%	6		-5000	1000	PMT	$909
4								
5	Pump A	13.32%	12		-7000	1500	PMT	$1,143
6	Pump B	13.32%	6		-5000	1000	PMT	$1,143
7		=B5					PMT difference	$0
8								
9								
10								
11								
12								
13								
14								

Goal Seek dialog box:
Set cell: $H$7
To value: 0
By changing cell: $B$5
OK   Cancel

FIGURE 7–7 Graphing present worth versus *i*.

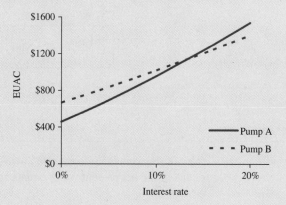

If Pump B's modified life of 9 years from Example 6–8 is used, then Pump B has a lower average cost and the EUAC at 7% falls to $684. The incremental rate of return for the difference between the two pumps falls to a very surprising −1.13%. This result happens because the average cost per year for Pump A (the more expensive and longer lived pump) is higher. This is true both for initial cost ($7000/12 for A is > $5000/9 for B) and for total costs ($7000 − $1500)/12 for A is > ($5000 − $1000)/9 for B).

XIRR

The tools of this chapter can solve for the rate of return—if the transaction dates are or can be approximated as uniformly spaced end-of-period cash flows. Using Excel's XIRR function, it is possible to solve for a rate of return with cash flows that occur on any set of dates. A daily interest rate is calculated, and then an effective annual interest rate is returned. This rate can be negative. There can be multiple sign changes in the cash flows, but there cannot be multiple roots (see Appendix 7A). This XIRR function is similar to the XNPV function that was used in Example 5–13.

The dates do not need to be in order, so it is possible to group cash flows that occur periodically together (for example, cash flows that occur every year on June 15, the first day of each month, or at the end of every other year). This is an easier way to build the cash flow table using copy and fill.

EXAMPLE 7–16

An engineer's employer offers a 401(k) that is invested in the stock market. The engineer deposited $1500 on September 13, 2016, and another $2200 on June 7, 2017. If the account is worth $3840 on August 7, 2017, what annual rate of return has the engineer's account earned?

SOLUTION

	A	B	C
1	Dates	Cash flows	
2	9/13/2016	-1500	
3	6/7/2017	-2200	
4	8/7/2017	3840	
5			
6	8.24%	=XIRR(B2:B4, A2:A4)	
7	= XIRR(Dates, Cash flows, *guess*)		

SUMMARY

Rate of return is the interest rate i at which the equivalent worth of benefits and costs are equal, or the net present worth equals zero.

There are a variety of ways of writing the cash flow equation in which the rate of return i may be the single unknown. Five of them are as follows:

$$\text{PW of benefits} - \text{PW of costs} = 0$$

$$\frac{\text{PW of benefits}}{\text{PW of costs}} = 1$$

$$\text{NPW} = 0$$

$$\text{EUAB} - \text{EUAC} = 0$$

$$\text{PW of costs} = \text{PW of benefits}$$

Rate of return analysis: Rate of return is the most frequently used measure in indus-try, as the resulting rate of return is readily understood. Also, the difficulties in selecting a suitable interest rate to use in present worth and annual cash flow analysis are avoided.

Criteria

Two Alternatives

Compute the incremental rate of return—ΔIRR—on the increment of *investment* between the alternatives. Then,

- if ΔIRR $\geq$ MARR, choose the higher-cost alternative, or,
- if ΔIRR $<$ MARR, choose the lower-cost alternative

When an increment of *borrowing* is examined, where ΔIRR is the incremental interest rate,

- if ΔIRR $\leq$ MARR, the increment is acceptable, or
- if ΔIRR $>$ MARR, the increment is not acceptable

Three or More Alternatives

See Chapter 8.

Looking Ahead

Rate of return is further described in Appendix 7A. This material concentrates on the dif-ficulties that occur with some cash flows series with multiple sign changes that may yield more than one root for the rate of return equation.

PROBLEMS

Rate of Return

7-1 (a) A mining firm makes annual deposits of $500,000
G into a reclamation fund for 30 years. If the firm must have $20 million when the mine is closed, what interest rate must the investment earn?

(b) The $20 million above is to be used to reclaim the negative impacts of the mine. List 6 to 10 potential environmental or community impacts that the fund might be used for.

7-2 An engineer invests $5800 at the end of every year
A for a 35-year career. If the engineer wants $1 million in savings at retirement, what interest rate must the investment earn?

7-3 An investment of $5000 in Biotech common stock proved to be very profitable. At the end of 5 years the stock was sold for $25,000. What was the rate of return on the investment?

7-4 The Diagonal Stamp Company, which sells used postage stamps to collectors, advertises that its average price has increased from $1 to $5 in the last 5 years. Thus, management states, investors who had purchased stamps from Diagonal 5 years ago would have received a 100% rate of return each year. What is the annual rate of return?

7-5 A woman went to the Beneficial Loan Company and borrowed $10,000. She must pay $323.53 at the end of each month for the next 60 months. What is the monthly interest rate she is paying? What effective annual interest rate is she paying?

7-6 Helen is buying a $12,375 car with a $3000 down payment, followed by 36 monthly payments of $325 each. The down payment is paid immediately, and the monthly payments are due at the end of each month. What nominal annual interest rate is Helen paying? What effective interest rate?

7-7 Your cousin Jeremy has asked you to bankroll his proposed business painting houses in the summer. He plans to operate the business for 5 years to pay his way through college. He needs $5000 to purchase an old pickup, some ladders, a paint sprayer, and some other equipment. He is promising to pay you $1500 at the end of each summer (for 5 years) in return. Calculate your annual rate of return. *Contributed by Paul R. McCright, University of South Florida*

7-8 Compute the rate of return for the following cash flow.

Year	Cash Flow
0	−$6400
1	0
2	1000
3	2000
4	3000
5	4000

7-9 Peter Minuit bought an island from the Manhattoes Indians in 1626 for $24 worth of glass beads and trinkets. The 1991 estimate of the value of land on this island was $12 billion.

(a) What rate of return would the Indians have received if they had retained title to the island rather than selling it for $24?

(b) What is your view of the ethics of this transaction? Do you believe deception was involved? In what circumstances do you view deception as ethical?

7-10 The student in Problem 5-6 wanted to buy a car costing $24,000 from a dealer offering 0% down and financing at 6% interest over 60 months. Her disposable income is $500 per month.

(a) What monthly interest rate can she afford? What effective annual rate is this?

(b) Insurance on this car will be $50 per month more than she had planned, which will leave her with only $450 per month for her car payment. Now what monthly interest rate can she afford? What effective annual rate is this?

7-11 You invest $2500 and in return receive two payments of $1800—one at the end of 2 years and the other at the end of 5 years. Calculate the resulting rate of return.

7-12 Compute the rate of return on the following investment.

Year	Cash Flow
0	−$10,000
1	0
2	3,000
3	3,000
4	3,000
5	3,000

7-13 Compute the rate of return for the following cash flow.

Year	Cash Flow
1–5	−$6,209
6–10	10,000

7-14 For the following diagram, compute the interest rate at which the costs are equivalent to the benefits.

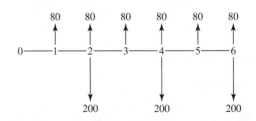

7-15 An investor has invested $250,000 in a new rental property. Her estimated annual costs are $6000 and annual revenues are $20,000. What rate of return per year will the investor make over a 30-year period ignoring the salvage value? If the property can be sold for $200,000 what is the rate of return? *Contributed by Hamed Kashani, Saeid Sadri, and Baabak Ashuri, Georgia Institute of Technology*

7-16 Installing an automated production system costing
A $278,000 is initially expected to save Zia Corporation $52,000 in expenses annually. If the system needs $5000 in operating and maintenance costs each year and has a salvage value of $25,000 at Year 10, what is the IRR of this system? If the company wants to earn at least 12% on all investments, should this system be purchased? *Contributed by Paul R. McCright, University of South Florida*

7-17 Compute the rate of return represented by the cash flow.

Year	Cash Flow
0	−$14,000
1	5,000
2	5,500
3	6,000
4	6,500

7-18 For the following diagram, compute the IRR.
A

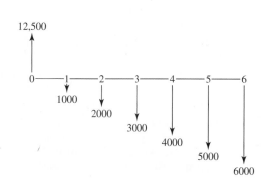

7-19 Switching to powder coating technology will reduce the emission of volatile organic carbons (VOCs) for a firm's production process. The initial cost is $200,000 with annual costs of $50,000 and savings of $90,000 in the first year. Savings are projected to increase by $3000 annually after Year 1. The salvage value 10 years from now is projected to be $30,000. What rate of return will the firm make on this investment?
Contributed by Hamed Kashani, Saeid Sadri, and Baabak Ashuri, Georgia Institute of Technology

7-20 To secure funding to convert their service vehicle
A fleet to hybrid technologies, MGL Industries is issu-
G ing *green bonds* to investors. You plan to buy $1 million of these bonds. This is how much you will raise by selling a vacation home that you acquired 25 years ago for $140,000.

(a) At what rate has the value of the vacation home increased?

(b) Write a short description of *green bonds*. How are they different from normal corporate, revenue, and debenture bonds? As an investor, would you buy *green bonds*? Explain.

7-21 For the following diagram, compute the rate of return on the $30,000 investment.

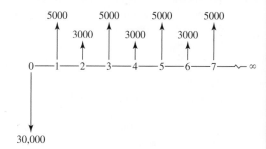

7-22 Consider the following cash flow:
A

Year	Cash Flow
0	−$4000
1	0
2	2000
3	1500
4	1000
5	500

Write one equation, with i as the only unknown, for the cash flow. Use no more than two single payment compound interest factors. Solve the equation for i.

7-23 You have just been elected into the "Society of Honorable Engineers." First-year dues are waived in honor of your election. Thus, your first-payment of $200 is due at the end of the year, and annual dues are expected to increase 3% annually. After 40 years of paying dues you become a life member and pay no more dues. Instead of paying annual dues, however, you can pay a one-time $2000 life membership fee.

(a) Show the equation for determining the rate of return for buying a life membership.

(b) What is the rate of return?

Contributed by D. P. Loucks, Cornell University

7-24 A bank proudly announces that it has changed its interest computation method to continuous compounding. Now $2000 left in the bank for 9 years will double to $4000. What nominal continuous interest rate is the bank paying? What effective interest rate is it paying?

Net Present Worth vs. *i* and Bonds

7-25 For Problem 7-7, graph the PW versus the interest rate for values from 0% to 50%. Is this the typical PW graph for an investment?

7-26 For Problem 7-8, graph the PW versus the interest rate for values from 0% to 50%. Is this the typical PW graph for an investment?

7-27 For Problem 7-18, graph the PW versus the interest rate for values from 0% to 50%. Is this the typical PW graph for an investment?

7-28 A well-known industrial firm has issued $1000 bonds with a 4% coupon interest rate paid semiannually. The bonds mature 20 years from now. From the financial pages of your newspaper you learn that the bonds may be purchased for $715 each ($710 for the bond plus a $5 sales commission). What nominal and effective annual rate of return would you receive if you purchased the bond now and held it to maturity 20 years from now?

7-29 A man buys a corporate bond from a bond brokerage house for $875. The bond has a face value of $1000 and a 4% coupon rate paid semiannually. If the bond will be paid off at the end of 12 years, what rate of return will the man receive?

7-30 An engineer bought a $1000 bond of an American airline for $875 just after an interest payment had been made. The bond paid a 6% coupon interest rate semiannually. What nominal rate of return did the engineer receive from the bond if he held it 13.5 years until its maturity?

7-31 Mildred can purchase a municipal bond with a par (face) value of $1000 that will mature in 10 years. The bond pays 6% interest compounded quarterly. If she can buy this bond for $1050, what rate of return will she earn? *Contributed by Paul R. McCright, University of South Florida*

7-32 Mike buys a corporate bond with a face value of $1000 for $900. The bond matures in 10 years and

pays a coupon interest rate of 6%. Interest is paid every quarter.

(a) Determine the effective rate of return if Mike holds the bond to maturity.

(b) What effective interest rate will Mike get if he keeps the bond for only 5 years and sells it for $950?

Contributed by Meenakshi Sundaram, Tennessee Tech University

7-33 An investor purchased a 5%, $1000 30-year bond for $850 with 22 years to maturity. The interest was payable quarterly. The bond was kept for only 9 years and sold for $950 immediately after the 36th interest payment was received. What nominal and effective rates of return per year were made on this investment? *Contributed by Meenakshi Sundaram, Tennessee Tech University*

7-34 A 9%, $10,000 bond that has interest payable semiannually sells for $8500. Determine what the maturity date should be so that the purchaser may enjoy a 12% nominal rate of return on this investment.
Contributed by Meenakshi Sundaram, Tennessee Tech University

7-35 ABC Corporation's recently issued bonds paying interest semiannually and maturing in 10 years. The face value of each bond is $1000, and 8% is the nominal interest rate.

(a) What is the effective interest rate an investor receives?

(b) If a 1.75% fee is deducted by the brokerage firm from the initial $1000, what is the effective annual interest rate paid by ABC Corporation?

7-36 ABC Corporation is issuing some *zero coupon bonds*, which pay no interest. At maturity in 20 years they pay a face value of $1000. The bonds are expected to sell for $311.80 when issued.

(a) What is the effective interest rate an investor receives?

(b) A 1% fee (based on the face value) is deducted by the brokerage firm from the initial revenue. What is the effective annual interest rate paid by ABC Corporation?

7-37 A zero coupon bond pays no interest—only its face value of $1000 at maturity. One such bond has a maturity of 18 years and an initial price of $130. What annual interest rate is earned if the bond is bought when issued and held to maturity?

Discounts and Fees

7-38 The cash price of a machine tool is $3500. The
Ⓐ dealer is willing to accept a $1200 down payment
and 24 end-of-month monthly payments of $110
each. At what effective interest rate are these terms
equivalent?

7-39 A local used car dealer calculates its "4%" financ-
Ⓔ ing as follows. If $3600 is borrowed to be
repaid over a 3-year period, the interest charge is
($3600)(0.04)(3 years) = $432. The $432 of interest
is deducted from the $3600 loan and the customer
has $3168 toward the cost of a car. The customer
must repay the loan with 1/36 of $3600, or $100,
monthly payments for 36 months.

(*a*) What effective annual interest rate is being
charged?

(*b*) What is your view of the ethics of this loan calculation? Do you believe deception is involved?
In what circumstances do you view deception as
ethical?

7-40 Jan purchased 100 shares of Peach Computer stock
Ⓐ for $18 per share, plus a $45 brokerage commission.
Every 6 months she received a dividend from Peach
of 50 cents per share. At the end of 2 years, just after
receiving the fourth dividend, she sold the stock for
$23 per share and paid a $58 brokerage commission
from the proceeds. What annual rate of return did she
receive on her investment?

7-41 A used car dealer advertises financing at 0% interest
over 3 years with monthly payments. You must pay
a processing fee of $500 at signing. The car you like
costs $9000.

(*a*) What is your effective annual interest rate?

(*b*) You believe that the dealer would accept $8200
if you paid cash. What effective annual interest
rate would you be paying, if you financed with
the dealer?

7-42 A new car dealer advertises financing at 0% inter-
Ⓐ est over 4 years with monthly payments or a $3000
rebate if you pay cash.

(*a*) The car you like costs $12,000. What effective
annual interest rate would you be paying if you
financed with the dealer?

(*b*) The car you like costs $18,000. What effective
annual interest rate would you be paying if you
financed with the dealer?

(*c*) The car you like costs $24,000. What effective
annual interest rate would you be paying if you
financed with the dealer?

7-43 A used car dealer advertises financing at 4% interest
over 3 years with monthly payments. You must pay
a processing fee of $500 at signing. The car you like
costs $9000.

(*a*) What is your effective annual interest rate?

(*b*) You believe that the dealer would accept $8200
if you paid cash. What effective annual interest
rate would be paying if you financed with the
dealer?

(*c*) Compare these answers with those for Problem
7-41. What can you say about what matters the
most for determining the effective interest rate?

7-44 Some laboratory equipment sells for $75,000. The
Ⓐ manufacturer offers financing at 8% with annual
payments for 4 years for up to $50,000 of the cost.
The salesman is willing to cut the price by 10% if
you pay cash. What is the interest rate you would
pay by financing?

7-45 A home mortgage with monthly payments for 30
years is available at 6% interest. The home you are
buying costs $120,000, and you have saved $12,000
to meet the requirement for a 10% down payment.
The lender charges "points" of 2% of the loan value
as a loan origination and processing fee. This fee is
added to the initial balance of the loan.

(*a*) What is your monthly payment?

(*b*) If you keep the mortgage until it is paid off in 30
years, what is your effective annual interest rate?

(*c*) If you move to a larger house in 10 years and
pay off the loan, what is your effective annual
interest rate?

(*d*) If you are transferred in 3 years and pay off the
loan, what is your effective annual interest rate?

Investments and Loans

7-46 An investor bought a one-acre lot on the outskirts
Ⓐ of a city for $9000 cash. Each year she paid $80 of
property taxes. At the end of 4 years, she sold the lot
for a net value of $15,000. What rate of return did
she receive on her investment?

7-47 A mine is for sale for $800,000. It is believed the
mine will produce a profit of $250,000 the first year,
but the profit will decline $25,000 a year after that,
eventually reaching zero, whereupon the mine will
be worthless. What rate of return would be earned
on the mine?

7-48 An apartment building in your neighborhood is for
Ⓐ sale for $140,000. The building has four units, which
are rented at $500 per month each. The tenants have

long-term leases that expire in 5 years. Maintenance and other expenses for care and upkeep are $8000 annually. A new university is being built in the vicinity and it is expected that the building could be sold for $160,000 after 5 years. What is the internal rate of return for this investment?

7-49 An engineering student is deciding whether to buy multiple term length parking permits or an annual permit. The annual parking permit costs $250 due August 15th, and the semester permits are $160 due August 15th and January 15th. What is the rate of return for buying the annual permit?

7-50 An engineering student is deciding whether to buy **G** multiple term length parking permits or an annual permit. Using the dates and costs for your university, find the rate of return for the incremental cost of the annual permit.

7-51 Fifteen families live in Willow Canyon. Although several water wells have been drilled, none has produced water. The residents take turns driving a water truck to a fill station in a nearby town. The water is hauled to a storage tank in Willow Canyon. Last year truck and water expenses totaled $6200. What rate of return would the Willow Canyon residents receive on a new water supply pipeline costing $100,000 that would replace the truck? The pipeline is considered to last

(a) Forever.

(b) 100 years.

(c) 50 years.

(d) Would you recommend that the pipeline be installed? Explain.

7-52 A new machine can be purchased today for **A** $300,000. The annual revenue from the machine is calculated to be $67,000, and the equipment will last 10 years. Expect the maintenance and operating costs to be $3000 a year and to increase $600 per year. The salvage value of the machine will be $20,000. What is the rate of return for this machine?

7-53 An insurance company is offering to sell an annuity for $20,000 cash. In return the firm will guarantee to pay the purchaser 20 annual end-of-year payments, with the first payment amounting to $1100. Subsequent payments will increase at a uniform 10% rate each year (second payment is $1210; third payment is $1331, etc.). What rate of return would someone who buys the annuity receive?

7-54 A popular magazine offers a lifetime subscription for **A** $1000. Such a subscription may be given as a gift to

an infant at birth (the parents can read it in those early years), or taken out by an individual for himself. Normally, the magazine costs $64.50 per year. Knowledgeable people say it probably will continue indefinitely at this $64.50 rate. What is the rate of return on a life subscription purchased for an infant?

7-55 A luxury car can be leased for $679 per month for 36 months. Terms are first month's lease payment, a $625 refundable security deposit, a consumer down payment of $3500, and an acquisition fee of $725 due at lease signing. Tax, license, title fees, and insurance extra. Option to purchase at lease end for $37,775 plus a fee of $350. Mileage charge of $0.20 per mile over 30,000 miles. Determine the interest rate (nominal and effective) for the lease.

(a) Use the MSRP of $64,025.

(b) You could buy the car for $58,000, if you arranged other financing.

7-56 An engineering firm can pay for its liability insur- **A** ance on an annual or quarterly basis. If paid quarterly, the insurance costs $18,000. If paid annually, the insurance costs $65,000. What are the quarterly rate of return and the nominal and effective interest rates for paying on an annual basis?

7-57 An engineering student must decide whether to pay for auto insurance on a monthly or an annual basis. If paid annually, the upfront annual cost is $2750. If paid monthly, the cost is $250 at the start of each month. What is the rate of return for buying the insurance on an annual basis expressed as an annual effective rate?

7-58 For your auto or home insurance, find out the cost of paying annually or on a shorter term. What is the rate of return for buying the insurance on an annual basis?

7-59 An investor bought 100 shares of Omega common stock for $14,000. He held the stock for 9 years. For the first 4 years he received annual end-of-year dividends of $800. For the next 4 years he received annual dividends of $400. He received no dividend for the ninth year. At the end of the ninth year he sold his stock for $6000. What rate of return did he receive on his investment?

7-60 One aspect of obtaining an engineering education is the prospect of improved future earnings in comparison to non-engineering graduates. Sharon Shay estimates that her engineering education has a $75,000 equivalent cost at graduation. She believes the benefits of her education will occur throughout 40 years of employment. She thinks that during the first 10 years out of college, her income will be higher than that of a non-engineering graduate by $20,000 per year. During the subsequent 10 years, she projects an annual income that is $30,000 per year higher. During the last 20 years of employment, she estimates an annual salary that is $50,000 above the level of the non-engineering graduate. If her estimates are correct, what rate of return will she receive as a result of her investment in an engineering education?

7-61 Upon graduation, every engineer must decide whether to go on to graduate school. Estimate the costs of going to the university full time to obtain a master of science degree. Then estimate the resulting costs and benefits. Combine the various consequences into a cash flow table and compute the rate of return. Nonfinancial benefits are probably relevant here too.

Incremental Analysis

7-62 If 7% is considered the minimum attractive rate of
(A) return, which alternative should be selected?

Year	A	B
0	−$20,000	−$28,000
1	8,000	11,000
2	8,000	11,000
3	8,000	11,000

7-63 If the MARR is 8%, which alternative should be selected?

Year	X	Y
0	−$5000	−$5000
1	−3000	2000
2	4000	2000
3	4000	2000
4	4000	2000

7-64 Alternatives A and B require investments of $10,310
(A) and $13,400, respectively. Their respective net annual cash inflows are $3,300 and $4,000. What is the rate of return for each alternative and for the incremental difference? If the interest rate is 10%, which alternative should be selected? *Contributed by Yasser Alhenawi, University of Evansville*

7-65 Two mutually exclusive alternatives are being considered. Both have lives of 10 years. Alternative *A* has a first cost of $10,000 and annual benefits of $4500. Alternative *B* costs $25,000 and has annual benefits of $8800.

If the minimum attractive rate of return is 6%, which alternative should be selected? Solve the problem by

(a) Present worth analysis
(b) Annual cash flow analysis
(c) Rate of return analysis

7-66 Two hazardous environment facilities are being
(A) evaluated, with the projected life of each facility being 10 years. The company uses a MARR of 15%. Using rate of return analysis, which alternative should be selected?

	Alt. *A*	Alt. *B*
First cost	$615,000	$300,000
O&M cost	10,000	25,000
Annual benefits	158,000	92,000
Salvage value	65,000	−5,000

7-67 The owner of a corner lot wants to find a use that will yield a desirable return on his investment. If the owner wants a minimum attractive rate of return on his investment of 15%, which of the two alternatives would you recommend? Both alternatives have a 20-year life and no salvage value.

	Build Fast Food Restaurant	Build Gas Station
First cost	$800,000	$1,200,000
Annual property taxes	30,000	50,000
Annual net income	150,000	200,000

7-68 A grocery distribution center is considering whether
(A) to invest in RFID or bar code technology to track its inventory within the warehouse and truck loading operations. The useful life of the RFID and bar code devices is projected to be 5 years with minimal or zero salvage value. The bar code investment cost is $100,000 and can be expected to save at least $50,000 in product theft and lost items annually. The RFID system is estimated to cost $200,000 and will save $30,000 the first year, with an increase of $15,000 annually after the first year. For a 5% MARR, should the manager invest in the RFID system or the bar code system? Analyze incrementally using rate of return. *Contributed by Oliver Hedgepeth, American Public University*

7-69 A state's department of transportation (DOT) is considering whether to buy or lease an RFID tracking

system for asphalt, concrete, and gravel trucks to be used in road paving. Purchasing the RFID system will cost $5000 per truck, with a salvage value of $1500 after the RFID system's useful life of 5 years. However, the DOT considering this purchase is also looking at leasing this same RFID system for an annual payment of $3500, which includes a full replacement warranty. Assuming that the MARR is 11% and on the basis of an internal rate of return analysis, which alternative would you advise the DOT to consider? Analyze incrementally using rate of return. The number of trucks used in a season varies from 5000 to 7500. Does this matter?
Contributed by Oliver Hedgepeth, American Public University

7-70 A contractor is considering whether to buy or lease (A) a new machine for her layout site work. Buying a new machine will cost $12,000 with a salvage value of $1200 after the machine's useful life of 8 years. On the other hand, leasing requires an annual lease payment of $3000, which occurs at the start of each year. The MARR is 15%. On the basis of an internal rate of return analysis, which alternative should the contractor be advised to accept?

7-71 A bulldozer can be purchased for $380,000 and used for 6 years, when its salvage value is 15% of the first cost. Alternatively, it can be leased for $60,000 a year. (Remember that lease payments occur at the start of the year.) The firm's interest rate is 12%.

(a) What is the interest rate for buying versus leasing? Which is the better choice?

(b) If the firm will receive $65,000 more each year than it spends on operating and maintenance costs, should the firm obtain the bulldozer? What is the rate of return for the bulldozer using the best financing plan?

7-72 A diesel generator for electrical power can be pur- (A) chased by a remote community for $480,000 and used for 10 years, when its salvage value is $50,000. Alternatively, it can be leased for $70,000 a year. (Remember that lease payments occur at the start of the year.) The community's interest rate is 8%.

(a) What is the interest rate for buying versus leasing? Which is the better choice?

(b) The community will spend $80,000 less each year for fuel and maintenance, than it currently spends on buying power. Should it obtain the generator? What is the rate of return for the generator using the best financing plan?

7-73 After 5 years of working for one employer, you accept a new job. Your retirement account total is $60,000. Half is "yours" and half is your employer's matching contributions. You now have two alternatives. (1) You may leave both contributions in the fund until retirement in 35 years, when you will receive its future value at 4% interest. (2) You may take out the total value of "your" contributions. You can do as you wish with the money you take out, but the other half will be lost as far as you are concerned. Which alternative is more attractive? Why?

7-74 Three mutually exclusive alternatives are being con- (A) sidered. All have a 10-year useful life. If the MARR is 25%, which alternative is preferred?

	A	B	C
Initial cost	$40,000	$50,000	$55,000
Uniform annual benefit	10,000	13,000	14,000

7-75 If the minimum attractive rate of return is 14%, which alternative should be selected?

Year	W	X	Y	Z
0	−$1000	−$500	−$1200	−$1500
1	350	165	420	500
2	350	165	420	500
3	350	165	420	500
4	350	165	420	500

Analysis Period

7-76 If the minimum attractive rate of return is 7%, which (A) alternative should be selected assuming identical replacement?

	A	B
First cost	$5000	$9200
Uniform annual benefit	1750	1850
Useful life, in years	4	8

7-77 Jean has decided it is time to buy a new battery for her car. Her choices are:

	Zappo	Kicko
First cost	$48	$78
Guarantee period, in months	12	24

Jean believes the batteries can be expected to last only for the guarantee period. She does not want to invest extra money in a battery unless she can expect a 40% rate of return. If she plans to keep her present car another 2 years, which battery should she buy?

7-78 Two investment opportunities are as follows:

Ⓐ

	A	B
First cost	$150	$100
Uniform annual benefit	25	22.25
End-of-useful-life salvage value	20	0
Useful life, in years	15	10

At the end of 10 years, Alt. *B* is not replaced. Thus, the comparison is 15 years of *A* versus 10 years of *B*. If the MARR is 10%, which alternative should be selected?

🖥️ Spreadsheets

7-79 The Southern Guru Copper Company operates a large mine in a South American country. A legislator in the National Assembly said in a speech that most of the capital for the mining operation was provided by loans from the World Bank; in fact, Southern Guru has only $1.5 million of its own money actually invested in the property. The cash flow for the mine is:

Year	Cash Flow
0	$-0.5M
1	0.9M
2	3.5M
3	3.9M
4	8.6M
5	4.3M
6	3.1M
7	1.2M

The legislator divided the $25.5 million total profit by the $1.5 million investment. This produced, he said, a 1700% rate of return on the investment. Southern Guru, claiming the actual rate of return is much lower, asks you to compute it.

7-80 A young engineer's starting salary is $52,000. The
Ⓐ engineer expects annual raises of 3%. The engineer will deposit 10% of the annual salary at the end of each year in a savings account that earns 4%. How much will the engineer have saved for starting a business after 15 years? We suggest that the spreadsheet include at least columns for the year, the year's salary, the year's deposit, and the year's cumulative savings.

7-81 A young engineer's starting salary is $75,000. The engineer expects annual raises of 2%. The engineer will deposit 15% of annual salary at the end of each year in an investment account that averages 6% interest. How much will the engineer have saved for retirement after 40 years?

7-82 A young engineer's starting salary is $55,000. The
Ⓐ engineer expects annual raises of 2%. The engineer will deposit a constant percentage of annual salary at the end of each year in a savings account that earns 5%. What percentage must be saved so that there will be $1 million in savings for retirement after 40 years? (*Hint*: Use GOAL SEEK: see last section of Chapter 8.)

7-83 Find the average starting engineer's salary for your discipline. Find and reference a source for the average annual raise you can expect. If you deposit 10% of your annual salary at the end of each year in a savings account that earns 4%, how much will you have saved for retirement after 40 years?

7-84 An engineer believes a firm is well managed with an exciting new technology. He purchased $4500 worth of its stock on March 14 and another $3500 on June 17. It is now November 15, and the stock is worth $8000. Use the XIRR function to find the IRR. If the stock is worth $7000, what is the IRR?

7-85 An investment has the following cash flows. Use the XIRR function to find the IRR.

Date	Cash Flow
12/11/2017	−$10,000
1/8/2018	50
7/2/2018	300
1/7/2019	400
7/1/2019	500
1/6/2020	10,700

7-86 A cold remedy's cash flows for one season's cycle are shown below. Use XIRR to find the IRR.

Date	Cash Flow (in $M)
6/1/2018	−$30
9/1/2018	12
11/1/2018	2
12/17/2018	6
1/7/2019	8
2/25/2019	5
4/1/2019	2

7-87 Use the XIRR function to find the IRR.

Date	Cash Flow
4/16/2018	−$5000
12/10/2018	1500
4/1/2019	1800
7/22/2019	2000
11/11/2019	1900
2/4/2020	1500
6/2/2020	1200

Minicases

7-88 Some lenders charge an up-front fee on a loan, which is subtracted from what the borrower receives. This is typically described as "points" (where one point equals 1% of the loan amount). The federal government requires that this be accounted for in the APR that discloses the loan's cost.

(*a*) A 5-year auto loan for $18,000 has monthly payments at a 9% nominal annual rate. If the borrower must pay a loan origination fee of 2 points, what is the true effective cost of the loan? What would the APR be?

(*b*) If the car is sold after 2 years and the loan is paid off, what is the effective interest rate and the APR?

(*c*) Graph the effective interest rate as the time to sell the car and pay off the loan varies from 1 to 5 years.

7-89 Some lenders charge an up-front fee on a loan, which is added to what the borrower owes. This is typically described as "points" (where one point equals 1% of the loan amount). The federal government requires that this be accounted for in the APR that discloses the loan's cost.

(*a*) A 30-year mortgage for $220,000 has monthly payments at a 6% nominal annual rate. If a borrower's loan origination fee is 3% (3 points) and it is added to the initial balance, what is the true effective cost of the loan? What would the APR be?

(*b*) If the house is sold after 6 years and the loan is paid off, what is the effective interest rate and the APR?

(*c*) Graph the effective interest rate as the time to sell the house and pay off the loan varies from 1 to 15 years.

APPENDIX 7A

Difficulties in Solving for an Interest Rate

After completing this chapter appendix, students should be able to:

- Describe why some projects' cash flows cannot be solved for a single positive interest rate.
- Identify the multiple roots if they exist.
- Evaluate whether the multiple roots are a problem for decision making.
- Use the *modified internal rate of return (MIRR)* methodology in multiple-root cases.

Example 7A–1 illustrates the situation.

EXAMPLE 7A–1	(Example 5–11 Revisited)

A piece of land may be purchased for $610,000 to be strip-mined for the underlying coal. Annual net income will be $200,000 for 10 years. At the end of 10 years, the surface of the land will be restored as required by a federal law on strip mining. The reclamation will cost $1.5 million more than the land's resale value after it is restored. Is this a desirable project, if the minimum attractive rate of return is 10%?

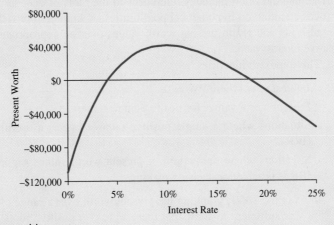

SOLUTION

In Example 5–11 the NPW was calculated to be $41,000. Since the NPW is positive, the project seems desirable. However, at interest rates below 4.07% or above 18.29% (see Figure 7A–1), the NPW is negative, indicating an undesirable project. This indicates that the project is undesirable at 4% and desirable at 10% and not at 20%. This does not make sense.

FIGURE 7A–1 Two positive roots.

The results warn us that NPW may not always be a proper criterion for judging whether or not an investment should be undertaken. In this example the disbursements ($610,000 + $1,500,000) exceed the benefits (10 × $200,000), which certainly does not portray a desirable investment. Thus Example 5–11 shows that NPW calculations in certain infrequent conditions can lead to unreliable results.

As outlined in this appendix, the problem is in the cash flows. It is *not* solved by relying on present worth as an economic measure over a confusing double root for the rate of return.

Example 7A–1 demonstrates that cash flows can have multiple solutions to the IRR equation (PW = 0). The example also demonstrates a common feature of such a problem— the cash flows fit neither the pattern of an investment nor the pattern of a loan. Both investments and loans have a single change in the sign of the cash flows over time. In contrast, Example 7A–1 has two sign changes in its series of cash flows.

In Example 7A–1 the open pit mine has a negative initial cash flow, which is followed by a series of positive cash flows—the pattern of an investment. However, the cash flows at the project's end follow the pattern of a loan—initial cash flows are positive and then the final cash flow for remediation is negative. Thus, Example 7A–1 is neither a loan nor an investment, but rather a combination. Thus it is impossible to know whether it is desirable to have a high interest rate for an investment or a low interest rate for a loan.

Two or more sign changes in the series of cash flows are the distinguishing feature of instances where multiple roots can but *may not* arise when solving for the rate of return for a set of cash flows.

Examples like the cash flow diagram with a first cost, uniform return, and final remediation expense have been thoroughly analyzed (Eschenbach, Baker, and Whittaker). Only problems with a very large final cost have double positive roots. More important, due to the uncertain remediation expenses being much larger than the initial costs, the sensitivity of the results means that the PW values seem also to be unreliable guides for decision making when there are two positive roots for the PW = 0 equation.

WHAT TO DO IF CASH FLOW DIAGRAM HAS TWO OR MORE SIGN CHANGES

Traditional coverage of multiple roots starts with Descartes' rule of signs for roots of a polynomial and often includes references to the tools of decades of research. This tends to focus attention on hypothetical possibilities. A simpler approach is to use the power of spreadsheets and graph present worth values over an appropriate range of negative and positive interest rates.

This approach identifies

- Interest rates where PW = 0.
- Present worth values for positive interest rates.
- Situations where a single positive interest rate is a useful internal rate of return (IRR).
- Situations where the pattern of present worth values suggest that neither PW nor IRR is reliable for decision making.

In Example 7A–1, graphing the present worth over a range of interest rates from 0% to 25% was sufficient to show that there were two positive roots of about 4% and 18%.

The values of present worth did not behave as investments or loans that this text has analyzed. As explained after Example 7A–1, this happened because the cash flows are neither an investment nor a loan, but a combination. The solution to Example 7A–1 noted that the total of the cash flows is −$110,000, which is not a sign of a good investment.[1]

In general, the range (−100%, 100%] includes the interest rates and present worth values that are useful for decision making. Note that present value calculations can only approach the lower limit of $i = -1$, since the denominator $1 + i$ equals 0 at that limit. For the suggested upper limit there do not seem to be occasions involving multiple roots where interest rates of over 100% are earned over an engineering project's life. Unrealistic cash flow examples with very high rates have been created and published.

As will be detailed in several examples, when multiple roots occur there is usually a negative root that can be ignored and a positive root that can be used. When this positive root is used as the IRR it implies the exact same decision as a PW measure that appears to be reliable. If there is only one positive root, then it is a valid and useful IRR.

If there is only one root and it is negative, then it is a valid IRR. One example of a valid negative IRR comes from a bad outcome, such as the IRR for a failed research and development project. Another example would be a privately sponsored student loan set up so that if a student volunteers for the Peace Corps, only half of the principal need be repaid.

If a project has a negative and a positive root for i, then, as stated for most projects only the positive root of i is used.

PROJECTS WITH MULTIPLE SIGN CHANGES

The most common case of two or more sign changes in cash flows is projects with a salvage cost which typically have two sign changes. This salvage cost can be large for environmental restoration at termination. Examples include pipelines, open-pit mines, and nuclear power plants. Example 7A–2 is representative.

EXAMPLE 7A–2

This project is representative of ones with a salvage cost. How many roots for the PW equation exist?

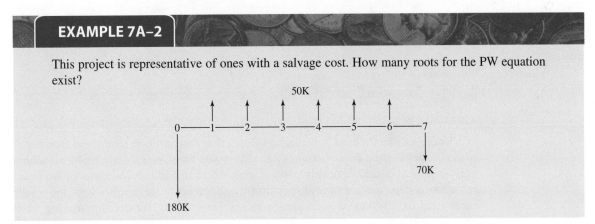

---

[1]All possible roots for combinations of *P*, *A*, and *F* are detailed in Eschenbach, Ted G., Elisha Baker, and John Whittaker, "Characterizing the Real Roots for *P*, *A*, and *F* with Applications to Environmental Remediation and Home Buying Problems," *The Engineering Economist*, Volume 52, Number 1, 2007, pp. 41–65.

SOLUTION

Figure 7A–2 shows the spreadsheet calculations and the graph of PW versus i. In this case, there is one positive root of 10.45%. The value can be used as an IRR. There is also a negative root of $i = -38.29\%$. This root is not useful. Few firms would accept a project with only a 10% rate of return.

	A	B	C	D	E	F	G
1	Year	Cash Flow		i	PW		
2	0	−180		−40%	−126.39	=$B$2+NPV(D2,$B$3:$B$9)	
3	1	50		−30%	219.99		
4	2	50		−20%	189.89		
5	3	50		−10%	114.49		
6	4	50		0%	50.00		
7	5	50		10%	1.84		
8	6	50		20%	−33.26		
9	7	−70		30%	−59.02		
10				40%	−78.24		
11	IRR	10.45%		50%	−92.88		
12	root	−38.29%					
13							
14							
15							
16							
17							
18							
19							
20							
21							
22							

FIGURE 7A–2 PW versus i for project with salvage cost.

Examples like the cash flow diagram with a first cost, uniform return, and final remediation expense have been thoroughly analyzed (see footnote 1). Only problems with a very large final cost have double positive roots. More importantly, because the uncertain remediation expenses are much larger than the initial costs, the sensitivity of the results means that the PW values seem also to be unreliable guides for decision making when there are two positive roots for the PW = 0 equation.

Many enhancement projects for existing mines and deposits have a pattern of two sign changes. Example 7A–3 describes an oil well in an existing field. The initial investment recovers more of the resource and speeds recovery of resources that would have been recovered eventually. The resources shifted for earlier recovery can lead to two sign changes.

In Example 7A–4, we consider staged construction, where three sign changes are common but a single root is also common.

Other examples with multiple sign changes and a single root can be found in comparisons of alternatives with unequal lives.[2]

EXAMPLE 7A–3

Adding an oil well to an existing field costs $4 million (4M). It will increase recovered oil by $3.5M, and it shifts $4.5M worth of production from Years 5, 6, and 7 to earlier years. Thus, the cash flows for Years 1 through 4 total $8M and Years 5 through 7 total −$4.5M. If the well is justified, one reason is that the oil is recovered sooner. How many roots for the PW equation exist? Is one useful as an IRR, and should the project be funded?

SOLUTION

The first step is to draw the cash flow diagram and count the number of sign changes. The following pattern is representative, although most wells have a longer life.

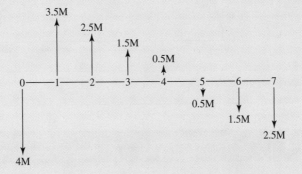

There are two sign changes, thus there may be multiple positive roots for the PW = 0 equation. The additional recovery corresponds to an investment, and the shifting of recovery to earlier years corresponds to a loan (positive cash flow now and negative later). Thus, the oil wells are neither an investment nor a loan; they are a combination of both.

Figure 7A–3 shows the spreadsheet calculations and the graph of PW versus i. In this case, there are positive roots at 4.73 and 37.20%. These roots are not useful. This project is a combination of an investment and a loan, so we don't even know whether we want a high rate or a low rate. If our interest rate is about 20%, then the project has a positive PW. However, small changes in the data can make for large changes in these results. At this stage no certain recommendation can be made.

It is useful to apply the modified internal rate of return described in the last section.

[2]Eschenbach, "Multiple Roots and the Subscription/Membership Problem," *The Engineering Economist* Volume 29, Number 3, Spring 1984, pp. 216–223.

	A	B	C	D	E	F	G
1	Year	Cash Flow		*i*	PW		
2	0	−4.00		0%	−0.50	=$B$2+NPV(D2,$B$3:$B$9)	
3	1	3.50		5%	0.02		
4	2	2.50		10%	0.28		
5	3	1.50		15%	0.37		
6	4	0.50		20%	0.36		
7	5	−0.50		25%	0.29		
8	6	−1.50		30%	0.19		
9	7	−2.50		35%	0.06		
10				40%	−0.08		
11	root	4.73%		45%	−0.22		
12	root	37.20%		50%	−0.36		

FIGURE 7A–3 PW versus *i* for oil well.

EXAMPLE 7A–4

A project has a first cost of $120,000. Net revenues begin at $30,000 in Year 1 and then increase by $2000 per year. In Year 5 the facility is expanded at a cost of $60,000 so that demand can continue to expand at $2000 per year. How many sign changes in the cash flows are there? How many roots for the PW equation are there? Is one a useful IRR, and what decision is recommended?

SOLUTION

The first step is to draw the cash flow diagram. Then counting the three sign changes is easy.

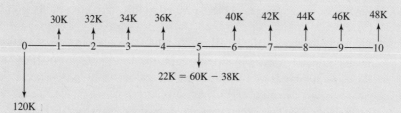

With three sign changes, there may be multiple roots for the PW = 0 equation.

Figure 7A–4 shows the spreadsheet calculations and the graph of PW versus *i*. In this case, there is one positive root of 21.69%. The value can be used as an IRR, and the project is attractive.

	A	B	C	D	E	F	G
1	Year	Cash Flow		*i*	PW		
2	0	−120		−80%	579,646,330	=$B$2+NPV(D2,$B$3:$B$12)	
3	1	30		−60%	735,723		
4	2	32		−40%	17,651		
5	3	34		−30%	1460		
6	4	36		0%	210		
7	5	−22		10%	73		
8	6	40		20%	7		
9	7	42		30%	−28		
10	8	44		40%	−48		
11	9	46		50%	−62		
12	10	48		60%	−71		
13				70%	−78		
14	IRR	21.69%		80%	−83		
15				90%	−87		
16							
17							
18							
19							
20							
21							
22							
23							
24							
25							
26							

FIGURE 7A-4 PW versus *i* for project with staged construction.

MODIFIED INTERNAL RATE OF RETURN (MIRR)

Two external rates of return can be used to ensure that the resulting equation is solvable for a unique rate of return—the MIRR. The MIRR is a measure of the attractiveness of the cash flows, but it is also a function of the two external rates of return.

The rates that are *external* to the project's cash flows are (1) the rate at which the organization normally invests and (2) the rate at which it normally borrows. These are external rates for investing, e_{inv}, and for financing, e_{fin}. Because profitable firms invest at higher rates than they borrow at, the rate for investing is generally higher than the rate for financing. Sometimes a single external rate is used for both, but this requires the questionable assumption that investing and financing happen at the same rate.

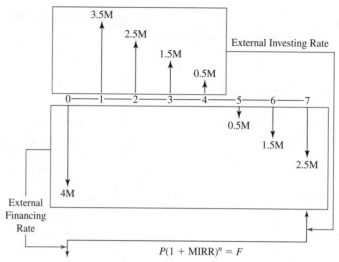

FIGURE 7A–5 MIRR for the oil well.

The approach is:

1. Combine cash flows in each period (t) into a single net receipt, R_t, or net expense, E_t.
2. Find the present worth of the expenses with the financing rate.
3. Find the future worth of the receipts with the investing rate.
4. Find the MIRR which makes the present and future worths equivalent.

The result is Equation 7A-1. This equation will have a unique root, since it has a single negative present worth and a single positive future worth. There is only one sign change in the resulting series.

$$(F/P, \text{MIRR}, n) \sum_t E_t (P/F, e_{\text{fin}}, t) = \sum_t R_t (F/P, e_{\text{inv}}, n - t) \qquad (7A\text{-}1)$$

There are other external rates of return, but the MIRR has historically been the most clearly defined. All of the external rates of return are affected by the assumed values for the investing and financing rates, so none are a *true* rate of return on the project's cash flow. The MIRR also has an Excel function, so it now can easily be used. Example 7A–5 illustrates the calculation, which is also summarized in Figure 7A–5.

EXAMPLE 7A–5   (Example 7A–3 Revisited)

Adding an oil well to an existing field had the cash flows summarized in Figure 7A–5. If the firm normally borrows money at 8% and invests at 15%, find the modified internal rate of return (MIRR).

SOLUTION

Figure 7A–6 shows the spreadsheet calculations.

	A	B	C	D	E	F	G	H	I
1	8%	external financing rate							
2	15%	external investing rate							
3	Year	0	1	2	3	4	5	6	7
4	Cash Flow	−4.00	3.50	2.50	1.50	0.50	−0.50	−1.50	−2.50
5	13.64%		Cell A5 contains = MIRR(B4:I4,A1,A2)						

FIGURE 7A–6 MIRR for oil well.

It is also possible to calculate the MIRR by hand. While more work, the process does clarify what the MIRR function is doing.

1. Each period's cash flow is already a single net receipt or expenditure.
2. Find the present worth of the expenses with the financing rate.

$$PW = -4M - 0.5M(P/F, 8\%, 5) - 1.5M(P/F, 8\%, 6) - 2.5M(P/F, 8\%, 7)$$
$$= -4M - 0.5M(0.6806) - 1.5M(0.6302) - 2.5M(0.5835) = -6.744M$$

3. Find the future worth of the receipts with the investing rate.

$$FW = 3.5M(F/P, 15\%, 6) + 2.5M(F/P, 15\%, 5) + 1.5M(F/P, 15\%, 4) + 0.5M(F/P, 15\%, 3)$$
$$= 3.5M(2.313) + 2.5M(2.011) + 1.5M(1.749) + 0.5M(1.521) = 16.507M$$

4. Find the MIRR that makes the present and future worths equivalent.

$$0 = (1 + MIRR)^n(PW) + FW$$
$$0 = (1 + MIRR)^7(-6.744M) + 16.507M$$
$$(1 + MIRR)^7 = 16.507M/6.744M = 2.448$$
$$(1 + MIRR) = 2.448^{1/7} = 1.1364$$
$$MIRR = 13.64\%$$

The MIRR does allow calculation of a rate of return for *any* set of cash flows. However, the result is only as realistic as the external rates that are used. The MIRR value can depend as much on the external rates that are used, as it does on the cash flows that it is describing.

SUMMARY

In cash flows with more than one sign change, we find that solving the cash flow equation can result in more than one root for the rate of return. Typical situations include a new oil well in an existing field, a project with a significant salvage cost, and staged construction.

In a sign change, successive nonzero values in the cash flow have different signs (that is, they change from + to −, or vice versa). Zero sign changes indicates there is no rate of return, as the cash flow is either all disbursements or all receipts.

One sign change is the usual situation, and a single positive rate of return generally results. There will be a negative rate of return whenever loan repayments are less than the loan or an investment fails to return benefits at least equal to the investment.

Multiple sign changes may result in multiple positive roots for i. When they occur, none of the positive multiple roots is a suitable measure of the project's economic desirability. The PW may also not be a reliable guide for decision making. If multiple positive roots are identified by graphing the present worth versus the interest rate, then the modified internal rate of return can be used to evaluate the project.

Graphing the present worth versus the interest rate ensures that the analyst recognizes that the cash flow has multiple sign changes. Otherwise a rate could be found and used that is not in fact a meaningful descriptor of the project. Graphing the present worth also identifies the much more common situation with multiple roots where one root is negative and the other positive. The negative root can be ignored for decision making. The positive root can be used for the IRR—with results identical to PW for decision making.

The modified internal rate of return (MIRR) relies on rates for investing and borrowing that are external to the project. The number of sign changes are reduced to one, ensuring that the MIRR can be found.

PROBLEMS

Unless the problem asks a different question or provides different data: (1) determine how many roots are possible and (2) graph the PW versus the interest rate to see whether multiple roots occur. If the root is a unique IRR, it is the project's rate of return. If there are multiple roots, then use an *external investing rate* of 12% and an *external borrowing rate* of 6%. Compute and use the MIRR as the project's rate of return.

7A-1 Given the following cash flow, determine the rate of return on the investment.

Year	Cash Flow
0	−$1000
1	3400
2	−5700
3	3800

7A-2 Find the rate of return for the following cash flow:
Ⓐ

Year	Cash Flow
0	−$15,000
1	10,000
2	−8,000
3	11,000
4	13,000

7A-3 Given the following cash flow, determine the rate of return on the investment.

Year	Cash Flow
0	−$5,000
1	3,500
2	−5,000
3	11,000

7A-4 Given the following cash flow, determine the rate of
Ⓐ return on the project.

Year	Cash Flow
0	−$5,000
1	20,000
2	−12,000
3	−3,000

7A-5 (a) Determine the rate of return on the investment
Ⓔ for the following cash flow.

Year	Cash Flow
0	−$1000
1	3600
2	−4280
3	1680

(b) If the firm's interest rate is 25%, is it ethical to recommend this investment?

7A-6 A firm invested $15,000 in a project that appeared (A) to have excellent potential. Unfortunately, there was a lengthy labor dispute in Year 3. Compute the project's rate of return.

Year	Cash Flow
0	−$15,000
1	10,000
2	6,000
3	−8,000
4	4,000
5	4,000
6	4,000

7A-7 Compute the rate of return for this project.

Year	Cash Flow
0	−$16,000
1	−8,000
2	11,000
3	13,000
4	−7,000
5	14,000

7A-8 Determine the rate of return on the investment on the (A) following cash flow.

Year	Cash Flow
0	−$3570
1–3	1000
4	−3170
5–8	1500

7A-9 What is the rate of return associated with this project?

Year	Cash Flow
0	−$210,000
1	88,000
2	68,000
3	62,000
4	−31,000
5	30,000
6	55,000
7	65,000

7A-10 A project has been in operation for 5 years. Calcu-(A) late the rate of return and state whether it has been an acceptable rate of return.

Year	Cash Flow
0	−$103,000
1	102,700
2	−87,000
3	94,500
4	−8,300
5	38,500

7A-11 Bill bought a vacation lot he saw advertised on television for $9000 down and monthly payments of $500. When he visited the lot, he found it was not something he wanted to own. After 40 months he was finally able to sell the lot. The new owner assumed the balance of the loan on the lot and paid Bill $19,000. What rate of return did Bill receive on his investment?

7A-12 The project, which had a projected life of 5 years, (A) was terminated early. Compute the interest rate.

Year	Cash Flow
0	−$5000
1	1500
2	1500

7A-13 Compute the rate of return on the investment.

Year	Cash Flow
0	−$5000
1	7500
2	1500
3	−6000

7A-14 Consider the following cash flow.

(A)
(E)

Year	Cash Flow
0	−$100
1	240
2	−143

(a) If the minimum attractive rate of return is 12%, should the project be undertaken?
(b) If the firm's interest rate is 25%, is it ethical to recommend this investment?

7A-15 Compute the rate of return on an investment having the following cash flow.

Year	Cash Flow
0	−$ 8,500
1	6,000
2–9	2,000
10	−18,000

7A-16 Consider the following situation. What is the rate of return?
Ⓐ

Year	Cash Flow
0	−$200
1	350
2	−100

7A-17 An investor is considering two mutually exclusive projects. She can obtain a 6% before-tax rate of return on external investments, but she requires a minimum attractive rate of return of 7% for these projects. Use a 10-year analysis period to compute the incremental rate of return from investing in Project A rather than Project B.

	Project A: Build Drive-Up Photo Shop	Project B: Buy Land in Hawaii
Initial capital investment	$58,500	$ 48,500
Net uniform annual income	6648	0
Salvage value 10 years hence	30,000	138,000
Computed rate of return	8%	11%

7A-18 Compute the rate of return on the investment on the following cash flow.
Ⓐ

Year	Cash Flow
0	−$1200
1–5	358
6	−200

7A-19 A problem often discussed in the engineering economy literature is the "oil-well pump problem." Pump 1 is a small pump; Pump 2 is a larger pump that costs more, will produce slightly more oil, and will produce it more rapidly. If the MARR is 20%, which pump should be selected? Assume that any temporary external investment of money earns 10% per year and that any temporary financing is done at 6%.

Year	Pump 1 ($000s)	Pump 2 ($000s)
0	−$100	−$110
1	70	115
3	70	30

7A-20 In January 2013, an investor bought a convertible debenture bond issued by the XLA Corporation. The bond cost $1000 and paid $60 per year interest in annual payments on December 31. Under the convertible feature of the bond, it could be converted into 20 shares of common stock by tendering the bond, together with $400 cash. The next business day after the investor received the December 31, 2015, interest payment, he submitted the bond together with $400 to the XLA Corporation. In return, he received the 20 shares of common stock. The common stock paid no dividends. On December 31, 2017, the investor sold the stock for $1740, terminating his 5-year investment in XLA Corporation. What rate of return did he receive?
Ⓐ

7A-21 An engineering firm is doing design work on a client's project. It has $40,000 in expenses at the beginning of each month from January 2018 through December 2018. The client has agreed to the following payment schedule, if the firm meets milestone delivery dates. Use the XIRR function to find the IRR.

Date	Income
5/14/2018	150,000
8/6/2018	140,000
10/22/2018	110,000
1/1/2019	140,000

CASES

The following case from *Cases in Engineering Economy* (www.oup.com/us/newnan) is suggested as matched with this chapter.

CASE 14 **Northern Gushers**

Incremental oil production investment with possible double root.

CHAPTER 8

CHOOSING THE BEST ALTERNATIVE

Coauthored with John Whittaker

Selecting the Best Pavement

The U.S. highway system is critical to meeting our mobility and economic needs. The goal is ensuring that users travel on pavements that are safe, smooth, quiet, durable, economical, and constructed of suitable materials—typically asphalt or concrete. Poor road surfaces are estimated to cost the average driver $324 annually in vehicle repairs and increase fuel consumption by about 2%.

In choosing between rigid concrete and flexible asphalt pavements, the three key questions are: (1) initial cost, (2) time to rehabilitation, and (3) cost of rehabilitation. A complete analysis would include the costs of the greenhouse gases (higher for asphalt), the ability to recycle (higher for asphalt), and the costs of traffic disruptions during rehabilitation (more frequent for asphalt). While government agencies must often focus on direct costs, the proper measure is life cycle and not initial costs.

Looking at the roads you drive on will indicate which pavement is generally preferred in your community. Often this is asphalt, since a thick asphalt structure can be a perpetual pavement with the only rehabilitation being milling of the surface followed by an asphalt overlay over a life of 60 years or more. Asphalt pavement often provides the smoothest, quietest ride with the greatest satisfaction for the motoring public.

On the other hand, asphalt pavement is also subject to rutting and shoving. Most of this damage is due to truck traffic, since the damage increases as a function of axle load to the 4th power. Deflection of flexible pavements under heavy truck loads can also increase fuel consumption. In northern climates, wear from studded tires also contributes to rutting. Thus in some cases rigid concrete pavement is a better choice than asphalt, which must be rehabilitated more frequently.

Historically, concrete pavements had design lives of 20 to 40 years and asphalt pavements had lives of 10 years. Today both can be designed for 50 to 60 years—if needed maintenance and rehabilitation occurs on schedule. In addition, combinations of pavements and overlays have increased the choices available to design engineers. ■ ■ ■

Contributed by Benedict N. Nwokolo, Grambling State University

QUESTIONS TO CONSIDER

1. Transportation agencies must choose the type of pavement to consider. Should the transportation agency be indifferent as to whether a rigid pavement or a flexible pavement be chosen? Why?

2. What engineering economic principles would you apply in the above problem if you are to choose from different types of pavement? What would you consider as the most effective measures for the public forum?

3. Considering the life cycle cited in the narrative, how can that be used in selecting a pavement type? How does a focus on initial vs. life-cycle costs affect the choice between asphalt and concrete pavements? Which pavement choice is made for the different types of road in your community?

4. What uncertainties are likely in estimating the (1) initial cost, (2) time to rehabilitation, and (3) cost of rehabilitation? Which are greater or smaller for asphalt or concrete choices?

5. If federal money is a significant share of a highway's initial cost, how might this distort the decision making of a state agency responsible for rehabilitation costs?

After Completing This Chapter...

The student should be able to:

- Use a *graphical technique* to visualize and solve problems involving mutually exclusive choices.

- Define *incremental analysis* and differentiate it from a standard present worth, annual worth, and internal rate of return analyses.

- Use spreadsheets to solve incremental analysis problems.

Key Words

choice table	incremental analysis	mutually exclusive
graphical approach		

INCREMENTAL ANALYSIS

This chapter was once titled *Incremental Analysis,* and it extended the **incremental analysis** presented in Chapter 7 to multiple alternatives—solely for the rate of return (IRR) measure. The approach relied on a series of numerical comparisons of challengers and defenders. That approach is presented in the last section of this chapter.

Recent editions use the more powerful and easier-to-understand approach of graphing each alternative's present worth (PW), equivalent uniform annual cost (EUAC), or equivalent uniform annual worth (EUAW). These graphs and spreadsheets support the calculation of incremental rates of return.

The graphical approach has the added benefit of focusing on the difference between alternatives. Often the difference is much smaller than the uncertainty in our estimated data. For ease of grading and instruction, we assume in this text that answers are exact, but in the real world uncertainty in the data must always be considered.

The engineering design process involves selection from competing alternatives. In engineering economy the words **mutually exclusive** alternatives are often used to emphasize that only one alternative may be implemented. Thus the objective is selecting the best of these mutually exclusive alternatives.

In earlier chapters we did this by maximizing PW, minimizing EUAC, or maximizing EUAW. We do the same here. In Chapter 7 we compared two alternatives incrementally to decide whether the IRR on the increment was acceptable. Any two alternatives can be compared by recognizing that:

[Higher-cost alternative] = [Lower-cost alternative] + [Increment between them]

When there are two alternatives, only a single incremental analysis is required. With more alternatives, a series of comparisons is required. Also, only by doing the analysis step by step can we determine which pairs must be compared. For example, if there are 4 alternatives, then 3 of 6 possible comparisons must be made. For 5 alternatives then 4 of 10 possible comparisons must be made. For N alternatives, $N-1$ comparisons must be made from $N(N-1)/2$ possibilities.

The **graphical approach** provides more information, and it is easier to understand and to present to others. It is best implemented with spreadsheets.

GRAPHICAL SOLUTIONS

Examples 8–1 and 8–2 illustrate why incremental or graphical analyses for the IRR criterion are required. They show that graphing makes it easy to choose the best alternative. Chapter 4 presented *xy* plots done with spreadsheets, and in Chapter 7 spreadsheets were used to graph the present worth versus the rate of return. In this chapter the present worth of each alternative is one *y* variable for graphs with multiple alternatives (or variables). In Chapter 9, one of the spreadsheet sections will present some of the ways that graphs can be customized for a better appearance.

EXAMPLE 8–1

The student engineering society is building a snack cart to raise money. Members must decide what capacity the cart should be able to serve. To serve 100 customers per hour costs $10,310, and to serve 150 customers per hour costs $13,400. The 50% increase in capacity is less than 50% of $10,310 because of economies of scale; but the increase in net revenue will be less than 50%, since the cart will not always be serving 150 customers per hour. The estimated net annual income for the lower capacity is $3300, and for the higher capacity it is $4000. After 5 years the cart is expected to have no salvage value. The engineering society is unsure of what interest rate to use to decide on the capacity. Make a recommendation.

SOLUTION

Since we do not know the interest rate, the easiest way to analyze the problem is to graph the PW of each alternative versus the interest rate, as in Figure 8–1. The Excel function[1]

$$= -\text{cost} + \text{PV}(\text{interest rate}, \text{life}, -\text{annual benefit})$$

is used to graph these two equations:

$$\text{PW}_{\text{low}} = -\$10,310 + \$3300(P/A, i, 5)$$
$$\text{PW}_{\text{high}} = -\$13,400 + \$4000(P/A, i, 5)$$

FIGURE 8–1 Maximizing PW to choose best alternative.

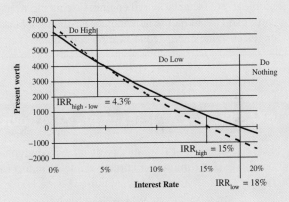

5-BUTTON SOLUTION

	A	B	C	D	E	F	G	H
	Alternative	*i*	*n*	*PMT*	*PV*	*FV*	Solve for	Answer
1								
2	Low		5	3300	-10,310	0	*i*	18.04%
3	High		5	4000	-13,400	0	*i*	15.03%
4	High − low		5	700	-3,090	0	*i*	4.30%

---

[1]The cost term in the equation has a minus sign because it is a cost and we are calculating a present worth. The annual benefit term has a minus sign because of the sign convention for the PV function.

We want to maximize the PW, so the choice between these alternatives is defined by where their curves intersect (the incremental rate of return of 4.3%), not by where those curves intersect the PW = 0 axis. Those internal rates of return of 15 and 18% are irrelevant to the choice between the low- and high-capacity carts.

Figure 8–1 also shows the main advantage of the graphical approach for real-world decision making. We can easily see that for interest rates between 0% and about 8%, there is little difference between the choices. In this region we are better off focusing our decision making on risk, benefits, or costs that we could not quantify, as well as on uncertainties in the data—since the PWs of our two alternatives are nearly the same. These topics are discussed in later chapters. Until then it is clearer if we analyze problems as though all numbers are known precisely.

From Figure 8–1, we see that if the interest rate is below 4.3%, the high-capacity cart has a higher PW. From 4.3 to 18% the lower-capacity cart has a higher PW. Above 18%, the third alternative of doing nothing is a better choice than building a low-capacity cart because the PW < 0. This can be summarized in a **choice table**:

Interest Rate	Best Choice
0% $\leq i \leq$ 4.3%	High capacity
4.3% $\leq i \leq$ 18.0%	Low capacity
18.0% $\leq i$	Do nothing

This example has been defined in terms of a student engineering society. But very similar problems are faced by a civil engineer sizing the weighing station for trucks carrying fill to a new earth dam, by an industrial engineer designing a package-handling station, and by a mechanical engineer sizing energy conservation equipment.

EXAMPLE 8–2 (Example 7–14 Revisited)

Solve Example 7–14 by means of an NPW graph. Two machines are being considered for purchase. If the minimum attractive rate of return (MARR) is 10%, which machine should be bought?

All $ values in 1000s	Machine X	Machine Y
Initial cost	$200	$700
Uniform annual benefit	95	120
End-of-useful-life salvage value	50	150
Useful life, in years	6	12

SOLUTION

Since the useful lives of the two alternatives are different, for an NPW analysis we must adjust them to the same analysis period. If the need seems continuous, then the "replace with an identical machine" assumption is reasonable and a 12-year analysis period can be used. The annual cash flows and the incremental cash flows are as follows:

End of Year	Cash Flows		Y − X
	Machine X	Machine Y	
0	−$200	−$700	−$500
1	95	120	25
2	95	120	25
3	95	120	25
4	95	120	25
5	95	120	25
6	−55	120	175
7	95	120	25
8	95	120	25
9	95	120	25
10	95	120	25
11	95	120	25
12	145	270	125

By means of a spreadsheet program, the NPWs are calculated for a range of interest rates and then plotted on an NPW graph (Figure 8–2).

For a MARR of 10%, Machine X is clearly the superior choice. In fact, as the graph clearly illustrates, Machine X is the correct choice for most values of MARR. The intersection point of the two graphs can be found by calculating ΔIRR, the rate of return on the incremental investment. From the spreadsheet IRR function applied to the incremental cash flows for $Y - X$:

$$\Delta IRR = 1.32\%$$

FIGURE 8–2 NPW graph.

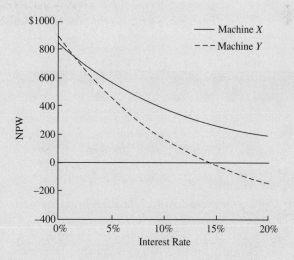

We can see from the graph that for MARR greater than 1.32%, Machine X is the right choice, and for MARR values less than 1.32%, Machine Y is the right choice.

Rate	NPW	
	Machine X	Machine Y
0%	$840.00	$890.00
1.32	752.24	752.24
2	710.89	687.31
4	604.26	519.90
6	515.57	380.61
8	441.26	263.90
10	378.56	165.44
12	325.30	81.83
14	279.77	10.37
16	240.58	−51.08
18	206.65	−104.23
20	177.10	−150.47

Examples 8–3 through 8–6 increase the number of alternatives being considered, but the same approach is used: graph the PW, EUAW, or EUAC, and for each possible interest rate choose the best alternative. Generally this means maximising the PW or EUAW, but for Example 8–4 the EUAC is minimized. Once the best choices have been identified from the graph, calculate the incremental rates of return.

Example 8–4 is a typical design problem where the most cost-effective solution must be chosen, but the dollar value of the benefit is not defined. For example, every building must have a roof, but deciding whether it should be metal, shingles, or a built-up membrane is a cost decision. No value is placed on a dry building; it is simply a requirement. In Example 8–4 having the pressure vessel is a requirement.

EXAMPLE 8–3   (Examples 7–14 and 8–2 Revisited)

In Example 8–2 Machine X was assumed to be "replaced with an identical machine" for a 12-year analysis period. In Chapter 6 this was shown to be equivalent to using EUAW as the comparison measure. Construct a choice table using EUAW.

SOLUTION

The first step is to construct the EUAW graph for Machines X and Y. If Figures 8–2 and 8–3 are compared, it is easy to see that there are only two differences. The first is the scale of the y-axis, where one is for NPW and the other is for EUAW. The second difference is the shape of the curves, where the EUAW curves are straighter than the NPW curves. The important similarity is that in both cases Machine Y is preferred for very low interest rates, and Machine X is preferred for other interest rates.

FIGURE 8–3 EUAW graph.

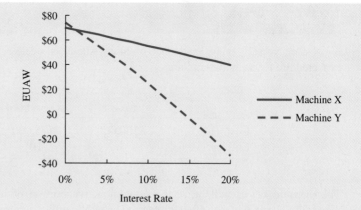

Finding the rate of return where Machines X and Y have the same EUAW is equivalent to finding the incremental IRR on $Y - X$. This was done for Example 8–2 by using the IRR spreadsheet function on the cash flows. Here we will use the GOAL SEEK function that was introduced in Example 7–15.

The 5-BUTTON SOLUTION starts by finding the EUAW for each machine at the 10% minimum attractive rate of return (MARR) that was used in Example 8–2. Note that this solution must add the annual benefit to the negative equivalent annual value of the initial cost and salvage value. To show this with the GOAL SEEK results, the 5-button calculations are duplicated in rows 9 to 16.

The GOAL SEEK function is found under DATA/WHAT-IF ANALYSIS. To use the GOAL SEEK function, the goal must be expressed in a single cell. Cell H16 simply calculates the difference in the EUAW values for each machine—the goal is a value of 0. GOAL SEEK varies a single cell to try to achieve the goal—in this case cell B10, the interest rate for Machine X. Cell B13 must contain =B10, so that both interest rates change at the same time.

When the lives of alternatives differ, this is the easiest way to find the breakeven interest rate. For the EUAWs of Machines X and Y the incremental IRR$_{Y-X}$ is 1.32%.

	A	B	C	D	E	F	G	H
1	Alternative	i	n	*PMT*	*PV*	*FV*	Solve for	Answer
2	Machine X	10%	6		-200	50	PMT	$39
3							Change sign	-39
4			annual benefit	95			EUAW	56
5	Machine Y	10%	12		-700	150	PMT	96
6							Change sign	-96
7			annual benefit	120			EUAW	24
8							Difference in EUAW	31
9								
10	Machine X	1.32%	6		-200	50	PMT	27
11							Change sign	-27
12			annual benefit	95			EUAW	68
13	Machine Y	1.32%	12		-700	150	PMT	52
14							Change sign	-52
15			annual benefit	120			EUAW	68
16							Difference in EUAW	0

EXAMPLE 8–4

A pressure vessel can be made out of brass, stainless steel, or titanium. The first cost and expected life for each material are:

	Brass	Stainless Steel	Titanium
Cost	$100,000	$175,000	$300,000
Life, in years	4	10	25

The pressure vessel will be in the nonradioactive portion of a nuclear power plant that is expected to have a life of 50 to 75 years. The public utility commission and the power company have not yet agreed on the interest rate to be used for decision making and rate setting. Build a choice table for the interest rates to show where each material is the best.

SOLUTION

The pressure vessel will be replaced repeatedly during the life of the facility, and each material has a different life. Thus, the best way to compare the materials is using EUAC (see Chapter 6). This assumes identical replacements.

Figure 8–4 graphs the EUAC for each alternative. In this case the best alternative at each interest rate is the material with the *lowest* EUAC. (We maximize worths, but minimize costs.) The Excel function is

$$= \text{PMT(interest rate, life, } -\text{first cost)}$$

The factor equation is

$$\text{EUAC} = \text{first cost}(A/P, \text{interest rate, life})$$

FIGURE 8–4 EUAC comparison of alternatives.

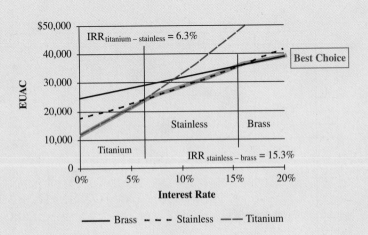

	A	B	C	D	E	F	G	H
1	Alternative	*i*	*n*	*PMT*	*PV*	*FV*	Solve for	Answer
2	Brass	15.3%	4		-100,000	0	PMT	$35,206
3	Stainless Steel	15.3%	10		-175,000	0	PMT	35,206
4	Titanium	15.3%	25		-300,000	0	PMT	47,114
5		15.3%		Difference in EUAW for brass & stainless				0
6		15.3%		Difference in EUAW for stainless & titanium				11,908
7								
8	Alternative	*i*	*n*	*PMT*	*PV*	*FV*	Solve for	Answer
9	Brass	6.3%	4		-100,000	0	PMT	29,043
10	Stainless Steel	6.3%	10		-175,000	0	PMT	24,091
11	Titanium	6.3%	25		-300,000	0	PMT	24,091
12		6.3%		Difference in EUAW for brass & stainless				-4,952
13		6.3%		Difference in EUAW for stainless & titanium				0

Because these alternatives have different-length lives, calculating the incremental IRRs is best done using the spreadsheet function GOAL SEEK. This spreadsheet is structured somewhat differently. In the top block of rows, each interest rate is set equal to the highlighted interest rate in cell B5. Copying the formula =$B$5 is easiest. In the lower block of rows, each interest rate is set equal to the highlighted interest rate in cell B13. Then two GOAL SEEKS are used to find the highlighted interest rates that make their respective differences in EUAW equal to zero.The choice table for each material is:

Interest Rate	Best Choice
0% $\leq i \leq 6.3\%$	Titanium
6.3% $\leq i \leq 15.3\%$	Stainless steel
15.3% $\leq i$	Brass

EXAMPLE 8–5

The following information refers to three mutually exclusive alternatives. The decision maker wishes to choose the right machine but is uncertain what MARR to use. Create a choice table that will help the decision maker to make the correct economic decision.

	Machine *X*	Machine *Y*	Machine *Z*
Initial cost	$2000	$7000	$4250
Uniform annual benefit	650	1100	1000
Useful life, in years	6	12	8

SOLUTION

In this example, the lives of the three alternatives are different. As we saw in Chapter 6 and Examples 8–3 and 8–4, when the service period is expected to be continuous and the assumption of identical replacement reasonable, we can assume a series of replacements and compare annual worth values just as we did with present worth values.

We will make these assumptions in this instance and plot EUAW. This was done in a spreadsheet using the Excel function = benefit + PMT (interest rate, life, initial cost), and the result is Figure 8–5.

The EUAW graph shows that for low values of MARR, Y is the correct choice. Then, as MARR increases, Z and X become the preferred machines. If the "do nothing" alternative is available, then it becomes the best choice for higher MARR values.

FIGURE 8–5 EUAW graph.

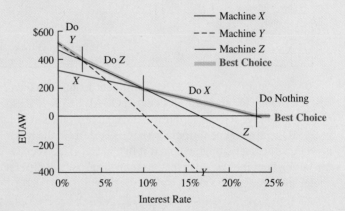

In this case, it is first necessary to solve for the EUAC of the initial cost. This is subtracted from the annual benefit of each machine to calculate the EUAWs. In the top block of rows, each interest rate is set equal to the highlighted interest rate in cell B5. Copying the formula =$B$5 is easiest. In the lower block of rows, each interest rate is set equal to the highlighted interest rate in cell B13. Then two GOAL SEEKS are used to find the highlighted interest rates that make their respective differences in EUAWs equal to zero. Notice that Figure 8–5 has been used to determine that finding the incremental or breakeven rates between Y & Z, Z & X, and X & doing nothing is necessary. The breakeven rate between X and doing nothing equals IRR_X, and it is found using the RATE function.

	A	B	C	D	E	F	G	H	I	J
1	Machine	i	n	PMT	PV	FV	Solve for	EUAC	Ann. Benefit	Benefit − cost
2	X	3.1%	6		-2000	0	PMT	371	650	279
3	Y	3.1%	12		-7000	0	PMT	709	1100	391
4	Z	3.1%	8		-4250	0	PMT	609	1000	391
5		3.1%					Difference in EUAW for	Y & Z		0
6		3.1%					Difference in EUAW for	Z & X		112
7										
8	Alternative	i	n	PMT	PV	FV	Solve for	EUAC	Ann. Benefit	Benefit − cost
9	X	10.8%	6		-2000	0	PMT	470	650	180
10	Y	10.8%	12		-7000	0	PMT	1068	1100	32
11	Z	10.8%	8		-4250	0	PMT	820	1000	180
12		10.8%					Difference in EUAW for	Y & Z		148
13		10.8%					Difference in EUAW for	Z & X		0
14										
15	Alternative	i	n	PMT	PV	FV	Solve for	IRR		
16	X		6	650	-2000	0	i	23.2%		

If		MARR ≥ 23.2%	do nothing
If	23.2% ≥ MARR ≥ 10.8%		choose X
If	10.8% ≥ MARR ≥ 3.1%		choose Z
If	3.1% ≥ MARR ≥ 0%		choose Y

If the "do nothing" alternative is not available, then the table reads

If		MARR ≥ 10.8%	choose X
If	10.8% ≥ MARR ≥ 3.1%		choose Z
If	3.1% ≥ MARR ≥ 0%		choose Y

The choice now is back in the decision maker's hands. There is still the need to determine MARR, but if the uncertainty was, for example, that MARR was some value in the range 12 to 18%, it can be seen that for this problem it doesn't matter. The answer is Machine X in any event. If, however, the uncertainty of MARR were in the 7 to 13% range, two things are clear. First, the decision maker would have to determine MARR with greater accuracy to maximize the EUAW. Second, there is little practical difference between the EUAWS for machines X and Y.

EXAMPLE 8–6

The following information is for five mutually exclusive alternatives that have 20-year useful lives. The decision maker may choose any one of the options or reject them all. Prepare a choice table.

	Alternatives				
	A	B	C	D	E
Cost	$4000	$2000	$6000	$1000	$9000
Uniform annual benefit	639	410	761	117	785

SOLUTION

Figure 8–6 is an NPW graph of the alternatives constructed by means of a spreadsheet.

FIGURE 8–6 NPW graph.

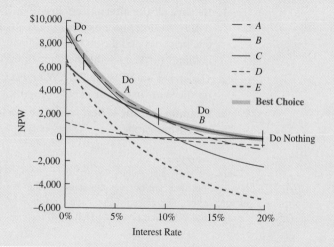

The graph clearly shows that Alternatives D and E are never part of the solution. They are dominated by the other three. The crossover points can either be read from the graph (if you have plotted it at a large enough scale) or found by calculating the ΔIRR of the intersecting curves.

Entering the data for the five alternatives also makes it easy to solve for the IRR of each. However, only IRR_B is useful. As the highest rate, it is the rate above which doing nothing is the best choice.[2] This can be seen from the right-hand side of Figure 8–6. Fortunately, the graph shows us that the rate is about 20%. Some spreadsheet packages fail to find rates that high, unless an *initial guess* is supplied. It turns out that the guess is almost exactly the IRR_B of 19.96%.

The figure also identifies the two incremental calculations that are needed—for $C - A$ and for $A - B$. The first value appears to be about 2% and the second a bit less than 10%. The rate on the incremental investment is also the rate at which the two alternatives have the same NPW. The three "useful" IRRs are left-indented for emphasis.

	A	B	C	D	E	F	G	H
	Alternative	i	n	PMT	PV	FV	Solve for	Answer
1								
2	A		20	639	-4000	0	i	15.0%
3	B		20	410	-2000	0	i	20.0%
4	C		20	761	-6000	0	i	11.2%
5	D		20	117	-1000	0	i	9.9%
6	E		20	785	-9000	0	i	6.0%
7	$C - A$		20	122	-2000	0	i	9.6%
8	$A - B$		20	229	-2000	0	i	2.0%

Placing these numbers in a choice table:

If	MARR ≥ 20%	do nothing
If	20% ≥ MARR ≥ 9.6%	select B
If	9.6% ≥ MARR ≥ 2%	select A
If	2% ≥ MARR ≥ 0%	select C

Examining Figure 8–6 also makes it clear that C is only slightly better than A for MARRs below 2%.

ELEMENTS IN COMPARING MUTUALLY EXCLUSIVE ALTERNATIVES

1. **Be sure all the alternatives are identified.** In textbook problems the alternatives will be well-defined, but real-life problems may be less clear. Before proceeding, one must have all the mutually exclusive alternatives tabulated, including the "do-nothing" or "keep doing the same thing" alternative, if appropriate.

2. **Construct an NPW or EUAW or EAC graph showing all alternatives plotted on the same axes.** This would be a difficult task were it not for spreadsheets.

3. **Examine the line of best values and determine which alternatives create it, and over what range.**

---

[2]There are analytical techniques for determining which incremental investments enter the solution set, but they are beyond the scope of this book and, in any case, are redundant in this era of spreadsheets. Interested readers can consult *Economic Analysis for Engineers and Managers* by Sprague and Whittaker (Prentice-Hall, 1987).

4. **Determine the changeover points** where the line of best values changes from one alternative to another. These can either be read directly off the graph or calculated, since they are the intersection points of the two curves and, what is more important and meaningful for engineering economy, they are **the ΔIRR of the incremental investment** between the two alternatives.

5. **Create a choice table** to present the information in compact and easily understandable form.

DOING A SET OF INCREMENTAL CHALLENGER–DEFENDER COMPARISONS

This chapter has focused on doing a graphical comparison of the PW of each alternative over a range of interest rates. We've calculated incremental interest rates, but we've looked at the curves to see which pair of intersecting PW curves we were analyzing. Before spreadsheets, this type of problem was solved by a series of challenger–defender comparisons, where defender $\equiv$ best alternative identified at this stage of the analysis, and challenger $\equiv$ next alternative being evaluated. Example 8–7 illustrates this approach.

The numerically based pairwise incremental comparisons is done at a single interest rate, and it does not show that "near" their intersection, two PW curves have "essentially" the same value. It also does not produce a choice table, since it is at a single interest rate. A choice table can be constructed, but it requires even more incremental comparisons.

EXAMPLE 8–7 (Example 8–6 Revisited)

For the alternatives in Example 8–6, conduct a pairwise incremental comparison. Which alternative is the best at an interest rate of 10%? Each of the five alternatives has no salvage value at the end of a 20-year useful life.

	Alternatives				
	A	B	C	D	E
Cost	$4000	$2000	$6000	$1000	$9000
Annual benefit	639	410	761	117	785

SOLUTION

1. The first step is to reorder the alternatives in order of increasing cost. This ensures that for each pairwise comparison the increment is an investment. In this example the order becomes D, B, A, C, E.

	Alternatives				
	D	B	A	C	E
Cost	$1000	$2000	$4000	$6000	$9000
Annual benefit	117	410	639	761	785

2. Calculate the IRR of the least expensive alternative to see if it is better than doing nothing at a MARR of 10%.

$$PW_D = 0 = -\$1000 + \$117(P/A, IRR_D, 20)$$

Solving this without a spreadsheet may take several tries and interpolation, but using the Excel function = RATE(life, annual benefit, –first cost) makes it easy and the answer exact:

$$IRR_D = 9.94\%$$

Because investment D earns less than 10%, doing nothing is preferred to doing D.

3. Doing nothing is still our *defender*, and the next Alternative, B, becomes the *challenger*. Calculate the IRR of Alternative B to see if it is better than doing nothing at a MARR of 10%.

$$PW_B = 0 = -\$2000 + \$410(P/A, IRR_B, 20)$$

Using the Excel function = RATE(life, annual benefit, –first cost), we find

$$IRR_B = 19.96\%$$

Because investment B earns more than 10%, Alternative B is preferred to doing nothing.

4. Alternative B is now our *defender,* and the next Alternative, A, becomes the *challenger*. This comparison must be made incrementally.

$$PW_{A-B} = 0 = -(\$4000 - \$2000) + (\$639 - \$410)(P/A, IRR_{A-B}, 20)$$

Using the Excel function = RATE(life, annual benefit, −first cost), we have

$$IRR_{A-B} = 9.63\%$$

Because the incremental investment $A - B$ earns less than 10%, Alternative B is preferred to doing Alternative A.

5. Alternative B is still our *defender,* and the next Alternative, C, becomes the *challenger*. This comparison must be made incrementally.

$$PW_{C-B} = 0 = -(\$6000 - \$2000) + (\$761 - \$410)(P/A, IRR_{C-B}, 20)$$

Using the Excel function = RATE(life, annual benefit, −first cost), we have

$$IRR_{C-B} = 6.08\%$$

Because the incremental investment $C - B$ earns less than 10%, Alternative B is preferred to doing Alternative C.

6. Alternative B is still our *defender,* and the final Alternative, E, becomes the *challenger*. This comparison must be made incrementally.

$$PW_{E-B} = 0 = -(\$9000 - \$2000) + (\$785 - \$410)(P/A, IRR_{E-B}, 20)$$

Using the Excel function = RATE(life, annual benefit, –first cost)

$$IRR_{E-B} = 0.67\%$$

Because the incremental investment $E - B$ earns less than 10%, Alternative B is preferred to doing Alternative E.

At an interest rate of 10%, Alternative *B* is the best choice. But only by referring to Figure 8–6 can we see that Alternative *B* is the best choice for all interest rates between 9.63% (where Alternative *A* is equally attractive) and 19.96% (where doing nothing is equally attractive). With other data, the curve for Alternative *B* might intersect at an incremental comparison we hadn't done. Also this pairwise comparison has not even compared Alternatives *A* and *C* so we cannot complete the choice table.

CHOOSING AN ANALYSIS METHOD

At this point, we have examined in detail the three major economic analysis techniques: present worth analysis, annual cash flow analysis, and rate of return analysis. A practical question is, which method should be used for a particular problem?

While the obvious answer is to use the method requiring the least computations, a number of factors may affect the decision.

1. Unless the MARR—minimum attractive rate of return (or minimum required interest rate for invested money)—is known, neither present worth analysis nor annual cash flow analysis is possible.
2. Present worth analysis and annual cash flow analysis often require far less computation than rate-of-return analysis.
3. In some situations, a rate-of-return analysis is easier to explain to people unfamiliar with economic analysis. At other times, an annual cash flow analysis may be easier to explain.
4. Business enterprises generally adopt one, or at most two, analysis techniques for broad categories of problems. If you work for a corporation and policy specifies the rate of return analysis, you would appear to have no choice in the matter.

Since one may not always be able to choose the analysis technique computationally best suited to the problem, this book illustrates how to use each of the three methods in all feasible situations. Ironically, the most difficult method to apply when using tabulated factors—rate-of-return analysis—is the one used most frequently by engineers in industry, which is one reason why spreadsheets and financial calculators are used so much in industry.

SUMMARY

For choosing from a set of mutually exclusive alternatives, the rate-of-return technique is more complex than the present worth or annual cash flow techniques. This results because in the latter two techniques the numbers can be compared directly, whereas with the rate of return it is necessary to consider the *increment of investment*. This is fairly straightforward if there are only two alternatives, but it becomes more and more complex as the number of alternatives increases.

A visual display of the problem can be created by using a spreadsheet to graph the economic value of the alternatives. The steps are as follows:

1. Be sure all the alternatives are identified.
2. Construct a NPW or EUAW (or EUAC) graph showing all alternatives plotted on the same axes.
3. Examine the line of maximum values (or minimum for the EUAC) and determine which alternatives create it, and over what range.
4. Determine the changeover points (ΔIRRs).
5. Create a choice table.

The graphical approach, where more values are calculated, is a more powerful one. By allowing the decision maker to see the range over which the choices are valid, it provides a form of sensitivity analysis. It also makes it clear that "close" to the changeover point the alternatives are very similar in value.

PROBLEMS

These problems are organized such that the (*a*) parts are best done with graphical analysis and as such are much more easily done with spreadsheets, and the (*b*) parts require numerical incremental analysis. Some problems include only one approach. Usually only (*b*) answers are included in Appendix E.

Two Action Alternatives

8-1 Including a do-nothing alternative, construct a choice table for interest rates from 0% to 100%.

Year	X	Y
0	−$1000	−$2000
1	1500	2800

8-2 Construct a choice table for interest rates from 0% to 100% for two mutually exclusive alternatives and the do-nothing alternative.

Year	Buy Y	Buy X
0	−$50.0	−$100.0
1-4	16.5	31.5

8-3 Consider three alternatives A, B, and "do-nothing."

(*a*) Construct a choice table for interest rates from 0% to 100%.

Year	A	B
0	−$10,000	−$15,000
1-5	3,200	4,500

(*b*) Step #4 of the decision-making process described in Chapter 1 is *Identify Feasible Alternatives*. Would you view it as ethical to not consider an alternative *C*, if you knew it was competitive but your boss asked you to leave it off the list? What would you do and why?

8-4 A paper mill is considering two types of pollution control equipment.

	Neutralization	Precipitation
Initial cost	$700,000	$500,000
Annual chemical cost	40,000	110,000
Salvage value	175,000	125,000
Useful life, in years	5	5

(*a*) Construct a choice table for interest rates from 0% to 100%.

(*b*) The firm wants a 12% rate of return on any avoidable increments of investment. Which equipment should be purchased?

8-5 A stockbroker has proposed two investments in low-rated corporate bonds paying high interest rates and selling at steep discounts (junk bonds). The bonds are rated as equally risky and both mature in 15 years.

Bond	Stated Value	Annual Interest Payment	Current Market Price with Commission
Gen Dev	$1000	$ 67	$480
RJR	1000	98	630

(*a*) Construct a choice table for interest rates from 0% to 100%.

(*b*) Which, if any, of the bonds should you buy if your MARR is 20%?

(*c*) Are there professional ethics standards for stockbrokers in the U.S.? What are some common ethical pitfalls?

8-6 A firm is considering two alternatives that have no salvage value.

	A	B
Initial cost	$10,700	$5500
Uniform annual benefits	2,100	1800
Useful life, in years	8	4

At the end of 4 years, another *B* may be purchased with the same cost, benefits, and so forth.

(*a*) Graph the EUAC or EUAW for the alternatives. Construct a choice table for interest rates from 0% to 100%.

(b) If the MARR is 10%, which alternative should be selected?

8-7 Don Garlits is a landscaper. He is considering the purchase of a new commercial lawn mower, either the Atlas or the Zippy. Graph the EUAC or EUAW for the alternatives. Construct a choice table for interest rates from 0% to 100%.

	Atlas	Zippy
Initial cost	$6700	$16,900
Annual O&M	1500	1,800
Annual benefit	4000	5,500
Salvage value	1000	3,500
Useful life, in years	3	6

8-8
Ⓐ Your cat's summer kitty-cottage needs a new roof. You feel a 15-year analysis period is in line with your cat's remaining lives. (There is no salvage value for old roofs.)

	Thatch	Slate
First cost	$200	$350
Annual upkeep	50	20
Service life, in years	3	5

(a) Graph the EUAC or EUAW for the alternatives. Construct a choice table for interest rates from 0% to 100%.
(b) Which roof should you choose if your MARR is 12%? What is the actual value of the IRR on the incremental cost?

8-9 The South End bookstore has an annual profit of $370,000. The owner may open a new bookstore by leasing an existing building for 5 years with an option to continue the lease for a second 5-year period. If he opens "The North End," it will take $1,200,000 of store fixtures and inventory. He believes that the two stores will have a combined profit of $560,000 a year after all the expenses of both stores have been paid.

The owner's economic analysis is based on a 5-year period. He will be able to recover $800,000 at the end of 5 years by selling the store fixtures and moving the inventory to The South End.

(a) Construct a choice table for interest rates from 0% to 100%.
(b) If The North End is opened, what rate of return can he expect?

8-10
Ⓐ
Ⓔ George is going to replace his car in 3 years when he graduates, but now he needs a radiator repair. The local shop has a used radiator, which will be guaranteed for 2 years, or they can install a new one, which is "guaranteed for as long as you own the car." The used radiator is $250 and the new one is $425. If George assumes the used radiator will last 3 years, but will need to be replaced so he can sell the car, which should he buy?

(a) Graph the EUAC or EUAW for the alternatives. Develop a choice table for interest rates from 0% to 50%.
(b) George's interest rate on his credit card is 20%. What should he do?
(c) Find the ASA auto repair ethics standards. How are the profession's ethical standards similar and different from those of your engineering discipline?

8-11
Ⓖ Using the current specifications, resurfacing a road will cost $1.5M initially, need $120K in annual maintenance, and need to be resurfaced every 10 years. A proposed new specification is expected to be more resistant to wear. The resurfacing cost will be $2.1M with $90K in annual maintenance and resurfacing every 15 years.

(a) Develop a choice table for interest rates from 0% to 25%.
(b) If the highway department's interest rate is 6%, which specification is preferred?
(c) How significant is the economic difference between the two specifications?
(d) Research the relationship between road surfacing and environmental impact. You may be surprised!

Multiple Alternatives

8-12
Ⓐ Each alternative has a 10-year useful life and no salvage value.

	A	B	C
Initial cost	$3000	$6000	$2000
Uniform annual benefits	410	980	350

(a) Construct a choice table for interest rates from 0% to 100%.
(b) If the MARR is 8%, which alternative should be selected?

8-13 The following three mutually exclusive alternatives have no salvage value after 5 years. Construct a choice table for interest rates from 0% to 100%.

	A	B	C
First cost	$2000	$3000	$6000
Uniform annual benefit	597	771	1652
Computed rate of return	15%	9%	11.7%

8-14 The following four mutually exclusive alternatives
Ⓐ have no salvage value after 10 years.

	A	B	C	D
First cost	$7500	$5000	$5000	$8500
Uniform annual benefit	1600	1200	1000	1700
Computed rate of return	16.8%	20.2%	15.1%	15.1%

(a) Construct a choice table for interest rates from
0% to 100%.

(b) Using 8% for the MARR, which alternative
should be selected?

8-15 Consider four mutually exclusive and a do-nothing
alternatives, each having an 10-year useful life:

	A	B	C	D
First cost	$1000	$800	$600	$500
Uniform annual benefit	125	120	100	125
Salvage value	750	500	250	0

(a) Construct a choice table for interest rates from
0% to 100%.

(b) If the minimum attractive rate of return is 8%,
which alternative should be selected?

8-16 Three mutually exclusive alternatives are being
Ⓐ considered.

	A	B	C
Initial investment	$50,000	$22,000	$15,000
Annual net income	5,093	2,077	1,643
Rate of return	8%	7%	9%

Each alternative has a 20-year useful life with no
salvage value.

(a) Construct a choice table for interest rates from
0% to 100%.

(b) If the minimum attractive rate of return is 7%,
which alternative should be selected?

8-17 Each alternative has a 10-year useful life and no salvage value. Construct a choice table for interest rates
from 0% to 100%, if doing nothing is allowed.

	A	B	C
Initial cost	$1500	$1000	$2035
Annual benefit for first 5 years	250	250	650
Annual benefit for subsequent 5 years	450	250	145

8-18 QZY, Inc. is evaluating new widget machines
Ⓐ offered by three companies. The chosen machine
will be used for 3 years.

	Company A	Company B	Company C
First cost	$15,000	$25,000	$20,000
Maintenance and operating	1,600	400	900
Annual benefit	8,000	13,000	9,000
Salvage value	3,000	6,000	4,500

(a) Construct a choice table for interest rates from
0% to 100%.

(b) MARR = 15%. From which company, if any,
should you buy the widget machine? Use rate of
return analysis.

8-19 Andrews Manufacturing offers three models for one
of its products to its customers. You have been asked
to analyze the choices from the customer's perspective. Which model should a customer choose if each
model has a life of 12 years? Doing nothing is an
alternative.

	Alternative		
	Deluxe	Regular	Economy
First cost	$220,000	$125,000	$75,000
Annual benefit	79,000	43,000	28,000
Maintenance and operating costs	38,000	13,000	8,000
Salvage value	16,000	6,900	3,000

(a) Construct a choice table for interest rates from
0% to 100%.

(b) MARR = 15%. Using incremental rate of return
analysis, which alternative, if any, should the
customer choose?

8-20 Wayward Airfreight, Inc. is considering a new automatic parcel sorter. Each choice has a 7-year life.

	SHIP-R	SORT-Of	U-SORT-M
First cost	$184,000	$235,000	$170,000
Salvage value	38,300	44,000	14,400
Annual benefit	75,300	89,000	68,000
Yearly O&M cost	19,000	19,000	12,000

(a) Construct a choice table for interest rates from
0% to 100%.
(b) Using a MARR of 15% and a rate of return
analysis, which alternative, if any, should be
selected?

8-21 A firm is considering the following alternatives, as
well as a fifth choice: do nothing. Each alternative
has a 5-year useful life.

	1	2	3	4
Initial cost	$100,000	$130,000	$200,000	$330,000
Uniform annual	26,380	38,780	47,480	91,550
Rate of return	10%	15%	6%	12%

(a) Construct a choice table for interest rates from
0% to 100%.
(b) The firm's minimum attractive rate of return is
8%. Which alternative should be selected?

8-22 Consider three mutually exclusive alternatives that
have a uniform annual benefit of $420. The analysis
period is 8 years. Assume identical replacements and
construct a choice table for interest rates from 0%
to 100%.
(a) Assume doing nothing is allowed.
(b) Assume A, B, or C must be chosen.

	A	B	C
Initial cost	$770	$1406	$2563
Useful life (years)	2	4	8
Rate of return	6.0%	7.5%	6.4%

8-23 Three mutually exclusive projects are being
considered:

	A	B	C
First cost	$1000	$2000	$3000
Uniform annual benefit	250	350	525
Salvage value	200	300	400
Useful life (years)	5	6	7

Assume identical replacements.
(a) Construct a choice table for interest rates from
0% to 100%.
(b) If 8% is the desired rate of return, which project
should be selected?

8-24 A business magazine is available for $58 for 1 year,
$108 for 2 years, $153 for 3 years, or $230 for 5
years. Assume you will read the magazine for at least
the next 5 years.
(a) For what interest rates do you prefer each payment plan?
(b) What is the environmental impact of glossy
magazines? How are they recycled?

8-25 Three office furniture firms that offer different payment plans have responded to a request for bids from
a state agency.

	Price	Payment Schedule
OfficeLess	$185,000	50% now, 25% in 6 months, 25% in 1 year
OfficeMore	182,000	50% now, 50% in 6 months
OfficeStation	180,000	100% in 1 year

(a) Develop a choice table for nominal interest rates
from 0% to 50%.
(b) If the agency's MARR is 10%, which vendor's
plan is preferred?

8-26 Consider the following alternatives:

	A	B	C
Initial cost	$10,000	$15,000	$20,000
Uniform annual benefit	1000	1762	5548
Useful life (years)	Infinite	20	5

Alternatives B and C are replaced at the end of
their useful lives with identical replacements. Use an
infinite analysis period.

(a) Construct a choice table for interest rates from 0% to 100%.

(b) At an 8% interest rate, which alternative is better?

8-27 Three mutually exclusive alternatives may replace the current equipment.

Year	A	B	C
0	−$20,000	−$24,000	−$25,000
1	10,000	10,000	5,000
2	5,000	10,000	5,000
3	10,000	8,000	5,000
4	5,000	5,000	25,000

(a) Construct a choice table for interest rates from 0% to 100%.

(b) If the MARR is 12%, which alternative should be selected?

8-28 Ⓐ A firm is considering three mutually exclusive alternatives as part of a production improvement program. The alternatives are as follows:

	A	B	C
Installed cost	$10,000	$15,000	$20,000
Uniform annual benefit	1,625	1,625	1,890
Useful life, in years	10	20	20

For each alternative, the salvage value at the end of useful life is zero. At the end of 10 years, Alt. A could be replaced by another A with identical cost and benefits.

(a) Construct a choice table for interest rates from 0% to 100%.

(b) The MARR is 6%. If the analysis period is 20 years, which alternative should be selected?

8-29 A new 10,000-square-meter warehouse next door to the Tyre Corporation is for sale for $450,000. The terms offered are $100,000 down with the balance being paid in 60 equal monthly payments based on 15% interest. It is estimated that the warehouse would have a resale value of $600,000 at the end of 5 years.

Tyre has the cash and could buy the warehouse but does not need all the warehouse space at this time. The Johnson Company has offered to lease half the new warehouse for $2500 a month. Modifying the space for two tenants will cost $12,000.

Tyre presently rents and uses 7000 square meters of warehouse space for $2700 a month. It has the option of reducing the rented space to 2000 square meters, in which case the monthly rent would be $1000 a month. Furthermore, Tyre could cease renting warehouse space entirely. Tom Clay, the Tyre Corp. plant engineer, is considering three alternatives:

1. Buy the new warehouse and lease half the space to the Johnson Company. In turn, the Tyre-rented space would be reduced to 2000 square meters.
2. Buy the new warehouse and cease renting any warehouse space.
3. Continue as is, with 7000 square meters of rented warehouse space.

Construct a choice table for interest rates from 0% to 100%.

8-30 Ⓐ Construct a choice table for interest rates from 0% to 100%. Similar alternatives will repeat indefinitely.

	Alternatives			
	A	B	C	D
Initial cost	$2000	$5000	$4000	$3000
Annual benefit	800	500	400	1300
Salvage value	2000	1500	1400	3000
Life, in years	5	6	7	4

8-31 One of these four mutually exclusive alternatives must be chosen. Each costs $13,000 and has no salvage value. Similar alternatives will repeat indefinitely.

Alternative	Annual Cost	Life (yrs)
A	$1000 first year; then increasing $575 per year	7
B	$100 first year; then increasing $500 per year	6
C	$2500	7
D	$4250 first year; then declining $500 per year	8

(a) Construct a choice table for interest rates from 0% to 100%.

(b) If the MARR is 8%, which alternative should be selected?

8-32 A more detailed examination of the situation in Problem 8-31 reveals that there are two additional mutually exclusive alternatives to be considered. Both have no salvage value after a useful life of 10 years.

Alternative	Initial Cost	Annual cost
E	$4880	$4880
F	8000	$2800 for first year; then increasing $200 per year

(a) Construct a choice table for interest rates from 0% to 100%.

(b) If the MARR remains at 8%, which one of the six alternatives should be selected?

8-33 The owner of a downtown parking lot has employed a civil engineering consulting firm to advise him on the economic feasibility of constructing an office building on the site. Betty Samuels, a newly hired civil engineer, has been assigned to make the analysis. She has assembled the following data:

Alternative	Total Investment*	Total Net Annual Revenue
Sell parking lot	$ 0	$ 0
Keep parking lot	200,000	22,000
Build 1-story building	400,000	60,000
Build 2-story building	555,000	72,000
Build 3-story building	750,000	100,000
Build 4-story building	875,000	105,000
Build 5-story building	1,000,000	120,000

* Includes the value of the land.

The analysis period is to be 15 years. For all alternatives, the property has an estimated resale (salvage) value at the end of 15 years equal to the present total investment.

(a) Construct a choice table for interest rates from 0% to 100%.

(b) If the MARR is 10%, what recommendation should Betty make?

8-34 A firm is considering moving its manufacturing plant **Ⓐ** from Chicago to a new location. The industrial engineering department was asked to identify the various alternatives together with the costs to relocate the plant and the benefits. The engineers examined six likely sites, together with the do-nothing alternative of keeping the plant at its present location. Their findings are summarized as follows:

Plant Location	First Cost ($000s)	Uniform Annual Benefit ($000s)
Denver	$300	$ 52
Dallas	550	137
San Antonio	450	117
Los Angeles	750	167
Cleveland	150	18
Atlanta	200	49
Chicago	0	0

The annual benefits are expected to be constant over the 8-year analysis period.

(a) Construct a choice table for interest rates from 0% to 100%.

(b) If the firm uses 10% annual interest in its economic analysis, where should the manufacturing plant be located?

8-35 An oil company plans to purchase a piece of vacant land on the corner of two busy streets for $70,000. On properties of this type, the company installs businesses of three different types. Each has an estimated useful life of 15 years. The salvage land for each is estimated to be the $70,000 land cost.

Plan	Cost*	Type of Business	Net Annual Income
A	$ 75,000	Conventional gas station	$23,300
B	230,000	Add automatic carwash	44,300
C	130,000	Add quick carwash	27,500

* Improvements cost does not include $70,000 for the land.

(a) Construct a choice table for interest rates from 0% to 100%.

(b) If the oil company expects a 10% rate of return on its investments, which plan (if any) should be selected?

8-36 *The Financial Advisor* is a weekly column in the **Ⓐ** local newspaper. Assume you must answer the following question. "I recently retired at age 65, and I have a tax-free retirement annuity coming due soon.

I have three options. I can receive (A) $30,976 now, (B) $359.60 per month for the rest of my life, or (C) $513.80 per month for the next 10 years. For option C if I die within 10 years, payments continue to my heirs. My interest rate is 9%. What should I do?" Ignore the timing of the monthly cash flows and assume that the payments are received at the end of year.

Contributed by D. P. Loucks, Cornell University

(a) Develop a choice table for life spans from age 66 to 100.

(b) If remaining life is 20 years and $i = 9\%$, use an incremental rate of return analysis to recommend which option should be chosen.

8-37 A firm must decide which of three alternatives to adopt to expand its capacity. The firm wishes a minimum annual profit of 20% of the initial cost of each separable increment of investment. Any money not invested in capacity expansion can be invested elsewhere for an annual yield of 20% of initial cost.

Alt.	Initial Cost	Annual Profit	Profit Rate
A	$100,000	$30,000	30%
B	300,000	66,000	22%
C	500,000	80,000	16%

Which alternative should be selected? Use a challenger–defender rate of return analysis.

8-38 The New England Soap Company is considering adding some processing equipment to the plant to aid in the removal of impurities from some raw materials. By adding the processing equipment, the firm can purchase lower-grade raw material at reduced cost and upgrade it for use in its products.

Four different pieces of processing equipment with 20-year lives are being considered:

	A	B	C	D
Initial investment	$10,000	$18,000	25,000	$30,000
Annual saving in materials costs	4,000	6,000	7,500	9,000
Annual operating cost	2,000	3,000	3,000	4,000

The company can obtain a 15% annual return on its investment in other projects and is willing to invest money on the processing equipment only as long as

it can obtain 15% annual return on each increment of money invested. Which one, if any, of the alternatives should be selected? Use a challenger–defender rate of return analysis.

Cash vs. Loan vs. Lease

8-39 Frequently we read in the newspaper that one should lease a car rather than buying it. For a typical 24-month lease on a car costing $9400, the monthly lease charge is about $267. At the end of the 24 months, the car is returned to the lease company (which owns the car). As an alternative, the same car could be bought with no down payment and 24 equal monthly payments, with interest at a 12% nominal annual percentage rate. At the end of 24 months the car is fully paid for. The car would then be worth about half its original cost.

(a) Over what range of nominal before-tax interest rates is leasing the preferred alternative?

(b) What are some of the reasons that would make leasing more desirable than is indicated in (a)?

8-40 *The Financial Advisor* is a weekly column in the local newspaper. Assume you must answer the following question. "I need a new car that I will keep for 5 years. I have three options. I can (A) pay $19,999 now, (B) make monthly payments for a 6% 5-year loan with 0% down, or (C) make lease payments of $299.00 per month for the next 5 years. The lease option also requires an up-front payment of $1000. What should I do?"

Assume that the number of miles driven matches the assumptions for the lease, and the vehicle's value after 5 years is $6500. Remember that lease payments are made at the beginning of the month, and the salvage value is received only if you own the vehicle.

(a) Develop a choice table for nominal interest rates from 0% to 50%. (You do not know what the reader's interest rate is.)

(b) If $i = 9\%$, use an incremental rate of return analysis to recommend which option should be chosen.

8-41 *The Financial Advisor* is a weekly column in the local newspaper. Assume you must answer the following question. "I need a new car that I will keep for 5 years. I have three options. I can (A) pay $25,999 now, (B) make monthly payments for a 9% 5-year loan with 0% down, or (C) make lease payments of $470 per month for the next 5 years.

The lease option also requires a security deposit of $1500. What should I do?"

Assume that the number of miles driven matches the assumptions for the lease, and the vehicle's value after 5 years is $7000. Remember that lease payments are made at the beginning of the month, and the salvage value is received only if you own the vehicle.

(a) Develop a choice table for nominal interest rates from 0% to 50%. (You do not know what the reader's interest rate is.)

(b) If i = 9%, use an incremental rate of return analysis to recommend which option should be chosen.

Minicases

8-42 Contact a car dealer and choose a car to evaluate a buy-versus-lease decision (keep it reasonable—no Lamborghinis). Tell the people at the dealership that you are a student working on an assignment. Be truthful and don't argue; if they don't want to help you, leave and find a friendlier dealer. For both buying and leasing, show all assumptions, costs, and calculations. Do not include the cost of maintenance, gasoline, oil, water, fluids, and other routine expenses in your calculations.

Determine the car's sales price (no need to negotiate) and the costs for sales tax, license, and fees. Estimate the "Blue Book value" in 5 years. Determine the monthly payment based on a 5-year loan at 9% interest. Assume that your down payment is large enough to cover only the sales tax, license, and fees. Calculate the equivalent uniform monthly cost of owning the car.

Identify the costs to lease the car (if available assume a 5-year lease). This includes the monthly lease payment, required down payments, and any return fees that are required. Calculate the equivalent uniform monthly cost of leasing the car.

The salesperson probably does not have the answers to many of these questions. Write a one- to two-page memo detailing the costs. Make a recommendation: Should you own or lease your car? Include nonfinancial items and potential financial items in your conclusions, such as driving habits and whether you are likely to drive more than the allowed number of miles dictated in the lease.

8-43 Develop the costs and benefits to compare owning a car versus depending on public transit, friends, and/or a bicycle. Place a monetary value on each advantage or disadvantage.

(a) Develop a choice table for interest rates between 0% and 25%.

(b) Since costs are likely to be similar for most people, how sensitive are your results to environmental concerns? What external costs did you omit? How does including them and increasing their importance change your results?

8-44 Develop the costs and benefits to compare owning a new car with one that is 2 years old. Place a monetary value on each advantage or disadvantage. Develop a choice table for interest rates between 0% and 25%.

8-45 For a vehicle that you or a friend owns, determine the number of miles driven per year. Find three alternative sets of 4 tires that differ in their tread warranty. Assume that the life of the tires equals the tread warranty divided by the number of miles driven per year. Compare the EUACs of the tires.

(a) For what interest rates is each choice the best?

(b) Develop a graph equivalent to Figure 8–4 to illustrate the results.

(c) For the interest rate that is in the "middle" of your range, how low and how high can the number of miles each year be without changing the best choice?

CASES

The following cases from *Cases in Engineering Economy* (www.oup.com/us/newnan) are suggested as matched with this chapter.

CASE 6 **Lease a Lot**

Compares leasing and ownership. Results show importance of separating financing and investing decisions.

CASE 10 **The Cutting Edge**

Make vs. buy and machine selection.

CASE 11 **Harbor Delivery Service**

Annual comparison of diesel versus gasoline engines.

OTHER ANALYSIS TECHNIQUES

Clean and Green

In 2016, the U.S. Green Building Council (USGBC) and U.S. Energy Information Administration reported that in the U.S., buildings account for 75% of electricity consumption, about 45% of energy use, almost 40% of CO_2 emissions, 40% of raw materials use, and 13% of potable water consumption. To address this huge impact on the environment, designers and engineers have made great progress in developing environment-friendly construction materials and building techniques. "Green" buildings offer numerous advantages, including improved worker productivity, reduced health and safety costs, improved indoor environmental quality, and reduced energy and maintenance costs.

Some still believe that green structures cost more to build, especially with respect to first costs. However, growing evidence is dispelling this myth, and what has been learned is that location and final finishing play a bigger role in cost differentiation of buildings. Increasingly, engineering, procurement, and construction groups are embracing sustainable design and construction and no longer see "green" as an added cost. To many, "green design" is simply "good design." Although documentation costs can be high, in particular in the pursuit of LEED certification (Leadership in Energy and Environmental Design), the USGBC estimates that as of May 2016, more than 15 billion square feet of building space is LEED-certified worldwide, and this is increasing by more than 10% annually. Furthermore, LEED is becoming increasingly international with more than 79,100 LEED projects in over 160 countries and territories.

Despite this mounting evidence in the support of building green, some commercial developers believe that environment-friendly features add to the expense of construction. And even with small additional costs in the range of 1–2%, they can make it harder for the builder to recoup its investment and break even on the project in the short term.

Furthermore, there is a general reluctance by many in this industry to be innovative and to embrace new ideas or new technologies.

Renters may like the idea of being cleaner and greener—but they may be unwilling to pay extra for it unless they can see tangible impact to their bottom line. ▪ ▪ ▪

Contributed by Kim LaScola Needy, University of Arkansas

QUESTIONS TO CONSIDER

1. Office leases frequently require building owners, rather than tenants, to pay heating and cooling costs. What effect might this have on the decision making of potential tenants who are considering renting space in green buildings?

2. Many environment-friendly buildings are architecturally distinctive and feature better quality materials and workmanship than traditional commercial structures. Environmental advocates hope these characteristics will help green buildings attract a rent "premium." How might these features make the buildings more attractive to tenants? Is it the green features or the higher quality materials and workmanship that add significantly to the perceived higher first costs of green buildings?

3. How can the costs and benefits of green buildings be economically validated by an independent party so that designers and engineers can make a fair assessment?

4. What ethical questions arise from government regulations intended to promote or require green building practices?

After Completing This Chapter...

You should be able to:

- Use future worth, benefit–cost ratio, payback period, and sensitivity analysis methods to solve engineering economy problems.

- Link the use of *future worth* analysis to the present worth and annual worth methods developed earlier.

- Mathematically develop the *benefit–cost ratio,* and use this model to select alternatives and make economic choices.

- Understand the concept of the *payback period* of an investment, and be able to calculate this quantity for prospective projects.

- Demonstrate a basic understanding of *sensitivity* and *breakeven analyses* and the use of these tools in an engineering economic analysis.

- Use a spreadsheet to perform *sensitivity* and *breakeven analyses.*

- Consider expected returns and risks when investing or saving. (Appendix 9A)

Key Words

benefit–cost ratio	liquidity	sensitivity analysis
breakeven chart	payback period	staged construction
discounted payback period	profitability	what-if analysis
future worth analysis		

Chapter 9 examines four topics:

- Future worth analysis
- Benefit–cost ratio or present worth index analysis
- Payback period
- Sensitivity, breakeven, and what-if analysis

Future worth analysis is very much like present worth analysis, dealing with *then* (future worth) rather than with *now* (present worth) situations.

Previously, we have written economic analysis relationships based on either

$$\text{PW of cost} = \text{PW of benefit} \quad \text{or} \quad \text{EUAC} = \text{EUAB}$$

Instead of writing it in this form, we could define these relationships as

$$\frac{\text{PW of benefit}}{\text{PW of cost}} = 1 \quad \text{or} \quad \frac{\text{EUAB}}{\text{EUAC}} = 1$$

When economic analysis is based on these ratios, the calculations are called benefit–cost ratio analysis. The PW ratio is also known as a present worth index.

Payback period is an approximate analysis technique, generally defined as the time required for cumulative benefits to equal cumulative costs.

Sensitivity describes how much a problem element must change to reverse a particular decision. Closely related is breakeven analysis, which determines the conditions under which two alternatives are equivalent. What-if analysis changes one or all variables to see how the economic value and recommended decision change. Thus, breakeven and what-if analysis are forms of sensitivity analysis.

FUTURE WORTH ANALYSIS

In present worth analysis, alternatives are compared in terms of their present consequences. In annual cash flow analysis, the comparison was in terms of equivalent uniform annual costs (or benefits). But the concept of resolving alternatives into comparable units is not restricted to a present or annual comparison. The comparison may be made at any point in time. In many situations we would like to know what the *future* situation will be, if we take some particular course of action *now*. This is called **future worth analysis.**

Example 9–1 illustrates how much impact seemingly small decisions can have on long-term objectives. *Happy Money* by Dunn and Norton notes that individuals can *decrease* spending and *increase* happiness by switching a daily habit to a weekly treat. A common variation of Example 9–1 defines a future worth goal in *n* years for an expensive vacation, a down payment on a house, a child's college fund, or for retirement. In these cases the question is how much must be saved each month to meet the goal. This type of savings and investing is the subject of Appendix 9A, which focuses on retirement because of its long horizon and importance. Appendix 9A can be applied to any future worth savings goal. Example 9–2 illustrates how to calculate the future worth at start-up of projects with multi-year construction periods.

EXAMPLE 9–1

Ron Jamison, a 20-year-old college student, smokes about a carton of cigarettes a week. He wonders how much money he could accumulate by age 65 if he quit smoking now and put his cigarette money into a savings account. Cigarettes cost $35 per carton. Ron expects that a savings account would earn 5% interest, compounded semiannually. Compute the future worth of Ron's savings at age 65.

Cigarette smoking is an unlikely choice for switching from a daily habit to a weekly treat. So assume that the weekly $35 is from a 5-day per week habit where this makes sense, and that there is a $2 per day alternative. Possible examples include lattes, breakfasts from a drive-through, and lunches not brought from home. What are the weekly savings and the future worth of the savings account?

TABLE SOLUTION

Stop and Save

$$\text{Semiannual saving} = (\$35/\text{carton})(26 \text{ weeks}) = \$910$$
$$FW = A(F/A, 2\tfrac{1}{2}\%, 90) = 910(329.2) = \$299{,}572$$

Daily Habit to Weekly Treat

A weekly $35 from a 5-day per week habit is $7 per day. Substituting a $2 per day alternative on 4 days per week would save $20 weekly ($= 4 \times \$5$). This would be $520 every 6 months.

$$FW = 520(F/A, 2\tfrac{1}{2}\%, 90) = 520(329.2) = \$171{,}184$$

5 BUTTON SOLUTION

	A	B	C	D	E	F	G	H	I
1	Problem	i	n	*PMT*	*PV*	*FV*	Solve for	Answer	Formula
2	Stop & save	2.5%	90	-910	0		FV	$299,530	=FV(B2,C2,D2,E2)
3									
4	Weekly treat	2.5%	90	-520	0		FV	$171,160	=FV(B2,C2,D2,E2)

EXAMPLE 9–2

An East Coast firm has decided to establish a second plant in Kansas City. There is a factory for sale for $850,000 that could be remodeled and used. As an alternative, the firm could buy vacant land for $85,000 and have a new plant constructed there. Either way, it will be 3 years before the firm will be able to get a plant into production. The timing of costs for the factory are:

Year	Construct New Plant		Remodel Available Factory	
0	Buy land	$ 85,000	Purchase factory	$ 850,000
1	Design	200,000	Design	250,000
2	Construction	1,200,000	Remodeling	250,000
3	Production equipment	200,000	Production equipment	250,000

If interest is 8%, which alternative has the lower equivalent cost when the firm begins production at the end of Year 3?

SOLUTION

New Plant

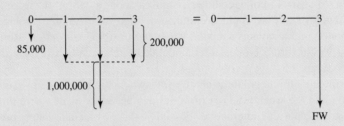

$$\text{FW of cost} = 85{,}000(F/P, 8\%, 3) + 200{,}000(F/A, 8\%, 3)$$
$$+ 1{,}000{,}000(F/P, 8\%, 1) = \$1{,}836{,}000$$

Remodel Available Factory

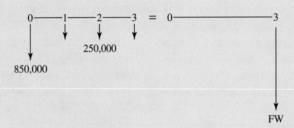

$$\text{FW of cost} = 850{,}000(F/P, 8\%, 3) + 250{,}000(F/A, 8\%, 3)$$
$$= \$1{,}882{,}000$$

The total cost of remodeling the available factory ($1,600,000) is smaller than the total cost of a new plant ($1,685,000). However, the timing of the expenditures is better with the new plant. The new plant is projected to have the smaller future worth of cost and thus is the preferred alternative.

BENEFIT–COST RATIO ANALYSIS

At a given minimum attractive rate of return (MARR), we would consider an alternative acceptable, provided that

$$\text{PW of benefits} - \text{PW of costs} \geq 0 \qquad \text{or} \qquad \text{EUAB} - \text{EUAC} \geq 0$$

These could also be stated as a ratio of benefits to costs, or

$$\textbf{Benefit–cost ratio } = \frac{\text{PW of benefit}}{\text{PW of costs}} = \frac{\text{EUAB}}{\text{EUAC}} \geq 1$$

TABLE 9–1 Benefit–Cost Ratio Analysis

Input/Output	Situation	Criterion
Neither input nor output fixed	Neither amount of money or other inputs nor amount of benefits or other outputs are fixed	*One alternative:* B/C $\geq$ 1 *Two or more alternatives:* Solve by incremental analysis of benefit–cost ratios
Fixed input	Amount of money or other input resources are fixed	Maximize B/C
Fixed output	Fixed task, benefit, or other output to be accomplished	Maximize B/C

Rather than using present worth or annual cash flow analysis to solve problems, we can base the calculations on the benefit–cost ratio, B/C. The criteria are presented in Table 9–1. In Table 9–1 the two special cases where maximizing the B/C ratio is correct are listed below the more common situation where incremental analysis is required. In Chapter 16 we will detail how this measure is applied in the public sector. Its use there is so pervasive that the term *present worth index* is sometimes used to distinguish private-sector applications.

Present worth, annual cash flow, rate of return, and benefit–cost ratio approaches *all* require a common horizon when comparing alternatives. Example 9–3 shows that incremental analysis is required for mutually exclusive alternatives to correctly apply the benefit–cost ratio, just as it was required to use rate of return. The intuitively appealing approach of maximizing the B/C ratio does not work. Example 9–4 applies incremental analysis to a six-alternative problem.

EXAMPLE 9–3   **(Examples 5–1, 6–5, and 7–13 Revisited)**

A firm is trying to decide which of two devices to install to reduce costs. Both devices have useful lives of 5 years and no salvage value. Device *A* costs $10,000 and can be expected to result in $3000 savings annually. Device *B* costs $13,500 and will provide cost savings of $3000 the first year, but savings will increase by $500 annually, making the second-year savings $3500, the third-year savings $4000, and so forth. With interest at 7%, which device should the firm purchase?

SOLUTION

We have used three types of analysis thus far to solve this problem: present worth in Example 5–1, annual cash flow in Example 6–5, and rate of return in Example 7–13. First we correctly analyze this incrementally, then we look at each device's benefit–cost ratio.

Incremental B–A

$$\text{PW of incremental cost} = \$3500$$

$$\text{PW of incremental benefits} = 500(P/G, 7\%, 5)$$

$$= 500(7.647) = \$3820$$

$$\frac{B}{C} = \frac{\text{PW of benefit}}{\text{PW of costs}} = \frac{3820}{3500} = 1.09$$

The increment is justified at the MARR of 7%. Device *B* should be purchased.

Device A

$$\text{PW of cost} = \$10,000$$

$$\text{PW of benefits} = 3000(P/A, 7\%, 5)$$

$$= 3000(4.100) = \$12,300$$

$$\frac{B}{C} = \frac{\text{PW of benefit}}{\text{PW of costs}} = \frac{12,300}{10,000} = 1.23$$

Device B

$$\text{PW of cost} = \$13,500$$

$$\text{PW of benefit} = 3000(P/A, 7\%, 5) + 500(P/G, 7\%, 5)$$

$$= 3000(4.100) + 500(7.647) = 12,300 + 3820 = 16,120$$

$$\frac{B}{C} = \frac{\text{PW of benefit}}{\text{PW of costs}} = \frac{16,120}{13,500} = 1.19$$

Maximizing the benefit–cost ratio indicates the wrong choice, Device *A*. Incremental analysis must be used.

EXAMPLE 9–4 | **(Examples 8–6 and 8–7 Revisited)**

Consider the five mutually exclusive alternatives from Examples 8–6 and 8–7 plus an additional alternative, *F*. They have 20-year useful lives and no salvage value. If the minimum attractive rate of return is 6%, which alternative should be selected?

	A	*B*	*C*	*D*	*E*	*F*
Cost	$4000	$2000	$6000	$1000	$9000	$10,000
PW of benefit	7330	4700	8730	1340	9000	9,500
$\dfrac{B}{C} = \dfrac{\text{PW of benefits}}{\text{PW of cost}}$	1.83	2.35	1.46	1.34	1.00	0.95

SOLUTION

Incremental analysis is needed to solve the problem. The steps in the solution are the same as the ones presented in Example 8–7 for incremental rate of return, except here the criterion is $\Delta B/\Delta C$, and the cutoff is 1, rather than ΔIRR with a cutoff of MARR.

1. Be sure all the alternatives are identified.
2. (Optional) Compute the B/C ratio for each alternative. Since there are alternatives for which B/C $\geq$ 1, we will discard any with B/C < 1. Discard Alt. *F*.
3. Arrange the remaining alternatives in ascending order of investment.

	D	B	A	C	E
Cost (= PW of cost)	$1000	$2000	$4000	$6000	$9000
PW of benefits	1340	4700	7330	8730	9000
B/C	1.34	2.35	1.83	1.46	1.00

	B − D Increment	A − B Increment	C − A Increment
ΔCost	$1000	$2000	$2000
Δ Benefits	3360	2630	1400
ΔB/C	3.36	1.32	0.70

4. For each increment of investment, if $\Delta B/\Delta C \geq 1$ the increment is attractive. If $\Delta B/\Delta C < 1$ the increment of investment is not desirable. The increment $B - D$ is desirable, so *B* is preferred to *D*. The increment $A - B$ is desirable. Thus, Alt. *A* is preferred. Increment $C - A$ is not attractive since $\Delta B/\Delta C = 0.70$. Now we compare *A* and *E*:

	E − A Increment
ΔCost	$5000
ΔBenefit	1670
ΔB/ΔC	0.33

The increment is undesirable. We choose Alt. *A* as the best of the six alternatives. [*Note:* The best alternative does not have the highest B/C ratio, nor is it the largest project with a B/C ratio > 1. Alternative *A* does have the largest difference between the PW of its benefits and costs (= $3330).]

Benefit–cost ratio analysis may be graphically represented. Figure 9–1 is a graph of Example 9–4. We see that *F* has a B/C < 1 and can be discarded. Alternative *D* is the starting point for examining the separable increments of investment. The slope of line *B–D* indicates a $\Delta B/\Delta C$ ratio of >1. This is also true for line *A–B*. Increment *C–A* has a slope much flatter than B/C = 1, indicating an undesirable increment of investment. Alternative *C* is therefore discarded and *A* retained. Increment *E–A* is similarly unattractive. Alternative *A* is therefore the best of the six alternatives.

Note three additional things about Figure 9–1: first, even if alternatives with B/C ratio < 1 had not been initially excluded, they would have been systematically eliminated in the

FIGURE 9–1 Benefit–cost ratio graph of Example 9–4.

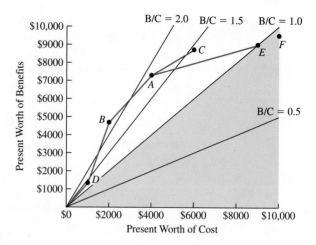

incremental analysis. Second, Alt. *B* had the highest B/C ratio (B/C = 2.35), but it is not the best of the six alternatives. We saw the same situation in rate of return analysis of three or more alternatives. The reason is the same in both analysis situations. We seek to maximize the *total* profit, not the profit rate.

Third and most important, the total profit or B − C for each alternative is the vertical distance from the point to the B/C = 1.0 line. *A* maximizes that distance.

Variations on the Theme of the Benefit–Cost Ratio

The basic benefit–cost ratio has been defined as placing *all* benefits in the numerator of the ratio and *all* costs in the denominator. One variation of the ratio considers the salvage values as reducing the costs rather than as increasing the benefits.

In the public sector, it is common to define the benefit–cost ratio so that the numerator includes all consequences to the users or the public and the denominator includes all consequences to the sponsor or government. For example, the numerator might include the positive benefits of improved highway traffic flow and the disbenefits of congestion during construction, since both accrue to the public or users. The denominator in this case includes consequences to the government, such as the costs of construction and the reduced maintenance cost for the new highway. Example 9–5 illustrates this for a public project.

In Example 9–6 exactly the same numbers are put in a private context. Here the benefit–cost ratio is typically called a present worth index. The calculation is modified so that the denominator is the project's first cost, and all other consequences are placed in the numerator. This formulation of the benefit–cost ratio emphasizes the "bang for the buck" of how much return is gained for each dollar of investment.

We will examine the public-sector application of the benefit–cost ratio in more detail in Chapter 16, and the present worth index will be used in Chapter 15. Here and in those chapters, the same standard applies for all versions of the ratio. Is the ratio ≥1? More importantly, if one version of the ratio is ≥1, then all versions are ≥1. As shown in Examples 9–5 and 9–6, the values of the ratios may differ, but whether they are above or below 1 and the recommended decisions do not change.

EXAMPLE 9–5

Traffic congestion on Riverview Boulevard has reached a point where something must be done. Two suggested plans have a life of 15 years, because that is the scheduled time for completion of the new Skyway Highway. After that time traffic will fall well below current levels.

Adding right-turn lanes at key intersections will cost $8.9M (million) with annual maintenance costs for signals and lane painting of $150,000. Added congestion during construction is a disbenefit of $900,000, but the reduced congestion after construction is an annual benefit of $1.6M. This benefit actually starts lower and increases over time, but for a simple initial analysis we are assuming a uniform annual benefit.

Adding a second left-turn lane at a few key intersections will cost an additional $3M with an added annual maintenance cost of $75,000. This construction is more disruptive, and the total disbenefit for congestion during construction is $2.1M. Upon completion, the total benefit for reduced congestion will be $2.2M annually.

Which alternative is preferred if the interest rate is 10%? Analyze using a government B/C ratio (public in numerator and government in denominator).

SOLUTION

Since something *must* be done and we have only two identified alternatives, we could simply analyze the difference between the alternatives to see which is better. But we are going to start by analyzing the less expensive "right-turns" alternative to check that this is a reasonable choice for what *must* be done.

The user consequences include an annual benefit for reduced congestion and a first "cost" that is the disbenefit of increased congestion during construction.

$$\text{PW-benefits}_{\text{right turns}} = -900,000 + 1,600,000(P/A, 10\%, 15)$$
$$= -900,000 + 1,600,000(7.606) = \$11.27\text{M}$$

The government costs include a first cost for construction and annual maintenance costs. Note that these are calculated as present *costs*.

$$\text{PW-costs}_{\text{right turns}} = 8,900,000 + 150,000(P/A, 10\%, 15)$$
$$= 8,900,000 + 150,000(7.606) = \$10.04\text{M}$$

The benefit–cost ratio for public divided by government consequences is

$$\text{B/C ratio} = \$11.27\text{M}/\$10.04\text{M} = 1.122$$

Thus, the right-turns-only alternative is better than doing nothing.

Now we evaluate the incremental investment for also doing the left-turn improvements. Because we are using a benefit–cost ratio, this evaluation must be done incrementally. The user consequences include an incremental annual benefit for reduced congestion and an incremental first "cost" that is the disbenefit of increased congestion during construction.

$$\text{PW-benefits}_{\text{left turns } - \text{ right turns}} = -2{,}100{,}000 - (-900{,}000) + (2{,}200{,}000 - 1{,}600{,}000)(P/A, 10\%, 15)$$
$$= -1{,}200{,}000 + 600{,}000(P/A, 10\%, 15)$$
$$= -1{,}200{,}000 + 600{,}000(7.606) = \$3.364\text{M}$$

The government costs include a first cost for construction and annual maintenance costs.

$$\text{PW-costs}_{\text{left turns } - \text{ right turns}} = 3{,}000{,}000 + 75{,}000(P/A, 10\%, 15)$$
$$= 3{,}000{,}000 + 75{,}000(7.606) = \$3.570\text{M}$$

The benefit–cost ratio for public divided by government consequences is

$$\text{B/C ratio} = \$3.364\text{M}/\$3.570\text{M} = 0.942$$

Thus, the right-turns-only alternative is better than adding the left-turn increment.

EXAMPLE 9–6

The industrial engineering department of Amalgamated Widgets is considering two alternatives for improving material flow in its factory. Both plans have a life of 15 years, because that is the estimated remaining life for the factory.

A minimal reconfiguration will cost \$8.9M (million) with annual maintenance costs of \$150,000. During construction there is a cost of \$900,000 for extra material movements and overtime, but more efficient movement of materials will save \$1.6M annually. The cost savings actually start lower and increase over time, but for a simple initial analysis we are assuming a uniform annual cost savings.

Reconfiguring a second part of the plant will cost an additional \$3M with an added annual maintenance cost of \$75,000. This construction is more disruptive, and the total cost for material movement and overtime congestion during construction is \$2.1M. Once complete, the total cost savings for more efficient movement of materials is \$2.2M annually.

Which alternative is preferred if the interest rate is 10%? Analyze using a present worth index (all consequences in Years 1 to n in numerator and all first costs in denominator).

SOLUTION

Since something *must* be done and we have only two identified alternatives, we could simply analyze the difference between the alternatives to see which is better. But we are going to start by analyzing the less expensive minimal reconfiguration alternative to check that this is a reasonable choice for what *must* be done.

The consequences in Years 1 to n include an annual cost savings for more efficient flow and annual maintenance costs.

$$\text{PW-benefits}_{\text{Years 1 to } n} = (1{,}600{,}000 - 150{,}000)(P/A, 10\%, 15)$$
$$= (1{,}600{,}000 - 150{,}000)(7.606) = \$11.03\text{M}$$

The first costs include a first cost for construction and the cost for disruption during construction.

$$\text{PW-costs} = 8{,}900{,}000 + 900{,}000 = \$9.8\text{M}$$

The present worth index is

$$\text{PW-index} = \$11.03\text{M}/\$9.8\text{M} = 1.125$$

Thus, the minimal reconfiguration is better than doing nothing.

Now we evaluate the incremental investment for also reconfiguring the second part of the plant. Because we are using a present worth index, this evaluation must be done incrementally. The annual consequences include an incremental annual cost savings and incremental maintenance costs.

$$\text{PW-benefits}_{\text{Years 1 to } n} = (600{,}000 - 75{,}000)(P/A, 10\%, 15)$$
$$= 525{,}000(7.606) = \$3.993\text{M}$$

There is a first cost for construction and for the associated disruption.

$$\text{PW-costs} = 3{,}000{,}000 + 1{,}200{,}000 = \$4.2\text{M}$$

The present worth index is

$$\text{PW-index} = \$3.993\text{M}/\$4.2\text{M} = 0.951$$

Thus, the minimal reconfiguration is better than reconfiguring the second part of the plant.

In Examples 9–5 and 9–6, the numbers that appeared in the numerator and denominator were changed, and the exact values of the B/C ratio and present worth index also changed. However, the conclusions did not. The ratios were above 1.0 for the minimal investment choice. The ratios were below 1.0 for the incremental investment. It was always best to make the minimal investment.

These examples demonstrate that present worth analysis and incremental benefit–cost ratio analysis lead to the same optimal decision. We saw in Chapter 8 that rate of return and present worth analysis led to identical decisions. Any of the exact analysis methods—present worth, annual cash flow, rate of return, or benefit–cost ratio—will lead to the same decision. Benefit–cost ratio analysis is extensively used in economic analysis at all levels of government.

PAYBACK PERIOD

Payback period measures the time required for the profit or other benefits from an investment to equal the cost of the investment. This is the general definition for payback period. Other definitions consider depreciation of the investment, interest, and income taxes; they, too, are simply called "payback period." We will limit our discussion to the simplest form.

> **Payback period** is the period of time required for the project's profit or other benefits to equal the project's cost.

The criterion in all situations is to minimize the payback period. The computation of payback period and its weaknesses relative to time value of money measures are illustrated in Examples 9–7 and 9-8.

EXAMPLE 9–7

The cash flows for two alternatives are as follows:

Year	A	B
0	−$1000	−$2783
1	200	1200
2	200	1200
3	1200	1200
4	1200	1200
5	1200	1200

You may assume the benefits occur throughout the year rather than just at the end of the year. Based on payback period, which alternative should be selected? For what interest rates is this the correct choice?

PAY BACK PERIOD SOLUTION

Because benefits occur throughout the year (like most engineering projects), fractional years have meaning.

Alternative A

Payback period is how long it takes for the profit or other benefits to equal the cost of the investment. In the first 2 years, only $400 of the $1000 cost is recovered. The remaining $600 cost is recovered in the first half of Year 3. Thus the payback period for Alt. A is 2.5 (= 2 + 600/1200) years.

Alternative B

Since the annual benefits are uniform, the payback period is simply

$$\$2783/\$1200 \text{ per year} = 2.3 \text{ years}$$

To minimize the payback period, choose Alt. B.

TIME VALUE OF MONEY SOLUTION

Previous chapters have shown that incremental analysis is needed to analyze mutually exclusive alternatives with rate of return. Since both alternatives have cash flows of $1200 in years 3 through 5, the incremental cash flows are zero, and the incremental analysis can focus on the cash flows for the first two years.

Year	B − A
0	−$1783
1	1000
2	1000

The incremental IRR is 8.0%, so for interest rates below 8% Alternative B will be preferred (matching the payback period solution). For interest rates above 8% Alternative A will be preferred. The easiest way to check this is to ask, which is preferred at 0% interest? By adding the cash flows we see that $PW_{B-A} = \$217 \, (= 2000 - 1783)$. So the PW of B is $217 higher than the PW of A at 0%. At 10% the PW of the increment is −$47, so A is preferred.

	A	B	C	D	E	F	G	H	I
1	Problem	i	n	PMT	PV	FV	Solve for	Answer	Formula
2	IRR_{B-A}		2	1000	-1783	0	RATE	8.011%	=RATE(C2,D2,E2,F2)
3	$PW_{10\%}$	10%	2	1000			PV	1736	
4								-1783	
5							$PW_{B-A\,10\%}$	-47	

EXAMPLE 9–8

A product will be phased out in 5 years, but it currently has a quality problem that is costing $4000 per year. Machines A and B will both solve the quality problem. Machine A costs only $10,000 but it will have no salvage value in 5 years. Machine B costs $15,000 but in 5 years it will have a salvage value of $9000. Based on payback period, which alternative should be selected? For what interest rates is this the correct choice?

PAYBACK PERIOD SOLUTION

As long as payback does not depend on the salvage value (usually true), the payback period for uniform cash flows is simply the first cost divided by the annual benefit. In this case:

$$\text{Payback}_A = \$10,000/(\$4000/\text{year}) = 2.5 \text{ years}$$

$$\text{Payback}_B = \$15,000/(\$4000/\text{year}) = 3.75 \text{ years}$$

So choosing to minimize the payback period would result in choosing Machine A.

TIME VALUE OF MONEY SOLUTION

Previous chapters have shown that incremental analysis is needed to analyze mutually exclusive alternatives with rate of return. Since both alternatives have cash flows of $4000 in years 1 through 5, the incremental analysis can focus on the cash flows for the first cost and the salvage value.

Year	B − A
0	−$5000
5	9000

	A	B	C	D	E	F	G	H	I
1	Problem	i	n	PMT	PV	FV	Solve for	Answer	Formula
2	IRR$_{B-A}$		5		-5000	9000	RATE	12.5%	=RATE(C2,D2,E2,F2)
3	PW$_{15\%}$	15%	5			9000	PV	4475	
4								-5000	
5							PW$_{B-A\ 15\%}$	-525	

The incremental IRR is 12.5%, so for interest rates below 12.5% Alternative *B* will be preferred (matching the payback period solution). For interest rates above 12.5% Alternative *A* will be preferred. The easiest way to check this is to ask, which is preferred at 0% interest? By adding the cash flows we see that at 0% PW$_{B-A}$ = $4000 (= 9000 − 5000). So the PW of *B* is $4000 higher than the PW of *A* at 0%. At 15% the PW of the increment is −$525, so *A* is preferred.

There are four important points to be understood about payback period calculations:

1. This is an approximate, rather than an exact, economic analysis calculation.
2. All costs and all profits or savings of the investment before payback are included *without* considering differences in their timing.
3. All the economic consequences beyond the payback period are completely ignored.
4. Being an approximate calculation, payback period may or may not select the correct alternative.

This last point—that payback period may select the *wrong* alternative—was illustrated by Examples 9–7 and 9–8. But if payback period calculations are approximate and may lead to selecting the wrong alternative for some or all interest rates, why are they used? First, the calculations can be readily made by people unfamiliar with economic analysis. Second, good projects and alternatives are usually clearly superior by all measures. Third, payback period is easily understood. Earlier we pointed out that this is also an advantage to rate of return. The vignette in Chapter 3 described a high payout project. The NPV was about $8.6M and the IRR was over 11,000%. Both of these can be confusing to management, but the idea that the project paid out in one day was very clear.

Moreover, payback period *does* measure **liquidity**—how long it will take for the cost of the investment to be recovered from its benefits. Firms are often very interested in this time period: a rapid return of invested capital means that the funds can be reused sooner for other purposes. But one must not confuse the *speed* of the return of the investment, as measured by the payback period, with **profitability**. They are two distinctly separate concepts.

There is a refinement of the payback period that does include interest—the **discounted payback period**. The discounted payback period is longer than the payback period, because the benefits must also cover the interest on the capital invested in the project. If the annual benefits are uniform and the salvage value is $0, then NPER can be used to calculate the discounted payback period—as long as that period is less than the alternative's life. If the salvage value becomes involved in achieving payback, then the horizon is the payback period.

The discounted payback period includes some consideration of the time value of money, so it is a better measure than payback period, but discounted payback period is still not a valid time value of money measure that includes all cash flows. In Example 9–9, there are clearly salvage values that would make either alternative the preferred choice. Payback periods, whether discounted or not, ignore cash flows that occur after payback.

EXAMPLE 9–9

Two alternatives have been identified. Alternative A has a first cost of $10,000 and benefits of $3000 annually. Alternative B has a first cost of $12,000 and benefits of $3500 annually. The salvage values of each alternative are currently unknown. Using discounted payback period, which alternative is preferred at an interest rate of 10%?

SOLUTION

Even if the salvage values were known, they must be ignored for the NPER calculation. If they are included, then NPER assumes that the salvage value occurs at time of payback—which may dramatically shorten the calculated time period.

	A	B	C	D	E	F	G	H	I
1	Exp. 9-9	i	n	PMT	PV	FV	Solve for	Answer	Formula
2	A	10%		3000	-10,000	0	NPER	4.25	=NPER(B2,D2,E2,F2)
3	B	10%		3500	-12,000	0	NPER	4.41	

Alternative A has a discounted payback period of 4.25 years versus 4.41 years for Alternative B. So A is preferred using the discounted payback period. Whether or not this is correct can only be determined by a valid time value of money computation that includes the unknown salvage values. We do not yet know which alternative has the better present worth at 10%.

From the discussion and the examples, we see that payback period can measure the speed of the return of the investment. This might be quite important, for example, for a company that is short of working capital or for a firm in an industry experiencing rapid changes in technology. Calculation of payback period alone, however, must not be confused with a careful economic analysis. Ignoring all cash flows after the payback period is seldom wise. We have shown that a short payback period does not always mean that the associated investment is desirable. Thus, payback period is not a suitable replacement for accurate economic analysis calculations.

SENSITIVITY AND BREAKEVEN ANALYSIS

Since many data gathered in solving a problem represent *projections* of future consequences, there may be considerable uncertainty regarding the data's accuracy. Since the goal is to make good decisions, an appropriate question is: To what extent do variations in the data affect my decision? When small variations in a particular estimate would change which alternative is selected, the decision is said to be **sensitive to the estimate.** To better evaluate the impact of any particular estimate, we compute "how much a particular estimate would need to change in order to change a particular decision." This is called **sensitivity analysis.** Chapter 8 compared economic values at different interest rates with graphs and choice tables. That was also a form of sensitivity analysis.

An analysis of the sensitivity of a problem's decision to its various parameters highlights the important aspects of that problem. For example, estimated annual maintenance and salvage values may vary substantially. Sensitivity analysis might indicate that a certain decision is insensitive to the salvage-value estimate over the full range of possible values. But, at the same time, we might find that the decision is sensitive to changes in the annual maintenance estimate. Under these circumstances, one should place greater emphasis on improving the annual maintenance estimate and less on the salvage-value estimate.

As indicated at the beginning of this chapter, breakeven analysis is a form of sensitivity analysis that is often presented as a **breakeven chart**. Another nomenclature that is sometimes used for the breakeven point is *point of indifference.* One application of these tools is **staged construction.** Should a facility be constructed now to meet its future full-scale requirement? Or should it be constructed in stages as the need for the increased capacity arises? What is the breakeven point on how soon the capacity is needed for this decision? Three examples are:

- Should we install a cable with 400 circuits now or a 200-circuit cable now and another 200-circuit cable later?
- A 10-cm water main is needed to serve a new area of homes. Should it be installed now, or should a 15-cm main be installed to ensure an adequate water supply to adjoining areas later, when other homes have been built?
- An industrial firm needs a new warehouse now and estimates that it will need to double its size in 4 years. The firm could have a warehouse built now and later enlarged, or the firm could have the largeer warehouse built right away.

Examples 9–10 and 9–11 illustrate sensitivity and breakeven analysis. These examples have focused on the breakeven project life, because that value of *n* is often one of the most uncertain values in an economic analysis. How long will that bridge, machine, or product function and meet the need?

EXAMPLE 9–10

Consider a project that may be constructed to full capacity now or may be constructed in two stages.

Construction Alternative	Costs
Two-stage construction	
Construct first stage now	$100,000
Construct second stage	120,000
n years from now	
Full-capacity construction	
Construct full capacity now	140,000

After 40 years all facilities will have zero salvage value. The annual cost of operation and maintenance is the same for both alternatives. With an 8% interest rate, what is the breakeven *n*?

5-BUTTON SOLUTION

As emphasized previously, the choice between two mutually exclusive alternatives can be made by analyzing the incremental difference between them. For rate of return and benefit–cost methods, incremental analysis is required. In this case, building the full capacity now costs $40,000 more than the initial cost of the two-stage alternative. This will save the $120,000 required whenever the second stage would be built. NPER can be used to solve for the life that gives these cash flows the same worth at an interest rate of 10%.

	A	B	C	D	E	F	G	H	I
1	Exp. 9-10	i	n	PMT	PV	FV	Solve for	Answer	Formula
2	Full–staged	8%		0	-40,000	120,000	NPER	14.27	=NPER(B2,D2,E2,F2)

Example 9–12 will revisit this example to create and improve the graph of each alternative's present worth.

TABLE SOLUTION

Since we are dealing with a common analysis period, the calculations may be either annual cost or present worth. Present worth calculations appear simpler and are used here. Incremental analysis with a simple factor equation matching the 5-BUTTON SOLUTION will be presented, but starting with a graphical approach (like Chapter 8) seems likely to support a better understanding.

Construct Full Capacity Now

$$\text{PW of cost} = \$140,000$$

Two-Stage Construction

If the first stage is to be constructed now and the second stage *n* years hence, compute the PW of cost for several values of *n* (years).

$$\text{PW of cost} = 100,000 + 120,000(P/F, 8\%, n)$$

$$n = 5 \qquad \text{PW} = 100,000 + 120,000(0.6806) = \$181,700$$
$$n = 10 \qquad \text{PW} = 100,000 + 120,000(0.4632) = \ \ 155,600$$

$$n = 20 \qquad PW = 100,000 + 120,000(0.2145) = 125,700$$
$$n = 30 \qquad PW = 100,000 + 120,000(0.0994) = 111,900$$

These data are plotted in the form of a breakeven chart in Figure 9–2.

FIGURE 9–2 Breakeven chart for Example 9–10.

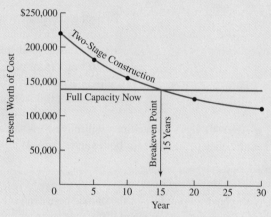

Age When Second Stage Constructed

In Figure 9–2 we see that the PW of cost for two-stage construction naturally decreases as the second stage is deferred. The one-stage construction (full capacity now) is unaffected by the *x*-axis variable and, hence, is a horizontal line.

The breakeven point on the graph is where both alternatives have the same PW. This is about 15 years. If the second stage were to be needed prior to Year 15, then one-stage construction has the lower cost. On the other hand, if the second stage would not be required until after 15 years, two-stage construction has the lower cost.

This breakeven point can be more accurately calculated by setting the two alternatives equal to each other.

$$PW = 140,000 = 100,000 + 120,000(P/F, 8\%, n)$$
$$(P/F, 8\%, n) = \frac{40,000}{120,000} = 0.3333$$

From the tables

$$n = 14 + (15 - 14)(0.3405 - 0.3333)/(0.3405 - 0.3152)$$
$$n = 14.3 \text{ years}$$

SENSITIVITY DISCUSSION

The decision on how to construct the project is sensitive to the age at which the second stage is needed *only* if the range of estimates includes 15 years. For example, if one estimated that the second-stage capacity would be needed between 5 and 10 years hence, the decision is insensitive to that estimate. For any value within that range, the decision does not change. But, if the second-stage capacity were to be needed sometime between, say, 12 and 18 years, the decision would be sensitive to the estimate of when the full capacity would be needed.

EXAMPLE 9–11 (Examples 9–5 and 9–6 Revisited)

In both Examples 9–5 (traffic congestion on Riverview Boulevard) and 9–6 (reconfiguring the plant of Amalgamated Widgets), the life of 15 years is clearly subject to some uncertainty. While holding the other data constant, analyze the sensitivity of the recommended decisions to the project life. Use the present worth measure, since it is the same for both examples.

SOLUTION

The two alternatives have the following present worth values.

$$PW_{\text{right turns or minimal}} = -900{,}000 - 8{,}900{,}000 + (1{,}600{,}000 - 150{,}000)(P/A, 10\%, n)$$
$$= -9{,}800{,}000 + 1{,}450{,}000(P/A, 10\%, n)$$
$$PW_{\text{left turns or 2}^{\text{nd}}\text{ part of plant}} = -2{,}100{,}000 - 8{,}900{,}000 - 3{,}000{,}000$$
$$+ (2{,}200{,}000 - 225{,}000)(P/A, 10\%, n)$$
$$= -14{,}000{,}000 + 1{,}975{,}000(P/A, 10\%, n)$$

These could be analyzed for breakeven values of n. However, it is easier to use the graphing technique for multiple alternatives that was presented in Chapter 8. Instead of using the interest rate for the x axis, use n.

As shown in Figure 9–3, the right-turn or minimal alternative is the best one for lives of 12 to 16 years. The left-turn increment or 2nd part of the plant, is the best choice for lives of 17 or more years. If the life is 11 years or less, doing nothing is better. To keep the graph readable, Figure 9–3 includes only Years 10 through 20.

FIGURE 9–3 Breakeven chart for Example 9–11.

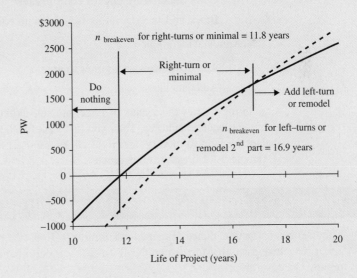

Breakeven points can be estimated from graphs or calculated with formulas or GOAL SEEK. Sensitivity analysis and breakeven point calculations can be very useful in identifying how different estimates affect the calculations. It must be recognized that these calculations assume that all parameters except one are held constant, and the sensitivity of the decision to that one variable is evaluated. The next section presents ways to modify your Excel chart to make it more effective.

GRAPHING WITH SPREADSHEETS FOR SENSITIVITY AND BREAKEVEN ANALYSIS

Chapter 4 introduced drawing xy plots with spreadsheets, and Chapter 8 relied on plots of PW, EUAW, and EUAC versus i. This section will present some of the spreadsheet tools and options that can make the xy plots more effective and attractive.

The spreadsheet tools and options can be used to:

- Modify the x or y axes

    Specify the minimum or maximum value

    Specify at what value the other axis intersects (default is 0)

- Match line types to data

    Use line types to distinguish one curve from another

    Use markers to show real data

    Use lines without markers to plot curves (straight segments or smooth curves)

- Match chart colors to how displayed

    Color defaults are fine for color computer screen

    Color defaults are OK for color printers

    Black-and-white printing is better with editing (use line types not colors)

- Annotate the graph

    Add text, arrows, and lines to graphs

    Add data labels

In most cases the menus of Excel are self-explanatory, so the main step is deciding what you want to achieve. Then you just look for the way to do it. Left clicks are used to select the item to modify, and right clicks are used to bring up the options for that item. Example 9–12 illustrates this process.

EXAMPLE 9–12   (Example 9–10 Revisited)

The staged construction choice described in Example 9–10 used a broad range of x values for the x axis. Create a graph that focuses on the 10- to 20-year period and is designed for printing in a report. The costs are:

Year	Full Capacity	Two Stages
0	$140,000	$100,000
n	0	120,000

SOLUTION

The first step is to create a table of values that shows the present worth of the costs for different values of $n \equiv$ the length of time until the second stage or full capacity is needed. Notice that the full capacity is calculated at $n = 0$. The only reason to calculate the corresponding value for staged construction is to see if the formula is properly entered, since building both stages at the same time will not really cost $220,000. The values for staged construction at 5, 10, 20, and 30 years check with the values in Example 9–10.

The next step is to select cells A8:C13, which includes the x values and two series of y values. Then the ChartWizard tool is selected. In the first step, the xy (scatter) plot is selected with the option of smoothed lines without markers. In Step 2 no action is required, since the cells A8:C13 were selected first. In Step 3 labels are added for the x and y axes. In Step 4 the chart is moved around on the worksheet page, so that it does not overlap with the data. The result is shown in Figure 9–4.

FIGURE 9–4

Automatic graph from spreadsheet.

	A	B	C	D
1	Time	Full Capacity	Two Stage	
2	0	140,000	100,000	
3	n		120,000	
4				
5	Life	40		
6	i	8%		
7			Present Worth of	
8	n	Full Capacity	Two Stage	
9	0	140,000	220,000	=$C$2+PV($B$6,A9,0,−$C$3)
10	5	140,000	181,670	
11	10	140,000	155,583	
12	20	140,000	125,746	
13	30	140,000	111,925	

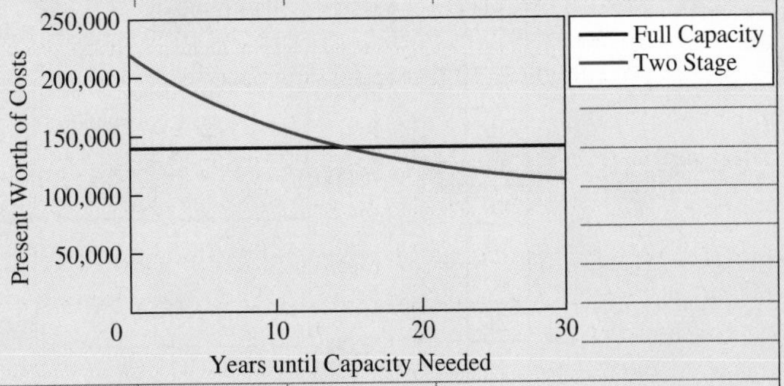

Our first step in cleaning up the graph is to delete the formula in cell C9, since two-stage construction will not be done at Time 0. We also delete the label in the adjacent cell, which explains the formula. Then we create a new label for cell C10. As shown in Appendix A, the easy way to create that label is to insert an apostrophe or space, as the first entry in cell C10. This converts the formula to a label that we can copy to D10. Then we delete the apostrophe or space in cell C10.

The axis scales must be modified to focus on the area of concern. Select the x axis and change the minimum from automatic to 10 and the maximum to 20. Select the y axis and change the minimum to 125,000 and the maximum to 160,000.

Left-click on the plot area to select it. Then right-click to bring up the options. Select Format Plot Area and change the area pattern to "none." This will eliminate the gray fill that made Figure 9–4 difficult to read.

Left-click on the two-stage curve to select it. Then right-click for the options. Format the data series using the Patterns tab. Change the line style from solid to dashed, the line color from automatic to black, and increase the line weight. Similarly, increase the line weight for the full-capacity line. Finally, select a grid line and change the line style to dotted. The result is far easier to read in black and white.

FIGURE 9–5
Spreadsheet of Figure 9–4 with improved graph.

	A	B	C	D
1	Time	Full Capacity	Two Stage	
2	0	140,000	100,000	
3	n		120,000	
4				
5	life	40		
6	i	8%		
7		Present worth of		
8	n	Full Capacity	Two Stage	
9	0	140,000		
10	5	140,000	181,670	=$C$2+PV($B$6,A10,0,−$C$3)
11	10	140,000	155,583	
12	20	140,000	125,746	
13	30	140,000	111,925	
14				

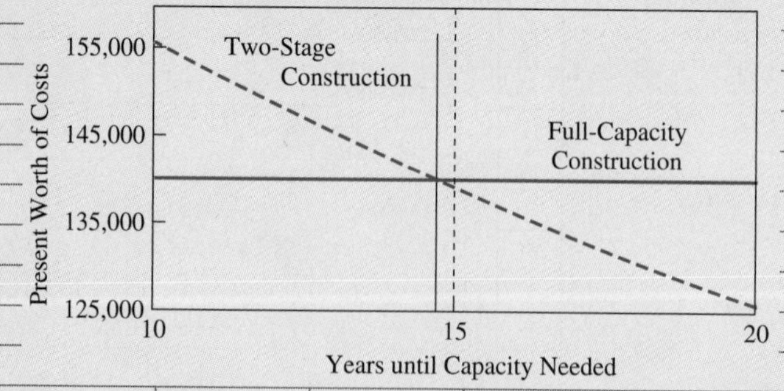

To further improve the graph, we can replace the legend with annotations on the graph. Left-click somewhere in the white area around the graph to select "chart area." Right-click and then choose the chart options on the menu. The legends tab will let us delete the legend by turning "show legend" off. Similarly, we can turn the *x*-axis gridlines on. The line style for these gridlines should be changed to match the *y*-axis gridlines. This allows us to see that the breakeven time is between 14 and 15 years.

To make the graph less busy, change the scale on the *x* axis so that the interval is 5 years rather than automatic. Also eliminate the gridlines for the *y* axis (by selecting the Chart Area, Chart Options, and Gridlines tabs). The graph size can be increased for easier reading as well. This may require specifying an interval of 10,000 for the scale of the *y* axis.

Finally, to add the labels for the full-capacity curve and the two-stage curve, find the toolbar for graphics, which is open when the chart is selected (probably along the bottom of the spreadsheet). Select the text box icon, and click on a location close to the two-stage chart. Type in the label for two-stage construction. Notice how including a return and a few spaces can shape the label to fit the slanted line. Add the label for full construction. Figure 9–5 is the result.

DOING WHAT- IF ANALYSIS WITH SPREADSHEETS

Breakeven charts change one variable at a time, while **what-if analysis** may change many of the variables in a problem. However, spreadsheets remain a very powerful tool for this form of sensitivity analysis. In Example 9–13 a project appears to be very promising. However, what-if analysis indicates that a believable scenario raises some questions about whether the project should be done.

EXAMPLE 9–13

You are an assistant to the vice president for manufacturing. The staff at one of the plants has recommended approval for a new product with a new assembly line to produce it. The VP believes that the numbers presented are too optimistic, and she has added a set of adjustments to the original estimates. Analyze the project's benefit–cost ratio or present worth index as originally submitted. Reanalyze the project, asking "What if the VP's adjustments are correct?"

	Initial Estimate	Adjustment
First cost	$70,000	+10%
Units/year	1,200	−20%
Net unit revenue	$ 25	−15%
Life, in years	8	−3
Interest rate	12%	None

> **SOLUTION**
>
> Figure 9–6 shows that the project has a 2.13 benefit–cost ratio with the initial estimates, but only a value of 0.96 with the what-if adjustments. Thus we need to determine which set of numbers is more realistic. Real-world experience suggests that in many organizations the initial estimates are too optimistic. Auditing of past projects is the best way to develop adjustments for future projects.

FIGURE 9–6 Spreadsheet for what-if analysis.

		A	B	C	D
1			Initial Estimate	Adjust-ment	Adjusted Values
2	First cost		$70,000	10%	$77,000
3	Units/year		1,200	−20%	960
4	Net unit revenue		$25	−15%	$21
5	Life (years)		8	−3	5
6	Interest rate		12%	none	12%
7					
8		Benefits	149,029		73,537
9		Cost	70,000		77,000
10		B/C Ratio	2.13		0.96
11					
12			=PV(B6,B5,-B3*B4)		

SUMMARY

In this chapter, we have looked at four new analysis techniques.

Future worth: When the comparison between alternatives will be made in the future, the calculation is called future worth. This is very similar to present worth, which is based on the present, rather than a future point in time.

Benefit–cost ratio analysis: This technique is based on the ratio of benefits to costs using either present worth or annual cash flow calculations. The method is graphically similar to present worth analysis. When neither input nor output is fixed, incremental benefit–cost ratios ($\Delta B/\Delta C$) are required. The method is similar in this respect to rate of return analysis. Benefit–cost ratio analysis is often used at the various levels of government.

Payback period: Here we define payback as the period of time required for the profit or other benefits of an investment to equal the cost of the investment. Although simple to use and understand, payback is a poor analysis technique for ranking alternatives. While it provides a measure of the liquidity or speed of the return of the investment, it is not an accurate measure of the profitability of an investment.

Sensitivity, breakeven, and what-if analysis: These techniques are used to see how sensitive a decision is to estimates for the various parameters. Breakeven analysis is done to locate conditions under which the alternatives are equivalent. This is often presented in the form of breakeven charts. Sensitivity analysis examines a range of values of some parameters to determine the effect on a particular decision. What-if analysis changes one or many estimates to see what results.

PROBLEMS

Future Worth

9-1 Pick a discretionary expense that you incur on a regular basis, such as buying premium coffees daily, buying fashion items monthly, buying sports tickets monthly, or going to movies weekly. Assume that you instead place the money in an investment account that earns 6% annually. After 40 years, how much is in the account?

9-2 (A) Sally deposited $100 a month in her savings account for 24 months. For the next 5 years she made no deposits. What is the future worth in Sally's savings account at the end of the 7 years, if the account earned 6% annual interest, compounded monthly?

9-3 A new engineer is considering investing in an individual retirement account (IRA) with a mutual fund that has an average annual return of 8%. What is the future worth of her IRA at age 70 if she makes annual investments of $2000 into the fund beginning on her 22nd birthday? Assume that the fund continues to earn an annual return of 8%.

9-4 (A) (G) You can buy a piece of vacant land for $30,000 cash. You plan to hold it for 15 years and then sell it at a profit. During this period, you would pay annual property taxes of $600. You would have no income from the property.

(*a*) Assuming that you want a 10% rate of return, at what net price would you have to sell the land 15 years hence?

(*b*) What is open space conservation and why is it important? What options would you have in selling your property with this in mind?

9-5 Compute the future worth for the following cash flows.

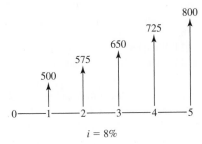

$i = 8\%$

9-6 (A) An individual who makes $32,000 per year anticipates retiring in 30 years. If his salary is increased by $600 each year and he deposits 10% of his yearly salary into a fund that earns 7% interest, what is the future worth at retirement?

9-7 For the following cash flows, compute the future worth.

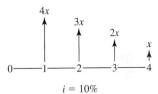

$i = 10\%$

9-8 (A) The interest rate is 9% per year and there are 48 compounding periods per year. The principal is $50,000. What is the future worth in 5 years?

9-9 A company deposits $10,000 in a bank at the beginning of each year for 20 years. The account earns 4% interest, compounded every 6 months. What is in the account at the end of 20 years?

9-10 (A) In the early 1980s, planners were examining alternate sites for a new London airport. At one potential site, the twelfth-century Norman church of St. Michaels, in the village of Stewkley, would have had to be demolished. The planners used the value of the fire insurance policy on the church—a few thousand pounds sterling—as the church's value.

An outraged antiquarian wrote to the London *Times* that an equally plausible computation would be to assume that the original cost of the church (estimated at 100 pounds sterling) be increased at the rate of 10% per year for 800 years. Based on his proposal, what would be the future worth of St. Michaels? (*Note:* There was great public objection to tearing down the church, and it was spared.)

9-11 If you invested $5000 in a 24-month bank certificate of deposit (CD) paying 3%, compounded monthly, what is the future worth of the CD when it matures?

9-12 For a 15% interest rate, compute the value of F so the following cash flows have a future worth of 0.

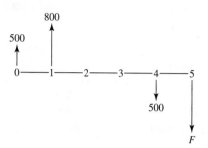

9-13 Calculate the present worth and the future worth of a series of 15 annual cash flows with the first cash flow equal to $15,000 and each successive cash flow increasing by $750. The interest rate is 6%.

9-14 A 20-year-old student decided to set aside $100 on his 21st birthday for investment. Each subsequent year through his 55th birthday, he plans to increase the investment on a $100 arithmetic gradient. He will not set aside additional money after his 55th birthday. If the student can achieve a 8% rate of return, what is the future worth of the investments on his 65th birthday?

9-15 For the following cash flows, compute the future worth.

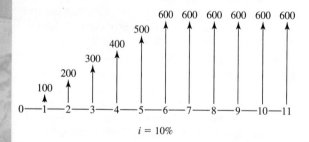

9-16 Stamp collecting has become an increasingly and popular hobby, stamps have been a good place to invest money over the last 10 years, as demand has caused resale prices to increase 18% each year. Suppose a collector purchased $100 worth of stamps 10 years ago and increased his purchases by $50 per year in each subsequent year. After 10 years of stamp collecting, what is the current worth of the stamp collection?

9-17 After receiving an inheritance of $50,000 on her 21st birthday, Katlyn deposited the inheritance in a savings account with an effective annual interest rate

of 3%. She decided to make regular deposits, beginning with $1000 on her 22nd birthday and increasing by $200 each year (i.e., $1200 on her 23rd birthday, $1400 on her 24th birthday, etc.). What was the future worth of Katlyn's deposits after her deposit on her 65th birthday?

9-18 Bill made a budget and planned to deposit $150 a month in a savings account, beginning September 1. He did this, but on the following January 1, he reduced the monthly deposits to $100. He made 18 deposits, four at $150 and 14 at $100. If the savings account paid 6% interest, compounded monthly, what was the future worth of his savings immediately after the last deposit?

9-19 Compute F so the following cash flows have a future worth of 0.

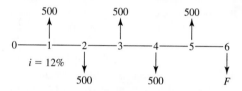

9-20 A family starts an education fund for their son Patrick when he is 8 years old, investing $150 on his eighth birthday, and increasing the yearly investment by $150 per year until Patrick is 18 years old. The fund pays 6% annual interest. What is the fund's future worth when Patrick is 18?

9-21 IPS Corp. will upgrade its package-labeling machinery. It costs $850,000 to buy the machinery and have it installed. Operation and maintenance costs, which are $10,000 per year for the first 3 years, increase by $1000 per year for the machine's 10-year life. The machinery has a salvage value of 10% of its initial cost. Interest is 25%. What is the future worth of cost of the machinery?

9-22 A bank account pays 3% interest with monthly compounding. A series of deposits started with a deposit of $5000 on January 1, 2007. Deposits in the series were to occur each 6 months. Each deposit in the series is for $150 less than the one before it. The last deposit in the series will be due on July 1, 2021. What is the future worth of the account on July 1, 2024, if the balance was zero before the first deposit and no withdrawals are made?

9-23 Jamal Brown is a 55-year-old engineer. According to mortality tables, a male at age 55 has an average life expectancy of 21 more years. Jamael has accumulated $200,000 toward his retirement. He is now adding $10,000 per year to his retirement fund. The fund earns 5% interest. Jamal will retire when he can obtain an annual income from his retirement fund of $100,000, assuming he lives to age 76. He will make no provision for a retirement income after age 76. What is the youngest age at which Jamal can retire?

9-24 An engineering graduate starts a new job at $60,000 per year. Her investments are deposited at the end of the year into a mutual fund that earns a nominal interest rate of 3% per year with semiannual compounding. How much money will be in the account immediately after she makes the last deposit?

(*a*) She makes $6000 annual deposits for the next 40 years.

(*b*) She makes the $6000 deposits for 10 years, then stops all investments for the next 10 years, and then resumes deposits of $10,000 per year for the next 20 years.

Contributed by Gillian Nicholls, Southeast Missouri State University

9-25 A company is considering buying a new bottle-capping machine. The initial cost of the machine is $1.2M and it has a 10-year life. Monthly maintenance costs are expected to be $2000 per month for the first 7 years and $2500 per month for the remaining years. The machine requires a major overhaul costing $100,000 at the end of the fifth year of service. Assume that all these costs occur at the end of the appropriate period.

(*a*) What is the future value of all the costs of owning and operating this machine if the nominal interest rate is 9% compounded monthly?

(*b*) Bottle caps are in the top 10 items found in beach cleanups. What are other items on such lists, and what are beach cities doing to reduce this debris?

9-26 The Association of General Contractors (AGC) is endowing a fund of $1 million for the Construction Engineering Technology Program at Grambling State University. The AGC established an escrow account in which 10 equal end-of-year deposits that earn 7% compound interest were to be made. After seven deposits, the Louisiana legislature revised laws relating to the licensing fees AGC can charge its members, and there was no deposit at the end of Year 8. What must the amount of the remaining equal end-of-year deposits be, to ensure that the $1 million is available on schedule for the Construction Engineering Technology Program?

9-27 A recent college graduate got a good job and began a savings account. He authorized the bank to automatically transfer $500 each month from his checking account to the savings account. The bank made the first withdrawal on July 1, 2015 and is instructed to make the last withdrawal on January 1, 2040. The bank pays a nominal interest rate of 6% and compounds twice a month. What is the future worth of the account on January 1, 2040?

9-28 A business executive is offered a management job at Generous Electric Company, which offers him a 5-year contract that calls for a salary of $62,000 per year, plus 600 shares of GE stock at the end of the 5 years. This executive is currently employed by Fearless Bus Company, which also has offered him a 5-year contract. It calls for a salary of $65,000, plus 100 shares of Fearless stock each year. The Fearless stock is currently worth $60 per share and pays an annual dividend of $2 per share. Assume end-of-year payments of salary and stock. Stock dividends begin one year after the stock is received. The executive believes that the value of the stock and the dividend will remain constant. If the executive considers 9% a suitable rate of return in this situation, what must the Generous Electric stock be worth per share to make the two offers equally attractive? Use the future worth analysis method in your comparison.

9-29 Jean invests $1000 in Year 1 in a socially responsible fund, and doubles the amount each year after that (so the investment is $1000, 2000, ...).

(*a*) If she does this for 10 years, and the investment pays 4% annual interest, what is the future worth of her investment?

(*b*) What are socialy/ethically responsible investment funds? How do they differ from other types of investments? Why do people invest in them?

Benefit–Cost Ratio

9-30 Each of the three mutually exclusive alternatives shown has a 5-year useful life. If the MARR is 10%, which alternative should be selected? Solve the problem by benefit–cost ratio analysis.

	A	*B*	*C*
Cost	$600.0	$500.0	$200.0
Uniform annual benefit	158.3	138.7	58.3

9-31 Consider three mutually exclusive alternatives, each with a 15-year useful life. If the MARR is 12%, which alternative should be selected? Solve the problem by benefit–cost ratio analysis.

	A	B	C
Cost	$800	$300	$150
Uniform annual benefit	130	60	35

9-32 An investor is considering buying some land for $100,000 and constructing an office building on it. Three different buildings are being analyzed.

	Building Height		
	2 Stories	5 Stories	10 Stories
Cost of building (excluding cost of land)	$400,000	$800,000	$2,100,000
Resale value* of land + building after 20-year horizon	200,000	300,000	400,000
Annual net rental income	70,000	105,000	256,000

*Resale value considered a reduction in cost—not a benefit.

Using benefit–cost ratio analysis and an 8% MARR, determine which alternative, if any, should be selected.

9-33 Using benefit–cost ratio analysis, determine which one of the three mutually exclusive alternatives should be selected. Each alternative has a 10-year useful life. Assume a 20% MARR.

	A	B	C
First cost	$560	$340	$120
Uniform annual benefit	140	100	40
Salvage value	40	0	0

9-34 A government agency is planning a new office building close to its current headquarters. Four proposed sites are to be evaluated. Any of these sites will save the agency $700,000 per year, since two of its current satellite offices will no longer need to be rented. The agency uses a 6% interest rate and assumes that the building and its benefits will last for 40 years. Based on a benefit–cost analysis what should the agency do?

	Site			
	A	B	C	D
Initial cost	$8.6M	$8.1M	$7.5M	$6.8M
Annual O&M	0.12M	0.155M	0.2M	0.3M

Contributed by Hamed Kashani, Saeid Sadri, and Baabak Ashuri, Georgia Institute of Technology

9-35 Using benefit–cost ratio analysis, a 10-year useful life, and a 25% MARR, determine which of the following mutually exclusive alternatives should be selected.

	A	B	C	D	E
Cost	$100	$200	$300	$400	$500
Annual benefit	37	60	83	137	150

9-36 Five mutually exclusive investment alternatives have been proposed. Based on benefit–cost ratio analysis, and a MARR of 15%, which alternative should be selected?

Year	A	B	C	D	E
0	−$200	−$100	−$125	−$150	−$225
1–5	68	25	42	52	68

9-37 Three mutually exclusive projects are being considered by Sesame Street Productions (SSP). SSP uses a MARR of 8%. SSP has heard about your excellent analysis skills and wants you to help them make a decision. Using a B/C analysis, which project do you recommend to SSP? Assume all benefits and costs repeat for Project A.

	Project A	Project B	Project C
Initial cost	$300	$450	$765
Annual benefits	$200	$190	$300
Project life (years)	2	4	4
B/C ratio	1.19	1.40	1.30

Contributed by Gana Natarajan, Oregon State University

9-38 Burns City may build a garbage incinerator on the outskirts of town. Environmental impact statements and safety planning/inspection will cost $25,000 (payable at start of construction). The annual upkeep and operating costs are expected to be $30,000. The new incinerator will save $12 each annually for 26,500 billed customers. Consultants have estimated an annual disbenefit to the surrounding area of $32,000. At the end of a 10-year useful life the incinerator will be dismantled at a cost of $45,000. Using benefit-cost ratio analysis, and assuming a cost of money of 4% what is the maximum that Burns City can pay to build the incinerator? *Contributed by Ed Wheeler, University of Tennessee, Martin*

9-39 Looking at Figure 9-1 another way to pick the best mutually exclusive alternative is to maximize the perpendicular distance from the alternative's point to the B/C = 1. Explain why this is true. *Contributed by Hector Medina, Liberty University*

9-40 Cornell has two options for upgrading their athletic facilities. The off-campus option costs only $20 million, but it will require frequent bus service to those facilities at an annual cost that starts at $300,000 and increases by 4% per year. (buses, drivers' and mechanics' salaries, maintenance, road wear, etc.). Improving the on-campus facilities will cost $50 million, but no extra transportation costs are required. Both options involve an estimated annual maintenance cost of $1 million for about 40 years before new facilities will again be needed. Using benefit–cost ratio analysis, determine which option is more economically efficient. Use an interest rate of 8% per year. *Contributed by D. P. Loucks, Cornell University*

9-41 A do-nothing and two mutually exclusive alternatives are being considered for reducing traffic congestion. User benefits come from reduced congestion once the project is complete, while user disbenefits are due to increased congestion during construction. The interest rate is 8%, and the life of each alternative is 15 years. Which alternative should be chosen?

Alternative	A	B
User benefits ($M/yr)	2.1	2.6
User disbenefits ($M)	1.2	2.1
First cost ($M)	6.9	9.9
Operations and maintenance ($M/yr)	0.75	0.825

(a) Use the benefit–cost ratio.
(b) Use the modified benefit–cost ratio.
(c) Use the public/government version of the B/C ratio.
(d) Assume these numbers apply to a private firm and use a present worth index.
(e) Are your recommendations for (a) through (d) consistent? Which measure gives the largest value? Why?

9-42 A school is overcrowded and there are three options. The do-nothing alternative corresponds to continuing to use modular classrooms. The school can be expanded, or a new school can be built to "split the load" between the schools. User benefits come from improvements in school performance for the expanded or new schools. If a new school is built, there are more benefits because more students will be able to walk to school, the average distance for those who ride the school buses will be shorter, and the schools will be smaller and more "student friendly." The disbenefits for the expanded school are due to the impact of the construction process during the school year. The interest rate is 8%, and the life of each alternative is 20 years. Which alternative should be chosen? What is the incremental ratio for the preferred alternative?

Alternative	Expand	New School
User benefits ($M/yr)	2.1	3.1
User disbenefits ($M)	0.8	0
First cost ($M)	8.8	10.4
Operations and maintenance ($M/yr)	0.95	1.7

(a) Use the benefit–cost ratio.
(b) Use the modified benefit–cost ratio.
(c) Use the public/government version of the B/C ratio.
(d) Assume these numbers apply to a private firm and use a present worth index.
(e) Are your recommendations for (a) through (d) consistent? Which measure gives the largest value? Why?
(f) Describe the ethical issues involved in the overcrowded schools dilemma in terms of stakeholders and impacts.

Payback Period and Exact Methods

9-43 A project has the following costs and benefits. What is the payback period?

Year	Costs	Benefits
0	$65,000	
1–2	15,000	
3	5,000	$50,000
4–10		$10,000 in each year

9-44 Able Plastics, an injection-molding firm, has
A negotiated a contract with a national chain of department stores. Plastic pencil boxes are to be produced for a 2-year period. If the firm invests $67,000 for special removal equipment to unload the completed pencil boxes from the molding machine, one machine operator can be eliminated saving $26,000 per year. The removal equipment has no salvage value and is not expected to be used after the 2-year production contract is completed. The equipment would be serviceable for about 15 years. What is the payback period? Should Able Plastics buy the removal equipment?

9-45 A car dealer leases a small computer with software for $5000 per year. As an alternative he could buy the computer for $7500 and lease the software for $3500 per year. Any time he would decide to switch to some other computer system he could cancel the software lease and sell the computer for $500.

(*a*) If he buys the computer and leases the software, what is the payback period?

(*b*) If he kept the computer and software for 8 years, what would be the benefit–cost ratio, based on an 5% interest rate?

9-46 Tom Sewel has gathered data on the relative costs
A of a solar water heater system and a conventional
G electric water heater. The data are based on statistics for a mid-American city and assume that during cloudy days an electric heating element in the solar heating system will provide the necessary heat.

The installed cost of a conventional electric water tank and heater is $200. A family of four uses an average of 300 liters of hot water a day, which takes $230 of electricity per year. The glass-lined tank has a 20-year guarantee. This is probably a reasonable estimate of its actual useful life.

The installed cost of two solar panels, a small electric pump, and a storage tank with auxiliary electric heating element is $1400. It will cost $60 a year for electricity to run the pump and heat water on cloudy days. The solar system will require $180 of maintenance work every 4 years. Neither the conventional electric water heater nor the solar water heater will have any salvage value at the end of its useful life.

Using Tom's data, what is the payback period if the solar water heater system is installed, rather than the conventional electric water heater?

9-47 A cannery is considering installing an automatic case-sealing machine to replace current hand methods. If they purchase the machine for $5000 in June, at the beginning of the canning season, they will save $500 per month for the 4 months each year that the plant is in operation. Maintenance costs of the case-sealing machine are expected to be negligible. The case-sealing machine is expected to be useful for five annual canning seasons and then will have no salvage value. What is the payback period? What is the nominal annual rate of return?

9-48 A large project requires an investment of $200 mil-
A lion. The construction will take 3 years: $30 million will be spent during the first year, $100 million during the second year, and $70 million during the third year of construction. Two project operation periods are being considered: 10 years with the expected net profit of $40 million per year and 20 years with the expected net profit of $32.5 million per year. For simplicity of calculations it is assumed that all cash flows occur at end of year. The company minimum required return on investment is 10%.
Calculate for each alternative:

(*a*) The payback period

(*b*) The total equivalent investment cost at the end of the construction period

(*c*) The equivalent uniform annual worth of the project (use the operation period of each alternative)

Which operation period should be chosen?

9-49 Two alternatives with identical benefits are being considered:

	A	B
Initial cost	$500	$800
Uniform annual cost	200	150
Useful life, in years	8	8

(a) Compute the payback period if Alt. B is purchased rather than Alt. A.

(b) Use a MARR of 10% and benefit–cost ratio analysis to identify the alternative that should be selected.

9-50 Consider four mutually exclusive alternatives:

	A	B	C	D
Cost	$75.0	$50.0	$15.0	$90.0
Uniform annual benefit	18.8	13.9	4.5	23.8

Each alternative has a 5-year useful life and no salvage value. The MARR is 10%. Which alternative should be selected, based on

(a) The payback period
(b) Future worth analysis
(c) Benefit–cost ratio analysis

9-51 Consider three alternatives:

	A	B	C
First cost	$50	$150	$110
Uniform annual benefit	30	45	45
Useful life, in years*	3	9	6

*At the end of its useful life, an identical alternative (with the same cost, benefits, and useful life) may be installed.

All the alternatives have no salvage value. If the MARR is 10%, which alternative should be selected?

(a) Solve the problem by payback period.
(b) Solve the problem by future worth analysis.
(c) Solve the problem by benefit–cost ratio analysis.
(d) If the answers in parts (a), (b), and (c) differ, explain why this is the case.

9-52 Consider three mutually exclusive alternatives. The MARR is 10%.

Year	X	Y	Z
0	−$100	−$50	−$50
1	25	16	21
2	25	16	21
3	25	16	21
4	25	16	21

(a) For Alt. X, compute the benefit–cost ratio.

(b) Based on the payback period, which alternative should be selected?

(c) Determine the preferred alternative based on an exact economic analysis method.

9-53 You are an investor who wants to make your investment back as quickly as possible. There are four potential projects that you can invest in. Which project should you choose?

	Project 1	Project 2	Project 3	Project 4
Initial Cost	$50,000	$60,000	$65,000	$125,000
Annual Revenues	8500	12,500	8,500	18,000
Length of Ownership	6	4	9	10

Contributed by Gana Natarajan, Oregon State University

9-54 Three mutually exclusive alternatives are being considered:

	A	B	C
Initial cost	$500	$400	$300
Benefit at end of the first year	200	200	200
Uniform benefit at end of subsequent years	100	125	100
Useful life, in years	6	5	4

At the end of its useful life, an alternative is *not* replaced. If the MARR is 10%, which alternative should be selected

(a) Based on the payback period?
(b) Based on benefit–cost ratio analysis?

9-55

Year	E	F	G	H
0	−$90	−$110	−$100	−$120
1	20	35	0	0
2	20	35	10	0
3	20	35	20	0
4	20	35	30	0
5	20	0	40	0
6	20	0	50	180

(a) Based on the payback period, which alternative is preferred?

(b) Based on future worth analysis, which of the four alternatives is preferred at 5% interest?

(c) Based on future worth analysis, which alternative is preferred at 20% interest?

(d) At 10% interest, what is the benefit–cost ratio for Alt. *G*?

9-56 A new piece of laboratory equipment costing $10,000 promises to save $4000 per year in materials and overtime pay. If the cost of money is 12% and projects must have a 3-year discounted payback period, should the equipment be purchased?

9-57 A new high-efficiency motor is being considered for a large compressor. The new motor will cost $20,000 but will save $8000 per year in electricity. If the firm's MARR is 15%, what is the discounted payback period?

9-58 Two alternative pumps are being compared. Pump A costs $4000, will save $1200 per year in operating and maintenance expenses, and is expected to last for 6 years. Pump B costs $4800, will save $1400 per year, and is also expected to last for 6 years. If the cost of capital is 16%, which has the better discounted payback period?

9-59 Two equipment investments are estimated as follows:

Year	A	B
0	−$15,000	−$18,000
1	5,000	6,500
2	5,000	6,500
3	5,000	6,500
4	5,000	6,500
5	5,000	6,500

Which investment has the better discounted payback period if *i* = 14%?

Sensitivity

9-60 If the MARR is 12%, compute the value of *X* that makes the two alternatives equally desirable.

	A	B
Cost	$800	$1000
Uniform annual benefit	230	230
Useful life, in years	5	X

9-61 Analyze Problem 9–60 again with the following changes:

(a) What if *B*'s first cost is $1200?

(b) What if *B*'s annual benefit is $280?

(c) What if the MARR is 10% annually?

(d) What if (a), (b), and (c) happen simultaneously?

9-62 Consider two alternatives:

	A	B
Cost	$500	$300
Uniform annual benefit	75	75
Useful life, in years	Infinity	X

Assume that Alt. *B* is not replaced at the end of its useful life. If the MARR is 10%, what must be the useful life of *B* to make Alternatives *A* and *B* equally desirable?

9-63 If the MARR is 5%, compute the value of *X* that makes the two alternatives equally desirable.

	A	B
Cost	$150	$ X
Uniform annual benefit	40	65
Salvage value	100	200
Useful life, in years	6	6

9-64 Ithaca is considering a new $50,000 snowplow that will save the city $600 per day of use compared to the existing one. It should last 10 years and have a resale value of $2000.

(a) To obtain a 12% rate of return what is the minimum number of days per year on average it will have to be used.

(b) Research the environmental impact of road salt. What are other options?

Contributed by D. P. Loucks, Cornell University

9-65 Victoria is choosing between a standard Honda Civic for $20,000 or a hybrid Civic for $25,000. She calculates her annual cost of ownership including payments but not including gasoline to be $5000 for the standard and $5800 for the hybrid. The standard Civic will cost Victoria 18¢/mile for gasoline, while the hybrid will cost her only 12¢/mile. How many miles must Victoria drive in a year before the hybrid vehicle becomes more cost efficient to her?

Contributed by Paul R. McCright, University of South Florida

9-66 Midwest Airlines flies a short nonstop with 137-passenger planes. Considering all the costs of owning each plane plus the salaries for their crews and the fuel costs and landing fees, the fixed cost for a single flight is $10,400. If the costs associated with each passenger (reservations cost, check-in cost, baggage handling cost, snack cost, etc.) total $48 per passenger and the average ticket price is $157 (before the various taxes are added), what percentage of seats must be filled for the flight to break even?
Contributed by Paul R. McCright, University of South Florida

9-67 Fence posts for a particular job cost $18.03 each to install, including the labor cost. They will last 10 years. If the posts are treated with a wood preservative, they can be expected to have a 15-year life. Assuming a 8% interest rate, how much could one afford to pay for the wood preservative treatment?

9-68 A piece of property is purchased for $10,000 and yields a $1000 yearly net profit. The property is sold after 5 years. What is its minimum price to break even with interest at 10%?

9-69 Analyze Problem 9–68 again with the following changes:

(*a*) What if the property is purchased for $12,000?

(*b*) What if the yearly net profit is $925?

(*c*) What if it is sold after 7 years?

(*d*) What if (*a*), (*b*), and (*c*) happen simultaneously?

9-70 Midwest Airlines (MWA) is planning to expand its fleet of jets to replace some old planes and to expand its routes. It has received a proposal to purchase 112 small jets over the next 4 years. What annual net revenue must each jet produce to break even on its operating cost? The analysis should be done by finding the EUAC for the 10-year planned ownership period. MWA has a MARR of 12%, purchases the jet for $22 million, has operating and maintenance costs of $3.2 million the first year, increasing 8% per year, and performs a major maintenance upgrade costing $4.5M at end of Year 5. Assume the plane has a salvage value at end of Year 10 of $13 million.
Contributed by Paul R. McCright, University of South Florida

9-71 Plan *A* requires a $30,000 investment now. Plan *B* requires an $28,700 investment now and an additional $10,000 investment at a later time. At 12%

interest, compute the breakeven point for the timing of the $10,000 investment.

9-72 A low-carbon-steel machine part, operating in a corrosive atmosphere, lasts 6 years and costs $350 installed. If the part is treated for corrosion resistance, it will cost $500 installed. How long must the treated part last to be the preferred alternative, assuming 10% interest?

9-73 Analyze Problem 9–72 again with the following changes:

(*a*) What if the installed cost of the corrosion-treated part is $600?

(*b*) What if the untreated part will last only 4 years?

(*c*) What if the MARR is 12% annually?

(*d*) What if (*a*), (*b*), and (*c*) happen simultaneously?

9-74 Tyrella Jackson is buying a used car. Alternative *A* is an American-built compact. It has an initial cost of $8900 and operating costs of 9¢/km, excluding depreciation. From resale statistics, Tyrella estimates the American car can be resold at the end of 3 years for $1700. Alternative *B* is a foreign-built Fiasco. Its initial cost is $8000, the operating cost, also excluding depreciation, is 8¢/km. How low could the resale value of the Fiasco be to provide equally economical transportation? Assume Tyrella will drive 12,000 km/year and considers 8% as an appropriate interest rate.

9-75 Analyze Problem 9–74 again with the following changes:

(*a*) What if the Fiasco is more reliable than expected, so that its operating cost is $0.075/km?

(*b*) What if Tyrella drives only 9000 km/year?

(*c*) What if Tyrella's interest rate is 6% annually?

(*d*) What if (*a*), (*b*), and (*c*) happen simultaneously?

9-76 A car company has decided to spend $150M on a museum for exhibiting its classic cars. Land can be purchased for $500,000. The museum building will require 30,000 square feet of general space, while each car displayed will require an additional 1000 square feet. The design and planning process will cost $100,000, which should be paid immediately. The construction of the building will cost $600 per square foot, and the building will be completed within the next 2 years, while the cost of construction will be distributed evenly between the 2 years of construction. All cars will be purchased during the second year of construction at an average cost of $120,000 per car. The annual operation of the museum will cost $2,000,000 plus $30,000 per car. If the funds are invested at 9% per year and the museum is to exist forever, how many cars can the trustees purchase? *Contributed by Hamed Kashani, Saeid Sadri, and Baabak Ashuri, Georgia Institute of Technology*

9-77 A road can be paved with either asphalt or concrete. Concrete costs $20,000/km and lasts 20 years. Assume the annual maintenance costs are $700 for concrete and $1000 for asphalt per kilometer per year. Use an interest rate of 5% per year.
Contributed by D. P. Loucks, Cornell University

(*a*) What is the maximum that should be spent for asphalt if it lasts only 10 years?

(*b*) Assume the asphalt road costs $8500 per kilometer. How long must it last to be the preferred alternative?

(*c*) Research and summarize conclusions of two scholarly articles on the environmental comparison of asphalt versus concrete.

9-78 Christina Cook studied the situation described in Problem 9-46 and decided that the solar system will *not* require the $180 of maintenance every 4 years. She believes future replacements of either the conventional electric water heater, or the solar water heater system can be made at the same costs and useful lives as the initial installation. Based on a 10% interest rate, what must be the useful life of the solar system to make it no more expensive than the electric water heater system?

9-79 A newspaper is considering buying locked vending machines to replace open newspaper racks in the downtown area. The vending machines cost $75 each. It is expected that the annual revenue from selling the same quantity of newspapers will increase $12 per vending machine. The useful life of the vending machine is unknown.

(*a*) To determine the sensitivity of rate of return to useful life, prepare a graph for rate of return versus useful life for lives up to 10 years.

(*b*) If the newspaper requires a 20% rate of return, what minimum useful life must it obtain from the vending machines?

(*c*) What would be the rate of return if the vending machines were to last indefinitely?

9-80 Rental equipment is for sale for $110,000. A prospective buyer estimates he would keep the equipment for 12 years and spend $6000 a year on maintaining it. Estimated annual net receipts from

equipment rentals would be $14,400. It is estimated the rental equipment could be sold for $80,000 at the end of 12 years. If the buyer wants a 7% rate of return on his investment, what is the maximum price he should pay for the equipment?

9-81 Neither of the following machines has any net salvage value.

	A	B
Original cost	$55,000	$75,000
Annual expenses		
Operation	9,500	7,200
Maintenance	5,000	3,000
Taxes and insurance	1,700	2,250

At what useful life are the machines equivalent if

(*a*) 10% interest is used in the computations?

(*b*) 0% interest is used in the computations?

9-82 Jane Chang is making plans for a summer vacation. She will take $1000 with her in the form of traveler's checks. From the newspaper, she finds that if she purchases the checks by May 31, she will not have to pay a service charge. That is, she will obtain $1000 worth of traveler's checks for $1000. But if she waits to buy the checks until just before starting her summer trip, she must pay a 1% service charge. (It will cost her $1010 for $1000 of traveler's checks.)

Jane can obtain a 13% interest rate, compounded weekly, on her money. How many weeks after May 31 can she begin her trip and still justify buying the traveler's checks on May 31?

9-83 A motor with a 200-horsepower output is needed in the factory for intermittent use. A Graybar motor costs $7000 and has an electrical efficiency of 90%. A Blueball motor costs $6000 and has an 85% efficiency. Neither motor would have any salvage value, since the cost to remove it would equal its scrap value. The annual maintenance cost for either motor is estimated at $500 per year. Electric power costs $0.12/kWh (1 hp = 0.746 kW). If an 18% interest rate is used in the calculations, what is the minimum number of hours the higher initial cost Graybar motor must be used each year to justify its purchase?

9-84 A machine costs $5240 and produces benefits of $1000 at the end of each year for 8 years. Assume an annual interest rate of 10%.

(*a*) What is the payback period (in years)?

(*b*) What is the breakeven point (in years)?

(c) Since the answers in (a) and (b) are different, which one is "correct"?

9-85 *The Financial Advisor* is a weekly column in the local newspaper. Assume you must answer the following question. "I recently retired at age 65, and I have a tax-free retirement annuity coming due soon. I have three options. I can receive (A) $30,976 now, (B) $359.60 per month for the rest of my life, or (C) $513.80 per month for the next 10 years. What should I do?" Ignore the timing of the monthly cash flows and assume that the payments are received at the end of year. Assume the 10-year annuity will continue to be paid to loved heirs if the person dies before the 10-year period is over.

Contributed by D. P. Loucks, Cornell University

(a) If $i = 6\%$, develop a choice table for lives from 5 to 30 years. (You do not know how long this person or other readers may live.)

(b) If $i = 10\%$, develop a choice table for lives from 5 to 30 years. (You do not know how long this person or other readers may live.)

(c) How does increasing the interest rate change your recommendations?

Minicases

9-86 A proposed steel mill may include a co-generation electrical plant. This plant will add $2.3M in first cost with net annual savings of $0.27M considering operating costs and electrical bills. The plant will have a $0.4M salvage value after 25 years. The firm uses an interest rate of 12% and present worth index (PWI) in its decision making.

The public utility offers a subsidy for co-generation facilities because it will not have to invest as much in new capacity. This subsidy is calculated as 20% of the co-generation facility's first cost, but it is paid annually. The utility calculates the subsidy using a benefit–cost ratio at 8% and a life of 20 years.

(a) Is the plant economically justifiable to the firm without the subsidy? What is the PWI?

(b) What is the annual subsidy?

(c) Is the plant economically justifiable to the firm with the subsidy? Now what is the PWI?

(d) How important is the difference in interest rates, and how does it affect these results?

(e) How important is the difference in horizons, and how does it affect these results?

(f) What is the "co" aspect of a co-generation power plant? What are the primary benefits of this system, and who accrues those benefits? Why aren't all power plants designed in this fashion?

9-87 Assume a cost improvement project has only a first cost of $100,000 and a monthly net savings, M. There is no salvage value. Graph the project's IRR for payback periods from 6 months to the project's life of N years. The firm accepts projects with a 2-year payback period or a 20% IRR. When are these standards consistent and when are they not?

(a) Assume that $N = 3$ years.

(b) Assume that $N = 5$ years.

(c) Assume that $N = 10$ years.

(d) What recommendation do you have for the firm about its project acceptance criteria?

9-88 The Louisiana Department of Transportation and Development (LaDOTD) in 2009 approved the feasibility analysis for upgrading 6 miles of US-167 South starting at the intersection with US-80 (California Avenue). An existing two-lane highway between is to be converted to a four-lane divided freeway. The proposed new freeway is projected to average 25,000 vehicles per day over the next 20 years. Truck volumes represent 6.25% of the total traffic. Annual maintenance on the existing highway is $1875 per lane mile. The existing accident rate is 5.725 per million vehicle miles (MVM). Capital improvement investment money can be secured at 6.25%. Which alternative is preferred (use benefit/cost ratio analysis)?

Plan 1: Add two adjacent lanes for $562,500 per mile. This will reduce auto travel time by 2.5 minutes and truck travel time by 1.25 minutes. It will reduce the accident rate to 3.125 per MVM. Annual maintenance is estimated to be $1560 per lane mile.

Plan 2: Make grade improvements while adding two adjacent lanes at a cost of $812,500 per mile. This would reduce auto and truck travel time by 3.75 minutes each. The accident rate is estimated to be 3.10 per MVM. Annual maintenance is estimated to be $1250 per lane mile.

Plan 3: Construct a new freeway on new alignment at a cost of $1,000,000 per mile. This would reduce auto travel time by 6.25 minutes and truck travel time by 5 minutes. Plan 3 is 0.5 miles longer than the others. The estimated accident rate is 3.00 per MVM. Annual maintenance is estimated to be $1250 per lane mile. Plan 3 abandons the existing highway with no salvage value.

Additional data:

Operating cost—autos:	15¢ per mile
Operating cost—trucks:	22.5¢ per mile
Time saving—autos:	3.75¢ per vehicle minute
Time saving—trucks:	18.75¢ per vehicle minute
Average accident cost:	$1500

Contributed by Benedict Nwokolo, Grambling State University

CASES

The following cases from *Cases in Engineering Economy* (www.oup.com/us/newnan) are suggested as matched with this chapter.

CASE 16   **Great White Hall**

Proposal comparison using B/C analysis for RFP with unclear specifications.

CASE 17   **A Free Lunch?**

Is proposal too good to be true from two perspectives? Realistic (unordered) statement of facts.

APPENDIX 9A

Investing for Retirement and Other Future Needs

Key Words

defined benefit	risk	volatility
defined contribution		

DEFINED CONTRIBUTION AND DEFINED BENEFIT PLANS

Firms and individuals often borrow to buy capital items like buildings/homes, land, and vehicles. The money is needed now, and often the item is some kind of security for the loan. However, pension funds, retirement accounts, college savings, and down payments are accumulated by saving and investing. Because pensions and retirement have the longest horizons, compound interest plays a larger role than in shorter horizon problems.

Planning for retirement has become even more important to individuals because of a shift in how it is handled. As recently as 2000, 60% of Fortune 500 firms offered **defined benefit** retirement plans—also called pensions. Today it is estimated that about 90% offer new employees only **defined contribution** plans. Many governmental units have also shifted from traditional pensions to defined contribution plans.

A defined benefit plan is overseen by a firm or a government body for its employees. The employer is responsible for managing the fund and ensuring that there is enough money to cover all obligations. In many cases, both the employee and the employer contribute to the fund. Most payouts involve a formula, whereby the employee can calculate his or her monthly or annual benefit, which continues until death. As an example, Equation 9A–1 is applied to a 30-year employee who earned $80,000 yearly in the 3 years before retirement.

$$\text{Annual benefit} = \text{average salary}_{\text{last 3 years}} \times 2\% \times \text{years}_{\text{employment}} \quad (9A-1)$$
$$= 80,000 \times 2\% \times 30 = \$48,000/\text{yr}$$

The benefit is predictable, and thus makes retirement planning much easier.

With a defined contribution plan, the employee designates the amount to be directed toward retirement and is responsible for its management. The employer may match part of an employee's contribution, and may offer a limited set of fund plans. Based on the choices available, employees select how much they deposit annually, for how many years, and what funds they invest in. Employees choose how safely or aggressively the funds are invested. The total in the account is the **defined contribution**; that total determines how much can be withdrawn and for how long. Thus, employees must also choose the withdrawal rate for the funds after they retire. Employees must estimate how long they will live.

Retirement planning has become more difficult and more necessary. Social security provides a safety net, but it is income replacement only for those who work half-time at the minimum wage. For those retiring in 2016 who contributed at the maximum taxed level for 35 years, the annual social security benefit is $31,668—much less than engineers typically earn.

WHAT RETURNS ARE REASONABLE TO EXPECT AND WHAT RISKS GO WITH THEM

Table 9A–1 summarizes the performance of three types of investments over the last 60 years: the U.S. stock market, U.S. government bonds with typical maturities of 10 to 30 years, and U.S. treasury bills (T-bills) with typical maturities of 1 to 6 months. Since these are market or nominal values, Table 9A–1 also includes the inflation rate (see Chapter 14 for more information). Over the past 60 years, stocks have returned about 7.3% over inflation, long-term treasury bonds about 2.5% above inflation, and T-bills about 0.8% above inflation. These are real rates of return.

The standard deviation and high and low values show that stocks have higher **risks** than bonds and much higher risks than T-bills. The annual returns vary much more for stocks, so stocks have a higher **volatility** than bonds or T-bills.

Note that the geometric mean is the correct average rather than the arithmetic average. The geometric rate of return is the interest rate that would be calculated using the present worth, the future value, and the number of years. It can also be calculated using Equation 9A–2.

$$\text{Geometric mean} = \left[\prod_{i=1}^{n} (1 + r_i) \right]^{1/n} - 1 \qquad (9A\text{-}2)$$

where r_i = return in year i

The arithmetic average of annual returns overstates expected returns (especially over long intervals with positive and negative annual values). A quick example is gaining 50% one year and losing 50% the other year in either order. The arithmetic average is 0% change. The geometric average using Equation 9A-2 is -13.4% $(= \sqrt{(0.5 \times 1.5)} - 1)$. You are not back where you started, as an arithmetic average of 0% would indicate. Instead, you have lost 25%, as the geometric average indicates. You have 75% $(= 1.5 \times 0.5)$ or $(= 0.866 \times 0.866)$ of what you started with.

The presence of a standard deviation demonstrates that returns vary from year to year. On average, the investments in Table 9A–1 will increase, but returns may be positive or

TABLE 9A–1 Returns and Standard Deviations for Investments and Inflation

	Common Stocks	Treasury Bonds	Treasury Bills	Inflation Rate
Geometric mean	11.0%	6.2%	4.5%	3.7%
Standard deviation	17.5%	10.8%	2.8%	2.9%
Maximum 1950–2012	52.6%	40.4%	11.6%	13.3%
Minimum 1950–2012	−35.5%	−12.2%	0.0%	−0.5%

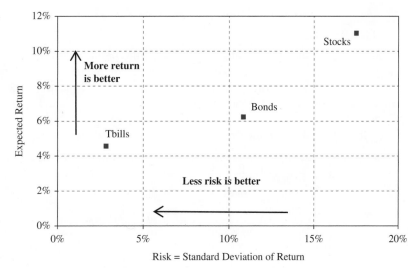

FIGURE 9A–1 Returns and risks for investments.

negative in any given year. All investments carry risk, and this risk is often characterized by the standard deviation. Investments that have a higher risk are expected to deliver a higher return in order to compensate for that risk.

Figure 9A–1 plots the geometric means and standard deviations of the three investment types. The arrows emphasize that both higher returns and lower risks are preferred. Thus the best mix of investments depends on how the investor evaluates the trade-off between risk and return. Chapter 10 includes more on probabilities in economic models. Appendix 10A explains why a combination of stocks and bonds is expected to perform better than either alone; this is also why a diversified portfolio of stocks is expected to perform better than one with only a few stocks.

EXAMPLE 9A–1

An engineer has just finished paying off a student loan and is ready to start saving for retirement. Her current annual salary is $63,000. She expects salary increases to exceed inflation, but to be safe she wants to assume that salary increases will be matched with inflation. She expects to work for another 35 years. If she invests 15% of her salary, how much can be expected to be in her account if she invests in T-bills, bonds, or stocks?

SOLUTION

The first step is to determine the interest rates. Since this problem is in constant-value dollars, like nearly all of the text (except for bonds) up to Chapter 14, the interest rates are the market rates minus inflation. From the first paragraph of this section, stocks have returned about 7.3% over inflation, long-term treasury bonds about 2.5% above inflation, and T-bills about 0.8% above inflation.

To make the solution more flexible, the spreadsheet starts with a data block.

	A	B	C	D	E	F	G	H
1	$63,000	Salary						
2	15%	% saved						
3								
4	Exp. 9A-1	i	n	PMT	PV	FV	Solve for	Answer
5	T-bills	0.8%	35	-9,450	0		FV	$379,957
6	Bonds	2.5%	35	-9,450	0		FV	$519,072
7	Stocks	7.3%	35	-9,450	0		FV	$1,394,949

EXAMPLE 9A–2   (Example 9A–1 Revisited)

The engineer in Example 9A–1 has decided that a retirement goal of $1 million is adequate. What fraction of her salary must be saved if she invests in T-bills, bonds, or stocks?

SOLUTION

The first step is to determine the annual deposits at the different interest rates. As in Example 9A–1, the interest rates are the market rates minus inflation: stocks have returned about 7.3% over inflation, long-term treasury bonds about 2.5% above inflation, and T-bills about 0.8% above inflation.

To make the solution more flexible, the spreadsheet starts with a data block.

	A	B	C	D	E	F	G	H	I
1	$63,000	Salary							
2									
3	Exp. 9A-2	i	n	PMT	PV	FV	Solve for	Answer	% salary
4	T-bills	0.8%	35		0	1,000,000	PMT	-$24,871	39.5%
5	Bonds	2.5%	35		0	1,000,000	PMT	-$18,206	28.9%
6	Stocks	7.3%	35		0	1,000,000	PMT	-$6,774	10.8%

An investor who chooses to invest safely before retirement should also invest safely after retirement. If the $1M is invested at a lower rate, then less can be withdrawn each year to live on. It is common to suggest that *all* investors should invest more safely once they are approaching retirement or are already retired.

EXAMPLE 9A–3	(Examples 9A–1 and 9A–2 Revisited)

The engineer in Examples 9A–1 and 9A–2 wants to know how many years she can live on her savings before they are exhausted. Assume the common guideline of retirement spending equals 80% of pre-retirement net income. Also assume the same investments and returns before and after retirement. How long after retirement before her $1 million is exhausted?

SOLUTION

The average returns were determined for the earlier examples: stocks have returned about 7.3% over inflation, long-term treasury bonds about 2.5% above inflation, and T-bills about 0.8% above inflation. The required savings at each rate were determined in Example 9A–2. Note that the annual amount spent before retirement is $63,000 \times (1 - \%_{saved})$. The expected spending level after retirement is 80% of that. So the spending level with T-bills is $30,503 (= (63,000 − 24,871) \times 80\%)$.

	A	B	C	D	E	F	G	H	I
1	$63,000	Salary							
2	80%	% retirement spending							
3	Exp. 9A-2	*i*	*n*	*PMT*	*PV*	*FV*	Solve for	Answer	% salary
4	T-bills	0.8%	35		0	1,000,000	PMT	-$24,871	39.5%
5	Bonds	2.5%	35		0	1,000,000	PMT	-$18,206	28.9%
6	Stocks	7.3%	35		0	1,000,000	PMT	-$6,774	10.8%
7									
8	Exp. 9A-3	*i*	*n*	*PMT*	*PV*	*FV*	Solve for	Answer	
9	T-bills	0.8%		30,503	1,000,000	0	NPER	38.2	
10	Bonds	2.5%		35,836	1,000,000	0	NPER	48.4	
11	Stocks	7.3%		44,980	1,000,000	0	NPER	#NUM!	

No answer is returned for stocks. The reason is that an annual return of 7.3% on $1 million is $73,000. This is more than the expected retirement spending. It is also more than the pre-retirement salary!

EXAMPLE 9A–4	(Example 9A–1 Revisited)

An engineer has just finished paying off a student loan and is ready to start saving for retirement. Her current annual salary is $63,000. Assume that her salary increases 2% faster than inflation. She expects to work for another 35 years. If she invests 15% of her salary, how much can be expected to be in her account if she invests in T-bills, bonds, or stocks?

SOLUTION

The interest rates were identified in Example 9A–1, but the annuity functions used in Examples 9A–1 through 9–3 cannot be used here. The annual deposit increases with the salary, so tables of cash flows must be created. Each year a deposit equal to a percentage of the income is made.

The balance in year t is that deposit plus $(1 + \text{return}) \times$ the previous year's balance. There is no interest in year 1 because the deposit is assumed to be made at the end of the year.

	A	B	C	D	E	F
1	$63,000	Salary				
2	15%	% saved				
3	2%	annual salary increase				
4				T-bills	Bonds	Stocks
5	Year	Salary	Deposit	0.8%	2.5%	7.3%
6	1	63,000	9,450	9,450	9,450	9,450
7	2	64,260	9,639	19,165	19,325	19,779
8	3	65,545	9,832	29,150	29,640	31,054
9	4	66,856	10,028	39,411	40,410	43,350
10	5	68,193	10,229	49,956	51,649	56,743
40	35	123,523	18,528	534,108	705,567	1,743,063
41						
42			=$C40+D39*(1+D$5)			
43					=$C40+F39*(1+F$5)	

PROBLEMS

For these problems assume that stocks, bonds, and T-bills return, respectively, 7.3%, 2.5%, and 0.8% over inflation.

9A-1 A 25-year-old engineer earning $65,000 per year wants to retire at age 55 with $2 million, and plans to invest in a stock fund.

(a) How much money must be invested each year?

(b) If the employer does a 100% match of retirement savings up to 3% of the employee's salary, how much money must each invest annually?

9A-2 An engineer earning $68,000 per year wants to retire in 35 years with $2 million and plans to invest in a treasury bond fund.

(a) How much money must be invested each year?

(b) If this person works 5 additional years, how much money must be invested each year?

(c) If the employer does a 100% match of retirement savings up to 4% of the employee's salary, and the engineer wants to retire in 35 years, how much money must each invest annually?

9A-3 A risk-averse 25-year-old engineer earning $70,000 per year wants to retire at age 65 with $2 million and plans to safely invest in treasury bills.

(a) How much money must be invested each year?

(b) If the employer does a 100% match of retirement savings up to 5% of the employee's salary, how much money must each invest annually?

9A-4 A new employee earning $60,000 annually has set a retirement goal of $1 million. She plans to work for 40 years, and will invest in stocks.

(a) What fraction of her salary must be saved?

(b) If the employer does a one-to-one match of retirement savings up to 3% of the employee's salary, what fraction of her salary must be saved?

9A-5 An employee earning $63,000 per year has set a goal of retiring in 40 years with $1 million. She will invest in treasury bonds.

(a) What fraction of her salary must be saved?

(b) If the employer does a one-to-one match of retirement savings up to 5% of the employee's salary, what fraction of her salary must be saved?

9A-6 A new engineer started her first job earning $60,000 annually, and wants to retire in 40 years with $1 million. She will invest in treasury bills.

(a) What fraction of her salary must be saved?

(b) If the employer does a one-to-one match of retirement savings up to 4% of the employee's salary, what fraction of her salary must be saved?

9A-7 A manager just retired at age 62, with his retirement savings of $800,000 invested in stocks. If he needs $60,000 per year for living expenses, how long will his savings last?

9A-8 (A) A manager retired at age 65, and has her retirement savings of $900,000 invested in treasury bonds. She needs $35,000 per year for living expenses, in addition to her social security benefit. How long will her investment last?

9A-9 A person worked for many years with the same company and has accumulated a retirement account of $900,000. If $30,000 per year needs to be withdrawn from this account, how long will the savings last if invested in treasury bills?

9A-10 (A) A new employee puts 4% of his salary of $60,000 into a retirement account, and his employer matches this, also putting 4% into the account. The money is invested in a diversified stock fund. His salary increases 3% per year.

(a) What is the value of the account after 10 years?

(b) What is the value of the employer's matching funds?

(c) How much will be in the account after 40 years?

9A-11 An engineer changed jobs and is signing up for benefits. The company 401(k) includes a low cost treasury bond fund. The engineer will put 3% of her salary of $70,000 into the account, and her employer will match half this amount. Her salary is expected to increase 2.8% per year.

(a) What is the value of the account after 10 years?

(b) If she expects to work for 30 years, how much will be in the account?

9A-12 (A) A long-term employee is nearing retirement and will adjust his retirement account. The account has $600,000 in it. The employee will put that money and all new money in a treasury bill (T-bill) account. He puts 5% of his $120,000 salary into the account, and his employer matches half of this amount. His salary increases 2% per year. What is the value of the account after 10 years?

9A-13 Mike just changed jobs, leaving a company after 6 years. He is fully vested, and can keep the money his employer deposited in his retirement account. His employer has been contributing $200 per month into a diversified stock fund.

(a) Using average market rates, how much money has accumulated in the account?

(b) If this money is "rolled over" into another retirement account with an 8% annual return, how much will this be worth after 30 years?

9A-14 (A) Monica became concerned about the stock market due to recent losses, and wants safe investments in treasury bonds. She moved $350,000 in her retirement account into an ETF specializing in government bonds, which average 2.5% above inflation. Her salary is currently $92,000 per year, and she expects raises of 2.0% each year. Her employer will match retirement savings up to 5% of her salary. If she wants $450,000 in her account in 5 years, what % of her salary does she need to save?

9A-15 Jorge was recently laid off, and may not be able to find another full-time job. Capital preservation is his primary concern, so he wants to invest in treasury bills. His retirement fund has $1.3 million. He can receive social security, so he will only need $38,000 per year from his savings to live well. Jorge is 64 years old. How long does he expect his savings to last? Should he look for a job or enjoy his retirement?

10

UNCERTAINTY IN FUTURE EVENTS

Video Game Development and Uncertainty

Today, most computing devices, from laptops to music players to cell phones, are being equipped with technologies that allow users to run a wide range of products. Consumers can download and install applications directly from the Internet, and firms rely on this capability to reach as wide an audience as possible. Unfortunately, developing products that run on multiple platforms adds a great deal of uncertainty to the product-development process.

The army has licensed several virtual software environment products to use for training, education, and outreach that are required to operate on many product architectures. The product developers must build the software environments so clients can use them across a wide range of different PC configurations and Internet speeds. In a closed environment, where every PC has the exact same configuration, testing is easy. However, software that is intended to be distributed to the public must be able to run on all common PC configurations.

One of the most successful projects has been America's Army. The series of video games was developed to help the U.S. Army with both public relations and recruitment. Since its first release in 2002, America's Army has had over forty versions released. Like most software development projects, each version of America's Army has placed time and budget constraints on the project manager. A typical software development cycle allows for minor tweaks and bug fixes prior to product release. However, there are times when a test reveals a critical issue that requires more resources to fix. Most of the time, the source of the critical issues, like an operating system patch that creates a major change in how applications access the system, lies outside the development team's control.

When patches and upgrades are announced and released, it is uncertain whether there will be problems and how significant they may be.

Complexity of the software can also affect uncertainty in the project. Some small projects, such as improving how a vehicle performs in the game, may take only a few weeks to complete. Other projects, such as adding a new training scenario, could take several months and require a dozen or more programmers. A large new training scenario might also require programmers to develop multiple terrains, different vehicles, and a range of new equipment types for the soldiers; all of these programming efforts require extensive testing and add to the uncertainty in budgeting and scheduling the project. ■ ■ ■

Contributed by Paul Componation, University of Texas at Arlington, and Marsha Berry, U.S. Army Research, Development & Engineering Command (AMRDEC), Software Engineering Directorate.

QUESTIONS TO CONSIDER

1. As a software developer how would you quantify the uncertainty in a specific software development effort?

2. How would you estimate the impact on your project's schedule and costs when you have to deal with uncertainty in your software testing program?

3. The number of new applications for software has increased dramatically over the past decade. If you are developing a new software product, how will you prepare it not only to meet all the applications today but also to deal with future unforeseen applications?

4. How do you economically justify projects whose payoff is in improved recruiting and training rather than revenue from product sales?

After Completing This Chapter...

You should be able to:

- Use a range of estimated variables to evaluate a project.
- Describe possible outcomes with probability distributions.
- Combine probability distributions for individual variables into joint probability distributions.
- Use expected values for economic decision making.
- Use economic decision trees to describe and solve more complex problems.
- Measure and consider risk when making economic decisions.
- Understand how simulation can be used to evaluate economic decisions.
- Understand why and how diversification reduces risk for investments and project portfolios. (Appendix 10A)

Key Words

beta distribution	expected value	real options
chance node	joint probability distribution	risk
decision node	most likely	scenario
decision tree	optimistic	simulation
discrete probability distribution	outcome node	standard deviation
dominated projects	pessimistic	statistically independent
efficient frontier	pruned branch	SUMPRODUCT

An assembly line is built after the engineering economic analysis has shown that the anticipated product demand will generate profits. A new motor, heat exchanger, or filtration unit is installed after analysis has shown that future cost savings will economically justify current costs. A new road, school, or other public facility is built after analysis has shown that the future demand and benefits justify the present cost to build. However, future performance of the assembly line, motor, and so on is uncertain, and demand for the product or public facility is more uncertain.

Engineering economic analysis is used to evaluate projects with long-term consequences when the time value of money matters. Thus, it must concern itself with future consequences; but describing the future accurately is not easy. In this chapter we consider the problem of evaluating the future. The easiest way to begin is to make a careful estimate and a breakeven analysis. Then we examine the possibility of predicting a range of possible outcomes. Finally, we consider what happens when the probabilities of the various outcomes are known or may be estimated. We will show that the tools of probability are quite useful for economic decision making.

ESTIMATES AND THEIR USE IN ECONOMIC ANALYSIS

Economic analysis requires evaluating the future consequences of an alternative. In practically every chapter of this book, there are cash flow tables and diagrams that describe precisely the costs and benefits for future years. We don't really believe that we can exactly foretell a future cost or benefit. Instead, our goal is to select a single value representing the *best* estimate that can be made.

Breakeven analysis, as shown in Examples 9–11, 12, and 13, is one means of examining the impact of the variability of some estimate on the outcome. It helps by answering the question, How much variability can a parameter have before the decision will be affected? While the preferred decision depends on whether the salvage value is above or below the breakeven value, the economic difference between the alternatives is small when the salvage value is "close" to breakeven.

What-if analysis, which was detailed in Example 9–15, is another way of creating and evaluating **scenarios** that describe future uncertainties. Breakeven and what-if analyses do not solve the basic problem of how to take the inherent variability of parameters into account in an economic analysis. This is the task of this chapter.

A RANGE OF ESTIMATES

It is usually more realistic to describe parameters with a range of possible values, rather than a single value. A range could include an **optimistic** estimate, the **most likely** estimate, and a **pessimistic** estimate. Then, the economic analysis can determine whether the decision is sensitive to the range of projected values. Sets of such optimistic, most likely, and pessimistic estimates are often used to form optimistic, most likely, and pessimistic **scenarios**.

EXAMPLE 10–1

A firm is considering an investment. The most likely data values were found during the feasibility study. Analyzing past data of similar projects shows that optimistic values for the first cost and the annual benefit are 5% better than most likely values. Pessimistic values are 15% worse.

The firm's most experienced project analyst has estimated the values for the useful life and salvage value. Note that 5% better and 15% worse is not $+5\%$ and -15% for cost. A lower cost is better and a higher cost is worse.

	Optimistic	Most Likely	Pessimistic
Cost	$950	$1000	$1150
Net annual benefit	210	200	170
Salvage value	100	0	0
Useful life, in years	12	10	8

Compute the rate of return for each estimate. If a 10% before-tax minimum attractive rate of return is required, is the investment justified under all three estimates? If it is justified only under some estimates, how can these results be used?

SOLUTION

	A	B	C	D	E	F	G	H
1	Scenario	i	n	*PMT*	*PV*	*FV*	Solve for	Answer
2	Optimistic		12	210	-950	100	i	19.8%
3	Most Likely		10	200	-1000	0	i	15.1%
4	Pessimistic		8	170	-1150	0	i	3.9%

To solve with tabulated factors, the equations for the respective cases are:

Optimistic

$$PW = 0 = -\$950 + 210(P/A, \text{IRR}_{\text{opt}}, 12) + 100(P/F, \text{IRR}_{\text{opt}}, 12)$$

Most Likely

$$PW = 0 = -\$1000 + 200(P/A, \text{IRR}_{\text{most likely}}, 10)$$

$$(P/A, \text{IRR}_{\text{most likely}}, 10) = 1000/200 = 5$$

Pessimistic

$$PW = 0 = -\$1150 + 170(P/A, \text{IRR}_{\text{pess}}, 8)$$

$$(P/A, \text{IRR}_{\text{pess}}, 8) = 1150/170 = 6.76$$

From the calculations we conclude that the rate of return for this investment is most likely to be 15.1%, but might range from 3.9% to 19.8%. The investment meets the 10% MARR criterion for two of the estimates. These estimates can be considered to be scenarios of what may happen with this project. Since one scenario indicates that the project is not attractive, we need to have a method of weighting the scenarios or considering how likely each is.

Example 10–1 made separate calculations for the sets of optimistic, most likely, and pessimistic values. The range of scenarios is useful. However, if there are more than a few uncertain variables, it is unlikely that all will prove to be optimistic (best case) or most likely or pessimistic (worst case). It is more likely that many parameters are the most likely values, while some are optimistic and some are pessimistic.

This can be addressed by using Equation 10-1 to calculate average or mean values for each parameter. Equation 10-1 puts four times the weight on the most likely value than on the other two. This equation has a long history of use in project management to estimate activity completion times. It is an approximation with the **beta distribution**.

$$\text{Mean value} = \frac{\text{Optimistic value} + 4(\text{Most likely value}) + \text{Pessimistic value}}{6} \qquad (10\text{-}1)$$

This approach is illustrated in Example 10–2.

EXAMPLE 10–2

Solve Example 10–1 by using Equation 10-1. Compute the resulting mean rate of return.

SOLUTION

Compute the mean for each parameter:

$$\text{Mean cost} = [950 + 4(1000) + 1150]/6 = \$1016.7$$

$$\text{Mean net annual benefit} = [210 + 4(200) + 170]/6 = \$196.7$$

$$\text{Mean useful life} = [12 + 4(10) + 8]/6 = 10.0$$

$$\text{Mean salvage life} = 100/6 = \$16.7$$

Compute the mean rate of return:

	A	B	C	D	E	F	G	H
1	Scenario	*weight*	*n*	*PMT*	*PV*	*FV*	Solve for	Answer
2	Optimistic	1	12	210	-950	100	*i*	19.8%
3	Most Likely	4	10	200	-1000	0	*i*	15.1%
4	Pessimistic	1	8	170	-1150	0	*i*	3.9%
5	weighted values		10	196.7	-1016.7	16.7	*i*	14.3%
6			=SUMPRODUCT($B$2:$B$4,C2:C4)/SUM($B$2:$B$4)					

Note that the **SUMPRODUCT** function is an easy way to multiply the set of weights by each of n, PMT, PV, and FV. The resulting 14.3% rate of return is lower than the return of the most likely case, because the pessimistic values for PV and FV are "further away" from the most likely values than the optimistic values are.

Example 10–1 gave a most likely rate of return (15.1%) that differed from the mean rate of return (14.3%) computed in Example 10–2. These values are different because the former is based exclusively on the most likely values and the mean considers other possible values.

In examining the data, we see that the pessimistic values are further away from the most likely values than are the optimistic values. This is common. For example, a savings of 10–20% may be the maximum possible, but a cost overrun can be 50%, 100%, or even more. This causes the weighted mean values to be less favorable than the most likely values. As a result, the mean rate of return is usually less than the rate of return based on the most likely values.

PROBABILITY

We all have used probabilities. For example, what is the probability of getting a "head" when flipping a coin? Using a model that assumes that the coin is fair, both the head and tail outcomes occur with a probability of 50%, or $1/2$. This probability is the likelihood of an event in a single trial. It also describes the long-run relative frequency of getting heads in many trials (out of 50 coin flips, we expect to average 25 heads).

Probabilities can also be based on data, expert judgment, or a combination of both. Past data on weather and climate, on project completion times and costs, and on highway traffic are combined with expert judgment to forecast future events. These examples can be important in engineering economy.

Another example based on long-run relative frequency is the PW of a flood-protection dam that depends on the probabilities of different-sized floods over many years. This would be based on data from past floods and would include many years of observation. An example of a single event that may be estimated by expert judgment is the probability of a successful outcome for a research and development project, which will determine its PW.

All the data in an engineering economy problem may have some level of uncertainty. However, small uncertainties may be ignored, so that more analysis can be done with the large uncertainties. For example, the price of an off-the-shelf piece of equipment may vary by only $\pm 5\%$. The price could be treated as a known or deterministic value. On the other hand, demand over the next 20 years will have more uncertainty. Demand should be analyzed as a random or stochastic variable. We should establish probabilities for different values of demand.

There are also logical or mathematical rules for probabilities. If an outcome can never happen, then the probability is 0. If an outcome will certainly happen, then the probability is 1, or 100%. This means that probabilities cannot be negative or greater than 1; in other words, they must be within the interval [0, 1], as indicated shortly in Equation 10-2.

Probabilities are defined so that the sum of probabilities for all possible outcomes is 1 or 100% (Equation 10-3). Summing the probability of 0.5 for a head and 0.5 for a tail leads to a total of 1 for the possible outcomes from the coin flip. An exploration well drilled in a potential oil field will have three outcomes (dry hole, noncommercial quantities, or commercial quantities) whose probabilities will sum to one.

Equations 10-2 and 10-3 can be used to check that probabilities are valid. If the probabilities for all but one outcome are known, the equations can be used to find the unknown probability for that outcome (see Example 10–3).

$$0 \leq \text{Probability} \leq 1 \qquad (10\text{-}2)$$

$$\sum_{j=1 \text{ to } K} P(\text{outcome}_j) = 1, \quad \text{where there are } K \text{ outcomes} \qquad (10\text{-}3)$$

In a probability course many probability distributions, such as the normal, uniform, and beta, are presented. These continuous distributions describe a large population of data. However, for engineering economy it is more common to use a **discrete probability distribution** with 2 to 5 outcomes—even though the 2 to 5 outcomes only represent or approximate the range of possibilities. Quite often 3 outcomes are used—optimistic, most likely, and pessimistic.

This is done for two reasons. First, the data often are estimated by expert judgment, so that using 7 to 10 outcomes would be false accuracy. Second, each outcome requires more analysis. In most cases the 2 to 5 outcomes represents the best trade-off between representing the range of possibilities and the amount of calculation required. Example 10–3 illustrates these calculations.

EXAMPLE 10–3

What are the probability distributions for the annual benefit and life for the following project?

The annual benefit's most likely value is $8000 with a probability of 60%. There is a 30% probability that it will be $5000 and the highest likely value is $10,000. A life of 6 years is twice as likely as a life of 9 years.

SOLUTION

For the annual benefit, probabilities are given for only two of the possible outcomes. The third value is found from the fact that the probabilities for the three outcomes must sum to 1 (Equation 10-3).

$$1 = P(\text{Benefit is \$5000}) + P(\text{Benefit is \$8000}) + P(\text{Benefit is \$10,000})$$
$$P(\text{Benefit is \$10,000}) = 1 - 0.6 - 0.3 = 0.1$$

The probability distribution can then be summarized in a table. The histogram or relative frequency diagram is Figure 10–1.

Annual benefit	$5000	$8000	$10,000
Probability	0.3	0.6	0.1

FIGURE 10–1 Probability distribution for annual benefit.

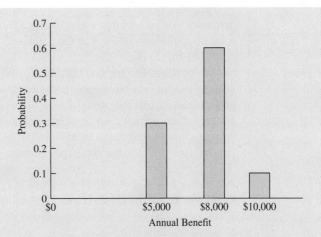

For the life's probability distribution, the problem statement tells us

$$P(\text{life is 6 years}) = 2P(\text{life is 9 years})$$

Equation 10-3 can be applied to write a second equation for the two unknown probabilities:

$$P(6) + P(9) = 1$$

Combining these, we write

$$2P(9) + P(9) = 1$$
$$P(9) = 1/3$$
$$P(6) = 2/3$$

The probability distribution for the life is $P(6) = 66.7\%$ and $P(9) = 33.3\%$.

JOINT PROBABILITY DISTRIBUTIONS

Example 10–3 constructed the probability distributions for a project's annual benefit and life. These examples show how likely each value is for the input data. We would like to construct a similar probability distribution for the project's present worth. This is the distribution that we can use to evaluate the project. That present worth depends on both input probability distributions, so we need to construct the **joint probability distribution** for the different combinations of their values.

For this introductory text, we assume that two random variables such as the annual benefit and life are unrelated or **statistically independent**. This means that the *joint* probability of a combined event (Event A defined on the first variable and Event B on the second variable) is the product of the probabilities for the two events. This is Equation 10-4:

$$\text{If A and B are independent, then } P(\text{A and B}) = P(\text{A}) \times P(\text{B}) \qquad (10\text{-}4)$$

For example, flipping a coin and rolling a die are statistically independent. Thus, the probability of {flipping a head and rolling a 4} equals the probability of a {heads} = 1/2 times the probability of a {4} = 1/6, for a joint probability = 1/12.

The number of outcomes in the joint distribution is the product of the number of outcomes in each variable's distribution. Thus, for the coin and the die, there are 2 times 6, or 12 combinations. Each of the 2 outcomes for the coin is combined with each of the 6 outcomes for the die.

Some variables are not statistically independent, and the calculation of their joint probability distribution is more complex. For example, a project with low revenues may be terminated early and one with high revenues may be kept operating as long as possible. In these cases annual cash flow and project life are not independent. While this type of relationship can sometimes be modeled with economic decision trees (covered later in this chapter), we will limit our coverage in this text to the simpler case of independent variables.

Example 10–4 uses the three values and probabilities for the annual benefit and the two values and probabilities for the life to construct the six possible combinations. Then the values and probabilities are constructed for the project's PW.

EXAMPLE 10–4

The project described in Example 10–3 has a first cost of $25,000. The firm uses an interest rate of 10%. Assume that the probability distributions for annual benefit and life are unrelated or statistically independent. Calculate the probability distribution for the PW.

SOLUTION

Since there are three outcomes for the annual benefit and two outcomes for the life, there are six combinations. The first four columns of the following table show the six combinations of life and annual benefit. The probabilities in columns 2 and 4 are multiplied to calculate the joint probabilities in column 5. For example, the probability of a low annual benefit and a short life is $0.3 \times 2/3$, which equals 0.2 or 20%.

The PW values include the $25,000 first cost and the results of each pair of annual benefit and life. For example, the PW for the combination of high benefit and long life is

$$\mathrm{PW}_{\$10,000,9} = -25,000 + 10,000(P/A, 10\%, 9) = -25,000 + 10,000(5.759) = \$32,590$$

Annual Benefit	Probability	Life (years)	Probability	Joint Probability	PW
$ 5,000	30%	6	66.7%	20.0%	−$ 3,224
8,000	60	6	66.7	40.0	+9,842
10,000	10	6	66.7	6.7	18,553
5,000	30	9	33.3	10.0	3,795
8,000	60	9	33.3	20.0	21,072
10,000	10	9	33.3	3.3	32,590
				100.0%	

Figure 10–2 shows the probabilities for the PW in the form of a histogram for relative frequency distribution, or probability distribution function.

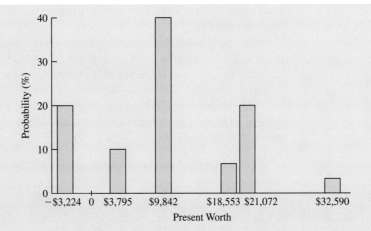

FIGURE 10–2 Probability distribution function for PW.

This probability distribution function shows that there is a 20% chance of having a negative PW. It also shows that there is a small (3.3%) chance of the PW being $32,590. The three values used to describe possible annual benefits for the project and the two values for life have been combined to describe the uncertainty in the project's PW.

Creating a distribution, as in Example 10–4, gives us a much better understanding of the possible PW values along with their probabilities. The three possibilities for the annual benefit and the two for the life are representative of the much broader set of possibilities that really exist. Optimistic, most likely, and pessimistic values are a good way to represent the uncertainty about a variable.

Similarly the six values for the PW represent the much broader set of possibilities. The 20% probability of a negative PW is one measure of risk that we will talk about later in the chapter.

Some problems, such as Examples 10–1 and 10–2, have so many variables or different outcomes that constructing the joint probability distribution is arithmetically burdensome. If the values in Equation 10-1 are treated as a discrete probability distribution function, the probabilities are $1/6, 2/3, 1/6$. With an optimistic, most likely, and pessimistic outcome for each of 4 variables, there are $3^4 = 81$ combinations. In Examples 10–1 and 10-2, the salvage value has only two distinct values, so there are still $3 \times 3 \times 3 \times 2 = 54$ combinations.

When the problem is important enough, the effort to construct the joint probability distribution is worthwhile. It gives the analyst and the decision maker a better understanding of what may happen. It is also needed to calculate measures of a project's risk. While spreadsheets can automate the arithmetic, simulation (described at the end of the chapter) can be a better choice when there are a large number of variables and combinations.

EXPECTED VALUE

For any probability distribution we can compute the **expected value (EV)** or weighted arithmetic average (mean). To calculate the EV, each outcome is weighted by its probability, and the results are summed. This is NOT the simple average or unweighted mean. When the class average on a test is computed, this is an unweighted mean. Each student's test has the same weight. This simple "average" is the one that is shown by the button $\bar{x}$ on many calculators.

The expected value is a weighted average, like a student's grade point average (GPA). To calculate a GPA, the grade in each class is weighted by the number of credits. For the expected value of a probability distribution, the weights are the probabilities.

This is described in Equation 10-5. We saw in Example 10–2 that these expected values can be used to compute a rate of return. They can also be used to calculate a present worth as in Example 10–5.

$$\text{Expected value} = \text{Outcome}_A \times P(A) + \text{Outcome}_B \times P(B) + \cdots \qquad (10\text{-}5)$$

EXAMPLE 10–5

The first cost of the project in Example 10–3 is $25,000. Use the expected values for annual benefits and life to estimate the present worth. Use an interest rate of 10%.

SOLUTION

$$\text{EV}_{\text{benefit}} = 5000(0.3) + 8000(0.6) + 10,000(0.1) = \$7300$$

$$\text{EV}_{\text{life}} = 6(2/3) + 9(1/3) = 7 \text{ years}$$

The PW using these values is

$$\text{PW(EV)} = -25,000 + 7300(P/A, 10\%, 7) = -25,000 + 6500(4.868) = \$10,536$$

[*Note:* This is the present worth of the expected values, PW(EV), not the expected value of the present worth, EV(PW). It is an easy value to calculate that approximates the EV(PW), which will be computed from the joint probability distribution found in Example 10–4.]

Example 10–5 is a simple way to approximate the project's expected PW. But the true expected value of the PW is somewhat different. To find it, we must use the joint probability distribution for benefit and life, and the resulting probability distribution function for PW that was derived in Example 10–4. Example 10–6 shows the expected value of the PW or the EV(PW).

EXAMPLE 10–6

Use the probability distribution function of the PW that was derived in Example 10–4 to calculate the EV(PW). Does this indicate an attractive project?

SOLUTION

The table from Example 10–4 can be reused with one additional column for the weighted values of the PW (= PW × probability). Then, the expected value of the PW is calculated by summing the column of present worth values that have been weighted by their probabilities.

Annual Benefit	Probability	Life (years)	Probability	Joint Probability	PW	PW × Joint Probability
$ 5,000	30%	6	66.7%	20.0%	−$ 3,224	−$ 645
8,000	60	6	66.7	40.0	9,842	3,937
10,000	10	6	66.7	6.7	18,553	1,237
5,000	30	9	33.3	10.0	3,795	380
8,000	60	9	33.3	20.0	21,072	4,214
10,000	10	9	33.3	3.3	32,590	1,086
				100.0%		EV(PW) = $10,209

With an expected PW of $10,209, this is an attractive project. While there is a 20% chance of a negative PW, the possible positive outcomes are larger and more likely. Having analyzed the project under uncertainty, we are much more knowledgeable about the potential result of the decision to proceed.

The $10,209 value is more accurate than the approximate value calculated in Example 10–5. The values differ because PW is a nonlinear function of the life. The more accurate value of $10,209 is lower because the annual benefit values for the longer life are discounted by $1/(1+i)$ for more years.

In Examples 10–5 and 10–6, the question was whether the project had a positive PW. With two or more alternatives, the criterion would have been to maximize the PW. With equivalent uniform annual costs (EUACs) the goal is to minimize the EUAC. Example 10–7 uses the criterion of minimizing the EV of the EUAC to choose the best height for a dam.

EXAMPLE 10–7

A dam is being considered to reduce river flooding. But if a dam is built, what height should it be? Increasing the dam's height will (1) reduce a flood's probability, (2) reduce the damage when floods occur, and (3) cost more. Which dam height minimizes the expected total annual cost? The state uses an interest rate of 5% for flood protection projects, and all the dams should last 50 years.

Dam Height (ft)	First Cost
No dam	$ 0
20	700,000
30	800,000
40	900,000

SOLUTION

The easiest way to solve this problem is to choose the dam height with the lowest equivalent uniform annual cost (EUAC). Calculating the EUAC of the first cost requires multiplying the first cost by $(A/P, 5\%, 50)$. For example, for the dam 20 ft high, this is $700,000(A/P, 5\%, 50) = \$38,344$.

Calculating the annual expected flood damage cost for each alternative is simplified because the term for the P(no flood) is zero, because the damages for no flood are $0. Thus we need to calculate only the term for flooding. This is done by multiplying the P(flood) times the damages if a flood happens. For example, the expected annual flood damage cost with no levee is $0.25 \times \$800,000$, or $200,000.

Then the EUAC of the first cost and the expected annual flood damage are added together to find the total EUAC for each height. The 30 ft dam is somewhat cheaper than the 40 ft dam.

Dam Height (ft)	EUAC of First Cost	Annual P (flood) > Height		Damages if Flood Occurs		Expected Annual Flood Damages	Total Expected EUAC
No dam	$ 0	0.25	×	$800,000	=	$200,000	$200,000
20	38,344	0.05	×	500,000	=	25,000	63,344
30	43,821	0.01	×	300,000	=	3000	46,821
40	49,299	0.002	×	200,000	=	400	49,699

ECONOMIC DECISION TREES

Some engineering projects are more complex, and evaluating them properly is correspondingly more complex. For example, consider a new product with potential sales volumes ranging from low to high. If the sales volume is low, then the product may be discontinued early in its potential life. On the other hand, if sales volume is high, additional capacity may be added to the assembly line and new product variations may be added. This can be modeled with a **decision tree**.

The following symbols are used to model decisions with decision trees:

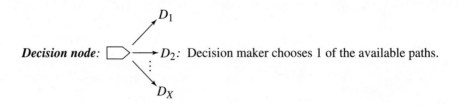

Decision node: D_2: Decision maker chooses 1 of the available paths.

Chance node: Represents a probabilistic (chance) event. Each possible outcome $(C_1, C_2, \ldots, C_Y)$ has a probability $(p_1, p_2, \ldots, p_y)$ associated with it.

Outcome node: →☐: Shows result for a particular path through the decision tree.

Pruned branch: ─∦→: The double hash mark indicates that a branch has been pruned because another branch has been chosen. This can happen only at decision nodes, not at chance nodes. The term "pruned" is chosen to correspond to the gardener's practice of trimming or pruning off branches to make a tree or bush healthier.

Figure 10–3 illustrates how decision nodes ▷, chance nodes ◯, and outcome nodes ☐ can be used to describe a problem's structure. Details such as the probabilities and costs can be added on the branches that link the nodes. With the branches from decision and chance nodes, the model becomes a decision tree.

Figure 10–3 illustrates that decision trees describe the problem by starting at the decision that must be made and then adding chance and decision nodes in the proper logical sequence. Thus describing the problem starts at the first step and goes forward in time with sequences of decision and chance nodes.

To make the decision, calculations begin with the final nodes in the tree. Since they are the final nodes, enough information is available to evaluate them. At decision nodes the criterion is either to maximize PW or to minimize EUAC. At chance nodes an expected value for PW or EUAC is calculated.

Once all nodes that branch from a node have been evaluated, the originating node can be evaluated. If the originating node is a decision node, choose the branch with the best PW or EUAC and place that value in the node. If the originating node is a chance node, calculate the expected value and place that value in the node. This process "rolls back" values from the terminal nodes in the tree to the initial decision. Example 10–8 illustrates this process.

EXAMPLE 10–8

What decision should be made on the new product summarized in Figure 10–3? What is the expected value of the product's PW? The firm uses an interest rate of 10% to evaluate projects. If the product is terminated after one year, the capital equipment has a salvage value of $550,000 for use with other new products. If the equipment is used for 8 years, the salvage value is $0.

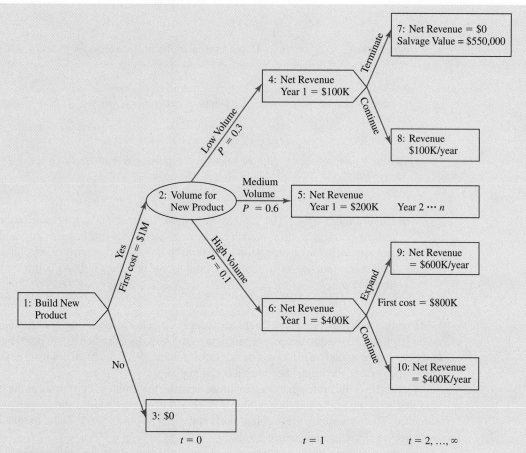

FIGURE 10–3 Economic decision tree for new product.

Evaluating decision trees is done by starting with the end outcome nodes and the decisions that lead to them. In this case the decisions are whether to terminate after 1 year if sales volume is low and whether to expand after 1 year if sales volume is high.

The decision to terminate the product depends on which is more valuable, the equipment's salvage value of $550,000 or the revenue of $100,000 per year for 7 more years. The worth (PW_1) of the salvage value is $550,000. The worth ($PW_1$) of the revenue stream at the end of Year 1 shown in node 8 is

$$PW_1 \text{ for node } 8 = 100,000(P/A, 10\%, 7)$$

$$= 100,000(4.868) = \$486,800$$

Thus, terminating the product and using the equipment for other products is better. We enter the two "present worth" values at the end of Year 1 in nodes 7 and 8. We make the *arc to node 7 bold* to indicate that it is our preferred choice at node 4. We use a *double hash mark* to show that we're *pruning the arc to node 8* to indicate that it has been rejected as an inferior choice at node 4.

The decision to expand at node 6 could be based on whether the $800,000 first cost for expansion can be justified based on increasing annual revenues for 7 years by $200,000 per year. However, this is difficult to show on the tree. It is easier to calculate the "present worth" values at the end of Year 1 for each of the two choices. The worth (PW_1) of node 9 (expand) is

$$PW_1 \text{ for node } 9 = -800,000 + 600,000(P/A, 10\%, 7)$$

$$= -800,000 + 600,000(4.868)$$

$$= \$2,120,800$$

The value of node 10 (continue without expanding) is

$$PW_1 \text{ for node } 10 = 400,000(P/A, 10\%, 7)$$

$$= 400,000(4.868)$$

$$= \$1,947,200$$

This is $173,600 less than the expansion node, so the expansion should happen if volume is high. Figure 10–4 summarizes what we know at this stage of the process.

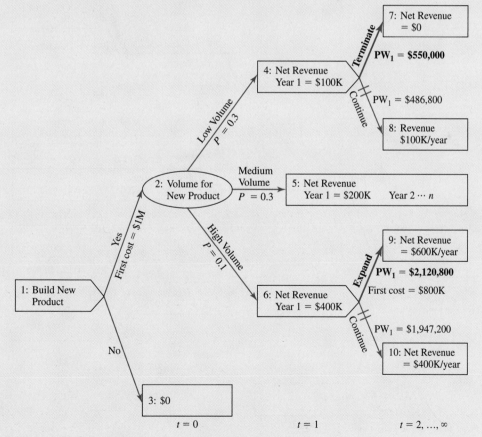

FIGURE 10–4 Partially solved decision tree for new product.

The next step is to calculate the PW (Time 0) at nodes 4, 5, and 6.

$$PW \text{ at node } 4 = (100,000 + 550,000)(P/F, 10\%, 1) = 650,000(0.9091) = \$590,915$$

$$PW \text{ at node } 5 = (200,000)(P/A, 10\%, 8) = 200,000(5.335) = \$1,067,000$$

$$PW \text{ at node } 6 = [400,000 - 800,000 + 600,000(P/A, 10\%, 7)] \, (P/F, 10\%, 1)$$

$$= [-400,000 + 600,000(4.868)](0.9091) = \$2,291,660$$

Now the expected value at node 2 can be calculated:

$$EV \text{ at node } 2 = 0.3(590,915) + 0.6(1,067,000) + 0.1(2,291,660) = \$1,046,640$$

The cost of selecting node 2 is \$1,000,000, so proceeding with the product has an expected PW of \$46,640. This is greater than the \$0 for not building the project. So the decision is to build. Figure 10–5 is the decision tree at the final stage.

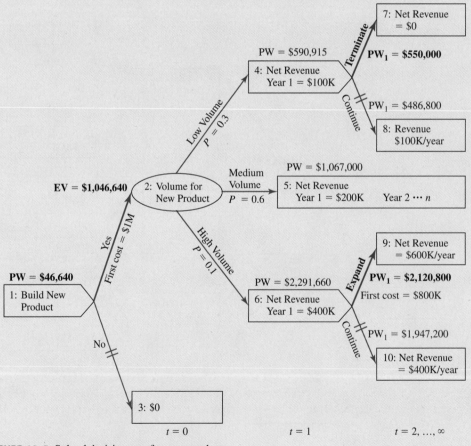

FIGURE 10–5 Solved decision tree for new product.

Example 10–8 is representative of many problems in engineering economy. The main criterion is maximizing PW or minimizing EUAC. However, as shown in Example 10–9, other criteria, such as risk, are used in addition to expected value.

EXAMPLE 10–9

Consider the economic evaluation of collision and comprehensive (fire, theft, etc.) insurance for a car. This insurance is typically required by lenders, but once the car has been paid for, this insurance is not required. (Liability insurance *is* a legal requirement.)

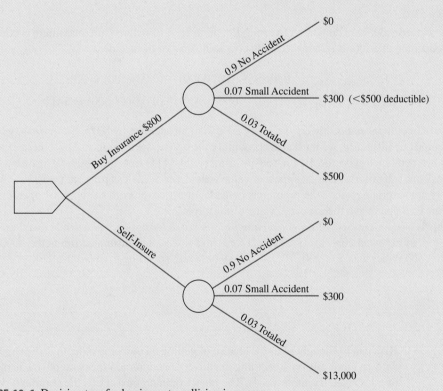

FIGURE 10-6 Decision tree for buying auto collision insurance.

Figure 10–6 begins with a decision node with two alternatives for the next year. Insurance will cost $800 per year with a $500 deductible if a loss occurs. The other option is to self-insure, which means to go without buying collision and comprehensive insurance. Then if a loss occurs, the owner must replace the vehicle with money from savings or a loan, or do without a vehicle until the owner can afford to replace it.

Three accident severities are used to represent the range of possibilities: a 90% chance of no accident, a 7% chance of a small accident (at a cost of $300, which is less than the deductible), and a 3% chance of totaling the $13,000 vehicle. Since our driving habits are likely to be the same with and without insurance, the accident probabilities are the same for both chance nodes.

Even though this is a text on engineering economy, we have simplified the problem and ignored the difference in timing of the cash flows. Insurance payments are made at the beginning of the covered period, and accident costs occur during the covered period. Since car insurance is usually paid semiannually, the results of the economic analysis are not changed significantly by the simplification. We focus on the new concepts of expected value, economic decision trees, and risk.

What are the expected values for each alternative, and what decision is recommended?

SOLUTION

The expected values are computed by using Equation 10-5. If insured, the maximum cost equals the deductible of $500. If self-insured, the cost is the cost of the accident.

$$EV_{\text{accident w/ins.}} = (0.9)(0) + (0.07)(300) + (0.03)(500) = \$36$$

$$EV_{\text{accident w/o ins.}} = (0.9)(0) + (0.07)(300) + (0.03)(13{,}000) = \$411$$

Thus, buying insurance lowers the expected cost of an accident by $375. To evaluate whether we should buy insurance, we must also account for the cost of the insurance. Thus, these expected costs are combined with the $0 for self-insuring (total $411) and the $800 for insuring (total $836). Thus self-insuring has an expected value cost that is $425 less per year (= $836 − $411). This is not surprising, since the premiums collected must cover both the costs of operating the insurance company and the expected value of the payouts.

This is also an example of *expected values alone not determining the decision*. Buying insurance has an expected cost that is $425 per year higher, but that insurance limits the maximum loss to $500 rather than $13,000. The $425 may be worth spending to avoid that risk.

RISK

Risk can be thought of as the chance of getting an outcome other than the expected value—with an emphasis on something negative. One common measure of risk is the probability of a loss (see Example 10–4). The other common measure is the **standard deviation** (σ), which measures the dispersion of outcomes about the expected value. For example, many students have used the normal distribution in other classes. The normal distribution has 68% of its probable outcomes within ±1 standard deviation of the mean and 95% within ±2 standard deviations of the mean.

Appendix 9A introduced risk as measured by the standard deviation. In this section the standard deviation is calculated. The next section will cover considering risk and return together. Appendix 10A shows how diversification reduces risk, with a focus on investing for retirement. Risk reduction through diversification is one reason that firms form portfolios of projects, products, and business lines.

Calculating the Standard Deviation

Mathematically, the standard deviation is defined as the square root of the variance. This term is defined as the weighted average of the squared difference between the outcomes of the random variable X and its mean. Thus the larger the difference between the mean and the values, the larger are the standard deviation and the variance. This is Equation 10-6:

$$\text{Standard deviation } (\sigma) = \sqrt{[\text{EV}(X - \text{mean})^2]} \qquad (10\text{-}6)$$

Squaring the differences between individual outcomes and the EV ensures that positive and negative deviations receive positive weights. Consequently, negative values for the standard deviation are impossible, and they instantly indicate arithmetic mistakes. The standard deviation equals 0 if only one outcome is possible. Otherwise, the standard deviation is positive.

This is not the standard deviation formula built into most calculators, just as the weighted average is not the simple average built into most calculators. The calculator formulas are for N equally likely data points from a randomly drawn sample, so that each probability is $1/N$. In economic analysis we will use a weighted average for the squared deviations since the outcomes may not be equally likely.

The second difference is that for calculations (by hand or the calculator), it is easier to use Equation 10-7, which is shown to be equivalent to Equation 10-6 in introductory probability and statistics texts.

$$\text{Standard deviation } (\sigma) = \sqrt{\{\text{EV}(X^2) - [\text{EV}(X)]^2\}} \qquad (10\text{-}7)$$

$$= \sqrt{\{\text{Outcome}_{\text{A}}^2 \times P(\text{A}) + \text{Outcome}_{\text{B}}^2 \times P(\text{B}) + \cdots - \text{expected value}^2\}} \qquad (10\text{-}7')$$

This equation is the square root of the difference between the average of the squares and the square of the average. The standard deviation is used instead of the *variance* because the standard deviation is measured in the same units as the expected value. The variance is measured in "squared dollars"—whatever they are.

The calculation of a standard deviation by itself is only a descriptive statistic of limited value. However, as shown in the next section on risk/return trade-offs, it is useful when the standard deviation of each alternative is calculated and these results are compared. But first, some examples of calculating the standard deviation.

EXAMPLE 10–10   **(Example 10–9 continued)**

Consider the economic evaluation of collision and comprehensive (fire, theft, etc.) insurance for an auto. One example was described in Figure 10–6. The probabilities and outcomes are summarized in the calculation of the expected values, which was done using Equation 10–5.

$$EV_{\text{accident w/ins.}} = (0.9)(0) + (0.07)(300) + (0.03)(500) = \$36$$

$$EV_{\text{accident w/o ins.}} = (0.9)(0) + (0.07)(300) + (0.03)(13{,}000) = \$411$$

Calculate the standard deviations for insuring and not insuring.

SOLUTION

The first step is to calculate the $EV(\text{outcome}^2)$ for each.

$$EV^2_{\text{accident w/ins.}} = (0.9)(0^2) + (0.07)(300^2) + (0.03)(500^2) = 13{,}800$$

$$EV^2_{\text{accident w/o ins.}} = (0.9)(0^2) + (0.07)(300^2) + (0.03)(13{,}000^2) = 5{,}076{,}300$$

Then the standard deviations can be calculated.

$$\sigma_{\text{w/ins.}} = \sqrt{EV^2_{\text{w/ins.}} - (EV_{\text{w/ins.}})^2}$$

$$= \sqrt{(13{,}800 - 36^2)} = \sqrt{12{,}504} = \$112$$

$$\sigma_{\text{w/o ins.}} = \sqrt{EV^2_{\text{w/o ins.}} - (EV_{\text{w/o ins.}})^2}$$

$$= \sqrt{(5{,}076{,}300 - 411^2)} = \sqrt{4{,}907{,}379} = \$2215$$

As described in Example 10–9, the expected value cost of insuring is $836 (= \$36 + \$800) and the expected value cost of self-insuring is $411. Thus the expected cost of not insuring is about half the cost of insuring. But the standard deviation of self-insuring is 20 times larger. It is clearly riskier.

Which choice is preferred depends on how much risk one is comfortable with.

As stated before, this is an example of *expected values alone not determining the decision*. Buying insurance has an expected cost that is $425 per year higher, but that insurance limits the maximum loss to $500 rather than $13,000. The $425 may be worth spending to avoid that risk.

EXAMPLE 10–11   (Example 10–4 continued)

Using the probability distribution for the PW from Example 10–4, calculate the PW's standard deviation.

SOLUTION

The following table adds a column for $PW^2 \times$ Probability to calculate the $EV(PW^2)$.

Annual Benefit	Probability	Life (years)	Probability	Joint Probability	PW	PW× Probability	PW²× Probability
$ 5,000	30%	6	66.7%	20.0%	−$ 3,224	−$ 645	2,079,480
8,000	60	6	66.7	40.0	9,842	3937	38,747,954
10,000	10	6	66.7	6.7	18,553	1237	22,950,061
5,000	30	9	33.3	10.0	3,795	380	1,442,100
8,000	60	9	33.3	20.0	21,072	4214	88,797,408
10,000	10	9	33.3	3.3	32,590	1086	35,392,740
						EV = $10,209	189,409,745

$$\text{Standard deviation} = \sqrt{\{EV(X^2) - [EV(X)]^2\}}$$

$$\sigma = \sqrt{\{189,405,745 - [10,209]^2\}} = \sqrt{85,182,064} = \$9229$$

For those with stronger backgrounds in probability than this chapter assumes, let us consider how the standard deviation in Example 10–11 depends on the assumption of independence between the variables. While exceptions exist, a positive statistical dependence between variables often increases the PW's standard deviation. Similarly, a negative statistical dependence between variables often decreases the PW's standard deviation.

RISK VERSUS RETURN

A graph of risk versus return is one way to consider these items together. Figure 10–7 in Example 10–12 illustrates the most common format. Risk measured by standard deviation is placed on the x axis, and return measured by expected value is placed on the y axis. This is usually done with internal rates of return of alternatives or projects.

EXAMPLE 10–12

A large firm is discontinuing an older product; thus some facilities are becoming available for other uses. The following table summarizes eight new projects that would use the facilities. Considering expected return and risk, which projects are good candidates? The firm believes it can earn 4% on a risk-free investment in government securities (labeled as Project F).

Project	IRR	Standard Deviation
1	13.1%	6.5%
2	12.0	3.9
3	7.5	1.5
4	6.5	3.5
5	9.4	8.0
6	16.3	10.0
7	15.1	7.0
8	15.3	9.4
F	4.0	0.0

SOLUTION

Answering the question is far easier if we use Figure 10–7. Since a larger expected return is better, we want to select projects that are as "high up" as possible. Since a lower risk is better, we want to select projects that are as "far left" as possible. The graph lets us examine the trade-off of accepting more risk for a higher return.

FIGURE 10–7 Risk-versus-return graph.

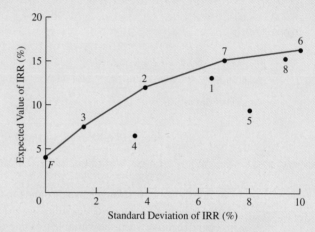

First, we can eliminate Projects 4 and 5. They are **dominated projects.** Dominated alternatives are no better than another alternative on all measures and inferior on at least one measure. Project 4 is dominated by Project 3, which has a higher expected return and a lower risk. Project 5 is dominated by Projects 1, 2, and 7. All three have a higher expected return and a lower risk.

Second, we look at the **efficient frontier.** This is the blue line in Figure 10–7 that connects Projects F, 3, 2, 7, and 6. Depending on the trade-off that we want to make between risk and return, any of these could be the best choice.

Project 1 appears to be inferior to Projects 2 and 7. Project 8 appears to be inferior to Projects 7 and 6. Projects 1 and 8 are inside and not on the efficient frontier.

There are models of risk and return that can allow us to choose between Projects F, 3, 2, 7, and 6; but those models are beyond what is covered here.

There is a simple rule of thumb for comparing a project's risk and return for which we would like to thank Joe Hartman. If the expected present worth is at least double the standard deviation of the present worth, then the project is relatively *safe*. For comparison, remember that a normal distribution has about 2.5% of its values less than 2 standard deviations below the mean.

Risk and Return as Multiple Objectives

To combine the effects of risk and return a single measure can be created by using the additive multiple objective model introduced in Chapter 1. Scores are calculated on a scale of 0 to 10 (worst to best). Since higher IRR values are better, Equation 10-8 is used for return. Since lower standard deviations are better, Equation 10-9 is used for risk. After scoring the alternatives, the next step is to estimate reasonable weights that add to 100% for the objectives, in this case return and risk. Then a total weighted score for each alternative can be calculated.

When higher values are better $\text{Score} = (\text{Value} - \text{minimum}) \times 10/\text{range}$ (10-8)

When lower values are better $\text{Score} = 10 - (\text{Value} - \text{minimum}) \times 10/\text{range}$
(10-9)

For examples with many alternatives like Example 10–13, the maximum and minimum values are calculated. If there are only a few alternatives, it is probably better to estimate theoretical or reasonably achievable maximums and minimums. Otherwise many, most, or even all scores will be 0 or 10.

EXAMPLE 10–13   (Example 10–12 Revisited)

For the projects in Example 10–12, apply a weighted multiple objective model using a weight of 60% for the IRR and a weight of 40% for the standard deviation. Which project is the best for this model?

SOLUTION

Whether by hand or spreadsheet, the first step is to calculate the minimum value, the maximum value, and the range of values for both the IRR and standard deviation data.

	IRR	Standard Deviation
Maximum	16.3%	10.0%
Minimum	4.0	0.0
Range	12.3	10.0

The solution will be by spreadsheet, but it is best to start with a hand calculation to ensure understanding and as a check on the later calculations. What are the scores for Project 1?

$$\text{IRR}_1 = (\text{Value} - \text{minimum}) \times 10/\text{range} \qquad \text{using (10-8)}$$
$$= (13.1\% - 4.0\%) \times 10/12.3\% = 7.4$$
$$\text{Standard deviation}_1 = 10 - (\text{Value} - \text{minimum}) \times 10/\text{range} \qquad \text{using (10-9)}$$
$$= 10 - (6.5\% - 0.0\%) \times 10/10.0\% = 3.5$$
$$\text{Score}_1 = 60\% \times 7.4 + 40\% \times 3.5 = 5.8$$

It is worth noting that the three highest-scoring projects, 7, 2, and 6, are all on the efficient frontier in Figure 10–7. However, projects 1 and 8, which are not on the efficient frontier, have a weighted score higher than Project 3, which is on the efficient frontier.

	A	B	C	D	E	F
1				60%	40%	Weight
2	Project	IRR	Std.Dev.	IRR score	SD score	Total
3	1	13.1%	6.5%	7.4	3.5	5.84
4	2	12.0%	3.9%	6.5	6.1	6.34
5	3	7.5%	1.5%	2.9	8.6	5.14
6	4	6.5%	3.5%	2.0	6.5	3.82
7	5	9.4%	8.0%	4.4	2.0	3.43
8	6	16.3%	10.0%	10.0	0.0	6.00
9	7	15.1%	7.0%	9.0	3.0	6.61
10	8	15.3%	9.4%	9.2	0.6	5.75
11	F	4.0%	0.0%	0.0	10.0	4.00
12	min	4.0%	0.0%			
13	max	16.3%	10.0%			
14	range	12.3%	10.0%			

SIMULATION

Simulation is a more advanced approach to considering risk in engineering economy problems. As such, the following discussion focuses on what it is. As the examples show, spreadsheet functions and add-in packages make simulation easier to use for economic analysis.

Economic **simulation** uses random sampling from the probability distributions of one or more variables to analyze an economic model for many iterations. For each iteration, all variables with a probability distribution are randomly sampled. These values are used to calculate the PW, IRR, or EUAC. Then the results of all iterations are combined to create a probability distribution for the PW, IRR, or EUAC.

Simulation can be done by hand, using a table of random numbers—if there are only a few random variables and iterations. However, results are more reliable as the number of iterations increases, so in practice this is usually computerized. This can be done in Excel using the RAND() function to generate random numbers, as shown in Example 10–14.

Because we were analyzing each possible outcome, the probability distributions earlier in this chapter (and in the end-of-chapter problems) used two or three discrete outcomes. This limited the number of combinations that we needed to consider. Simulation makes it easy to use continuous probability distributions like the uniform, normal, exponential, log normal, binomial, and triangular. Examples 10–14 and 10–15 use the normal and the discrete uniform distributions.

EXAMPLE 10–14

ShipM4U is considering installing a new, more accurate scale, which will reduce the error in computing postage charges and save $250 a year. The scale's useful life is believed to be uniformly distributed over 12, 13, 14, 15, and 16 years. The initial cost of the scale is estimated to be normally distributed with a mean of $1500 and a standard deviation of $150.

Use Excel to simulate 25 random samples of the problem and compute the rate of return for each sample. Construct a graph of rate of return versus frequency of occurrence.

SOLUTION

FIGURE 10-8 Excel spreadsheet for simulation
($N = 25$).

	A	B	C	D
1	250	Annual Savings		
2		Life	First Cost	
3	Min	12	1500	Mean
4	Max	16	150	Std dev
5				
6	Iteration			IRR
7	1	12	1277	16.4%
8	2	15	1546	13.9%
9	3	12	1523	12.4%
10	4	16	1628	13.3%
11	5	14	1401	15.5%
12	6	12	1341	15.2%
13	7	12	1683	10.2%
14	8	14	1193	19.2%
15	9	15	1728	11.7%
16	10	12	1500	12.7%
17	11	16	1415	16.0%
18	12	12	1610	11.2%
19	13	15	1434	15.4%
20	14	12	1335	15.4%
21	15	14	1468	14.5%
22	16	13	1469	13.9%
23	17	14	1409	15.3%
24	18	15	1484	14.7%
25	19	14	1594	12.8%
26	20	15	1342	16.8%
27	21	14	1309	17.0%
28	22	12	1541	12.1%
29	23	16	1564	14.0%
30	24	13	1590	12.2%
31	25	16	1311	17.7%
32				
33	Mean	14	1468	14.4%
34	Std dev	2	135	2.2%

This problem is simple enough to allow us to construct a table with each iteration's values of the life and the first cost. From these values and the annual savings of $250, the IRR for each iteration can be calculated using the RATE function. These are shown in Figure 10–8. The IRR values are summarized in a relative frequency diagram in Figure 10–9.

Note: Each time Excel recalculates the spreadsheet, different values for all the random numbers are generated. Thus the results depend on the set of random numbers, and your results will be different if you create this spreadsheet.

Note for students who have had a course in probability and statistics: Creating the random values for life and first cost is done as follows. Select a random number in [0, 1] using Excel's RAND function. This is the value of the cumulative distribution function for the variable. Convert this to the variable's value by using an inverse function from Excel, or build the inverse function. For the discrete uniform life, the function is = min life + INT(range * RAND()). For the normally distributed first cost, the functions is = NORMINV(RAND(), mean, standard deviation).

FIGURE 10–9
Graph of IRR values.

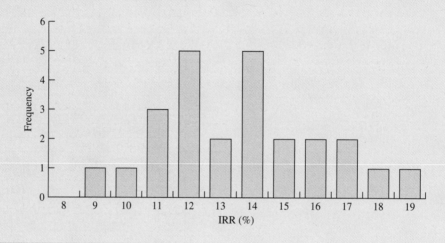

Stand-alone simulation programs and commercial spreadsheet add-in packages such as @Risk and Crystal Ball provide probability distribution functions to use for each input variable. In Example 10–15 the functions RiskUniform and RiskNormal are used. The packages also collect values for the output variables, such as the IRR for Example 10–15. In other problems the PW or EUAC could be collected. These values form a probability distribution for the PW, IRR, or EUAC. From this distribution the simulation package can calculate the expected return, P(loss), and the standard deviation of the return.

Example 10–15 uses @Risk to simulate 1000 iterations of PW for the data in Example 10–14. A simulation package makes it easy to do more iterations. More important, accurate models can be built because it is much easier to use different probability distributions and parameters. Because the models are easier to build, they are less likely to contain errors.

EXAMPLE 10–15 | (Example 10–14 Revisited)

Consider the scale described in Example 10–14. Generate 1000 iterations and construct a frequency distribution for the scale's rate of return.

SOLUTION

The first IRR (cell A7) of 14.01% that is computed in Figure 10–10 is based on the average life and the average first cost. The second IRR (cell A11) of 14.01% is computed by @Risk using the average of each distribution. The cell content in A11 is the RATE formula, referencing cells that are input cells to the simulation.

There are two input cells that contain variables that will change with each iteration. Cell A9 is the value of life, which uses a uniform distribution between 12 and 16, showing a static value of 14. Cell A10 is the first cost, using a normal distribution with a mean of −1500 and a standard deviation of 150, showing a static value of −1500. @Risk will use the static values to compute the value in the output cell (A11). Performing a simulation of 1000 iterations provides the graph, showing the distribution of the IRR. The graph will change each time the simulation is performed. The graph in Figure 10–10 with 1000 iterations is much smoother than Figure 10–9, the graph from Example 10–14, where 25 iterations were done.

FIGURE 10–10 Simulation spreadsheet for Examples 10–14 and 10–15.

	A	B	C	D	E	F
1	−1500	average first cost				
2	150	standard deviation of first cost				
3	12	minimum value of life				
4	16	maximum value of life				
5	250	annual benefit				
6	IRR computed using averages					
7	14.01%	= RATE ((A3+A4)/2,A5,A1)				
8	Simulation					
9	14	=RiskUniform(12,16,RiskStatic(14))				
10	−1500	=RiskNormal(-1500,150,RiskStatic(-1500))				
11	14.01%	=RiskOutput()+RATE(A9,A5,A10)				
12						
13		**Distribution for IRR (cell A11)**				
14		14.07% = mean value for IRR from 1000 iterations				
15		2.15% = standard deviation from 1000 iterations				
16						
17						
18						
19						
20						
21						
22						
23						
24						
25						
26						
27						
28						

REAL OPTIONS

Real options analysis is another approach for evaluating projects with significant future uncertainties. It is derived from the theory underlying financial options in the stock market. Rather than buying or selling a share of stock, a financial option gives you the right to buy or sell the stock at a given price for a given time period. Since the option costs less than the stock, a $1000 investment in options has more potential gain than the same investment in the stock. It also has more risk. For engineering projects, a real option is an alternative that places a value on the ability to delay or change the decision in the future.

For an intuitive understanding of why real options make sense, consider the example of a lease for gas and oil development rights. Many such projects are never developed because they are not "economic." Nevertheless, it may make sense to renew the lease for a noneconomic project because of the possibility that the project will *become* economically profitable. For example, prices may rise, or pipelines and facilities may be built to support a nearby tract. The lease is an option to pursue the project in the future.

The models and assumptions of real options are beyond the scope of this text. Interested readers are referred to the third edition of the advanced text *Economic Analysis of Industrial Projects* by Eschenbach, Lewis, Hartman, and Bussey.

These options may include delaying, abandoning, expanding, shrinking, changing, and replicating a project—with different models for different actions. However, some suggested guidelines may help you decide when you should consider applying real options to your project analysis.

- The value of exercising an option cannot be negative, since you can choose not to exercise it.
- An option to delay a project is valuable when the project is not currently economic but may become economic. Thus, if a project's present worth is positive or very negative, a delay option is unlikely to be worthwhile.
- Unlike projects, options become more valuable as risk (described as volatility) increases. For projects, we want more return but less risk. However, options are more likely to be worthwhile and to result in action if the risk increases.
- Volatility is the key parameter in real option valuation, but it is difficult to calculate. In addition, much of the risk associated with a project may not be *actionable* volatility, and only actionable volatility can create value for a real option [Lewis, Eschenbach, and Hartman, 2008].
- Many published examples of real option analysis ignore the cost of waiting, but this is rarely appropriate for engineering projects [Eschenbach, Lewis, and Hartman, 2009]. For example, if an R&D project, a new product, or a new building is delayed, then it will later face more competition and during the period of delay there are no revenues or cost savings since the project is not yet operating.
- In many cases decision trees and simulation are required to properly develop and describe the real option, and by then the real option analysis may add little value to the decision-making process.

SUMMARY

Estimating the future is required for economic analysis, and there are several ways to do this. Precise estimates will not ordinarily be exactly correct, but they are considered to be the best single values to represent what we think will happen.

A simple way to represent uncertainty is through a range of estimates for each variable, such as optimistic, most likely, and pessimistic. The full range of prospective results may be examined by using the optimistic values to solve the problem and then using the pessimistic values. Solving the problem with the most likely values is a good single value estimate. However, the extremes with all optimistic values or all pessimistic values are less likely—it is more likely that a mix of optimistic, most likely, and pessimistic values will occur.

One approach taken from project management uses the following weighted values instead of a range of estimates.

Estimate	Optimistic	Most Likely	Pessimistic
Relative weight	1	4	1

The most commonly used approach for decision making relies on **expected values.** Here, known or estimated probabilities for future events are used as weights for the corresponding outcomes.

$$\text{Expected value} = \text{Outcome}_A \times \text{Probability}_A + \text{Outcome}_B \times \text{Probability}_B + \cdots$$

Expected value is the most useful and the most frequently used technique for estimating a project's attractiveness.

However, risk as measured by standard deviation and the probability of a loss is also important in evaluating projects. Since projects with higher expected returns also frequently have higher risk, evaluating the trade-offs between risk and return is useful in decision making.

More complicated problems can be summarized and analyzed by using decision trees, which allow logical evaluation of problems with sequential chance, decision, and outcome nodes.

Where the elements of an economic analysis are stated in terms of probability distributions, a repetitive analysis of a random sample is often done. This simulation-based approach relies on the premise that a random sampling of increasing size becomes a better and better estimate of the possible outcomes. The large number of computations means that simulation is usually computerized.

PROBLEMS

Range of Estimates

10-1 Telephone poles exemplify items that have varying useful lives. Telephone poles, once installed in a location, remain in useful service until one of a variety of events occur.

(*a*) Name three reasons why a telephone pole might be removed from useful service at a particular location.

(*b*) You are to estimate the total useful life of telephone poles. If the pole is removed from an original location while it is still serviceable, it will be installed elsewhere. Estimate the optimistic life, most likely life, and pessimistic life for telephone poles. What percentage of all telephone poles would you expect to have a total useful life greater than your estimated optimistic life?

(*c*) What is an environmental life cycle assessment (LCA)? How do treated wood, metal, and concrete poles compare?

10-2 The purchase of a used pickup for $14,000 is being considered. Records for other vehicles show that costs for oil, tires, and repairs about equal the cost for fuel.

Fuel costs are $1500 per year if the truck is driven 10,000 miles. The salvage value after 5 years of use drops about 10¢ per mile. Find the equivalent uniform annual cost if the interest rate is 6%. How much does this change if the annual mileage is 15,000? 5000?

10-3 For the data in Problem 10-2 assume that the 5000, 10,000, and 15,000 mileage values are, respectively, pessimistic, most likely, and optimistic estimates. Use a weighted estimate to calculate the equivalent annual cost.

10-4 A heat exchanger is being installed as part of a plant modernization program. It costs $80,000, including installation, and is expected to reduce the overall plant fuel cost by $20,000 per year. Estimates of the useful life of the heat exchanger range from an optimistic 12 years to a pessimistic 4 years. The most likely value is 5 years. Assume the heat exchanger has no salvage value at the end of its useful life.

(a) Determine the pessimistic, most likely, and optimistic rates of return.

(b) Use the range of estimates to compute the mean life and determine the estimated before-tax rate of return.

10-5 A new engineer is evaluating whether to use a higher-voltage transmission line. It will cost $250,000 more initially, but it will reduce transmission losses. The optimistic, most likely, and pessimistic projections for annual savings are $25,000, $20,000, and $13,000. The interest rate is 6%, and the transmission line should have a life of 30 years.

(a) What is the present worth for each estimated value?

(b) Use the range of estimates to compute the mean annual savings, and then determine the present worth.

(c) Does the answer to (b) match the present worth for the most likely value? Why or why not?

10-6 A new 2-lane road is needed in a part of town that is growing. At some point the road will need 4 lanes to handle the anticipated traffic. If the city's optimistic estimate of growth is used, the expansion will be needed in 4 years. For the most likely and pessimistic estimates, the expansion will be needed in 8 and 15 years, respectively. The expansion will cost $5.4 million. Use an interest rate of 6%.

(a) What is the PW for each scenario, and what is the range of values?

(b) Use Equation 10-1 to find the mean value of the expansion's PW.

Probabilities

10-7 When a pair of dice are tossed, the results may be any whole number from 2 through 12. In the game of craps one can win by tossing either a 7 or an 11 on the first roll. What is the probability of doing this? (*Hint:* There are 36 ways that a pair of six-sided dice can be tossed. What portion of them result in either a 7 or an 11?)

10-8 The construction time for a bridge depends on the weather. The project is expected to take 250 days if the weather is dry and hot. If the weather is damp and cool, the project is expected to take 350 days. Otherwise, it is expected to take 300 days. Historical data suggest that the probability of cool, damp weather is 40% and that of dry, hot weather is 10%. Find the project's probability distribution.

10-9 Over the last 10 years, the hurdle or discount rate for projects from the firm's research and development division has been 18% twice, 20% four times, and 25% the rest of the time. There is no recognizable pattern. Calculate the probability distribution for next year's discount rate.

10-10 A new product's sales and profits are uncertain. The marketing department has predicted that sales might be as high as 10,000 units per year with a probability of 10%. The most likely value is 7000 units annually. The pessimistic value is estimated to be 5000 units annually with a probability of 20%. Manufacturing and marketing together have estimated the most likely unit profit to be $32. The pessimistic value of $24 has a probability of 0.3, and the optimistic value of $38 has a probability of 0.2. Construct the probability distributions for sales and unit profits.

10-11 A road between Fairbanks and Nome, Alaska, will have a most likely construction cost of $7 million per mile. Doubling this cost is considered to have a probability of 30%, and cutting it by 25% is considered to have a probability of 10%. The state's interest rate is 4%, and the road should last 25 years before major reconstruction. What is the probability distribution of the equivalent annual construction cost per mile?

10-12 Al took a midterm examination in physics and received a score of 65. The mean was 60 and the standard deviation was 20. Bill received a score of 15 in mathematics, where the exam mean was 12 and the standard deviation was 4. Which student ranked higher in his class? Explain.

10-13 You recently had an auto accident that was your fault. If you have another accident or receive a another moving violation within the next 3 years, you will become part of the "assigned risk" pool, and you will pay an extra $1500 per year for insurance.

(a) If the probability of an accident or moving violation is 20% per year, what is the probability distribution of your "extra" insurance payments over the next 4 years? Assume that insurance is purchased annually and that violations register at the end of the year—just in time to affect next year's insurance premium.

(b) Would it be ethically questionable to pay out of your own pocket for a fender bender to avoid having it reported to your insurance company?

Joint Probabilities

10-14 For the data in Problem 10-10, construct the probability distribution for the annual profit. Assume that the sales and unit profits are statistically independent.

10-15 A project has a life of 10 years, and no salvage value. The firm uses an interest rate of 12% to evaluate engineering projects. The project has an uncertain first cost and net revenue.

First Cost	P	Net Revenue	P
$300,000	0.2	$ 70,000	0.3
400,000	0.5	90,000	0.5
600,000	0.3	100,000	0.2

(a) What is the joint probability distribution for first cost and net revenue?

(b) Define optimistic, most likely, and pessimistic scenarios by using both optimistic, both most likely, and both pessimistic estimates. What is the present worth for each scenario?

10-16 A robot has just been installed at a cost of $81,000. It will have no salvage value at the end of its useful life.

Savings per Year	Probability	Useful Life (years)	Probability
$18,000	0.2	12	1/6
20,000	0.7	5	2/3
22,000	0.1	4	1/6

(a) What is the joint probability distribution for savings per year and useful life?

(b) Define optimistic, most likely, and pessimistic scenarios by using both optimistic, both most likely, and both pessimistic estimates. What is the rate of return for each scenario?

10-17 Modifying an assembly line has a first cost of $150,000, and its salvage value is $0. The firm's interest rate is 20%. The savings shown in the table depend on whether the assembly line runs one, two, or three shifts and on whether the product is made for 3 or 5 years.

Shifts/ day	Savings/ year	Probability	Useful Life (years)	Probability
1	$15,000	0.3	3	0.6
2	30,000	0.5	5	0.4
3	45,000	0.2		

(a) Give the joint probability distribution for savings per year and the useful life.

(b) Define optimistic, most likely, and pessimistic scenarios by using both optimistic, both most likely, and both pessimistic estimates. Use a life of 4 years as the most likely value. What is the present worth for each scenario?

Expected Value

10-18 For the data in Problem 10-8, compute the project's expected completion time.

10-19 For the data in Problem 10-9, compute the expected value for the next year's discount rate.

10-20 A man wants to decide whether to invest $1000 in a friend's speculative venture. He will do so if he thinks he can get his money back in one year. He believes the probabilities of the various outcomes at the end of one year are as follows:

Result	Probability
$2000 (double his money)	0.3
1500	0.1
1000	0.2
500	0.3
0 (lose everything)	0.1

What would be his expected outcome if he invests the $1000?

10-21 For the data in Problem 10-11 calculate the expected value of the equivalent annual construction cost per mile.

10-22 Two instructors announced that they "grade on the curve," that is, give a fixed percentage of each of the various letter grades to each of their classes. Their curves are as follows:

Grade	Instructor A	Instructor B
A	10%	15%
B	15	15
C	45	30
D	15	20
F	15	20

If a random student came to you and said that his object was to enroll in the class in which he could expect the higher grade point average, which instructor would you recommend?

10-23 Annual savings due to an energy efficiency project
(G) have a most likely value of $30,000. The high estimate of $40,000 has a probability of .25, and the low estimate of $20,000 has a probability of .35.

(a) What is the expected value for the annual savings?

(b) What types of tax incentives are available to firms for green projects?

10-24 In the New Jersey and Nevada gaming casinos, craps
(A) is a popular gambling game. One of the many bets available is the "Hard-way 8." A $1 bet in this fashion will win the player $4 if in the game the pair of dice come up 4 and 4 before one of the other ways of totaling 8. For a $1 bet, what is the expected result?

10-25 A man went to Atlantic City with $2000 and placed 100 bets of $20 each, one after another, on the same number on the roulette wheel. There are 38 numbers on the wheel, and the gaming casino pays 35 times the amount bet if the ball drops into the bettor's numbered slot in the roulette wheel. In addition, the bettor receives back the original $20 bet. Estimate how much money the man is expected to win or lose in Atlantic City.

10-26 Assume that the pessimistic and optimistic estimates
(A) in Problem 10-6 have 40% and 20% probabilities, respectively.

(a) What is the expected PW the expansion costs?

(b) What is the expected number of years until the expansion?

(c) What is PW of the expansion cost using the expected number of years until the expansion?

(d) Do your answers to (a) and (c) match? If not, why not?

10-27 For the data in Problems 10-2 and 10-3, assume that the optimistic probability is 20%, the most likely is 50%, and the pessimistic is 30%.

(a) What is the expected value of the equivalent uniform annual cost?

(b) Compute the expected value for the number of miles, and the corresponding equivalent uniform annual cost.

(c) Do the answers to (a) and (b) match? Why or why not?

10-28 If your interest rate is 8%, what is the expected
(A) value of the present worth of the "extra" insurance payments in Problem 10-13?

10-29 For the data in Problem 10-5, assume that the optimistic probability is 20%, the most likely is 50%, and the pessimistic is 30%.

(a) What is the expected value of the present worth?

(b) Compute the expected value for annual savings, and the corresponding present worth.

(c) Do the answers to (a) and (b) match? Why or why not?

10-30 An industrial park is being planned for a tract of land
(A) near the river. To prevent flood damage to the industrial buildings that will be built on this low-lying land, an earthen embankment can be constructed. The height of the embankment will be determined by an economic analysis of the costs and benefits. The following data have been gathered.

Embankment Height	
Above Roadway (m)	Initial Cost
2.0	$100,000
2.5	165,000
3.0	300,000
3.5	400,000
4.0	550,000

Flood Level Above Roadway (m)	Average Frequency That Flood Level Will Exceed Height in Col. 1
2.0	Once in 3 years
2.5	Once in 8 years
3.0	Once in 25 years
3.5	Once in 50 years
4.0	Once in 100 years

The embankment can be expected to last 50 years and will require no maintenance. Whenever the flood water flows over the embankment, $300,000 of damage occurs. Should the embankment be built? If so, to which of the five heights above the roadway? A 12% rate of return is required.

10-31 An energy efficiency project has a first cost of
(G) $400,000, a life of 10 years, and no salvage value. Assume that the interest rate is 10%. The most likely value for annual savings is $50,000. The optimistic value for annual savings is $80,000 with a probability of 0.2. The pessimistic value is $40,000 with a probability of 0.25.

(a) What is the expected annual savings and the expected PW?

(b) Compute the PW for the pessimistic, most likely, and optimistic estimates of the annual savings. What is the expected PW?

(c) Do the answers for the expected PW match? Why or why not?

10-32 The MSU football team has 10 games scheduled for next season. The business manager wishes to estimate how much money the team can be expected to have left over after paying the season's expenses, including any postseason "bowl game" expenses. From records for the past season and estimates by informed people, the business manager has assembled the following data:

Situation	Probability	Situation	Net Income
Regular season		Regular season	
Win 3 games	0.10	Win 5 or	$250,000
Win 4 games	0.15	fewer	
Win 5 games	0.20	games	
Win 6 games	0.15	Win 6 to 8	400,000
Win 7 games	0.15	games	
Win 8 games	0.10		
Win 9 games	0.07	Win 9 or 10	600,000
Win 10 games	0.03	games	
Postseason		Postseason	Additional
Bowl game	0.10	Bowl game	income of
			$100,000

What is the expected net income for the team next season?

10-33 For the data in Problem 10-4, assume that the optimistic probability is 15%, the most likely is 80%, and the pessimistic is 5%.

(a) What is the expected value of the rate of return?

(b) Compute the expected value for the life, and the corresponding rate of return.

(c) Do the answers to (a) and (b) match? Why or why not?

EV Joint Distribution

10-34 For the data in Problem 10-10, calculate the expected value of sales and unit profits. For the data in Problem 10-14, calculate the expected value of annual profit. Are these results consistent?

10-35 The energy efficiency project described in Problem 10-23 has a first cost of $150,000, a life of 10 years, and no salvage value. Assume that the interest rate is 8%.

(a) What is the equivalent uniform annual worth for the expected annual savings?

(b) Compute the equivalent uniform annual worth for the pessimistic, most likely, and optimistic estimates of the annual savings. What is the expected value of the equivalent uniform annual worth?

(c) Do the answers to (a) and (b) match? Why or why not?

10-36 Should the project in Problem 10-15 be undertaken if the firm uses an expected value of present worth to evaluate engineering projects?

(a) Compute the PW for each combination of first cost and revenue and the corresponding expected worth.

(b) What are the expected first cost, expected net revenue, and corresponding present worth of the expected values?

(c) Do the answers for (a) and (b) match? Why or why not?

10-37 A new engineer is evaluating whether to use a larger-diameter pipe for a water line. It will cost $600,000 more initially, but it will reduce pumping costs. The optimistic, most likely, and pessimistic projections for annual savings are $80,000, $50,000, and $7500, with respective probabilities of 25%, 45%, and 30%. The interest rate is 8%, and the water line should have a life of 50 years.

(a) What is the PW for each estimated value? What is the expected PW?

(b) Compute the expected annual savings and expected PW.

(c) Do the answers for the expected PW match? Why or why not?

10-38 For the data in Problem 10-16:

(a) What are the expected savings per year, life, and corresponding rate of return for the expected values?

(b) Compute the rate of return for each combination of savings per year and life. What is the expected rate of return?

(c) Do the answers for (a) and (b) match? Why or why not?

10-39 For the data in Problem 10-17:

(a) What are the expected savings per year, life, and corresponding present worth for the expected values?

(b) Compute the present worth for each combination of savings per year and life. What is the expected present worth?

(c) Do the answers for (a) and (b) match? Why or why not?

Decision Trees

10-40 The tree in Figure P10-40 has probabilities after each chance node and PW values for each terminal node. What decision should be made? What is the expected value?

FIGURE P10-40

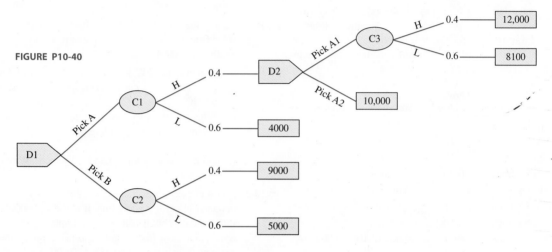

10-41 The tree in Figure P10-41 has probabilities after each chance node and PW values for each terminal node. What decision should be made? What is the expected value?

FIGURE P10-41

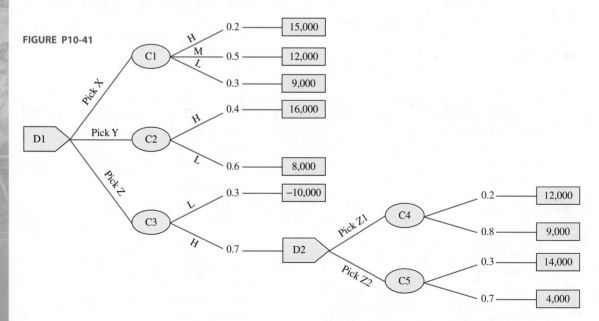

10-42 A decision has been made to perform certain repairs
Ⓐ on the outlet works of a small dam. For a particular 36-inch gate valve, there are three available alternatives:

A. Leave the valve as it is.

B. Repair the valve.

C. Replace the valve.

If the valve is left as it is, the probability of a failure of the valve seats, over the life of the project, is 60%; the probability of failure of the valve stem is 50%; and of failure of the valve body is 40%.

If the valve is repaired, the probability of a failure of the seats, over the life of the project, is 40%; of failure of the stem is 30%; and of failure of the body is 20%. If the valve is replaced, the probability of a failure of the seats, over the life of the project, is 30%; of failure of the stem is 20%; and of failure of the body is 10%.

The present worth of cost of future repairs and service disruption of a failure of the seats is $10,000; the present worth of cost of a failure of the stem is $20,000; the present worth of cost of a failure of the body is $30,000. The cost of repairing the valve now is $10,000; and of replacing it is $20,000. If the criterion is to minimize expected costs, which alternative is best?

10-43 A factory building is located in an area subject to occasional flooding by a nearby river. You have been brought in as a consultant to determine whether flood-proofing of the building is economically justified. The alternatives are as follows:

A. Do nothing. Damage in a moderate flood is $10,000 and in a severe flood, $25,000.

B. Alter the factory building at a cost of $15,000 to withstand moderate flooding without damage and to withstand severe flooding with $10,000 damages.

C. Alter the factory building at a cost of $20,000 to withstand a severe flood without damage.

In any year the probability of flooding is as follows: 0.70, no flooding of the river; 0.20, moderate flooding; and 0.10, severe flooding. If interest is 15% and a 15-year analysis period is used, what do you recommend?

10-44 Five years ago a dam was constructed to impound
Ⓐ irrigation water and to provide flood protection
Ⓖ for the area below the dam. Last winter a 100-year flood caused extensive damage both to the

dam and to the surrounding area. This was not surprising, since the dam was designed for a 50-year flood.

The cost to repair the dam now will be $250,000. Damage in the valley below amounts to $750,000. If the spillway is redesigned at a cost of $250,000 and the dam is repaired for another $250,000, the dam may be expected to withstand a 100-year flood without sustaining damage. However, the storage capacity of the dam will not be increased and the probability of damage to the surrounding area below the dam will be unchanged. A second dam can be constructed up the river from the existing dam for $1 million. The capacity of the second dam would be more than adequate to provide the desired flood protection. If the second dam is built, redesign of the existing dam spillway will not be necessary, but the $250,000 of repairs must be done.

The development in the area below the dam is expected to be complete in 10 years. A new 100-year flood in the meantime would cause a $1 million loss. After 10 years the loss would be $2 million. In addition, there would be $250,000 of spillway damage if the spillway is not redesigned. A 50-year flood is also likely to cause about $200,000 of damage, but the spillway would be adequate. Similarly, a 25-year flood would cause about $50,000 of damage.

There are three alternatives: (1) repair the existing dam for $250,000 but make no other alterations, (2) repair the existing dam ($250,000) and redesign the spillway to take a 100-year flood ($250,000), and (3) repair the existing dam ($250,000) and build the second dam ($1 million). Based on an expected annual cash flow analysis, and a 7% interest rate, which alternative should be selected? Draw a decision tree to clearly describe the problem.

10-45 In Problems 10-17 and 10-39, how much is it worth
🖥 to the firm to be able to extend the product's life by 3 years, at a cost of $50,000, at the end of the product's initial useful life?

Risk

10-46 An engineer decided to make a careful analysis of
Ⓐ the cost of fire insurance for his $200,000 home.

From a fire rating bureau he found the following risk of fire loss in any year.

Outcome	Probability
No fire loss	0.986
$ 10,000 fire loss	0.010
40,000 fire loss	0.003
200,000 fire loss	0.001

(a) Compute his expected fire loss in any year.

(b) He finds that the expected fire loss in any year is less than the $550 annual cost of fire insurance. In fact, an insurance agent explains that this is always true. Nevertheless, the engineer buys fire insurance. Explain why this is or is not a logical decision.

10-47 For the data in Problems 10-9 and 10-19, compute the standard deviation of the interest rate.

10-48 The Graham Telephone Company may invest in new switching equipment. There are three possible outcomes, having net present worth of $6570, $8590, and $9730. The outcomes have probabilities of 0.3, 0.5, and 0.2, respectively. Calculate the expected return and risk measured by the standard deviation associated with this proposal.

10-49 For the data in Problems 10-11 and 10-21, compute the standard deviation of the equivalent annual cost per mile.

10-50 What is your risk associated with Problem 10-28?

10-51 For the data in Problem 10-31, compute the standard deviation of the present worth.

10-52 For the data in Problem 10-37, compute the standard deviation of the present worth.

10-53 A new machine will cost $50,000. The machine is expected to last 10 years and have no salvage value. If the interest rate is 15%, determine the return and the risk associated with the purchase.

P	0.3	0.4	0.3
Annual savings	$9000	$10,500	$12,000

10-54 A new product's chief uncertainty is its annual net revenue. So far, $35,000 has been spent on development, but an additional $30,000 is required to finish development. The firm's interest rate is 10%.

(a) What is the expected PW for deciding whether to proceed?

(b) Find the P(loss) and the standard deviation for proceeding.

	State		
	Bad	**OK**	**Great**
Probability	0.3	0.5	0.2
Net revenue	−$15,000	$15,000	$20,000
Life, in years	5	5	10

10-55 (a) In Problem 10-54, how much is it worth to the firm to terminate the product after 1 year if the net revenues are negative?

(b) How much does the ability to terminate early change the P(loss) and the standard deviation?

10-56 Measure the risk for Problems 10-15 and 10-36 using the P(loss), range of PW values, and standard deviation of the PWs.

10-57 (a) In Problems 10-17 and 10-39, describe the risk using the P(loss) and standard deviation of the PWs.

(b) How much do the answers change if the possible life extension in Problem 10-45 is allowed?

Risk Versus Return

10-58 A firm wants to select one new research and development project. The following table summarizes six possibilities. Considering expected return and risk, which projects are good candidates? The firm believes it can earn 5% on a risk-free investment in government securities (labeled as Project F).

Project	IRR	Standard Deviation
1	15.8%	6.5%
2	12.0	4.1
3	10.4	6.3
4	12.1	5.1
5	14.2	8.0
6	18.5	10.0
F	5.0	0.0

10-59 A firm is choosing a new product. The following table summarizes six new potential products. Considering expected return and risk, which products are good candidates? The firm believes it can earn 4% on a risk-free investment in government securities (labeled as Product F).

Product	IRR	Standard Deviation
1	10.4%	3.2%
2	9.8	2.3
3	6.0	1.6
4	12.1	3.6
5	12.2	8.0
6	13.8	6.5
F	4.0	0.0

Risk and Return as Multiple Objectives

10-60 Which project in Problem 10-58 should be selected if the rate of return has a weight of 70% and the standard deviation has a weight of 30%? How much difference in the weighted score is there between the best and the worst of projects 1 to 6?

10-61 Which project in Problem 10-58 should be selected if the rate of return has a weight of 40% and the standard deviation has a weight of 60%? How much difference in the weighted score is there between the best and the worst of projects 1 to 6?

10-62 Which product in Problem 10-59 should be selected if the rate of return has a weight of 75% and the standard deviation has a weight of 25%? How much difference in the weighted score is there between the best and the worst of projects 1 to 6?

10-63 Which product in Problem 10-59 should be selected if the rate of return has a weight of 30% and the standard deviation has a weight of 70%? How much difference in the weighted score is there between the best and the worst of projects 1 to 6?

Simulation

10-64 A project's first cost is $25,000, and it has no salvage value. The interest rate for evaluation is 7%. The project's life is from a discrete uniform distribution that takes on the values 7, 8, 9, and 10. The annual benefit is normally distributed with a mean of $4400 and a standard deviation of $1000. Using Excel's RAND function, simulate 25 iterations. What are the expected value and standard deviation of the present worth?

10-65 A factory's power bill is $55,000 a year. The first cost of a small geothermal power plant is normally distributed with a mean of $150,000 and a standard deviation of $50,000. The power plant has no salvage value. The interest rate for evaluation is 8%. The project's life is from a discrete uniform distribution that takes on the values 3, 4, 5, 6, and 7. (The life is relatively short due to corrosion.) The annual operating cost is expected to be about $10,000 per year. Using Excel's RAND function,

simulate 25 iterations. What are the expected value and standard deviation of the present worth?

CASES

The following cases from *Cases in Engineering Economy* (www.oup.com/us/newnan) are suggested as matched with this chapter.

CASE 13 **Guaranteed Return**

Risk versus expected return and choice of interest rate.

CASE 18 **Gravity-Free High**

New product with development stages and probability of failure.

CASE 19 **Crummy Castings**

Decision tree problem with modest ambiguity. Options create more challenging problems.

CASE 21 **Glowing in the Dark**

Expected return and variance for facility of minimal size, of minimal size plus preparation for later expansion, and of the full size now.

CASE 22 **City Car**

Decision tree analysis. Includes discussion of strategy and risk. Some assumptions must be made.

CASE 23 **Washing Away**

Levee height and probability of flood damage.

CASE 24 **Sinkemfast**

Decision tree with assumptions required for realistic comparisons. Info supports creation of new, better alternatives.

CASE 39 **Uncertain Demand at WM[3]**

Includes inflation, taxes, and uncertainty.

CASE 44 **Sunnyside—Up or Not?**

Uncertain growth rates over 30 years and setting utility rates.

CASE 50 **Capital Planning Consultants**

Capital budgeting including mutually exclusive alternatives. Includes uncertainties in first cost, annual benefit, and lives.

APPENDIX 10A

Diversification Reduces Risk

Key Words

capital allocation line   diversification

PORTFOLIOS OF STOCKS AND BONDS

Appendix 9A described the expected returns and risks for stocks, treasury bonds, and treasury bills (T-bills). The returns listed were for portfolios of only stocks and portfolios of only bonds. Figure 10A–1 adds portfolios of combined stocks and bonds to Figure 9A–1. It shows that due to **diversification**, portfolios of bonds and stocks can offer better combinations of risk and return than either alone. We cannot draw a curve for more than two investments (funds or individual bonds and stocks), but the reduction of risk through diversification applies.

Table 10A–1 details the numbers, but this effect is easier to see in Figure 10A–1. For example, a portfolio that is invested 50% in each of stocks and bonds has the same risk (10.8%) as investing only in bonds. However, the 50/50 portfolio's expected return is 2.4% higher (= 8.6% − 6.2%).

The availability and performance of the 50/50 portfolio reduces the attractiveness of the all-stock portfolio. From Table 10A–1, the stock portfolio's risk is 6.7% (= 17.5% − 10.8%) higher, but its expected return is now only 2.4% higher than the 50/50 portfolio. The *efficient frontier* is the set of choices that should be considered. Other available choices are a worse combination of risk and return.

If portfolios of a low-risk stock and a higher-risk stock were formed, the performance would follow a similar curve. The math behind this is based on Equations 10A-1 and 10A-2. The expected return of a portfolio is simply a weighted average of the returns for the two choices. However, the standard deviation or risk of the portfolio is reduced, as long as the two choices are not perfectly correlated (ρ is the correlation coefficient in Equation 10A-2).

$$\text{Expected return} \qquad E(R_P) = w_1 E(R_1) + w_2 E(R_2) \tag{10A-1}$$

$$\text{Standard deviation} \qquad \sigma_P = \sqrt{(w_1^2 \sigma_1^2 + w_2^2 \sigma_2^2 + 2w_1 w_2 \rho_{12} \sigma_1 \sigma_2)} \tag{10A-2}$$

The equations can be generalized for portfolios of tens, hundreds, or even thousands of stocks. The reduction in risk through diversification is routine for large investors; thus an investor who buys only one or a few stocks has a much higher risk than the buyers and sellers that set the market price.

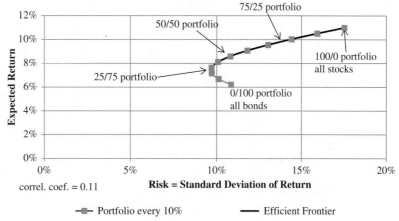

FIGURE 10A-1 Risk and return for portfolios of stocks and bonds.

TABLE 10A-1 Risk and Return Data for Portfolios of Stocks and Bonds

Treasury		Portfolio	
Stock	**Bond**	**Std Dev**	**Return**
100%	0%	17.5%	11.0%
75%	25%	13.7%	9.8%
50%	50%	10.8%	8.6%
25%	75%	9.6%	7.4%
0%	100%	10.8%	6.2%

The implication for individuals with a defined contribution retirement program is that the best strategy is to take advantage of all available company match dollars and invest in a diverse set of firms and bonds through an exchange traded fund (ETF) or a low-cost mutual fund. It is difficult for individuals to compete with institutional investors (who have millions to spend on how to invest billions). The key to long-term success seems to be keeping transaction costs down and investing regularly. Fortunately, ETFs offer the individual investor a way to hold very well diversified portfolios of stocks and/or bonds that are designed to match major market indices. We also note that these ETFs include returns to investors (dividends on stocks and interest on bonds) that go beyond the simple price indices. Both price and total return versions of the Dow Jones and the S&P 500 indices are available, but only total return versions should be used. We note that the S&P 500 and the Russell 1000 are generally regarded as better measures than the Dow Jones, as they include more firms and they weight them by their size, rather than simply averaging the stock market prices of 30 firms, as does the Dow Jones (with a divisor calculated to maintain historical continuity).

MAKING THE TRADE-OFF BETWEEN RISK AND RETURN

If U.S. Treasury bill data is added to Figure 10A–1, there is a more complete invest-ment model with a *better* efficient frontier. T-bills typically have maturities of 3, 6, or 12 months. Because these are issued by the U.S. government and have short maturities, they are generally considered to be free of default risk. The new efficient frontier line in Figure 10A–2 connects the T-bill point with the point of tangency to the efficient frontier of the stock/bond portfolio. The point of tangency is typically near the 50/50 portfolio of stocks and bonds. While points of tangency and line slopes could be mathematically calculated, we suggest that the accuracy of the data is better matched to simply eyeballing the line as it is added. The efficient frontier is commonly called the **capital allocation line** because it describes the risk/return relationship that drives how capital is allocated. The risk and return combination that is preferred by an investor determines how that investor's capital should be allocated between T-bills, bonds, and stocks.

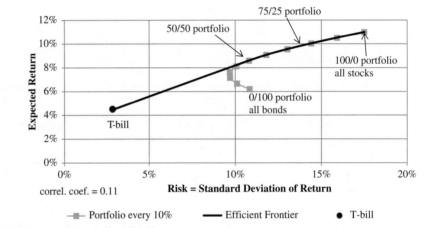

FIGURE 10A–2 Capital allocation line.

EXAMPLE 10A–1   (Example 9A–1 Revisited)

An engineer has just finished paying off a student loan and is ready to start saving for retirement. Her current annual salary is $63,000. She expects salary increases to exceed inflation, but to be safe she wants to assume that salary increases will be matched with inflation. She expects to work for another 35 years. If she invests 15% of her salary, how much can be expected to be in her account if she invests only in bonds or in a 50/50 portfolio of stocks and bonds?

SOLUTION

The first step is to recognize that both choices have a standard deviation of 10.8%—the same level of risk. Next we determine the expected returns or interest rates. Since this problem is in constant-value dollars, as is all of the text up to Chapter 14, the interest rates are the market rates minus inflation. Long-term treasury bonds have returned about 2.5% above the average inflation rate of 3.7% (Table 9A–1). For the 50/50 portfolio we must subtract average inflation from the 8.6% expected return shown in Tables A–1 and A–2. Thus the rate of return for the 50/50 portfolio is 4.9%.

To make the solution more flexible, the spreadsheet starts with a data block.

	A	B	C	D	E	F	G	H
1	$63,000	Salary						
2	15%	% saved						
3								
4	Exp. 10A-1	*i*	*n*	*PMT*	*PV*	*FV*	Solve for	Answer
5	Bonds	2.5%	35	-9450	0		FV	$519,072
6	50/50	4.9%	35	-9450	0		FV	$836,054

TABLE 10A–2 Risk and Return Data for Capital Allocation Line

	Treasury		Portfolio	
T-bill	Stock	Bond	Std Dev	Return
	100%	0%	17.5%	11.0%
	90%	10%	15.9%	10.5%
	80%	20%	14.4%	10.0%
	70%	30%	13.0%	9.6%
	60%	40%	11.8%	9.1%
0%	50%	50%	10.8%	8.6%
20%	40%	40%	9.2%	7.8%
40%	30%	30%	7.6%	7.0%
60%	20%	20%	6.0%	6.2%
80%	10%	10%	4.4%	5.3%
100%	0%	0%	2.8%	4.5%

EXAMPLE 10A–2   (Example 10A–1 Revisited)

The engineer in Example 10A–1 has decided that a retirement goal of $1 million is adequate. What fraction of her salary must be saved if she invests only in bonds or in a 50/50 portfolio of stocks and bonds?

SOLUTION

The first step is to determine the annual deposits at the different interest rates. As in Example 10A–1, the interest rates are the market rates minus inflation: long-term treasury bonds about 2.5% above inflation and a 50/50 portfolio about 4.9% above inflation.

To make the solution more flexible, the spreadsheet starts with a data block.

	A	B	C	D	E	F	G	H	I
1	$63,000	Salary							
2									
3	Exp. 10A-2	i	n	*PMT*	*PV*	*FV*	Solve for	Answer	% salary
4	Bonds	2.5%	35		0	1,000,000	PMT	-$18,206	28.9%
5	50/50	4.9%	35		0	1,000,000	PMT	-$11,303	17.9%

An investor who chooses to invest safely before retirement should also invest safely after retirement. If the $1M is invested at a lower rate, then less can be drawn out each year to live on. It is common to suggest that *all* investors should invest more safely once they are approaching retirement or are already retired.

PROBLEMS

As in the examples, assume constant value dollars and use real rates of return above inflation.

10A-1 A 25-year-old engineer earning $65,000 per year wants to retire at age 55 with $2 million, and plans to invest in a fund made up of 60% stocks and 40% bonds.

(a) How much money must be invested each year?

(b) If the employer does a 100% match of retirement savings up to 3% of the employee's salary, how much money must each invest annually?

10A-2 An engineer earning $68,000 per year wants to retire in 35 years with $2 million and plans to invest in a fund containing 40% stocks, 40% bonds, and 20% treasury bills.

(a) How much money must be invested each year?

(b) If this person works 5 additional years, how much money must be invested each year?

(c) If the employer does a 100% match of retirement savings up to 4% of the employee's salary, and the engineer wants to retire in 35 years, how much money must each invest annually?

10A-3 A new employee earning $60,000 annually set a retirement goal of $2 million. She plans to work for 40 years, and will invest in a fund having a portfolio of 80% stocks and 20% bonds.

(a) What fraction of her salary must be saved?

(b) If the employer does a one-to-one match of retirement savings up to 3% of the employee's salary, what fraction of her salary must be saved?

10A-4 An employee earning $63,000 per year set a goal of retiring in 40 years with $1 million. She will invest in a fund investing 70% in stocks and 30% in treasury bonds.

(a) What fraction of her salary must be saved?

(b) If the employer does a one-to-one match of retirement savings up to 5% of the employee's salary, what fraction of her salary must be saved?

10A-5 A manager just retired at age 62, with his retirement savings of $800,000 invested in a portfolio of 90% stocks and 10% bonds. If he continues with this portfolio, and needs $85,000 per year for living expenses, how long will his savings last?

10A-6 A manager retired at age 65, and has her retirement savings of $600,000 invested in a mutual fund composed of 75% treasury bonds and 25% stocks. She needs $45,000 per year for living expenses, in addition to her social security benefit. How long will her investment last if there is no inflation?

10A-7 A new employee puts 4% of his salary of $60,000 into a retirement account, and his employer matches this, also putting 4% into the account. The money is invested into a diversified fund of 50% stocks and 50% bonds. His salary increases 3% per year.

(*a*) What is the value of the account after 10 years?

(*b*) What is the value of the employer's matching funds after 10 years?

(*c*) How much will be in the account after 40 years?

10A-8 An engineer changed jobs and is signing up for benefits. The company 401(k) includes a low-cost fund that invests 40% in stocks, 40% in bonds, and 20% in T-bills. The engineer will put 3% of her salary of $70,000 into the account, and her employer will match half this amount. Her salary is expected to increase 2.8% per year.

(*a*) What is the value of the account after 10 years?

(*b*) If she expects to work for 30 years, how much will be in the account?

10A-9 Monica became concerned about the stock market due to recent losses, and moved $350,000 into a fund of 60% T-bills, 20% stocks, and 20% bonds. Her salary is currently $92,000 per year, and she expects raises of 2.0% each year. Her employer will match retirement savings up to 5% of her salary. If she wants $500,000 in her account in 5 years, what percentage of her salary does she need to save?

10A-10 Jorge was recently laid off, and may not be able to find another full-time job. Capital preservation is his primary concern, so he wants to invest in a portfolio of 80% treasury bills, 10% stocks, and 10% bonds. His retirement fund has $1.0 million. He can receive social security, so he will only need $55,000 per year from his savings to live well. Jorge is 64 years old. How long does he expect his savings to last? Should he look for a job or enjoy his retirement?

10A-11 Discuss the ethical considerations of a firm's retirement savings plan. The firm will match retirement savings up to 10% of an employee's salary, but only if the employee selects the option to invest in the firm's stock. The employee's contribution can be withdrawn or invested in another fund at any time. However, the matching funds have a 5-year vesting period, and the match goes away if the employee's contribution is shifted out of the firm's stock before vesting.

CHAPTER 11

DEPRECIATION

Depreciation and Intangible Property

The U.S. government allows firms to subtract many business expenses from their gross income in determining taxes due. This process is relatively straightforward for some expenses, such as labor and materials, which are "consumed" in the process of producing goods and services. This chapter on depreciation presents what happens when a business purchases a piece of durable equipment, such as a forklift, crane, or computer, which will be used over many years. This equipment is not directly consumed but does deteriorate with time and is clearly a business expense. It does not make sense for firms to be able to subtract the entire equipment cost immediately, when full value has not yet been realized from the equipment and payment may not even be fully complete. Therefore, the government has devised depreciation rules that allow firms to recoup durable equipment and other durable property value over time, much like other business expenses.

Firms can also apply depreciation to durable intangible assets, such as patents, trademarks, or even the estimated value of customer relationships. Like durable equipment, these types of durable intangible property provide value to the business over time, rather than being consumed during production, and also degrade in value or usefulness over time. For instance, patents and most customer relationships have a limited life span. While trademarks do not, the goods and services they are associated with are not generally expected to have indefinite appeal. When depreciation is applied to an intangible asset, this process is typically referred to as *amortization*.

While amortization of intangible assets may seem like it would be a minor concept for most businesses, the value of a Coca-Cola or Nike brand, a major drug or hardware patent, or the customer base of an acquired firm can be in the millions or billions of dollars. For instance, Apple and Samsung have been engaged in an ongoing and highly publicized

multinational legal battle over patents, trademarks, and other intangible assets with damages sought totaling in the billions of dollars. While not every firm will have intangible assets worth quite this much, intangible assets are a critical property class in many firms. Recent data indicate that intangible property accounts for around 80% of the total market value of the "typical" U.S. firm. For example, intangible property comprised about 78% of the market value of Alphabet, Inc. (the parent company of Google) circa 2015. Thus, all firms should consider their intangible property in investment decisions, including correctly evaluating tax implications over time through the application of the appropriate amortization procedures. ▪ ▪ ▪

Contributed by Jennifer A. Cross, Texas Tech University

QUESTIONS TO CONSIDER

1. Besides Apple and Samsung, what other examples of legal battles over intangible assets can you identify?

2. If you had to develop a method for amortizing an intangible asset such as a patent, how would you go about doing this? What sorts of parameters would you need to consider in developing this method?

3. Does the percentage of market value tied to intangible assets in U.S. firms surprise you? Why or why not? Do you think this percentage differs in other countries? Why or why not?

After Completing This Chapter...

The student should be able to:

- Describe depreciation, deterioration, and obsolescence.

- Distinguish various types of depreciable property and differentiate between depreciation expenses and other business expenses.

- Use classic depreciation methods to calculate the *annual depreciation charge* and *book value* over the asset's life.

- Explain the differences between the classic depreciation methods and the modified accelerated cost recovery system (MACRS).

- Use MACRS to calculate allowable *annual depreciation charge* and *book value* over the asset's life for various cost bases, property classes, and recovery periods.

- Fully account for *capital gains/losses, ordinary losses,* and *depreciation recapture* due to the disposal of a depreciated business asset.

- Use the *units of production* and *depletion* depreciation methods as needed in engineering economic analysis problems.

- Use spreadsheets to calculate depreciation.

Key Words

annual depreciation	depreciation recapture	percentage depletion
asset depreciation range (ADR)	deteriorating	personal property
book value	double declining balance	real property
capital gain	expensed item	recovery period
cost basis	income statement	straight-line depreciation
cost depletion	intangible property	sum-of-years'-digits
declining balance depreciation	loss	depreciation
depletion	loss on disposal	tangible property
depreciable life	MACRS	unit-of-production
depreciation	obsolescence	depreciation

We have so far dealt with a variety of economic analysis problems and many techniques for their solution. In the process we have avoided income taxes, which are an important element in many private sector economic analyses. Now, we move to more realistic—and more complex—situations.

Governments tax individuals and businesses to support their processes—lawmaking, domestic and foreign economic policy making, even the making and issuing of money itself. The omnipresence of taxes requires that they be included in economic analyses, which means that we must understand the *way* taxes are imposed. For capital equipment, knowledge about depreciation is required to compute income taxes. Chapter 11 examines depreciation, and Chapter 12 illustrates how depreciation and other effects are incorporated in income tax computations. The goal is to support decision making on engineering projects, not to support final tax calculations.

The focus of this chapter is computing depreciation in order to find after-tax cash flows in Chapter 12. However, depreciation is also part of computing the firm's value. Many people do not realize that firms often use different depreciation methods for the two tasks. For tax purposes firms want the tax deductions as soon as possible but for valuing the firm a classic depreciation method that better approximates the item's value is often chosen.

INCOME, DEPRECIATION, AND CASH FLOW

The role of depreciation is most easily understood by starting with a firm's **income statement** or profit and loss statement. Revenue or sales may be a single line item, or it may be broken out into products and services. As shown in Figure 11–1, costs are often broken down into far more detail. All of the revenues and costs shown are cash flows—except for depreciation.

Depreciation is not a cash flow; it is an accounting entry that allocates a portion of the cost of machines, buildings, etc., in each year. However, because depreciation is one of the costs subtracted from revenue to determine taxable income, it does change the cash that flows when taxes are paid.

Now some items—wages, materials, and the like—are paid for shortly after they are used. Other items—like office rent and insurance premiums—are paid for monthly, quarterly, or yearly. But most items are paid within the period of a year and so it is reasonable to compare revenues for the year with expenses for the year.

However, capital assets—such as land, buildings, equipment, machinery—last longer than one year and typically cost more. It would be misleading to charge for all of their costs

Simplified Income Statement for XYZ Company for
year ending December 31, 2017   (all amounts in $M)

Revenue	*Sales of products and services*	**Total Revenue**	**$184**
Costs	**Cost of Goods Sold**		
	Wages, materials, and utilities		110
	Depreciation		35
	Selling, Administration, and Financing Costs		19
		Total Costs	**$164**
		Net Income Before Taxes	**20**
Taxes			7
Profit			13

FIGURE 11–1 The Income Statement.

at the time of purchase. Instead, their costs are typically depreciated over time, except for land, which doesn't wear out and thus can't be depreciated.

Our presentation of depreciation methods begins with the classic time-based depreciation techniques. It then covers the MACRS approach that is required in the U.S. for tax purposes. Finally it covers units-of-production and depletion methods that are based on physical quantities or costs.

BASIC ASPECTS OF DEPRECIATION

The word *depreciation* is defined as a "decrease in value." This is somewhat ambiguous because *value* has several meanings. In economic analysis, value may refer to either *market value* or *value to the owner.* For example, an assembly line is far more valuable to the manufacturing firm that it was designed for, than it is to a used equipment market. Thus, we now have two definitions of depreciation: a decrease in value to the market or a decrease to the owner.

Deterioration and Obsolescence

A machine may depreciate because it is **deteriorating**, or wearing out and no longer performing its function as well as when it was new. Many kinds of machinery require increased maintenance as they age, reflecting a slow but continuing failure of individual parts. Anyone who has worked to maintain a car has observed deterioration due to failure of individual parts (such as fan belts, mufflers, and batteries) and the wear on mechanical components. In other types of equipment, the quality of output may decline due to wear on components and resulting poorer mating of parts.

Depreciation is also caused by **obsolescence.** A machine that is in excellent working condition, and serving a needed purpose, may still be obsolete. Newer models are more capable with new features. Generations of computers have followed this pattern. The continuing stream of newer models makes older ones obsolete.

The accounting profession defines depreciation in yet another way, as allocating an asset's cost over its **depreciable life.** Thus, we now have *three distinct definitions of depreciation:*

1. Decline in market value of an asset.
2. Decline in value of an asset to its owner.
3. Systematic allocation of an asset's cost over its depreciable life.

Depreciation and Expenses

It is the third (accountant's) definition that is used to compute depreciation for business assets. Business costs are generally either **expensed** or *depreciated.* Expensed items, such as labor, utilities, materials, and insurance, are part of regular business operations, and for tax purposes they are subtracted from business revenues when they occur. Expensed costs reduce income taxes because businesses are able to *write off* their full amount when they occur. Section 179 of the tax code allows profitable *small* businesses to expense some capital asset purchases.

In contrast, business costs due to capital assets (buildings, forklifts, chemical plants, etc.) are not fully written off when they occur. Capital assets lose value gradually and must be written off or *depreciated* over an extended period. For instance, consider an injection-molding machine used to produce the plastic beverage cups found at sporting events. The plastic pellets melted into the cup shape lose their value as raw material directly after manufacturing. The raw material cost for production material (plastic pellets) is expensed immediately. On the other hand, the injection-molding machine itself will lose value over time, and thus its costs (purchase price and installation expenses) are written off (or depreciated) over its **depreciable life** or **recovery period.** This is often different from the asset's useful or most economic life. Depreciable life is determined by the depreciation method used to spread out the cost—depreciated assets of many types operate well beyond their depreciable life.

Depreciation is a *noncash* cost that requires no exchange of dollars. Companies do not write a check to someone to *pay* their depreciation expenses. Rather, these are business expenses that are allowed by the government to offset the loss in value of business assets. Remember, the company has paid for assets up front; depreciation is simply a way to claim these "business expenses" over time. As shown in Figure 11–1, depreciation deductions reduce the taxable income of businesses and thus reduce the amount of taxes paid. Since taxes are cash flows, depreciation must be considered in after-tax economic analyses.

In general, business assets can be depreciated only if they meet the following basic requirements:

1. The property must be used for business purposes to produce income.
2. The property must have a useful life that can be determined, and this life must be longer than one year.
3. The property must be an asset that decays, gets used up, wears out, becomes obsolete, or loses value to the owner from natural causes.

EXAMPLE 11–1

Consider the costs that are incurred by a local pizza business. Identify each cost as either *expensed* or *depreciated* and describe why that classification applies.

- Cost for pizza dough and toppings
- Cost to pay wages for janitor
- Cost of a new baking oven

- Cost of new delivery van
- Cost of furnishings in dining room
- Utility costs for soda refrigerator

SOLUTION

Cost Item	Type of Cost	Why
Pizza dough and toppings	Expensed	Life < 1 year; lose value immediately
New delivery van	Depreciated	Meets 3 requirements for depreciation
Wages for janitor	Expensed	Life < 1 year; lose value immediately
Furnishings in dining room	Depreciated	Meet 3 requirements for depreciation
New baking oven	Depreciated	Meets 3 requirements for depreciation
Utilities for soda refrigerator	Expensed	Life < 1 year; lose value immediately

Types of Property

The rules for depreciation are linked to the classification of business property as either tangible or intangible. Tangible property is further classified as either real or personal.

Tangible property can be seen, touched, and felt.

Real property includes land, buildings, and all things growing on, built upon, constructed on, or attached to the land.

Personal property includes equipment, furnishings, vehicles, office machinery, and anything that is tangible excluding assets defined as *real property*.

Intangible property is all property that has value to the owner but cannot be directly seen or touched. Examples include patents, copyrights, trademarks, trade names, and franchises.

Many different types of property that wear out, decay, or lose value can be depreciated as business assets. This wide range includes copy machines, helicopters, buildings, interior furnishings, production equipment, and computer networks. Almost all tangible property can be depreciated.

One important and notable exception is land, which is *never* depreciated. Land does not wear out, lose value, or have a determinable useful life and thus does not qualify as a depreciable property. Rather than decreasing in value, most land becomes more valuable as time passes. In addition to the land itself, expenses for clearing, grading, preparing, planting, and landscaping are not generally depreciated because they have no fixed useful life. Other tangible property that *cannot* be depreciated includes factory inventory, containers considered as inventory, and leased property. The leased property exception highlights the fact that only the owner of property may claim depreciation expenses.

Tangible properties used in *both* business and personal activities, such as a vehicle used in a consulting engineering firm that is also used to take one's kids to school, can be depreciated, but only in proportion to the use for business purposes.

Depreciation Calculation Fundamentals

To understand the complexities of depreciation, the first step is to examine the fundamentals of depreciation calculations. Figure 11–2 illustrates the general depreciation problem of allocating the total depreciation charges over the asset's depreciable life. The vertical axis is labeled **book value.** At time zero the curve of book value starts at the **cost basis** (= the first cost plus installation cost). Over time, the book value declines to the salvage value. Thus, at any point in time:

Book value = Cost basis − Depreciation charges made to date

Looked at another way, book value is the asset's remaining unallocated cost.

In Figure 11–2, *book value* goes from a value of B at time zero in the recovery period to a value of S at the end of Year 5. Thus, book value is a *dynamic* variable that changes over an asset's recovery period. The equation used to calculate an asset's book value over time is

$$BV_t = \text{Cost basis} - \sum_{j=1}^{t} d_j \qquad (11\text{-}1)$$

where BV_t = book value of the depreciated asset at the end of time t

Cost basis = B = dollar amount that is being depreciated; this includes the asset's purchase price as well as any other costs necessary to make the asset "ready for use"

$\sum_{j=1}^{t} d_j$ = sum of depreciation deductions taken from time 0 to time t, where d_j is the depreciation deduction in Year j

Equation 11-2 shows that year-to-year depreciation charges reduce an asset's book value over its life. The following section describes methods that are or have been allowed under federal tax law for quantifying these yearly depreciation deductions.

FIGURE 11–2 General depreciation.

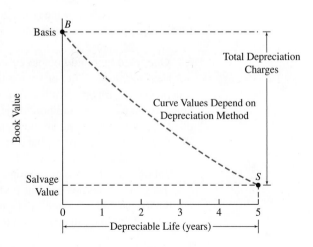

TIME-BASED DEPRECIATION METHODS

Classic methods: These methods include the *straight-line, sum-of-the-years'-digits,* and *declining balance* methods. Each method requires estimates of an asset's useful life and salvage value. Most countries allow or require that one or more of the classic techniques be used for tax purposes, and multinational firms often have to pay taxes of different kinds in multiple jurisdictions. In addition, as noted earlier, depreciation is also part of computing the firm's value. U.S. firms often use the classic depreciation methods to value the firm, as they better approximate the item's value.

MACRS and the U.S. tax code: In 1981 Congress created the accelerated cost recovery system (ACRS). The ACRS method had four key features: (1) property class lives were created, and all depreciated assets were assigned to one particular category; (2) the need to estimate salvage values was eliminated because all assets were *fully* depreciated over their recovery period; (3) shorter recovery periods were used to calculate annual depreciation, which *accelerated* the write-off of capital costs more quickly than did the classic methods—thus the name; and (4) the tables were based on straight-line and declining balance methods.

The modified accelerated cost recovery system (**MACRS**) was implemented in 1986. The modified system (1) expanded the number of property classes and (2) included a half-year convention for the first and final years.

In this chapter, our primary focus is to describe the MACRS depreciation method, because we must calculate after-tax cash flows in Chapter 12 to decide what projects have the most economic value. However, it is useful to first describe three classic depreciation methods. These methods are used in many countries, and MACRS is based on two of them.

Straight-Line Depreciation

The simplest and best known depreciation method is **straight-line depreciation.** To calculate the constant **annual depreciation charge,** the total amount to be depreciated, $B - S$, is divided by the depreciable life, in years, N:[1]

$$\text{Annual depreciation charge} = d_t = \frac{B - S}{N} \tag{11-2}$$

EXAMPLE 11–2

Consider the following (in $1000):

Cost of the asset, B	$900
Depreciable life, in years, N	5
Salvage value, S	$70

Compute the straight-line depreciation schedule.

---

[1]N is used for the depreciation period because it may be shorter than n, the horizon (or project life).

SOLUTION

$$\textbf{Annual depreciation charge} = d_t = \frac{B - S}{N} = \frac{900 - 70}{5} = \$166$$

Year	Depreciation for Year t ($1000)	Sum of Depreciation Charges Up to Year t ($1000)	Book Value at the End of Year t ($1000)
t	d_t	$\displaystyle\sum_{j=1}^{t} d_j$	$BV_t = B - \displaystyle\sum_{j=1}^{t} d_j$
1	$166	$166	$900 - 166 = 734$
2	166	332	$900 - 332 = 568$
3	166	498	$900 - 498 = 402$
4	166	664	$900 - 664 = 236$
5	166	830	$900 - 830 = \ 70 = S$

This situation is illustrated in Figure 11–3. Notice the constant $166,000 d_t each year for 5 years and that the asset has been depreciated down to a book value of $70,000, which was the estimated salvage value.

FIGURE 11–3 Straight-line depreciation.

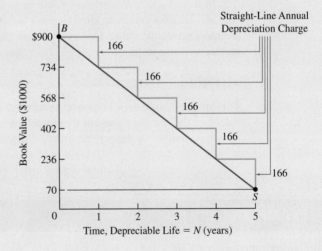

The straight-line (SL) method is often used for intangible property. For example, Veronica's firm bought a patent in April that was not acquired as part of acquiring a business. She paid $6800 for this patent and must use the straight-line method to depreciate it over 17 years with no salvage value. The annual depreciation is $400 (= $6800/17). Since the patent was purchased in April, the first year's deduction must be prorated over the 9 months of ownership. This year the deduction is $300 (= $400 × 9/12), and then next year she can begin taking the full $400 per year.

Sum-of-Years'-Digits Depreciation

Another method for allocating an asset's cost *minus* salvage value *over* its depreciable life is called **sum-of-years'-digits (SOYD) depreciation.** This method results in larger-than-straight-line depreciation charges during an asset's early years and smaller charges as the asset nears the end of its depreciable life. Each year, the depreciation charge equals a fraction of the total amount to be depreciated $(B-S)$. The denominator of the fraction is the sum of the years' digits. For example, if the depreciable life is 5 years, $1+2+3+4+5 = 15 = SOYD$. Then $5/15, 4/15, 3/15, 2/15$, and $1/15$ are the fractions from Year 1 to Year 5. Each year the depreciation charge shrinks by $1/15$ of $B - S$. Because this change is the same every year, SOYD depreciation can be modeled as an arithmetic gradient, G. The equations can also be written as

$$\left(\begin{array}{l}\text{Sum-of-years'-digits}\\ \text{depreciation charge for}\\ \text{any year}\end{array}\right) = \frac{\left(\begin{array}{l}\text{Remaining depreciable life}\\ \text{at beginning of year}\end{array}\right)}{\left(\begin{array}{l}\text{Sum of years' digits}\\ \text{for total depreciable life}\end{array}\right)}\text{(Total amount depreciated)}$$

$$d_t = \frac{N-t+1}{SOYD}(B-S) \tag{11-3}$$

where d_t = depreciation charge in any year t
  N = number of years in depreciable life
  $SOYD$ = sum of years' digits, calculated as $N(N+1)/2 = SOYD$
  B = cost of the asset made ready for use
  S = estimated salvage value after depreciable life

EXAMPLE 11–3

Compute the SOYD depreciation schedule for the situation in Example 11.2 (in $1000):

Cost of the asset, B	$900
Depreciable life, in years, N	5
Salvage value, S	$70

SOLUTION

$$SOYD = \frac{5 \times 6}{2} = 15$$

Thus,

$$d_1 = \frac{5-1+1}{15}(900-70) = 277$$

$$d_2 = \frac{5-2+1}{15}(900-70) = 221$$

$$d_3 = \frac{5-3+1}{15}(900-70) = 166$$

$$d_4 = \frac{5-4+1}{15}(900-70) = 111$$

$$d_5 = \frac{5-5+1}{15}(900-70) = 55$$

Year	Depreciation for Year t ($1000)	Sum of Depreciation Charges Up to Year t ($1000)	Book Value at End of Year t ($1000)
t	d_t	$\sum_{j=1}^{t} d_j$	$BV_t = B - \sum_{j=1}^{t} d_j$
1	$277	$277	$900 - 277 = 623$
2	221	498	$900 - 498 = 402$
3	166	664	$900 - 664 = 236$
4	111	775	$900 - 775 = 125$
5	55	830	$900 - 830 = 70 = S$

These data are plotted in Figure 11–4.

FIGURE 11–4 Sum-of-years'-digits depreciation.

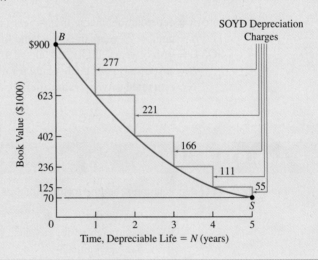

Declining Balance Depreciation

Declining balance depreciation applies a *constant depreciation rate* to the property's declining book value. Two rates were commonly used before the 1981 and 1986 tax revisions, and they are used to compute MACRS depreciation percentages. These are 150 and 200% of the straight-line rate. Since 200% is twice the straight-line rate, it is called

double declining balance, or DDB; the general equation is

$$\text{Double declining balance} \quad d_t = \frac{2}{N}(\text{Book value}_{t-1}) \quad \quad (11\text{-}4a)$$

Since book value equals cost *minus* depreciation charges to date,

$$\text{DDB} \quad d_t = \frac{2}{N}(\text{Cost} - \text{Depreciation charges to date})$$

or

$$d_t = \frac{2}{N}\left(B - \sum_{j=1}^{t-1} d_j\right) \quad \quad (11\text{-}4b)$$

EXAMPLE 11–4

Compute the DDB depreciation schedule for the situations in Examples 11–2 and 11–3 ($1000):

Cost of the asset, B	$900
Depreciable life, in years, N	5
Salvage value, S	$70

SOLUTION

Year t	Depreciation for Year t Using Equation 11-4a ($1000) d_t	Sum of Depreciation Charges Up to Year t ($1000) $\sum_{j=1}^{t} d_j$	Book Value at End of Year t ($1000) $BV_t = B - \sum_{j=1}^{t} d_j$
1	$(2/5)900 = 360$	$360	$900 - 360 = 540$
2	$(2/5)540 = 216$	576	$900 - 576 = 324$
3	$(2/5)324 = 130$	706	$900 - 706 = 194$
4	$(2/5)194 = 78$	784	$900 - 784 = 116$
5	$(2/5)116 = 46$	830	$900 - 830 = 70 = S$

Figure 11–5 illustrates the situation.

FIGURE 11–5 Declining balance depreciation.

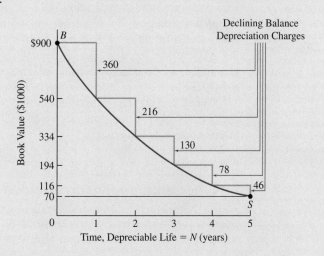

The final salvage value of $70,000 for Examples 11–2, 11–3, and 11-4 was chosen to match the ending value for the double declining balance method. This does not normally happen. If the final salvage value of Example 11–4 had not been $70,000, the double declining balance method would have had to be modified. One modification stops further depreciation once the book value has come to equal the salvage value—this prevents taking too much depreciation. The other modification would switch from declining balance depreciation to straight line—this ensures taking enough depreciation.

These modifications are not detailed here because (1) MACRS has been the legally appropriate system since 1986 and (2) MACRS incorporates the shift from declining balance to straight-line depreciation.

MODIFIED ACCELERATED COST RECOVERY SYSTEM (MACRS)

The modified accelerated cost recovery system (MACRS) depreciation method, introduced by the Tax Reform Act of 1986, was continued with the Taxpayer Relief Act of 1997. Three major advantages of MACRS are that (1) the "property class lives" are less than the "actual useful lives," (2) salvage values are assumed to be zero, and (3) tables of annual percentages simplify computations.

The definition of the MACRS classes of depreciable property is based on work by the U.S. Treasury Department. In 1971 the Treasury published guidelines for about 100 broad asset classes. For each class there was a lower limit, a midpoint, and an upper limit of useful life, called the **asset depreciation range (ADR).** The ADR midpoint lives were somewhat shorter than the actual average useful lives. These guidelines have been incorporated into MACRS so that the property class lives are again shorter than the ADR midpoint lives.

Use of MACRS focuses on the general depreciation system (GDS), which is based on declining balance with switch to straight-line depreciation. The alternative depreciation system (ADS) provides for a longer period of recovery and uses straight-line depreciation. Thus it is much less economically attractive. Under law, ADS must be used for (1) any tangible property used primarily outside the U.S., (2) any property that is tax exempt or financed by tax-exempt bonds, and (3) farming property placed in service when uniform capitalization rules are not applied. The ADS may also be *elected* for property that can be depreciated using the GDS system. However, once ADS has been elected for an asset, it is not possible to switch back to the GDS system. Because the ADS makes the depreciation deductions less valuable, unless ADS is specifically mentioned, subsequent discussion assumes the GDS system when reference is made to MACRS.

Once a property has been determined to be eligible for depreciation, the next step is to calculate its depreciation deductions over its life. The following information is required to calculate these deductions:

- The asset's cost basis.
- The asset's *property class* and *recovery period*.
- The asset's placed-in-service date.

Cost Basis and Placed-in-Service Date

The cost basis, B, is the cost to obtain and place the asset in service fit for use. However, for real property the basis may also include certain fees and charges that the buyer pays as part of the purchase. Examples of such fees include legal and recording fees, abstract fees, survey charges, transfer taxes, title insurance, and amounts that the seller owes that you pay (back taxes, interest, sales commissions, etc.).

Depreciation for a business asset begins when the asset is *placed in service* for a business purpose. If an asset is purchased and used in a personal context, depreciation may not be taken. If that asset is later used in business for income-producing activity, depreciation may begin with the change in usage.

Property Class and Recovery Period

Each depreciated asset is placed in a *MACRS property class,* which defines the **recovery period** and the depreciation percentage for each year. Historically the IRS assigned each type of depreciable asset a *class life* or an *asset depreciation range*. With MACRS, asset class lives have been pooled together in the *property classes*. Table 11–1 lists the class lives and GDS and ADS property classes for several example depreciable assets. Table 11–2 lists the MACRS GDS property classes.

The MACRS GDS property classes are described in more detail in Table 11–2. The proper MACRS property class can be found several different ways. Of the five approaches listed, the first one that works should be used.

1. Property class given in problem.
2. Asset is named in Table 11–2.
3. IRS tables or Table 11–1.
4. Class life.
5. Seven-year property for "all other property not assigned to another class."

Once the MACRS property class is known, as well as the placed-in-service date and the cost basis, the year-to-year depreciation deductions can be calculated for GDS assets over their depreciable life using

$$d_t = B \times r_t \tag{11-5}$$

where d_t = depreciation deduction in year t
B = cost basis being depreciated
r_t = appropriate MACRS percentage rate

Percentage Tables

The IRS has prepared tables to assist in calculating depreciation charges when MACRS GDS depreciation is used. Table 11–3 gives the yearly depreciation percentages (r_t) that are used for the six personal property classes (3-, 5-, 7-, 10-, 15-, and 20-year property classes), and Table 11–4 gives the percentages for nonresidential real property. Notice that the values are given in *percentages*—thus, for example, the value of 33.33% (given in Table 11–3 for Year 1 for a 3-year MACRS GDS property) is 0.3333.

TABLE 11–1 **Example Class Lives and MACRS Property Classes***

IRS Asset Class	Asset Description	Class Life (years) ADR	MACRS Property Class (years)	
			GDS	ADS
00.11	Office furniture, fixtures, and equipment	10	7	10
00.12	Information systems: computers/peripheral	6	5	6
00.22	Automobiles, taxis	3	5	5
00.241	Light general-purpose trucks	4	5	5
00.25	Railroad cars and locomotives	15	7	15
00.40	Industrial steam and electric distribution	22	15	22
01.11	Cotton gin assets	12	7	12
01.21	Cattle, breeding or dairy	7	5	7
13.00	Offshore drilling assets	7.5	5	7.5
13.30	Petroleum refining assets	16	10	16
15.00	Construction assets	6	5	6
20.10	Manufacture of grain and grain mill products	17	10	17
22.2	Manufacture of yarn, thread, and woven fabric	11	7	11
24.10	Cutting of timber	6	5	6
32.20	Manufacture of cement	20	15	20
37.11	Manufacture of motor vehicles	12	7	12
48.11	Telephone communications assets and buildings	24	15	24
48.2	Radio and television broadcasting equipment	6	5	6
49.12	Electric utility nuclear production plant	20	15	20
49.13	Electric utility steam production plant	28	20	28
49.23	Natural gas production plant	14	7	14
50.00	Municipal wastewater treatment plant	24	15	24
80.00	Theme and amusement park assets	12.5	7	12.5

*See Table B.1, *Table of Class Lives and Recovery Periods* in IRS Publication 946, *How to Depreciate Property* (www.irs.gov) for a complete list of depreciable properties under MACRS.

Notice in Table 11–3 that the depreciation percentages continue for *one year beyond* the property class life. For example, a MACRS 10-year property has an r_t value of 3.28% in Year 11. This is due to the *half-year convention* that also halves the percentage for the first year. The half-year convention assumes that all assets are placed in service at the midpoint of the first year.

Another characteristic of the MACRS percentage tables is that the r_t values in any column sum to 100%. This means that assets depreciated using MACRS are *fully depreciated* at the end of the recovery period. This assumes a salvage value of zero. This is a departure from the classic methods, where an estimated salvage value was considered.

Where MACRS Percentage Rates (r_t) Come From

This section describes the connection between classic depreciation methods and the MACRS percentages that are shown in Table 11–3. Before ACRS and MACRS, the most common depreciation method was declining balance with a switch to straight line. That combined method is used for MACRS with three further assumptions.

TABLE 11–2 MACRS GDS Property Classes

Property Class	Personal Property (all property except real estate)
3-year property	Special handling devices for food and beverage manufacture Special tools for the manufacture of finished plastic products, fabricated metal products, and motor vehicles Property with ADR class life of 4 years or less
5-year property	Automobiles* and trucks Aircraft (of non-air-transport companies) Equipment used in research and experimentation Computers Petroleum drilling equipment Property with ADR class life of more than 4 years and less than 10 years
7-year property	All other property not assigned to another class Office furniture, fixtures, and equipment Property with ADR class life of 10 years or more and less than 16 years
10-year property	Assets used in petroleum refining and certain food products Vessels and water transportation equipment Property with ADR class life of 16 years or more and less than 20 years
15-year property	Telephone distribution plants Municipal sewage treatment plants Property with ADR class life of 20 years or more and less than 25 years
20-year property	Municipal sewers Property with ADR class life of 25 years or more

Property Class	Real Property (real estate)
27.5 years	Residential rental property (does not include hotels and motels)
39 years	Nonresidential real property

*The depreciation deduction for passenger automobiles, trucks, and vans is limited and determined by year placed in service. For 2016, limits on passenger automobiles were $3560 for the first tax year (if special depreciation allowance is not elected), $5700 for Year 2, $3350 for Year 3, and $2075 in later years. For more detail, see details in IRS Publication 946, *How to Depreciate Property* (www.irs.gov). Special tables apply to pure electric vehicles.

1. Salvage values are assumed to be zero for all assets.
2. The first and last years of the recovery period are each assumed to be *half-year*.
3. The declining balance rate is 200% for 3-, 5-, 7-, and 10-year property and 150% for 15- and 20-year property.

As shown in Example 11–5, the MACRS percentage rates can be derived from these rules and the declining balance and straight-line methods. However, it is obviously much easier to simply use the r_t values from Tables 11–3 and 11–4.

TABLE 11–3 MACRS Depreciation for Personal Property: Half-Year Convention

Recovery Year	Applicable Percentage for Property Class					
	3-Year Property	5-Year Property	7-Year Property	10-Year Property	15-Year Property	20-Year Property
1	33.33	20.00	14.29	10.00	5.00	3.750
2	44.45	32.00	24.49	18.00	9.50	7.219
3	14.81*	19.20	17.49	14.40	8.55	6.677
4	7.41	11.52*	12.49	11.52	7.70	6.177
5		11.52	8.93*	9.22	6.93	5.713
6		5.76	8.92	7.37	6.23	5.285
7			8.93	6.55*	5.90*	4.888
8			4.46	6.55	5.90	4.522
9				6.56	5.91	4.462*
10				6.55	5.90	4.461
11				3.28	5.91	4.462
12					5.90	4.461
13					5.91	4.462
14					5.90	4.461
15					5.91	4.462
16					2.95	4.461
17						4.462
18						4.461
19						4.462
20						4.461
21						2.231

Computation method

- The 3-, 5-, 7-, and 10-year classes use 200% and the 15- and 20-year classes use 150% declining balance depreciation.

- All classes convert to straight-line depreciation in the optimal year, shown with asterisk (*).

- A half-year of depreciation is allowed in the first and last recovery years.

- If more than 40% of the year's MACRS property is placed in service in the last 3 months, then a midquarter convention must be used with depreciation tables that are not shown here.

EXAMPLE 11–5

Consider a 5-year MACRS property asset with an installed and "made ready for use" cost basis of $100. (*Note*: The $100 value used here is for illustration purposes in developing the rates. One would not normally depreciate an asset with a cost basis of only $100.) Develop the MACRS percentage rates (r_t) for the asset based on the underlying depreciation methods.

SOLUTION

To develop the 5-year MACRS property percentage rates, we use the 200% declining balance method, switching over to straight line at the optimal point. Since the assumed salvage value is zero, the entire cost basis of $100 is depreciated. Also the $100 basis mimics the 100% that is used in Table 11–3.

Let's explain the accompanying table year by year. In Year 1 the basis is $100 − 0$, and the d_t values are halved for the initial half-year assumption. Double declining balance has a rate of 40% for 5 years $(= 2/5)$. This is larger than straight line for Year 1. So one-half of the 40% is used for Year 1. The rest of the declining balance computations are simply 40%×(basis minus the cumulative depreciation).

In Year 2 there are 4.5 years remaining for straight line, so 4.5 is the denominator for dividing the remaining $80 in book value. Similarly in Year 3 there are 3.5 years remaining. In Year 4 the DDB and SL calculations happen to be identical, so the switch from DDB to SL can be done in either Year 4 or Year 5. Once we know that the SL depreciation is 11.52 at the switch point, then the only further calculation is to halve that for the last year.

Notice that the DDB calculations get smaller every year, so that at some point the straight-line calculations lead to faster depreciation. This point is the optimal switch point, and it is built into Table 11–3 for MACRs.

Year	DDB Calculation	SL Calculation	MACRS r_t (%) Rates	Cumulative Depreciation (%)
1	$(1/2)(2/5)(100 - 0) = \mathbf{20.00}$	$^1/_2(100 - 0)/5 = 10.00$	20.00 (DDB)	20.00
2	$(2/5)(100 - 20.00) = \mathbf{32.00}$	$(100 - 20)/4.5 = 17.78$	32.00 (DDB)	52.00
3	$(2/5)(100 - 52.00) = \mathbf{19.20}$	$(100 - 52)/3.5 = 13.71$	19.20 (DDB)	71.20
4	$(2/5)(100 - 71.20) = \mathbf{11.52}$	$(100 - 71.20)/2.5 = \mathbf{11.52}$	11.52 (either)	82.72
5		$\mathbf{11.52}$	11.52 (SL)	94.24
6		$(^1/_2)(11.52) = \mathbf{5.76}$	5.76 (SL)	100.00

The values given in this example match the r_t percentage rates given in Table 11–3 for a 5-year MACRS property.

TABLE 11–4 **MACRS Depreciation for Real Property (real estate)***

Recovery Year	Recovery Percentages for Nonresidential Real Property (month placed in service)											
	1	2	3	4	5	6	7	8	9	10	11	12
1	2.461	2.247	2.033	1.819	1.605	1.391	1.177	0.963	0.749	0.535	0.321	0.107
2–39						2.564			[all months]			
40	0.107	0.321	0.535	0.749	0.963	1.177	1.391	1.605	1.819	2.033	2.247	2.461

*The useful life is 39 years for nonresidential real property. Depreciation is straight line using the midmonth convention. Thus a property placed in service in January would be allowed $11^1/_2$ months depreciation for recovery Year 1.

MACRS Method Examples

Remember the key points in using MACRS: (1) what type of asset you have, and whether it qualifies as depreciable property, (2) the amount you are depreciating [cost basis], and (3) when you are placing the asset in service. Let's look at several examples of using MACRS to calculate both depreciation deductions and book values.

EXAMPLE 11–6

Use the MACRS GDS method to calculate the yearly depreciation allowances and book values for a firm that has purchased $150,000 worth of office equipment that qualifies as depreciable property. The equipment is estimated to have a salvage (market) value of $30,000 (20% of the original cost) after the end of its depreciable life.

SOLUTION

1. The assets qualify as depreciable property.
2. The cost basis is given as $150,000.
3. The assets are being placed in service in Year 1 of our analysis.
4. MACRS GDS applies.
5. The salvage value is not used with MACRS to calculate depreciation or book value.

Office equipment is listed in Table 11–2 as a 7-year property. We now use the MACRS GDS 7-year property percentages from Table 11–3 and Equation 11-5 to calculate the year-to-year depreciation allowances. We use Equation 11-1 to calculate the book value of the asset.

Year, t	MACRS, r_t		Cost Basis	d_t	Cumulative d_t	$BV_t = B - Cum.d_t$
1	14.29%	×	$150,000	$ 21,435	$ 21,435	$128,565
2	24.49		150,000	36,735	58,170	91,830
3	17.49		150,000	26,235	84,405	65,595
4	12.49		150,000	18,735	103,140	46,860
5	8.93		150,000	13,395	116,535	33,465
6	8.92		150,000	13,380	129,915	20,085
7	8.93		150,000	13,395	143,310	6,690
8	4.46		150,000	6,690	150,000	0
	100.00%			$150,000		

Notice in this example several aspects of the MACRS depreciation method: (1) the sum of the r_t values is 100.00%, (2) this 7-year MACRS GDS property is depreciated over 8 years (=property class life + 1), and (3) the book value after 8 years is $0.

EXAMPLE 11–7

Investors in the JMJ Group purchased a hotel resort in April. The group paid $20 million for the hotel resort and $5 million for the grounds surrounding the resort. The group sold the resort

5 years later in August. Calculate the depreciation deductions for Years 1 through 6. What was the book value at the time the resort was sold?

SOLUTION

Hotels are nonresidential real property and are depreciated over a 39-year life. Table 11–4 lists the percentages for each year. In this case the cost basis is $20 million, and the $5 million paid for the land is not depreciated. JMJ's depreciation is calculated as follows:

Year 1 (obtained in April) $d_1 = 20,000,000(1.819\%) = \$363,800$
Year 2 $d_2 = 20,000,000(2.564\%) =\ \ 512,800$
Year 3 $d_3 = 20,000,000(2.564\%) =\ \ 512,800$
Year 4 $d_4 = 20,000,000(2.564\%) =\ \ 512,800$
Year 5 $d_5 = 20,000,000(2.564\%) =\ \ 512,800$
Year 6 (disposed of in August) $d_6 = 20,000,000(1.605\%) =\ \ 321,000$

Thus the hotel's book value when it was sold was

$$BV_6 = B - (d_1 + d_2 + d_3 + d_4 + d_5 + d_6)$$
$$= 20,000,000 - (2,736,000) = \$17,264,000$$

The value of the land has not changed in terms of book value.

Comparing MACRS and Classic Methods

In Examples 11–2 through 11–4 we used the *straight-line, sum-of-the-years'-digits,* and *declining balance* depreciation methods to illustrate how the book value of an asset that cost $900,000 and had a salvage value of $70,000 changed over its 5-year depreciation life. Figures 11–2 through 11–4 provided a graphical view of book value over the 5-year depreciation period using these methods. Example 11–8 compares the MACRS GDS depreciation method directly against the classic methods.

EXAMPLE 11–8

Consider the equipment that was purchased in Example 11–6. Calculate the asset's depreciation deductions and book values over its depreciable life for MACRS and the classic methods.

SOLUTION

Table 11–5 and Figure 11–6 compare MACRS and classic depreciation methods. MACRS depreciation is the most *accelerated* or fastest depreciation method—remember its name is the modified *accelerated* cost recovery system. The book value drops fastest and furthest with MACRS, thus the present worth is the largest for the MACRS depreciation deductions.

Depreciation deductions *benefit* a firm after taxes because they reduce taxable income and taxes. The time value of money ensures that it is better to take these deductions as soon as possible. In general, MACRS, which allocates larger deductions earlier in the depreciable life, provides more economic benefits than classic methods.

TABLE 11–5 Comparison of MACRS and Classic Methods for Asset in Example 11–6

Year, t	MACRS		Straight Line		Double Declining		Sum-of-Years' Digits	
	d_t	BV_t	d_t	BV_t	d_t	BV_t	d_t	BV_t
1	21,435	128,565	12,000	138,000	30,000	120,000	21,818	128,182
2	36,735	91,830	12,000	126,000	24,000	96,000	19,636	108,545
3	26,235	65,595	12,000	114,000	19,200	76,800	17,455	91,091
4	18,735	46,860	12,000	102,000	15,360	61,440	15,273	75,818
5	13,395	33,465	12,000	90,000	12,288	49,152	13,091	62,727
6	13,380	20,085	12,000	78,000	9,830	39,322	10,909	51,818
7	13,395	6,690	12,000	66,000	7,864	31,457	8,727	43,091
8	6,690	0	12,000	54,000	1,457	30,000	6,545	36,545
9	0	0	12,000	42,000	0	30,000	4,364	32,182
10	0	0	12,000	30,000	0	30,000	2,182	30,000
PW (10%)	$108,217		$73,734		$89,918		$84,118	

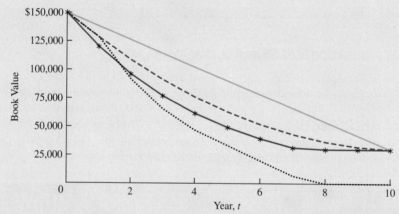

FIGURE 11–6 Comparing MACRS (·············) and classic depreciation methods: double declining balance (——*——), sum of the years' digits (– – – –), and straight line (———).

When computing cash flows and income taxes, the MACRS depreciation schedule results in the highest PW and thus it will minimize the PW of taxes. However, for valuing the business the straight-line deductions have the lowest PW, thus maximizing the firm's reported value.

DEPRECIATION AND ASSET DISPOSAL

When a depreciated asset is disposed of, the key question is, Which is larger, the asset's *book value, BV*, or the asset's *market value, MV*? If the book value is lower than the market value, then excess depreciation will be recaptured and taxed. On the other hand, if the book value is higher than the market value, there is a **loss on the disposal**. In either case, the level of taxes owed changes.

Depreciation recapture *(ordinary gains)*: Depreciation recapture, also called ordinary gains, is necessary when an asset is sold for more than an asset's current book value. If more than the original cost basis is received, only the amount up to the original cost basis is recaptured depreciation. Since MACRS assumes $S = 0$ for its annual calculations, MACRS often has recaptured depreciation at disposal.

Losses: A *loss* occurs when less than book value is received for a depreciated asset. In the accounting records we've disposed of an asset for a dollar amount less than its book value, which is a loss.

Capital gains: Capital gains occur when more than the asset's original cost basis is received for it. The excess over the original cost basis is the *capital gain*. As described in Chapter 12, the tax rate on such gains is sometimes lower than the rate on ordinary income, but this depends on how long the investment has been held ("short," ≤ 1 year; "long," ≥ 1 year). In most engineering economic analyses capital gains are very uncommon because business and production equipment and facilities almost always *lose* value over time. Capital gains are much more likely to occur for nondepreciated assets like stocks, bonds, real estate, jewelry, art, and collectibles.

The relationship between depreciation recapture, loss, and capital gain is illustrated in Figure 11–7. Each case given is at a point in time in the life of the depreciated asset, where the original cost basis is $10,000 and the book value is $5000. Case (a) represents depreciation recapture (ordinary gain), Case (b) represents a loss, and in Case (c) both recaptured depreciation and a capital gain are present.

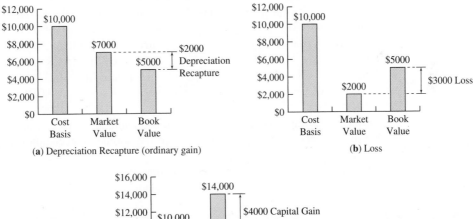

(a) Depreciation Recapture (ordinary gain)

(b) Loss

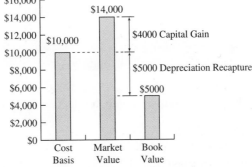

(c) Capital Gain and Depreciation Recapture

FIGURE 11–7 Recaptured depreciation, loss on sale, and capital gain.

EXAMPLE 11–9

Consider an asset with a cost basis of $10,000 that has been depreciated using the MACRS method. This asset is a 3-year MACRS property. What is the gain or loss if the asset is disposed of after 5 years of operation for (**a**) $7000, (**b**) $0, and (**c**) a cost of $2000?

SOLUTION

To find *gain* or *loss* at disposal we compare *market* and *book value*. Since MACRS depreciates to a salvage value of 0, and 5 years is greater than the recovery period, the book value equals $0.

   (**a**) Recaptured depreciation = $7000, since the book value is $7000 higher.

   (**b**) Since market value equals salvage value, there is no recaptured depreciation or loss.

   (**c**) Since the money is paid for disposal, this is less than the book value, and there is a loss of $2000.

This general method for calculating recaptured depreciation or loss applies to all of the depreciation methods described in this chapter.

If the asset is in the middle of its depreciable life, then recaptured depreciation and losses are calculated in a similar manner—compare the *market* and *book values* at the time of disposal. However, in computing the book value with MACRS depreciation, a special rule must be applied for assets disposed of before the end of the recovery period. The rule is to *take one half of the allowable depreciation deduction for that year.* This rule assumes that disposals take place on average halfway through the year. Thus for a 5-year asset disposed of in the middle of Year 4, the rate allowed for MACRS depreciation is half of 11.52% or 5.76%. If the asset is disposed of in Year 6, it is already past the recovery period, and a half-year assumption has already been built into the MACRS schedule. Thus, the full r_6 is taken.

However, Example 11–10 illustrates that economic analyses will arrive at the same taxable income whether 0%, 50%, or 100% of the normal depreciation is claimed in the year of disposal.

Thus, correct economic analyses can assume the year of disposal is *just like* every other year and claim 100% of that year's depreciation.

EXAMPLE 11–10

Consider the $10,000 asset in Example 11–9. Using MACRS and a 3-year recovery period, calculate the effect of disposal if this asset is sold during Year 2 for $5000 and

   1. 50% depreciation is claimed.

   2. 0% depreciation is claimed.

   3. 100% depreciation is claimed.

SOLUTION

The first effect of the disposal is a before-tax cash flow of $5000. This is not affected by the amount of depreciation claimed. The second effect of the disposal is the total deduction from taxable income for that year.

In every case

$$MV_2 = \$5000 \text{ (market value in year 2)}$$

$$BV_1 = 10{,}000 - 10{,}000r_1$$

$$= 10{,}000 - 10{,}000 \times 0.3333$$

$$= 6667$$

1. If 50% depreciation is claimed in year 2:

$$D_2 = 10{,}000(r_2/2) = 10{,}000\,(0.4445/2)$$

$$= \$2222.50$$

$$BV_2 = BV_1 - D_2 = 6667 - 2222.50$$

$$= \$4444.50$$

This is less than MV_2, so excess depreciation must be recaptured.

$$\text{Recaptured depreciation} = 5000 - 4444.50 = 555.50$$

$$\text{Total deduction from taxable income} = D_2 - \text{Recaptured depreciation}$$

$$= 2222.50 - 555.50 = \$1667$$

2. If 0% depreciation is claimed in year 2: No depreciation, but total deduction from taxable income is the loss because the market value of $5000 is $1667 less than the book values of $6667.

3. If 100% depreciation is claimed in year 2:

$$D_2 = 10{,}000\,r_2 = 10{,}000 \times 0.4445 = \$4450$$

$$BV_2 = BV_1 - D_2 = 6667 - 4450 = \$2217 < MV_2 \Rightarrow \text{depreciation recapture}$$

$$\text{Recapturred depreciation} = 5000 - 2217 = \$2783$$

$$\text{Total deduction from taxable income} = D_2 - \text{recapture}$$

$$= 4450 - 2783 = \$1667$$

In all three cases, the total deduction from taxable income is $1667. The first approach follows the tax language. The second, with 0% claimed, is probably the easiest for hand calculations. The third approach with 100% claimed is probably the easiest for spreadsheet calculations, because it treats the year of disposal like any other year.

UNIT-OF-PRODUCTION DEPRECIATION

At times the recovery of depreciation on a particular asset is more closely related to use than to time. In these few situations (and they are rare), the **unit-of-production (UOP) depreciation** in any year is

$$\text{UOP depreciation in any year} = \frac{\text{Production for year}}{\text{Total lifetime production for asset}}(B - S) \qquad (11\text{-}6)$$

This method might be useful for machinery that processes natural resources if the resources will be exhausted before the machinery wears out. Historically, this method was sometimes used for construction equipment that had very heavy use in some years and very light use in others. It is not considered an acceptable method for general use in depreciating industrial equipment.

EXAMPLE 11–11

For numerical similarity with previous examples, assume that equipment costing $900,000 has been purchased for use in a sand and gravel pit. The pit will operate for 5 years, while a nearby airport is being reconstructed and paved. Then the pit will be shut down, and the equipment removed and sold for $70,000. Compute the unit-of-production (UOP) depreciation schedule if the airport reconstruction schedule calls for 40,000 m^3 of sand and gravel as follows:

Year	Required Sand and Gravel (m^3)
1	4,000
2	8,000
3	16,000
4	8,000
5	4,000

SOLUTION

The cost basis, B, is $900,000. The salvage value, S, is $70,000. The total lifetime production for the asset is 40,000 m^3 of sand and gravel. From the airport reconstruction schedule, the first-year UOP depreciation would be

$$\text{First-year UOP depreciation} = \frac{4000 \text{ m}^3}{40,000 \text{ m}^3}(900,000 - 70,000) = \$83,000$$

Similar calculations for the subsequent 4 years give the complete depreciation schedule:

Year	UOP Depreciation (in $1000)
1	$ 83
2	166
3	332
4	166
5	83
	$830

It should be noted that the actual unit-of-production depreciation charge in any year is based on the actual production for the year rather than the scheduled production.

DEPLETION

Depletion is the exhaustion of natural resources as a result of their removal. Since depletion covers such things as mineral properties, oil and gas wells, and standing timber, removal may take the form of digging up metallic or nonmetallic minerals, producing petroleum or natural gas from wells, or cutting down trees.

Depletion is recognized for income tax purposes for the same reason depreciation is—capital investment is being consumed or used up. Thus a portion of the gross income should be considered to be a return of the capital investment. The calculation of the depletion allowance is different from depreciation because there are two distinct methods of calculating depletion: *cost depletion* and *percentage depletion*. Except for standing timber and most oil and gas wells, depletion is calculated by both methods and the larger value is taken as depletion for the year. For standing timber and most oil and gas wells, only cost depletion is permissible.

Cost Depletion

Depreciation relied on an asset's cost, depreciable life, and salvage value to apportion the cost *minus* salvage value *over* the depreciable life. In some cases, where the asset is used at fluctuating rates, we might use the unit-of-production (UOP) method of depreciation. For mines, oil wells, and standing timber, fluctuating production rates are the usual situation. Thus, **cost depletion** is computed like unit-of-production depreciation using:

1. Property cost, less cost for land.
2. Estimated number of recoverable units (tons of ore, cubic meters of gravel, barrels of oil, million cubic feet of natural gas, thousand board-feet of timber, etc.).
3. Salvage value, if any, of the property.

EXAMPLE 11–12

A small lumber company bought a tract of timber for $35,000, of which $5000 was the land's value and $30,000 was the value of the estimated 1.5 million board-feet of standing timber. The first year, the company cut 100,000 board-feet of standing timber. What was the year's depletion allowance?

SOLUTION

$$\text{Depletion allowance per 1000 board-ft} = \frac{\$35{,}000 - \$5000}{1500 \times 1000 \text{ board-ft}}$$

$$= \$20 \text{ per 1000 board-ft}$$

The depletion allowance for the year would be

$$100{,}000 \text{ board-ft} \times \$20 \text{ per 1000 board-ft} = \$2000$$

Percentage Depletion

Percentage depletion is an alternate method for mineral property. The allowance is a certain percentage of the property's gross income during the year. This is an entirely different concept from depreciation. Unlike depreciation, which allocates cost *over* useful life, the **percentage depletion** allowance (see Table 11–6) is based on the property's gross income.

TABLE 11–6 Percentage Depletion Allowances for Selected Deposits

Deposits	Rate
Sulfur, uranium, and, if from deposits in the U.S., asbestos, lead ore, zinc ore, nickel ore, and mica	22%
Gold, silver, copper, iron ore, and certain oil shale, if from deposits in the U.S.	15%
Borax, granite, limestone, marble, mollusk shells, potash, slate, soapstone, and carbon dioxide produced from a well	14%
Coal, lignite, and sodium chloride	10%
Clay and shale used or sold for use in making sewer pipe or bricks or used or sold for use as sintered or burned lightweight aggregates	$7\frac{1}{2}$%
Clay used or sold for use in making drainage and roofing tile, flower pots, and kindred products, and gravel, sand, and stone (other than stone used or sold for use by a mine owner or operator as dimension or ornamental stone)	5%

Source: Internal Revenue Service, Publication 535, Chapter 9. Section 613(b) of the Internal Revenue Code gives a complete list of minerals and their percentage depletion rates.

Since percentage depletion is computed on the *income* rather than the property's cost, the total depletion *may exceed the cost of the property*. In computing the *allowable percentage depletion* on a property in any year, the *percentage depletion allowance* cannot exceed 50% of the property's taxable income computed without the depletion deduction. The percentage depletion calculations are illustrated by Example 11–13.

EXAMPLE 11–13

A coal mine has a gross income of $250,000. Mining expenses equal $210,000. Compute the allowable percentage depletion deduction.

SOLUTION

From Table 11–6, coal has a 10% depletion allowance based on gross mining income. The allowable percentage depletion deduction is also limited to a maximum of 50% of taxable income.

Computed Percentage Depletion

Gross income from mine	$250,000
Depletion percentage	× 10%
Computed percentage depletion	$ 25,000

Taxable Income Limitation

Gross income from mine	$250,000
Less: Expenses other than depletion	−210,000
Taxable income from mine	40,000
Deduction limitation	× 50%
Taxable income limitation	$20,000

Since the taxable income limitation ($20,000) is less than the computed percentage depletion ($25,000), the allowable percentage depletion deduction is $20,000. If the cost depletion were higher, it could be claimed instead.

As previously stated, on mineral property the depletion deduction can be based on either cost or percentage depletion. Each year, depletion is computed by cost and percentage depletion methods, and the allowable depletion deduction is the larger of the two amounts.

SPREADSHEETS AND DEPRECIATION

The spreadsheet functions for straight-line, double declining balance, and sum-of-years'-digits depreciation are listed in Table 11–7. Because these techniques are simple and were replaced for tax purposes by MACRS in 1986, they are not covered in detail here. All three

TABLE 11–7 Spreadsheet Functions for Depreciation

Depreciation Technique	Excel
Straight line	SLN(cost, salvage, life)
Double declining balance	DDB(cost, salvage, life, period, factor)
Sum of years' digits	SYD(cost, salvage, life, period)
MACRS	VDB(cost, salvage, life, start_period, end_period, factor, no_switch)

functions include parameters for *cost* (initial book value), *salvage* (final salvage value), and *life* (depreciation period). Both DDB and SYD change depreciation amounts every year, so they include a parameter to pick the *period* (year). Finally, DDB includes a *factor*. The default value is 2 for 200% or double declining balance, but another commonly used value is 1.5 for 150%.

Using VDB for MACRS

The Excel function VDB is a flexible or variable declining balance method. It includes the ability to specify the starting and ending periods, rather than simply a year. It also includes an optional no_switch for problems where a switch from declining balance to straight-line depreciation is *not* desired.

To use VDB to calculate MACRS depreciation, the following are true.

1. Salvage $= 0$, since MACRS assumes no salvage value.
2. Life $=$ recovery period of 3, 5, 7, 10, 15, or 20 years.
3. First period runs from 0 to 0.5, 2nd period from 0.5 to 1.5, 3rd from 1.5 to 2.5, t^{th} from $t - 1.5$ to $t - 0.5$, and last period from *life* $- 0.5$ to *life*.
4. Factor $= 2$ for recovery periods of 3, 5, 7, or 10 years and $= 1.5$ for recovery periods of 15 or 20 years.
5. Since MACRS includes a switch to straight line, no_switch can be omitted.

The start_period and end_period arguments are from $t - 1.5$ to $t - 0.5$, because MACRS uses a half-year convention for the first year. Thus the first year has 0 to 0.5 year of depreciation, and the second year starts where the first year stops. When one is writing the Excel function, either the first and last periods must be edited individually, or start_period must be defined with a minimum of 0 and end_period with a maximum of life. This prevents the calculation of depreciation from -0.5 to 0 and from *life* to *life* $+ 0.5$.

The results of using the VDB function match Table 11–4, except that the VDB function has more significant digits rather than being rounded to 2 decimals. Example 11–14 illustrates the use of the VDB function.

EXAMPLE 11–14   (Example 11–6 Revisited)

Return to the data of Example 11–6 which had $150,000 of office equipment, which is 7-year MACRS property. Use VDB to compute the depreciation amounts.

SOLUTION

	A	B	C	D	E	F	G
1	150,000	First Cost					
2	0	Salvage					
3	7	Life = recovery period					
4	200%	Factor					
5							
6	Period	Depreciation					
7	1	$21,428.57	=VDB($A$1,0,$A$3,MAX(0,A7−1.5),MIN($A$3,A7−0.5),$A$4)				
8	2	$36,734.69	or (cost, salvage, life, max(0, t−1.5), min (life, t−.5), factor)				
9	3	$26,239.07					
10	4	$18,742.19					
11	5	$13,387.28					
12	6	$13,387.28					
13	7	$13,387.28					
14	8	$6,693.64					
15		$150,000	= Sum				

FIGURE 11-8 Using VDB to calculate MACRS depreciation.

The spreadsheet in Figure 11–8 defines the start_period with a minimum of 0 and the end_period with a maximum of life. Remember that this is the *depreciable* life or recovery period. Thus this formula could be used for any year of any recovery schedule. Notice that the VDB formula uses the value 0 for the salvage value, rather than referring to the data cell for the salvage value. MACRS assumes a salvage value of zero—no matter what the value truly is.

SUMMARY

From the perspective of engineering economy, depreciation matters even though it is *not* a cash flow. It is part of determining taxable income and the cash flow of taxes.

Depreciation is part of computing income taxes in economic analysis. There are three distinct definitions of depreciation:

1. Decline in asset's market value.
2. Decline in asset's value to its owner.
3. Allocating the asset's cost *less* its salvage value *over* its recovery period or depreciable life.

While the first two definitions are used in valuing an asset or a firm, it is the third definition that is used in tax computations and thus the focus of this chapter. Book value is the remaining unallocated cost of an asset, or

$$\text{Book value} = \text{Asset cost} - \text{Depreciation charges made to date}$$

This chapter describes how depreciable assets are *written off* (or claimed as a business expense) over a period of years instead of *expensed* in a single period (like wages, material costs, etc.). The depreciation methods described include the classic methods: *straight line, sum of the years' digits,* and *declining balance.* These methods required estimating the asset's salvage value and depreciable life.

The current tax law specifies use of the modified accelerated capital recovery system (MACRS). This chapter has focused on the general depreciation system (GDS) with limited discussion of the less attractive alternative depreciation system (ADS). MACRS (GDS) specifies faster *recovery periods* and a salvage value of zero, so it is generally economically more attractive than the classic methods. While historical from a U.S. tax perspective, the classic methods are often used in the U.S. when valuing assets and firms and internationally for both valuation and tax purposes.

The MACRS system is the current tax law, and it assumes a salvage value of zero. This is in contrast with historical methods, which ensured the final book value would equal the predicted salvage value. Thus, when one is using MACRS it is often necessary to consider recaptured depreciation. This is the excess of salvage value over book value, and it is taxed as ordinary income. Similarly, losses on sale or disposal are taxed as ordinary income.

Unit-of-production (UOP) depreciation relies on usage to quantify the loss in value. UOP is appropriate for assets that lose value based on the number of units produced, the tons of gravel moved, and so on (vs. number of years in service). However, this method is not considered to be acceptable for most business assets.

Depletion is the exhaustion of natural resources like minerals, oil and gas, and standing timber. The owners of the natural resources are consuming their investments as the natural resources are removed and sold. Cost depletion is computed based on the fraction of the resource that is removed or sold. For minerals and some oil and gas wells, an alternate calculation called percentage depletion is allowed. Percentage depletion is based on income, so the total allowable depletion deductions may *exceed* the invested cost.

Integrating depreciation schedules with cash flows often involves a lot of arithmetic. Thus, the tool of spreadsheets can be quite helpful. The functions for the classic methods, straight line, sum of the years' digits, and declining balance, are straightforward. Rather than individually entering MACRS percentages into the spreadsheet, the function VDB can be used to calculate MACRS depreciation percentages.

PROBLEMS

Assume that the depreciation methods listed in each problem can be used for tax and valuation purposes.

Depreciation Schedules

11-1 For an asset that fits into the MACRS "all property not assigned to another class" designation, show in a table the depreciation and book value over the asset's 10-year life of use. The cost basis of the asset is $20,000.

11-2 (A) A large profitable corporation bought a small jet plane for use by the firm's executives in January. The plane cost $1.5 million and, for depreciation purposes, is assumed to have $400,000 salvage value at the end of 5 years. Compute the MACRS depreciation schedule.

11-3 Hillsborough Architecture and Engineering, Inc. has purchased a blueprint printing machine for $25,000. This printer falls under the MACRS category "Office Equipment" and thus has a 7-year MACRS class life. Prepare a depreciation table.
Contributed by Paul R. McCright, University of South Florida

11-4 (A) If Hillsborough Architecture and Engineering (in Problem 11-3) purchases a new office building in May for $5.7M, determine the allowable depreciation for each year. (*Note*: This asset is non-residential real property.)
Contributed by Paul R. McCright, University of South Florida

11-5 A new machine tool is being purchased for $25,000 and is expected to have a zero salvage value at the end of its 5-year useful life. Compute its DDB depreciation schedule. Assume any remaining depreciation is claimed in the last year.

11-6 (A) A company that manufactures food and beverages in the vending industry has purchased some handling equipment that cost $75,000 and will be depreciated using MACRS GDS; the class life of the asset is 4 years. Show in a table the yearly depreciation amount and book value of the asset over its depreciation life.

11-7 A depreciable asset costs $50,000 and has an estimated salvage value of $5000 at the end of its 6-year depreciable life. Compute the depreciation schedule for this asset by both SOYD depreciation and DDB depreciation.

11-8 (A) Gamma Cruise, Inc. purchased a new tender (a small motorboat) for $35,000. Its salvage value is $7500 after its useful life of 5 years. Calculate the depreciation schedule using (*a*) MACRS and (*b*) SOYD methods.

11-9 (G) A $5 million oil drilling rig has a 5-year depreciable life and a $250,000 salvage value at the end of that time.
(*a*) Determine which one of the following methods provides the preferred depreciation schedule: DDB or SOYD.
(*b*) Show the depreciation schedule for the preferred method.
(*c*) Search for new oil rig technologies and describe three that improve environmental impact.

11-10 (A) The RX Drug Company has just purchased a capsulating machine for $76,000. The plant engineer estimates the machine has a useful life of 5 years and no salvage value. Compute the depreciation schedule using:
(*a*) Straight-line depreciation
(*b*) Sum-of-years'-digits depreciation
(*c*) Double declining balance depreciation (assume any remaining depreciation is claimed in the last year)

11-11 Some special handling devices can be obtained for $20,000. At the end of 5 years, they can be sold for $2000. Compute the depreciation schedule for the devices using the following methods:
(*a*) Straight-line depreciation
(*b*) Sum-of-years'-digits depreciation
(*c*) Double declining balance depreciation
(*d*) MACRS depreciation

11-12 (A) The company treasurer must determine the best depreciation method for office furniture that costs $50,000 and has a zero salvage value at the end of a 10-year depreciable life. Compute the depreciation schedule using:
(*a*) Straight line
(*b*) Double declining balance
(*c*) Sum-of-years'-digits
(*d*) Modified accelerated cost recovery system

11-13 The Acme Chemical Processing Company paid $50,000 for research equipment, which it believes will have zero salvage value at the end of its 5-year life. Compute the depreciation schedule using:
(*a*) Straight line
(*b*) Sum-of-years'-digits

(c) Double declining balance

(d) Modified accelerated cost recovery system

(e) What is the U.S. EPA's *Presidential Green Chemistry Challenge*? What impact has this initiative had on green chemical processing?

11-14 Units-of-production depreciation is being used for a machine that, based on usage, has an allowable depreciation charge of $6500 the first year and increasing by $1000 each year until complete depreciation. If the machine's cost basis is $110,000, set up a depreciation schedule that shows depreciation charge and book value over the machine's 10-year useful life.

11-15 Consider a $6500 piece of machinery, with a 5-year depreciable life and an estimated $1200 salvage value. The projected utilization of the machinery when it was purchased, and its actual production to date, are as follows:

Year	Projected Production (tons)	Actual Production (tons)
1	3500	3000
2	4000	5000
3	4500	[Not
4	5000	yet
5	5500	known]

Compute the depreciation schedule using:

(a) Straight line

(b) Sum-of-years'-digits

(c) Double declining balance

(d) Unit of production (for first 2 years only)

(e) Modified accelerated cost recovery system

11-16 Al Jafar Jewel Co. purchased a crystal extraction machine for $50,000 that has an estimated salvage value of $10,000 at the end of its 8-year useful life. Compute the depreciation schedule using:

(a) MACRS depreciation

(b) Straight-line depreciation

(c) Sum-of-years' digits (SOYD) depreciation

(d) Double declining balance depreciation

11-17 The depreciation schedule for a machine has been arrived at by several methods. The estimated salvage value of the equipment at the end of its 6-year useful life is $600. Identify the resulting depreciation schedules.

Year	A	B	C	D
1	$2114	$2000	$1600	$1233
2	1762	1500	2560	1233
3	1410	1125	1536	1233
4	1057	844	922	1233
5	705	633	922	1233
6	352	475	460	1233

11-18 Consider five depreciation schedules:

Year	A	B	C	D	E
1	$45.00	$35.00	$29.00	$58.00	$43.50
2	36.00	20.00	46.40	34.80	30.45
3	27.00	30.00	27.84	20.88	21.32
4	18.00	30.00	16.70	12.53	14.92
5	9.00	20.00	16.70	7.52	10.44
6			8.36		

They are based on the same initial cost, useful life, and salvage value. Identify each schedule as one of the following:

- Straight-line depreciation
- Sum-of-years'-digits depreciation
- 150% declining balance depreciation
- Double declining balance depreciation
- Unit-of-production depreciation
- Modified accelerated cost recovery system

11-19 The depreciation schedule for an asset, with a salvage value of $90 at the end of the recovery period, has been computed by several methods. Identify the depreciation method used for each schedule.

Year	A	B	C	D	E
1	$323.3	$212.0	$424.0	$194.0	$107.0
2	258.7	339.2	254.4	194.0	216.0
3	194.0	203.5	152.6	194.0	324.0
4	129.3	122.1	91.6	194.0	216.0
5	64.7	122.1	47.4	194.0	107.0
6		61.1			
	970.0	1060.0	970.0	970.0	970.0

11-20 A heavy construction firm has been awarded a contract to build a large concrete dam. It is expected that a total of 8 years will be required to complete the work. The firm will buy $600,000 worth of special equipment for the job. During the preparation of the job cost estimate, the following utilization schedule was computed for the special equipment:

Year	Utilization (hr/yr)	Year	Utilization (hr/yr)
1	6000	5	800
2	4000	6	800
3	4000	7	2200
4	1600	8	2200

At the end of the job, it is estimated that the equipment can be sold at auction for $60,000.

(a) Compute the sum-of-years'-digits depreciation schedule.

(b) Compute the unit-of-production depreciation schedule.

11-21 A profitable company making earthmoving equipment is considering an investment of $150,000 on equipment that will have a 5-year useful life and a $50,000 salvage value. Use a spreadsheet function to compute the MACRS depreciation schedule. Show the total depreciation taken (=sum( )) as well as the PW of the depreciation charges discounted at the MARR%.

11-22 A custom-built production machine is being depreciated using the units-of-production method. The machine costs $65,000 and is expected to produce 1.5 million units, after which it will have a $5000 salvage value. In the first 2 years of operation the machine was used to produce 140,000 units each year. In the 3rd and 4th years, production went up to 400,000 units. After that time annual production returned to 135,000 units. Use a spreadsheet to develop a depreciation schedule showing the machine's depreciation allowance and book value over its depreciable life.

11-23 You are equipping an office. The total office equipment will have a first cost of 2.0M and a salvage value of $200,000. You expect the equipment will last 10 years. Use a spreadsheet function to compute the MACRS depreciation schedule.

11-24 Office equipment whose initial cost is $100,000 has an estimated actual life of 6 years, with an estimated salvage value of $10,000. Prepare tables listing the annual costs of depreciation and the book value at the end of each 6 years, based on straight-line, sum-of-years'-digits, and MACRS depreciation. Use spreadsheet functions for the depreciation methods.

Comparing Depreciation Methods

11-25 The XYZ Block Company purchased a new office computer and other depreciable computer hardware for $12,000. During the third year, the computer is declared obsolete and is donated to the local community college. Using an interest rate of 10%, calculate the PW of the depreciation deductions. Assume that no salvage value was initially declared and that the machine was expected to last 5 years.

(a) Straight-line depreciation
(b) Sum-of-years'-digits depreciation
(c) MACRS depreciation
(d) Double declining balance depreciation
(e) Which method is preferred for determining the firm's taxes?
(f) Which method is preferred for determining the firm's value?
(g) Is using two accounting methods ethical?

11-26 Some equipment that costs $1000 has a 5-year depreciable life and an estimated $50 salvage value at the end of that time. You have been assigned to determine whether to use straight-line or SOYD depreciation. If a 10% interest rate is appropriate, which is the preferred depreciation method for this profitable corporation? Use a spreadsheet to show your computations of the difference in present worths.

11-27 The FOURX Corp. has purchased $50,000 of experimental equipment. The anticipated salvage value is $5000 at the end of its 5-year depreciable life. This profitable corporation is considering two methods of depreciation: sum-of-years'-digits and double declining balance. If it uses 10% interest in its comparison, which method do you recommend? Use a spreadsheet to develop your solution.

11-28 The White Swan Talc Company paid $120,000 for
A mining equipment for a small talc mine. The min-
G ing engineer's report indicates the mine contains 40,000 cubic meters of commercial-quality talc. The company plans to mine all the talc in the next 5 years as follows:

Year	Talc Production (m^3)
1	15,000
2	11,000
3	4,000
4	6,000
5	4,000

At the end of 5 years, the mine will be exhausted and the mining equipment will be worthless. The company accountant must now decide whether to use sum-of-years'-digits depreciation or unit-of-production depreciation. The company considers 15% to be an appropriate time value of money.

(a) Which would you recommend? How much better is the present worth for the recommended choice?

(b) What is talc and how is it used? As the softest mineral, are there special health and environmental issues/risks that are present in the mining, processing, and use of talc?

11-29 For the data in Problem 11-21, if money is worth 10%, which one of the following three methods of depreciation would be preferable?

(a) Straight-line method

(b) Double declining balance method

(c) MACRS method

11-30 TELCO Corp. has leased some industrial land near its plant. It is building a small warehouse on the site at a cost of $250,000. The building will be ready for use January 1. The lease will expire 15 years after the building is occupied. The warehouse will belong at that time to the landowner, with the result that there will be no salvage value to TELCO. The warehouse is to be depreciated either by MACRS or SOYD depreciation. If 10% interest is appropriate, which depreciation method should be selected?

11-31 A small used delivery van can be purchased for $20,000. At the end of its useful life (8 years), the van can be sold for $3000. Determine the PW of the depreciation schedule based on 15% interest using:

(a) Straight-line depreciation

(b) Sum-of-years'-digits depreciation

(c) MACRS depreciation

(d) Double declining balance depreciation

11-32 Loretta Livermore Labs purchased R&D equipment
A costing $200,000. The interest rate is 5%, salvage
E value is $20,000, and expected life is 10 years. Compute the PW of the depreciation deductions assuming:

(a) Straight-line depreciation

(b) Sum-of-years'-digits depreciation

(c) MACRS depreciation

(d) Double declining balance depreciation

(e) Which method is preferred for determining the firm's taxes?

(f) Which method is preferred for determining the firm's value?

(g) Is using two accounting methods ethical?

Depreciation and Book Value

11-33 Explain in your own words the difference between capital gains and ordinary gains. In addition, explain why it is important to our analysis as engineering economists. Do we see capital gains much in industry-based economic analyses or in our personal lives?

11-34 On July 1, Nancy paid $600,000 for a commercial
A building and an additional $150,000 for the land on which it stands. Four years later, also on July 1, she sold the property for $850,000. Compute the modified accelerated cost recovery system depreciation for each of the five calendar years during which she had the property.

11-35 To meet increased delivery demands, Mary Moo Dairy just purchased 15 new delivery trucks. Each truck cost $30,000 and has an expected life of 4 years. The trucks can each be sold for $5000 after 4 years. Using MACRS depreciation (*a*) determine the depreciation allowance for Year 2 for the fleet, and (*b*) determine the book value of the fleet at the end of Year 3.

11-36 Mary, Medhi, Marcos, and Marguerite have purchased a small warehouse in St. Pete Beach. If they paid $745,000 for the unit in September, how much will their depreciation be in the 15th year? (*Note:* This asset is nonresidential real property.)
Ⓐ
Contributed by Paul R. McCright, University of South Florida

11-37 A minicomputer purchased in 2016 costing $12,000 has no salvage value after 4 years. What is the MACRS depreciation allowance for Year 3 and book value at the end of that year?

11-38 The MACRS depreciation percentages for 7-year personal property are given in Table 11–3. Make the necessary computations to determine if the percentages shown are correct.

11-39 The MACRS depreciation percentages for 10-year personal property are given in Table 11–3. Make the necessary computations to determine if the percentages shown are correct.

11-40 A group of investors has formed SandInn Corporation to purchase a small hotel. The price is $200,000 for the land and $800,000 for the hotel building. If the purchase takes place in June, compute the MACRS depreciation for the first three calendar years. Then assume the hotel is sold in June of the fourth year, and compute the MACRS depreciation in that year also.
Ⓐ

11-41 Mr. Donald Spade bought a backhoe in January. It cost $70,000 and is to be depreciated using MACRS. Donald's accountant pointed out that under a special tax rule which applies when the value of property placed in service in the last 3 months of the tax year exceeds 40% of the total value of property placed in service during the tax year, the backhoe and all property that year would be subject to the midquarter convention. This assumes that all property is placed in service at the midpoint of its quarter. Compute Donald's MACRS depreciation for the backhoe's first year.

11-42 For its fabricated metal products, the Able Corp. is paying $10,000 for special tools that have a 4-year useful life and no salvage value. Compute the depreciation charge for the *second* year by each of the following methods:
Ⓐ
(*a*) DDB
(*b*) Sum-of-years'-digits
(*c*) Modified accelerated cost recovery system

11-43 Global Fitters, an international clothing company, has purchased material handling equipment that cost $100,000 and a salvage value of $18,000 after 10 years. Determine the book value of the equipment after 3 years using:
Ⓔ
(*a*) MACRS depreciation
(*b*) Straight-line depreciation
(*c*) Sum-of-years'-digits (SOYD) depreciation
(*d*) 150% declining balance depreciation
(*e*) Global Fitters uses low-cost labor in emerging world economies to manufacture its products. List three potential ethical issues that are associated with the use of this labor pool.

11-44 A pump in an ethylene production plant costs $15,000. After 9 years, the salvage value is declared at $0.
Ⓐ
(*a*) Determine depreciation charge and book value for Year 9 using straight-line, sum-of-years'-digits, and 7-year MACRS depreciation.
(*b*) Find the PW of each depreciation schedule if the interest rate is 5%.

11-45 A used drill press costs $60,000, and delivery and installation charges add $5000. The salvage value after 10 years is $10,000. Compute the accumulated depreciation through Year 5 using
(*a*) 7-year MACRS depreciation
(*b*) Straight-line depreciation
(*c*) Sum-of-years'-digits (SOYD) depreciation
(*d*) Double declining balance depreciation

11-46 Metal Stampings, Inc., can purchase a new forging machine for $100,000. After 20 years of use the forge should have a salvage value of $15,000. What depreciation is allowed for this asset in Year 3 for
Ⓐ
(*a*) MACRS depreciation?
(*b*) straight-line depreciation?
(*c*) double declining balance depreciation?

11-47 Muddy Meadows Earthmoving can purchase a bulldozer for $150,000. After 7 years of use, the bulldozer should have a salvage value of $50,000. What depreciation is allowed for this asset in Year 4 for

(a) MACRS depreciation?
(b) straight-line depreciation?
(c) sum-of-years'-digits (SOYD) depreciation?
(d) 150% declining balance depreciation?

11-48 An asset costs $150,000 and has a salvage value
Ⓐ of $15,000 after 10 years. What is the depreciation charge for the 4th year, and what is the book value at the end of the 8th year with

(a) MACRS depreciation?
(b) straight-line depreciation?
(c) sum-of-years'-digits (SOYD) depreciation?
(d) double declining balance depreciation?

11-49 A precision five-axis CNC milling machine costs $200,000, and it will be scrapped after 10 years. Compute the book value and depreciation for the first 3 years using

(a) MACRS depreciation
(b) straight-line depreciation
(c) sum-of-years'-digits (SOYD) depreciation
(d) double declining balance depreciation
(e) 150% declining balance depreciation

11-50 Use MACRS GDS depreciation for each of the
Ⓐ assets, 1–3, to calculate the following items, (a)–(c).
1. A light general-purpose truck used by a delivery business, cost = $17,000

2. Production equipment used by a Detroit automaker to produce vehicles, cost = $30,000

3. Cement production facilities used by a construction firm, cost = $130,000

(a) The MACRS GDS property class
(b) The depreciation deduction for Year 3
(c) The book value of the asset after 6 years

11-51 Sarah Jarala recently purchased an asset that she intends to use for business purposes in her small Iceland Tourism business. The asset has MACRS class life of 5 years. Sarah purchased the asset for $85,000 and uses a *salvage value for tax purposes of $15,000* (when applicable). Also, the ADR life of the asset is 8 years.

(a) Using MACRS depreciation, what is the book value after 2 years?
(b) Using MACRS depreciation, what is the book value after 4 years?

(c) Using MACRS depreciation, what is the sum of the depreciation charges through the 5th year?
(d) Using MACRS depreciation, what is the depreciation for the 6th year?
(e) Using MACRS depreciation, what is the book value after 8 years?
(f) Using straight-line depreciation (with no half-year convention), what is the book value after the 3rd year?
(g) Using straight-line depreciation (with no half-year convention), what is the book value after the 8th year?

11-52 A company is considering buying a new piece of machinery. A 10% interest rate will be used in the computations. Two models of the machine are available.

	Machine I	Machine II
Initial cost	$80,000	$100,000
End-of-useful-life salvage value, S	20,000	25,000
Annual operating cost	18,000	15,000 first 10 years
		20,000 thereafter
Useful life, in years	20	25

(a) Determine which machine should be purchased, based on equivalent uniform annual cost.
(b) What is the capitalized cost of Machine I?
(c) Machine I is purchased and a fund is set up to replace Machine I at the end of 20 years. Compute the required uniform annual deposit.
(d) Machine I will produce an annual saving of material of $28,000. What is the rate of return if Machine I is installed?
(e) What will be the book value of Machine I after 2 years, based on sum-of-years'-digits depreciation?
(f) What will be the book value of Machine II after 3 years, based on double declining balance depreciation?

(g) Assuming that Machine II is in the 7-year property class, what would be the MACRS depreciation in the third year?

Gain/Loss on Disposal

11-53 Equipment costing $20,000 that is a MACRS 5-year property is disposed of during the second year for $15,000. Calculate any depreciation recapture, ordinary losses, or capital gains associated with disposal of the equipment.

11-54 An asset with an 8-year ADR class life costs $50,000
(A) and was purchased on January 1, 2001. Calculate any depreciation recapture, ordinary losses, or capital gains associated with selling the equipment on December 31, 2003, for $15,000, $25,000, and $60,000. Consider two cases of depreciation for the problem: if MACRS GDS is used, and if straight-line depreciation over the ADR class life is used with a $10,000 salvage value.

11-55 A purchased machine cost $320,000 with delivery and installation charges amounting to $30,000. The declared salvage value was $50,000. Early in Year 3, the company changed its product mix and found that it no longer needed the machine. One of its competitors agreed to buy the machine for $180,000. Determine the loss, gain, or recapture of MACRS depreciation on the sale. The ADR is 12 years for this machine.

11-56 O'Leary Engineering Corp. has been depreciating a
(A) $50,000 machine for the last 3 years. The asset was just sold for 60% of its first cost. What is the size of the recaptured depreciation or loss at disposal using the following depreciation methods?

(a) Sum-of-years'-digits with $N = 8$ and $S = 2000$
(b) Straight-line depreciation with $N = 8$ and $S = 2000$
(c) MACRS GDS depreciation, classified as a 7-year property

11-57 A $150,000 asset has been depreciated with the straight-line method over an 10-year life. The estimated salvage value was $30,000. At the end of the 7th year the asset was sold for $38,000. From a tax perspective, what is happening at the time of disposal, and what is the dollar amount?

11-58 A numerically controlled milling machine was pur-
(A) chased for $95,000 four years ago. The estimated salvage value was $15,000 after 15 years. What is

the machine's book value after 5 years of depreciation? If the machine is sold for $20,000 early in Year 7, how much gain on sale or recaptured depreciation is there? Assume

(a) 7-year MACRS depreciation
(b) Straight-line depreciation
(c) Sum-of-years'-digits (SOYD) depreciation
(d) 150% declining balance depreciation

11-59 A computer costs $3500 and its salvage value in 5
(G) years is negligible. What is the book value after 3 years? If the machine is sold for $1500 in Year 5, how much gain or recaptured depreciation is there? Assume

(a) MACRS depreciation
(b) Straight-line depreciation
(c) Sum-of-years'-digits (SOYD) depreciation
(d) Double declining balance depreciation
(e) There are two primary important considerations when disposing of old computers—one environmental and one personal. What are they, and how can you lessen the effect of each?

11-60 A belt-conveyor purchased for $140,000 has ship-
(A) ping and installation costs of $20,000. It was expected to last 6 years, when it would be sold for $25,000 after paying $5000 for dismantling. Instead, it lasted 4 years, and several workers were permitted to take it apart on their own time for reassembly at a private technical school. How much gain, loss, or recaptured depreciation is there? Assume

(a) 7-year MACRS depreciation
(b) Straight-line depreciation
(c) Sum-of-years'-digits (SOYD) depreciation
(d) 150% declining balance depreciation

Depletion and Unit-of-Production

11-61 A piece of machinery has a cost basis of $50,000. Its salvage value will be $5000 after 9000 hours of operation. With units-of-production depreciation, what is the allowable depreciation rate per hour? What is the book value after 4000 hours of operation?

11-62 When a major highway was to be constructed nearby,
(A) a farmer realized that a dry streambed running through his property might be a valuable source of sand and gravel. He shipped samples to a testing laboratory and learned that the material met the requirements for certain low-grade fill material. The farmer contacted the highway construction contractor, who

offered 65¢ per cubic meter for 45,000 cubic meters of sand and gravel. The contractor would build a haul road and would use his own equipment. All activity would take place during a single summer.

The farmer hired an engineering student for $2500 to count the truckloads of material hauled away. The farmer estimated that 2 acres of streambed had been stripped of the sand and gravel. The 640-acre farm had cost him $300 per acre, and the farmer felt the property had not changed in value. He knew that there had been no use for the sand and gravel prior to the construction of the highway, and he could foresee no future use for any of the remaining 50,000 cubic meters of sand and gravel.

Determine the farmer's depletion allowance.

11-63 During the construction of a highway bypass, earth-moving equipment costing $40,000 was purchased for use in transporting fill from the borrow pit. At the end of the 4-year project, the equipment will be sold for $20,000. The schedule for moving fill calls for a total of 100,000 cubic feet during the project. In the first year, 40% of the total fill is required; in the second year, 30%; in the third year, 25%; and in the final year, the remaining 5%. Determine the units-of-production depreciation schedule for the equipment.

11-64 Mr. H. Salt purchased an $\frac{1}{8}$ interest in a producing oil well for $45,000. Recoverable oil reserves for the well were estimated at that time at 15,000 barrels, $\frac{1}{8}$ of which represented Mr. Salt's share of the reserves. During the subsequent year, Mr. Salt received $12,000 as his $\frac{1}{8}$ share of the gross income from the sale of 1000 barrels of oil. From this amount, he had to pay $3000 as his share of the expense of producing the oil. Compute Mr. Salt's depletion allowance for the year.

11-65 The Piney Copper Company purchased an ore-bearing tract of land for $10.0M. The geologist for Piney estimated the recoverable copper reserves to be 500,000 tons. During the first year, 50,000 tons were mined and 40,000 tons were sold for $5.0M. Expenses (not including depletion allowances) were $3.0M.

(a) What are the percentage depletion and the cost depletion allowances?

(b) Copper mining and production are subject to high levels of regulation, including control of air and water quality as well as materials handling and disposal practices. What are the primary environmental risks, and how has regulation

lessened those risks? What are the most significant health risks, and how have these been lessened?

11-66 American Pulp Corp. (APC) has entered into a contract to harvest timber for $450,000. The total estimated available harvest is 150 million board-feet.

(a) What is the depletion allowance for Years 1 to 3, if 42, 45, and 35 million board-feet are harvested by APC in those years?

(b) After 3 years, the total available harvest for the original tract was reestimated at 180 million board-feet. Compute the depletion allowances for Years 4 and beyond.

11-67 An automated assembly line is purchased for $2,500,000. The company has decided to use units-of-production depreciation. At the end of 8 years, the line will be scrapped for an estimated $500,000. Using the following information, determine the depreciation schedule for the assembly line.

Year	Production Level
1	5,000 units
2	10,000 units
3	15,000 units
4	15,000 units
5	20,000 units
6	20,000 units
7	10,000 units
8	5,000 units

11-68 Western Carolina Coal Co. expects to produce 125,000 tons of coal annually for 15 years. The deposit cost $3M to acquire; the annual gross revenues are expected to be $9.50 per ton, and the net revenues are expected to be $4.25 per ton.

(a) Compute the annual depletion on a cost basis.

(b) Compute the annual depletion on a percentage basis.

(c) What are some of the primary environmental impacts of both surface and underground mining? What health risks do underground miners face? What is being done to make mining safer and more environmentally friendly?

11-69 Eastern Gravel expects to produce 60,000 tons of gravel annually for 5 years. The deposit cost $150K to acquire; the annual gross revenues are expected to be $9 per ton, and the net revenues are expected to be $4 per ton.

(*a*) Compute the annual depletion on a cost basis.

(*b*) Compute the annual depletion on a percentage basis.

11-70 A 2500-acre tract of timber is purchased by the Houser Paper Company for $1,200,000. The acquisitions department at Houser estimates-the land will be worth $275 per acre once the timber is cleared. The materials department estimates that a total of 5 million board-feet of timber is available from the tract. The harvest schedule calls for equal amounts of the timber to be harvested each year for 5 years. Determine the depletion allowance for each year.

11-71 Mining recently began on a new deposit of 10 million metric tons of ore (2% nickel and 4% copper). Annual production of 350,000 metric tons begins this year. The market price of nickel is $3.75 per pound and $0.65 for copper. Mining operation costs are expected to be $0.50 per pound. XYZ Mining Company paid $600 million for the deposits.

What is the maximum depletion allowance each year for the mine?

11-72 The Red River oil field will become less productive each year. Rojas Brothers is a small company that owns Red River, which is eligible for percentage depletion. Red River costs $2.5M to acquire, and it will be produced over 15 years. Initial production costs are $4 per barrel, and the wellhead value is $10 per barrel. The first year's production is 90,000 barrels, which will decrease by 6000 barrels per year.

(*a*) Compute the annual depletion (each year may be cost-based or percentage-based).

(*b*) What is the PW at $i = 12\%$ of the depletion schedule?

CASE

The following case from *Cases in Engineering Economy* (www.oup.com/us/newnan) is suggested as matched with this chapter.

CASE **26 Molehill & Mountain Movers**
    Compare depreciation methods with option for inflation.

INCOME TAXES FOR CORPORATIONS

On with the Wind

For decades, environmental activists and community leaders have bemoaned American dependence on foreign oil. One solution is to rely more on renewable sources of energy, such as wind and solar power. However, while the sun and the wind may be "free," wind and solar farms require a substantial investment.

In the late 1980s, wind-generated power cost roughly twice as much to produce as energy from conventional sources. The reasons included wind turbines are typically located far from cities making grid connections expensive, wind turbines typically spin only about 30% of the time, and manufacturing costs were high. The Alta Wind Energy Center in Kern County, CA, is the largest wind farm in the U.S. (1548 MW as of 2016, with a planned size of 3000 MW). By 2016, wind-generated power contributed 31% of Iowa's power, at least 15% of the total electricity supply in 7 more states, and at least 5% of total supply in 13 more states. Overall, wind-generated power supplied nearly 5% of total electricity supply in the U.S. The American Wind Energy Association (AWEA) reports that many modern, state-of-the-art wind plants can now produce power for less than 5¢ per kilowatt-hour, making them competitive with conventional sources.

These lower costs have been driven by advances in wind turbine technology and the increased scale of a sector increasing by 25% annually. But it was helped significantly by a federal production tax credit contained in the Energy Policy Act of 1992. This statute allowed utilities and other electricity suppliers a "production tax credit" of 1.5¢ per kilowatt-hour (by 2012 adjusted to 2.2¢ to account for the effects of inflation).

But what the government giveth, the government can take away. When the production tax credit (PTC) briefly expired at the end of 2001, an estimated $3 billion worth of wind projects were suspended, and hundreds of workers were laid off. Fortunately for the industry, the credit has been subsequently extended several times. The extension is

currently scheduled to continue at 70% of its present value in 2017, 60% in 2018, and 40% in 2019. ■ ■ ■

Revised by Eva Andrijcic, Rose-Hulman Institute of Technology

QUESTIONS TO CONSIDER

1. Opponents of the wind energy tax credit argue that the market should determine the future of alternative energy sources. Supporters argue that government has historically supported the oil and coal industries through subsidies and that alternative energy reduces what society pays for health and the environment. Which side has the stronger argument in your view? Why?

2. Clearly wind energy has both costs and benefits as an alternate energy technology, and government may have some role in enabling its development and promoting investments in it. Develop a table of the costs and benefits of wind energy technology. Who would be the winners and losers if our nation dramatically increased the fraction of our energy from wind sources? What ethical issues arise? Can we solve any of the ethical issues that you've identified?

3. Developing a wind power project takes many years and requires the commitment of large sums of investment capital before the project begins to return a profit. What is the effect on investment when the wind power production tax credit is allowed to expire or is extended for periods of only a few years?

4. Has the production tax credit for wind energy changed yet again?

After Completing This Chapter...

The student should be able to:

- Calculate *taxes due* or *taxes owed* for corporations.
- Understand the incremental nature of corporate tax rates used for calculating taxes on income.
- Calculate a combined income tax rate for state and federal income taxes and select an appropriate tax rate for engineering economic analyses.
- Utilize an *after-tax tax table* to find the after-tax cash flows for a prospective investment project.
- Calculate after-tax measures of merit, such as present worth, annual worth, payback period, internal rate of return, and benefit–cost ratio, from developed after-tax cash flows.
- Evaluate investment alternatives on an after-tax basis including asset disposal.
- Use spreadsheets for solving after-tax economic analysis problems.
- Calculate personal incomes taxes and make choices about student loans, retirement accounts, insurance, and personal budgeting. (Appendix 12A)

Key Words

after-tax cash flow table	combined incremental tax rate	Section 179 deduction
bonus depreciation	expensed	taxable income
capital expenditures	incremental tax rates	
capital gain	investment tax credit	

As Benjamin Franklin said, two things are inevitable: death and taxes. There are many types of taxes and structures for taxation in the U.S., including sales taxes, gasoline taxes, property taxes, and state and federal income taxes. In this chapter we will concentrate our attention on federal income taxes for corporations. Income taxes are part of most real problems and often have a substantial impact that must be considered.

First, we must understand the way in which taxes are imposed. Chapter 11 concerning depreciation is an integral part of this analysis, so the principles covered there must be well understood. Then, having understood the mechanism, we will see how federal income taxes affect our economic analysis.

A PARTNER IN THE BUSINESS

Probably the most straightforward way to understand the role of federal income taxes is to consider the U.S. government as a partner in every business activity. As a partner, the government shares in the profits from every successful venture. In a somewhat more complex way, the government shares in the losses of unprofitable ventures too. The tax laws are complex, and it is not our purpose to fully explain them. Instead, we will examine the fundamental concepts of federal income tax for corporations—we emphasize at the start that there are exceptions and variations to almost every statement we shall make!

CALCULATION OF TAXABLE INCOME

At the mention of income taxes, one can visualize dozens of elaborate and complex calculations. There is some truth to that vision, for there can be complexities in computing income taxes. Yet incomes taxes are just another type of disbursement or cash outflow that affects profitability. Our economic analysis calculations in prior chapters have dealt with all sorts of disbursements: operating costs, maintenance, labor and materials, and so forth. Now we simply add one more prospective disbursement to the list—income taxes.

Classification of Business Expenditures

When an individual or a firm operates a business, there are three distinct types of business expenditure:

1. For depreciable assets.
2. For nondepreciable assets.
3. All other business expenditures.

Expenditures for Depreciable Assets: When facilities or productive equipment with useful lives in excess of one year are acquired, the firm will normally recover the investment through depreciation charges. Chapter 11 detailed how to allocate an asset's cost over its useful life.

Expenditures for Nondepreciable Assets: Land is considered to be a nondepreciable asset, for there is no finite life associated with it. Other nondepreciable assets are properties *not* used either in a trade, in a business, or for the production of income. The final category of nondepreciable assets comprises those subject to *depletion,* rather than *depreciation.* Since business firms generally acquire assets for use in the business, their only nondepreciable assets normally are land and assets subject to depletion.

All Other Business Expenditures: This category is probably the largest of all, for it includes all the ordinary and necessary expenditures of operating a business. Labor costs, materials, all direct and indirect costs, and facilities and productive equipment with a useful life of one year or less are part of routine expenditures. They are charged as a business expense—*expensed*—when they occur.

Business expenditures in the first two categories—that is, for either depreciable or nondepreciable assets—are called **capital expenditures.** In the accounting records of the firm, they are **capitalized**; all ordinary and necessary expenditures in the third category are **expensed.**

Taxable Income of Business Firms

The starting point in computing a firm's taxable income is *gross income.* All ordinary and necessary expenses to conduct the business—*except* capital expenditures—are deducted from gross income. Capital expenditures may *not* be deducted from gross income. Except for land, business capital expenditures are allowed on a period-by-period basis through depreciation or depletion charges.

For business firms, taxable income is computed as follows:

$$\textbf{Taxable income} = \text{Gross income}$$
$$- \text{All expenditures except capital expenditures}$$
$$- \text{Depreciation and depletion charges} \qquad (12\text{-}1)$$

Because of the treatment of capital expenditures for tax purposes, the taxable income of a firm may be quite different from the actual cash flows.

EXAMPLE 12–1

During a 3-year period, a firm had the following cash flows (in millions of dollars):

	Year 1	Year 2	Year 3
Gross income from sales	$200	$200	$200
Purchase of special tooling (useful life: 3 years)	−60	0	0
All other expenditures	−140	−140	−140
Cash flows for the year	$ 0	$ 60	$ 60

Compute the taxable income for each of the 3 years.

The cash flows for each year would suggest that Year 1 was a poor one, while Years 2 and 3 were very profitable. A closer look reveals that the firm's cash flows were adversely affected in Year 1 by the purchase of special tooling. Since the special tooling has a 3-year useful life, it is a capital expenditure with its cost allocated over the useful life. If we assume that straight-line depreciation applies with no salvage value, we use Equation 11-2 to find the annual charge:

$$\text{Annual depreciation charge} = \frac{B - S}{N} = \frac{60 - 0}{3} = \$20 \text{ million}$$

Applying Equation 12-1, we write

$$\text{Taxable income} = 200 - 140 - 20 = \$40 \text{ million}$$

In each of the 3 years, the taxable income is $40 million.

An examination of the cash flows and the taxable income in Example 12–1 indicates that taxable income is a better indicator of the firm's annual performance.

INCOME TAX RATES

Corporate Tax Rates

Income tax rates in the U.S. are based on an incremental scale—for different levels of taxable income different tax rates apply. These rates applied to different levels of taxable income are called *marginal* or **incremental tax rates**. For corporations in the U.S. the top incremental tax rate has changed somewhat over time—from the 1950's to mid-1980's that rate was near 50% and since then has been under 40%. Figure 12–1 and Table 12–1

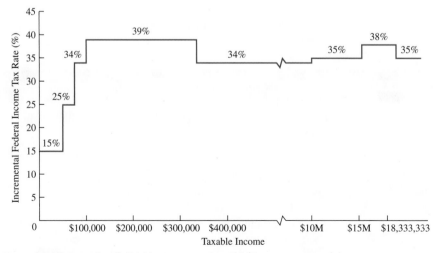

FIGURE 12–1 Corporation federal income tax rates (2016).

TABLE 12–1 Corporate Income Tax Rates

Taxable Income	Tax Rate	Corporate Income Tax
Not over $50,000	15%	15% over $0
$50,000–75,000	25%	$7,500 + 25% over $50,000
$75,000–100,000	34%	$13,750 + 34% over $75,000
$100,000–335,000	39%*	$22,250 + 39% over $100,000
$335,000–10 million	34%	$113,900 + 34% over $335,000
$10 million–15 million	35%	$3,400,000 + 35% over $10 million
$15 million–18,333,333	38%	$5,150,000 + 38% over $15 million
$\geq$$18,333,333	35%†	Flat rate at 35%

*The extra 5% from $100,000 to $335,000 was chosen so that firms in the $335,000 to $10 million bracket pay a flat 34% tax rate. $[(0.39 - 0.34)](335,000 - 100,000) = (0.34 - 0.15)(50,000) + (0.34 - 0.25)(75,000 - 50,000)]$ so tax = 0.34 (tax income) in $335,000 to $10 million bracket.
†*Personal service corporations* are subject to a 35% flat tax.

demonstrate the current corporate tax rates. The data in Figure 12–1 can be presented in schedule form (Table 12–1).

EXAMPLE 12–2

The French Chemical Corporation was formed to produce household bleach. The firm bought land for $220,000, had a $900,000 factory building erected, and installed $650,000 worth of chemical and packaging equipment. The plant was completed and operations begun on April 1. The gross income for the calendar year was $450,000. Supplies and all operating expenses, excluding the capital expenditures, were $100,000. The firm will use modified accelerated cost recovery system (MACRS) depreciation.

(a) What is the first-year depreciation charge?
(b) What is the first-year taxable income?
(c) How much will the corporation pay in federal income taxes for the year?

SOLUTION TO PART a

MACRS depreciation: Chemical equipment is personal property. From Table 11–2, it is in the "7-year, all other property" class.

$$\text{First-year depreciation(equipment)} = \$650,000 \times 14.29\% = \$92,885$$

The building is in the 39-year real property class. Since it was placed in service on April 1, the first-year depreciation is

$$\text{First-year depreciation(building)} = \$900,000 \times 1.819\% = \$16,371 \quad \text{(see Table 11–4)}$$

The land is a nondepreciable asset. Thus we have

$$\text{Total first-year MACRS depreciation} = \$92,885 + \$16,371 = \$109,256$$

SOLUTION TO PART b

Taxable income = Gross income

— All expenditures except capital expenditures

— Depreciation and depletion charges

$$= \$450{,}000 - 100{,}000 - 109{,}256 = \$240{,}744$$

SOLUTION TO PART c

Federal income tax $= \$22{,}250 + 39\%(240{,}744 - 100{,}000)$

$$= \$77{,}140$$

Combined Federal and State Income Taxes

In addition to federal income taxes, most corporations pay state income taxes. It would be convenient if we could derive a single tax rate to represent both the state and federal incremental tax rates. In the computation of taxable income for federal taxes, the amount of state taxes paid is one of the allowable itemized deductions. Federal income taxes are not, however, generally deductible in the computation of state taxable income. Therefore, the state income tax is applied to a *larger* taxable income than is the federal income tax rate. As a result, the combined incremental tax rate will not be the sum of two tax rates.

For an increment of income (ΔIncome) and tax rate on incremental income (ΔTax rate):

State income taxes $= (\Delta$State tax rate$)(\Delta$Income$)$

Federal taxable income $= (\Delta$Income$)(1 - \Delta$State tax rate$)$

Federal income taxes $= (\Delta$Federal tax rate$)(\Delta$Income$) \times (1 - \Delta$State tax rate$)$

The total of state and federal income taxes is

$$[\Delta\text{State tax rate} + (\Delta\text{Federal tax rate})(1 - \Delta\text{State tax rate})](\Delta\text{Income})$$

The term in the brackets gives the combined incremental tax rate.

Combined incremental tax rate

$$= \Delta\text{State tax rate} + (\Delta\text{Federal tax rate})(1 - \Delta\text{State tax rate}) \quad \text{(12-2)}$$

EXAMPLE 12–3

A small start-up engineering firm has taxable income that places it in the 34% federal income tax bracket and at the 12% state incremental tax rate. The firm can earn an extra $25,000 on a small job at the end of the tax year. What will the combined federal and state income tax rate be on the additional income?

SOLUTION

Use Equation 12-2 to find the combined incremental tax rate: $0.12 + 0.34(1 - 0.12) = 41.9\%$.

Selecting an Income Tax Rate for Economy Studies

Since income tax rates vary with the level of taxable income one must decide which tax rate to use in a particular situation. The simple answer is that the tax rate to use is the incremental tax rate that applies to the change in taxable income projected in the economic analysis. If a corporation has a taxable income of $8 million and can increase its income by $2 million, what tax rate should be used for this additional income? From Table 12–1, we see that the entire $2 million falls within the 34% tax bracket—thus 34% would be used on this incremental income.

Now suppose the corporation has a great opportunity to double its income of $8 million. In this case what tax rate should be used for this additional income? From Table 12–1 we see that the additional $8 million (taking the firm from a taxable income of $8 million to $16 million) would be subject to three different incremental rates. The first $2 million would be taxed at 34%, the next $5 million taxed at 35%, and the last $1 million would be taxed at 38%. For larger corporations, the federal incremental tax rate is 35%. In addition, there may be up to a 12 to 15% state tax.

ECONOMIC ANALYSIS TAKING INCOME TAXES INTO ACCOUNT

An important step in economic analysis has been to resolve the consequences of alternatives into a cash flow. Because income taxes have been ignored, the result has been a *before-tax cash flow*. This before-tax cash flow is an essential component in economic analysis that also considers the consequences of income tax. The principal elements in an *after-tax analysis* are as follows:

- Before-tax cash flow
- Depreciation
- Taxable income (Before-tax cash flow − Depreciation)
- Income taxes (Taxable income × Incremental tax rate)
- After-tax cash flow (Before-tax cash flow − Income taxes)

These elements are usually arranged to form an **after-tax cash flow table**. This is illustrated by Example 12–4.

EXAMPLE 12–4

A medium-sized profitable corporation may buy a $15,000 used pickup truck for use by the shipping and receiving department. During the truck's 5-year useful life, it is estimated the firm will save $4000 per year after all the costs of owning and operating the truck have been paid. Truck salvage value is estimated at $4500. So that this can be solved using an equation or annuity function rather than a full spreadsheet, assume that straight-line depreciation is used.

(a) What is the before-tax rate of return?

(b) What is the after-tax rate of return on this capital expenditure? Assume straight-line depreciation.

SOLUTION TO PART a

For a before-tax rate of return, we must first compute the before-tax cash flow.

Year	Before-Tax Cash Flow
0	−$15,000
1	4,000
2	4,000
3	4,000
4	4,000
5	{ 4,000 / 4,500 }

Solve for the before-tax rate of return, IRR_{BT}:

$$15{,}000 = 4000(P/A, i, 5) + 4500(P/F, i, 5)$$

Try $i = 15\%$:

$$15{,}000 \doteq 4000(3.352) + 4500(0.4972) \doteq 13{,}408 + 2237 = 15{,}645$$

Since i is slightly low, try $i = 18\%$:

$$15{,}000 \doteq 4000(3.127) + 4500(0.4371) \doteq 12{,}508 + 1967 = 14{,}475$$

$$IRR_{BT} = 15\% + 3\% \left(\frac{15{,}645 - 15{,}000}{15{,}645 - 14{,}475} \right) = 15\% + 3\%(0.55) = 16.7\%$$

5-BUTTON SOLUTION

	A	B	C	D	E	F	G	H
1	Problem	i	m	PMT	PV	FV	Solve for	Answer
2	Exp. 12-4a		5	4000	−15,000	4500	i	16.61%

SOLUTION TO PART b

For an after-tax rate of return, we must set up an after-tax cash flow table (Table 12–2). The starting point is the before-tax cash flow. Then we will need the depreciation schedule for the truck:

$$\text{Straight-line depreciation} = \frac{B - S}{N} = \frac{15{,}000 - 4500}{5} = \$2100 \text{ per year}$$

Taxable income is the before-tax cash flow *minus* depreciation. For this medium-sized profitable corporation, let's assume an incremental federal income tax rate of 34%. Therefore income taxes are 34% of taxable income. Finally, the after-tax cash flow equals the before-tax cash flow *minus* income taxes. These data are used to compute Table 12–2.

TABLE 12–2 After-Tax Cash Flow Table for Example 12–4*

Year	(a) Before-Tax Cash Flow	(b) Straight-Line Depreciation	(c) Δ(Taxable Income) (a)−(b)	(d) 34% Income Taxes −0.34 (c)	(e) After-Tax Cash Flow (a)+(d)†
0	−$15,000				−$15,000
1	4,000	$2100	$1900	−$646	3,354
2	4,000	2100	1900	−646	3,354
3	4,000	2100	1900	−646	3,354
4	4,000	2100	1900	−646	3,354
5	{ 4,000 / 4,500 }	2100	1900	−646	{ 3,354 / 4,500 }

*Sign convention for income taxes: a minus (−) represents a disbursement of money to pay income taxes; a plus (+) represents the receipt of money by a decrease in the tax liability.
†The after-tax cash flow is the before-tax cash flow minus income taxes. Based on the income tax sign convention, this is accomplished by *adding* columns (**a**) and (**d**).

5-BUTTON SOLUTION

We now use the after-tax cash flows to find the after-tax IRR.

	A	B	C	D	E	F	G	H
1	Problem	i	m	PMT	PV	FV	Solve for	Answer
2	Exp. 12-4b		5	3354	−15,000	4500	i	11.15%

The calculations required to compute the after-tax rate of return in Example 12–4 were certainly more elaborate than those for the before-tax rate of return. It must be emphasized, however, that often the after-tax rate of return is the key value, since income taxes are a major disbursement that cannot be ignored.

EXAMPLE 12–5

An analysis of a firm's sales activities indicates that a number of profitable sales are lost each year because the firm cannot deliver some of its products quickly enough. By investing an additional $100,000 in inventory, it is believed that the firm will realize $10,000 more in before-tax profits in the first year. In the second year, before-tax extra profit will be $17,500. Profits for subsequent years are expected to continue to increase on a $7500-per-year gradient. The investment in the additional inventory may be recovered at the end of a 4-year analysis period simply by selling and not replenishing the inventory. Compute:

(a) The before-tax rate of return.
(b) The after-tax rate of return assuming an incremental tax rate of 39%.

SOLUTION

Inventory is not considered to be a depreciable asset. Therefore, the investment in additional inventory is not depreciated. The after-tax cash flow table for the problem is presented in Table 12–3. The sign convention described in the first footnote to Table 12–2 applies.

TABLE 12–3 After-Tax Cash Flow Table for Example 12–5

Year	(a) Before-Tax Cash Flow	(b) Depreciation	(c) △(Taxable Income) (a)−(b)	(d) 39% Income Taxes (c) x−(0.39)	(e) After-Tax Cash Flow (a) + (d)
0	−$20,000				−$20,000
1	1,000	—	$1000	−$390	610
2	1,500	—	1500	−585	915
3	2,000	—	2000	−780	1,220
4	{ 2,500 20,000	—	2500	−975	{ 1,525 20,000

SOLUTION TO PART a

Use the following equation to calculate the before-tax rate of return:

$$20{,}000 = 1000(P/A, i, 4) + 500(P/G, i, 4) + 20{,}000(P/F, i, 4)$$

Try $i = 8\%$:

$$20{,}000 \neq 1000(3.312) + 500(4.650) + 20{,}000(0.7350)$$

$$\neq 3312 + 2325 + 14{,}700 = 20{,}337$$

Since i is too low, try $i = 10\%$:

$$20{,}000 \neq 1000(3.170) + 500(4.378) + 20{,}000(0.6830)$$

$$\neq 3170 + 2189 + 13{,}660 = 19{,}019$$

Now we interpolate between these values.

$$\text{Before-tax rate of return } = 8\% + 2\%\left(\frac{20{,}337 - 20{,}000}{20{,}337 - 19{,}019}\right) = 8.5\%$$

SOLUTION TO PART b

For a before-tax cash flow gradient of $500, the resulting after-tax cash flow gradient is $(1 - 0.39)(500) = \$305$.

$$20{,}000 = 610(P/A, i, 4) + 305(P/G, i, 4) + 20{,}000(P/F, i, 4)$$

Try $i = 5\%$:

$$20,000 \doteq 610(3.546) + 305(5.103) + 20,000(0.8227) \doteq 20,173$$

Since i is too low, try $i = 6\%$:

$$20,000 \doteq 610(3.465) + 300(4.4945) + 20,000(0.7921) \doteq 19,304$$

$$\text{After-tax rate of return} = 5\% + 1\% \left(\frac{20,173 - 20,000}{20,173 - 19,304} \right) = 5.2\%$$

Section 179 Deduction

Not all assets held for more than one year and used in business are depreciated over time. There is an exception as part of the U.S. tax law called a **Section 179 deduction** (named after the section of code) meant primarily to spur capital investment by small businesses. For the 2016 tax year, Section 179 allows businesses to immediately deduct (*expense*) up to $500,000 of the cost of qualifying assets in the year of purchase. Above $2 million in purchases, businesses must reduce this limit dollar-for-dollar. Thus, qualifying Section 179 deductions reduce taxable income and thus taxes paid for the year of purchase.

Investment Tax Credit

Another tax policy technique used historically in the U.S. to promote capital investments is the **investment tax credit (ITC)**. Under the ITC, businesses are able to deduct a percentage of the purchase price of equipment as a *tax credit*. Depending on the provisions of the ITC law in effect, this credit might be subtracted from the asset's basis for depreciation, or perhaps not. The Tax Reform Act of 1986 eliminated ITC for most assets, although credits are allowed in some specialized cases such as historic building preservation and in the development of alternate energy sources. It is likely, however, that the general ITC will appear at some future time.

Bonus Depreciation

Another policy tool used to promote capital investment and/or provide economic relief is **bonus depreciation**. For qualified assets this is additional depreciation in year 1. When available, bonus depreciation has varied from 30% to 100%, with 50% being the most common. Bonus depreciation is available for assets with a MACRS class life of 20 years or less and for some assets with longer lives. As shown in Example 12–6, all forms of depreciation may be available.

EXAMPLE 12–6

An organic foods company engaged in the farm-to-table market purchased $800,000 of new 7-year MACRS equipment. This equipment and firm meet the requirements for full Section 179 and 50% bonus depreciation deductions. What is the total deduction from taxable income for the first year? Assume that a 10% investment tax credit applies to the $800,000 investment. What are the equipment's time-0 and year-1 after-tax cash flows for the capital expenditure if the firm has a 34% tax rate?

SOLUTION

Deductions from taxable income in year 1:

- Section 179: The firm claims the $500,000 limit, which reduces the equipment's cost basis to $300,000 (= $800,000 − $500,000).
- Bonus depreciation: At a 50% rate, an additional $150,000 (= $300,000 × 0.50) is claimed, which lowers the equipment's cost basis to $150,000 (= $300,000 − $150,000).
- Standard depreciation: For 7-year MACRS equipment, the first-year deduction is $21,435 (= $150,000 × 0.1429). The remaining years of MACRS depreciation will use a cost basis of $150,000.

The deductions total $671,435 (= $500,000 + $150,000 + $21,435) or 84% of the investment cost in the first year. This will reduce income taxes by $228,288 (= 0.34 × $671,435). The 10% investment tax credit reduces the company's taxes by another $80,000 (= $800,000 × 0.10). The after-tax cash flow at time 0 is −$800,000 and the tax reduction in year 1 is $308,288 (= $228,288 + $80,000).

THE AFTER-TAX RATE OF RETURN

Estimating a Project's After-Tax IRR

There is no shortcut method for computing the after-tax rate of return from the before-tax rate of return. One possible exception to this statement is in the situation of nondepreciable assets. In this special case, we have

After-tax rate of return = (1 − Incremental tax rate) × (Before-tax rate of return)

For Example 12–5, we could estimate the after-tax rate of return from the before-tax rate of return as follows:

$$\text{After-tax rate of return} = (1 - 0.39)(8.5\%) = 5.2\%$$

This value agrees with the value computed in Example 12–5(b).

This relationship may be helpful for selecting a trial after-tax rate of return when the before-tax rate of return is known. It must be emphasized, however, this relationship is only a rough approximation in almost all situations.

Firm's After-tax Minimum Attractive Rate of Return

While a project's IRR can sometimes be estimated by applying the incremental tax rate to the before-tax rate of return, the firm's after-tax MARR cannot be estimated this way. The reason is that virtually all firms are financed by a mix of debt and equity, and only the interest payments on debt are tax deductible. The dividends paid on stock and other returns to equity holders are not tax deductible.

This will be demonstrated in Example 15–1 where the pre-tax MARR is 9.8%, and the after-tax MARR is 8.52%. Even though the tax rate is 40%, the tax shield only applies to the 40% of the capital structure that is debt. If the tax shield is incorrectly applied to the pre-tax MARR, the after-tax MARR is incorrectly estimated as 5.88% (= 9.8% × (1 − 0.40)).

CAPITAL GAINS AND LOSSES FOR NONDEPRECIATED ASSETS

When a nondepreciated capital asset is sold or exchanged, appropriate entries are made in the firm's accounting records. If the selling price of the capital asset exceeds the original cost basis, the excess is called a **capital gain.** If the selling price is less than the original cost basis, the difference is a **capital loss.** Examples of nondepreciated assets include stocks, land, art, and collectibles.

$$\text{Capital} \begin{Bmatrix} \text{Gain} \\ \text{Loss} \end{Bmatrix} = \text{Selling price} - \text{Original cost basis}$$

It is not uncommon for capital gains tax rates to be different from those for ordinary income—historically these have been set at a *lower* rate. This is in contrast to recaptured depreciation, which is taxed at the same rate as other (ordinary) income. The tax treatment of capital gains and losses for nondepreciated assets is shown in Table 12–4.

TABLE 12–4   Tax Treatment of Capital Gains and Losses for Corporations

Capital gain	Taxed as ordinary income.
Capital loss	Corporations may deduct capital losses only to the extent of capital gains. Any capital loss in the current year that exceeds capital gains can be carried back 3 years and forward for up to 5 years.

AFTER-TAX CASH FLOWS AND SPREADSHEETS

Realistic after-tax analyses require spreadsheets. Even if costs and revenues are the same every year, MACRS depreciation percentages are not. The steps for calculating an after-tax internal rate of return are illustrated in Example 12–7. Because some cash flows are taxed and some are not, the spreadsheet is easier to build if these two types are separated. Spreadsheet construction is easier, as well, if recaptured depreciation or other gain/loss on disposal or sale are tabulated separately.

Taxes are considered even if only the costs of a project are known. The firm that does an engineering project must generate profits—or go out of business. Even if a firm has an unprofitable year, the tax law includes carry-forward and -backward provisions to transfer deductions to profitable years. The depreciation and revenues in Example 12–7 result in a negative taxable income for Year 2. Thus the *positive cash flow* due to taxes that is shown in Year 2 of Example 12–7 really represents tax savings for the firm.

EXAMPLE 12–7

Return to the data of Example 12–4, where the used truck had a first cost of $15,000, a salvage value after 5 years of $4500, and savings of $4000 per year. Use MACRS depreciation and calculate the after-tax rate of return.

SOLUTION

Under MACRS, vehicles have a 5-year recovery period. Thus the MACRS depreciation can be calculated by using a VDB function (see "Spreadsheets and Depreciation" at the end of Chapter 11) or by lookup in Table 11–3.

The depreciation in Year 5 has been halved, since it is the year of disposal, and the 50% in year of disposal matches IRS regulations. However, as demonstrated in Example 11–10, claiming 0% or 100% of the final year's depreciation has the same taxable income and ATCF. In this case both D13 and E13 would be $864 smaller for 0% or $864 bigger for 100%.

	A	B	C	D	E	F	G	H
1	15,000	First Cost						
2	4000	Annual Benefit						
3	5	Recovery Period						
4	4500	Salvage Value						
5	0.34	Tax Rate						
6								
7	Year	Untaxed BTCF	Taxed BTCF	MACRS	Recaptured Depreciation	Tax Income	Tax	ATCF
8	0	−15,000						−15,000.0
9	1		4000	3000		1000	340.0	3660.0
10	2		4000	4800		−800	−272.0	4272.0
11	3		4000	2880		1120	380.8	3619.2
12	4		4000	1728		2272	772.5	3227.5
13	5	4500	4000	864	2772	5908	2008.7	6491.3
14		Cum. Depr.=		13,272			IRR=	11.96%
15								
16						=Taxed BTCF − MACRS + Recapt.		
17					=Salvage − BookValue			

FIGURE 12–2 Spreadsheet for after-tax IRR calculation.

Note from Figure 12–2 that using MACRS rather than straight-line depreciation increases the after-tax IRR from 11.15% to 11.96%. This is due solely to the faster write-off that is allowed under MACRS.

SUMMARY

Since income taxes are part of most problems, no realistic economic analysis can ignore their consequences. Income taxes make the U.S. government a partner in every business venture. Thus the government benefits from all profitable ventures and shares in the losses of unprofitable ventures.

For corporations, taxable income equals gross income *minus* all ordinary and necessary expenditures (except capital expenditures) and depreciation and depletion charges. The income tax computation (whether for an individual or a corporation) is relatively simple, with rates ranging from 10 to 39%. The proper rate to use in an economic analysis is the incremental tax rate applicable to the increment of taxable income being considered.

Most corporations pay state income taxes in addition to federal income taxes. Since state income taxes are an allowable deduction in computing federal taxable income, it follows that the taxable income for the federal computation is lower than the state taxable income.

Combined state and federal incremental tax rate

$$= \Delta \text{State tax rate} + (\Delta \text{Federal tax rate})(1 - \Delta \text{State tax rate})$$

To introduce the effect of income taxes into an economic analysis, the starting point is a before-tax cash flow. Then the depreciation schedule is deducted from appropriate parts of the before-tax cash flow to obtain taxable income. Income taxes are obtained by multiplying taxable income by the proper tax rate. Before-tax cash flow less income taxes equals the after-tax cash flow. This data is all captured in an after-tax cash flow table.

Governments can encourage corporate capital investment using tax policy, giving companies an incentive to make investments now versus deferring or not investing at all. Section 179 deductions, bonus depreciation, and investment tax credits are examples of such policies.

When dealing with nondepreciable assets, there is a nominal relationship between before-tax and after-tax rate of return. It is

$$\text{After-tax rate of return} = (1 - \Delta \text{Tax rate})(\text{Before-tax rate of return})$$

There is no simple relationship between before-tax and after-tax rate of return in the more usual case of investments involving depreciable assets.

PROBLEMS

These problems can be solved by hand, but most will be solved much more easily with a spreadsheet.

Corporate Taxes

12-1 From the perspective of taxes paid, describe the difference between capital asset costs and other costs-of-operation that are expensed.

12-2 (*a*) If permitted to choose between depreciating a cost over several years versus expensing it in a single year, which would you choose for your company? What factors might come into play in your recommendation?

(*b*) Firms can reduce the taxes they pay in the U.S. by setting internal transfer prices so the "profit" is earned in countries with low tax rates or by selling themselves to an international firm.

What are the ethical pros and cons of these practices?

12-3 Earth Powered Oil Company has purchased green engineering technology equipment for algae farms that turn sunlight into automotive biofuel. Two sets of equipment, each costing $200,000, are needed. One set is being depreciated using MACRS depreciation and the other is being depreciated with sum-of-the-years'-digits depreciation with zero salvage value. Assume the company pays taxes annually and the tax rate is constant. Does the firm save more on income taxes with either one of the equipment sets? If so, which one?

12-4 What is the (1) marginal and (2) average tax rate paid for a firm with taxable income of
 (a) $30,000?
 (b) $70,000?
 (c) $280,000?
 (d) $8 million?
 (e) $45 million?

12-5 What is the (1) marginal and (2) average tax rate paid for a firm with taxable income of
 (a) $15,000?
 (b) $60,000?
 (c) $150,000?
 (d) $2.6 million?
 (e) $20 million?

12-6 Compute income taxes owed for a firm with the following data:

Gross income from sales	$ 20 million
Accumulated business expenses	$ 5.5 million
Depreciation charges on assets	$ 3.5 million

12-7 A firm's annual revenues are $850,000. Its expenses for the year are $615,000, and it claims $135,000 in depreciation expenses. What does it pay in taxes, and what is its after-tax income?

12-8 A major industrialized state has a state corporate tax rate of 9.6% of taxable income. If a corporation has a state taxable income of $275,000, what is the total state and federal income tax it must pay? Also, compute its combined incremental state and federal income tax rate.

12-9 A company wants to set up a new office in a country where the corporate tax rate is: 15% of first $50,000 profits, 25% of next $25,000, 34% of next $25,000, and 39% of everything over $100,000. Executives estimate that they will have gross revenues of $500,000, total costs of $300,000, $30,000 in allowable tax deductions, and a one-time business start-up credit of $8000. What is taxable income for the first year, and how much should the company expect to pay in taxes?

12-10 To increase its market share, Sole Brother Inc. decided to borrow $50,000 for more advertising for its shoe retail line. The loan is to be paid in four equal annual payments with 15% interest. The loan is discounted 12 points. The first 6 "points" are an additional interest charge of 6% of the loan, deducted immediately from what is received from the $50,000 loan. Another 6 points or $3000 of additional interest is deducted as four $750 additional annual interest payments. What is the after-tax interest rate on this loan, if the firm's combined tax rate is 40%?

12-11 Concepcion Industries paid $308,000 in federal taxes last year. If business expenses and depreciation charges were $345,000, what were their gross sales for the year?

Straight-Line Depreciation

12-12 A salad oil bottling plant can either buy caps for the glass bottles at 5¢ each or install $500,000 worth of plastic molding equipment and manufacture the caps at the plant. The manufacturing engineer estimates the material, labor, and other costs would be 3¢ per cap.
 (a) If 11 million caps per year are needed and the molding equipment is installed, what is the payback period?
 (b) The plastic molding equipment would be depreciated by straight-line depreciation using a 6-year useful life and no salvage value. Assuming a combined 40% income tax rate, what is the after-tax payback period, and what is the after-tax rate of return?

12-13 A computer-controlled milling machine will cost Ajax Manufacturing $65,000 to purchase plus $4700 to install.
 (a) If the machine would have a salvage value of $6600 at EOY 20, how much could Ajax charge annually to depreciation of this equipment? Ajax uses straight-line depreciation.
 (b) What is the book value of the machine at EOY 3?

(c) Ajax Manufacturing earns a net profit before tax (also called a taxable income) of $28,800,000. How much tax would Ajax owe for this year?

Contributed by Paul R. McCright, University of South Florida

12-14 The effective combined tax rate in a firm is 40%. An outlay of $2 million for certain new assets is under consideration. Over the next 9 years, these assets will be responsible for annual receipts of $650,000 and annual disbursements (other than for income taxes) of $225,000. After this time, they will be used only for stand-by purposes with no future excess of receipts over disbursements.

(a) What is the prospective rate of return before income taxes?

(b) What is the prospective rate of return after taxes if straight-line depreciation can be used to write off these assets for tax purposes in 9 years?

(c) What is the prospective rate of return after taxes if it is assumed that these assets must be written off for tax purposes over the next 20 years, using straight-line depreciation?

12-15 A project using passive heating/cooling design concepts to reduce energy costs requires an investment of $125,000 in equipment (straight-line depreciation with a 10-year depreciable life and $0 salvage value), and $30,000 in labor (not depreciable). At the end of 10 years, the project will be terminated. Assuming a combined tax rate of 34% and after-tax MARR of 15%, determine the project's after-tax present worth.

12-16 Florida Construction Equipment Rentals (FCER) purchases a new 10,000-pound-rated crane for rental to its customers. This crane costs $1,125,000 and is expected to last for 25 years, at which time it will have an expected salvage value of $147,000. FCER earns $195,000 before-tax cash flow each year in rental income from this crane, and its total taxable income each year is between $10M and $15M. If FCER uses straight-line depreciation and a MARR of 15%, what is the present worth of the after-tax cash flow for this equipment? Should the company invest in this crane? *Contributed by Paul R. McCright, University of South Florida*

12-17 A firm manufactures padded shipping bags. A cardboard carton should contain 100 bags, but machine operators fill the cardboard cartons by eye, so a carton may contain anywhere from 98 to 123 bags (average = 105.5 bags). Each padded bag costs $0.03.

Management realizes that they are giving away 5 1/2% of their output by overfilling the cartons. One solution is to automate the filling of shipping cartons. This should reduce the average quantity of bags per carton to 100.3, with almost no cartons containing fewer than 100 bags.

The equipment would cost $18,600 and straight-line depreciation with a 10-year depreciable life and a $3600 salvage value would be used. The equipment costs $16,000 annually to operate. 200,000 cartons will be filled each year. This large profitable corporation has a 40% combined federal-plus-state incremental tax rate. Assume a 10-year study period for the analysis and an after-tax MARR of 15%.

Compute:

(a) The after-tax present worth

(b) The after-tax internal rate of return

(c) The after-tax simple payback period

12-18 ACDC Company is considering the installation of a new machine that costs $150,000. The machine is expected to lead to net income of $44,000 per year for the next 5 years. Using straight-line depreciation, $0 salvage value, and an effective income tax rate of 40%, determine the after-tax rate of return for this investment. If the company's after-tax MARR rate is 12%, would this be a good investment or not?

Contributed by Mukasa Ssemakula, Wayne State University

12-19 An old duplex was bought for $200,000 cash. Both sides were rented for $2500 per month. The total annual expenses for property taxes, repairs, gardening, and so forth are estimated at $200 per month. For tax purposes, straight-line depreciation over a 20-year remaining life with no salvage value is used. Of the total $200,000 cost of the property $50,000 is the value of the lot. Assume 38% incremental income tax bracket (combined state and federal taxes) applies throughout the 20 years.

If the property is held for 20 years, what after-tax rate of return can be expected?

(a) Assume the building and the lot can be sold for the lot's $50,000 estimated value.

(b) A more optimistic estimate of the future value of the building and the lot is that the property can be sold for $150,000 at the end of 20 years.

Declining Balance Depreciation

12-20 A firm is considering the following investment
 (A) project:

Year	Before-Tax Cash Flow (thousands)
0	−$1000
1	500
2	340
3	244
4	100
5	$\begin{cases} 100 \\ 125 \text{ Salvage value} \end{cases}$

The project has a 5-year useful life with a $125,000
salvage value, as shown. Double declining balance
depreciation will be used, assuming the $125,000
salvage value. The combined income tax rate is
34%. If the firm requires a 10% after-tax rate of
return, should the project be undertaken?

12-21 The Shellout Corp. owns a piece of petroleum
drilling equipment that costs $300,000 and will be
depreciated over 10 years by double declining balance depreciation. There is a combined 45% tax
rate. Shellout will lease the equipment to others and
each year receive $165,000 in rent. At the end of 5
years, the firm will sell the equipment for $80,000.
What is the after-tax rate of return Shellout will
receive from this equipment investment?

12-22 An automaker is buying some special tools for
 (A) $100,000. The tools are being depreciated by dou-
ble declining balance depreciation using a 4-year
depreciable life and a $6250 salvage value. It is
expected the tools will actually be kept in service
for 6 years and then sold for $6250. The before-tax
benefit of owning the tools is as follows:

Year	Before-Tax Cash Flow
1	$30,000
2	30,000
3	35,000
4	40,000
5	10,000
6	10,000
	6,250 Selling price

Compute the after-tax rate of return for this
investment situation, assuming a 46% incremental
tax rate.

12-23 This is the continuation of Problem 12-22. Instead
of paying $100,000 cash for the tools, the
corporation will pay $20,000 now and borrow the
remaining $80,000. The depreciation schedule will
remain unchanged. The loan will be repaid by 4
equal end-of-year payments of $25,240.
  Prepare an expanded cash flow table that takes
into account both the special tools and the loan.

(a) Compute the after-tax rate of return for the
tools, taking into account the $80,000 loan.
(b) Explain why the rate of return obtained in part
(a) is different from the rate of return obtained
in Problem 12-22.

12-24 A firm may invest in equipment that will be depreci-
 (A) ated by double declining balance depreciation with
conversion to straight-line depreciation in year 5.
For depreciation purposes a $700,000 salvage value
at the end of 6 years is assumed. But the actual value
is thought to be $1,000,000, and it is this sum that
is shown in the before-tax cash flow.

Year	Before-Tax Cash Flow (in $1000)
0	−$12,000
1	1,727
2	2,414
3	2,872
4	3,177
5	3,358
6	1,997
	1,000 Salvage value

If the firm wants a 9% after-tax rate of return and
its combined incremental income tax rate is 34%,
determine by annual cash flow analysis whether the
investment is desirable.

SOYD Depreciation

12-25 A firm has invested $60,000 in machinery with
a 5-year useful life. The machinery will have no
salvage value, as the cost to remove it will equal
its scrap value. The uniform annual benefits from
the machinery are $15,000. For a combined 45%

income tax rate, and sum-of-years'-digits depreciation, compute the after-tax rate of return.

12-26 A farmer bought a new harvester for $120,000. The
Ⓐ harvester's operating expenses averaged $10,000 per year but the harvester saved $40,000 per year in labor costs. It was depreciated over a life of 5 years using the SOYD method, assuming a salvage value of $30,000. The farmer sold the harvester for only $10,000 at the end of the fifth year. Given an income tax rate of 30% and a MARR rate of 5% per year, determine the after-tax net present worth of this investment. *Contributed by Mukasa Ssemakula, Wayne State University*

12-27 A firm has invested $400,000 in car-washing
Ⓖ equipment. They will depreciate the equipment by sum-of-years'-digits depreciation, assuming a $50,000 salvage value at the end of the 5-year useful life. The firm is expected to have a before-tax cash flow, after meeting all expenses of operation (except depreciation), of $165,000 per year. The firm's combined corporate tax rate is 40%.

(*a*) If the projected income is correct, and the equipment can be sold for $100,000 at the end of 5 years, what after-tax rate of return would the corporation receive from this venture?

(*b*) Summarize the environmental impact of commercial car washing. Include how wash water drainage is directed, how discharge sources are treated (or not), the impacts of sediment, detergents, waxes, and heavy metals, and local/state policies/practices.

12-28 A mining corporation purchased $120,000 of
Ⓐ production machinery and depreciated it using SOYD depreciation, a 5-year depreciable life, and zero salvage value. The corporation is a profitable one that has a 34% combined incremental tax rate.

At the end of 5 years the mining company changed its method of operation and sold the production machinery for $40,000. During the 5 years the machinery was used, it reduced mine operating costs by $32,000 a year, before taxes. If the company MARR is 12% after taxes, was the investment in the machinery a satisfactory one?

12-29 Zeon, a large, profitable corporation, is considering adding some automatic equipment to its production facilities. An investment of $120,000 will produce an annual benefit of $40,000. If the firm uses sum-of-years'-digits depreciation, an 8-year useful life, and $12,000 salvage value, will it obtain the desired 12% after-tax rate of return? Assume that the equipment can be sold for its $12,000 salvage

value at the end of the 8 years. Also assume a 40% income tax rate for state and federal taxes combined.

MACRS Depreciation

12-30 A special power tool for plastic products costs
Ⓐ $400,000 and has a 4-year useful life, no salvage value, and a 2-year before-tax payback period. Assume uniform annual end-of-year benefits.

(*a*) Compute the before-tax rate of return.

(*b*) Compute the after-tax rate of return, based on MACRS depreciation and a 34% combined corporate income tax rate.

12-31 (*a*) The BVM Corp., a construction company, pur-
Ⓖ chased a used hybrid electric pickup truck for $30,000 and used MACRS depreciation in the income tax return. During the time the company had the truck, they estimated that it saved $9500 a year. At the end of 4 years, BVM sold the truck for $9000. The combined federal and state income tax rate for BVM is 40%. Compute the after-tax rate of return for the truck.

(*b*) Hybrid electric vehicles (HEVs), such as the truck purchased by BVM Corp., are recognized for their fuel efficiency and earth-friendliness. However, no major technology is without negative environmental impacts. Summarize the key environmental issues, such as greenhouse emissions in use, life-cycle impacts of battery sources (such as lithium-ion and nickel metal hydride), vehicle production impacts, and other life-cycle impacts.

12-32 A chemical company bought a small vessel for
Ⓐ $550,000; it is to be depreciated by MACRS depreciation. When requirements changed suddenly, the chemical company leased the vessel to an oil company for 6 years at $100,000 per year. The lease also provided that the oil company could buy the vessel at the end of 6 years for $350,000. At the end of the 6 years, the oil company exercised its option and bought the vessel. The chemical company has a 34% combined incremental tax rate. Compute its after-tax rate of return on the vessel.

12-33 Xon, a small oil equipment company, purchased
Ⓔ a new petroleum drilling rig for $2,000,000. Xon will depreciate it using MACRS depreciation. The drilling rig has been leased to a firm, which will pay Xon $750,000 per year for 8 years. After 8 years the drilling rig will belong to the firm. Xon has a 40% combined incremental tax rate and a 20% after-tax MARR

(a) Does the investment appear to be satisfactory?

(b) Some claim that the coal and/or oil industries are inherently unsustainable and harmful to the environment. Develop a short position summary to support and challenge this perspective.

12-34 The profitable Palmer Golf Cart Corp. is
A considering investing $300,000 in special tools for some of the plastic golf cart components. The present golf cart model will continue to be manufactured and sold for 5 years, after which a new cart design will be needed, together with a different set of special tools.

    The saving in manufacturing costs, owing to the special tools, is estimated to be $150,000 per year for 5 years. Assume MACRS depreciation for the special tools and a 39% combined income tax rate.

(a) What is the after-tax payback period for this investment?

(b) If the company wants a 12% after-tax rate of return, is this a desirable investment?

12-35 Granny's Butter and Egg Business is such that she pays an effective tax rate of 35%. Granny is considering the purchase of a new Turbo Churn for $25,000. This churn is a special handling device for food manufacture and has an estimated life of 4 years and a salvage value of $5000. The new churn is expected to increase net income by $8000 per year for each of the 4 years of use. If Granny works with an after-tax MARR of 10% and uses 3-year MACRS depreciation, should she buy the churn?

12-36 An engineer is working on the layout of a new
A research and experimentation facility. Two plant operators will be required. If, however, an additional $100,000 of instrumentation and remote controls were added, the plant could be run by a single operator. The total before-tax cost of each plant operator is projected to be $35,000 per year. The instrumentation and controls will be depreciated by means of the modified accelerated cost recovery system (MACRS).

    If this corporation (34% combined corporate tax rate) invests in the additional instrumentation and controls, how long will it take for the after-tax benefits to equal the $100,000 cost? In other words, what is the after-tax payback period?

12-37 A corporation with a 39% combined income tax rate is considering the following investment in research equipment.

Year	Before-Tax Cash Flow
0	−$7,500,000
1	650,000
2	950,000
3	2,750,000
4	1,900,000
5	800,000
6	450,000

Prepare an after-tax cash flow table assuming MACRS depreciation.

(a) What is the before-tax rate of return?

(b) What is the after-tax rate of return?

12-38 Specialty Machining, Inc. bought a new multi turret
A turning center for $250,000. The machine generated new revenue of $80,000 per year. Operating costs for the machine averaged $10,000 per year. Following IRS regulations, the machine was depreciated using the MACRS method, with a recovery period of 7 years. The center was sold for $75,000 after 5 years of service. The company uses an after-tax MARR rate of 12% and is in the 35% tax bracket. Determine the after-tax net present worth of this asset over the 5-year service period. *Contributed by Mukasa Ssemakula, Wayne State University*

12-39 ABC Co. is contemplating an $18,000 investment
G in a methane gas generator. They estimate gross income will be $4500 the first year and increase by $500 each year over the next 10 years. Expenses of $300 the first year would increase by $250 each year over the next 10 years. ABC will depreciate the generator by MACRS depreciation, assuming a 7-year property class. A 10-year-old methane generator has no market value. The combined income tax rate is 45%. (Remember that recaptured depreciation is taxed at the same 45% rate.)

(a) Construct the after-tax cash flow for the 10-year project life.

(b) Determine the after-tax rate of return on this investment. ABC Co. thinks it should be at least 8%.

(c) If ABC Co. could sell the generator for $10,000 at the end of the fifth year, what would the rate of return be?

12-40 Fleet Fleet rental car company purchased 10 new
Ⓐ cars for a total cost of $180,000. The cars generated
income of $150,000 per year and incurred operating
expenses of $60,000 per year. The company uses
MACRS depreciation and its marginal tax rate is
40% (*Note:* Per IRS regulations, cars have a class
life of 5 years). The 10 cars were sold at the end
of the third year for a total of $75,000. Assuming
a MARR of 10% and using NPW, determine if
this was a good investment on an after-tax basis.
*Contributed by Mukasa Ssemakula, Wayne State
University*

12-41 Mid-America Shipping is considering purchasing
a new barge for use on its Ohio River routes. The
new barge will cost $13.2 million and is expected
to generate an income of $7.5 million the first year
(growing $1M each year), with additional expenses
of $2.6 million the first year (growing $400,000
per year). If Mid-America uses MACRS, is in the
38% tax bracket, and has a MARR of 12%, what
is the present worth of the first 4 years of after-tax
cash flows from this barge? Would you recommend
that Mid-America purchase this barge? Does your
answer change at 5 or 6 or 7 or . . . years? *Contributed
by Paul R. McCright, University of South Florida*

12-42 An investor bought a racehorse for $1 million. The
horse's average winnings were $700,000 per year
and expenses averaged $200,000 per year. The
horse was retired after 3 years, at which time it was
sold to a breeder for $175,000. Assuming MACRS
depreciation, a class life of 3 years, and an income
tax rate of 40%, determine the investor's after-tax
rate of return on this investment. *Contributed by
Mukasa Ssemakula, Wayne State University*

12-43 Refer to Problem 12-31. To help pay for the pickup
truck, the BVM Corp. obtained a $20,000 loan
from the truck dealer, payable in four end-of-year
payments of $5000 plus 10% interest on the loan
balance each year.

(*a*) Compute the after-tax rate of return for the
truck together with the loan. Note that the interest on the loan is tax deductible, but the $5000
principal payments are not.

(*b*) Why is the after-tax rate of return computed
in part (*a*) different from the rate obtained in
Problem 12-31?

(*c*) Assume now that the loan is paid off as a
uniform payment loan at 10% per year. Compute the after-tax rate of return with this loan
arrangement.

12-44 A house and lot are for sale for $210,000. $40,000 is
Ⓐ the value of the land. On January 1 Bonnie buys the
house to rent out. After 5 years, she expects to sell
the house and land on December 31 for $225,000.
Total annual expenses (maintenance, property
taxes, insurance, etc.) are expected to be $5000 per
year. The house would be depreciated by MACRS
depreciation using a 27.5-year straight-line rate
with midmonth convention for rental property. For
depreciation, a salvage value of zero was used.
Bonnie wants a 15% after-tax rate of return on
her investment. Assume that Bonnie's incremental
income tax rate is 28% each year and the capital
gains are taxed at 15%. Determine the following:

(*a*) The annual depreciation

(*b*) The capital gain (loss) resulting from the sale
of the house

(*c*) The annual rent Bonnie must charge to produce
an after-tax rate of return of 15%.

12-45 Tampa Electric Company (TECO) is planning a
major upgrade in its computerized demand management system. In order to accommodate this
upgrade, a building will be constructed on land
already owned by the company. The building is
estimated to cost $1.8M and will be opened in
August of this year. The computer equipment
for the building will cost $2.75M, and all office
equipment will cost $225,000. Annual expenses
for operating this facility (labor, materials, insurance, energy, etc.) are expected to be $325,000
for the rest of this year. Use of the new demand
management system is expected to decrease fuel
and other costs for the company by $1.8M this
year. If the company expects to earn 9% on its
investments, is in the 35% tax bracket, and uses a
20-year planning horizon, determine the estimated
after-tax cash flow from this project for this
year. *Contributed by Paul R. McCright, University
of South Florida*

12-46 A profitable wood products corporation is
Ⓐ considering buying a parcel of land for $50,000,
building a small factory building at a cost of
$200,000, and equipping it with $150,000 of
MACRS 5-year-class machinery.

If the project is undertaken, MACRS depreciation will be used. Assume the plant is put in service
October 1. The before-tax net annual benefit from
the project is estimated at $70,000 per year. The
analysis period is to be 5 years, and planners assume

the sale of the total property (land, building, and machinery) at the end of 5 years, also on October 1, for $328,000. Compute the after-tax cash flow based on a 34% combined income tax rate. If the corporation's criterion is a 15% after-tax rate of return, should it proceed with the project?

12-47 One January Geraldo Adair bought a small house and lot for $99,700. He estimated that $9700 of this amount represented the land's value. He rented the house for $6500 a year during the 4 years he owned the house. Expenses for property taxes, maintenance, and so forth were $500 per year. For tax purposes the house was depreciated by MACRS depreciation (27.5-year straight-line depreciation with a midmonth convention is used for rental property). At the end of 4 years the property was sold for $105,000. His incremental state and federal combined tax rate is 24%. What after-tax rate of return did Geraldo obtain on his investment?

12-48 Ms. Sami Jones, a successful businessperson, is
A considering erecting a small building on a commercial lot. A local furniture company is willing to lease the building for $9000 per year, paid at the end of each year. It is a net lease, which means the furniture company must also pay the property taxes, fire insurance, and all other annual costs. The furniture company will require a 5-year lease with an option to buy the building and land on which it stands for $125,000 after 5 years. Ms. Jones could have the building constructed for $82,000. She could sell the commercial lot now for $30,000, the same price she paid for it. She would depreciate the commercial building by modified accelerated cost recovery system (MACRS) depreciation. Ms. Jones believes that at the end of the 5-year lease she could easily sell the property for $125,000. What is the after-tax present worth of this 5-year venture if Ms. Jones has a 25% tax rate on ordinary income and 15% on capital gains? She uses a 10% after-tax MARR.

12-49 Vanguard Solar Systems is building a new manu-
G facturing facility to be used for the production of solar panels. Vanguard uses a MARR of 15%, and MACRS depreciation on its assets and is in the 35% tax bracket. The building will cost $2.75 million, and the equipment (3-year property) will cost $1.55 million plus installation costs of $135,000. O&M costs are expected to be $1.3 million the first year, increasing 6% annually.
Contributed by Paul R. McCright, University of South Florida

(*a*) If the facility opens in the month of March and sales are as shown, determine the present worth of the after-tax cash flows for the first 5 years of operation.

Year	Sales
1	$2,100,000
2	3,200,000
3	3,800,000
4	4,500,000
5	5,300,000

(*b*) Photovoltaic solar panels are an emerging energy source in Europe and the U.S., yet most are being manufactured in China. What are the market effects, environmental impacts, and ethical issues related to production to China?

Solving for Unknowns

12-50 A store owner, Jing Lang, believes his business
A has suffered from the lack of adequate customer parking space. He may buy an old building and lot next to his store. He would demolish the old building and make off-street parking on the lot. Jing estimates that the new parking would increase his before-income-tax profit by $7000 per year. It would cost $2500 to demolish the old building.

Mr. Lang's accountant advised that both costs (the property and demolishing the old building) would be considered to comprise the total value of the land for tax purposes, and it would not be depreciable.

Mr. Lang would spend an additional $3000 right away to put a light gravel surface on the lot. He believes this may be charged as an operating expense immediately. His combined state and federal incremental income tax rate will average 40%. If Jing wants a 15% after-tax rate of return from this project, how much could he pay to purchase the adjoining land with the old building?

Assume that the analysis period is 10 years and that the parking lot could always be sold to recover the costs of buying the property and demolishing the old building.

12-51 The management of a private hospital is considering automating some back office functions. This would replace five personnel that currently cover three shifts per day, 365 days per year. Each person earns $35,000 per year. Company-paid benefits and overhead are 45% of wages. Money costs 8% after income taxes. Combined federal and state income taxes are 40%. Annual property taxes and maintenance are 2¹/₂ and 4% of investment, respectively. Depreciation is 15-year straight line. Disregarding inflation, how large an investment in the automation project can be economically justified?

12-52 (A) A house and lot are for sale for $155,000. It is estimated that $45,000 is the land's value and $110,000 is the value of the house. The net rental income would be $12,000 per year after taking all expenses, except depreciation, into account. The house would be depreciated by straight-line depreciation using a 27.5-year depreciable life and zero salvage value.

Mary Silva, the prospective purchaser, wants a 10% after-tax rate of return on her investment after considering both annual income taxes and a capital gain when she sells the house and lot. At what price would she have to sell the house at the end of 10 years to achieve her objective? Assume that Mary has an incremental income tax rate of 28% in each of the 10 years and a capital gain rate of 20%.

12-53 A contractor has to choose one of the following alternatives in performing earthmoving contracts:

A. Buy a heavy-duty truck for $35,000. Salvage value is expected to be $8000 at the end of the vehicle's 7-year depreciable life. Maintenance is $2500 per year. Daily operating expenses are $200.

B. Hire a similar unit for $550 per day.

Based on a 10% after-tax rate of return, how many days per year must the truck be used to justify its purchase? Base your calculations on straight-line depreciation and a 40% income tax rate.

12-54 (A) A large profitable company, in the 40% combined federal/state tax bracket, is considering the purchase of a new piece of equipment that will yield benefits of $10,000 in Year 1, $15,000 in Year 2, $20,000 in Year 3, and $20,000 in Year 4. The equipment is to be depreciated using 5-year MACRS depreciation starting in the year of purchase (Year 0). It is expected that the equipment will be sold at the end of Year 4 at 20% of its purchase price. What is the maximum equipment

purchase price the company can pay if its after-tax MARR is 10%?

12-55 The Able Corporation is considering the installation of a small electronic testing device for use in conjunction with a government contract the firm has just won. The testing device will cost $20,000 and will have an estimated salvage value of $5000 in 5 years when the government contract is finished. The firm will depreciate the instrument by MACRS using 5 years as the class life. Assume that Able pays 40% federal and state corporate income taxes and uses 8% *after-tax* in economic analysis. What minimum equal annual benefit must Able obtain *before taxes* in each of the 5 years to justify purchasing the electronic testing device?

12-56 (A) A sales engineer has the following alternatives to consider in touring his sales territory.

A. Buy a 2-year old used car for $14,500. Salvage value is expected to be about $5000 after 3 more years. Maintenance and insurance cost is $1000 in the first year and increases at the rate of $500/year in subsequent years. Daily operating expenses are $50/day.

B. Rent a similar car for $80/day.

Based on a 12% after-tax rate of return, how many days per year must he use the car to justify its purchase? You may assume that this sales engineer is in the 28% incremental tax bracket. Use MACRS depreciation.

12-57 (G) For many firms, environmental regulation and oversight are converting external "social costs" to internal "real costs." TDE Industries is considering investing in equipment to reduce emissions. Without the investment they expect to incur fines and fees of $750,000 per year. TDE uses a 40% combined tax rate, an after-tax MARR of 20%, and will depreciate such equipment as a 5-year MACRS property.

(a) Management has asked you to determine the maximum amount they should pay for equipment to eliminate these fines. Use a 6-year study period.

(b) What argument would you add to the economic analysis to justify the investment?

(c) If your manager decides to purchase $1M in equipment, what level of decreased fees would justify the investment? Use a 6-year study period.

Section 179, Bonus Depreciation, and Tax Credits

12-58 Christopher wants to add a solar photovoltaic system to his home. He plans to install a 2-kW system and has received a quote from an installer who will install this unit for $19,750. The federal government will give him a tax credit of 30% of the cost, and the state will give a 10% tax credit. State law requires the utility company to buy back all excess power generated by the system. Christopher's annual power bill is estimated to be $2000, and this will be eliminated by the solar system. Christopher expects to receive a check from the power company for $600 each year for his excess production. If the tax credits are received at EOY 1 and Christopher receives a $2000 savings plus a $600 income at the end of each year, use present worth to determine if the system pays for itself in 8 years. Assume that Christopher earns 3% on all investments. *Contributed by Paul R. McCright, University of South Florida*

12-59 Rework Problem 12-32 assuming that a $250,000 Section 179 deduction and a 50% bonus depreciation deduction apply.

12-60 Rework Problem 12-33 assuming that a $350,000 Section 179 deduction applies.

12-61 Rework Problem 12-35 assuming that a 50% bonus appreciation deduction applies.

12-62 Rework Problem 12-38 assuming that a $100,000 Section 179 deduction applies.

12-63 Rework Problem 12-12 assuming that a 10% Investment Tax Credit (ITC) that does not reduce the depreciable basis applies.

12-64 Rework Problem 12-14 assuming that a 10% Investment Tax Credit (ITC) that does not reduce the depreciable basis applies.

12-65 Rework Problem 12-27 assuming that a 10% Investment Tax Credit (ITC) that does reduce the depreciable basis applies.

12-66 Rework Problem 12-28 assuming that a 10% Investment Tax Credit (ITC) that does reduce the depreciable basis applies.

Multiple Alternatives

12-67 A small-business corporation is considering whether to replace some equipment in the plant. An analysis indicates there are five alternatives in addition to the do-nothing option, Alt. *A*. The alternatives have a 5-year useful life with no salvage value. Straight-line depreciation would be used.

Alternatives	Cost (thousands)	Before-Tax Uniform Annual Benefits (thousands)
A	$ 0	$ 0
B	25	8
C	10	5
D	5	2
E	15	5

The firm has a combined federal and state income tax rate of 38%. If the corporation expects a 8% after-tax rate of return for any new investments, which alternative should be selected?

12-68 A corporation with $7 million in annual taxable income is considering two alternatives:
Ⓐ

Year	Before-Tax Cash Flow ($1000) Alt. 1	Before-Tax Cash Flow ($1000) Alt. 2
0	−$10,000	−$20,000
1–10	4,500	4,500
11–20	0	4,500

Both alternatives will be depreciated by straight-line depreciation assuming a 10-year depreciable life and no salvage value. Neither alternative is to be replaced at the end of its useful life. If the corporation has a minimum attractive rate of return of 10% *after taxes,* which alternative should it choose? Solve the problem by:

(*a*) Present worth analysis

(*b*) Annual cash flow analysis

(*c*) Rate of return analysis

(*d*) Future worth analysis

(*e*) Benefit–cost ratio analysis

12-69 A plant can be purchased for $1,000,000 or it can be leased for $200,000 per year. The annual income is expected to be $800,000 with the annual operating cost of $200,000. The resale value of the plant is estimated to be $400,000 at the end of its 10-year life. The company's combined federal and state income tax rate is 40%. A straight-line depreciation can be used over the 10 years with the full first-year depreciation rate.

(a) If the company uses the after-tax minimum attractive rate of return of 10%, should it lease or purchase the plant?

(b) What is the breakeven rate of return of purchase versus lease?

12-70 Two mutually exclusive alternatives are being considered by a profitable corporation with an annual taxable income between $5 million and $10 million.

Before-Tax Cash Flow ($1,000)

Year	Alt. A	Alt. B
0	−$3000	−$5000
1	1000	1000
2	1000	1200
3	1000	1400
4	1000	2600
5	1000	2800

Both alternatives have a 5-year useful and depreciable life and no salvage value. Alternative A would be depreciated by sum-of-years'-digits depreciation, and Alt. B by straight-line depreciation. If the MARR is 10% after taxes, which alternative should be selected?

12-71 If a firm's after-tax minimum attractive rate of return is 12% and its combined incremental income tax rate is 42%, which alternative should be selected?

	Alt. A	Alt. B
Initial cost	$11,000	$33,000
Uniform annual benefit	3,000	9,000
End-of-depreciable-life salvage value	2,000	3,000
Depreciation method	MACRS 5-year	MACRS 5-year
End-of-useful-life salvage value obtained	2,000	5,000
Depreciable life, in years	5	5
Useful life, in years	5	5

12-72 Use the after-tax IRR method to evaluate the following three alternatives for MACRS 3-year property, and offer a recommendation. The after-tax MARR is 25%, the project life is 5 years, and the firm has a combined incremental tax rate of 45%.

Alt.	First Cost	Annual Costs	Salvage Value
A	$14,000	$2500	$ 5,000
B	18,000	1000	10,000
C	10,000	5000	0

12-73 VML Industries has need of specialized yarn manufacturing equipment for operations over the next 3 years. The firm could buy the machinery for $95,000 and depreciate it using MACRS. Annual maintenance would be $7500, and it would have a salvage value of $25,000 after 3 years. Another alternative would be to lease the same machine for $45,000 per year on an "all costs" inclusive lease (maintenance costs included in lease payment). These lease payments are due at the beginning of each year. VML Industries uses an after-tax MARR of 18% and a combined tax rate of 40%. Do an after-tax present worth analysis to determine which option is preferred.

12-74 Padre Pio owns a small business and has taxable income of $150,000. He is considering four mutually exclusive alternative models of machinery. Which machine should be selected on an after-tax basis? The after-tax MARR is 15%. Assume that each machine is MACRS 5-year property and can be sold for a market value that is 25% of the purchase cost, and the project life is 10 years.

Model	I	II	III	IV
First cost	$9000	$8000	$7500	$6200
Annual costs	25	200	300	600

12-75 LoTech Welding can purchase a machine for $175,000 and depreciate it as 5-year MACRS property. Annual maintenance would be $9800, and its salvage value after 8 years is $15,000. The machine can also be leased for $35,000 per year on an "all costs" inclusive lease (maintenance costs included). Lease payments are due at the beginning of each year, and they are tax-deductible. The firm's combined tax rate for state and federal income taxes is 40%. If the firm's after-tax interest rate is 9%, which alternative has the lower EAC and by how much?

12-76 For Problem 12-75, assume that Section 179 depreciation must be considered as this is LoTech's only major equipment investment this year. Also assume that the firm is profitable. How much lower is the annual cost of purchasing the machine?

12-77 For Problem 12-75, assume that the machine will be financed with a 4-year loan whose interest rate is 12%.

  (*a*) Graph the EAC for the purchase for financing fractions ranging from 0% to 100%.

  (*b*) Assume that 50% bonus depreciation is also available. Graph the EAC for the purchase for financing fractions ranging from 0% to 100%.

  (*c*) Assume that Section 179 depreciation must be considered, as this is LoTech's only major equipment investment this year. Also assume that the firm is profitable. Graph the EAC for the purchase for financing fractions ranging from 0% to 100%.

CASES

The following cases from *Cases in Engineering Economy* (www.oup.com/us/newnan) are suggested as matched with this chapter.

CASE 26   Molehill & Mountain Movers
  Compares depreciation methods with option for inflation.

CASE 28   Olives in Your Backyard
  Emphasizes taxes and sensitivity analysis.

CASE 29   New Fangled Manufacturing
  Emphasizes taxes and sensitivity analysis.

CASE 36   Brown's Nursery (Part A)
  After-tax analysis of expansion opportunity.

CASE 38   West Muskegon Machining and Manufacturing
  More complex inflation and tax problem with sunk cost and leverage.

CASE 53   Problems in Pasta Land
  Long case statement. Includes taxes and limited uncertainty.

APPENDIX 12A
Taxes and Personal Financial Decision Making

Key Words

adjusted gross income	gross income	Stafford loans
buy term invest difference	individual retirement account	taxable income
cash value benefit	insurance	401(k) plan
death benefit	liability insurance	403(b) plan
deductible	needs/savings/wants	529 plan
defined benefit	Perkins loans	
defined contribution	premiums	

In this appendix our goal is to focus on decisions involving personal income taxes and likely financial scenarios that students will face. We first describe how taxes are computed for individuals. As is the case with corporations, individuals should consider decisions that will be affected by taxes on an *after-tax* basis. We detail several example applications of this knowledge to common personal finance scenarios including student loans and retirement accounts. Lastly, we provide details on decisions involving insurance and personal budgeting.

INCOME TAXES FOR INDIVIDUALS

Calculating Taxable Income

The amount of federal income taxes to be paid by individuals depends on taxable income and the income tax rates. Therefore, our first concern is the definition of **taxable income.** To begin, one must compute his or her **gross income:**

$$\text{Gross income} = \text{Wages, salary, etc.} + \text{Interest income} + \text{Dividends} + \text{Capital gains}$$
$$+ \text{Unemployment compensation} + \text{Other income}$$

From gross income, we subtract any allowable retirement plan contributions and other adjustments. The result is **adjusted gross income (AGI).** From adjusted gross income, individuals may deduct the following items:

1. **Personal Exemptions.** One exemption ($4000 for 2015 returns) is provided for each person who depends on the gross income for his or her living.
2. **Itemized Deductions.** Some of these are:
   (a) Excessive medical and dental expenses (exceeding 10% of AGI)
   (b) State and local income, real estate and personal property tax
   (c) Home mortgage interest
   (d) Charitable contributions (limited to 50, 30, or 20% of AGI based on assets contributed)

(e) Casualty and theft losses (exceeding $100 + 10% of AGI)

(f) Job expenses and certain miscellaneous deductions (some categories must exceed 2% of AGI)

3. **Standard Deduction**. Each taxpayer may either itemize his or her deductions, or instead take a standard deduction as follows:

(a) Single taxpayers, $6300 (for 2015 returns)

(b) Married taxpayers filing a joint return, $12,600 (for 2015 returns)

The result is **taxable income.**

For individuals, taxable income is computed as follows:

$$\text{Adjusted gross income} = \text{Gross income} - \text{Adjustments}$$

$$\begin{aligned}\textbf{Taxable income} = {} &\text{Adjusted gross income} \\ &- \text{Personal exemption(s)} \\ &- \text{Itemized deductions or standard deduction}\end{aligned} \qquad (12\text{-}1A)$$

Individual Tax Rates

Income tax rates for individuals are modified every year to adjust for inflation and on average about every 4 years (since 1960) for policy reasons. For example, since 1960 the maximum federal income rate for individuals has fallen from over 90% to the current level of 39.6%. There have been 4 increases and 5 decreases since 1960.

Two other schedules (not shown here) are applicable to unmarried individuals with dependent relatives ("head of household"), and married taxpayers filing separately.

EXAMPLE 12A–1

An unmarried self-supporting student earned $10,000 in the summer plus another $6000 during the rest of the year. When she files an income tax return, she will be allowed one exemption (for herself). She estimates that she spent $1000 on allowable itemized deductions. How much income tax will she pay?

$$\textbf{Adjusted gross income} = 10{,}000 + 6000 = \$16{,}000$$

$$\begin{aligned}\textbf{Taxable income} = {} &\text{Adjusted gross income} \\ &- \text{Deduction for one exemption (\$4000)} \\ &- \text{Standard deduction (\$6300)} \\ = {} &16{,}000 - 4000 - 6300 = \$5700\end{aligned}$$

$$\textbf{Federal income tax} = 10\% \,(5700) = \$570 \quad \text{(tax rate from Table 12A–1a)}$$

SOLUTION

The $1000 in itemized deductions is less than the standard deduction, so that higher value is claimed. In filing her tax return the student will compare the $570 to any amount withheld. Any withholding above $570 will be refunded; any amount below is the balance that she owes.

TABLE 12A–1a 2015 Tax Rate Schedules Schedule X–a If Filing Status Is Single

Taxable Income	Tax Rate
$0 to $9225	10%
$9226 to $37,450	$922.50 plus 15% of the amount over $9,225
$37,451 to $90,750	$5156.25 plus 25% of the amount over $37,450
$90,751 to $189,300	$18,481.25 plus 28% of the amount over $90,750
$189,301 to $411,500	$46,075.25 plus 33% of the amount over $189,300
$411,501 to $413,200	$119,401.25 plus 35% of the amount over $411,500
$413,201 or more	$119,996.25 plus 39.6% of the amount over $413,200

TABLE 12A–1b 2015 Tax Rates Schedule Y-1—If Your Filing Status Is Married Filing Jointly or Qualifying Window(er)

Taxable Income	Tax Rate
$0 to $18,450	10%
$18,451 to $74,900	$1845.00 plus 15% of the amount over $18,450
$74,901 to $151,200	$10,312.50 plus 25% of the amount over $74,900
$151,201 to $230,450	$29,387.50 plus 28% of the amount over $151,200
$230,451 to $411,500	$51,577.50 plus 33% of the amount over $230,450
$411,501 to $464,850	$111,324.00 plus 35% of the amount over $411,500
$464,851 or more	$129,996.50 plus 39.6% of the amount over $464,850

Two other schedules (not shown here) are applicable to unmarried individuals with dependent relatives ("head of household"), and married taxpayers filing separately.

EXAMPLE 12A–2

Two grad students can be claimed as dependents on their respective parental tax returns for this year—if they get married on January 2 rather than December 28. Each grad student has an income of $9000. The parents of each grad student file a joint return using the standard deduction and a combined income of $110,000. What are the net tax consequences of the marriage date decision?

SOLUTION

Since we have assumed that the students and families have matching incomes, we need only analyze one. If each student files a return as a single individual and is not claimed as a dependent on their parents' return, then using 2015 numbers the sum of the personal exemption and the standard deduction is $10,300 (= $4000 + $6300). Since this is more than the income of $9000, the tax savings would only be the tax rate times the potential taxable income or $900 (= 10% × $9000).

If the student is claimed as a dependent on the parental return, the student cannot claim the standard deduction and the standard deduction on the parental return does not change. However, the personal exemption is more valuable on the parental return. In this case the taxable income is $85,400 (= 110,000 − 3× 4000 − 12,600), which in the 25% bracket for joint returns. Thus, claiming the student as a dependent saves $1000 (= 25% × 4000).

If the students marry on December 28 and file a joint return for 2015, then their gross income, exemptions, and standard deduction all double. So their taxable income and tax are both $0.

Postponing the wedding until January and allowing the parents to claim the students as dependents for one more year would save $100 for each student. For smaller student incomes the savings would be greater. For student incomes above $10,300, their independent returns can fully use the exemption and standard deduction. For higher income parental incomes the tax savings for the extra exemption can be as high as −$1584 (= 39.6% × 4000).

While these differences have been fairly small, larger differences can result from deductions for student loan interest or credits or deductions for tuition payments.

Combined Federal and State Income Taxes

Individuals, like corporations, are often subject to both federal and state income taxes. When state taxes are deductible Equation 12-2 applies. However, if the standard deduction is used then Equation 12A-1 applies.

$$\textbf{Combined incremental tax rate} = \Delta\text{State tax rate} + \Delta\text{Federal tax rate} \qquad (12\text{A-}1)$$

EXAMPLE 12A–3

A single engineer's expected taxable income for 2015 is expected to be $92,000. Instead of taking a planned end-of-year vacation, the engineer could accept assignment to a project with a year-end deadline. The expected overtime would increase earnings by $8000. The marginal rate for state income tax at either total income is 8%. What is the engineer's combined marginal tax rate?

1. If deductions are itemized.
2. If the standard deduction is used.

SOLUTION

In either case the marginal tax rate at the federal level is 28%.

1. If deductions are itemized then Eq. 12-2 applies.

$$\text{Combined rate} = 0.08 + 0.28(1 - 0.08) = 0.08 + 0.258 = 33.8\%$$

2. If the standard deduction is used then Eq. 12A-1 applies.

$$\text{Combined rate} = 0.08 + 0.28 = 36\%$$

Capital Gains/Losses for Individuals

As with corporations, individual taxpayers incur capital gains or losses when selling houses, land, stocks, art, jewelry, and other collectables. Depending on tax bracket, type of asset, and duration of ownership, the capital gain tax rate can range from 0% to 39.6%. The most important capital gain provision for many is the exemption of up to a $500,000 gain from sale of a principal residence. Because this Appendix is only an introduction to personal income taxes, the regulations for the sale of a house and capital gains tax rates are not detailed here.

Tax Credits vs. Tax Deductions

This chapter and appendix have already included examples of *tax credits* and *deductions from taxable income*. Comparing the choices available in "writing off" a portion of college tuition expenses allows us to directly compare the value of a tax deduction vs. a tax credit. Table 12A–2 summarizes tuition-related federal tax deductions and credits for single filers. Income levels are doubled for joint filers, and so is the education expense deduction. The credits are not changed.

TABLE 12A–2 **Federal tax deductions and credits for single filers (2015)**

Provision for Education Expenses	Modified Adjusted Gross Income (MAGI)	Details
Education Expenses	$0–65,000	$4000
Deduction	$65,000–80,000	$2000
American Opportunity Tax Credit	$0–80,000	100% of first $2000 + 25% of next $2000
	$80,000–90,000	Phase out bracket
Lifetime Learning Credit	$0–65,000	$2000

The reduction in allowable values is proportional to percentage of phase out bracket. Income levels and the Education Expenses Deduction are doubled for joint filers.

Student Loan Interest Deduction

The interest on most consumer loans is not tax deductible, but interest on student loans (up to the limits shown in Table 12A–3) and home mortgages is tax deductible.

TABLE 12A–3 Federal tax deduction for student loan interest (2015)

Modified Adjusted Gross Income (MAGI)	Details
$ 0–65,000 (single) $ 0–130,000 (joint)	$2500 or total paid
$ 65,000–80,000 (single) $ 130,000–160,000 (joint)	Phase out bracket

EXAMPLE 12A–4

A freshman's single-parent family has a modified adjusted gross income of $35,000 and tuition expenses of $10,000 after scholarships. The student started in January 2015 so all of the tuition was paid in 2015. Which of the three deductions and credits described in Table 12A–2 is the most valuable?

SOLUTION

The family files a single not a joint return, so the limits in Table 12A–2 apply. The family's income is within the brackets where the full value of each deduction or credit is available. Assume that the freshman is an only child and that $6000 in itemized deductions is claimed. Then the taxable income is $21,000 ($= 35,000 - 2 \times 4000 - 6000$), and the family is in the 15% tax bracket.

Only $4000 of the tuition payments is deductible under the education expenses deduction. Since deductions reduce taxable income, the reduction in taxes would be $600 ($= 4000 \times 0.15$).

While deductions reduce *taxable income*, credits reduce *income taxes*. Thus, both credits are much more attractive than the deduction. The American Opportunity Credit at $2500 allows $500 ($= 25\%$ of next 2000) more than the Lifetime Learning Credit, which is limited to $2000.

EXAMPLE 12A–5

A recent engineering graduate is single and earning $55,000 annually with a government labs group. What is the after-tax cost of the $3200 in student loan interest that the engineer paid last year?

SOLUTION

Only $2500 of the interest is deductible from taxable income. With $55,000 in income, the engineer will be in a 25% tax bracket. Thus, there is a reduced tax bill (savings) of $625 ($= 2500 \times .25$). The after-tax cost of the interest is $2575 ($= 3200 - 625$).

EXAMPLE 12A–6

Two engineers are married and filing a joint return. Last year their MAGI was $150,000. What is the amount of allowable deduction if they paid $800 interest on their student loans? If they paid $2750 in student loan interest?

SOLUTION

For 2015 the maximum deduction is $2500, with phase out for joint filers with MAGI between $130,000 and 160,000. Their MAGI is 2/3 of the way through the phase out bracket, so in each case the allowable deduction is reduced by 2/3 to 1/3 of the maximum deduction. For the $2750 in interest the deduction is calculated using the $2500 limit.

$$\text{Deduction}_{\$800} = \$800 \times 1/3 = \$267$$
$$\text{Deduction}_{\$2750} = \$2500 \times 1/3 = \$833$$

STUDENT LOANS

If you are borrowing money to attend college, you are part of a very large group. In 2015 over 20 million people attended colleges or universities in the U.S., and over 70% borrowed from the federal government or private sources. An estimated 40 million current and former students owe over $1.2 trillion collectively—with an average student debt of about $35,000.

This is not an overview of the different programs with their rates, limits, and regulations. Instead Examples 12A–7 and 8 illustrate the economic analysis that is part of better decision making. Some of that decision making is easy. Subsidized student loans with low rates of interest are better than loans with more expensive rates. Other decisions, such as the choice between working more, borrowing less, and perhaps taking longer to graduate, are more complex. On-the-job learning, building a resume, and balancing life and school are difficult to include in an economic analysis—but they may be the decision-making drivers.

EXAMPLE 12A–7

An undergraduate is comparing the financial aid packages offered by two institutions. Both packages include $20,000 in Stafford loans over 4 years but one is subsidized and one is not. If the student pays back the loans in 5 years after graduation, how much interest is paid on each loan? To simplify the calculations for an easier to follow example assume annual tuition and loan payments and ignore reductions in taxable income. To further simplify the calculations, ignore the loan fees (see Chapter 7 to include). Assume there is one loan for 4 years rather than a separate loan at a possibly different rate for each year. Assume interest rates are 3.4% for subsidized loans and 6.8% for unsubsidized loans.

5-BUTTON SOLUTION

For the subsidized Stafford loan the interest is paid by the government while the student is in school (0% compounding for the student). Thus for the subsidized loan only $20,000 is owed at the end of 4 years (the government pays the interest), and thus the payment is calculated at $4417. The total interest paid in the 5 years after graduation is $2085.

	A	B	C	D	E	F	G	H	I	J
1	Alternative	i	n	PMT	PV	FV	Solve for		Total Paid	Interest Paid
2	Subsidized	3.4%	5		-20,000	0	PMT	4,417	22,085	2,085
3				begin						
4	Unsubsidized	6.8%	4	5,000	0		FV	-23,639		
5		6.8%	5		-23,639	0	PMT	5,735	28,673	8,673

For the unsubsidized loan the amount due at graduation is $23,639 (18% higher than $20,000). The payments after graduation of $5735 are 30% higher than the $4417. The total interest paid is $8673, which is more than 4 times the interest paid for the subsidized scenario.

Example 12A–7 illustrates that even if student loans are paid off quickly, rates and subsidies can dramatically change what must be repaid. Many students owe more and take much longer to repay the loan, which can interfere with their ability to buy a house, save for retirement, or even take vacations. It is also worth noting that student loans can be discharged in bankruptcy only if undue hardship can be proven.

EXAMPLE 12A–8

Find the effective rate of interest the student pays for the subsidized loan in Example 12A–7, assuming the loan rate is 3.4%.

SPREADSHEET SOLUTION

The timing of the $5000 tuition payments and the loan payment amounts and timing do not change. The future value of the tuition payments and the present value of the loan payments are equal, and the interest rate is the unknown. For visibility and convenience the total payment value for the subsidized loan is entered to establish a goal. Then GOAL SEEK is used to find the effective interest rate of 1.82%.

	A	B	C	D	E	F	G	H	I	J
1	Alternative	i	n	PMT	PV	FV	Solve for		Total Paid	
2				begin						
3	Subsidized	1.82%	4	5,000	0		FV	-20,928		
4		1.82%	5		-20,928	0	PMT	4,417	22,085	=C4*H4
5		=B3			=H3		Subsidized Exp. 12A-7		22,085	
6								difference	0	

EXAMPLE 12A–9

A student has borrowed the maximum permitted for 2015–16 of $27,500 in Perkins loans (available to students with exceptional need). The rate on the loan is 5% and the government has paid the interest while the student has been in school. To simplify the calculations assume annual tuition and loan payments. What are the differences in the payment amount and the total paid if the student pays the loan back in 5 years and in 20 years? Is the availability of the student loan interest likely to reduce the after-tax cost of the loan?

5-BUTTON SOLUTION

The first year the interest is 5% of the loan amount or $1375. This is less than the maximum deduction for student loan interest, so it all would be deductible. However, it is also *much* less than the standard deduction. Few students buy a home right out of school, and student loan debt is forcing some to postpone it longer than they would like. It is mortgage interest and property taxes that often make itemizing deductions attractive.

	A	B	C	D	E	F	G	H	I	J
1	Alternative	*i*	*n*	*PMT*	*PV*	*FV*	Solve for	Annual Payment	Total Paid	Interest Paid
2	5-year	5%	5		-27,500	0	PMT	6352	31,759	4259
3	20-year	5%	20		-27,500	0	PMT	2207	44,133	16,633

Using the longer repayment term reduces the payment by 65%, but it increases the total paid by 39% and the interest paid by 391%.

RETIREMENT ACCOUNTS

Appendix 9A introduced **defined benefit** and **defined contribution** retirement plans. The shift to defined contribution plans by firms and governments means that students graduating today must pay more attention than their parents' generation to saving enough for retirement. Examples 12A–9 and 10 illustrate a few of the options for retirement savings. Good planning starts with understanding the power of compound interest. It includes the analytical techniques used here. It also includes keeping up with changing investment details as your personal situation also evolves so that you can make the best choices.

401(k) plans are available to employees of corporations. Similar retirement plans exist for employees of other types of organizations (403(b) for public education and some nonprofit employees, 401(a) plans for some government workers, 457(b) and (f) plans for other government, non-government, and nonprofit employees).

These plans allow qualified employees to direct part of each paycheck into a designated retirement account. Human resource (HR) departments have staff and resources dedicated to explaining the options and establishing these accounts. When choosing a plan, any fees paid by the employee will need to be considered. (In some cases expected returns before fees are quoted.)

Dollars are directed into 401(k) accounts on a before-tax basis. Deposits are subtracted from salary and wages when reporting taxable income so taxes are also reduced. In 2015 the federal limit (the section 402(g) limit) on pre-tax dollar investments and/or expenses was $18,000 per individual ($24,000 if over 50, via a "catch-up contribution").

EXAMPLE 12A–10

A 25-year-old with a dual BS/MS engineering degree accepts employment with a national firm which has a required retirement age of 65. For simplicity assume that the engineer stays through retirement and that the average annual salary over the next 40 years is $96,000. The firm offers a *$1 firm-for-$2 employee 401(k) matching plan* (up to a maximum employee contribution of 6%). The chosen plan offers a 3% expected rate of return after subtracting fees. What will the engineer have in the plan at retirement?

SOLUTION

Pre-tax monthly salary	$= \$96,000/12$	$= \$8000$
Monthly contributions from employee	$= \$8000 \times 6\%$	$= \$480$
Monthly contributions from company	$= \$8000 \times 3\%$	$= \$240$ (1-for-2 match)
Total monthly contribution	$= \$480 + 240$	$= \$720$
Future value in 40 years?	$= \$720(F/A, 0.25\%, 480)^1$	$= \$666,763$
	$^1 (F/A, 3\%/12, 40 \times 12)$	

Without the company match the accumulated amount in the account would have been 1/3 less. All distributions from this account to individuals will be taxable income. But marginal tax rates after retirement should be much lower.

EXAMPLE 12A–11

For the data in Example 12A–10 compute the after-tax difference in income with and without the 401(k) deduction. Assume a single taxpayer and standard exemptions and deductions.

SOLUTION

	Choose 401(k) Plan	No 401(k) Plan	Notes
Wages	$90,240	$96,000	$96,000*.06 = $5760 contribution
Personal exemption	4,000	4,000	
Std. deduction	6,300	6,300	
Taxable income	79,940	85,700	marginal tax levels is 25% for both
Taxes (from table)	15,779	17,219	
After-tax wages	74,461	78,781	= Wages – taxes
			Difference is $4320

Tax computations:

$$401(k) \; \$15,779 = 5156.25 + (.25)(79,940 - 37,450)$$

$$\text{No } 401(k) \; \$17,219 = 5156.25 + (.25)(85,700 - 37,450)$$

By choosing the 401(k) plan the engineer is putting $8640 ($5760 employee + 2880 company match) into the account and losing only $4320 in after-tax wages.

INSURANCE

Insurance helps individuals manage risk. Policyholders pay monthly or annual **premiums** to insure their life, home, vehicle, and so on. Renters insurance may cover your possessions and your potential liability if someone is hurt in your apartment. For trip insurance or an extended warranty, the purchase is for *a* trip or *a* product, where the insurance company pays in the event of an insured loss. With insurance, the benefit payment is often reduced by a **deductible** that the individual is responsible for. Because the insurance company has expenses to operate, to market, and to limit fraudulent claims, the payout is on average less than the premiums collected—sometimes less than 50% and sometimes higher than 85%. This means that for a properly rated individual, the policy's expected value is negative.

So how do you decide which types of insurance you want? Sometimes you may not have a choice. If you finance your vehicle or purchase a home with a mortgage, the lender will require coverage to protect itself. If your employer does not offer medical insurance, you may be required by the government to purchase it. In most states, if you license a vehicle you are required to have **liability insurance**. Once you decide to purchase insurance, in some cases you will also have to choose how much coverage to carry.

For example, most students and engineers starting out have few assets, so bankruptcy may be an alternative to higher liability insurance levels; however, they may have young children and a non-working spouse, so life insurance may be critical. In contrast, an engineer approaching retirement may have substantial assets to protect and no dependents.

Automobile Insurance

Buying a vehicle usually means an insurance policy must be chosen. Coverage and deductible limits, state requirements, and optional choices can quickly overwhelm a car buyer. Policy buyers must choose property damage and bodily injury liability levels, medical payments protection/personal injury protection levels, collision versus comprehensive coverage, uninsured motorist protection, roadside assistance, and rental car options.

EXAMPLE 12A–12

The state of New Hampshire is the one state that does not require drivers to carry liability insurance. Assume you just received your dream job in the Granite State. Should you purchase optional automotive insurance?

SOLUTION

The answer is that it depends! What can you afford and what is your risk tolerance? Let's start by looking at the financial outcomes of three scenarios:

1. No insurance,* no accident:

   - Annual insurance premium = $0   **Total Annual Expense = $0**

2. No insurance, 1 accident: Assume the driver is responsible for 1 two-car accident with these costs:

   - Own car is totaled   = $5000 net replacement value
   - Damage to other car   = 4000
   - Own medical bills   = 2500
   - Other driver, medical   = 8000
   - Property damage   = 1500 replace traffic post/sign

Due others	=	**$13,500**
Cost to self	=	**7,500**
Total	=	**$21,000**

3. Insurance, 1 accident: Assume the driver has the minimum liability coverage policy for a driver in NH** and collision coverage with a $500 deductible and causes the scenario 2 accident:

   - Annual insurance premium = $150/month (estimate)
   - Deductible expenses   = $500

   Total Annual Expense = $2300

Scenario 2 shows the risk of not having at least minimum auto insurance. But there are other costs that are harder to quantify that are part of choosing insurance types and levels. Scenario 2 could easily have included legal costs that are part of the insurance firm's expertise and cost of doing business, but that would be expensive in time and money for you to buy.

Less expensive coverage without collision insurance for the Scenario 2 accident would have meant having to replace the $5000 vehicle. Do you have the funds to do so? Do you need the car to get to work, or just for weekend activities? How much is the time that you would need to spend resolving issues with the insurance company worth to you?

*Currently 17 of the 49 states that require liability minimums also require uninsured motorists coverage.
**2016 NH minimums for those electing to insure their vehicles are 25/50/25 in $1000 for bodily liability per person and per accident, and property per accident.

Most policyholders are willing to trade off the expense of a periodic premium for the peace of mind that insurance provides, as well as protection against potentially catastrophic financial circumstances. To reduce premiums: make firms compete for your business, consider bundling different insurance coverages with the same company; make one annual payment versus monthly payments; increase your deductible levels; and ask for available low mileage, good driver, and student discounts.

Life Insurance

Life insurance policies provide a **death benefit** that is paid to the survivors (beneficiaries) when the insured person dies. This benefit is generally not subject to income tax, but there may be estate taxes. Some types of life insurance also offer a **cash value benefit** to policy-holders. The policy's cash portion can be withdrawn or borrowed against, often after some initial accumulation period.

EXAMPLE 12A–13

A recent engineering graduate started her new job with a required session in the firm's human resources (HR) office. She needs to explore the *life insurance* options that were discussed. What should she know to assist her choice? Her options are:

(1) Do not purchase any additional life insurance.

(2) Purchase additional *term* life insurance.

(3) Purchase a *permanent* life insurance policy.

SOLUTION

(1) As allowed in IRS Code Section 79, her employer provides a $50,000 term life policy to all employees free of charge and not subject to income tax.

(2) *Term* life insurance policies provide a death benefit for the specified 5-, 10-, 20-, . . . year term as long as premiums are paid on time. The younger and healthier the insured and the shorter the term, the lower the premium. Once the term ends, some policies can be renewed at a higher cost and others require a new policy.

(3) *Permanent* life insurance comes in many types that include whole, ordinary, variable, universal, index-universal, and variable universal life. Most have both a cash value and a death benefit. The cash value accumulation acts as a tax-free investment instrument over the policy's life. In many cases the policyholder has the right to keep the policy for his or her *whole* life until death, when the death benefit is paid.

There is no one right answer. Life insurance is a complex decision based on one's life circumstances, goals, and perspectives. Decisions are often influenced by number and age of dependents and one's desire to provide financial security for survivors.

EXAMPLE 12A–14   **(Example 12A–13 Revisited)**

The engineer in Example 12A–13 recently returned from an investment seminar that recommended the **Buy Term and Invest the Difference** (BTID) strategy. Should she use this approach?

SOLUTION

Permanent life policies offer more than death benefits for a defined term, and thus the premiums are higher than for term life policies. If term insurance is purchased, this *difference* in premiums can be invested.

The main disadvantage of BTID is that the insured must actually *invest* the difference (versus spend it!). In addition, investments in permanent life policies grow tax free and other BTID investments might not. Advantages of the BTID strategy include control of the investment portfolio, higher potential investment gains, lower investment fees, greater access to funds, and the fact that the investor's nest egg (cash value) passes on to beneficiaries after death—rather than being kept by the insurance company, as is the case for permanent life insurance policies.

The biggest advantage of the BTID strategy may be that the amount of insurance purchased can be better matched to the insured's need (which varies over the insured's life) to provide for dependents in the case of the insured's death.

PERSONAL BUDGETING

At its essence, money management employs a simple concept: *live within your means.* Yet, this concept is challenged by the abundance of heavily marketed consumer choices, as well as the spending habits of many people. A simple way to organize a personal budget divides spending into:

1. **Needs:** spending on the basics such as housing, groceries, transportation, insurances, energy/utilities, clothing, and other essentials. This includes loan payments that must be made. Since income taxes are usually withheld by the employer, it is common to base other calculations on after-tax income that has subtracted the withholding.

2. **Savings:** building an emergency fund (target 6 months of needs), paying off high-interest loans faster, and saving for the future—retirement, home down payment, a new vehicle, or major travel.

3. **Wants:** discretionary spending on entertainment, meals out, clothing beyond basics, hobbies, recreational travel, and the like.

A common guideline (the 50/20/30 rule) allocates after-tax income as 50% or less for needs, 20% or more for savings, and 30% or less for wants. These percentages are not always achievable. When income is low, all spending may be on needs, but even then the guidelines represent a goal as income increases.

Recognizing and naming types of spending forces individuals to be thoughtful about where their money goes each month. This text has many examples of how a few dollars here and a few dollars there add up to significant sums. Be aware of where your expenses occur and prioritize your spending goals. It is incredibly easy to let wants become needs. For example, taking out a loan to buy a more expensive vehicle or incurring credit card debt on electronics, meals out, or travel converts a want into a required payment. Be intentional about your spending.

EXAMPLE 12A–15

Dave Parish is a mid-career civil engineer with an annual after-tax income of $90,000. His monthly expenses are:

Category	Monthly expense	Category	Monthly expense
House payment	$1250	Groceries	$400
Gasoline (automobile)	180	Utilities (gas, water, electric)	250
Communications (cable, phone, Internet)	255	Car payment	535
Restaurants	350	Hobbies	175
Home essentials	230	All savings	1550
Insurance and taxes	565	Vacations	550
Child support	1000	Entertainment	210

What percentage of Dave's expenses fit into each of the three categories? What adjustments should he consider if his goal is 50/20/30%?

SOLUTION

Categorizing Dave's annual expenses:

Category	Expenses	Total $/year	Percentage of budget
Needs	Home payment, groceries, gas, utilities, car, home, insurances/taxes, child support	$52,920	58.8%
Savings	Retirement account	18,600	20.7
Wants	Communications, restaurants, hobbies, vacations, entertainment	18,480	20.5

Like many of us, Dave's monthly budget is heavier than he'd prefer in the needs category. However, he is meeting his goal for savings! Looking into the future, he could have more money for wants if he keeps his car after it is paid off. That removes the car loan payment from the needs category. In general, this budget could be right for Dave at this life stage.

PROBLEMS

Income Taxes

12A-1 Miriam Anne is a single taxpayer. What federal taxes does she pay if her taxable income is

(a) $5000,

(b) $20,000,

(c) $95,000,

(d) $350,000,

(e) $1 million?

12A-2 (A) John Adams has a $95,000 adjusted gross income from Apple Corp. and allowable itemized deductions of $7200. Mary Eve has a $75,000 adjusted gross income and $3000 of allowable itemized deductions. Compute the total tax they would pay as unmarried individuals. Then compute their tax as a married couple filing a joint return.

12A-3 An unmarried taxpayer with no dependents expects an adjusted gross income of $87,000 in a given year. His nonbusiness deductions are expected to be $7000.

(a) What will his federal income tax be?

(b) He is considering an additional activity expected to increase his adjusted gross income. If this increase should be $15,000 and there should be no change in nonbusiness deductions or exemptions, what will be the increase in his federal income tax?

12A-4 (A) Bill Jackson had a total taxable income of $3000. Bill's employer wants him to work another month during the summer, but Bill had planned to spend the month hiking. If an additional month's work would increase Bill's taxable income by $2000, how much more money would he have after paying the income tax?

12A-5 Amara and her husband, Mosi, are both employed. Amara will have an adjusted gross income this year of $130,000. Mosi has an adjusted gross income of $5000 a month. Amara and Mosi have agreed that Mosi should continue working only until the federal income tax on their joint income tax return becomes $30,700. On what date should Mosi quit his job?

12A-6 (A) A married couple filing jointly have a combined total adjusted gross income of $110,000. They have computed that their allowable itemized deductions are $5000. Compute their federal income tax.

12A-7 An unmarried individual in California with a taxable income of about $75,000 has a federal incremental tax rate of 25% and a state incremental tax rate of 9.3%. What is his combined incremental tax rate?

(a) Assume standard deduction taken.

(b) Assume deductions are itemized.

12A-8 (A) Veronica Marie has an income that places her in the 25% federal and 6.5% state incremental tax brackets. What is her combined tax on a $1000 honorarium that she received recently for a speech?

(a) Assume standard deduction taken.

(b) Assume deductions are itemized.

12A-9 Given the following data, compute your combined income tax rate (CTR) assuming you deduct allowable expenses on your income tax forms: a before-tax MARR of 5%, an inflation rate of 3%, a federal income tax rate of 28%, a state income tax rate of 6%, a local city income tax of 3%, and a capital gains tax rate of 15%, as applicable.
Contributed by D. P. Loucks, Cornell University

12A-10 (A) A $10,000 commercial bond that has a 6% bond rate and matures in 5 years can be purchased for $11,000. Interest is paid at the end of each year for the next 5 years. Find the annual after-tax rate of return of this investment. Assume a 35% tax rate applies.
Contributed by D. P. Loucks, Cornell University

12A-11 Jane Shay operates a management consulting business. The business has been successful and now produces a taxable income of $100,000 per year after all "ordinary and necessary" expenses and depreciation have been deducted. At present the business is operated as a proprietorship; that is, Jane pays personal federal income tax on the entire $100,000. For tax purposes, it is as if she had a job that pays her a $100,000 salary per year.

As an alternative, Jane is considering incorporating the business. If she does, she will pay herself a salary of $40,000 a year from the corporation. The corporation will then pay taxes on the remaining $60,000 and retain the balance of the money as a corporate asset. Thus Jane's two alternatives are to operate the business as a proprietorship or as a corporation. Jane is single and has $3500 of itemized personal deductions. Which alternative will result in a smaller total payment of taxes to the government?

12A-12 Gains from non-business-related investment assets held by individuals that increase in value are subject to taxes. What distinguishes assets taxed at the ordinary rate from those taxes at the capital gains rate?

12A-13 Juan DeBaptist purchased $10,000 in corporate stock on June 1 and sold the stock when its value reached $13,000 on October 26. Ignoring stock transaction fees, what federal taxes did Juan pay on this stock investment if his taxable income is $90,000? Assume a capital gains tax rate of 15%.

12A-14 A married couple (filing jointly) bought an antique armoire at an estate sale, then sold it 6 months later for twice what they paid for it. If their federal taxable income is $80,000 and they paid $750 in taxes on this transaction, how much did they pay for the armoire? Assume capital gains are taxed at a rate of 15%.

12A-15 An investor bought investment property at the beach for $35,000 per acre. Twenty years later she sold the 100-acre lot to a developer for a profit, and paid $1.05 million in taxes as a result of the sale. If capital gains are taxed at 15% and her marginal tax bracket is 35%, what was the price paid by the developer for the lot?

12A-16 You recently bought a mini-supercomputer for $10,000 to allow for tracking and analysis of real-time changes in stock and bond prices. Assume you plan on spending half your time tending to the stock market with this computer and the other half as personal use. Also assume you can depreciate your computer by 20% per year over 5 years (straight line rate). How much tax savings will you have in each of those 5 years, if any? Use a tax rate of 28%.
Contributed by D. P. Loucks, Cornell University

Student Loans

12A-17 If you have a choice between a $3000 deduction and a $3000 tax credit, explain which you would choose and why.

12A-18 An independent student has a modified adjusted gross income of $25,000 and qualifying educational expenses of $18,000. If the American Opportunity Tax Credit is used, what is the amount of the tax credit?

12A-19 Joint filing parents have $30,000 in educational expenses as the first of their three children goes off to college. Assuming the family has a modified adjusted gross income of $80,000, what effect would each of the education deductions/credits listed in Table 12A-2 have on the family's taxes?

12A-20 A single MS engineer started a new job at $70,000 and paid $3000 in interest on her student loans last year. If she uses the federal tax deduction on student loan interest paid, what tax savings does this represent?

12A-21 Federal subsidized Stafford student loans historically had interest calculated at 3.4%, and the government pays the interest while the student in school. What are a student's annual payments if she borrows $2500 per year in subsidized Stafford loans for 4 years and then pays off the loan in 3 years after graduation?

12A-22 Referring to Problem 12A-21 above, compare the total amount paid by the student over the term of school and repaying the loan if the borrowing was a non-subsidized 8% bank loan.

Retirement

12A-23 An engineering professor has contributed $300 per month into her 403b tax-deferred retirement account for the past 24 years. She is now eligible to retire and wants to know her after-tax account balance (she is contemplating a lump sum distribution) if it has earned 0.5% per month over the 24 years. Assume that she is currently in the 25% marginal tax bracket.

12A-24 A newly hired engineer signed up for the 401k plan at her new job. She decided to contribute the maximum amount allowed (assume that the annual federal limit is $18,000 of tax-exempted income) and to take advantage of the match offered by her company ($1 firm-for-$3 employee). What is the size of total monthly contribution to her retirement?

12A-25 Mike just changed jobs, leaving a company after 6 years. He is fully vested, and thus can keep the money his employer deposited in his retirement account. His employer has been contributing $200 per month into a diversified stock fund.

(a) Using average market rates of 7.3%, how much money has accumulated in the account?

(b) If this money is "rolled over" into another retirement account with an 8% annual return today, how much will this be worth after 30 years?

Insurance

12A-26 The required automobile insurance varies by state. Investigate the laws in your state and identify the minimum required coverage for bodily injury per person and per accident, and property per accident.

12A-27 A newly hired professional bought a used car to **E** get to the new job. Automobile insurance is expensive, so he decided to not get any. It is against the law to drive without insurance, but he figures that he won't be fined if he isn't caught. If he is in an accident, he might get sued, but he doesn't have anything so he has nothing to lose. Comment on the ethics of this situation.

12A-28 A $500,000 whole life policy is available for $5150 **A** per year, payable at the beginning of the year. If 10% of this amount pays for insurance, and 90% goes into a savings plan, what is the cash value of the policy after 10 years? The insurance company guarantees a rate of 2.4% per year.

12A-29 A newly hired employee has a choice regarding life insurance. A $250,000 whole life policy is available for an annual premium of $1180, payable at the beginning of each year. At the end of 10 years, the policy has a cash value of $11,600. A term life policy for the same amount is available for an annual premium of $140, with the premium increasing 1% each year. There is no cash value to the term policy.

(a) Which policy would you choose, and why?

(b) If the whole life policy were cashed in after 10 years, what is the rate of return on the savings portion of the whole life policy?

Personal Budgeting

12A-30 Your neighbors have a household after-tax income **A** of $70,000. Their monthly expenses are:

Category	Monthly expense	Category	Monthly expense
House payment	$1500	Groceries	$650
Gasoline	200	Utilities	250
Phone/Internet/TV	250	Car payment	450
Eating out	200	Savings	1000
Home maintenance	233	Vacation &	500
Insurance and taxes	600	entertainment	

(a) What percentage of the family's expenses fit into the *Needs* category?

(b) What percentage of the family's expenses fit into the *Savings* category?

(c) What percentage of the family's expenses fit into the *Wants* category?

(d) They are paying $1050 per year for fire insurance and $3990 per year in taxes on the house as part of their house payment. What is the monthly mortgage payment?

(e) If the family pays 15% of their total income in taxes (combined federal and state), what is their before-tax income?

12A-31 Your neighbors have a household after-tax income of $120,000. Their monthly expenses are:

Category	Monthly expense	Category	Monthly expense
House payment	$2100	Groceries	$550
Gasoline	200	Utilities	250
Phone/Internet/TV	250	Car payment	750
Eating out	800	Savings	900
Home maintenance	333	Vacation &	800
Insurance and taxes	800	entertainment	

(a) What percentage of the family's expenses fit into the *Needs* category?

(b) What percentage of the family's expenses fit into the *Savings* category?

(c) What percentage of the family's expenses fit into the *Wants* category?

(d) They are paying $1600 per year for fire insurance and $17,500 in mortgage payments as part of their house payment. What is their annual property tax that is paid as a part of their house payment?

(e) If the family pays 18.5% of their total income in taxes (combined federal and state), what is their before-tax income?

12A-32 Many people graduate from college with large student loan debt. What category of a personal budget does this debt fall in? Why?

12A-33 A couple wants to purchase a $260,000 house, and they have the required 20% down payment and money for other closing costs. The bank is offering a 30-year mortgage at 4.625% interest, compounded monthly. The couple has an annual after-tax income of $55,000 and other debts totaling $650 per month. *Contributed by Kate Abel, Stevens Institute of Technology*

(*a*) If the maximum debt-to-income ratio (total monthly debt divided by after-tax monthly income) is 43%, can the couple afford to purchase the home?

(*b*) If the couple lives in the house for 30 years, what is the total amount paid for the house, including down payment, principal, and interest?

12A-34 A couple wants to purchase a $170,000 house,
Ⓐ and they have enough saved for a 5% down payment and money for other closing costs. The bank is offering a 30-year mortgage at 5.35% interest, compounded monthly. The couple has an annual after-tax income of $85,000 and other debts totaling $850 per month. Because their down payment is less than 20%, they are required to pay for private mortgage insurance, which costs 1% of the loan amount each year. *Contributed by Kate Abel, Stevens Institute of Technology*

(*a*) If the maximum debt-to-income ratio (total monthly debt divided by after-tax monthly income) is 43%, can the couple afford to purchase the home?

(*b*) If the couple lives in the house for 30 years, what is the total amount paid for the house,

including down payment, principal, interest, and private mortgage insurance?

12A-35 The couple in Problem 12A-33 has only enough money for a 10% down payment and other closing costs. Thus the bank is offering a loan at a higher rate of 5.15% and requiring private mortgage insurance, which costs 1% of the loan amount each year. *Contributed by Kate Abel, Stevens Institute of Technology*

(*a*) If the maximum debt-to-income ratio (total monthly debt divided by after-tax monthly income) is still 43%, can the couple afford to purchase the home?

(*b*) If the couple lives in the house for 30 years, what is the total amount paid for the house, including down payment, principal, interest, and private mortgage insurance?

12A-36 Why is how long you expect to live in a house important to the choice between buying and renting a home? *Contributed by Kate Abel, Stevens Institute of Technology*

(*a*) Can you find credible guidelines for how long you should expect to own a home to justify buying it? What are some of the central concepts?

(*b*) What are the key factors that would lengthen or shorten this period?

(*c*) If a couple will be moving in 3 years, what will be the key factors in the decision of whether or not to buy now?

REPLACEMENT ANALYSIS

Aging Bridges

Mike Siegel/The Seattle Times. Copyright 2013, Seattle Times Company.
Used with Permission.

On May 23, 2013, a 160-foot span of the Skagit River Bridge on I-5 north of Seattle collapsed moments after upper bridge supports were struck by a tractor-trailer with an oversized load. The truck made it safely across, but two other vehicles fell into the water 24 feet below. Three people were rescued without major injuries. The bridge was constructed in 1955 and designed for an expected life of 50 years.

The Skagit River Bridge is rated by the Federal Highway Administration (FHWA) as *functionally obsolete*—it is not designed to today's standards, but it is not necessarily unsafe. The steel through-truss bridge has a fracture critical design, which means that the failure of a single element could cause collapse. There are about 18,000 fracture critical bridges throughout the United States, built mostly between the mid-1950s and late 1970s. Modern construction methods are much more resilient to damage.

In 2007 the I-35W bridge carrying traffic over the Mississippi River between Minneapolis and St. Paul collapsed suddenly during rush hour, killing 13 people and injuring 145. The Minnesota bridge, completed in 1967, was also a fracture critical bridge and was classified as *structurally deficient* by the FHWA. Structural deficiency indicates that the bridge has one or more defects in its support structure or deck and therefore requires maintenance, repair, and eventual rehabilitation or replacement.

The nation's 611,845 bridges have an average age of 43 years, and almost 23% are rated as either structurally deficient, functionally obsolete, or both. The FHWA calculates that more than 30% of U.S. bridges exceed their 50-year design life. The required fiscal investment for reconstruction and renovation poses a significant challenge for federal, state, and local governments—but some progress is being made. ■ ■ ■

Contributed by Letitia M. Pohl, University of Arkansas

1. Decisions on how to allocate funding to upgrade and replace deficient bridges are influenced by both economic and non-economic factors. List three of each.

2. The Skagit River Bridge carries an estimated 71,000 vehicles a day and is a main commercial route between the United States and Canada. How would you calculate the economic impact of the catastrophic failure of the bridge? Compare the economic impact to commuters versus commercial traffic.

3. What factors should be considered when engineers determine whether to either rehabilitate or replace a deficient bridge?

4. The Federal Highway Administration released $1 million in federal emergency funding to the state of Washington the day after the I-5 bridge collapse and almost a month later allocated $15.6 million in federal funding to help rebuild the bridge. Discuss the ethical dilemma of state and local governments that have aging infrastructure to repair before tragedy strikes, but insufficient funding to make the repairs.

5. Both of the bridges described here were routinely inspected and deemed safe for use. Discuss how this inability to predict structural failures complicates the job of transportation officials.

After Completing This Chapter...

The student should be able to:

- Recast an equipment reinvestment decision as a *challenger vs. defender* analysis.

- Use the *replacement analysis decision map* to select the correct economic analysis technique to apply.

- Calculate the *minimum cost life* of economic challengers.

- Incorporate concepts such as *repeatability assumption for replacement analysis* and *marginal cost data for the defender* to select the correct economic analysis techniques.

- Perform replacement problems on an after-tax basis, utilizing the *defender sign change procedure* when appropriate.

- Use spreadsheets for solving before-tax and after-tax replacement analysis problems.

Key Words

cash flow approach	marginal costs	replacement analysis
challenger	minimum cost life	replacement repeatability assumptions
defender	obsolescence	
deterioration	opportunity first cost	

Up to this point in our economic analysis we have considered the evaluation and selection of *new* alternatives. Which new car or production machine should we buy? What new material handling system or ceramic grinder should we install? More frequently, however, economic analysis weighs *existing* versus *new* facilities. For most engineers, the problem is less likely to be one of building a new plant; rather, the goal is more often keeping a present plant operating economically. We are not choosing between new ways to perform the desired task. Instead, we have equipment performing the task, and the question is: Should the existing equipment be retained or replaced? This adversarial situation has given rise to the terms **defender** and **challenger.** The defender is the existing equipment; the challenger is the best available replacement equipment. Economically evaluating the existing defender and its challengers is the domain of **replacement analysis.**

THE REPLACEMENT PROBLEM

Replacement of an existing asset may be appropriate due to obsolescence and deterioration due to aging. In each case, the ability of a previously implemented business asset to produce a desired output is challenged. For cases of obsolescence and aging, it may be economical to replace the existing asset. We define each situation.

Obsolescence: occurs when an asset's technology is surpassed by newer and/or different technologies. Obsolete assets may need to be replaced with newer, more technologically advanced ones.

Deterioration due to aging: costs for maintenance often increase as production machinery and other business assets age. Aging equipment often has a greater risk of breakdowns. Planned replacements can be scheduled to minimize the time and cost of disruptions. Unplanned replacements can be very costly or even, as with an airplane engine, potentially catastrophic.

There are variations of the replacement problem: An existing asset may be abandoned or retired, augmented by a new asset but kept in service, or overhauled to reduce its operating and maintenance costs. These variations are most easily considered as potential new challengers.

Replacement problems are normally analyzed by looking only at the *costs* of the existing and replacement assets. Since the assets typically perform the same function, the value of using the vehicle, machine, or other equipment can be ignored. If the new asset has new features or better performance, these can be included as a cost savings.

Alternatives in a replacement problem almost always have *different lives*. This is because an existing asset will often be kept for at most a few years longer, while the potential replacements may have lives of any length. Because the alternatives have differing lives, economic comparisons often use annual values—annual marginal costs and EUACs. We can calculate present costs, but only as a step in calculating EUAC values.

In industry, as in government, expenditures are normally monitored by means of *annual budgets.* One important facet of a budget is the allocation of money for new capital expenditures, either new facilities or replacement and upgrading of existing facilities.

Replacement analysis may recommend that certain equipment be replaced, with the cost included in the capital expenditures budget. Even if no recommendation to replace now is made, such a recommendation may be made in a year or more. At *some* point, existing equipment will be replaced, either when it is no longer necessary or when better equipment is available. Thus, the question is not *if* the defender will be replaced, but *when* it will be replaced. This leads us to the first question in the defender–challenger comparison:

Should we replace the defender now, or shall we keep it for one or more additional years?

If we do decide to keep the asset for another year, we will often reanalyze the problem next year. The operating environment and costs may change, or new challengers with lower costs or better performance may emerge.

REPLACEMENT ANALYSIS DECISION MAP

Figure 13–1 is a basic decision map for conducting a replacement analysis.

Looking at the map, we can see there are three *replacement analysis techniques* that can be used under different circumstances. The correct replacement analysis technique depends on the data available for the alternatives and how the data behave over time.

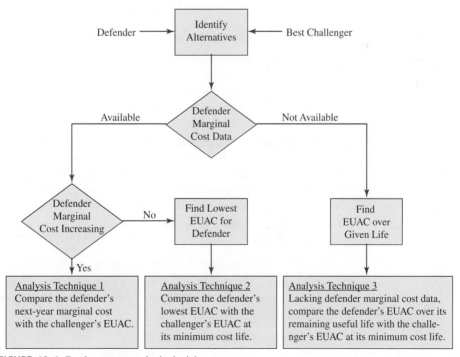

FIGURE 13–1 Replacement analysis decision map.

Each of the replacement analysis techniques typically applies to a specific replacement scenario. In all scenarios, the challenger's minimum EUAC is compared with a specific economic measure of the defender.

- When an existing asset nears the end of its economic life, then technique 1 is typically used because the defender's marginal costs are increasing. We evaluate: Is the marginal cost of using the existing asset for another year lower than the challenger's EUAC?
- When the existing asset is relatively new, but needs have changed or a challenger with better performance has become available, then technique 2 is typically used. We evaluate: Does the defender or the challenger have a lower EUAC for the best life of each?
- When the existing asset will be replaced now or left in place until the next major plant overhaul, then technique 3 is typically used. We evaluate: Is the EUAC of the defender over its life until the next overhaul better than the challenger's EUAC? This would also apply for the duration of a multi-year construction project.

Technique 1 is the core of what is normally meant by the term *replacement problem.*

- The defender is an aging asset whose past purchase, installation, repair, and maintenance costs are all *sunk* costs. Its realizable salvage value is a market price minus removal and sales costs. This value is usually only a small fraction of its initial cost. Costs for repair and maintenance are rising, as are the risks of equipment breakdowns, which can cause unscheduled production shutdowns (and are often the largest cost).
- There may be one or many potential challengers, but the best life for each that minimizes its EUAC is chosen. The best challenger is compared with the defender.

Two of these techniques are applications of previous chapters.

- Technique 2 is the comparison of mutually exclusive alternatives over an indefinite horizon. In Chapter 6, equivalent annual comparisons were used for alternatives with lives that might or might not match. In this case, the EUACs of both the challenger and defender are calculated over different possible lives, identifying the time, if any, when the challenger becomes more cost effective.
- In technique 2, the defender's life that minimizes its EUAC **cannot** be assumed to be the defender's best life, as other time periods may still be better than the best challenger. Even if the challenger is a used asset of the same age, condition, and price, costs for buying and installing it must be included. Any used asset is best considered as just another challenger.
- Technique 3 is the comparison of mutually exclusive alternatives over a selected project horizon.

By looking at the replacement analysis map, we see that the first step is to identify the alternatives. Again, in replacement analysis we are interested in comparing the previously implemented asset (the *defender*) against the best current available *challenger.*

If the defender proves more economical, it will be retained. If the challenger proves more economical, it will be installed.

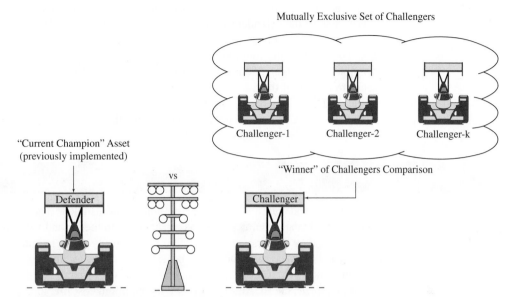

FIGURE 13–2 Defender–challenger comparison.

In this comparison the defender is being evaluated against a challenger that has been selected from a set of mutually exclusive competing challengers. Figure 13–2 illustrates this concept as a drag race between the defender and a challenger. The challenger that is competing against the defender has emerged from an earlier competition among a set of potential challengers. Any of the methods previously discussed in this text for evaluating sets of mutually exclusive alternatives could be used to identify the "best" challenger to race against the defender.

All scenarios and decision map techniques select the challenger in the same way— they choose the challenger with the lowest EUAC. Thus we will start by analyzing what life minimizes a challenger's EUAC. Learning how to analyze this will also illustrate the underlying structure that drives replacement analysis. If assets are kept only a few years, then the EUAC is high because capital costs are spread over a few years. If assets are kept for too long, then the EUAC is high because maintenance and repair costs and the cost of unplanned replacement have become large.

This chapter, like most real world economic analyses, assumes that the year-to-year costs have "smooth" curves or patterns. Real world policies recognize unexpected irregularities in cost by modifying these analyses with practical policies. An example might be to repair an asset until two-thirds of the economic life has been reached, and then replace it the first time a major repair is needed after that.

MINIMUM COST LIFE OF A NEW ASSET—THE CHALLENGER

The **minimum cost life** of any new asset is the number of years at which the equivalent uniform annual cost (EUAC) of ownership is minimized. This minimum cost life is often shorter than either the asset's physical or useful life, because of increasing operating and maintenance costs in the later years of asset ownership. The challenger asset selected to "race" against the defender (in Figure 13–2) is the one having the lowest minimum annual cost of all the competing mutually exclusive challengers.

To calculate the minimum cost life of an asset, we determine the EUAC for each possible life less than or equal to the useful life. As illustrated in Example 13–1, the EUAC tends to be high if the asset is kept only a few years; then it decreases to some minimum value, and then increases again as the asset ages. By identifying the number of years at which the EUAC is a minimum and then keeping the asset for that number of years, we are minimizing the yearly cost of ownership.

EXAMPLE 13–1

A machine costs $7500 and has no salvage value after it is installed. The manufacturer's warranty will pay the first year's maintenance and repair costs. In the second year, maintenance costs will be $900, and they will increase on a $900 arithmetic gradient in subsequent years. Also, operating expenses for the machinery will be $500 the first year and will increase on a $400 arithmetic gradient in the following years. If interest is 8%, compute the machinery's economic life that minimizes the EUAC. That is, find its minimum cost life.

SOLUTION

	If Retired at the End of Year n			
Year, n	EUAC of Capital Recovery Costs: $7500(A/P, 8\%, n)$	EUAC of Maintenance and Repair Costs: $900(A/G, 8\%, n)$	EUAC of Operating Costs: $500 + 400(A/G, 8\%, n)$	EUAC Total
1	$8100	$ 0	$ 500	$8600
2	4206	433	692	5331
3	2910	854	880	4644
4	2264	1264	1062	4589 ←
5	1878	1661	1238	4779
6	1622	2048	1410	5081
7	1440	2425	1578	5443
8	1305	2789	1740	5834
9	1200	3142	1896	6239
10	1117	3484	2048	6650
11	1050	3816	2196	7063
12	995	4136	2338	7470
13	948	4446	2476	7871
14	909	4746	2609	8265
15	876	5035	2738	8648

The total EUAC data are plotted in Figure 13–3. From either the tabulation or the figure, we see that the machinery's minimum cost life is 4 years, with a minimum EUAC of $4589 for each of those 4 years.

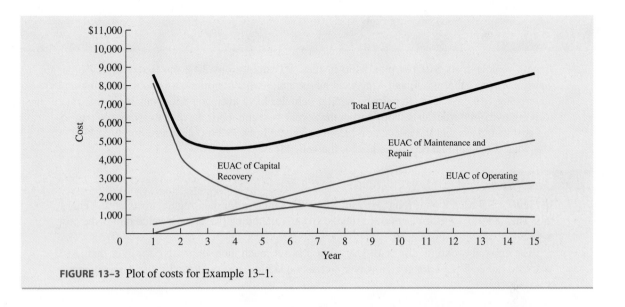

FIGURE 13-3 Plot of costs for Example 13–1.

Looking at Figure 13–3 a bit more closely, we see the effects of each of the individual cost components on total EUAC (capital recovery, maintenance/repair, and operating expense EUACs) and how they behave over time. The total EUAC curve of most assets tends to follow this concave "bathtub" shape—high for short lives due to capital recovery costs and high for long lives due to increased maintenance/repair and operating expenses. The minimum EUAC occurs somewhere between these high points.

Like many pieces of installed equipment, the machine considered in Example 13–1 had no salvage value. However, some assets, like your car, can easily be sold for a value that depends on the car's age and condition. Another possible complication is that repair costs may be reduced in early years by a warranty. The resulting cost curves will look like Figure 13–3, but the calculations are more work and a spreadsheet is very helpful. Example 13–2 illustrates finding the minimum cost life for a new vehicle using spreadsheets.

Example 13–2 illustrates an important fact about replacement problems—there are almost always costs to acquire and install an asset, and there are costs to remove and sell used assets. These costs increase the initial cost of an asset and reduce its net salvage value. Because these costs increase an asset's capital cost, the installation and removal costs may increase the minimum cost life. Equations 13–1 and 13–2 define the PW_t through year t and the $EUAC_t$ for a life of t years. Note that it is easier to include the installation and buying costs in CF_0 (cash flow at time zero) and the PWt calculation. It is easier to include the removal and selling costs in the net salvage value and the $EUAC_t$ calculation. The life with the lowest $EUAC_t$ is the minimum cost life.

$$PW_t = CF_0 + \sum_{j=0}^{t} cost_t/(1+i)^j \qquad (13\text{-}1)$$

$$PW_t = CF_0 + NPV(i, CF_1:CF_t) \qquad (13\text{-}1')$$

$$EUAC_t = PW_t(A/P, i, t) - (SV_t - \text{removal costs})(A/F, i, t) \qquad (13\text{-}2)$$

$$EUAC_t = -PMT(n, i, PW_t, -(SV_t - \text{removal costs})) \qquad (13\text{-}2')$$

EXAMPLE 13–2

A new vehicle costs $19,999 plus $400 in fees. Its value drops 30% the first year, 20% per year for Years 2 through 4, and 15% each additional year. When the car is sold, detailing and advertising will cost $250. Repairs on similar vehicles have averaged $50 annually in lost time (driving to/from the dealer's shop) during the 3-year warranty period. After the warranty period, the cost of repairs and the associated inconvenience climbs at $400 annually. If the MARR is 8%, what is the optimal economic life for the vehicle?

SOLUTION

In Figure 13–4 each PW$_t$ is found by using the NPV function for the irregular costs from Years 1 to t. Then a PMT function is used to find the EUAC$_t$ including the salvage value minus the cost to sell.

The optimal economic life is 10 years, but it is not much more expensive for lives that are 3 years shorter or 5 years longer. However, keeping a vehicle for less than 5 years is significantly more expensive.

	A	B	C	D	E	F
1	$19,999	first cost				
2	$400	cost to buy				
3	$250	cost to sell				
4	30%	salvage value drop yr 1				
5	20%	salvage value drop yr 2-4				
6	15%	salvage value drop yr 5+				
7	$50	repair during 3 yr warranty				
8	$400	gradient for repair after 3 yr warranty				
9	8%	interest rate				
10	year	cost	PW$_t$ without SV	salvage value w/o cost to sell	EUAC$_t$	
11	0	$20,399				
12	1	50	-$20,445	$13,999	$8,332	=PMT($A$9,A12,C12,D12-$A$3)
13	2	50	-20,488	11,199	6,225	
14	3	50	-20,528	8,960	5,283	
15	4	450	-20,859	7,168	4,762	
16	5	850	-21,437	6,092	4,373	
17	6	1250	-22,225	5,179	4,136	
18	7	1650	-23,188	4,402	3,988	
19	8	2050	-24,295	3,742	3,899	
20	9	2450	-25,521	3,180	3,851	
21	10	2850	-26,841	2,703	3,831	economic life & min cost
22	11	3250	-28,235	2,298	3,832	
23	12	3650	-29,684	1,953	3,849	
24	13	4050	-31,173	1,660	3,879	
25	14	4450	-32,688	1,411	3,917	
26	15	4850	-34,217	1,199	3,963	
27						
28	=-$B$11-NPV($A$9,$B$12:B26)					
29	= year 0 + NPV(i, B column)					

FIGURE 13–4 Spreadsheet for vehicle's optimal economic life.

Example 13–2 shows that even with complicated cost structures, it is still relatively easy to find the minimum cost EUAC with the power of spreadsheets.

Example 13–2 also illustrates the assumptions common in replacement analysis.

- Salvage values decline over time.
- Annual costs for operations, maintenance, risk of breakdowns, and so on increase over time.

MARGINAL COST CALCULATIONS

The decision map (see Figure 13–2) starts with two questions regarding marginal costs. *Do we have marginal cost data for the defender?* and *Are the defender's marginal costs increasing on a year-to-year basis?* Let us first define marginal cost and then discuss why it is important to answer these two questions.

Marginal costs, as opposed to an EUAC, are the year-by-year costs of keeping an asset. Therefore, the "period" of any yearly marginal cost of ownership is always *1 year*. In our analysis, marginal cost is compared with EUAC, which is an end-of-year cash flow. Therefore, the marginal cost is also calculated as end-of-year cash flow. The marginal cost of ownership for any year in an asset's life is the cost for *that year only*. To calculate an asset's yearly marginal cost of ownership, it is necessary to have estimates of an asset's net salvage value each year, as well as yearly expenses. Example 13–3 illustrates total marginal costs for all years of an asset.

Choosing to keep an asset for a given year ensures that the marginal cost for that year is incurred. At the beginning of the year, the net salvage value from the previous year is "invested" by keeping the asset. Then, at the end of the year, the new net salvage value is available. Other costs for that year are assumed to be end-of-year cash flows. This is illustrated in Figure 13–5 and formalized in Equation 13-3.

$$\text{MC}_t = \text{Net SV}_{t-1}(1 + i) - \text{Net SV}_t + \text{Cost}_t \tag{13-3}$$

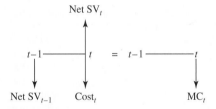

Let us first apply this to year 11 of Example 13–2. The salvage value at the end of year 10 (cell D21) is $2703. Subtracting the $250 cost to sell means than $2453 is forgone if the vehicle is kept for year 11. The net salvage value at the end of year 11 is $2048 (=2298 − 250). The yearly cost for year 11 is $3250 (cell B22). Applying Equation 13-3:

$$\text{MC}_{11} = 2703(1 + 0.08) - 2048 + 3250 = \$3906$$

The minimum EUAC ($3831) for Example 13–2 is at a 10-year life. The marginal cost for year 11 is higher, which is why the EUAC for 11 years is slightly higher. If we were to compute all of the marginal costs in Example 13–2, we would prove that the marginal cost for year 11 is the *first* marginal cost to exceed the EUAC for the previous year.

Example 13–3 illustrates a second way to calculate marginal costs using Equation 13-4. This calculates capital costs as a loss in net salvage value plus interest on the capital tied up for the year.

$$\text{MC}_t = (\text{Net SV}_{t-1} - \text{Net SV}_t) + \text{Net SV}_{t-1}(i) + \text{Cost}_t \qquad (13\text{-}4)$$

EXAMPLE 13–3

A new piece of production machinery has the following costs.

Investment cost $= \$25,000$

Annual operating and maintenance cost $= \$2000$ in Year 1 and then increasing by
$500 per year

Annual cost for risk of breakdown $= \$5000$ per year for 3 years, then
increasing by $1500 per year

Useful life $= 7$ years

MARR $= 15\%$

Calculate the marginal cost of keeping this asset over its useful life.

SOLUTION

From the problem data we can easily find the marginal costs for O&M and risk of breakdowns. However, to calculate the marginal capital recovery cost, we need estimates of each year's market or salvage value.

Year	Market Value
1	$18,000
2	13,000
3	9,000
4	6,000
5	4,000
6	3,000
7	2,500

We can now calculate the machinery's *marginal cost* (year-to-year cost of ownership) over its 7-year useful life.

Year, n	Loss in Market Value in Year n	Interest in Year n	O&M Cost in Year n	Cost of Breakdown Risk in Year n	Total Marginal Cost in Year n
1	$25,000 - 18,000 = \$7000$	$25,000(0.15) = \$3750$	$2000	$ 5,000	$17,750
2	$18,000 - 13,000 = 5000$	$18,000(0.15) = 2700$	2500	5,000	15,200
3	$13,000 - 9,000 = 4000$	$13,000(0.15) = 1950$	3000	5,000	13,950
4	$9,000 - 6,000 = 3000$	$9,000(0.15) = 1350$	3500	6,500	14,350
5	$6,000 - 4,000 = 2000$	$6,000(0.15) = 900$	4000	8,000	14,900
6	$4,000 - 3,000 = 1000$	$4,000(0.15) = 600$	4500	9,500	15,600
7	$3,000 - 2,500 = 500$	$3,000(0.15) = 450$	5000	11,000	16,950

Notice that each year's total marginal cost includes loss in market value, interest, O&M cost, and cost for risk of breakdowns. For example, the Year-5 marginal cost of \$14,900 is calculated as $2000 + 900 + 4000 + 8000$.

Note that in year 1 the marginal cost and the EUAC are the same. In each later year the marginal cost *is* the cost to keep the asset for one more year. The pattern in Example 13–3 is typical. In early years the marginal costs are decreasing because the capital costs are being spread over more years. In later years the marginal costs are increasing because yearly costs are increasing.

Do We Have Marginal Cost Data for the Defender?

Our decision map indicates that it is necessary to know whether marginal cost data are available for the defender asset to determine the appropriate replacement technique to use. Usually in engineering economic problems, annual savings and expenses are given for all alternatives. However, as in Example 13–3, it is also necessary to estimate year-to-year salvage values to calculate total marginal costs. If the total marginal costs for the defender can be calculated, and if the data are increasing from year to year, then *replacement analysis technique 1* should be used to compare the defender to the challenger. For Example 13–3 this would include years 4 to 7.

Are These Marginal Costs Increasing?

We have seen that it is important to know whether the marginal cost for the defender is increasing from year to year. This is determined by inspecting the total marginal cost of ownership of the defender over its remaining life. In most replacement analyses the defender is nearing the end of its economic life. The question usually is, Should we replace it now, next year, or perhaps the year after? In the early years of an asset's life we rarely analyze whether it is time for replacement. Thus, the defender's marginal costs are usually increasing, as shown in Example 13–4.

EXAMPLE 13-4

An asset purchased 5 years ago for $75,000 can be sold today for $15,000. Operating expenses will be $10,000 this year, but these will increase by $1500 per year. It is estimated that the asset's market value will decrease by $1000 per year over the next 5 years. If the MARR used by the company is 15%, calculate the total marginal cost of ownership of this old asset (that is, the defender) for each of the next 5 years.

SOLUTION

We calculate the total marginal cost of maintaining the old asset for the next 5-year period as follows:

Year, n	Loss in Market Value in Year n	Interest in Year n	Operating Cost in Year n	Marginal Cost in Year n
1	$15,000 - 14,000 = $1000	$15,000(0.15) = $2250	$10,000	$13,250
2	14,000 - 13,000 = 1000	14,000(0.15) = 2100	11,500	14,600
3	13,000 - 12,000 = 1000	13,000(0.15) = 1950	13,000	15,950
4	12,000 - 11,000 = 1000	12,000(0.15) = 1800	14,500	17,300
5	11,000 - 10,000 = 1000	11,000(0.15) = 1650	16,000	18,650

We can see that marginal costs increase in each subsequent year of ownership. When the condition of increasing marginal costs for the defender has been met, then the defender–challenger comparison should be made with *replacement analysis technique 1*.

REPLACEMENT ANALYSIS TECHNIQUE 1: DEFENDER MARGINAL COSTS CAN BE COMPUTED AND ARE INCREASING

When our first method of analyzing the defender asset against the best available challenger is used, the basic comparison involves *the defender's marginal cost data and the challenger's minimum cost life data.*

When the defender's marginal cost is increasing from year to year, we will maintain that defender as long as the marginal cost of keeping it one more year is less than the challenger's minimum EUAC. Thus our decision rule is as follows:

> *Maintain the defender as long as the marginal cost of ownership for one more year is less than the challenger's minimum EUAC. When the defender's marginal cost becomes greater than the challenger's minimum EUAC, then replace the defender with the challenger.*

One can see that this technique assumes that the current best challenger, with its minimum EUAC, will be available and unchanged in the future. However, it is easy to update a replacement analysis when marginal costs for the defender change or when there is a change in the cost and/or performance of available challengers. Example 13–5 illustrates the use of this technique for comparing defender and challenger assets.

EXAMPLE 13–5

Taking the machinery in Example 13–3 as the *challenger* and the machinery in Example 13–4 as the *defender,* use *replacement analysis technique 1* to determine when, if at all, a replacement decision should be made.

SOLUTION

Replacement analysis technique 1 should be used only in the condition of increasing marginal costs for the defender. Since these marginal costs are increasing for the defender (from Example 13–4), we can proceed by comparing defender marginal costs against the minimum EUAC of the challenger asset. In Example 13–3 we calculated only the marginal costs of the challenger; thus it is necessary to calculate the challenger's minimum EUAC. The EUAC of keeping this asset for each year of its useful life is worked out as follows.

Year, n	Challenger Total Marginal Cost in Year n	Present Cost if Kept Through Year n (PC_n)		EUAC if Kept Through Year n
1	$17,750	$[17,750(P/F,15\%,1)]$	$\times$ $(A/P,15\%,1) =$	$17,750
2	15,200	$PC_1 + 15,200(P/F,15\%,2)$	$\times$ $(A/P,15\%,2) =$	16,560
3	13,950	$PC_2 + 13,950(P/F,15\%,3)$	$\times$ $(A/P,15\%,3) =$	15,810
4	14,350	$PC_3 + 14,350(P/F,15\%,4)$	$\times$ $(A/P,15\%,4) =$	15,520
5	14,900	$PC_4 + 14,900(P/F,15\%,5)$	$\times$ $(A/P,15\%,5) =$	15,430 $\leftarrow$
6	15,600	$PC_5 + 15,600(P/F,15\%,6)$	$\times$ $(A/P,15\%,6) =$	15,450
7	16,950	$PC_6 + 16,950(P/F,15\%,7)$	$\times$ $(A/P,15\%,7) =$	15,580

A minimum EUAC of $15,430 is attained for the challenger at Year 5, which is the challenger's *minimum cost life.* We proceed by comparing this value against the *marginal* costs of the defender from Example 13–4:

Year, n	Defender Total Marginal Cost in Year n	Challenger Minimum EUAC	Comparison Result and Recommendation
1	$13,250	$15,430	Since $13,250 is *less than* $15,430, keep defender.
2	14,600	15,430	Since $14,600 is *less than* $15,430, keep defender.
3	15,950	15,430	Since $15,950 is *greater than* $15,430, replace defender.
4	17,300		
5	18,650		

Based on the data given for the challenger and for the defender, we would keep the defender for 2 more years and then replace it with the challenger because at that point the defender's marginal cost of another year of ownership would be greater than the challenger's minimum EUAC.

We previously noted that the marginal cost in year 11 of Example 13–2 exceeded the EUAC for the minimum cost life of 10 years. In Example 13–5 the challenger's minimum cost life is 5 years with an EUAC of $15,430. In Example 13–3 the marginal cost for this challenger in year 6 was calculated to be $15,600. Year 6 is the year where it is less costly to buy a new asset to use for 5 years than to extend the existing asset's life for a sixth year.

REPLACEMENT REPEATABILITY ASSUMPTIONS

The decision to use the challenger's minimum EUAC reflects two assumptions: the best challenger will be available "with the same minimum EUAC" in the future; and the period of needed service is indefinitely long. In other words, we assume that once the decision has been made to replace, there will be an indefinite cycle of replacement with the current best challenger asset. These assumptions must be satisfied for our calculations to be correct into an indefinite future. However, because the near future is economically more important than the distant future, and because our analysis is done with the *best data currently available*, our results and recommendations are robust or stable for reasonable changes in the estimated data.

The repeatability assumptions together are much like the repeatability assumptions that allowed us to use the annual cost method to compare competing alternatives with different useful lives. Taken together, we call these the **replacement repeatability assumptions.** They allow us to greatly simplify comparing the defender and the challenger.

Stated formally, these two assumptions are:

1. The currently available best challenger will continue to be available in subsequent years and will be unchanged in its economic costs. When the defender is ultimately replaced, it will be replaced with this challenger. Any challengers put into service will also be replaced with the same currently available challenger.
2. The period of needed service of the asset is indefinitely long. Thus the challenger asset, once put into service, will continuously replace itself in repeating cycles.

If these two assumptions are satisfied completely, then our calculations are exact. Often, however, future challengers represent further improvements so that Assumption 1 is not satisfied. While the calculations are no longer exact, the repeatability assumptions allow us to make the best decision we can with the data we have.

If the defender's marginal cost is increasing, once it rises above the challenger's minimum EUAC, it will continue to be greater. Under the repeatability assumptions, we would never want to incur a defender's marginal cost that was greater than the challenger's minimum EUAC. Thus, we use replacement analysis technique 1 when the defender's marginal costs are increasing.

REPLACEMENT ANALYSIS TECHNIQUE 2: DEFENDER MARGINAL COSTS CAN BE COMPUTED AND ARE NOT INCREASING

If the defender's marginal costs do not increase, we have no guarantee that *replacement analysis technique 1* will produce the alternative that is of the greatest economic advantage. Consider the new asset in Example 13–3, which has marginal costs that begin at a high of $17,750, then *decrease* over the next years to a low of $13,950, and then *increase* thereafter to $16,950 in Year 7. If evaluated *one year after implementation*, the asset would not have increasing marginal costs. Defenders in their early stages typically do not fit the requirements of *replacement analysis technique 1*. In the situation graphed in Figure 13–3, such defender assets would be in the downward slope of a concave marginal cost curve.

Example 13–6 details why *replacement analysis technique 1* cannot be applied when defenders do not have consistently increasing marginal cost curves. Instead we apply *replacement analysis technique 2*. That is, we calculate the defender's minimum EUAC to see whether the replacement should occur immediately. If not, as shown in Example 13–6, the replacement occurs after the defender's minimum cost life when the marginal costs are increasing. Then *replacement analysis technique 1* applies again.

The minimum EUAC of the defender for 3 years is $14,618, which is less than that of the challenger's minimum EUAC of $15,430. Thus, under the replacement repeatability assumptions, we will keep the defender for at least 3 years. We must still decide how much longer.

EXAMPLE 13–6

Let us look again at the defender and challenger assets in Example 13–5. This time let us arbitrarily change the defender's marginal costs for its 5-year useful life. Now when, if at all, should the defender be replaced with the challenger?

Year, n	Defender Total Marginal Cost in Year n
1	$16,000
2	14,000
3	13,500
4	15,300
5	17,500

SOLUTION

In this case the defender's total marginal costs are *not* consistently increasing from year to year. However, if we ignore this fact and apply *replacement analysis technique 1*, the recommendation would be to replace the defender now, because the defender's marginal cost for the first year ($16,000) is greater than the minimum EUAC of the challenger ($15,430). This would be the wrong choice.

Since the defender's marginal costs are below the challenger's minimum EUAC in the second through fourth years, we must calculate the EUAC of keeping the defender asset in each of its remaining 5 years, at $i = 15\%$.

Year, n	Present Cost if Kept n Years (PC_n)		EUAC if Kept n Years
1	$16,000(P/F, 15\%, 1)$	$\times$	$(A/P, 15\%, 1) = \$16,000$
2	$PC_1 + 14,000(P/F, 15\%, 2)$	$\times$	$(A/P, 15\%, 2) = 15,070$
3	$PC_2 + 13,500(P/F, 15\%, 3)$	$\times$	$(A/P, 15\%, 3) = 14,618$
4	$PC_3 + 15,300(P/F, 15\%, 4)$	$\times$	$(A/P, 15\%, 4) = 14,754$
5	$PC_4 + 17,500(P/F, 15\%, 5)$	$\times$	$(A/P, 15\%, 5) = 15,162$

The defender's EUAC begins to rise in Year 4, because the marginal costs are increasing, and because they are above the defender's minimum EUAC. Thus, we can use *replacement analysis technique 1* for Year 4 and later. The defender's marginal cost in Year 4 is $15,300, which is $130 below the challenger's minimum EUAC of $15,430. Since the defender's marginal cost of $17,500 is higher in Year 5, we replace it with the new challenger at the end of Year 4. Notice that we did *not* keep the defender for its minimum cost life of 3 years, we kept it for 4 years.

If the challenger's minimum EUAC were less than the defender's minimum EUAC of $14,618, then the defender would be immediately replaced.

Example 13–6 illustrates several potentially confusing points about replacement analysis.

- If the defender's marginal cost data is not increasing, the defender's minimum EUAC must be calculated.
- If the defender's minimum EUAC exceeds the challenger's minimum EUAC, then replace immediately. If the defender's minimum EUAC is lower than the challenger's minimum EUAC, then under the replacement repeatability assumptions the defender will be kept *at least* the number of years for its minimum EUAC.
- After this number of years, then replace when the defender's increasing marginal cost exceeds the challenger's minimum EUAC.

The problem statement for Example 13–7 uses Equation 13-3 to calculate the defender's marginal costs. Then the solution to Example 13–7 details the calculation of the minimum EUAC when the defender's data is presented as costs and salvage values in each year rather than as marginal costs. Notice that this is calculated the same way as the minimum cost life was calculated for new assets—the challengers.

Now to apply *replacement analysis technique 2* to Example 13–7, we ask: Is the challenger's minimum EUAC higher or lower than the defender's minimum EUAC of $1058? If the challenger's minimum EUAC is lower, then we replace the defender now.

EXAMPLE 13-7

A 5-year-old machine, whose current market value is $5000, is being analyzed to determine its minimum EUAC at a 10% interest rate. Salvage value and maintenance estimates and the corresponding marginal costs are given in the following table.

	Data		Calculating Marginal Costs		
Year	Salvage Value	O&M Cost	$S_{t-1}(1+i)$	$-S_t$	Marginal Cost
0	$5000				
1	4000	$ 0	$5500	$-4000	$1500
2	3500	100	4400	-3500	1000
3	3000	200	3850	-3000	1050
4	2500	300	3300	-2500	1100
5	2000	400	2750	-2000	1150
6	2000	500	2200	-2000	700
7	2000	600	2200	-2000	800
8	2000	700	2200	-2000	900
9	2000	800	2200	-2000	1000
10	2000	900	2200	-2000	1100
11	2000	1000	2200	-2000	1200

SOLUTION

Because the marginal costs have a complex, nonincreasing pattern, we must calculate the defender's minimum EUAC.

			If Retired at End of Year n		
Years Kept, n	Salvage Value (S) at End of Year n	Maintenance Cost for Year	EUAC of Capital Recovery $(P-S) \times$ $(A/P, 10\%, n) + Si$	EUAC of Maintenance $100(A/G, 10\%, n)$	Total EUAC
0	$P = $5000				
1	4000	$ 0	$1100 + 400	$ 0	$1500
2	3500	100	864 + 350	48	1262
3	3000	200	804 + 300	94	1198
4	2500	300	789 + 250	138	1177
5	2000	400	791 + 200	181	1172
6	2000	500	689 + 200	222	1111
7	2000	600	616 + 200	262	1078
8	2000	700	562 + 200	300	1062
9	2000	800	521 + 200	337	1058 ←
10	2000	900	488 + 200	372	1060
11	2000	1000	462 + 200	406	1068

A minimum EUAC of $1058 is computed at Year 9 for the existing machine. Notice that the EUAC begins to increase with n when the marginal cost in Year 10 exceeds the EUAC for 9 years.

Under the repeatability assumptions, if the challenger's minimum EUAC is higher, we would keep the defender at least 9 years. Replacement would occur in Year 10 or later when the defender's marginal costs exceed the challenger's minimum EUAC. Allowing for better challengers, we may replace the defender whenever a new challenger has an EUAC that is lower than $1058.

Example 13–8 illustrates the common situation of a current defender that may be kept if overhauled. This can also be analyzed as a potential new challenger.

EXAMPLE 13–8

We must decide whether existing (defender) equipment in an industrial plant should be replaced. A $4000 overhaul must be done now if the equipment is to be retained in service. Maintenance is $1800 in each of the next 2 years, after which it increases by $1000 each year. The defender has no present or future salvage value. The equipment described in Example 13–1 is the challenger (EUAC = $4589). Should the defender be kept or replaced if the interest rate is 8%?

FIGURE 13–5 Overhaul and maintenance costs for the defender in Example 13–7.

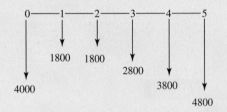

SOLUTION

The first step is to determine the defender's lowest EUAC. The pattern of overhaul and maintenance costs (Figure 13–5) suggests that if the overhaul is done, the equipment should be kept for several years. The computation is as follows:

	If Retired at End of Year *n*		
Year, *n*	EUAC of Overhaul $4000(A/P, 8%, n)	EUAC of Maintenance $1800 + $1000 Gradient from Year 3 on	Total EUAC
1	$4320	$1800	$6120
2	2243	1800	4043
3	1552	1800 + 308*	3660 ←
4	1208	1800 + 683†	3691
5	1002	1800 + 1079	3881

*For the first 3 years, the maintenance is $1800, $1800, and $2800. Thus, EUAC $= 1800 + 1000(A/F, 8\%, 3) = 1800 + 308$.

†EUAC $= 1800 + 1000(P/G, 8\%, 3)(P/F, 8\%, 1)(A/P, 8\%, 4) = 1800 + 683$.

The lowest EUAC of the overhauled defender is $3660. In Example 13–1, the challenger's minimum cost life was 4 years with an EUAC of $4589. If we assume the equipment is needed for at least 4 years, the overhauled defender's EUAC ($3660) is less than the challenger's EUAC ($4589). Overhaul the defender.

If the defender's and challenger's cost data do not change, we can use *replacement analysis technique 1* to determine when the overhauled defender should be replaced. We know from the minimum EUAC calculation that the defender should be kept at least 3 years. Is this the best life? The following table computes the marginal cost to answer this question.

Year, n	Overhaul Cost	Maintenance Cost	Marginal Cost to Extend Service
0	$4000	$ 0	
1	0	1800	$6120 = 4000(1.08) + 1800
2	0	1800	1800
3	0	2800	2800
4	0	3800	3800
5	0	4800	4800

Year 5 is the first year after Year 3, which has the overhauled defender's lowest EUAC (in which the $4800 marginal cost exceeds the challenger's $4589 minimum EUAC). Thus, the overhauled defender should be kept 4 more years if costs do not change. (Note that if the defender can be overhauled again after 3 or 4 years, that might be an even better choice.)

REPLACEMENT ANALYSIS TECHNIQUE 3: WHEN DEFENDER MARGINAL COST DATA ARE NOT AVAILABLE

In our third case, we simply compare the defender's *EUAC over its stated useful life and the challenger's minimum EUAC*. Pick the EUAC that is lower.

If the defender's marginal cost data is not known and cannot be estimated, it is impossible to apply *replacement analysis techniques 1 or 2* to decide *when* the defender should be replaced. Instead we must assume that the defender's stated useful life is the only one to consider. From a student problem-solving perspective, the defender in Example 13–8 might be described as follows.

The defender can be overhauled for $4800 to extend its life for 5 years. Maintenance costs will average $3000 per year, and there will be no salvage value. In this case the only possibility is to compare the defender's EUAC for a 5-year life with the best challenger.

In the real world, the most likely scenario for this approach involves a facility-wide overhaul every 3, 5, 10, etc. years. Pipelines and many process plants, such as refineries, chemical plants, and steel mills, must shut down to do major maintenance. All equipment is overhauled or replaced with a new challenger as needed, and the facility is expected to operate until the next maintenance shutdown.

The defender's EUAC over its remaining useful life is compared with the challenger's EUAC at its minimum cost life, and the lower cost is chosen. However, in making this basic comparison an often complicating factor is deciding what first cost to assign to the challenger and the defender.

COMPLICATIONS IN REPLACEMENT ANALYSIS

Defining Defender and Challenger First Costs

Because the defender is already in service, analysts often misunderstand what first cost to assign it. Example 13–9 demonstrates this problem.

EXAMPLE 13–9

A model SK-30 was purchased 2 years ago for $1600; it has been depreciated by straight-line depreciation using a 4-year life and zero salvage value. Because of recent innovations, the current price of the SK-30 is $995. An equipment firm has offered a trade-in allowance of $350 for the SK-30 on a new $1200 model EL-40. Some discussion revealed that without a trade-in, the EL-40 can be purchased for $1050. Thus, the originally quoted price of the EL-40 was overstated to allow a larger trade-in allowance. The true current market value of the SK-30 is probably only $200. In a replacement analysis, what value should be assigned to the SK-30?

SOLUTION

In the example, five different dollar amounts relating to the SK-30 have been outlined:

1. *Original cost:* It cost $1600 2 years ago.
2. *Present cost:* It now sells for $995.
3. *Book value:* The original cost less 2 years of depreciation is $1600 - \frac{2}{4}(1600 - 0) = \$800$.
4. *Trade-in value:* The offer was $350.
5. *Market value:* The estimate was $200.

We know that an economic analysis is based on the current situation, not on the past. We refer to past costs as *sunk* costs to emphasize this. These costs cannot be altered, and they are not relevant. (There is one exception: past costs may affect present or future income taxes.)

We want to use actual cash flows for each alternative. Here the question is, What value should be used in an economic analysis for the SK-30? The relevant cost is the equipment's present market value of $200. Neither the original cost, the present cost, the book value, nor the trade-in value is relevant.

At first glance, an asset's trade-in value would appear to be a suitable present value for the equipment. Often the trade-in price is inflated *along with* the price for the new item. (This practice is so common in new-car showrooms that the term *overtrade* is used to describe the excessive portion of the trade-in allowance. The buyer is also quoted a higher price for the new car.) Distorting the defender's present value, or the challenger's price can be serious because these distortions do not cancel out in an economic analysis.

Example 13–9 illustrated that of the several different values that can be assigned to the defender, the correct one is the present market value. If a trade-in value is obtained, care should be taken to ensure that it actually represents a fair market value.

Determining the value for the challenger's installed cost should be less difficult. In such cases the first cost is usually made up of purchase price, sales tax, installation costs, and other items that occur initially on a one-time basis if the challenger is selected. These values are usually rather straightforward to obtain. One aspect to consider in assigning a first cost to the challenger is the defender's potential market or salvage value. One must not arbitrarily subtract the defender's disposition value from the challenger's first cost, for this practice can lead to an incorrect analysis.

As described in Example 13–9, the correct first cost to assign to the defender SK-30 is its $200 current market value. This value represents the present economic benefit that we would be *forgoing* to keep the defender. This can be called our **opportunity first cost.** If, instead of assuming that this is the defender's *opportunity cost*, we assume it is a *cash benefit* to the challenger, a potential error arises.

The error lies in incorrect use of a *cash flow* perspective when the lives of the challenger and the defender are not equal, which is usually the case. Subtracting the defender's salvage value from the challenger's first cost is called the **cash flow approach**. From this perspective, keeping the defender in place causes $0 in cash to flow, but selecting the challenger causes the cash flow now to equal the challenger's first cost minus the defender's salvage value.

If the lives of the defender and the challenger are the same, then the cash flow perspective will lead to the correct answer. However, it is normally the case that the defender is an aging asset with a relatively short horizon of possible lives and the challenger is a new asset with a longer life. The *opportunity cost* perspective will always lead to the correct answer, so it is the one that should be used.

For example, consider the SK-30 and EL-40 from Example 13–9. It is reasonable to assume that the 2-year-old SK-30 has 3 years of life left and that the new EL-40 would have a 5-year life. Assume that neither will have any salvage value at the end of its life. Compare the difference in their annual capital costs with the correct *opportunity cost* perspective and the incorrect *cash flow* perspective.

SK-30		**EL-40**	
Market value	$200	First cost	$1050
Remaining life	3 years	Useful life	5 years

Looking at this from an *opportunity cost* perspective, the annual cost comparison of the first costs is

$$\text{Annualized first cost}_{SK\text{-}30} = \$200(A/P, 10\%, 3) = \$80$$

$$\text{Annualized first cost}_{EL\text{-}40} = \$1050(A/P, 10\%, 5) = \$277$$

The *difference* in annualized first cost between the SK-30 and EL-40 is

$$\text{AFC}_{EL\text{-}40} - \text{AFC}_{SK\text{-}30} = \$277 - \$80 = \$197$$

Now using an incorrect *cash flow* perspective to look at the first costs, we can calculate the *difference* due to first cost between the SK-30 and EL-40.

$$\text{Annualized first cost}_{\text{SK-30}} = \$0(A/P, 10\%, 3) = \$0$$
$$\text{Annualized first cost}_{\text{EL-40}} = (\$1050 - 200)(A/P, 10\%, 5) = \$224$$
$$\text{AFC}_{\text{EL-40}} - \text{AFC}_{\text{SK-30}} = \$224 - \$0 = \$224$$

When the defender's remaining life (3 years) differs from the challenger's useful life (5 years), the two perspectives yield different results. The correct difference of $197 is shown by using the *opportunity cost* approach, and an inaccurate difference of $224 is obtained if the *cash flow* perspective is used. From the opportunity cost perspective, the $200 is spread out over 3 years as a cost to the defender, and in the cash flow perspective, the opportunity cost is spread out over 5 years as a benefit to the challenger. Spreading the $200 over 3 years in one case and 5 years in the other case does not produce equivalent annualized amounts. Because of the difference in the lives of the assets, the annualized $200 opportunity cost for the defender cannot be called an equivalent benefit to the challenger.

In the case of unequal lives, the correct method is to assign the defender's current market value as its Time-0 opportunity costs, rather than subtracting this amount from the challenger's first cost. Because the cash flow approach yields an incorrect value when challenger and defender have unequal lives, the *opportunity cost* approach for assigning a first cost to the challenger and defender assets should *always* be used.

REPEATABILITY ASSUMPTIONS NOT ACCEPTABLE

Under certain circumstances, the repeatability assumptions described earlier may not apply. Then replacement analysis techniques 1, 2, and 3 may not be valid. For instance, there may be a specific study period instead of an indefinite need for the asset. For example, consider the case of phasing out production after a certain number of years—perhaps a person who is about to retire is closing down a business and selling all the assets. Another example is production equipment such as molds and dies that are no longer needed when a new model with new shapes is introduced. Yet another is a construction camp that may be needed for only a year or two or three.

This specific study period could potentially be any number of years relative to the lives of the defender and the challenger, such as equal to the defender's life, equal to the challenger's life, less than the defender's life, greater than the challenger's life, or somewhere between the lives of the defender and challenger. The analyst must be explicit about the challenger's and defender's economic costs and benefits, as well as residual or salvage values at the end of the specific study period. In this case the repeatability replacement assumptions do not apply; costs must be analyzed over the study period. The analysis techniques in the decision map also may not apply when future challengers are not assumed to be identical to the current best challenger. This concept is discussed in the next section.

FIGURE 13–6 Two possible ways the EUAC of future challengers may decline.

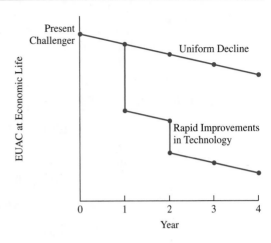

A Closer Look at Future Challengers

We defined the challenger as the best available alternative to replace the defender. But over time, the best available alternative can indeed change. And given the trend in our technological society, it seems likely that future challengers will be better than the present challenger. If so, the prospect of improved future challengers may affect the present decision between the defender and the challenger.

Figure 13–6 illustrates two possible estimates of future challengers. In many technological areas it seems likely that the equivalent uniform annual costs associated with future challengers will decrease by a constant amount each year. In other fields, however, a rapidly changing technology will produce a sudden and substantially improved challenger—with decreased costs or increased benefits. The uniform decline curve of Figure 13–6 reflects the assumption that each future challenger has a minimum EUAC that is a fixed amount less than the previous year's challenger. This assumption, of course, is only one of many possible assumptions that could be made regarding future challengers.

If future challengers will be better than the present challenger, what impact will this have on an analysis now? The prospect of better future challengers may make it more desirable to retain the defender and to reject the present challenger. By keeping the defender for now, we may be able to replace it later by a better future challenger. Or, to state it another way, the present challenger may be made less desirable by the prospect of improved future challengers.

As engineering economic analysts, we must familiarize ourselves with potential technological advances in assets targeted for replacement. This part of the decision process is much like the search for all available alternatives, from which we select the best. Upon finding out more about what alternatives and technologies are emerging, we will be better able to understand the repercussions of investing in the current best available challenger. Selecting the current best challenger asset can be particularly risky when (1) the costs are very high and/or (2) the challenger's useful minimum cost life is relatively long (5–10 years or more). When one or both of these conditions exist, it may be better to keep or even augment the defender asset until better future challengers emerge.

There are, of course, many assumptions that *could* be made regarding future challengers. However, if the replacement repeatability assumptions do not hold, the analysis needed becomes more complicated.

AFTER-TAX REPLACEMENT ANALYSIS

As described in Chapter 12, an after-tax analysis provides greater realism and insight. Tax effects can alter before-tax recommendations. After-tax effects may influence calculations in the defender–challenger comparisons discussed earlier. Consequently, one should always perform or check these analyses on an after-tax basis.

Marginal Costs on an After-Tax Basis

Marginal costs on an after-tax basis represent the cost that would be incurred through ownership of the defender *in each year*. On an after-tax basis we must consider the effects of ordinary taxes as well as gains and losses due to asset disposal. Consider Example 13–10.

EXAMPLE 13–10

Refer to Example 13–3, where we calculated the before-tax marginal costs for a new piece of production machinery. Calculate the asset's after-tax marginal costs considering this additional information.

- Depreciation is by the straight-line method, with $S = \$0$ and $n = 5$ years, so $d_t = (\$25,000 - \$0)/5 = \$5000$.
- Ordinary income, recaptured depreciation, and losses on sales are taxed at a rate of 40%.
- The after-tax MARR is 10%.

Some classes skip or have not yet covered Chapter 10's explanation of expected value. Thus, the expected cost for risk of breakdowns is described here as an insurance cost.

SOLUTION

The after-tax marginal cost of ownership will involve the following elements: incurred or forgone loss or recaptured depreciation, interest on invested capital, tax savings due to depreciation, and annual after-tax operating/maintenance and insurance. Figure 13–7 shows example cash flows for the marginal cost detailed in Table 13–1.

As a refresher of the recaptured depreciation calculations in Chapter 12:

The market value in Year $0 = 25,000$.

The market value decreases to $18,000 at Year 1.

The book value at Year $1 = 25,000 - 5000 = \$20,000$.

So loss on depreciation $= 20,000 - 18,000 = \$2000$.

This results in a tax savings of $(2000)(0.4) = \$800$.

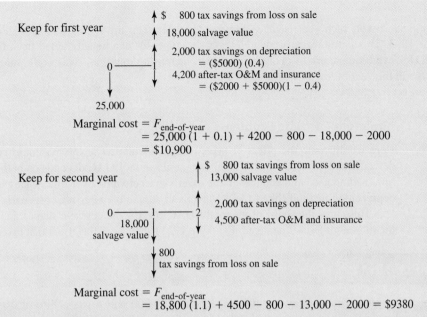

Keep for first year

$ 800 tax savings from loss on sale
18,000 salvage value
2,000 tax savings on depreciation
= ($5000) (0.4)
4,200 after-tax O&M and insurance
= ($2000 + $5000)(1 − 0.4)

0 ——— 1

25,000

$$\text{Marginal cost} = F_{\text{end-of-year}}$$
$$= 25,000\,(1 + 0.1) + 4200 - 800 - 18,000 - 2000$$
$$= \$10,900$$

Keep for second year

$ 800 tax savings from loss on sale
13,000 salvage value
2,000 tax savings on depreciation
4,500 after-tax O&M and insurance

0 ——— 1 ——— 2
18,000
salvage value

800
tax savings from loss on sale

$$\text{Marginal cost} = F_{\text{end-of-year}}$$
$$= 18,800\,(1.1) + 4500 - 800 - 13,000 - 2000 = \$9380$$

FIGURE 13–7 Cash flow diagrams and calculations for marginal cost.

TABLE 13–1 Marginal Costs of Ownership

Year	Market Value	Book Value	Recaptured Depr. or Loss	Taxes or Tax Savings	After-Tax Market Value
0	$25,000	$25,000			$25,000
1	18,000	20,000	−$2000	−$800	18,800
2	13,000	15,000	−2000	−800	13,800
3	9,000	10,000	−1000	−400	9,400
4	6,000	5,000	1000	400	5,600
5	4,000	0	4000	1600	2,400
6	3,000		3000	1200	1,800
7	2,500		2500	1000	1,500

	Col. B	Col. C	Col. D		Col. F	= C + D + F − B
	After-Tax	Beg. Yr	Tax Savings	O&M and	After-Tax	Marginal
Year	Market Value	Value × (1 + i)	from Depr. Deduct.	Insurance Cost	Annual Expense	Cost
0	$25,000					
1	18,800	$27,500	−$2000	$ 7,000	$4200	$10,900
2	13,800	20,680	−2000	7,500	4500	9,380
3	9,400	15,180	−2000	8,000	4800	8,580
4	5,600	10,340	−2000	10,000	6000	8,740
5	2,400	6,160	−2000	12,000	7200	8,960
6	1,800	2,640	0	14,000	8400	9,240
7	1,500	1,980	0	16,000	9600	10,080

The marginal cost in each year is much lower after taxes than the pretax numbers shown in Example 13–3. This is true because depreciation and expenses can be subtracted from taxable income. However, the pattern of declining and then increasing marginal costs is the same, and Year 3 is still the year of lowest marginal costs.

Minimum Cost Life Problems

Here we illustrate the effect of taxes on the calculation of the minimum cost life of the defender and the challenger. The after-tax minimum EUAC depends on both the depreciation method used and changes in the asset's market value over time. Using an accelerated depreciation method (like MACRS) tends to reduce the after-tax costs early in the asset's life. This alters the shape of the total EUAC curve—the concave shape can be shifted and the minimum EUAC changed. Example 13–11 illustrates the effect that taxes can have.

EXAMPLE 13–11

Some new production machinery has a first cost of $100,000 and a useful life of 10 years. Its estimated operating and maintenance (O&M) costs are $10,000 the first year, which will increase annually by $4000. The asset's before-tax market value will be $50,000 at the end of the first year and then will decrease by $5000 annually. This property is a 7-year MACRS property. The company uses a 6% after-tax MARR and is subject to a combined federal/state tax rate of 40%.

Calculate the after-tax cash flows.

SOLUTION

To find this new production machinery's minimum cost life, we first find the after-tax cash flow (ATCF) effect of the O&M costs and depreciation (Table 13–2). Then, we find the ATCFs of disposal if the equipment is sold in each of the 10 years (Table 13–3). Finally in the closing section on spreadsheets, we combine these two ATCFs (in Figure 13–8) and choose the minimum cost life.

In Table 13–2, the O&M expense starts at $10,000 and increases at $4000 per year. The depreciation entries equal the 7-year r_t MACRS depreciation values given in Table 11–3 multiplied by the $100,000 first cost. The taxable income, which is simply the O&M costs minus the depreciation values, is then multiplied by minus the tax rate to determine the tax savings. The O&M expense plus the tax savings is the Table 13–2 portion of the total ATCF.

Regarding the market value data in this problem, it should be pointed out that the initial decrease of $50,000 in Year 1 is not uncommon. This is especially true for custom-built equipment for a particular and unique application at a specific plant. Such equipment would not be valuable to others in the marketplace. Also, the $100,000 first cost (cost basis) could have included costs due to installation, facility modifications, or removal of old equipment. The $50,000 is realistic for the market value of one-year-old equipment.

The next step is to determine the ATCFs that would occur in each possible year of disposal. (The ATCF for Year 0 is easy; it is −$100,000.) For example, as shown in Table 13–3, in Year 1 there is a $35,710 loss as the book value exceeds the market value. The tax savings from this loss are added to the salvage (market) value to determine the ATCF (*if the asset is disposed of during Year 1*).

TABLE 13–2 ATCF for O&M and Depreciation for Example 13–11

Year, t	O&M Expense	MACRS Depreciation, d_t	Taxable Income	Tax Savings (at 40%)	O&M Depreciation ATCF
1	−$10,000	$14,290	−$24,290	$ 9,716	−$284
2	−14,000	24,490	−38,490	15,396	1,396
3	−18,000	17,490	−35,490	14,196	−3,804
4	−22,000	12,490	−34,490	13,796	−8,204
5	−26,000	8,930	−34,930	13,972	−12,028
6	−30,000	8,920	−38,920	15,568	−14,432
7	−34,000	8,930	−42,930	17,172	−16,828
8	−38,000	4,460	−42,460	16,984	−21,016
9	−42,000	0	−42,000	16,800	−25,200
10	−46,000	0	−46,000	18,400	−27,600

TABLE 13–3 ATCF in Year of Disposal for Example 13–11

Year, t	Market Value	Book Value	Gain or Loss	Gain/Loss Tax (at 40%)	ATCF if Disposed of
1	$50,000	$85,710	−$35,710	$14,284	$64,284
2	45,000	61,220	−16,220	6,488	51,488
3	40,000	43,730	−3,730	1,492	41,492
4	35,000	31,240	3,760	−1,504	33,496
5	30,000	22,310	7,690	−3,076	26,924
6	25,000	13,390	11,610	−4,644	20,356
7	20,000	4,460	15,540	−6,216	13,784
8	15,000	0	15,000	−6,000	9,000
9	10,000	0	10,000	−4,000	6,000
10	5,000	0	5,000	−2,000	3,000

These tables assume that depreciation is taken during the year of disposal and then calculates the recaptured depreciation (gain) or loss on the year-end book value.

Spreadsheets and After-tax Replacement Analysis

Spreadsheets are very useful in nearly all after-tax calculations. However, they are absolutely required for optimal life calculations in after-tax situations, because MACRS is the tax law, and after-tax cash flows are different in every year. Thus, the NPV function and the PMT function are both needed to find the minimum EUAC after taxes. Figure 13–8 illustrates the calculation of the minimum cost life for Example 13–11.

In Figure 13–8, the NPV finds the present worth of the irregular cash flows from Period 1 through Period t for $t = 1$ to life. Then PMT can be used to find the EUAC over each potential life. Before-tax replacement analysis was done this way in Example 13–2. The spreadsheet block function NPV is used to find the PW of cash flows from Period 1 to Period t. Note that the cell for Period 1 is an absolute address and the cell for period t is a relative address. This allows the formula to be copied.

	A	B	C	D	E	F
1		**Table 13-2**		**Table 13-3**	6%	Interest Rate
2		O&M & Depr.	PW_t	if disposed of		
3	Year	ATCF	without SV	ATCF	$EUAC_t$	
4	0		–$100,000			
5	1	−284	−100,268	$64,284	$42,000	=PMT($E$1,A5,C5,D5)
6	2	1,396	−99,025	51,488	29,018	
7	3	−3,804	−102,219	41,492	25,208	
8	4	−8,204	−108,718	33,496	23,718	
9	5	−12,028	−117,706	26,924	23,167	
10	**6**	**−14,432**	**−127,880**	**20,356**	**23,088**	**optimal life**
11	7	−16,828	−139,071	13,784	23,270	
12	8	−21,016	−152,257	9,000	23,610	
13	9	−25,200	−167,173	6,000	24,056	
14	10	−27,600	−182,585	3,000	24,580	=PMT($E$1,A14,C14,D14)
15						
16		=$C$4+NPV($E$1,$B$5:B14)				
17		= year 0 + NPV(*i*, B column)				

FIGURE 13–8 Spreadsheet for life with minimum after-tax cost.

SUMMARY

The question in selecting new equipment is, *Which machine will be more economical?* But when there is an existing machine (called the **defender**), the question is, *Shall we replace it now, or shall we keep it for one or more years?* When replacement is indicated, the best available replacement equipment (called the **challenger**), will be acquired. When we already have equipment, there is a tendency to use past or sunk costs in the replacement analysis. But only present and future costs are relevant.

This chapter has presented three distinctly different **replacement analysis techniques.** All use the simplifying **replacement repeatability assumptions.** These state that the defender will ultimately be replaced by the current best challenger (as will any challengers implemented in the future), and that we have an indefinite need for the asset's service.

In the usual case, marginal cost data are both available and increasing on a year-to-year basis, and **replacement analysis technique 1** compares *the defender's marginal cost data against the challenger's minimum EUAC.* We keep the defender as long as its marginal cost is less than the challenger's minimum EUAC.

When marginal cost data are available for the defender but are not increasing from year to year, **replacement analysis technique 2** compares *the defender's lowest EUAC against the challenger's minimum EUAC.* If the challenger's EUAC is less, select this asset in place of the defender today. If the defender's lowest EUAC is smaller, we do not replace it yet. If the cost data for the challenger and the defender do not change, replace the defender after the life that minimizes its EUAC when its marginal cost data exceed the challenger's minimum EUAC.

In the case of no marginal cost data being available for the defender, **replacement analysis technique 3** compares *the defender's EUAC over its stated life, against the* *challenger's minimum EUAC.* As in the case of replacement analysis technique 2, select the alternative that has the smallest EUAC. An important concept when calculating the EUAC of both defender and challenger is the first cost to be assigned to each alternative for calculation purposes. When the lives of the two alternatives match, either an **opportunity cost** or a **cash flow approach** may be used. However, in the more common case of different useful lives, only the opportunity cost approach accurately assigns first costs to the defender and challenger.

It is important when performing engineering economic analyses to include the effects of taxes. This is much easier to accomplish with spreadsheets. Spreadsheets also make it easy to compute the **optimal economic life** of vehicles and equipment—even when this includes complex patterns of declining salvage values, warranty periods, and increasing repair costs.

Replacement analyses are vastly important, yet often ignored by companies as they invest in equipment and facilities. Investments in business and personal assets should not be forgotten once an initial economic evaluation has produced a "buy" recommendation. It is important to continue to evaluate assets over their respective life cycles to ensure that invested monies continue to yield the greatest benefit. Replacement analyses help us to ensure this.

PROBLEMS

Replacement Problems

13-1 Typically there are two alternatives in a replacement analysis. One alternative is to replace the defender now. The other alternative is which one of the following?

    A. Keep the defender for its remaining useful life.

    B. Keep the defender for another year and then reexamine the situation.

    C. Keep the defender until there is an improved challenger that is better than the present challenger.

    D. The answer to this question depends on the data available for the defender and challenger as well as the assumptions made.

13-2 The defender's economic life can be found if certain estimates about the defender can be made. Assuming those estimates prove to be exactly correct, one can accurately predict the year when the defender should be replaced, even if nothing is known about the challenger. True or false? Explain.

13-3 A proposal has been made to replace a large heat exchanger (3 years ago, the initial cost was $85,000) with a new, more efficient unit at a cost of $120,000. The existing heat exchanger is being depreciated by the MACRS method. Its present book value is $20,400, but its scrap value just equals the cost to remove it from the plant. In preparing the before-tax

economic analysis, should the $20,400 book value of the old heat exchanger be

    A. *added* to the cost of the new exchanger?

    B. *subtracted* from the cost of the new exchanger?

    C. *ignored* in this before-tax economic analysis?

13-4 Which one of the following is the proper dollar value of defender equipment to use in replacement analysis?

    A. Original cost

    B. Present market value if sold

    C. Present trade-in value

    D. Present book value

    E. Present market value if sold minus removal and selling expenses

13-5 Consider the following data for a defender asset. What is the correct replacement analysis technique to compare this asset against a competing challenger? How is this method used? That is, what comparison is made, and how do we choose?

Year, n	BTCF in Year n (marginal costs)
1	−$2000
2	−2250
3	−2500
4	−3000
5	−3750

13-6 Describe an example in a replacement analysis sce-
E nario where the replacement is being considered
due to

(a) reduced performance of the existing equipment.
(b) altered requirements.
(c) obsolescence of the existing equipment.
(d) risk of catastrophic failure or unplanned replacement of the existing equipment.
(e) a previous equipment choice that was incorrectly analyzed by your boss. What ethical and
pragmatic challenges exist in this case?

13-7 A pulpwood-forming machine was purchased and
installed 6 years ago for $50,000. The declared sal-
vage value was $5000, with a useful life of 10 years.
The machine can be replaced with a more efficient
model that costs $90,000, including installation. The
old machine can be sold on the open market for
$25,000. The cost to remove the old machine is $4000.
Which are the relevant costs for the old machine?

13-8 A $25,000 machine that has been used for one year
A has a salvage value of $16,000 now, which will drop
by $4000 per year. The maintenance costs for the
next 4 years are $1250, $1450, $1750, and $2250.
When the machine is sold, it will cost $2000 to
remove and sell. When the machine was purchased,
the estimated salvage value in 5 years was $3500.
What are the relevant costs for the machine?

13-9 A drill press was purchased 4 years ago for $40,000.
Its estimated salvage value after 7 years was $5000.
The press can be sold for $15,000 today, or for
$12,000, $9000, or $6000 at the ends of each of the
next 3 years. The annual operating and maintenance
cost for the next 3 years will be $2700 for this year
and then will increase by $800 per year. What are the
relevant cash flows for this machine?

Challenger Minimum EUAC

13-10 A machine has a first cost of $50,000. Its market
A value declines by 20% annually. The operating and
maintenance costs start at $3500 per year and climb
by $2000 per year. The firm's MARR is 9%. Find the
minimum EUAC for this machine and its economic
life.

13-11 A machine has a first cost of $24,000. Its market
value declines by 20% annually. The repair costs are
covered by the warranty in Year 1, and then they
increase $900 per year. The firm's MARR is 12%.
Find the minimum EUAC for this machine and its
economic life.

13-12 A vehicle has a first cost of $25,000. Its market value
A declines by 15% annually. It is used by a firm that
estimates the effect of older vehicles on the firm's
image. A new car has no "image cost." But the image
cost of older vehicles climbs by $800 per year. The
firm's MARR is 10%. Find the minimum EUAC for
this vehicle and its economic life.

13-13 The Clap Chemical Company needs a large insulated
stainless steel tank to expand its plant. A recently
closed brewery has offered to sell their tank for
$15,000 delivered. The price is so low that Clap
believes it can sell the tank at any future time and
recover its $15,000 investment.

Installing the tank will cost $9000 and remov-
ing it will cost $5000. The outside of the tank is
covered with heavy insulation that requires considerable maintenance. This will cost $3500 in year 1
and increase by $1000 per year.

(a) Based on a 12% before-tax MARR, what life of
the insulated tank has the lowest EUAC?
(b) When the insulated tank is replaced by another
tank is the replacement's economic life likely to
be shorter or larger? Explain.

13-14 An electric oil pump's first cost is $45,000, and the
A interest rate is 10%. The pump's end-of-year salvage
values over the next 5 years are $42K, $40K, $38K,
$32K, and $26K. Determine the pump's economic
life.

13-15 A chemical process in your plant leaves scale
deposits on the inside of pipes. The scale cannot be
removed, but increasing the pumping pressure maintains flow through the narrower diameter. The pipe
costs $25 per foot to install, and it has no salvage
value when it is removed. The pumping costs are
$8.50 per foot of pipe initially, and they increase
annually by $5 per year starting in Year 2. What
is the economic life of the pipe if the interest rate
is 15%?

13-16 A $40,000 machine will be purchased by a company
A whose interest rate is 12%. The installation cost is
$10K, and removal costs are insignificant. What is
its economic life if its salvage values and O&M costs
are as follows?

Year	1	2	3	4	5
S	$35K	$30K	$25K	$20K	$15K
O&M	$8K	$14K	$20K	$26K	$32K

13-17 A $50,000 machine will be purchased by a company whose interest rate is 10%. The installation cost is $8K, and removal costs are insignificant. What is its economic life if its salvage values and O&M costs are as follows?

Year	1	2 and later
S	$35K	drop $5K per year
O&M	$8K	up $5K per year

13-18 A $20,000 machine will be purchased by a company whose interest rate is 10%. It will cost $5000 to install, but its removal costs are insignificant. What is its economic life if its salvage values and O&M costs are as follows?

Year	1	2	3	4	5
S	$16K	$13K	$11K	$10K	$9.5K
O&M	$5K	$8K	$11K	$14K	$17K

13-19 An injection-molding machine has a first cost of $1,050,000 and a salvage value of $225,000 in any year. The maintenance and operating cost is $235,000 with an annual gradient of $75,000. The MARR is 10%. What is the most economic life?

13-20 The plant manager may purchase a piece of unusual machinery for $10,000. Its resale value after 1 year is estimated to be $3000. Because the device is sought by antique collectors, resale value is rising $500 per year.

The maintenance cost is $300 per year for each of the first 3 years, and then it is expected to double each year. Thus the fourth-year maintenance will be $600; the fifth-year maintenance, $1200, and so on. Based on a 15% before-tax MARR, what life of this machinery has the lowest EUAC?

13-21 J&E Fine Wines recently purchased a new grape press for $150,000. The annual operating and maintenance costs for the press are estimated to be $7500 the first year. These costs are expected to increase by $2200 each year after the first. The market value is expected to decrease by $25,000 each year to a value of zero. Installation and removal of a press each cost $5000. Using an interest rate of 8%, determine the economic life of the press.

13-22 In a replacement analysis problem, the following facts are known:

Initial cost	$12,000
Annual maintenance	Year
None	1–3
$2000	4–5
$4500	6
+$2500/yr	7–10

Salvage value in any year is zero. Assume a 10% interest rate and ignore income taxes. Compute the life for this challenger having the lowest EUAC.

13-23 Mytown's street department repaves a street every 8 years. Potholes cost $15,000 per mile beginning at the end of Year 3 after construction or repaving. The cost to fix potholes generally increases by $15,000 each year. Repaving costs are $200,000 per mile. Mytown uses an interest rate of 8%. What is the EUAC for Mytown's policy? What is the EUAC for the optimal policy? What is the optimal policy?

13-24 A 2000-pound, counterbalanced, propane forklift can be purchased for $30,000. Due to the intended service use, the forklift's market value drops 20% of its prior year's value in Years 1 and 2 and then declines by 15% until Year 10 when it will have a scrap/market value of $1000. Maintenance of the forklift is $400 per year during Years 1 and 2 while the warranty is in place. In Year 3 it jumps to $750 and increases $200 per year thereafter. What is the optimal life of the forklift using $i = 10\%$?

13-25 Demonstrate how one would calculate the economic life of a truck costing $40,000 initially, and at the end of this and each following year (y) costing OMR_y in operation, maintenance and repair costs. The truck is depreciated using the straight-line method over 4 years (i.e., $40,000/4 = D_y$). Its salvage value each year equals its book value. Develop an expression to show how to determine the truck's economic life— that is, the year when the truck's uniform equivalent annual cost is a minimum.
Contributed by D. P. Loucks, Cornell University

13-26 A 2000-pound, counterbalanced, electric forklift can be purchased for $25,000 plus $3000 for the charger and $3000 for a battery. The forklift's market value is 10% less for each of its first 6 years of service. After this period the market value declines at the rate of 7.5% for the next 6 years.

The battery has a life of 4 years and a salvage value of $300. The charger has a 12-year life and a $100 salvage value. The charger's market value declines 20% per year of use. The battery's market value declines by 30% of its purchase price each year. Maintenance of the charger and battery are minimal. The battery will most likely not work with a replacement forklift.

Maintenance of the forklift is $200 per year during Years 1 and 2 while the warranty is in place. In Year 3 it jumps to $600 and increases $50 per year thereafter. What is the optimal ownership policy using $i = 10\%$?

13-27 Your firm is in need of a machine that costs $5000. During the first year the maintenance costs are estimated to be $400. The maintenance costs are expected to increase by $150 each year up to a total of $1750 at the end of Year 10. The machine can be depreciated over 5 years using the straight-line method. Assume each year's depreciation is a known amount D_y. There is no salvage value. Develop an expression to show how you would find the most economical useful life of this machine on a before-tax basis.

Contributed by D. P. Loucks, Cornell University

13-28 Bill's father read that each year a car's value declines by 10%. He also read that a new car's value declines by 10% as it is driven off the dealer's lot. Maintenance costs and the costs of "car problems" are only $100 per year during the 2-year warranty period. Then they jump to $600 per year, with an annual increase of $400 per year.

Bill's dad wants to keep his annual cost of car ownership low. The car he prefers cost $22,000 new, and he uses an interest rate of 10%. For this car, the new vehicle warranty is transferrable.

(*a*) If he buys the car new, what is the minimum cost life? What is the minimum EUAC?

(*b*) If he buys the car after it is 2 years old, what is the minimum cost life? What is the minimum EUAC?

(*c*) If he buys the car after it is 4 years old, what is the minimum cost life? What is the minimum EUAC?

(*d*) If he buys the car after it is 6 years old, what is the minimum cost life? What is the minimum EUAC?

(*e*) What strategy do you recommend? Why?

Replacement Technique 1

13-29 In Problem 13-9, what is the marginal cost for each year?

13-30 In Problem 13-8, what is the marginal cost for each year?

13-31 In Problem 13-28, what is the marginal cost for each year?

13-32 SHOJ Enterprises has asked you to look at the following data. The interest rate is 10%.

Year, n	Marginal Cost Data Defender	EUAC if Kept n Years Challenger
1	$3000	$6500
2	3150	4150
3	3400	3200
4	3800	3100
5	4250	3150
6	4950	3400

(*a*) What is the *challenger's* economic life?

(*b*) When, if at all, should we replace the *defender* with the *challenger*?

13-33 Should NewTech's pollution testing system be replaced this year? The system has a salvage value now of $5000, which will fall to $4000 by the end of the year. The cost of lower productivity linked to the older system is $3000 this year. NewTech uses an interest rate of 15%. What is the cost advantage of the best system? A potential new system costs $12,000 and has the following salvage values and lost productivity for each year.

Year	S	Lost Productivity
0	$12,000	
1	9,000	$ 0
2	7,000	1000
3	5,000	2000
4	3,000	3000

13-34 Five years ago, Thomas Martin installed production machinery that had a first cost of $25,000. At that time initial yearly costs were estimated at $1250, increasing by $500 each year. The market value of this machinery each year would be 90% of the previous year's value. There is a new machine available now that has a first cost of $27,900 and no yearly costs over its 5-year minimum cost life. If Thomas Martin uses an 8% before-tax MARR, when, if at all, should he replace the existing machinery with the new unit?

13-35 Consider Problem 13-34 involving Thomas Martin. When, if at all, should the old machinery be replaced with the new, given the following changes in the data. The old machine retains only 70% of its value in the market from year to year. The yearly costs of the old machine were $3000 in Year 1 and increase at 10% thereafter.

13-36 In evaluating projects, LeadTech's engineers use a rate of 15%. One year ago a robotic transfer machine was installed at a cost of $38,000. At the time, a 10-year life was estimated, but the machine has had a downtime rate of 28%, which is unacceptably high. A $12,000 upgrade should fix the problem, or a labor-intensive process costing $3500 in direct labor per year can be substituted. The plant estimates indirect plant expenses at 60% of direct labor, and it allocates front office overhead at 40% of plant expenses (direct and indirect). The robot has a value in other uses of $15,000. What is the difference between the EUACs for upgrading and switching to the labor-intensive process?

13-37 Mary O'Leary's company ships fine wool garments from County Cork, Ireland. Five years ago she purchased some new automated packing equipment having a first cost of $125,000. The annual costs for operating, maintenance, and insurance, as well as market value data for each year of the equipment's 10-year useful life are as follows.

| Year | Annual Costs in Year n for | | | Market Value |
n	Operating	Maintenance	Insurance	in Year n
1	$16,000	$ 5,000	$17,000	$80,000
2	20,000	10,000	16,000	78,000
3	24,000	15,000	15,000	76,000
4	28,000	20,000	14,000	74,000
5	32,000	25,000	12,000	72,000
6	36,000	30,000	11,000	70,000
7	40,000	35,000	10,000	68,000
8	44,000	40,000	10,000	66,000
9	48,000	45,000	10,000	64,000
10	52,000	50,000	10,000	62,000

Now Mary is looking at the remaining 5 years of her investment in this equipment. What is the marginal cost for each of the remaining 5 years? When, if at all, should Mary replace this packing equipment with a new challenger that has a minimum EUAC of $110,000?

13-38 Eight years ago, the Blank Block Building Company installed an automated conveyor system for $38,000. When the conveyor is replaced, the net cost of removal will be $2500. The minimum EUAC of a new conveyor is $6500. When should the conveyor be replaced if BBB's MARR is 12%? The O&M costs for the next 5 years are $5K, $6K, $7K, $8K, and $9K.

13-39 Big-J Construction Company, Inc. is conducting a routine periodic review of existing field equipment. They use a MARR of 20%. This includes a replacement evaluation of a paving machine now in use. The machine was purchased 3 years ago for $200,000. The paver's current market value is $120,000, and yearly operating and maintenance costs are as follows.

Year, n	Operating Cost in Year n	Maintenance Cost in Year n	Market Value if Sold in Year n
1	$15,000	$ 9,000	$85,000
2	15,000	10,000	65,000
3	17,000	12,000	50,000
4	20,000	18,000	40,000
5	25,000	20,000	35,000
6	30,000	25,000	30,000
7	35,000	30,000	25,000

Data for a new paving machine have been analyzed. Its most economic life is at 8 years, with a minimum EUAC of $62,000. When should the existing paving machine be replaced?

Replacement Technique 2

13-40 A new $100,000 bottling machine has just been installed in a plant. It will have no salvage value when it is removed. The plant manager has asked you to estimate the machine's economic service life, ignoring income taxes. He estimates that the annual maintenance cost will be constant at $2500 per year. What service life will result in the lowest equivalent uniform annual cost?

13-41 The total marginal cost data for a grouping of defender machines over the next five years is:

Year	1	2	3	4	5
Total Marginal Cost ($1000s)	120	100	50	80	250

(a) What replacement technique is recommended if there is a challenger asset being considered to replace the machine grouping? Why is this technique chosen?

(b) What is the economic life and minimum EUAC of the machine grouping? Use $i = 12\%$.

(c) If the challenger asset has a minimum EUAC of $110,000, what is your recommendation?

13-42 VMIC Corp. has asked you to look at the following data. The interest rate is 10%.

Year, n	Marginal Cost Data Defender	EUAC If Kept n Years Challenger
1	$2500	$6500
2	2400	4200
3	2300	3000
4	2550	2650
5	2900	2700
6	3400	2800
7	4000	3000

(a) What is the *challenger's* minimum cost life?

(b) What is the *defender's* lowest EUAC?

(c) When, if at all, should we replace the *defender* with the *challenger*?

13-43 The existing business asset has the tabulated total marginal cost data. What is the maximum first cost for a challenger for it to be preferred over the defender today? With no salvage value and no differential annual costs from the defender, the challenger's minimum cost life is 5 years. Use $i = 20\%$.

Year	1	2	3	4
Total Marginal Cost ($1000s)	300	250	100	400

Replacement Technique 3

13-44 You are considering the purchase of a new high-efficiency machine to replace older machines now. The new machine can replace four of the older machines, each with a current market value of $600. The new machine will cost $5000 and will save the equivalent of 10,000 kWh of electricity per year. After a period of 10 years, neither option (new or old) will have any market value. If you use a before-tax MARR of 25% and pay $0.075 per kilowatt-hour, would you replace the old machines today with the new one?

13-45 A professor of engineering economics owns an older car. In the past 12 months, he has paid $2000 to replace the transmission, bought two new tires for $160, and installed a CD player for $110. He wants to keep the car for 2 more years because he invested money 3 years ago in a 5-year certificate of deposit, which is earmarked to pay for his dream machine, a red European sports car. Today the old car's engine failed.

The professor has two alternatives. He can have the engine overhauled at a cost of $1800 and then most likely have to pay another $800 per year for the next 2 years for maintenance. The car will have no salvage value at that time.

Alternatively, a colleague offered to make the professor a $5000 loan to buy another used car. He must pay the loan back in two equal installments of $2500 due at the end of Year 1 and Year 2, and at the end of the second year he must give the colleague the car. What interest rate is the professor paying on the loan from his colleague, if the vehicle will be worth $3000 after 2 years? Is this an ethical interest rate?

The "new" used car has an expected annual maintenance cost of $300. If the professor selects this alternative, he can sell his current vehicle to a junkyard for $500. Interest is 6%. Using present worth analysis, which alternative should he select and why?

13-46 Sacramento Cab Company owns several taxis that were purchased for $25,000 each 4 years ago. The cabs' current market value is $12,000 each, and if they are kept for another 6 years they can be sold for $2000 per cab. The annual maintenance cost per cab is $1000 per year. Sacramento Cab has been approached about a leasing plan that would replace the cabs. The leasing plan calls for payments of $6000 per year. The annual maintenance cost for each leased cab is $750 per year. Should the cabs be replaced if the interest rate is 10%?

13-47 The local telephone company purchased four special pole hole diggers 8 years ago for $14,000 each. Owing to an increased workload, additional machines will soon be required.

Recently an improved model of the digger was announced. The new machines have a higher production rate and lower maintenance expense than the old machines but will cost $32,000 each with a service life of 8 years and salvage value of $750 each. The four original diggers have an immediate salvage of $2000 each and an estimated salvage value of $500 each 8 years hence. The average annual maintenance expense of each old machine is about $1500, compared with $600 each for the new machines.

The workload would require three additional new machines if the old machines continue in service. However, if the old machines were all retired from service, the workload could be carried by six new machines with an annual savings of $12,000 in operation costs. A training program to prepare employees to run the machines will be necessary at an estimated cost of $1200 per new machine. If the MARR is 8% before taxes, what should the company do?

13-48 JMJ Inc. bought a manufacturing line 5 years ago
Ⓐ for $35M (million). At that time it was estimated to have a service life of 10 years and salvage value at the end of its service life of $10M. JMJ's CFO recently proposed to replace the old line with a modern line expected to last 15 years and cost $95M. This line will provide $5M savings in annual O&M costs, increase revenues by $2M, and have a $15M salvage value. The seller of the new line is willing to accept the old line as a trade-in for its current fair market value, which is $12M. The CFO estimates that if the old line is kept for 5 more years, its salvage value will be $6M. If the MARR is 8% per year, should the company keep the old line or replace it with the new line?
Contributed by Hamed Kashani, Saeid Sadri, and Baabak Ashuri, Georgia Institute of Technology

13-49 A couple bought their house 10 years ago for $165,000. At the time of purchase, they made a $35,000 down payment, and the balance was financed by a 30-year mortgage with monthly payments of $988.35. They expect to live in this house for 20 years, after which time they plan to sell the house and move to another state.

Alternatively, they can sell the house now and live in a rental unit for the next 20 years. The house can be sold now for $210,000, from which a 6% real estate commission and $110,000 remaining loan balance and miscellaneous expenses will be deducted.

If they stay in the house, the house can be sold after 20 years for $320,000, from which a 6% real estate commission and $10,000 miscellaneous expenses will be deducted. A comparable rental unit rents for $960 payable at the beginning of every month. No security deposit will be required of them to rent the unit, and the rent will not increase if they maintain a good payment record. They use an interest rate of 0.5% per month for analyzing this financial opportunity. Should they stay in the house or should they sell it and move into a rental unit?
Contributed by Hamed Kashani, Saeid Sadri, and Baabak Ashuri, Georgia Institute of Technology

13-50 A used car can be kept for two more years and then
Ⓐ sold for an estimated value of $3000, or it can be sold now for $7500. The average annual maintenance cost over the past 7 years has been $500 per year. However, if the car is kept for two more years, this cost is expected to be $1800 the first year and $2000 the second year. As an alternative, a new car can be purchased for $22,000 and be used for 4 years, after which it will be sold for $8,000. The new car will be under warranty the first 4 years, and no extra maintenance cost will be incurred during those years. If the MARR is 15% per year, what is the better option?
Contributed by Hamed Kashani, Saeid Sadri, and Baabak Ashuri, Georgia Institute of Technology

13-51 The Quick Manufacturing Company, a large profitable corporation, may replace a production machine tool. A new machine would cost $37,000, have an 8-year useful life, and have no salvage value. For tax purposes, 3-year MACRS depreciation would be used. The existing machine tool cost $40,000 4 years ago, and it has been completely depreciated. The tool could be sold now to a used equipment dealer for $10,000 or be kept in service for another 8 years. It would then have no salvage value. The new machine tool would save about $9000 per year in operating costs compared to the existing machine.

Assume a 40% combined state and federal tax rate. Compute the **before**-tax rate of return on the replacement proposal of installing the new machine rather than keeping the existing machine.

After-Tax Replacement

13-52 Fifteen years ago the Acme Manufacturing Company
Ⓐ bought a propane-powered forklift truck for $4800. The company depreciated the forklift using straight-line depreciation, a 12-year life, and zero salvage

value. Estimated end-of-year maintenance costs for the next 10 years are as follows:

Year	Maintenance Cost
1	$ 400
2	600
3	800
4	1000
5–10	1400/year

The old forklift has no present or future net salvage value (scrap value equals disposal costs). A modern unit can be purchased for $6500. It has an economic life equal to its 10-year depreciable life. Straight-line depreciation will be employed, with zero salvage value at the end of the 10-year depreciable life. Maintenance on the new forklift is estimated to be a constant $50 per year for the next 10 years, after which maintenance is expected to increase sharply. Should Acme Manufacturing keep its old forklift truck for the present or replace it now with a new one? The firm expects an 8% after-tax rate of return on its investments. Assume a 40% combined state-and-federal tax rate.

13-53 Machine *A* has been completely overhauled for $9000 and is expected to last another 12 years. The $9000 was treated as an expense for tax purposes last year. Machine *A* can be sold now for $30,000 net after selling expenses, but will have no salvage value 12 years hence. It was bought new 9 years ago for $54,000 and has been depreciated since then by straight-line depreciation using a 12-year depreciable life.

Because less output is now required, Machine *A* can be replaced with a smaller machine: Machine *B* costs $42,000, has an anticipated life of 12 years, and would reduce operating costs $2500 per year. It would be depreciated by straight-line depreciation with a 12-year depreciable life and no salvage value.

The income tax rate is 40%. Compare the after-tax annual costs and decide whether Machine *A* should be retained or replaced by Machine *B*. Use a 10% after-tax rate of return.

13-54 A new employee at CLL Engineering Consulting Inc., you are asked to join a team performing an economic analysis for a client. The client has a combined federal/state tax rate of 45% on ordinary income, depreciation recapture, and losses.

Defender: This asset was placed in service 7 years ago. At that time the $50,000 cost basis was set up on a straight-line depreciation schedule with an estimated salvage value of $15,000 over its 10-year ADR life. This asset has a present market value of $30,000.

Challenger: The new asset has a first cost of $85,000 and will be depreciated by MACRS depreciation over its 10-year class life. This asset qualifies for a 10% investment tax credit.

(*a*) Your task is to find the Time-0 ATCFs.
(*b*) How would your calculations change if the present market value of the *defender* is $25,500?
(*c*) How would your calculations change if the present market value of the *defender* is $18,000?

13-55 State the advantages and disadvantages with respect to after-tax benefits of the following options for a major equipment unit:

A. Buy new equipment.
B. Trade in and buy a similar, rebuilt equipment from the manufacturer.
C. Have the manufacturer rebuild your equipment with all new available options.
D. Have the manufacturer rebuild your equipment to the original specifications.
E. Buy used equipment.

13-56 A firm is concerned about the condition of some of its plant machinery. Bill James, a newly hired engineer, reviewed the situation and identified five feasible, mutually exclusive alternatives.

Alternative A: Spend $44,000 now repairing various items. The $44,000 can be charged as a current operating expense (rather than capitalized) and deducted from other taxable income immediately. These repairs will keep the plant functioning for 7 years with current operating costs.

Alternative B: Spend $49,000 to buy general-purpose equipment. Depreciation would be 5-year MACRS. The equipment has no salvage value after 7 years. The new equipment will reduce annual operating costs by $7000.

Alternative C: Spend $56,000 to buy new specialized equipment. This equipment would be depreciated by 5-year MACRS. This equipment would reduce annual operating costs by $12,000. It will have no salvage value.

Alternative D: This is the "do nothing" alternative, with annual operating costs $8000 above the present level.

This profitable firm pays 40% corporate income taxes and uses a 10% after-tax rate of return. Which alternative should the firm adopt?

13-57 The Ajax Corporation purchased a pressure tank 8 years ago for $60,000. It is being depreciated by SOYD depreciation, assuming a 10-year depreciable life and a $7000 salvage value. The tank needs to be reconditioned now at a cost of $35,000. If this is done, it is estimated the tank will last for 10 more years and have a $10,000 salvage value at the end of the 10 years.

On the other hand, the existing tank could be sold now for $10,000 and a new tank purchased for $85,000. The new tank would be depreciated by SOYD depreciation. Its estimated actual salvage value after 10 years would be $15,000. In addition, the new tank would save $7000 per year in maintenance costs, compared to the reconditioned tank.

Based on a 15% after-tax rate of return, determine whether the existing tank should be reconditioned or a new one purchased.

13-58 BC Junction purchased some embroidering
(A) equipment for their Denver facility 3 years ago for $15,000. This equipment qualified as MACRS 5-year property. Maintenance costs are estimated to be $1000 this next year and will increase by $1000 per year thereafter. The market (salvage) value for the equipment is $10,000 at the end of this year and declines by $1000 per year in the future. If BC Junction has an after-tax MARR of 30%, a marginal tax rate of 45% on ordinary income, depreciation recapture, and losses, what after-tax life of this previously purchased equipment has the lowest EUAC?

13-59 Compute the after-tax rate of return on the replacement proposal for Problem 13-51.

13-60 Reconsider the acquisition of packing equipment for Mary O'Leary's business, as described in Problem 13-37. Given the data tabulated there use an after-tax MARR of 25% and a tax rate of 35% on ordinary income to evaluate the investment. Determine the lowest after-tax EUAC of the equipment at its initial purchase.

CASES

The following cases from *Cases in Engineering Economy* (www.oup.com/us/newnan) are suggested as matched with this chapter.

CASE 27 To Use or Not to Use?

Focus is treatment of sunk costs. More complicated than most. Some discoveries in the data gathering process. Solution uses equation rather than cash flow table.

CASE 31 Freeflight Superdiscs

Inflation and sensitivity analysis for three alternatives. Includes taxes.

CASE 32 Mr. Speedy

Includes two memos using different inappropriate financial comparisons. Choose optimal life for replacement of vehicles.

CASE 33 Piping Plus

Data from case intro, three memos, income statement, and balance sheet. Computer improvement in a professional services firm. Assumptions will lead to an instructive variety of results.

CASE 34 R&D Device at EBP

Equipment replacement cost comparison with unequal lives. Continuing demand requires careful analysis statement by the student because of detailed cost data. Before or after taxes.

INFLATION AND PRICE CHANGE

Price Trends in Solar Technologies

In spite of inflation increasing most prices, the price of solar power is declining. Also in contrast to fossil fuels, the cost of generating electricity from solar energy is driven by the infrastructure costs instead of the cost of the natural resource. Therefore the costs and prices are more stable, particularly for large-scale electricity generation.

Historically, solar technologies have had high upfront infrastructure costs but low operating costs. The SunShot program launched in 2011 by the U.S. Department of Energy (DOE) seeks to make solar energy economically competitive by 2020. To achieve this goal, the cost of solar power will have to be reduced by roughly 75% relative to 2010 prices. The U.S. DOE has reported that cumulative adoption of solar technologies has increased over tenfold since 2008. It is expected that achieving the price reduction set by the SunShot initiative could lead to solar representing 14% of the electricity demand in the U.S. by 2030 and 27% by 2050.

According to a study by the National Renewable Energy Laboratory and the Lawrence Berkeley National Laboratory, reported prices of both residential and commercial photovoltaic systems decreased on average from 6% to 12% annually between 1998 and 2014. At the end of 2014, photovoltaic prices ranged from $4.27/W for residential systems with a median installed capacity of 6kW to $2.08/W for utility-scale systems with a median of 14MW. The average cost for solar power is dominated by the utility-scale systems. Costs for utility scale systems are projected to fall $1.00/W to $1.75/W by 2020. It is expected that within the next two decades, the cost of solar technologies will be lower than the costs of conventional fossil fuel electricity technologies. These lower electricity prices can decrease inflation and contribute to increased economic activity and growth.

Interestingly, the deployment of solar technologies continues to increase despite the 2015 plunge in oil prices. Solar was the third most added electricity-generation capacity in 2015, after wind and natural gas. However, the share of solar compared to the total U.S. electricity capacity in 2015 was 2%, while its generation share was only 0.9%. So it remains to be seen if the goal of the SunShot initiative will be met. ■ ■ ■
Contributed by Ona Egbue, University of Minnesota Duluth

QUESTIONS TO CONSIDER

1. Deflation decreases the general prices of goods and services and increases purchasing power. How does this differ from the declining price of solar technologies?

2. Currently, most electricity generation in the U.S. is from fossil fuels, including petroleum, coal, and natural gas. In what ways has the 2015 decline in oil prices affected the adoption of solar technologies?

3. Use Internet resources to compare the changes in the prices of renewable energy technologies including wind, solar, and geothermal over the last 5 years. Which of these three technologies has achieved the greatest cost decline during that period?

4. In addition to cost, what are other barriers to the adoption of photovoltaic systems?

After Completing This Chapter...

The student should be able to:

- Describe inflation, explain how it happens, and list its effects on purchasing power.

- Define real and actual dollars and interest rates.

- Conduct constant dollar and nominal dollar analyses.

- Define and use composite and commodity-specific price indexes.

- Develop and use cash flows that inflate at different interest rates and cash flows subject to different interest rates per period.

- Incorporate the effects of inflation in before-tax and after-tax calculations.

- Develop spreadsheets to incorporate the effects of inflation and price change.

Key Words

actual dollar	demand-pull inflation	nominal dollars
base year	exchange rate	price index
composite cost index	inflation rate	purchasing power
constant value dollars	market interest rate	real dollar
cost-push inflation	money supply	real interest rate
deflation		

Thus far we have used constant-value dollars in our analyses, thus they were unaffected by inflation or price change. However, this is not always valid or realistic. In this chapter we develop several key concepts and illustrate how inflation and price changes may be explicitly modeled.

MEANING AND EFFECT OF INFLATION

Inflation is an important concept because the purchasing power of money used in most world economies rarely stays constant. Rather, over time the amount of goods and services that can be bought with a fixed amount of money tends to change. Inflation causes money to lose **purchasing power.** That is, when prices inflate we can buy less with the same amount of money. *Inflation makes future dollars less valuable than present dollars.* Think about examples in your own life, or for an even starker comparison, ask your grandparents how much a loaf of bread or a new car cost 50 years ago. Then compare those prices with what you would pay today for the same items. This exercise will reveal the effect of inflation: as time passes, goods and services cost more, and more monetary units are needed to buy the same goods and services.

Because of inflation, dollars in one period of time are not equivalent to dollars in another. We know that engineering economic analysis requires that comparisons be made on an equivalent basis. So, it is important for us to be able to incorporate the effects of inflation.

When the purchasing power of a monetary unit *increases* rather than decreases as time passes, the result is **deflation.** Deflation, very rare in the modern world, nonetheless can exist. Deflation has the opposite effect of inflation—one can buy *more* with money in future years than can be bought today. Thus, deflation makes future dollars more valuable than current dollars.

How Does Inflation Happen?

Economists generally believe that inflation depends on the following, either in isolation or in combination.

Money supply: The amount of money in our national economy has an effect on its purchasing power. If there is too much money in the system (the Federal Reserve controls the flow of money) versus goods and services to purchase with that money, the value of dollars tends to decrease. When there are fewer dollars in the system, they become more valuable. The Federal Reserve attempts to influence economic growth and employment by controlling the amount of money in the system.

Exchange rates: The strength of the dollar in world markets affects the profitability of international companies. Prices may be adjusted to compensate for the dollar's relative strength or weakness in the world market. As corporations' profits are weakened or eliminated in some markets owing to fluctuations in exchange rates, prices may be raised in other markets to compensate.

Cost–push inflation: This cause of inflation develops as producers of goods and services "push" their increasing operating costs along to the customer through higher prices. These operating costs include fabrication/manufacturing, marketing, and sales.

Demand–pull inflation: This cause is realized when consumers spend money freely on goods and services. As more and more people demand certain goods and services, the prices of those goods and services will rise (demand exceeding supply).

A further consideration in analyzing how inflation works is the usually different rates at which prices and wages rise. Do workers benefit if, as their wages increase, the prices of goods and services increase? To determine the net effect of differing rates of inflation, we must be able to make comparisons and understand costs and benefits from an equivalent perspective. In this chapter we will learn how to make such comparisons.

Definitions for Considering Inflation in Engineering Economy

The following definitions are used throughout this chapter to illustrate how inflation and price change affect two quantities: interest rates and cash flows.

Inflation rate (f): The inflation rate captures the effect of goods and services costing more—a decrease in the purchasing power of dollars. More money is required to buy a good or service whose price has inflated. The inflation rate is measured as the annual rate of increase in the number of dollars needed to pay for the same amount of goods and services.

Real interest rate (i'): This interest rate measures the "real" value of money excluding the effect of inflation. Because it does not include inflation, it is sometimes called the *inflation-free interest rate.*

Market interest rate (i): This is the rate of interest that one obtains in the general marketplace. For instance, the interest rates on passbook savings, checking plus, and certificates of deposit quoted by banks are all market rates. The lending interest rate for autos and boats is also a market rate. This rate is sometimes called the *combined interest* rate because it incorporates the effect of both real interest **and** inflation. We can view i as follows:

Market interest rate	**has in it**	**"Real" value of money**	**and**	**Effect of inflation**

The mathematical relationship between the inflation, real and market interest rates is given as

$$i = i' + f + i'f \tag{14-1}$$

This is the first point where we have defined a real interest rate and a market or combined interest rate. This naturally leads to the question of what meaning should be attached to the interest rate i, which is found throughout the text. In fact, both meanings have been used.

- In problems about savings accounts and loans, the interest rate is usually a market rate.
- In problems about engineering projects where costs and benefits are often estimated as $\$x$ per year, the interest rate is a real rate.

EXAMPLE 14–1

Suppose a professional golfer wants to invest some recent golf winnings in her hometown bank for one year. Currently, the bank is paying a rate of 5.5% *compounded annually*. Assume inflation is expected to be 2% per year. Identify i, f, and i'. Repeat for inflation of 8% per year.

SOLUTION

If Inflation Is 2% per Year

The bank is paying a *market rate* (i). The *inflation rate* (f) is given. What then is the *real interest rate* (i')?

$$i = 5.5\%, \qquad f = 2\%, \qquad i' = ?$$

Solving for i' in Equation 14-1, we have

$$i = i' + f + i'f$$
$$i - f = i'(1 + f)$$
$$i' = (i - f)/(1 + f)$$
$$= (0.055 - 0.02)/(1 + 0.02) = 0.034 \quad \text{or} \quad \textbf{3.4\% per year}$$

This means that the golfer will have 3.4% **more** purchasing power than she had a year ago. At the end of the year she can buy 3.4% more goods and services than she could have at the beginning of the year. For example, assume she was buying golf balls that cost $5 each and that she had invested $1000.

At the *beginning* of the year she could buy

$$\text{Number of balls purchased today} = \frac{\text{Dollars today available to buy balls}}{\text{Cost of balls today}}$$

$$= 1000/\$5 = 200 \text{ golf balls}$$

At the *end* of the year she could buy

$$\text{Number of balls bought at end of year} = \frac{\text{Dollars available for purchase at end of year}}{\text{Cost per ball at end of year}}$$

$$= \frac{(\$1000)(F/P, 5.5\%, 1)}{(\$5)(1 + 0.02)^1} = \frac{\$1055}{\$5.10} = 207 \text{ golf balls}$$

The golfer can, after one year, buy 3.4% more golf balls than she could before. With rounding, this is 207 balls.

If Inflation Is 8%

As with the lower inflation rate, we would solve for i':

$$i' = (i - f)/(1 + f)$$
$$= (0.055 - 0.08)/(1 + 0.08)$$
$$= -0.023 \quad \text{or} \quad \textbf{-2.3\% per year}$$

In this case the real growth in money has *decreased* by 2.3%, so that the golfer can now buy 2.3% fewer balls with the money she had invested. Even though she has more money year-end, it is worth less, so she can purchase less.

Regardless of how inflation behaves over the year, the bank will pay the golfer $1055 at the end of the year. However, as we have seen, inflation can greatly affect the "real" growth of dollars over time. In a presidential speech, inflation has been called "that thief" because it steals real purchasing power from our dollars.

Let us continue the discussion of inflation by focusing on cash flows. We define dollars of two types:

Actual dollars ($A\$$):   This is the type of dollar that we ordinarily think of when we think of money. They circulate in our economy and are used for investments and payments. We can touch these dollars and often keep them in our purses and wallets—they are "actual" and exist physically. Sometimes they are called *inflated dollars* because they carry any inflation that has reduced their worth. These are also the dollars shown or **nominal dollars** on paychecks, credit card receipts, and normal financial transactions.

Real dollars ($R\$$):   This type of dollar is a bit harder to define. Real dollars are always expressed in terms of some constant purchasing power "base" year, for example, 2017-based dollars. Real dollars are sometimes called **constant value dollars or constant purchasing power dollars,** and because they have been adjusted for the effects of inflation, they are also known as *inflation-free dollars.*

Having defined *market, inflation,* and *real interest rates* as well as *actual* and *real dollars,* let us describe how these quantities relate. Figure 14–1 illustrates the relationship between these quantities.

Figure 14–1 illustrates the following principles:

When dealing with actual dollars ($A\$$), use a market interest rate (i), and when discounting $A\$$ over time, also use i.

When dealing with real dollars ($R\$$), use a real interest rate (i'), and when discounting $R\$$ over time, also use i'.

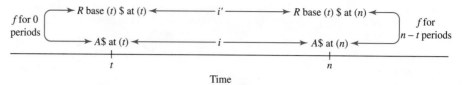

FIGURE 14-1 Relationship between i, f, i', $A\$$, and $R\$$.

Figure 14–1 shows the relationships between $A\$$ and $R\$$ that occur *at the same period of time*. Actual and real dollars are related by the *inflation rate*, in this case, over the period of years defined by $n - t$. To translate between dollars of one type to dollars of the other ($A\$$ to $R\$$ or $R\$$ to $A\$$), use the inflation rate for the right number of periods. The following example illustrates many of these relationships.

EXAMPLE 14–2

When the university's stadium was completed in 1965, the total cost was $1.2 million. At that time a wealthy alumnus gifted the university with $1.2 million to be used for a future replacement. University administrators are now considering building the new facility in the year 2020. Assume that:

- Inflation is 6.0% per year from 1965 to 2020.
- In 1965 the university invested the gift at a market interest rate of 8.0% per year.

(a) Define i, i', f, and $A\$$.
(b) How many actual dollars in the year 2020 will the gift be worth?
(c) How much would the actual dollars in 2020 be in terms of 1965 *purchasing power*?
(d) How much better or worse should the new stadium be?

SOLUTION TO PART a

Since 6.0% is the inflation rate (f) and 8.0% is the market interest rate (i), we can write

$$i' = (0.08 - 0.06)/(1 + 0.06) = 0.01887, \quad \text{or} \quad 1.887\%$$

The building's cost in 1965 was $1,200,000, which were the actual dollars ($A\$$) spent in 1965.

SOLUTION TO PART b

From Figure 14–1 we are going from *actual dollars at t, in 1965*, to *actual dollars at n, in 2020*. To do so, we use the *market interest rate* and compound this amount forward 55 years, as illustrated in Figure 14–2.

$$\text{Actual dollars in 2020} = \text{Actual dollars in 1965} \ (F/P, i, 55 \ \text{years})$$

$$= \$1,200,000(F/P, 8\%, 55)$$

$$= \$82,701,600$$

FIGURE 14–2 Compounding A$ in 1965 to A$ in 2020.

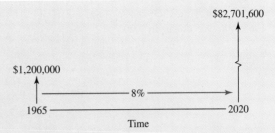

SOLUTION TO PART c

Now we want to determine the amount of *real 1965 dollars that occur in the year 2020,* which are equivalent to the $82.7 million from the solution to part **b**. Let us solve this problem two ways.

1. Translate *actual dollars in the year 2020 to real 1965 dollars in the year 2020.* From Figure 14–1 we can use the inflation rate to **strip 55 years of inflation** from the actual dollars. We do this by using the P/F factor for 55 years at the inflation rate. We are not physically moving the dollars in time; rather, we are simply removing inflation from these dollars one year at a time—the P/F factor does that for us. This is illustrated in the following equation and Figure 14–3.

$$\text{Real 1965 dollars in 2020} = (\text{Actual dollars in 2020})(P/F, f, 55)$$
$$= (\$82,701,600)(P/F, 6\%, 55)$$
$$= \$3,357,000$$

FIGURE 14–3 Translation of A$ in 2020 to R 1965-based dollars in 2020.

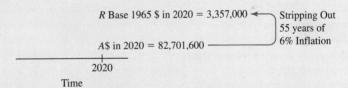

2. Translate *real 1965 dollars in 1965 to real 1965 dollars in 2020.* The $1.2 million can also be said to be *real 1965 dollars that circulated in 1965.* So, let us translate those real dollars from 1965 to the year 2020 (Figure 14–4). Since they are *real dollars,* we use the *real interest rate.*

$$\text{Real 1965 dollars in 2020} = (\text{Real 1965 dollars in 1965})(P/F, i', 55)$$
$$= (\$1,200,000)(F/P, 1.887\%, 55)$$
$$= \$3,355,000$$

Note: The answers differ due to rounding the market interest rate to 1.887% versus carrying it out to more significant digits. The difference due to this rounding is less than 0.1%. If i' and the factors have enough digits, the answers to the two parts would be identical.

FIGURE 14–4 Translation of R 1965 dollars in 1965 to R 1965 dollars in 2020.

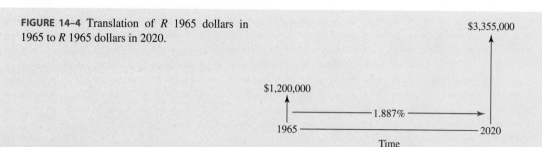

$3,355,000

$1,200,000

1.887%

1965 ——————————————— 2020

Time

SOLUTION TO PART d

Assuming that construction costs increased at the rate of 6% per year, then the amount available for the project *in terms of 1965 dollars* is almost $3.4 million. This means that the new stadium will be about 3.4/1.2 or approximately 2.8 times "better" than the original one using *real dollars*.

EXAMPLE 14–3

In 1924 Mr. O'Leary buried $1000 worth of quarters in his backyard. Over the years he had always thought that the money would be a nice nest egg to give to his first grandchild. His first granddaughter, Gabrielle, arrived in 1994. From 1924 to 1994, inflation averaged 4.5%, the stock market increased an average of 15% per year, and investments in government bonds averaged 6.5% return per year. What was the relative purchasing power of the jar of quarters that Mr. O'Leary gave to his granddaughter Gabrielle at birth? What might have been a better choice for his "backyard investment"?

SOLUTION

Mr. O'Leary's $1000 in quarters was *actual dollars* both in 1924 **and** in 1994.

To obtain the *real 1924 dollar equivalent* of the $1000 that Gabrielle received in 1994, we would **strip 70 years of inflation out of those dollars.** As it turned out, Gabrielle's grandfather gave her $45.90 worth of 1924 purchasing power. Because inflation has "stolen" purchasing power from his stash of quarters during the 70-year period, Mr. O'Leary gave his grandaughter much less than the amount he first spaded underground. This loss of purchasing power caused by inflation can be calculated as follows:

$$\text{Real 1924 dollars in 1994} = (\text{Actual dollars in 1994})(P/F, f, 1994 - 1924)$$

$$= \$1000(P/F, 4.5\%, 70) = \$45.90$$

On the other hand, if Mr. O'Leary had put his $1000 in the stock market in 1924, he would have made baby Gabrielle an instant multimillionaire by giving her $17,735,000. We calculate this as follows:

$$\text{Actual dollars in 1994} = (\text{Actual dollars in 1924})(F/P, i, 1994 - 1924)$$

$$= \$1000(F/P, 15\%, 70) = \$17,735,000$$

At the time of Gabrielle's birth that \$17.7 million translates to \$814,069 in 1924 purchasing power. This is quite a bit different from the \$45.90 in 1924 purchasing power calculated for the unearthed jar of quarters.

$$\text{Real 1924 dollars in 1994} = \$17,735,000(P/F, 4.5\%, 70) = \$814,069$$

Mr. O'Leary was never a risk taker, so it is doubtful he would have chosen the stock market for his future grandchild's nest egg. If he had chosen government bonds instead of his backyard, by 1994 the investment would have grown to \$59,076 (actual dollars)—the equivalent of \$2712 in 1924 purchasing power.

$$\text{Actual dollars in 1994} = (\text{Actual dollars in 1924})(F/P, i, 1994 - 1924)$$

$$= \$1000(F/P, 6\%, 70) = \$59,076$$

$$\text{Real 1924 dollars in 1994} = \$59,076(P/F, 4.5\%, 70) = \$2712$$

Obviously, either option would have been better than the choice Mr. O'Leary made. This example illustrates the effects of inflation and purchasing power, as well as the power of compound interest. However, in Mr. O'Leary's defense, if the country had experienced 70 years of *deflation* instead of *inflation,* he might have had the last laugh!

There are in general two ways to approach an economic analysis problem after the effects of inflation have been recognized. The first is to ignore these effects in conducting the analysis, as we've done so far in the text.

Ignoring inflation in the analysis: Use **real dollars** and a **real interest rate** that is adjusted for inflation.

The second approach is to systematically include the effects of inflation, as studied in this chapter.

Incorporating inflation in the analysis: Use a **market interest rate** and **actual dollars** that include inflation.

Since inflation is so common, why do many economic analyses of engineering projects and most of this text choose *not* to explicitly address inflation? This question is best answered by referring to the many examples and problems that contain statements like "Operations and maintenance costs are expected to be \$30,000 annually for the equipment's 20-year life."

Does such a statement mean that accounting records for the next 20 years will show constant costs? Obviously not. Instead, it means that in constant-dollar terms the O&M costs are not expected to increase. In real dollars, O&M costs are uniform. In actual or inflated dollars, we will pay more each year, but each of those dollars will be worth less.

Most costs and benefits in the real world and in this text have prices that increase at about the same rate of inflation as the economy as a whole. In most analyses these inflation increases are addressed by simply stating everything in real dollar terms and using a real interest rate.

There are specific items, such as computers and depreciation deductions, where inflation is clearly expected to differ from the general rate of inflation. It is for these cases, that this chapter is included in this text.

ANALYSIS IN CONSTANT DOLLARS VERSUS THEN-CURRENT DOLLARS

Performing an analysis requires that we distinguish cash flows as being either constant dollars (real dollars, expressed in terms of some purchasing power base) or then-current dollars (actual dollars that are then-current when they occur). As previously stated, constant (real) dollars require the use of a *real interest rate* for discounting; then-current dollars require a *market (or combined) interest rate*. We must not mix these two dollar types when performing an analysis. If both types are stated in the problem, one type must be converted to the other, so that a consistent comparison can be made.

One of the challenges in inflation analysis is that cash flows are stated in a variety of ways. The three most common are (1) then-current or actual dollars, (2) today's or time-0 dollars, and (3) year-1 dollars. This is combined with the normal assumption that first costs occur at time 0 and other costs occur at the end of the year. Very careful reading of the examples, homework problems, and test questions is recommended. If a simple statement of constant-value dollars is all you are given, the normal assumption is that they are time-0 dollars and that the values have been chosen to match normal time 0 and end-of-period assumptions.

EXAMPLE 14–4

The Waygate Corporation is interested in contracting out a testing function. Two firms have submitted bids. Waygate believes that both companies will be able to deliver equivalent services for the 5-year period. Determine which one Waygate should choose if the corporate MARR (investment market rate) is 25% and price inflation is assumed to be 3.5% per year over the next 5 years.

Company Alpha costs: Costs will be $150,000 in year-1 dollars the first year and will increase at a rate of 5% over the 5-year period.

Company Beta costs: Costs will be a constant $150,000 per year in terms of today's dollars over the 5-year period.

SOLUTION

The costs for each of the two alternatives are as follows:

Year	Alpha Then-Current Costs	Beta Time-0 Dollar Costs
1	$150,000 \times 1.05^0 = \$150,000$	$150,000
2	$150,000 \times 1.05^1 = \ \ 157,500$	150,000
3	$150,000 \times 1.05^2 = \ \ 165,375$	150,000
4	$150,000 \times 1.05^3 = \ \ 173,644$	150,000
5	$150,000 \times 1.05^4 = \ \ 182,326$	150,000

We inflate (or escalate) the stated yearly cost given by Company Alpha by 5% per year to obtain the then-current (actual) dollars each year.

Using a Constant Dollar Analysis

Here we must convert the then-current costs given by Company Alpha to constant today-based dollars. We do this by stripping the number of years of general inflation from each year's cost using $(P/F, f, n)$ or $(1 + f)^{-n}$.

Year	Alpha Time-0 Dollar Costs	Beta Time-0 Dollar Costs
1	$\$150,000 \times 1.035^{-1} = \$144,928$	$150,000
2	$157,500 \times 1.035^{-2} = 147,028$	150,000
3	$165,375 \times 1.035^{-3} = 149,159$	150,000
4	$173,644 \times 1.035^{-4} = 151,321$	150,000
5	$182,326 \times 1.035^{-5} = 153,514$	150,000

We use the *real interest rate* (i') calculated from Equation 14-1 to calculate the present worth of costs for each alternative:

$$i' = (i - f)/(1 + f) = (0.25 - 0.035)/(1 + 0.035) = 0.208$$

$$\text{PW of cost (Alpha)} = 144{,}928(P/F, 20.8\%, 1) + 147{,}028(P/F, 20.8\%, 2)$$
$$+ \ 149{,}159(P/F, 20.8\%, 3) + 151{,}321(P/F, 20.8\%, 4)$$
$$+ \ 153{,}514(P/F, 20.8\%, 5) = \$436{,}000$$

$$\text{PW of cost (Beta)} = \$150{,}000(P/A, 20.8\%, 5) = \$150{,}000(2.9387) = \$441{,}000$$

Using a Then-Current Dollar Analysis

Here we must convert the constant dollar costs of Company Beta to then-current dollars. We do this by using $(F/P, f, n)$ or $(1 + f)^n$ to "add in" the appropriate number of years of general inflation to each year's cost.

Year	Alpha Then-Current Costs	Beta Then-Current Costs
1	$\$150,000 \times 1.05^{0} = \$150,000$	$\$150,000 \times 1.035^{1} = \$155,250$
2	$150,000 \times 1.05^{1} = 157,500$	$150,000 \times 1.035^{2} = 160,684$
3	$150,000 \times 1.05^{2} = 165,375$	$150,000 \times 1.035^{3} = 166,308$
4	$150,000 \times 1.05^{3} = 173,644$	$150,000 \times 1.035^{4} = 172,128$
5	$150,000 \times 1.05^{4} = 182,326$	$150,000 \times 1.035^{5} = 178,153$

Calculate the present worth of costs for each alternative using the *market interest rate* (i).

$$\text{PW of cost (Alpha)} = 150{,}000(P/F, 25\%, 1) + 157{,}500(P/F, 25\%, 2)$$
$$+ \ 165{,}375(P/F, 25\%, 3) + 173{,}644(P/F, 25\%, 4)$$
$$+ \ 182{,}326(P/F, 25\%, 5) = \$436{,}000$$

PW of cost (Beta) $= 155{,}250(P/F, 25\%, 1) + 160{,}684(P/F, 25\%, 2)$

$$+\ 166{,}308(P/F, 25\%, 3) + 172{,}128(P/F, 25\%, 4)$$

$$+\ 178{,}153(P/F, 25\%, 5) = \$441{,}000$$

Using either a constant dollar or then-current dollar analysis, Waygate should chose Company Alpha's offer, which has the lower present worth of costs. There may, of course, be intangible elements in the decision that are more important than a 1% difference in the costs.

Example 14–5 confirms that analyses done in constant-value dollars with real interest rates will lead to the same results as analyses done in nominal dollars with market interest rates—if both are done correctly. It also illustrates another very important point. When costs are estimated to be uniform over time, this is virtually always consistent with assuming constant-value dollars. The only common exception is bonds, where the interest payments are uniform and stated in nominal dollar terms.

EXAMPLE 14–5

A new heat exchanger will cost $220,000, and it will save $50,000 annually. After 10 years it will have no salvage value. The firm's real interest rate is 15%. If inflation is 5%, is this project worth doing? Analyze in actual and constant value dollars.

SOLUTION

While not stated explicitly, the $50,000 in annual savings should be assumed to be constant value dollars and a constant amount of energy. Only coincidentally would decreases in volume exactly match increases in prices. It is also assumed that the constant value dollar is a time-0 dollar and that the values are chosen to match the beginning and end-of-period assumptions.

Thus, one way to analyze the problem is to use the real interest rate of 15% with the constant value dollar annual savings of $50,000. Inflation has been correctly included by using constant value dollars and a real rate.

Alternatively, the inflation rate of 5% can be explicitly included. The $50,000 is treated as a time 0 value and inflation is explicitly applied. Now the interest rate must also include inflation. Using Equation 14-1 we find that the appropriate interest rate is 20.75%.

$$i = i' + f + i'f$$
$$i = 0.15 + 0.05 + 0.15 \times 0.05$$
$$i = 0.2075 = 20.75\%$$

	A	B	C	D
1	$220,000	First cost ($M)		
2	$50,000	Annual savings		
3	5%	Inflation rate		
4	Interest rate	15%	20.75%	
5	Year	Constant value $s	Actual $s	
6	0	-$220,000	-$220,000	
7	1	50,000	52,500	=$A$2*(1+$A$3)^A7
8	2	50,000	55,125	
9	3	50,000	57,881	
10	4	50,000	60,775	
11	5	50,000	63,814	
12	6	50,000	67,005	
13	7	50,000	70,355	
14	8	50,000	73,873	
15	9	50,000	77,566	
16	10	50,000	81,445	
17	Present worth	$30,938	$30,938	=C6+NPV(C4,C7:C16)

Both approaches calculate the same present worth of $30,938. The project is attractive.

Inflation and Uniform Flow Equivalence

Can we convert the PWs in Example 14–5 to meaningful EUAWs? For the constant dollar case, the answer is obviously yes. The $6165 calculated below is a constant value dollar EUAW. However, the $7568 that is calculated for the actual dollar case is a time and discount rate weighted average of a non-uniform series of actual dollars that is hard to correctly interpret. Thus analyses that state conclusions as an equivalent uniform annual worth or cost may only make sense in conjunction with a constant value dollar assumption.

	A	B	C	D	E	F	G	H
1	Alternative	*i*	*n*	*PMT*	*PV*	*FV*	Solve for	Answer
2	Constant $	15%	10		30,938	0	PMT	6165
3	Actual $	20.75%	10		30,938	0	PMT	~~7568~~ ??

PRICE CHANGE WITH INDEXES

We have already described the effects that inflation can have on money over time. Also, several definitions and relationships regarding dollars and interest rates have been given. We have seen that it is not correct to compare the benefits of an investment in 2010-based dollars with costs in 2018-based dollars. This is like comparing apples and oranges. Such comparisons of benefits and costs can be meaningful only if a standard purchasing power base of money is used. So we ask, "How do I know what inflation rate to use in my studies? and How can we measure price changes over time?"

What Is a Price Index?

Price indexes (introduced in Chapter 2) describe the relative price fluctuation of goods and services. They provide a *historical* record of prices over time. Price indexes are tracked for *specific commodities* as well as for *bundles (composites) of commodities.* As such, price indexes can be used to measure historical price changes for individual cost items (like labor and material costs) as well as general costs (like consumer products). We use *past* price fluctuations to predict *future* prices.

Table 14–1 lists the historic prices of sending a first-class letter in the U.S. from 1970 to 2016. The cost is given both in terms of dollars (cents) and as measured by a fictitious price index that we could call the letter cost index (LCI).

Notice two important aspects of the LCI. First, as with all cost or price indexes, there is a **base year,** which is assigned a value of 100. Our LCI has a base year of 1970—thus for 1970, LCI = 100. Values for subsequent years are stated in relation to the 1970 value. Second, the LCI changes only when the cost of first-class postage changes. In years when this quantity does not change, the LCI is not affected. These general observations apply to all price indexes.

TABLE 14–1 Historic Prices of First-Class Mail, 1970–2010, and Letter Cost Index

Year, n	Cost of First-Class Mail	LCI	Annual Increase for n	Year, n	Cost of First-Class Mail	LCI	Annual Increase for n
1970	$0.06	100		1994	0.29	483	0%
1971	0.08	133	33.33%	1995	0.32	533	10.34
1972	0.08	133	0	1996	0.32	533	0
1973	0.08	133	0	1997	0.32	533	0
1974	0.10	166	25.00	1998	0.33	550	3.13
1975	0.13	216	30.00	1999	0.33	550	0
1976	0.13	216	0	2000	0.33	550	0
1977	0.13	216	0	2001	0.34	567	3.03
1978	0.15	250	15.74	2002	0.37	617	8.82
1979	0.15	250	0	2003	0.37	617	0
1980	0.15	250	0	2004	0.37	617	0
1981	0.20	333	33.33	2005	0.37	617	0
1982	0.20	333	0	2006	0.39	650	5.41
1983	0.20	333	0	2007	0.41	683	5.13
1984	0.20	333	0	2008	0.42	700	2.44
1985	0.22	367	10.00	2009	0.42	700	0
1986	0.22	367	0	2010	0.44	733	4.76
1987	0.22	367	0	2011	0.45	750	2.27
1988	0.25	417	13.64	2012	0.45	750	0
1989	0.25	417	0	2013	0.46	767	2.27
1990	0.25	417	0	2014	0.49	817	6.52
1991	0.29	483	16.00	2015	0.49	817	0
1992	0.29	483	0	2016	0.47	783	−4.08
1993	0.29	483	0				

In general, engineering economists are the "users" of cost indexes such as our hypothetical LCI. That is, cost indexes are calculated or tabulated by some other party, and our interest is in assessing what the index tells us about the historical prices and how these may affect our estimate of future costs. However, we should understand how the LCI in Table 14–1 was calculated.

In Table 14–1, the LCI is assigned a value of 100 because 1970 serves as our base year. In the following years the LCI is calculated on a year-to-year basis based on the annual percentage increase in first-class mail. Equation 14-2 illustrates the arithmetic used.

$$\text{LCI}_n = \frac{\text{cost } (n)}{\text{cost } 1970} \times 100 \tag{14-2}$$

For example, consider the LCI for the year 2010. We calculate the LCI as follows.

$$\text{LCI}_{2010} = \frac{0.44}{0.06} \times 100 = 733$$

As mentioned, engineering economists often use cost indexes to project future cash flows. As such, our first job is to use a cost index to **calculate** the *year-to-year* percentage increase (or *inflation*) of prices tracked by an index. We can use Equation 14-3:

$$\text{Annual percentage increase}_n = \frac{\text{Index}_n - \text{Index}_{n-1}}{\text{Index}_{n-1}} \times 100\% \tag{14-3}$$

To illustrate, let us look at the percent change from 2012 to 2013 for the LCI.

$$\text{Annual percentage increase}_{2013} = \frac{767 - 750}{750} \times 100\% = 2.27\%$$

For 2013 the price of mailing a first-class letter increased by 2.27% over the previous year.

An engineering economist often wants to know how a particular cost quantity changes over time. Often we are interested in calculating the *average* rate of price increases over a period of time. For instance, we might want to know the average yearly increase in postal prices from 2000 to 2010. If we generalize Equation 14-3 to calculate the percent change from 2000 to 2010, we obtain

$$\% \text{ Increase}_{2000 \text{ to } 2010} = \frac{700 - 550}{550} \times 100\% = 273\%$$

How do we use this to obtain the *average* rate of increase over those 10 years? Should we divide 273% by 10 years ($273/10 = 27.3\%$)? Of course not! Inflation, like interest, compounds. Such a simple division treats inflation like simple interest—without compounding. So the question remains: How do we calculate an *equivalent average rate of increase* in postage rates over a period of time? If we think of the index numbers as cash flows, we have

$$P = 550, \qquad F = 700, \qquad n = 10, \text{ years}, \qquad i = ?$$
$$\text{Using } F = P(1 + i)^n$$
$$700 = 500(1 + i)^{10}$$
$$i = (700/550)^{1/10} - 1 = 0.0244 = 2.44\%$$

We can use a cost index to calculate the average rate of increase over any period of years, which should provide insight into how prices may behave in the future.

Composite Versus Commodity Indexes

Cost indexes come in two types: commodity-specific indexes and composite indexes. Commodity-specific indexes measure the historical change in price for specific items—such as construction labor or iron ore. Commodity indexes, like our letter cost index, are useful when an economic analysis includes individual cost items that are tracked by such indexes. For example, if we need to estimate the direct-labor cost portion of a construction project, we could use an index that tracks the inflation, or escalation, of labor costs. The U.S. Departments of Commerce and Labor track many cost quantities through the Department of Economic Analysis and Bureau of Labor Statistics. Example 14–6 uses data from the California Construction Cost Index (CCCI) to demonstrate using a commodity index. This data is compiled from the *Engineering News-Record*.

EXAMPLE 14–6

In January 2016 bids were opened for a new building in Los Angeles. The low bid and the final construction cost were $52.5 million. Another building of the same size, quality, and purpose is planned with a bid opening in January 2018. Estimate the new building's low bid and cost using the CCCI.

SOLUTION

In January 2016 the California Construction Cost Index (CCCI) had a value of 6106 and in January 2000 the value was 3746. If we wanted a cost estimate for January 2016, we could simply use the ratio of these values and Equation 14-2. But we want a value for January 2018, which is outside our data set. (This is true for all future estimates.)

The solution is to estimate the average annual rate of increase and then to apply this to the January 2016 cost.

$$F = 6106, P = 3746, n = 16, \text{ find } f$$

$$F = P(1+f)^n$$

$$f = (6106/3746)^{1/16} - 1 = 3.10\% \text{ per year}$$

Now we can apply the inflation rate for $n = 2$ years to the building's cost in 2016.

$$F = 52.5 \text{ million } \times 1.0310^2 = \$55.8 \text{ million in } 2018$$

Composite cost indexes do not track historical prices for individual classes of items. Instead, they measure the historical prices of *bundles* or *market baskets* of assets. Examples of composite indexes include the *Consumer Price Index* (CPI) and the *Producer Price Index* (PPI). The CPI measures prices for consumers in the U.S. marketplace, and each PPI measures prices for categories of producers in the U.S. economy.

The CPI, an index calculated by the Bureau of Labor Statistics, tracks the cost of a standard *bundle of consumer goods* from year to year. This "consumer bundle" or "market basket" includes housing, clothing, food, transportation, and entertainment. The CPI enjoys popular identification as the "inflation" indicator. Table 14–2 gives yearly index values and annual percent increases in the CPI. Figure 14–5 charts the CPI inflation rate for the same period.

TABLE 14–2 CPI Index Values and Yearly Percentage Increases, 1955–2015

Year	CPI Value*	CPI Increase	Year	CPI Value*	CPI Increase
1955	26.8	–	1994	148.2	2.6%
1960	29.6	–	1995	152.4	2.8
1965	31.5	–	1996	156.9	2.9
1970	38.8	–	1997	160.5	2.3
1975	53.8	9.1%	1998	163.0	1.6
1976	56.9	5.8	1999	166.6	2.2
1977	60.6	6.5	2000	172.2	3.4
1978	65.2	7.6	2001	177.1	2.8
1979	72.6	11.3	2002	179.9	1.6
1980	82.4	13.5	2003	184.0	2.3
1981	90.9	10.3	2004	188.9	2.7
1982	96.5	6.2	2005	195.3	3.4
1983	99.6	3.2	2006	201.6	3.2
1984	103.9	4.3	2007	207.3	2.8
1985	107.6	3.6	2008	215.3	3.8
1986	109.6	1.9	2009	214.5	−0.4
1987	113.6	3.6	2010	218.1	1.6
1988	118.3	4.1	2011	224.9	3.2
1989	124.0	4.8	2012	229.6	2.1
1990	130.7	5.4	2013	233.0	1.5
1991	136.2	4.2	2014	236.7	1.6
1992	140.3	3.0	2015	237.0	0.1
1993	144.5	3.0			

*Reference base: 1982–1984 = 100.

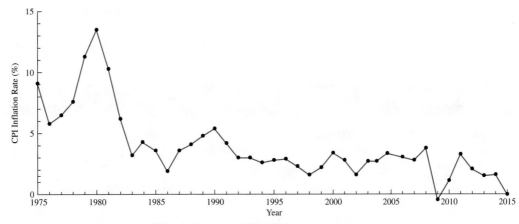

FIGURE 14–5 CPI inflation rate: 1975–2015.

Composite indexes are used the same way as commodity-specific indexes. That is, we can pick a single value from the table if we are interested in measuring the historic price for a single year, or we can calculate an *average inflation rate* or *average rate of price increase* as measured by the index over a time period extending several years.

How to Use Price Indexes in Engineering Economic Analysis

One may question the usefulness of *historical* data (as provided by price indexes) when engineering economic analysis deals with economic effects projected to occur in the *future*. However, historical index data are often better predictors of future prices than official government predictions, which may be influenced by political considerations. The engineering economist can use *average historical percentage increases (or decreases)* from commodity-specific and composite indexes, along with data from market analyses and other sources, to estimate future costs and benefits.

When the estimated quantities are items that are tracked by commodity specific indexes, then those indexes should be used to calculate *average historical percentage increases (or decreases)*. If no commodity-specific indexes are kept, use an appropriate composite index to make this calculation.

For example, to estimate electric usage costs for a turret lathe over a 5-year period, you would first want to refer to a commodity-specific index for electric power in your area. If such an index does not exist, you might use a specific index for a very closely related commodity—perhaps an index of electric power costs nationally. In the absence of such substitute or related commodity indexes, you could use a composite index for national energy prices. The key point is that you should try to identify and use the price index that most closely relates to the quantity being estimated.

CASH FLOWS THAT INFLATE AT DIFFERENT RATES

Engineering economic analysis requires the estimation of various parameters. Over time, it is not uncommon for these parameters to *inflate* or *increase* (or even decrease) at different rates. For instance, one parameter might *increase* 5% per year and another 15% per year, and a third might *decrease* 3.5% per year. Since we are looking at the behavior of cash flows over time, we must have a way of handling this effect.

In Example 14–7 several commodity prices change at different rates. By using the respective individual inflation rates, the *actual dollar* amounts for each commodity are obtained in each year. Then we use a market interest rate to discount these actual dollar amounts.

EXAMPLE 14–7

On your first assignment as an engineer, your boss asks you to develop the utility cost estimate for a new manufacturing facility. After some research you define the problem as finding the present worth of utility costs given the following data:

- Your company uses a minimum attractive rate of return (MARR) = 35% as *i* (not adjusted for inflation.)
- The project has a useful life of 25 years.

- Utilities to be estimated include electricity, water, and natural gas.
- The 35-year historical data reveal:

  Electricity costs increase at 8.5% per year

  Water costs increase at 5.5% per year

  Natural gas costs increase at 6.5% per year

- First-year estimates of the utility costs (in today's dollars) are as follows:

  Electricity will cost $55,000

  Water will cost $18,000

  Natural gas will cost $38,000

SOLUTION

For this problem we will take each of the utilities used in our manufacturing facility and inflate them independently at their respective historical annual rates. Once we have these actual dollar amounts ($A\$$), we can total them and then discount each year's total at 35% back to the present.

The formulas shown calculate each cost using the appropriate inflation rate—calculated from time 0. The formulas can also be constructed by applying the inflation rate from one year to the next. This approach is better if inflation rates are changing over time.

$$CF_{t+1} = CF_t(1 + f_{commodity})$$

Year	Electricity			Water			Natural Gas			Total
1	$55,000(1.085)^0$	$=$	$55,000	$18,000(1.055)^0$	$=$	$18,000	$38,000(1.065)^0$	$=$	$38,000	$111,000
2	$55,000(1.085)^1$	$=$	59,675	$18,000(1.055)^1$	$=$	18,990	$38,000(1.065)^1$	$=$	40,470	119,135
3	$55,000(1.085)^2$	$=$	64,747	$18,000(1.055)^2$	$=$	20,034	$38,000(1.065)^2$	$=$	43,101	127,882
4	$55,000(1.085)^3$	$=$	70,251	$18,000(1.055)^3$	$=$	21,136	$38,000(1.065)^3$	$=$	45,902	137,289
5	$55,000(1.085)^4$	$=$	76,222	$18,000(1.055)^4$	$=$	22,299	$38,000(1.065)^4$	$=$	48,886	147,407
6	$55,000(1.085)^5$	$=$	82,701	$18,000(1.055)^5$	$=$	23,525	$38,000(1.065)^5$	$=$	52,063	158,290
7	$55,000(1.085)^6$	$=$	89,731	$18,000(1.055)^6$	$=$	24,819	$38,000(1.065)^6$	$=$	55,447	169,997
8	$55,000(1.085)^7$	$=$	97,358	$18,000(1.055)^7$	$=$	26,184	$38,000(1.065)^7$	$=$	59,051	182,594
·	·			·			·			·
·	·			·			·			·
·	·			·			·			·
24	$55,000(1.085)^{23}$	$=$	359,126	$18,000(1.055)^{23}$	$=$	61,671	$38,000(1.065)^{23}$	$=$	161,743	582,539
25	$55,000(1.085)^{24}$	$=$	389,652	$18,000(1.055)^{24}$	$=$	65,063	$38,000(1.065)^{24}$	$=$	172,256	626,970

The present worth of the total yearly utility costs is

$$PW = 111,000(P/F, 35\%, 1) + 119,135(P/F, 35\%, 2) + \cdots + 626,970(P/F, 35\%, 25)$$
$$= \$5,540,000$$

DIFFERENT INFLATION RATES PER PERIOD

In this section we address the situation of inflation rates that are changing over the study period. Rather than different inflation rates for different cash flows, in Example 14–8 the *inflation rate* for the same cash flow is changing over time. A method for handling this situation is much like that of the preceding section. We can simply apply the inflation rates in the years in which they are projected to occur. We would do this for each cash flow. Once we have all these actual dollar amounts, we can use the market interest rate to apply PW or other measures of merit.

Example 14–8 provides another example of how the effect of changes in inflation rates over time can affect an analysis.

EXAMPLE 14–8

If general price inflation is estimated to be 5% for the next 5 years, 7.5% for the 3 years after that, and 3% the following 5 years, at what market interest rate (i) would you have to invest your money to maintain a real purchasing power growth rate (i') of 10% during those years?

SOLUTION

In Years 1–5 you must invest at $0.10 + 0.050 + (0.10)(0.050) = 0.1150 = 11.50\%$ per year.

In Years 6–8 you must invest at $0.10 + 0.075 + (0.10)(0.075) = 0.1825 = 18.25\%$ per year.

In Years 9–13 you must invest at $0.10 + 0.030 + (0.10)(0.030) = 0.1330 = 13.30\%$ per year.

Note: Most interest-bearing investments have fixed, up-front rates that the investor well understands going in. On the other hand, inflation is not quantified, and its effect on real return is not measured until the end of the year. Therefore the real investment return (i') may not turn out to be what was originally required.

INFLATION EFFECT ON AFTER-TAX CALCULATIONS

Earlier we noted the impact of inflation on before-tax calculations. If the future benefits keep up with the rate of inflation, the rate of return will not be adversely affected by the inflation. Unfortunately, we are not so lucky when we consider a situation with income taxes, as illustrated by Example 14–9. The value of the depreciation deduction is diminished by inflation.

EXAMPLE 14–9

A $12,000 investment with no salvage value will return annual benefits for 6 years. Assume straight-line depreciation and a 46% income tax rate. Solve for both before- and after-tax rates of return for two situations:

1. *No inflation:* the annual benefits are constant at $2918 per year.
2. *Inflation equal to 5%:* the benefits from the investment increase at this same rate, so that they continue to be the equivalent of $2918 in Year-0 dollars. Assume the $2918 is a time-0 value.

The benefit schedules are as follows:

Year	Annual Benefit for Both Situations (Year-0 dollars)	No Inflation, Actual Dollars Received	5% Inflation Factor*	5% Inflation, Actual Dollars Received
1	$2918	$2918	1.05^1	$3064
2	2918	2918	1.05^2	3217
3	2918	2918	1.05^3	3378
4	2918	2918	1.05^4	3547
5	2918	2918	1.05^5	3724
6	2918	2918	1.05^6	3910

*May be read from the 5% compound interest table as $(F/P, 5\%, n)$.

SOLUTIONS

Before-Tax Rate of Return

Since both situations (no inflation and 5% inflation) have an annual benefit, stated in Year-0 dollars of $2918, they have the same before-tax rate of return.

$$PW \text{ of cost} = PW \text{ of benefit}$$

$$12,000 = 2918(P/A, i, 6), \qquad (P/A, i, 6) = \frac{12,000}{2918} = 4.11$$

The before-tax rate of return equals 12%.

After-Tax Rate of Return, No Inflation

Year	Before-Tax Cash Flow	Straight-Line Depreciation	Taxable Income	46% Income Taxes	Actual Dollars, and Year-0 Dollars, After-Tax Cash Flow
0	−$12,000				−$12,000
1–6	2,918	$2000	$918	−$422	2,496

$$PW \text{ of cost} = PW \text{ of benefit}$$

$$12,000 = 2496(P/A, i, 6), \qquad (P/A, i, 6) = \frac{12,000}{2496} = 4.81$$

The after-tax rate of return equals 6.7%.

After-Tax Rate of Return, 5% Inflation

Year	Before-Tax Cash Flow	Straight-Line Depreciation	Taxable Income	46% Income Taxes	Actual Dollars, After-Tax Cash Flow
0	−$12,000				−$12,000
1	3,064	$2000	$1064	−$489	2,575
2	3,217	2000	1217	−560	2,657
3	3,378	2000	1378	−634	2,744
4	3,547	2000	1547	−712	2,835
5	3,724	2000	1724	−793	2,931
6	3,910	2000	1910	−879	3,031

Converting to Year-0 Dollars and Solving for Rate of Return

Year	Actual Dollars, After-Tax Cash Flow	Conversion Factor		Year-0 Dollars, After-Tax Cash Flow	Present Worth at 4%	Present Worth at 5%
0	−$12,000			−$12,000	−$12,000	−$12,000
1	2,575 ×	1.05^{-1} =		2,452	2,358	2,335
2	2,657 ×	1.05^{-2} =		2,410	2,228	2,186
3	2,744 ×	1.05^{-3} =		2,370	2,107	2,047
4	2,835 ×	1.05^{-4} =		2,332	1,993	1,919
5	2,931 ×	1.05^{-5} =		2,297	1,888	1,800
6	3,031 ×	1.05^{-6} =		2,262	1,788	1,688
					+362	−25

Linear interpolation between 4 and 5%:

$$\text{After-tax rates of return} = 4\% + 1\%[362/(362 + 25)] = 4.9\%$$

From Example 14–9, we see that the before-tax rate of return for both situations (no inflation and 5% inflation) is the same. Equal before-tax rates of return are expected because the benefits in the inflation situation increased in proportion to the inflation. No special calculations are needed in before-tax calculations when future benefits are expected to respond to inflation or deflation rates.

The after-tax calculations illustrate that equal before-tax rates of return do not produce equal after-tax rates of return considering inflation.

	Rate of Return	
Situation	Before Taxes	After Taxes
No inflation	12%	6.7%
5% inflation	12	4.9

Inflation reduces the after-tax rate of return, even though the benefits increase at the same rate as the inflation. A review of the cash flow table reveals that while benefits increase, the depreciation schedule does not. Thus, the inflation results in increased taxable income and, hence, larger income tax payments.

The result is that while the after-tax cash flow in actual dollars increases, the augmented amount is not high enough to offset *both* inflation and increased income taxes. The Year-0-dollar after-tax cash flow is smaller with inflation than the Year-0-dollar after-tax cash flow without inflation.

USING SPREADSHEETS FOR INFLATION CALCULATIONS

Spreadsheets are the perfect tool for incorporating consideration of inflation into analyses of economic problems. For example, next year's labor costs are likely to be estimated as equal to this year's costs times $(1 + f)$, where f is the inflation rate. Thus each year's value is different, so we can't use factors for uniform flows, A. Also the formulas that link different years are easy to write. As a result, problems that are tedious to do by hand are easy with a spreadsheet.

Example 14–10 illustrates two different ways to write the equation for inflating costs. In Figure 14–6 the labor costs (column B) are calculated by applying the inflation rate to the previous year's labor costs. The transportation costs (column D) vary each year, so each year's costs are calculated by applying the inflation rate for the number of years since time-0. Example 14–11 illustrates that inflation reduces the after-tax rate of return because inflation makes the depreciation deduction less valuable.

EXAMPLE 14–10

Two costs for construction of a small, remote mine are for labor and transportation. Labor costs are expected to be $350,000 the first year, with inflation of 6% annually. Unit transportation costs are expected to inflate at 5% annually, but the volume of material being moved changes each year. In Time-0 dollars, the transportation costs are estimated to be $40,000, $60,000, $50,000, and $30,000 in Years 1 through 4. The inflation rate for the value of the dollar is 3%. If the firm uses an i' of 7%, what is the equivalent annual cost in Time-0 dollars for this 4-year project?

SOLUTION

The data for labor costs can be stated so that no inflation needs to be applied in Year 1: the cost is $350,000. In contrast, the transportation costs for Year 1 are determined by multiplying $40,000 by 1.05 ($ = 1 + f$).

Also in later years the labor cost$_t$= labor cost$_{t-1}(1 + f)$, while each transportation cost must be computed as the Time-0 value times $(1 + f)^t$. In Figure 14–6, the numbers in the Year-0 (or real) dollar column equal the values in the actual dollars column divided by $(1.03)^t$.

	A	B	C	D	E	F	G	H
1						7%	Inflation-Free Interest	
2	Inflation Rate	6%		5%		3%		
3			Transportation Costs		Total	Total		
4	Year	Labor Costs	Year 0 $s	Actual $s	Actual $s	Real $s		
5	1	120,000	40,000	42,000	162,000	157,282	= E5/(1+$F$2)^A5	
6	2	127,200	60,000	66,150	193,350	182,251		
7	3	134,832	50,000	57,881	192,713	176,360		
8	4	142,922	30,000	36,465	179,387	159,383		
9						$571,732	= NPV(F1,F5:F8)	
10					=B8+D8	$168,791	= −PMT(F1,4,F9)	
11	=B7*(1+$B$2)		=C8*(1+$D$2)^A8					

FIGURE 14–6 Spreadsheet for inflation.

The equivalent annual cost equals $168,791 in Time-0 dollars. As noted earlier the meaning of an EUAC in actual dollars in unclear.

EXAMPLE 14–11   (Example 14–9 Revisited)

For the data of Example 14–9, calculate the IRR with and without inflation with MACRS depreciation. How are the results affected by inflation by comparison with the earlier results.

SOLUTION

Most of the formulas for this spreadsheet are given in rows 11 and 12 of Figure 14–7 for the data in Year 6. The benefits received are computed from the base value in cell B5. The depreciation is the MACRS percentage times the $12,000 spent in Year 0. This value is not influenced by inflation, so the depreciation deduction is less valuable as inflation increases. The tax paid equals the tax rate times the taxable income, which equals dollars received minus the depreciation charge. Then ATCF (after-tax cash flow) equals the before-tax cash flow minus the tax paid.

In Figure 14–7, notice that in Year 2 the depreciation charge is large enough to cause this project to pay "negative" tax. For a firm, this means that the deduction on this project will be used to offset income from other projects.

	A	B	C	D	E	F	G	H
1		0%	= Inflation Rate		46%	= Tax Rate		
2			Actual $s	MACRS	Actual $s	Actual $s	Actual $s	Real $s	
3		Year	Received	Deprec. %	Deprec.	Tax	ATCF	ATCF	
4		0	−12000				−12000	−12000	
5		1	2918	20.00%	2400	238	2680	2680	= F5/(1+$B$1)^A5
6		2	2918	32.00%	3840	−424	3342	3342	
7		3	2918	19.20%	2304	282	2636	2636	
8		4	2918	11.52%	1382	706	2212	2212	
9		5	2918	11.52%	1382	706	2212	2212	
10		6	2918	5.76%	691	1024	1894	1894	
11	Formulas			= −$B$4*C10		=B10−E10			
12	for Yr 6	=$B$5*(1+$B$1)^A10		=(B10−D10)*$E$1			7.29%	= IRR	
13									
14			5%	= Inflation Rate		46%	= Tax Rate		
15			Actual $s	MACRS	Actual $s	Actual $s	Actual $s	Real $s	
16		Year	Received	Deprec. %	Deprec.	Tax	ATCF	ATCF	
17		0	−12000				−12000	−12000	
18		1	3064	20.00%	2400	305	2759	2627	
19		2	3217	32.00%	3840	−287	3504	3178	
20		3	3378	19.20%	2304	494	2884	2491	
21		4	3547	11.52%	1382	996	2551	2099	
22		5	3724	11.52%	1382	1077	2647	2074	
23		6	3910	5.76%	691	1481	2430	1813	
24								5.68%	= IRR

FIGURE 14–7 After-tax IRRs with MACRS and inflation.

The IRRs are higher in this example (7.29% without inflation vs. 6.7% with straight-line depreciation in Example 14–9, and 5.68% with inflation vs. 4.9%) because MACRS supports faster depreciation, so the depreciation deductions are more valuable. Also because the depreciation is faster, the results are affected somewhat less by inflation. Specifically, 5% inflation lowers the IRR by 1.6% with MACRS and by 1.8% with straight-line depreciation.

SUMMARY

Inflation is characterized by rising prices for goods and services, whereas deflation produces a fall in prices. An inflationary trend makes future dollars have less **purchasing power** than present dollars. Deflation has the opposite effect. If money is borrowed over a period of time in which deflation is occurring, then debt will be repaid with dollars that have **more** purchasing power than those originally borrowed. Inflation and deflation have opposite effects on the purchasing power of a monetary unit over time.

To distinguish and account for the effect of inflation in our engineering economic analysis, we define *inflation, real,* and *market* interest rates. These interest rates are related by the following expression:

$$i = i' + f + i'f$$

Each rate applies in a different circumstance, and it is important to apply the correct rate to the correct circumstance. Cash flows are expressed in terms of either *actual* or *real dollars*. The *market interest* rate should be used with *actual dollars* and the *real interest rate* should be used with *real dollars*. Virtually always when costs are estimated to be uniform over time, this is consistent with assuming constant-value dollars and a real interest rate must be used. This may be the easiest way to correctly account for inflation. In addition, when results are stated as an EUAW, this may only be meaningful in constant value dollar terms.

The different cash flows in our analysis may inflate or change at different interest rates when we look over the life cycle of the investment. Also, a single cash flow may inflate or deflate at different rates over time. These two circumstances are handled easily by applying the correct inflation rates to each cash flow over the study period to obtain the actual dollar amounts occurring in each year. After the actual dollar quantities have been calculated, the analysis proceeds using the market interest rate to calculate the measure of merit of interest.

Historical price change for single commodities and bundles of commodities are tracked with price indexes. The Consumer Price Index (CPI) is an example of a composite index formed by a bundle of consumer goods. The CPI serves as a surrogate for general inflation in our economy. Indexes can be used to calculate the *average annual increase* (or decrease) of the costs and benefits in our analysis. The historical data provide valuable information about how economic quantities may behave in the future over the long run.

The effect of inflation on the computed rate of return for an investment depends on how future benefits respond to the inflation. Usually the costs and benefits increase at the same rate as inflation, so the before-tax rate of return will not be adversely affected by the inflation. This outcome is not found when an after-tax analysis is made because the allowable depreciation schedule does not increase. The result will be increased taxable income and income tax payments, which reduce the available after-tax benefits and, therefore, the after-tax rate of return. The important conclusion is that estimates of future inflation or deflation may be important in evaluating capital expenditure proposals.

PROBLEMS

Meaning and Effect

14-1 Define inflation in terms of the purchasing power of dollars.

14-2 Define and describe the relationships between the following: actual and real dollars, inflation, and real and market (combined) interest rates.

14-3 How does inflation happen?

Ⓔ (*a*) Describe a few circumstances that cause prices in an economy to increase.

(*b*) Is it common for governments with large debts to promote inflation, in order to pay off their debts more easily? Is this ethical?

14-4 Is it necessary to account for inflation in an engineer-

Ⓐ ing economy study? What are the two approaches for handling inflation in such analyses?

14-5 In Chapters 5 (Present Worth Analysis) and 6 (Annual Cash Flow Analysis) it is assumed that prices are stable and a machine purchased today for $5000 can be replaced for the same amount many years hence. In fact, prices have generally been rising, so the stable price assumption tends to be incorrect. Under what circumstances is it correct to use the "stable price" assumption when prices actually are changing?

14-6 An economist has predicted 3% inflation during the

Ⓐ next 15 years. How much will an item that presently sells for $100 cost in 15 years?

14-7 Explain how high inflation in a booming real estate market could benefit an engineer who sells a home 5 years after she buys it.

14-8 A newspaper reports that in the last 10 years, prices

Ⓐ have increased a total of 65%. This is equivalent to what annual inflation rate, compounded annually?

14-9 An investor wants a real rate of return i' of 6% per year. If the expected annual inflation rate for the next several years is 2.5%, what interest rate i should be used in project analysis calculations?

14-10 A man wishes to set aside some money for his daugh-

Ⓐ ter's college education. His goal is to have a bank savings account containing an amount equivalent to $20,000 in today's dollars at the girl's 18th birthday. The estimated inflation rate is 8%. If the bank pays 5% compounded annually, what lump sum should he deposit today on the child's 4th birthday?

14-11 A man bought a 5% tax-free municipal bond. It

Ⓔ cost $1000 and will pay $50 interest each year for 20 years. At maturity the bond returns the original $1000.

(*a*) If there is 2% annual inflation, what real rate of return will the investor receive?

(*b*) What ethical questions are linked to governments issuing negative interest rate bonds?

14-12 An automaker has a car that gets 10 kilometers per

Ⓐ liter of gasoline. Gas prices will increase 3.5% per

Ⓖ year, compounded annually, for the next 10 years. The manufacturer believes that the fuel consumption for its new cars should decline as fuel prices increase to keep the fuel costs constant.

(*a*) To achieve this, what must be the fuel rating, in kilometers per liter, of the cars 10 years from now?

(*b*) What limit to average kilometers per liter is likely?

14-13 Inflation has been a reality for the general economy of the U.S. in many years. Given this assumption, calculate the number of years it will take for the purchasing power of today's dollars to equal *one-third* of their present value. Assume that inflation will average 2.5% per year.

14-14 Sally Johnson loaned a friend $10,000 at 15%

Ⓐ interest, compounded annually. The loan will be paid in four equal end-of-year payments. Sally expects the inflation rate to be 3%. After taking inflation into account, what rate of return is Sally receiving on the loan? Compute your answer to the nearest 0.1%.

14-15 If inflation averages 2.5% each year from 2015 to 2025, what is the purchasing power in 2015 dollars of $75,000 in 2025? *Contributed by Paul R. McCright, University of South Florida*

14-16 If inflation is currently 3.35% and a bank is lending
(A) money at 6.65% interest, what is the real interest rate the bank is earning on its loans? *Contributed by Paul R. McCright, University of South Florida*

14-17 A South American country has had a high rate of inflation. Recently, its exchange rate was 15 cruzados per dollar; that is, one dollar will buy 15 cruzados. It is likely that the country will continue to experience a 25% inflation rate and that the U.S. will continue at a 3% inflation rate. Assume that the exchange rate will vary the same as the inflation. In this situation, one dollar will buy how many cruzados 5 years from now?

14-18 An economist has predicted that for the next 5 years,
(A) the U.S. will have an 2% annual inflation rate, followed by 5 years at a 4% inflation rate. This is equivalent to what average price change per year for the entire 10-year period?

14-19 An economist has predicted that during the next 12 years, prices in the U.S. will increase 55%. He expects a further increase of 25% in the subsequent 8 years. Compute the annual inflation rate, f, for the entire 20-year period.

14-20 You may pay $15,000 for an annuity that pays $2500
(A) per year for the next 10 years. You want a real rate of return of 5%, and you estimate inflation will average 6% per year. Should you buy the annuity?

14-21 A homebuilder's advertising had the caption, "Inflation to Continue for Many Years." The ad explained that if one buys a home now for $297,000, and construction cost inflation continues at 7%, the home will be worth $819,400 in 15 years. Thus, by buying a new home now, one can realize a profit of $522,400 in 15 years. Do you find this logic persuasive? Explain.

14-22 The average cost of a certain model car was
(A) $18,000 ten years ago. This year the average cost is $30,000.
    (a) Calculate the average monthly inflation rate (f_m) for this model.
    (b) Given the monthly rate f_m, what is the effective annual rate, f, of inflation for this model?

(c) Estimate what these will sell for 10 years from now, expressed in today's dollars.

Contributed by D. P. Loucks, Cornell University

14-23 You were recently looking at the historical prices paid for homes in a neighborhood that interests you. Calculate on a year-to-year basis how home prices in this neighborhood have inflated (*a–e* in the table).

Year	Average Home Price	Inflation Rate for That Year
5 years ago	$265,000	
4 years ago	267,000	(a)
3 years ago	272,000	(b)
2 years ago	280,000	(c)
Last year	283,000	(d)
This year	288,000	(e)

(f) What is your estimate of the inflation rate for next year?

14-24 ShaNey saw that the campus bookstore is having
(A) a special on pads of computation paper normally priced at $3.75 a pad, now on sale for $3.50 a pad. This sale is unusual and ShaNey assumes the paper will not be put on sale again. On the other hand, she expects that there will be no increase in the $3.75 regular price, even though the inflation rate is 0.75% every 3 months. ShaNey believes that competition in the paper industry will keep wholesale and retail prices constant. She uses a pad of computation paper every 3 months. ShaNey considers 9% a suitable minimum attractive rate of return. ShaNey will buy one pad of paper for her immediate needs. How many extra pads of computation paper should she buy?

14-25 Samantha receives a starting salary offer of $60,000 for Year 1. If inflation is 3% each year, what must her salary be to have the same purchasing power in Year 10? Year 20? Year 30? Year 40?

14-26 Inflation is 2.5%. If $1000 is invested in an account
(A) paying 6% compounded semiannually, what is the Year-0 dollar value of the account at the end of the 10 years?

14-27 Assume that Samantha (Problem 14-25) receives an annual 5% raise. How much more, in Year-1 dollars, is her salary in Year 10? Year 20? Year 30? Year 40?

14-28 Assume your salary is $55,000 in 2015 and $160,000
(A) in 2045. If inflation has averaged 2% per year, what

is the real or differential inflation rate of salary increases?

14-29 In the 1920 Sears Roebuck catalog, an oak chest of drawers cost $8 plus freight. In 1990 this same chest of drawers, in good condition, was $1200. If the average rate of inflation over that 70-year period was 3%, what was the average yearly rate of appreciation, adjusted for inflation?

14-30 Assume general inflation is 3.5% per year. What
Ⓐ is the price tag in 10 years for an item that has an inflation rate of 5.5% that costs $500 today?

14-31 The expected rise in prices due to inflation over the next 6 years is expected to be 30%. Determine the average annual inflation rate over the 6-year period.

14-32 The price of a HeeHaw Model BR549 computer is
Ⓐ presently $2200. If deflation of 2% per quarter is expected on this computer, what will its price be in nominal dollars at the end of 1 year? If inflation is 4.5% per year, what will the price be in Year-0 dollars?

14-33 You place $4000 into an account paying 8% compounded annually. Inflation is 2.5% during each of the next 3 years. What is the account's value at the end of the 3 years in Year-0 dollars?

14-34 Felix Jones, a recent engineering graduate, expects
Ⓐ a starting salary of $65,000 per year. His future employer has averaged 5% per year in salary increases for the last several years. If inflation is estimated to be 4% per year for the next 3 years, how much, in Year-1 dollars, will Felix be earning each year? What is the real growth rate in Felix's salary?

14-35 If $25,000 is deposited in a 5% savings account and inflation is 3%, what is the value of the account at the end of Year 20 in Year-0 dollars? If the time value of money is 4%, what is the present worth?

14-36 The cost of garbage pickup in Green Gulch is
Ⓐ $4,500,000 for Year 1. The population is increasing
Ⓖ at 6%, the nominal cost per ton is increasing at 5%, and the general inflation rate is estimated at 4%.

(a) Estimate the cost in Year 4 in Year-1 dollars and in nominal dollars.

(b) Reference a data source for trends in volume of garbage per person. How does including this change your answer?

14-37 The following series from the *Historical Statistics of the United States* can be combined with data in Table 14–2 to construct a long-term measure of inflation.

What is the average inflation rate from _____ to 2015?

(a) 1955
(b) 1940
(c) 1910
(d) 1800
(e) 1779

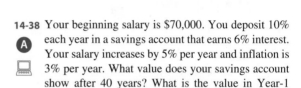

1910 = 100 (base)

Year	1779	1785	1800	1803	1830	1850	1864	1880
Index	226	92	129	118	91	84	193	100

1927 = 100 (base)

Year	1890	1910	1920	1921	1932	1940	1950	1955
Index	56	90	154	98	65	79	162	174

14-38 Your beginning salary is $70,000. You deposit 10%
Ⓐ each year in a savings account that earns 6% interest.
💻 Your salary increases by 5% per year and inflation is 3% per year. What value does your savings account show after 40 years? What is the value in Year-1 dollars?

14-39 The market for widgets is increasing by 15% per
💻 year from current profits of $2,000,000. Investing in a design change will allow the profit per widget to stay steady; otherwise the price will drop by 3% per year. If inflation in the economy is 4%, what is the present worth in Year-1 dollars of the savings over the next 5 years? 10 years? The interest rate is 10%.

14-40 A homeowner is considering an upgrade of a
Ⓐ fuel-oil-based furnace to a natural gas unit. The
💻 investment in the fixed equipment, such as a new boiler, will be $2500 installed. The cost of the natural
Ⓖ gas will average $60 per month over the year, instead of the $145 per month that the fuel oil costs. If funds cost 9% per year and cost inflation in fossil fuels will be 3% per year, how long will it take to recover the initial investment? Solve on a monthly basis.

14-41 Enrollment at City University is increasing 3% per
💻 year, its cost per credit hour is increasing 8% per year, and state funds are decreasing by 4% per year. State funds currently pay a third of the costs for City U., while tuition pays the rest. What annual increase in tuition is required?

14-42 Joan earns a salary of $110,000 per year, and she expects to receive increases at a rate of 4% per year for the next 30 years. She is purchasing a home for

$380,000 at 7% for 30 years (under a special veterans, preference loan with 0% down). She expects the home to appreciate at a rate of 3% per year. She will also save 10% of her gross salary in savings certificates that earn 5% per year. Assume that her payments and deposits are made annually. If inflation is assumed to have a constant 5% rate, what is the value (in Year-1 dollars) of each of Joan's two investments at the end of the 30-year period? Use a before-tax analysis.

Before-Tax Problems

14-43 Using a market interest rate of 6% and an inflation rate of 2.5%, calculate the future equivalent in Year 15 of $10,000 today:

(a) Having today's purchasing power.
(b) Having then-current purchasing power.

14-44 Auntie Frannie wants to help with tuition for her nieces to attend a private school. She intends to send a check for $10,000 at the end of each of the next 8 years.

(a) If general price inflation, as well as tuition price inflation, is expected to average 4% per year for those 8 years, calculate the present worth of the gifts. Assume that the real interest rate will be 3% per year.
(b) If Auntie Frannie wants her gifts to keep pace with inflation, what would be the present worth of her gifts? Again assume inflation is 4% and the real interest rate is 3%.

14-45 Ima Luckygirl recently found out that her grandfather had passed away and left her his Rocky Mountain Gold savings account. The only deposit was 75 years ago, when Ima's grandfather deposited $5000. If the account has earned an average rate of 10% and inflation has been 4%, answer the following:

(a) How much money is now in the account in *actual dollars*?
(b) Express the answer to part (a) in terms of the purchasing power of dollars from 75 years ago.

14-46 The ABC Block Company anticipates receiving $50,000 per year from its investments (with no change) over the next 10 years. If ABC's interest rate is 8% and the inflation rate is 3%, determine the present value of the cash flows.

14-47 Pollution control equipment must be purchased to remove the suspended organic material from liquid being discharged from a vegetable packing plant. Two alternative pieces of equipment are available

that would accomplish the task. A Filterco unit costs $7000 and has a 5-year useful life. A Duro unit, on the other hand, costs $10,000 but will have a 10-year useful life.

With inflation, equipment costs are rising at 5% per year, compounded annually, so when the Filterco unit needed to be replaced, the cost would be much more than $7000. Based on a 10-year analysis period, and a 20% minimum attractive rate of return, which pollution control equipment should be purchased?

14-48 A 30-year mortgage for $100,000 has been issued. The interest rate is 10% and payments are made annually. If your time value of money is 12%, what is the PW of the payments in Year-1 dollars if inflation is 0%? 3%? 6%? 9%?

14-49 A group of students decided to lease and run a gasoline service station. The lease is for 10 years. Almost immediately the students were confronted with the need to alter the gasoline pumps to read in liters. The Dayton Company has a conversion kit available for $3000 that may be expected to last 10 years. The firm also sells a $1000 conversion kit that has a 5-year useful life. The students believe that any money not invested in the conversion kits may be invested elsewhere at a 10% interest rate. Income tax consequences are to be ignored in this problem.

(a) Assuming that future replacement kits cost the same as today, which alternative should be selected?
(b) If one assumes a 4% inflation rate, which alternative should be selected?

14-50 If McDonnell Manufacturing has a MARR of 20%, inflation is 2.75%, and the company uses present worth analysis with a planning horizon of 15 years in making economic decisions, which of the following alternatives would be preferred?

	Alternative A	Alternative B
Initial costs	$236,000	$345,000
Annual operating costs	64,000	38,000
Annual maintenance costs	4,000	5,000
Salvage value (EOY 20)	23,000	51,000

Contributed by Paul R. McCright, University of South Florida

14-51 Due to competition from a new polycarbon, revenues for the mainstay product of Toys-R-Polycarbon are declining at 7.5% per year. Revenues will be $5M for this year. The product will be discontinued at the end of Year 6. If the firm's interest rate is 20%, calculate the PW of the revenue stream.

14-52 The City of Columbia is trying to attract a new man-
Ⓐ ufacturing business. It has offered to install and operate a water pumping plant to provide service to the proposed plant site. This would cost $50,000 now, plus $5000 per year in operating costs for the next 10 years, all measured in Year-0 dollars.

To reimburse the city, the new business must pay a fixed uniform annual fee, A, at the end of each year for 10 years. In addition, it is to pay the city $50,000 at the end of 10 years. It has been agreed that the city should receive a 3% rate of return, after taking an inflation rate, f, of 7% into account.

Determine the amount of the uniform annual fee.

14-53 The Wildwood Widget Company needs a milling machine for its new assembly line. The machine presently costs $85,000, but has a cost inflation rate of 2%. Widget will not need to purchase the machine for 3 years. If general inflation is expected to be 4% per year during those 3 years, determine the price of the machine. What is the present worth of the machinery if the market rate of interest for Widget is 25%?

14-54 As a recent graduate, you are considering
Ⓐ employment offers from three different companies. However, in an effort to confuse you and perhaps make their offers seem better, each company has used a different *purchasing power base* for expressing your annual salary over the next 5 years. If you expect inflation to be 6% for the next 5 years and your personal (real) MARR is 8%, which plan would you choose?

Company A: A constant $50,000 per year in terms of today's purchasing power.
Company B: $45,000 the first year, with increases of $2500 per year thereafter.
Company C: A constant $65,000 per year in terms of Year-5-based purchasing power.

14-55 Bob has lost his job and had to move back in with his mother. She agreed to let Bob have his old room back on the condition that he pay her $250 rent per month, and an additional $1500 every other year to pay for her biannual jaunt to Florida. Since he is down on his luck, she will allow him to pay his rent at the end of the year. If inflation is 3% and Bob's interest rate is 5%, how much is the present cost (in Year-1 dollars) for a 5-year contract with mom? (*Note:* Mom's trips are in Years 2 and 4).

14-56 A firm is having a large piece of equipment
Ⓐ overhauled. It expects that the machine will be needed for the next 12 years. The firm has an 8% minimum attractive rate of return. The contractor has suggested three alternatives:

A. A complete overhaul for $6000 that should permit 12 years of operation.
B. A major overhaul for $4500 that can be expected to provide 8 years of service. At the end of 8 years, a minor overhaul would be needed.
C. A minor overhaul now. At the end of 4 and 8 years, additional minor overhauls would be needed.

If minor overhauls cost $2500, which alternative should the firm select? If minor overhauls, which now cost $2500, increase in cost at +5% per year, but other costs remain unchanged, which alternative should the firm select?

14-57 Sam bought a house for $150,000 with some creative financing. The bank, which agreed to lend Sam $120,000 for 6 years at 15% interest, took a first mortgage on the house. The Joneses, who sold Sam the house, agreed to lend Sam the remaining $30,000 for 6 years at 12% interest. They received a second mortgage on the house. Thus Sam became the owner without putting up any cash. Sam pays $1500 a month on the first mortgage and $300 a month on the second mortgage. In both cases these are "interest only" loans, and the principal is due at the end of the loan.

Sam rented the house to Justin and Shannon, but after paying the taxes, insurance, and so on, he had only $800 left, so he was forced to put up $1000 a month to make the monthly mortgage payments. At the end of 3 years, Sam sold the house for $205,000. After paying off the two loans and the real estate broker, he had $40,365 left. After taking an 8% inflation rate into account, what was his before-tax rate of return?

Indexes

14-58 What is the Consumer Price Index (CPI)? What is the difference between commodity-specific and composite price indexes? Can each be used in engineering economic analysis?

14-59 A composite price index for the cost of vegetarian foods called *Eggs, Artichokes, and Tofu* (EAT) was 350 ten years ago and has averaged an annual increase of 5% since. Calculate the current value of the index.

14-60 From the data in Table 14–1 in the text, calculate
Ⓐ the *overall rate change* of first-class postage as

CASES

The following cases from *Cases in Engineering Economy* (www.oup.com/us/newnan) are suggested as matched with this chapter.

CASE 4 **Northern Windows**

Tax affected energy conservation project for a home with simplifying assumptions made clear. Includes inflation.

CASE 37 **Brown's Nursery (Part B)**

Adds inflation to Case 36.

CASE 38 **West Muskegon Machining and Manufacturing**

More complex inflation and tax problem with sunk cost and leverage.

CASE 39 **Uncertain Demand at WM³**

Includes inflation, taxes, and uncertainty.

CASE 41 **Freeflight Superdiscs**

Inflation and sensitivity analysis for three alternatives. Includes taxes.

SELECTION OF A MINIMUM ATTRACTIVE RATE OF RETURN

What's the Rate of Return on a Dam?

The broadly defined "energy" industry has often done well in recent years even when the U.S. and world economies have not. It's now quite evident that humans are becoming increasingly dependent on various forms of energy to power our world. Some forms, especially oil and coal, pose environmental threats. Thus government and businesses alike are scrambling to come up with "green" solutions to the world's energy needs. *But is the economic investment in renewable forms of energy (solar, wind, water) worth the risks? Do private investors need to reconsider their rates of return when evaluating risky investments?*

Consider Tajikistan, a former Soviet republic with a population of about 8 million people. While the country is lacking in most natural resources and has few major industries (cotton and aluminum), it is the source of over 40% of Central Asia's water. Tajikistan's government is relying on hydroelectric power for the country's economic and political future. The Rogun Dam is an unfinished dam across the Vakhsh River in southern Tajikistan with a planned capacity of 3600 megawatts of power. Construction began in 1976 but was halted after the collapse of the Soviet Union. In 2010, hoping to revitalize this project and complete the dam, Tajikistan made a public offering of bonds worth $1.4 billion. Because this is an extremely expensive undertaking, the country must rely heavily on foreign investment, and the investors are likely to expect significant rates of returns.

Upstream and downstream countries have different perspectives on how dam projects will affect their own economies and water supplies. The more hydroelectric projects Tajikistan develops, the more control they will have over water in the region; thus political tensions are a significant issue. The Nurek Dam, downstream on the Vakhsh River, is one of the world's highest. Its associated 9-unit power plant provides about 3000 megawatts of power, and the water impounded irrigates about 1.6 million acres of farmland.

Uzbekistan, Tajikistan's downstream neighbor, strongly opposed the Rogun Dam. In addition, Tajikistan is in an earthquake zone, and any dam projects must be built to withstand significant seismic shocks, not to mention the effects that these projects will have on the plant and animal life in the region. Dam construction halted in 2012. Tajikistan had been unable to raise additional required capital, and a World Bank–funded study was evaluating the dam's potential economic, environmental, and sociopolitical impact. In July 2014 the World Bank gave the green light, and Tajikistan was able to raise additional funds. Construction continues, despite Uzbekistan's continuing vehement objections and recent earthquakes in the Rogun region. This kind of investment is very risky for any business or government. ■ ■ ■

Contributed by Karen M. Bursic, University of Pittsburgh

QUESTIONS TO CONSIDER

1. What are the advantages of relying on hydroelectric power in a country such as Tajikistan?

2. What are the disadvantages of relying on hydroelectric power in a country such as Tajikistan? What are the possible consequences of new dam construction?

3. How would you balance the benefits against the costs in order to determine whether a new hydroelectric generating facility should be built in a given location?

4. What are some of the ethical issues that arise when one is considering investment in foreign countries?

5. What risks are associated with the Rogun Dam for Tajikistan? For its neighboring countries downstream? For foreign investors?

6. How would a private company determine the MARR to be used in evaluating such a project?

After Completing This Chapter...

The student should be able to:

- Define various sources of capital and the costs of those funds to the firm.

- Discuss the impact of inflation and the cost of borrowed money.

- Select a firm's MARR based on the opportunity cost approach for analyzing investments.

- Adjust the firm's MARR to account for risk and uncertainty.

- Use spreadsheets to develop cumulative investments and the opportunity cost of capital.

Key Words

capital budgeting	equity	risk
composite value	opportunity cost	treasury stock
cost of capital	ownership	uncertainty
debt	prime rate	weighted average cost of capital

Selecting the interest rate or minimum attractive rate of return that is suitable for use in a particular situation is complex, and no single answer is always appropriate. A discussion of what interest rate to use must inevitably begin by examining the sources of capital, followed by looking at the prospective investment opportunities and risk. Only in this way can an interest rate or minimum attractive rate of return (MARR) be chosen intelligently.

MARR FOR INDIVIDUALS

This chapter focuses on business firms, but the concepts that apply to firms also make sense for individuals. For example, the *minimum attractive rate of return* is the highest rate of several possibilities in both cases. For individuals the *cost of capital* becomes the highest interest rate loan or credit card balance that the individual is paying interest on. If a credit card payment has been late, these rates may reach nearly 30% in some cases! On the other hand, if credit cards are paid off every month, the interest rate on them is typically 0%. Interest rates on vehicle, student, or home loans may be the highest interest rates for an individual.

The *opportunity cost* for investing for most individuals is the expected rate of return on the individual's investment portfolio. Historical returns above inflation were calculated in Appendix 9A for stocks, long-term treasury bonds, and treasury bills (T-bills). Then Appendix 10A showed that portfolios can reduce risk by including a mix of bonds, stocks, and T-bills. Other common investments include corporate bonds, which offer a higher return than treasury bonds, but at a higher risk of default—bankrupt firms may make no or only partial payment on their bonds. Some individuals earn higher or possibly lower returns by investing in a small business, real estate, or high-growth firms.

SOURCES OF CAPITAL

In broad terms there are three sources of capital available to a firm: money generated from the firm's operation, borrowed money, and money from selling stock.

Money Generated from the Firm's Operations

A major source of capital investment money is retained profits from the firm's operation. Overall, industrial firms retain about half of their profits and pay out the rest to stockholders. In addition to profit, the firm generates money equal to the annual depreciation charges on existing capital assets. In other words, a profitable firm will generate money equal to its depreciation charges *plus* its retained profits. Even a firm that earns zero profit will still generate money from operations equal to its depreciation charges. (A firm with a loss, of course, will have still less funds.)

External Sources of Money

When a firm requires money for a few weeks or months, it typically borrows from banks. Longer-term unsecured loans (of, say, 1–4 years) may also be arranged through banks. While banks undoubtedly finance a lot of capital expenditures, regular bank loans cannot be considered a source of permanent financing.

Longer-term financing is done by selling bonds (see Chapter 5) to banks, insurance firms, pension funds, and the public. A wide variety of bonds exist, but most are interest-only loans, where interest is paid at a *coupon* rate every 6 months or once a year and the principal is due at the bond's maturity. Common maturities are 10 to 30 years, although some extend to 100 years and a few even longer. Chapters 5 and 7 include examples of how to calculate the interest rates.

A firm can also raise funds by issuing new stock (shares of ownership in the firm). Many firms have also repurchased their own stock in the past, which is called **treasury stock**. Another way firms can raise funds is to sell this treasury stock.

One of the finance questions each firm must address is maintaining an appropriate balance between **debt** (loans and bonds) and **equity** (stock and retained earnings). The debt has a maturity date, and there are legal obligations to repay it unless the firm declares bankruptcy. On the other hand, stockholders expect a higher rate of return to compensate them for the risks of ownership. Those who are interested in models used to calculate the cost of equity capital are referred to *The Economic Analysis of Industrial Projects*, 3rd edition, by Eschenbach, Lewis, Hartman, and Bussey, published by Oxford University Press.

Choice of Sources of Funds

Choosing the source of funds for capital expenditures is a decision for the firm's top executives, and it may require approval of the board of directors. When internal operations generate adequate funds for the desired capital expenditures, external sources of money are not likely to be used. But when the internal sources are inadequate, external sources must be employed or the capital expenditures will have to be deferred or canceled.

COST OF FUNDS

Cost of Borrowed Money

A first step in deciding on a minimum attractive rate of return might be to determine the interest rate at which money can be borrowed. Longer-term loans or bonds may be obtained from banks, insurance companies, or the variety of places in which substantial amounts of money accumulates (for example, the sovereign wealth funds of the oil-producing nations).

A large, profitable corporation might be able to borrow money at the **prime rate,** that is, the interest rate that banks charge their best and most sought-after customers. All other firms are charged an interest rate that is higher by one-half to several percentage points. In addition to the firm's financial strength and ability to repay the debt, the interest rate will depend on the debt's duration and on whether the debt has collateral or is unsecured.

Cost of Capital

Another relevant interest rate is the **cost of capital.** This is also called the **weighted average cost of capital (WACC)**. This is the rate from *all* sources of funds in the firm's overall capitalization. The mechanics for computing the cost of capital or WACC are given in Example 15–1.

EXAMPLE 15–1

For a particular firm, the purchasers of common stock require an 11% rate of return, bonds are sold at a 7% interest rate, and bank loans are available at 9%. Compute the cost of capital or WACC for the following capital structure:

		Rate of Return
$ 20 million	Bank loan	9%
20	Bonds	7
60	Common stock and retained earnings	11
$100 million		

SOLUTION

The weighted cost of capital "weights" the return on each source of capital by the fraction of the total capital it represents. In this case, 20% of the total capital is from the bank loan, 20% of the capital is from bonds, and 60% of the capital is from equity sources—that is, common stock and retained earnings.

$$WACC_{before\text{-}taxes} = (0.2)(9\%) + (0.2)(7\%) + (0.6)(11\%)$$
$$= 1.8\% + 1.4\% + 6.6\% = 9.8\%$$

Note that since this is an *average*, the result must be between the lowest rate of return (7%) and the largest (11%). Since it is a *weighted* average, the return with the largest weight (60%) has the most impact on the final average. We recommend some "mental" arithmetic to approximate the expected answer, which will help catch any errors with your calculator.

The cost of capital is also computed after considering that interest payments on debt, like bank loans and bonds, are tax-deductible business expenses. Thus,

$$After\text{-}tax\ interest\ cost = (Before\text{-}tax\ interest\ cost) \times (1 - Tax\ rate)$$

If we assume that the firm pays 40% income taxes, the computations become

Bank loan   After-tax interest cost $= 9\%(1 - 0.40) = 5.4\%$

Bonds   After-tax interest cost $= 7\%(1 - 0.40) = 4.2\%$

Dividends paid on the ownership in the firm (common stock + retained earnings) are not tax deductible. The cost of debt capital can also be computed by dividing the total amount of interest by the total amount of debt capital. Combining the three components, the after-tax interest cost for the $100 million of capital is

$20 million (5.4%) + $20 million (4.2%) + $60 million (11%) = $8.52 million

$$\text{WACC}_{\text{after-taxes}} = \frac{\$8.52 \text{ million}}{\$100 \text{ million}} = 8.52\%$$

In practical situations, the cost of capital is often difficult to compute. The fluctuation in the price of common stock, for example, makes it difficult to pick a cost, and because of the fluctuating prospects of the firm, it is even more difficult to estimate the future benefits that purchasers of the stock might expect to receive. Given the fluctuating costs and prospects of future benefits, what rate of return do stockholders require? There is no precise answer, but we can obtain an approximate answer. As described in the next section, inflation complicates the task of finding the *real* interest rate.

Inflation and the Cost of Borrowed Money

As inflation varies, what is its effect on the cost of borrowed money? A widely held view has been that interest rates on long-term borrowing, like 20-year Treasury bonds, will be about 3% more than the inflation rate. For borrowers this is the real—that is, after-inflation—cost of money, and for lenders the real return on loans. If inflation rates were to increase, it would follow that borrowing rates would also increase. All this suggests a rational and orderly situation, about as we might expect.

FIGURE 15–1 The real interest rate. The interest rate on 20-year Treasury bonds *minus* the inflation rate, f, as measured by changes in the Consumer Price Index.

Unfortunately, things have not worked out this way. Figure 15–1 shows that the real interest rate has not always been about 3% and, in fact, there have been long periods during which the real interest rate was negative. Can this be possible? Would anyone invest money at an interest rate several percentage points below the inflation rate? Well, consider this: when the U.S. inflation rate was 12%, savings banks were paying 5¹/2% on regular passbook deposits—and there was a lot of money in those accounts. While there must be a relationship between interest rates and inflation, Figure 15–1 suggests that it is complex.

The relationship between the inflation rate and the rate of return on investments is quite important, because inflation reduces the real rate of return (as shown in Appendix 9A and Chapter 14). In addition, many interest rates are reported without adjusting for inflation. For example, interest rates on loans (auto, home, and student) and investment returns (savings accounts, etc.) are all stated without adjusting for inflation.

INVESTMENT OPPORTUNITIES

An industrial firm has larger amounts of money, which allows investments that are unavailable to individuals, with their more limited investment funds. More important, however, is the fact that a firm conducts a business, which itself offers many investment opportunities. While exceptions can be found, a good generalization is that the opportunities for investment of money within the firm are superior to the investment opportunities outside the firm. Consider the available investment opportunities for a particular firm as outlined in Table 15–1. The cumulative investment required for all projects at or above a given rate of return is given in Figure 15–2.

Figure 15–2 illustrates that a firm may have a broad range of investment opportunities available at varying rates of return and with varying lives and uncertainties. It may take some study and searching to identify the better investment projects available to a firm. Typically, the good projects will almost certainly require more money than the firm budgets for capital investment projects.

Opportunity Cost

There are two aspects of investing that are basically independent. One factor is the source and quantity of money available for capital investment projects. The other aspect is the firm's investment opportunities.

Investment opportunities typically exceed the available money supply. Thus some investment opportunities can be selected and others must be rejected. Obviously, we want to ensure that *all the selected projects are better than the best rejected project.* To do this, we must know something about the rate of return on the best rejected project. The best rejected project is the best opportunity forgone, and this in turn is called the **opportunity cost.**

Opportunity cost = Cost of the best opportunity forgone

= Rate of return on the best rejected project

TABLE 15–1 A Firm's Available Investment Opportunities

Project Number	Project	Cost ($\times 10^3$)	Estimated Rate of Return
	Investment Related to Current Operations		
1	New equipment to reduce labor costs	$150	30%
2	Other new equipment to reduce labor costs	50	45
3	Overhaul particular machine to reduce material costs	50	38
4	New test equipment to reduce defective products produced	100	40
	New Operations		
5	Manufacture parts that previously had been purchased	200	35
6	Further processing of products previously sold in semifinished form	100	28
7	Further processing of other products	200	18
	New Production Facilities		
8	Relocate production to new plant	250	25
	External Investments		
9	Investment in a different industry	300	20
10	Other investment in a different industry	300	10
11	Overseas investment	400	15
12	Purchase of Treasury bills	Unlimited	0.8

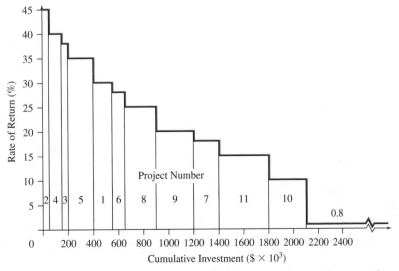

FIGURE 15–2 Cumulative investment required for all projects at or above a given rate of return.

If one could predict the opportunity cost for some future period (like the next 12 months), this rate of return could be one way to judge whether to accept or reject any proposed capital expenditure. Examples 15–2 and 15–3 illustrate this.

EXAMPLE 15–2

Consider the situation represented by Table 15–1 and Figure 15–2. For a capital expenditure budget of $1.2 million ($1.2 × 10^6$), what is the opportunity cost?

SOLUTION

From Figure 15–2 we see that the eight projects with a rate of return of 20% or more require a cumulative investment of $1.2 ($× 10^6$). We would take on these projects and reject the other four (7, 11, 10, and 12) with rates of return of 18% or less. The best rejected project is 7, and it has an 18% rate of return. Thus the opportunity cost is 18%.

EXAMPLE 15–3

Nine independent projects are being considered. Figure 15–3 may be prepared from the following data.

Project	Cost (thousands)	Uniform Annual Benefit (thousands)	Useful Life (years)	Salvage Value (thousands)	Computed Rate of Return
1	$100	$23.85	10	$0	20%
2	200	39.85	10	0	15
3	50	34.72	2	0	25
4	100	20.00	6	100	20
5	100	20.00	10	100	20
6	100	18.00	10	100	18
7	300	94.64	4	0	10
8	300	47.40	10	100	12
9	50	7.00	10	50	14

If a capital budget of $650,000 is available, what is the opportunity cost of capital? With this model, which projects should be selected?

SOLUTION

Looking at the nine projects, we see that some are expected to produce a larger rate of return than others. It is natural that if we are to select from among them, we will pick those with a higher rate of return. When the projects are arrayed by rate of return, as in Figure 15–3, Project 2 is the last one funded. With Project 2 the cumulative first cost of $650,000 matches the budget. Thus, the opportunity cost of capital is 14% from Project 9, the highest ranked unfunded project. This model implies funding Projects 3, 1, 4, 5, 6, and 2.

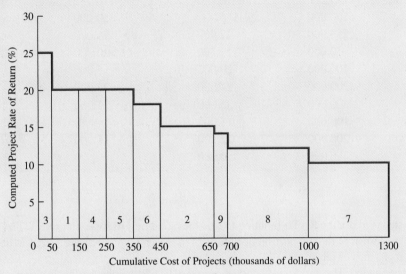

FIGURE 15–3 Cumulative cost of projects versus rate of return.

Spreadsheets can be used to sort the projects by rate of return and then calculate the cumulative first cost. This is accomplished through the following steps.

1. Enter or calculate each project's rate of return.
2. Select the data to be sorted. Do *not* include headings, but do include all information on the row that goes with each project.
3. Select the SORT tool (found in the menu under DATA), identify the rate of return column as the first key, and a sort order of descending. Also ensure that row sorting is selected. Sort.
4. Add a column for the cumulative first cost. This column is compared with the capital limit to identify the opportunity cost of capital and which projects should be funded.

Example 15–4 illustrates these steps. Example 15–5 outlines the more complicated case of independent projects with mutually exclusive alternatives.

EXAMPLE 15-4

A firm has a budget of $800,000 for projects this year. Which of the following projects should be accepted? What is the opportunity cost of capital?

Project	First Cost	Annual Benefit	Salvage Value	Life (years)
A	$200,000	$25,000	$50,000	15
B	250,000	47,000	−25,000	10
C	150,000	17,500	20,000	15
D	100,000	20,000	15,000	10
E	200,000	24,000	25,000	20
F	300,000	35,000	15,000	15
G	100,000	18,000	0	10
H	200,000	22,500	15,000	20
I	350,000	50,000	0	25

SOLUTION

The first step is to use the RATE function to find the rate of return for each project. The results of this step are shown in the top portion of Figure 15–4. Next the projects are sorted in descending order by their rates of return. Finally, the cumulative first cost is computed. Projects D, I, B, and G should be funded. The opportunity cost of capital is 10.6% the rate for the first project rejected.

	A	B	C	D	E	F	G	H
1	Project	First Cost	Annual Benefit	Salvage Value	Life	IRR		
2	A	200,000	25,000	50,000	15	10.2%	=RATE(E2,C2,−B2,D2)	
3	B	250,000	47,000	−25,000	10	12.8%		
4	C	150,000	17,000	20,000	15	8.6%		
5	D	100,000	20,000	15,000	10	16.0%		
6	E	200,000	24,000	25,000	20	10.6%		
7	F	300,000	35,000	15,000	15	8.2%		
8	G	100,000	18,000	0	10	12.4%		
9	H	200,000	22,500	15,000	20	9.6%		
10	I	350,000	50,000	0	25	13.7%		
11	Projects Sorted by IRR						Cumulative First Cost	
12	D	100,000	20,000	15,000	10	16.0%	100,000	
13	I	350,000	50,000	0	25	13.7%	450,000	
14	B	250,000	47,000	−25,000	10	12.8%	700,000	
15	G	100,000	18,000	0	10	12.4%	800,000	
16	E	200,000	24,000	25,000	20	10.6%	1,000,000	
17	A	200,000	25,000	50,000	15	10.2%	1,200,000	
18	H	200,000	22,500	15,000	20	9.6%	1,400,000	
19	C	150,000	17,500	20,000	15	8.6%	1,550,000	
20	F	300,000	35,000	15,000	15	8.2%	1,850,000	

FIGURE 15-4 Spreadsheet for finding opportunity cost of capital.

In Example 15–5, each of the more expensive mutually exclusive alternatives (1*B*, 1*C*, and 3*B*) has lower rates of return than the less expensive alternatives (1*A* and 3*A*). This is common because many projects exhibit decreasing returns to scale. As in Example 15–5, when it occurs, then it makes the ranking of increments easy to use. Sometimes more expensive mutually exclusive alternatives have higher rates of return; then several combinations or more advanced techniques must be tried.

EXAMPLE 15–5

A company is preparing its capital budget for next year. The amount has been set at $250,000 by the board of directors. Rank the following project proposals for the board's consideration and recommend which should be funded.

Project Proposals	Cost (thousands)	Uniform Annual Benefit (thousands)	Salvage Value (thousands)	Useful Life (years)
Proposal 1				
Alt. *A*	$100	$23.85	$0	10
Alt. *B*	150	32.20	0	10
Alt. *C*	200	39.85	0	10
Proposal 2	50	14.92	0	5
Proposal 3				
Alt. *A*	100	18.69	25	10
Alt. *B*	150	19.42	125	10

SOLUTION

For project proposals with two or more alternatives, incremental rate of return analysis is required.

Combination of Alternatives	Cost ($1000s)	Uniform Annual Benefit ($1000s)	Salvage Value ($1000s)	Rate of Return	Incremental Analysis Cost ($1000s)	Incremental Analysis Uniform Annual Benefit ($1000s)	Incremental Analysis Salvage Value ($1000s)	Incremental Analysis Rate of Return
Proposal 1								
A	$100	$23.85	$0	20.0%				
B − *A*					$50	8.35	$0	10.6%
B	150	32.20	0	17.0				
C − *B*					50	7.65	0	8.6
C − *A*					100	16.00	0	9.6
C	200	39.85	0	15.0				
Proposal 2	50	14.92	0	15.0				
Proposal 3								
A	100	18.69	25	15.0				
B − *A*					50	0.73	100	8.3
B	150	19.42	125	12.0				

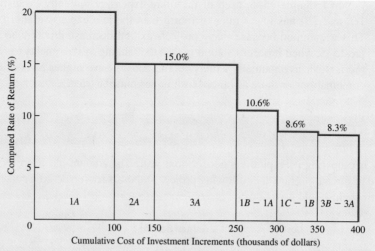

FIGURE 15–5 Cumulative cost versus incremental rate of return.

The various separable increments of investment may be ranked by the rate of return. They are plotted in a graph of cumulative cost versus rate of return in Figure 15–5. The ranking of projects by rate of return gives the following:

Project	Rate of Return
1A	20.0%
2	15.0%
3A	15.0%
1B in place of 1A	10.6%
1C in place of 1B	8.6%
3B in place of 3A	8.3%

For a budget of $250,000 the selected projects are 1A, 2, and 3A. Note that if a budget of $300,000 were available, 1B would replace 1A, making the proper set of projects 1B, 2, and 3A. At a budget of $400,000 1C would replace 1B; and 3B would replace 3A, making the selected projects 1C, 2, and 3B.

SELECTING A MINIMUM ATTRACTIVE RATE OF RETURN

Focusing on the three concepts on the cost of money (the cost of borrowed money, the cost of capital, and opportunity cost), which, if any, of these values should be used as the minimum attractive rate of return (MARR) in economic analyses?

Fundamentally, we know that unless the benefits of a project exceed its cost, we cannot add to the profitability of the firm. A lower boundary for the minimum attractive rate of return must be the cost of the money invested in the project. It would be unwise, for example, to borrow money at 8% and invest it in a project yielding a 6% rate of return.

Further, we know that no firm has an unlimited ability to borrow money. Bankers—and others who evaluate the limits of a firm's ability to borrow money—look at both the profitability of the firm and the relationship between the components in the firm's capital structure. This means that increases in a firm's debt will usually be matched by proportionate increases in equity (stock and retained earnings from operations). This maintains an acceptable ratio between **ownership** (equity) and debt. In other words, borrowing for a particular investment project is only a block of money from the overall capital structure of the firm. This suggests that the MARR should not be less than the cost of capital. Finally, we know that the MARR should not be less than the rate of return on the best opportunity forgone. Stated simply,

> **Minimum attractive rate of return should be greater than or equal to the largest of: cost of borrowed money, cost of capital, or opportunity cost.**

ADJUSTING MARR TO ACCOUNT FOR RISK AND UNCERTAINTY

We know from our study of estimating the future that what actually occurs is often different from the estimate. When we are fortunate enough to be able to assign probabilities to a set of possible future outcomes, we call this a **risk** situation. We saw in Chapter 10 that techniques like expected value and simulation may be used when the probabilities are known.

Uncertainty is the term used to describe the condition when the probabilities are *not* known. Thus, if the probabilities of future outcomes are known, we have *risk*, and if the probabilities are unknown, we have *uncertainty*. With uncertainty, adjustments for risk are more subjective.

In projects accompanied by normal business risk and uncertainty, the MARR is used without adjustment. For projects with greater than average risk or uncertainty, most firms increase the MARR. As reported in Block (2005), the percentage of firms using risk-adjusted rates varied from 66% in retail to 82% in health care. Some of the percentages for other industries are 70% for manufacturing, 73% for energy, and 78% for technology firms. Table 15–2 lists an example of risk-adjusted MARRs in manufacturing.

Some firms use the same rates for all divisions and groups. Other firms vary the rates by division for strategic reasons. There are even cases when a project-specific rate based on that project's financing may be justified. For example, a firm or joint venture may be founded to develop a specific mine, pipeline, or other resource development project.

However, as shown in Example 15–6, risk-adjusted rates may not work well. A preferable way deals explicitly with the probabilities using the techniques from Chapter 10. When the interest rate (MARR) used in economic analysis calculations is raised to adjust for risk or uncertainty, greater emphasis is placed on immediate or short-term results and less emphasis on longer-term results.

TABLE 15–2 Example Risk-Adjusted MARR Values in Manufacturing

Rate (%)	Applies to:
6	Equipment replacement
8	New equipment
10	New product in normal market
12	New product in related market
16	New product in new market
20	New product in foreign market

EXAMPLE 15–6

Consider the two following alternatives. The MARR of Alt. *B* has been raised from 10% to 15% to take into account the greater risk and uncertainty associated with that alternative. What is the impact of this change of MARR on the decision?

Year	Alt. *A*	Alt. *B*
0	−$80	−$80
1–10	10	13.86
11–20	20	10

SOLUTION

		NPW		
Year	Alt. *A*	At 14.05%	At 10%	At 15%
0	−$80	−$80.00	−$80.00	−$80.00
1–10	10	52.05	61.45	50.19
11–20	20	27.95	47.38	24.81
		0	+28.83	−5.00

		NPW		
Year	Alt. *B*	At 15.48%	At 10%	At 15%
0	−$80	−$80.00	−$80.00	−$80.00
1–10	13.86	68.31	85.14	69.56
11–20	10	11.99	23.69	12.41
		0	+28.83	+1.97

Computations at MARR of 10% Ignoring Risk and Uncertainty

Both alternatives have the same positive NPW (+ $28.83) at a MARR of 10%. Also, the differences in the benefits schedules (*A*−*B*) produce a 10% incremental rate of return. (This calculation is not shown here.) This must be true if NPW for the two alternatives is to remain constant at a MARR of 10%.

Considering Risk and Uncertainty with MARR of 10%

At 10%, the alternatives are equally desirable. Since Alt. *B* is believed to have greater risk and uncertainty, a logical conclusion is to select Alt. *A* rather than *B*.

Increase MARR to 15%

At a MARR of 15%, Alt. *A* has a negative NPW and Alt. *B* has a positive NPW. Alternative *B* is preferred under these circumstances.

Conclusion

Based on a business-risk MARR of 10%, the two alternatives are equivalent. Recognizing some greater risk of failure for Alt. *B* makes *A* the preferred alternative. If the MARR is increased to 15%, to add a margin of safety against risk and uncertainty, the computed decision is to select *B*. Since Alt. *B* has been shown to be less desirable than *A*, the decision, based on a MARR of 15%, may be an unfortunate one. The difficulty is that the same risk adjustment (increase the MARR by 5%) is applied to both alternatives even though they have different amounts of risk.

The conclusion to be drawn from Example 15–6 is that increasing the MARR to compensate for risk and uncertainty is only an approximate technique and may not always achieve the desired result. Nevertheless, it is common practice in industry to adjust the MARR upward to compensate for increased risk and uncertainty.

REPRESENTATIVE VALUES OF MARR USED IN INDUSTRY

We argued that the minimum attractive rate of return should be established at the highest one of the following: cost of borrowed money, cost of capital, or the opportunity cost.

The cost of borrowed money will vary from enterprise to enterprise, with the lowest rate being the prime interest rate. The prime rate may change several times in a year; it is widely reported in newspapers and business publications. As we pointed out, the interest rate for firms that do not qualify for the prime interest rate may be 1/2 to several percentage points higher.

The cost of capital of a firm is an elusive value. There is no widely accepted way to compute it; we know that as a **composite value** for the capital structure of the firm, it conventionally is higher than the cost of borrowed money. The cost of capital must consider the market valuation of the shares (common stock, etc.) of the firm, which may fluctuate widely, depending on future earnings prospects of the firm. We cannot generalize on representative costs of capital.

Somewhat related to cost of capital is the computation of the return on total capital (long-term debt, capital stock, and retained earnings) actually achieved by firms. *Fortune* magazine, among others, does an annual analysis of the rate of return on total capital. The after-tax rate of return on total capital for individual firms ranges from 0% to about 40% and averages 8%. *Business Week* does a periodic survey of corporate performance. This magazine reports an after-tax rate of return on common stock and retained earnings. We would expect the values to be higher than the rate of return on total capital, and this is the case. The after-tax return on common stock and retained earnings ranges from 0% to about 65% with an average of 14%.

Higher values for the MARR are used by firms that are short of capital, such as high-technology start-ups. They are also used in industries, such as petroleum and mining, where volatile prices increase the risk of poor returns for projects. Rates of 25 to 30% are relatively common, and even higher rates are sometimes used. For companies with more normal levels of risk, rates of 12 to 15% are more typical.

Note that the values of MARR given earlier are approximations. But the values quoted appear to be opportunity costs, rather than cost of borrowed money or cost of capital. This indicates that firms cannot or do not obtain money to fund projects whose anticipated rates of return are nearer to the cost of borrowed money or cost of capital. One reason that firms operate as they do is that they can focus limited resources of people, management, and time on a smaller number of good projects.

One cannot leave this section without noting that the MARR used by enterprises is much higher than can be obtained by individuals. (Where can you get a 30% after-tax rate of return without excessive risk?) The reason appears to be that businesses are not obliged to compete with the thousands of individuals in any region seeking a place to invest $2000 with safety, whereas the number of people who could or would want to invest $500,000 in a business is far smaller. This diminished competition, combined with a higher risk, appears to explain at least some of the difference.

CAPITAL BUDGETING OR SELECTING THE BEST PROJECTS

The opportunity cost of capital approach of ranking projects by their rate of return introduced a new type of problem. Up to that point we'd been analyzing mutually exclusive alternatives, where only one could be chosen. Engineering design problems are this type of problem, where younger engineers use engineering economy to choose the best alternative design.

At higher levels in the organization, engineering economy is applied to solve a different problem. For example, 30 projects have passed initial screening and are being proposed for funding. Every one of the 30 meets the MARR. The firm can afford to invest in only some of them. So, which ones should be chosen and how? This is called the **capital budgeting** problem.

Examples 15–2 and 15–3 applied the opportunity cost of capital approach to the capital budgeting problem. Firms use this approach as a starting point to rank the projects from best to worst. In some cases the ranking by rate of return is used to make the decision.

More often, managers then meet and decide which projects will be funded by obtaining a consensus, or a decision by the highest-ranking manager, which will modify the rate of return ranking. At this meeting, business units argue for a larger share of the capital budget, as do plants in the same business, groups at the same plants, and individuals within the groups. Some considerations, such as strategy, necessity, and the availability and capability of particular resources and people, are difficult to represent in the project's *numbers*, which are the subject of economic analysis.

Other firms rank projects using a benefit–cost ratio or present worth index. As shown in Example 15–7, the present worth index is the NPW divided by the cost's present value.

Anyone who has ever bought firecrackers probably used the practical ranking criterion of "biggest bang for the buck" in making a selection. This same criterion—stated more elegantly—is used by some firms to rank independent projects.

> Rank independent projects according to their value of net present worth divided by the present worth of cost. The appropriate interest rate is MARR (as a reasonable estimate of the cutoff rate of return).

Some consider the present worth index to be a better measure, but this can be true only if PW is applied at the correct interest rate. It is more common for firms to simply rank on the rate of return. If independent projects can be ranked in their order of desirability, then the selection of projects to be included in a capital budget is a simple task. One may proceed down the list of ranked projects until the capital budget has been exhausted. The only difficulty with this scheme occurs, occasionally, when the capital budget is more than enough for n projects but too little for $n + 1$ projects.

EXAMPLE 15–7

Rank the following nine independent projects in their order of desirability, based on a 14.5% minimum attractive rate of return.

Project	Cost (thousands)	Uniform Annual Benefit (thousands)	Useful Life (years)	Salvage Value (thousands)	Computed Rate of Return	Computed NPW at 14.5% (thousands)	Computed NPW/Cost (thousands)
1	$100	$23.85	10	$0	20%	$22.01	0.2201
2	200	39.85	10	0	15	3.87	0.0194
3	50	34.72	2	0	25	6.81	0.1362
4	100	20.00	6	100	20	21.10	0.2110
5	100	20.00	10	100	20	28.14	0.2814
6	100	18.00	10	100	18	17.91	0.1791
7	300	94.64	4	0	10	−27.05	−0.0902
8	300	47.40	10	100	12	−31.69	−0.1056
9	50	7.00	10	50	14	−1.28	−0.0256

SOLUTION

Ranked NPW/PW of cost, the projects are listed as follows:

Project	NPW/PW of Cost	Rate of Return
5	0.2814	20%
1	0.2201	20
4	0.2110	20
6	0.1791	18
3	0.1362	25
2	0.0194	15
9	−0.0256	14
7	−0.0902	10
8	−0.1056	12

With a 14.5% MARR, Projects 1 to 6 are recommended for funding and 7 to 9 are not. However, they are ranked in a different order by the present worth index and by the rate of return approaches. For example, Project 3 has the highest ranking for the rate of return and is fifth by the present worth index.

In Example 15–7, suppose the capital budget is $550,000. This is more than enough for the top five projects (sum = $450,000) but not enough for the top six projects (sum = $650,000). When we have this situation, it may not be possible to say with certainty that the best use of a capital budget of $550,000 is to fund the top five projects. There may be some other set of projects that makes better use of the available $550,000. While some trial-and-error computations may indicate the proper set of projects, more elaborate techniques are needed to prove optimality.

As a practical matter, a capital budget total or line items have some flexibility. If in Example 15–7 the capital budget is $550,000, then a careful examination of Project 2 will dictate whether to it should be funded. If so, then the budget may be expanded or some or all projects may be approved with a lower funding level. Or perhaps Project 2 can be started in this budget year and finished next year.

SUMMARY

There are three general sources of capital available to a firm. The most important one is money generated from the firm's operations. This has two components: there is the portion of profit that is retained in the business; in addition, funds equal to its depreciation charges are available for reinvestment.

The two other sources of capital are from outside the firm's operations.

Debt: borrowed as loans from banks, insurance companies, and so forth.

   Longer-term borrowing: from selling bonds.

Equity: sale of equity securities like common or preferred stock.

Retained profits and cash equal to depreciation charges are the primary sources of investment capital for most firms, and the only sources for many enterprises.

In selecting a value of MARR, three values are frequently considered:

1. Cost of borrowed money.
2. Cost of capital. This is a composite cost of the components of the firm's overall capitalization.
3. Opportunity cost, which is the rate of return on the best investment project that is rejected.

The MARR should be equal to the highest one of these three values.

When there is a risk aspect to the problem (probabilities are known or reasonably estimated), this can be handled by techniques like expected value and simulation. Where there is uncertainty (probabilities of the various outcomes are not known), there are analytical techniques, but they are less satisfactory. A method commonly used to adjust for risk and uncertainty is to increase the MARR. This method can distort the time-value-of-money relationship. The effect is to discount longer-term consequences more heavily than short-term consequences, which may or may not be desirable. Prior to this chapter we had assumed that all worthwhile projects are approved and implemented. But industrial firms, like individuals and governments, are typically faced with more good projects than can be funded with the money available. The task is to select the best projects and reject, or at least delay, the rest.

Capital may be rationed among competing investment opportunities by either rate of return or present worth methods. The results may not always be the same for these two methods in many practical situations.

If projects are ranked by rate of return, a proper procedure is to go down the list until the capital budget has been exhausted. The rate of return at this point is the cutoff rate of return. This procedure gives the best group of projects, but does not necessarily have them in the proper priority order.

It has been shown in earlier chapters that the usual business objective is to maximize NPW, and this is not necessarily the same as maximizing rate of return. One suitable procedure is to use the ratio (NPW/PW of cost) to rank the projects, letting the MARR equal the cutoff rate of return (which is the opportunity cost of capital). This present worth ranking method will order the projects so that, for a limited capital budget, NPW will be maximized. The MARR must equal the cutoff rate of return for the rate of return and present worth methods to yield compatible results.

PROBLEMS

Cost of Funds

15-1 Assume you have $2000 available for investment for a 5-year period. You wish to *invest* the money—not just spend it on fun things. There are obviously many alternatives available. You should be willing to assume a modest amount of risk of loss of some or all of the money if this is necessary, but not a great amount of risk (no investments in poker games or at horse races). How would you invest the money? What is your minimum attractive rate of return? Explain.

15-2 There are many venture capital syndicates that consist of a few (say, eight or ten) wealthy people who combine to make investments in small and (hopefully) growing businesses. Typically, the investors hire a young investment manager (often an engineer with an MBA) who seeks and analyzes investment opportunities for the group. Would you estimate that the MARR sought by this group is more or less than 12%? Explain.

15-3 Determine the current interest rate on the following securities, and explain why the interest rates are different for these different bonds.

(*a*) U.S. Treasury bond due in 10 years

(*b*) General obligation bond of a municipal district, city, or a state due in 10 years

(*c*) Corporate debenture bond of a U.S. industrial firm due in 10 years

15-4 A firm's stockholders expect a 15% rate of return, and there is $12M in common stock and retained earnings. The firm has $5M in loans at an average rate of 7%. The firm has raised $8M by selling bonds at an average rate of 6%. What is the firm's cost of capital:

(*a*) Before taxes?

(*b*) After taxes with a tax rate of 35%?

15-5 An engineering firm has borrowed $1.0M at 6%. The stockholders have invested another $1.5M. The firm's retained earnings total $1.2M. The return on equity is estimated to be 12%. What is the firm's cost of capital:

(*a*) Before taxes?

(*b*) After taxes with a tax rate of 40%?

15-6 A small engineering firm has borrowed $125,000 at 8%. The partners have invested another $75,000. If the partners require a 12% rate of return, what is the firm's cost of capital:

(*a*) Before taxes?

(*b*) After taxes with a tax rate of 30%?

15-7 A public university system wants to apply the concept of the WACC to developing its interest rate for analyzing capital projects. It has an endowment of $850 million which is earning 6.3% interest. It is paying 4.5% interest on $300 million in bonds. It believes that $120 million in general funds from the taxpayers should be assigned an interest rate of 13%.

What is the university's cost of capital? Note that only new bonds or the interest on the endowment is available to fund capital projects.

15-8 A firm's stockholders expect an 15% rate of return, and there is $22M in common stock and retained earnings. The firm has $9M in loans at an average rate of 8%. The firm has raised $14M by selling bonds at an average rate of 4%. What is the firm's cost of capital:

(*a*) Before taxes?

(*b*) After taxes with a tax rate of 38%?

15-9 A firm has 40,000 shares whose current price is $80.75. Those stockholders expect a return of 15%. The firm has a 2-year loan of $900,000 at 6.4%. It has issued 12,500 bonds with a face value of 1000, 15 years left to maturity, semiannual compounding, a coupon interest rate of 6%, and a current price of $1090. Using market values for debt and equity, what is the firm's cost of capital:

(*a*) Before taxes?

(*b*) After taxes with a tax rate of 40%?

15-10 A firm has 60,000 shares whose current price is $45.90. Those stockholders expect a return of 14%. The firm has a 3-year loan of $1,900,000 at 7.3%. It has issued 22,000 bonds with a face value of 1000, 20 years left to maturity, semiannual compounding, a coupon interest rate of 7%, and a current price of $925. Using market values for debt and equity, what is the firm's cost of capital:

(*c*) Before taxes?

(*d*) After taxes with a tax rate of 34%?

15-11 Some countries have initiated negative interest rates on short-term bonds and savings accounts to spur economic growth. What is the impact of such policies on people trying to save money for their future?

15-12 Payday loan companies make short-term loans to people with no credit or poor credit. These loans often need to be repaid the next time a person receives a paycheck. Fees and interest are added, and effective interest rates often exceed 100% on an annualized basis. Investigate the payday loan industry.

(*a*) Are these legal in all states?

(*b*) What is the legal limit on the interest rate in the state where you are in?

(*c*) Are the high interest rates justified?

Inflation

15-13 What is the interest rate on a 2-year certificate of deposit at a bank or credit union in your area? What is the most recent value of the Consumer Price Index (CPI)? If inflation matches that rate, what is the real rate of return on the 2-year CD? Include references for the sources of your data.

15-14 What is the interest rate on a 4- or 5-year new car loan at a bank or credit union in your area? What is the most recent value of the Consumer Price Index (CPI)? If inflation matches that rate, what is the real interest rate you would pay on the car loan? Include references for the sources of your data.

15-15 Over the last 10 years, what has the inflation rate been? Compare this with the rate of return on the "Dow" average over the same period. What has the real rate of return for investing in this mix of stocks been? Include references for the sources of your data.

15-16 Over the last 10 years, what has the inflation rate been? Compare this with the rate of return on the NASDAQ average over the same period. What has the real rate of return for investing in this mix of stocks been? Include references for the sources of your data.

15-17 Per the effect referenced in Figure 15-1, calculate the real rate of investment for a country that has the following data. Find (a), (b), (c), and (d). Then graph these results as in the figure.

Year	Rate on 20-Year Bonds	General Inflation Rate	Real Rate on Investment
0	15%	9.52%	5%
1	10%	22.22%	(a)
2	12%	1.82%	(b)
3	18%	12.38%	(c)
4	5%	10.53%	(d)

15-18 The long-term borrowing rate has historically been approximately 3% higher than general inflation. What would the Long Term Borrowing Index (LTBI) be in 2017 if its index base is the year 2000, and inflation since 2000 has averaged 2.3%?

Opportunity Cost of Capital

15-19 A factory has a $100,000 capital budget. Determine which project(s) should be funded and the opportunity cost of capital.

Project	First Cost	Annual Benefits	Life (years)	Salvage Value
A	$50,000	$13,500	5	$5000
B	50,000	9,000	10	0
C	50,000	13,250	5	1000
D	50,000	9,575	8	6000

15-20 Chips USA is considering the following projects to improve their production process. Chips have a short life, so a 3-year horizon is used in evaluation. Which projects should be done if the budget is $70,000? What is the opportunity cost of capital?

Project	First Cost	Benefit
1	$20,000	$11,000
2	30,000	14,000
3	10,000	6,000
4	5,000	2,400
5	25,000	13,000
6	15,000	7,000
7	40,000	21,000

15-21 A city has identified a series of projects to replace incandescent light bulbs with LED bulbs in traffic lights, office buildings, and other uses. Which projects should be done if the budget is (a) $500,000, and (b) $1.2 million? What is the opportunity cost of capital?

Project	First Cost	Annual Benefit	Life (years)
1	$200,000	$50,000	15
2	300,000	70,000	10
3	100,000	40,000	5
4	50,000	12,500	10
5	250,000	75,000	5
6	150,000	32,000	20
7	400,000	125,000	5

15-22 The Winthrop Company has decided to fund $2.0 million in new project proposals for the coming budget year. Which projects should be funded? What is the opportunity cost of capital?

Project	First Cost	Annual Benefit	Life (years)
A	$150,000	$44,290	5
B	200,000	61,730	5
C	300,000	98,780	5
D	250,000	62,610	6
E	400,000	119,330	6
F	500,000	115,500	6
G	350,000	67,940	10
H	600,000	126,920	10
I	750,000	140,580	10

15-23 For the following projects what is the opportunity cost of capital if the budget is (a) $60,000, and (b) $120,000? If Project 4 has an external environmental cost of $1000 annually that is included, (c) how does this change the answer to (a)? (d) How does this change the answer to (b)?

Project	Life (years)	First Cost	Annual Benefit	Salvage Value
1	20	$20,000	$4000	
2	20	20,000	3200	$20,000
3	30	20,000	3300	10,000
4	15	20,000	4500	
5	25	20,000	4500	−20,000
6	10	20,000	5800	
7	15	20,000	4000	10,000

Risk-Adjusted MARR

15-24 Use the example risk-adjusted interest rates for manufacturing projects in Table 15–2. Assume Project B in Problem 15-19 is for new equipment. What is the interest rate for evaluating this project? Should it be done?

15-25 Use the example risk-adjusted interest rates for manufacturing projects in Table 15–2. Assume Project E in Problem 15-22 is a new product in an normal market. What is the interest rate for evaluating this project? Should it be done?

15-26 Use the example risk-adjusted interest rates for manufacturing projects in Table 15–2. Assume Project 1 in Problem 15-23 is a new product in a foreign market. What is the interest rate for evaluating this project? Should it be done?

15-27 Use the example risk-adjusted interest rates for manufacturing projects in Table 15–2 and the project data in Problem 15-22. Assume that each project is to launch a new product. New products for the normal market will have a life of 10 years. New products for a related market will have a life of 6 years. New products for a new market will have a life of 5 years. Which new products are recommended for funding? What is the total cost of these projects?

Capital Budgeting

15-28 Each of the following 10 independent projects has a 10-year life and no salvage value.

Project	Cost (thousands)	Uniform Annual Benefits (thousands)	Rate of Return
1	$ 50	$10.3	16%
2	150	32.2	17
3	100	17.7	12
4	300	48.8	10
5	50	11.9	20
6	200	38.3	14
7	50	10.0	15
8	200	36.9	13
9	50	11.5	19
10	100	22.3	18

The projects have been proposed by the staff of the Ace Card Company. The MARR of Ace has been 12% for several years.

(a) If there is ample money available, what projects should Ace approve?
(b) Rank-order all the acceptable projects according to desirability using NPW/cost.
(c) If only $450,000 is available, which projects should be approved?
(d) Are the results the same if the projects are ranked on IRR? What is the opportunity cost of capital?

15-29 Ten capital spending proposals have been made to the budget committee as the members prepare the annual budget for their firm. Each independent project has a 5-year life and no salvage value.

Project	Initial Cost (thousands)	Uniform Annual Benefit (thousands)	Rate of Return
A	$10	$2.98	15%
B	15	5.58	25
C	5	1.53	16
D	20	5.55	12
E	15	4.37	14
F	30	9.81	19
G	25	7.81	17
H	10	3.49	22
I	5	1.67	20
J	10	3.20	18

(a) Based on a MARR of 14%, which projects should be approved?
(b) Rank-order all the projects according to desirability.
(c) If only $70,000 is available, which projects should be approved?
(d) Are the results the same if the projects are ranked on IRR? What is the opportunity cost of capital?

15-30 At Miami Products, four project proposals (three with mutually exclusive alternatives) are being considered. All the alternatives have a 10-year useful life and no salvage value.

Project Proposal	Cost (thousands)	Uniform Annual Benefits (thousands)	Rate of Return
Project 1			
Alt. A	$25	$4.61	13%
Alt. B	50	9.96	15
Alt. C	10	2.39	20
Project 2			
Alt. A	20	4.14	16
Alt. B	35	6.71	14
Project 3			
Alt. A	25	5.56	18
Alt. B	10	2.15	17
Project 4	10	1.70	11

(a) Use rate of return methods to determine which set of projects should be undertaken if the capital budget is limited to about $100,000.

(b) For a budget of about $100,000, what interest rate should be used in rationing capital by present worth methods? (Limit your answer to a value for which there is a compound interest table available in Appendix C).

(c) Using the interest rate determined in part (b), rank-order the eight different investment opportunities by means of the present worth method.

(d) For a budget of about $100,000 and the ranking in part (c), which of the investment opportunities should be selected?

Minicases

15-31 A financier has a staff of three people whose job it is to examine possible business ventures for him. Periodically they present their findings concerning business opportunities. On a particular occasion, they presented the following investment opportunities:

Project A: This is a project for the use of commercial land the financier already owns. There are three mutually exclusive alternatives.

- *A1.* Sell the land for $500,000.

- *A2.* Lease the property for a car-washing business. An annual income, after all costs (property taxes, etc.) of $98,700 would be received at the end of each year for 20 years. At the end of the 20 years, it is believed that the property could be sold for $750,000.

- *A3.* Construct an office building on the land. The building will cost $4.5 million to construct and will not produce any net income for the first 2 years. The probabilities of various levels of rental income, after all expenses, for the subsequent 18 years are as follows:

Annual Rental Income	Probability
$1,000,000	0.1
1,100,000	0.3
1,200,000	0.4
1,900,000	0.2

The property (building and land) probably can be sold for $3 million at the end of 20 years.

Project B: An insurance company is seeking to borrow money for 90 days at 13¾% per annum, compounded continuously.

Project C: A financier owns a manufacturing company. The firm desires additional working capital to allow it to increase its inventories of raw materials and finished products. An investment of $2 million will allow the company to obtain sales that in the past the company had to forgo. The additional capital will increase company profits by $500,000 a year. The financier can recover this additional investment by ordering the company to reduce its inventories and to return the $2 million. For planning purposes, assume the additional investment will be returned at the end of 10 years.

Project D: The owners of *Sunrise* magazine are seeking a loan of $500,000 for 10 years at a 16% interest rate.

Project E: The Galveston Bank has indicated a willingness to accept a deposit of any sum of money over $100,000, for any desired duration, at a 14.06% interest rate, compounded monthly. It seems likely that this interest rate will be available from Galveston, or some other bank, for the next several years.

Project F: A car rental firm is seeking a loan of $2 million to expand its fleet. The firm offers to repay the loan by paying $1 million at the end of Year 1 and $1,604,800 at the end of Year 2.

- If there is $4 million available for investment now (or $4.5 million if the Project A land is sold), which projects should be selected? What is the MARR in this situation?

- If there is $9 million available for investment now (or $9.5 million if the Project A land is sold), which projects should be selected?

15-32 The Raleigh Soap Company has been offered a 5-year contract to manufacture and package a leading brand of soap for Taker Bros. It is understood that the contract will not be extended past the 5 years because Taker Bros. plans to build its own plant nearby. The contract calls for 10,000 metric tons (one metric ton equals 1000 kg) of soap a year. Raleigh normally produces 12,000 metric tons of soap a year, so production for the 5-year period would be increased to 22,000 metric tons. Raleigh must decide what changes, if any, to make to accommodate this increased production. Five projects are under consideration.

Project 1: Increase liquid storage capacity. Raleigh has been forced to buy caustic soda in tank truck quantities owing to inadequate storage capacity. If another liquid caustic soda tank is installed to hold 1000 cubic meters, the caustic soda may be purchased in railroad tank car quantities at a more favorable price. The result would be a saving of 0.1¢ per kilogram of soap. The tank, which would cost $83,400, has no net salvage value.

Project 2: Acquire another sulfonation unit. The present capacity of the plant is limited by the sulfonation unit. The additional 12,000 metric tons of soap cannot be produced without an additional sulfonation unit. Another unit can be installed for $320,000.

Project 3: Expand the packaging department. With the new contract, the packaging department must either work two 8-hour shifts or have another packaging line installed. If the two-shift operation is used, a 20% wage premium must be paid for the second shift. This premium would amount to $35,000 a year. The second packaging line could be installed for $150,000. It would have a $42,000 salvage value at the end of 5 years.

Project 4: Build a new warehouse. The existing warehouse will be inadequate for the greater production. It is estimated that 400 square meters of additional warehouse is needed. A new warehouse can be built on a lot beside the existing warehouse for $225,000, including the land. The annual taxes, insurance, and other ownership costs would be $5000 a year. It is believed the warehouse could be sold at the end of 5 years for $200,000.

Project 5: Lease a warehouse. An alternative to building an additional warehouse would be to lease warehouse space. A suitable warehouse one mile away could be leased for $15,000 per year. The $15,000 includes taxes, insurance, and so forth. The annual cost of moving

materials to this more remote warehouse would be $34,000 a year.

The contract offered by Taker Bros. is a favorable one, which Raleigh Soap plans to accept. Raleigh management has set a 15% before-tax minimum attractive rate of return as the criterion for any of the projects. Which projects should be undertaken?

15-31 Mike Moore's microbrewery is considering production of a new ale called Mike's Honey Harvest Brew. To produce this new offering, Mike is considering two independent projects. Each of these projects has two mutually exclusive alternatives, and each alternative has a useful life of 10 years and no salvage value. Mike's MARR is 8%. Information regarding the projects and alternatives is given in the following table.

Project/Alternative	Cost	Annual Benefit
Project 1. Purchase new fermenting tanks		
Alt. A: 5000-gallon tank	$ 5,000	$1192
Alt. B: 15,000-gallon tank	10,000	1992
Project 2. Purchase bottle filler and capper		
Alt. A: 2500-bottle/hour machine	15,000	3337
Alt. B: 5000-bottle/hour machine	25,000	4425

Use incremental rate of return analysis to complete the following worksheet.

Proj./Alt.	Cost, P	Annual Benefit, A	(A/P, i, 10)	IRR
1A	$ 5,000	$1192	0.2385	20%
1B–1A	5,000	800	0.1601	
2A	15,000	3337		
2B–2A	10,000			

Use this information to determine:

(a) which projects should be funded if only $15,000 is available.

(b) the cutoff rate of return if only $15,000 is available.

(c) which projects should be funded if $25,000 is available.

CASES

The following cases from *Cases in Engineering Economy* (www.oup.com/us/newnan) are suggested as matched with this chapter.

CASE 46 **Aero Tech**

Case budgeting with memos for three different approaches to ranking.

CASE 47 **Bigstate Highway Department**

Capital budgeting including mutually exclusive alternatives. Includes uncertain value of a life.

CASE 48 **Dot Puff Project Selection**

Public-sector capital budgeting with added constraint on man-years required.

CASE 49 **The Arbitrator**

Includes two memos of suggested solutions for four projects with nine alternatives. Data presented in text discussion rather than tabulated.

CASE 50 **Capital Planning Consultants**

Capital budgeting including mutually exclusive alternatives. Includes uncertainties in first cost, annual benefit, and lives.

CASE 51 **Refrigerator Magnet Company**

Capital budgeting including mutually exclusive alternatives.

CHAPTER 16

ECONOMIC ANALYSIS IN THE PUBLIC SECTOR

From Waste to Power and Money

One human activity that contributes to global warming is the dumping of organic material into landfills. This material decomposes over time, producing a significant amount of landfill gas (LFG). About 50% of LFG is methane, which traps heat in the atmosphere with an effect that is twenty times greater than carbon dioxide. Landfills that do not control the escape of landfill gas are the largest single human source of methane emissions in the United States, as well as a source of air pollution and odors.

But humans must produce waste to live—it is one of the basic laws of thermodynamics. That is bad news all around—except that using engineering principles, bringing together various "constituencies," and evaluating alternatives with engineering economy can bring benefits out of almost every one of these issues.

Landfill gas (LFG) can be an asset if it is extracted and properly utilized. LFG has about half of the heating value of natural gas. As a result, more and more owners of municipal solid waste landfills extract this gas and use it to generate electricity. Such electricity is used directly for their power needs or is sold to the general power grid.

Extracting and using a landfill to generate electricity is a "6 win" situation.

1. It benefits the environment by reducing unwanted gas emission;
2. It adds electrical power to the grid;
3. It produces cash flow for the landfill owner;
4. It lowers the use of non-renewable fossil fuels;
5. It reduces financial costs to the local population; and
6. It reduces the hazardous, noxious, and odorous gases for those nearby and downwind. ■ ■ ■

Contributed by William R. Truran, U.S. Department of Defense, and Peter A. Cerenzio, Cerenzio & Panaro Consulting Engineers

QUESTIONS TO CONSIDER

1. How can this topic best be summarized for a non-technical public audience? This is an important "skill" of an engineer—conveying sometimes complex and confusing scientific issues to a less informed, less scientific, and less interested audience with a short attention span.

2. What engineering economic principles would you apply in analyzing this application? Which measures would be most effective for the public audience? The important measures, and the metrics used to describe them, may be considered "key performance indicators."

3. This waste to power application cannot be answered with a simple equation with a closed-ended answer that is either "correct" or "wrong." This problem requires the cooperation of several (at least) agencies and communities of interest. Name some of these stakeholders and identify how their objectives are aligned or in conflict.

4. The Enron example—which is a "poster child" of what can go wrong in business—like this vignette is related to electric power generation. Discuss ways in which some of the stakeholders may be "shady" in such situations. Discuss how you—as an engineer—could not only affect factual answers but could project perhaps an image that may not be real. As an engineer, you can be honored for (or guilty of) your representations.

After Completing This Chapter...
The student should be able to:

- Distinguish the unique objective and viewpoint of public decisions.
- Explain methods for determining the interest rates for evaluating public projects.
- Use the benefit–cost ratio to analyze projects.
- Distinguish between the conventional and modified versions of the benefit–cost ratio.
- Use an incremental benefit–cost ratio to evaluate a set of mutually exclusive projects.
- Discuss the impact of financing, duration, quantifying and valuing consequences, and politics in public investment analysis.

581

Key Words

benefit–cost ratio	incremental benefit–cost	promote the general
conventional B/C ratio	ratio	welfare
design to cost	modified B/C ratio	revenue bonds
disbenefits	net benefits to the users	taxpayer
general obligation	project duration	opportunity cost
municipal bonds	project financing	viewpoint
government opportunity cost	project politics	

Federal, state, and local governments, port authorities, school districts, government agencies, and other public organizations make investment decisions. For these decision-making bodies, economic analysis is complicated by several factors that do not affect firms in the private sector. These factors include the overall purpose of investment, the viewpoint for analysis, and how to select the interest rate. Other factors include project financing sources, expected project duration, quantifying and valuing benefits and disbenefits, effects of politics, beneficiaries of investment, and the multipurpose nature of investments. The overall mission in the public sector is the same as that in the private sector—to make prudent investment decisions that promote the organization's overall objectives.

The primary economic measure used in the public sector is the benefit–cost (B/C) ratio, which was introduced in Chapter 9. This measure is calculated as a ratio of the equivalent worth of the project's benefits to the equivalent worth of the project's costs. If the B/C ratio is *greater than 1.0*, the project under evaluation is accepted; if not, it is rejected. The B/C ratio is used to evaluate both single investments and sets of mutually exclusive projects (where the incremental B/C ratio is used). The uncertainties of quantifying cash flows, long project lives, and low interest rates all tend to lessen the reliability of public sector engineering economic analysis. There are two versions of the B/C ratio: *conventional* and *modified*. Both provide consistent recommendations to decision makers for single investment decisions and for decisions involving sets of mutually exclusive alternatives. The B/C ratio is a widely used and accepted measure in government economic analysis and decision making.

INVESTMENT OBJECTIVE

Organizations exist to promote the overall goals of those they serve. In private sector firms, investment decisions are based on increasing the firm's wealth and economic stability. Beneficiaries of investments generally are clearly identified as the firm's owners and/or stockholders.

In the public sector, the purpose of investment decisions is sometimes ambiguous. In the United States of America, the Preamble to the Constitution establishes the overall theme:

> We the People of the United States, in order to form a more perfect Union, establish justice, insure domestic tranquillity, provide for the common defense, *promote the general welfare* [italics supplied], and secure the blessings of liberty to ourselves and our posterity, do ordain and establish this Constitution for the United States of America.

The catch phrase **promote the general welfare** serves as a guideline for public decision making. But what does this phrase mean? At best it is a general guideline; at worst it is a vague slogan that can be used to justify any action. Projects some citizens want may be opposed by other citizens. In government economic analysis, it is not always easy to distinguish which investments promote the "general welfare" and which do not.

Consider the case of a dam construction project to provide water, electricity, flood control, and recreational facilities. Such a project might seem to be advantageous for a region's entire population. But on closer inspection, decision makers must consider that the dam will require the loss of land upstream due to backed-up water. Farmers will lose pastures or cropland, and nature lovers will lose canyon lands. Or perhaps the land to be lost is a pivotal breeding ground for protected species, and environmentalists will oppose the project. The project may also have a negative impact on towns, cities, and states downstream. How will it affect their water supply? Thus, a project initially deemed to have many benefits may also have many conflicts. Projects' conflicting aspects are characteristic of public-sector investment and decision making.

Public investment decisions are more difficult than those in the private sector owing to the many people, organizations, and political units potentially affected. Opposition to a proposal is more likely in public investment decisions than in those made by private-sector companies because for every group that benefits from a particular project, there is usually an opposing group. Many conflicts in opinion arise when the project involves the use of public lands, including industrial parks, housing projects, business districts, roadways, sewage and power facilities, and landfills. Opposition may be based on the belief that development of *any* kind is bad or that the proposed development should not be near "our" homes, schools, or businesses.

Consider the decision that a small town might face when considering whether to establish a municipal rose garden, seemingly a beneficial public investment with no adverse consequences. However, an economic analysis of the project must consider *all* effects of the project, including potential unforeseen outcomes. Where will visitors park their vehicles? Will increased travel around the park necessitate new traffic lights and signage? Will traffic and visitors to the park increase noise levels for adjacent homes? Will special varieties of roses create a disease hazard for local gardens? Will the garden require high levels of fertilizers and insecticides, and where will these substances wind up after they have been applied? Clearly, many issues must be addressed. Our simple rose garden illustrates how effects on *all parties involved* must be identified, even for projects that seem very useful.

The **Flood Control Act of 1936** specified that waterway improvements for flood control could be made as long as "the benefits *to whomsoever they accrue* [italics supplied] are in excess of the estimated costs." Perhaps the overall general objective of investment decision analysis in government should be a dual one: to promote the general welfare and to ensure that the value to those who can potentially benefit exceeds the overall costs to those who do not benefit.

VIEWPOINT FOR ANALYSIS

When governmental bodies do economic analysis, an important concern is the proper **viewpoint** of the analysis. A look at industry will help to explain how the viewpoint from which an analysis is conducted influences the final recommendation. A firm pays its costs and counts *its* benefits. Thus, both the costs and benefits are measured from the firm's perspective.

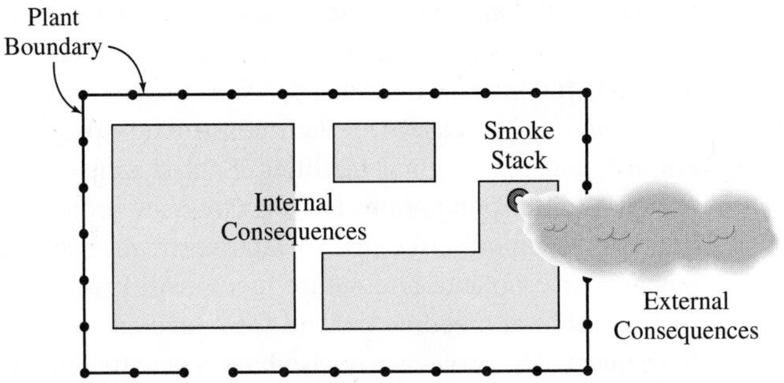

FIGURE 16–1 Internal and external consequences of an industrial plant.

The discussion of internal and external costs in Chapter 2 demonstrated one approach to expanding the firm's perspective that allows some consideration of the external consequences depicted in Figure 16–1.

The council members of a small town that levies taxes can be expected to take the "viewpoint of the town" in making decisions: unless it can be shown that the money from taxes can be used *effectively*, the town council is unlikely to spend it. But what happens when the money is contributed to the town by the federal government, as in "revenue sharing" or by means of some other federal grant? Often the federal government pays a share of project costs varying from 10 to 90%. Example 16–1 illustrates the viewpoint problem that is created.

EXAMPLE 16–1

A municipal project will cost $1 million. The federal government will pay 60% of the cost if the project is undertaken. Although the original economic analysis showed that the PW of benefits was $1.5 million, a subsequent detailed analysis by the city engineer indicates a more realistic estimate of the PW of benefits is $750,000. The city council must decide whether to proceed with the project. What would you advise?

SOLUTION

From the viewpoint of the city, the project is still a good one. If the city puts up 40% of the cost ($400,000) it will receive all the benefits ($750,000). On the other hand, from an *overall* viewpoint, the revised estimate of $750,000 of benefits does not justify the $1 million expenditure. This illustrates the dilemma caused by varying viewpoints. For economic efficiency, one does not want to encourage the expenditure of money, regardless of the source, unless the benefits at least equal the costs.

Possible viewpoints that may be taken include those of an individual, a business firm or corporation, a town or city district, a city, a state, a nation, or a group of nations. The proper approach is to *take a viewpoint at least as broad as the larger of those who pay the costs and those who receive the benefits*. When the costs and benefits are totally confined to a town, for example, the town's viewpoint seems to be an appropriate basis for the analysis. But when the costs or the benefits are spread beyond the proposed viewpoint, then the viewpoint should be enlarged to this broader population.

Other than investments in defense and social programs, most of the benefits provided by government projects are realized at a regional or local level. Projects, such as dams for electricity, flood control, and recreation, as well as transportation facilities, such as roads, bridges, and harbors, all benefit most those in the region in which they are constructed. Even smaller-scale projects, such as the municipal rose garden, although funded by public monies at a local or state level, provide most benefit to those nearby. It is important to adopt an appropriate and *consistent* viewpoint and to designate all the costs and benefits that arise from the prospective investment from that perspective. To use different perspectives when quantifying costs and benefits could greatly skew the results of the analysis and subsequent decision, and thus different perspectives are inappropriate.

SELECTING AN INTEREST RATE

Several factors, not present for private-sector firms, influence selecting an interest rate for economic analysis in the government sector. Recall that for private-sector firms the overall objective is wealth maximization, and the interest rate is selected consistent with this goal. Most firms use *cost of capital* or *opportunity cost* concepts when setting an interest rate. How to set an interest rate is less clear-cut for public projects. Possibilities include no interest rate, cost of capital concepts, and an opportunity cost concept.

No Time-Value-of-Money Concept

In government, monies are obtained through taxation and spent about as quickly as they are obtained. Often, there is little time delay between collecting money from taxpayers and spending it. (Remember that the federal government and many states collect taxes every paycheck in the form of withholding tax.) The collection of taxes, like their disbursement, although based on an annual budget, is actually a continuous process. Using this line of reasoning, some would argue that there is little or no time lag between collecting and spending tax dollars. Thus, they would advocate the use of a 0% interest rate for economic analysis of public projects. Not surprisingly, this viewpoint is most often expressed by people who are *not* engineering economists and who are pushing a marginal project.

Cost of Capital Concept

Another approach in determining interest rates in public investments is that most levels of government (federal, state, and local) borrow money for capital expenditures in addition to collecting taxes. Where money is borrowed for a specific project, one line of reasoning is to use an interest rate equal to the *cost of borrowed money*. This argument is less valid for state and local governments than for private firms because the federal government, through the income tax laws, subsidizes state and local bonded debt. If a state, county, city, or special assessment district raises money by selling bonds, the interest paid on these bonds is exempt from federal taxes. In this way the federal government is *subsidizing* the debt, thereby encouraging investors to purchase such bonds.

These bonds, called municipal bonds, can be either general obligation or revenue bonds. **General obligation municipal bonds** pay interest and are retired (paid off) through taxes raised by the issuing government unit. A school district may use property taxes it receives to finance bond debt for construction of new language labs. **Revenue bonds** are not supported by the taxing authority of the government unit; rather, they are supported by revenues earned by the project being funded. As an example, a city could use toll revenues from a new bridge to retire debt on revenue bonds sold for the bridge's construction.

For those who purchase municipal bonds, the instruments' tax-free status means that the expected return on this investment is somewhat less than that required of fully taxed bond investments (of similar risk). As a rough estimate, when fully taxed bonds yield an 8% interest rate, municipal bonds might make interest payments at a rate of 6%. The difference of 2% is due to the preferred treatment for federal taxation and the federal subsidy on tax-free bonds. This can skew the *cost of capital* approach by lowering the apparent cost of long-term bonds. The true cost of the bonds should include the federal subsidy, which is paid by all taxpayers.

Opportunity Cost Concept

Opportunity cost, which is related to the interest rate on the best opportunity forgone, may be based on either the government's or the taxpayers' opportunity cost. The **government opportunity cost** is the interest rate of the best prospective project for which funding is not available. One disadvantage of the government opportunity cost concept is that different government agencies and subdivisions have different opportunities. Therefore, units could use different interest rates for economic analysis, and a project that is rejected in one branch might be accepted in another. Differing interest rates lead to inconsistent evaluation and decision making across government.

Dollars used for public investments are generally gathered by taxing the citizenry. The concept of **taxpayer opportunity cost** suggests that a correct interest rate for evaluating public investments is that which the *taxpayer* could have received if the government had not collected those dollars through taxation. This philosophy holds that through taxation the government is taking away the taxpayers' opportunity to use the same dollars for investment. The interest rate that the government requires should not be less than what the taxpayer would have received. This concept argues that it is not economically desirable to take money from a taxpayer with a 12% opportunity cost, for example, and invest it in a government project yielding 4%.

The most widely followed standard is found in the Office of Management and Budget (OMB) A94 directive, which stipulates that a 7% interest rate be used in economic analysis for a wide range of federal projects.

Recommended Concept

The general rule of thumb in setting an interest rate for government investments has been to select the *largest* of the cost of capital, the government opportunity cost, or the taxpayer opportunity cost interest rates. However, as is the case in the private sector, there is no hard and fast rule, universally applied in all decision circumstances. Setting an interest rate for use in economic analysis is at the discretion of the government entity performing the analysis.

THE BENEFIT–COST RATIO

The **benefit–cost ratio** was described briefly in Chapter 9. This method is used almost exclusively in public investment analysis, and because of the magnitude of the amount of public dollars committed each year through such analysis, the benefit–cost ratio deserves our attention and understanding.

One of the primary reasons for using the benefit–cost ratio (B/C ratio) in public decision making is its simplicity. The ratio is formed by calculating the equivalent worth of the project's benefits divided by the equivalent worth of the project's costs. The benefit–cost ratio can be shown as follows:

$$\text{B/C ratio} = \frac{\text{Equivalent worth of net benefits}}{\text{Equivalent worth of costs}}$$

$$= \frac{\text{PW benefits}}{\text{PW costs}} = \frac{\text{FW benefits}}{\text{FW costs}} = \frac{\text{AW benefits}}{\text{AW costs}}$$

Notice that *any* of the equivalent worth methods (present, future, and annual) can be used to calculate this ratio. Each formulation of the ratio will produce an identical result, as illustrated in Example 16–2.

EXAMPLE 16–2

Demonstrate that for this highway expansion project, the same B/C ratio is obtained using the present, future, and annual worth formulations.

Initial costs of expansion	$1,500,000
Annual costs for operating/maintenance	65,000
Annual savings and benefits to travelers	225,000
Residual value of benefits after horizon	300,000
Useful life of investment	30 years
Interest rate	8%

SOLUTION

Using Present Worth

$$\text{PW benefits} = 225,000(P/A, 8\%, 30) + 300,000(P/F, 8\%, 30) = \$2,563,000$$

$$\text{PW costs} = 1,500,000 + 65,000(P/A, 8\%, 30) = \$2,232,000$$

5-BUTTON SOLUTION

	A	B	C	D	E	F	G	H	I
1	Problem	*i*	*n*	*PMT*	*PV*	*FV*	Solve for	Answer	change sign
2	Benefits	8.00%	30	225,000		300,000	PV	-$2,562,814	$2,562,814
3	Costs	8.00%	30	65,000		0	PV	-$731,756	$731,756
4					1,500,000			Total costs	$2,231,756
5								B/C ratio	1.15

Using Future Worth

$$\text{FW benefits} = 225,000(F/A, 8\%, 30) + 300,000 = \$25,790,000$$

$$\text{FW costs} = 1,500,000(F/P, 8\%, 30) + 65,000(F/A, 8\%, 30) = \$22,460,000$$

Using Annual Worth

$$\text{AW benefits} = 225{,}000 + 300{,}000(A/F, 8\%, 30) = \$227{,}600$$

$$\text{AW costs} = 1{,}500{,}000(A/P, 8\%, 30) + 65{,}000 = \$198{,}200$$

$$\text{B/C ratio} = \frac{2{,}563{,}000}{2{,}232{,}000} = \frac{25{,}790{,}000}{22{,}460{,}000} = \frac{227{,}600}{198{,}200} = 1.15$$

One can see that the ratio provided by each of these methods produces the same result: 1.15.

When one is using the B/C ratio, the decision rule is:

If the B/C ratio is > 1.0, then invest.

If the B/C ratio is < 1.0, then do not invest.

Cases of a B/C ratio equal to 1.0 are analogous to the case of a calculated net present worth of \$0 or an IRR analysis that yields $i = \text{MARR}\%$. In other words, the decision measure is at the breakeven criteria. In such cases a detailed analysis of the input variables and their estimates is necessary, and one should consider the merits of other available opportunities for the targeted funds. But, if the B/C ratio is clearly greater than or less than 1.0, the recommendation is clear.

The B/C ratio is a numerator/denominator relationship between the equivalent worths (EW) of *benefits* and *costs:*

$$\text{B/C ratio} = \frac{\text{EW of net benefits to whomsoever they may accrue}}{\text{EW of costs to the sponsors of the project}}$$

The numerator and denominator aspects of the ratio are sometimes interpreted and used in different fashions. For instance, the **conventional B/C ratio** (see Example 16–2) can be restated as follows:

$$\text{Conventional B/C ratio} = \frac{\text{EW of net benefits}}{\text{EW of initial costs} + \text{EW of operating and maintenance costs}}$$

However, there is also the **modified B/C ratio,** which subtracts the *annual operating and maintenance costs* in the numerator, rather than adding them as a cost in the denominator. This *modified B/C ratio* is mathematically similar to the present worth index defined in Chapter 9. The ratio becomes

$$\text{Modified B/C ratio} = \frac{\text{EW of net benefits} - \text{EW of operating and maintenance costs}}{\text{EW of initial costs}}$$

For decision making, the two versions of the benefit–cost ratio will produce the same recommendation on whether to *invest* or *not invest* in the project being considered. The *numeric values for the B/C ratio* will not be the same, but the recommendation will be. This is illustrated in Example 16–3.

EXAMPLE 16–3	(Example 16–2 Revisited)

Consider the highway expansion project from Example 16–2. Let us use the present worth formulation of *conventional* and *modified* versions to calculate the B/C ratio.

SOLUTION

Using the Conventional B/C Ratio

$$\text{B/C ratio} = \frac{225,000(P/A, 8\%, 30) + 300,000(P/F, 8\%, 30)}{1,500,000 + 65,000(P/A, 8\%, 30)} = 1.15$$

Using the Modified B/C Ratio

$$\text{B/C ratio} = \frac{225,000(P/A, 8\%, 30) + 300,000(P/F, 8\%, 30) - 65,000(P/A, 8\%, 30)}{1,500,000} = 1.22$$

We could calculate these same values using a spreadsheet, as in Figure 16–2. Figure 16–2 replicates these results with a 5-button solution.

	A	B	C	D	E	F	G	H
1	Problem	*i*	*n*	*PMT*	*PV*	*FV*	Solve for	Answer
2	Conventional							
3	Benefits	8.00%	30	225,000		300,000	PV	$2,562,814
4	Costs	8.00%	30	65,000		0	PV	$731,756
5					1,500,000		Total cost	$2,231,756
6							B/C ratio	1.15
7	Modified							
8	Benefits	8.00%	30	160,000		300,000	PV	$1,831,059
9	Costs	8.00%	30		1,500,000		Total cost	$1,500,000
10							B/C ratio	1.22

FIGURE 16–2

Whether the conventional or the modified ratio is used, the recommendation is to invest in the highway expansion project. The ratios are not identical in magnitude (1.15 vs. 1.22), but the decision is the same.

It is important when one is using the conventional and modified B/C ratios not to directly compare the magnitudes of the two versions. Evaluating a project with one version may produce a higher ratio than is produced with the other version, but this does not imply that the project is somehow better.

The **net benefits to the users** of government projects are the difference between the expected *benefits* from investment minus the expected *disbenefits.* Disbenefits are the negative effects of government projects felt by some individuals or groups. For example, consider the U.S. National Park System. Development projects by the skiing or lumber industries might provide enormous benefits to the recreation or construction sectors while creating simultaneous disbenefits for environmental groups.

INCREMENTAL BENEFIT–COST ANALYSIS

In Chapter 9 we demonstrated using the **incremental benefit–cost ratio** in economic decision analysis. As with the internal rate of return (IRR) measure, the incremental B/C ratio should be used in comparing *mutually exclusive alternatives*. Incremental B/C ratio analysis is consistent with maximizing the present worth of the alternatives. As with the incremental IRR method, it is *not* proper to simply calculate the B/C ratio for each alternative and choose the one with the highest value. Rather, an *incremental* approach is called for.

Elements of the Incremental Benefit–Cost Ratio Method

1. *Identify all relevant alternatives.* Decision rules or models can recommend a *best* course of action *only* from the set of identified alternatives. If a better alternative exists but is not considered, then it will never be selected, and when available, the solution will be suboptimal. For benefit–cost ratio problems, the "do-nothing" option is always the "base case" from which the incremental methodology proceeds.

2. *(Optional) Calculate the B/C ratio of each alternative.* Once the individual B/C ratios have been calculated, the alternatives with a ratio *less than 1.0* are eliminated from further consideration. This step gets the "poor performers" out of the way before the incremental procedure is initiated. This step may be omitted, however, because the incremental analysis method will eliminate the subpar alternatives in due time.

Note: There is a case where this step *must* be skipped. If doing nothing is not an alternative and if *all* alternatives have a B/C ratio less than 1.0, then the best of the unattractive alternatives will be selected through incremental analysis.

3. *Rank-order the projects.* The alternatives must be ordered from smallest to largest size of the *denominator of the B/C ratio.* (The rank order will be the same regardless of whether one uses the present worth, annual worth, or future worth of costs to calculate the *denominator.*) When available, the "do-nothing" alternative always becomes the first on the ordered list.

4. *Identify the increment under consideration.* The first increment considered is always that of going from the lowest cost alternative (when it is available, this is the do-nothing option) to the next higher cost alternative. As the analysis proceeds, any identified increment is always in reference to some previously justified alternative.

5. *Calculate the B/C ratio for the incremental cash flows.* First, calculate the *incremental benefits* and the *incremental costs.* This is done by finding the cash flows that represent the difference (Δ) between the two alternatives under consideration. For two alternatives X and Y, where X is the defender, or base case, and Y is the challenger, the increment can be written as ($X \rightarrow Y$) to signify *going from X to Y* or as ($Y - X$) to signify the *cash flows of Y minus cash flows of X.* Both modes identify the incremental costs (ΔC) and benefits (ΔB) of investing in Alternative Y, where X is a previously justified (or base) alternative. The *incremental B/C ratio* equals ΔB$/\Delta$C.

6. *Use the incremental B/C ratio to decide which alternative is better.* If the incremental B/C ratio ($\Delta B/\Delta C$) calculated in Step 5 is greater than 1.0, then the increment is desirable or justified; if the ratio is less than 1.0, it is not desirable, or is not justified. If an increment is accepted, the alternative (the challenger) associated with that increment of investment becomes the base from which the next increment is formed. When the increment is not justified, the alternative associated with the additional increment is rejected and the previously justified alternative (the defender) is maintained as the base for formation of the next increment.

7. *Iterate to Step 4 until all increments (projects) have been considered.* The incremental method requires that the entire list of ranked feasible alternatives be evaluated. All pairwise comparisons are made such that the additional increment being considered is examined with respect to a previously justified alternative. The incremental method continues until all alternatives have been evaluated.

8. *Select the best alternative from the set of mutually exclusive competing projects.* After all alternatives (and associated increments) have been considered, the incremental B/C ratio method calls for selecting the alternative *associated with the last justified increment.* This assures that a maximum investment is made such that each ratio of equivalent worth of incremental benefits to equivalent worth of incremental costs is greater than 1.0. (A common error in applying the incremental B/C method is selecting the alternative with the *largest* incremental B/C ratio, which is inconsistent with the objective of maximizing investment size with incremental B/C ratios above 1.0.)

Both the conventional and modified versions of the B/C ratio can be used with the incremental B/C ratio methodology just described, but the two versions should not be mixed in the same problem. Such an approach could cause confusion and errors. Instead, *one* of the two versions should be *consistently* used throughout the analysis.

In Examples 16–4 and 16–5 the incremental B/C ratio (conventional and modified version) is used to evaluate a set of mutually exclusive alternatives.

EXAMPLE 16–4

A midwestern industrial state may construct and operate two coal-burning power plants and a distribution network to provide electricity to several state-owned properties. The following costs and benefits have been identified.

Primary costs: Construction of the power plant facilities; cost of installing the power distribution network; life-cycle maintenance and operating costs.

Primary benefits: Elimination of payments to the current electricity provider; creation of jobs for construction, operation, and maintenance of the facilities and distribution network; revenue from selling excess power to utility companies; increased employment at coal mines in the state.

There have been four competing designs identified for the power plants. Each has a life of 45 years. Use the *conventional* B/C ratio with an interest rate of 8% to recommend a course of action.

	Values ($\times$ \$10^4) for Competing Design Alternatives			
	I	II	III	IV
Project costs				
Plant construction cost	\$12,500	\$11,000	\$12,500	\$16,800
Annual operating and maintenance cost	120	480	450	145
Project benefits				
Annual savings from utility payments	580	700	950	1,300
Revenue from overcapacity	700	550	200	250
Annual effect of jobs created	400	750	150	500

SOLUTION

Alternatives I through IV and the do-nothing alternative are *mutually exclusive* choices because one and only one will be selected. Therefore, an incremental B/C ratio method is used to obtain the solution.

Step 1 *Identify alternatives.* The alternatives are do nothing and designs I, II, III, and IV.

Step 2 *Calculate B/C ratio for each alternative.*

B/C ratio (I)   = $(580 + 700 + 400)\,(P/A, 8\%, 45)/[12,500 + 120\,(P/A, 8\%, 45)] = 1.46$
B/C ratio (II)   = $(700 + 550 + 750)\,(P/A, 8\%, 45)/[11,000 + 480\,(P/A, 8\%, 45)] = 1.44$
B/C ratio (III) = $(200 + 950 + 150)\,(P/A, 8\%, 45)/[12,500 + 325\,(P/A, 8\%, 45)] = 0.96$
B/C ratio (IV) = $(1300 + 250 + 500)(P/A, 8\%, 45)/[16,800 + 145(P/A, 8\%, 45)] = 1.34$

Alternatives I, II, and IV all have B/C ratios greater than 1.0 and thus merit further consideration. Alternative III does not meet the acceptability criterion and could be eliminated from further consideration. However, to illustrate that Step 2 is optional, all four design alternatives will be analyzed incrementally.

Step 3 *Rank-order projects.* Here we calculate the PW of costs for each alternative. The denominator of the *conventional* B/C ratio includes first cost and annual O&M costs, so the PW of costs for the alternatives are:

PW costs (I)   = $12,500 + 120(P/A, 8\%, 45) = \$13,953$
PW costs (III) = $12,500 + 325(P/A, 8\%, 45) = \$16,435$
PW costs (II) = $11,000 + 480(P/A, 8\%, 45) = \$16,812$
PW costs (IV) = $16,800 + 145(P/A, 8\%, 45) = \$18,556$

The rank order from low to high value of the B/C ratio *denominator* is as follows: do nothing, I, III, II, IV.

Step 4 *Identify increment under consideration.*

Step 5 *Calculate B/C ratio.*

Step 6 *Which alternative is better?*

Step 4	Increment Under Consideration	1st Iteration (Do Nothing → I)	2nd Iteration (I → III)	3rd Iteration (I → II)	4th Iteration (II → IV)
	ΔPlant construction cost	$12,500	$ 0	$−1500	$ 5800
	ΔAnnual O&M cost	120	205	360	−335
	PW of ΔCosts	13,953	2482	2859	1744
	ΔAnnual utility payment savings	580	370	120	600
	ΔAnnual overcapacity revenue	700	500	−150	−300
	ΔAnnual benefits of new jobs	400	−250	350	−250
	PW of ΔBenefits	20,342	−4601	3875	605
Step 5	ΔB/C ratio (PW ΔB)/(PW ΔC)	1.46	−1.15	1.36	0.35
Step 6	Is increment justified?	Yes	No	Yes	No

As an example of these calculations, consider the third increment (I → II).

ΔPlant construction cost		$= 11,000 - 12,500 = -\$1500$
ΔAnnual O&M cost		$= 480 - 120 = \$360$
PW of ΔCosts		$= -1500 + 360(P/A, 8\%, 45) = \$2859$
	or	$= 16,812 - 13,953 = \$2859$
ΔAnnual utility payment savings		$= 700 - 580 = \$120$
ΔAnnual overcapacity revenue		$= 550 - 700 = -\$150$
ΔAnnual benefits of new jobs		$= 750 - 400 = \$350$
PW of ΔBenefits		$= (120 - 150 + 350)(P/A, 8\%, 45)$
		$= \$3875$
ΔB/C ratio (PW ΔB)/(PW ΔC)		$= 3875/2859 = 1.36$

The analysis in the table proceeded as follows: do nothing to Alternative I was justified ($\Delta B/C$ ratio $= 1.46$), Alternative I became the new base; Alternative I to Alternative III was not justified (ΔB/C ratio $= -1.15$), Alternative I remained base; Alternative I to Alternative II was justified (ΔB/C ratio $= 1.36$), Alternative II became the base; Alternative II to Alternative IV was not justified (ΔB/C ratio $= 0.35$).

Step 7 *Select best alternative.* Alternative II became the recommended power plant design alternative because it is the one associated with the last justified increment. Notice that Alternative III did not affect the recommendation and was eliminated through the incremental method. Notice also that the first increment considered (do nothing → I) was not selected even though it had the *largest* ΔB/C ratio (1.45). The alternative associated with the *last justified increment* (in this case, Alt. II) should be selected.

EXAMPLE 16–5   (Example 16–4 Revisited)

Let us reconsider Example 16–4, this time using the modified B/C ratio to analyze the alternatives. Again we will use the present worth method.

SOLUTION

Here we use the modified B/C ratio.

Step 1 *Identify alternatives.* The alternatives are still do nothing and designs I, II, III, and IV.

Step 2 *Calculate modified B/C ratio for each alternative.*

$$\text{B/C ratio (I)} = (580 + 700 + 400 - 120)\,(P/A, 8\%, 45)/12{,}500 = 1.51$$
$$\text{B/C ratio (II)} = (700 + 550 + 750 - 480)\,(P/A, 8\%, 45)/11{,}000 = 1.67$$
$$\text{B/C ratio (III)} = (200 + 950 + 150 - 325)\,(P/A, 8\%, 45)/12{,}500 = 0.95$$
$$\text{B/C ratio (IV)} = (1300 + 250 + 500 - 145)\,(P/A, 8\%, 45)/16{,}800 = 1.37$$

Again Alternative III can be eliminated from further consideration because its B/C ratio is *less than 1.0*. In this case we will eliminate it. The remaining alternatives are do nothing and alternative designs I, II, and IV.

Step 3 *Rank-order projects.* The PW of costs for each alternative:

$$\text{PW Costs (I)} = \$12{,}500$$
$$\text{PW Costs (II)} = \$11{,}000$$
$$\text{PW Costs (IV)} = \$16{,}800$$

The correct rank order is now do nothing, II, I, IV. Notice that the *modified* B/C ratio produces an order of comparison different from that yielded by the *conventional* version in Example 16–4.

Step 4 *Identify increment being considered.*

Step 5 *Calculate B/C ratio.*

Step 6 *Which alternative is better?*

		1st Iteration	2nd Iteration	3rd Iteration
Step 4	**Incremental Effects**	**(Do Nothing → II)**	**(II → I)**	**(II → IV)**
	ΔPlant construction cost	$11,000	$ 1500	$ 5800
	PW of ΔCosts	11,000	1500	5800
	ΔAnnual utility payment savings	700	−120	600
	ΔAnnual overcapacity revenue	550	150	−300
	ΔAnnual benefits of new jobs	750	−350	−250
	ΔAnnual O&M disbenefit	480	−360	−335
	PW of ΔBenefits	18,405	484	4662
Step 5	ΔB/C ratio (PW ΔB)/(PW ΔC)	1.67	0.32	0.80
Step 6	Is increment justified?	Yes	No	No

As an example of the calculations in the foregoing table, consider the third increment (II → IV).

ΔPlant construction cost	$= 16{,}800 - 11{,}000 = \$5800$
PW of ΔCosts	$= \$5800$
ΔAnnual utility payment savings	$= 1300 - 700 = \$600$

ΔAnnual overcapacity revenue $= 250 - 550 = -\$300$

ΔAnnual benefits of new jobs $= 500 - 750 = -\$250$

ΔAnnual O&M disbenefit $= 145 - 480 = -\$335$

PW of ΔBenefits $= (600 - 300 - 250 + 335)(P/A, 8\%, 45) = \$4662$

ΔB/C ratio, (PW ΔB)/(PW ΔC) $= 4662/5800 = 0.80$

When the modified version of the B/C ratio is used, Alt. II emerges as the recommended power plant design—just as it did when we used the conventional B/C ratio.

NOTE: A useful exercise for the student would be to develop a spreadsheet that can be used to solve this problem. Try it out!

OTHER EFFECTS OF PUBLIC PROJECTS

Four areas remain that merit discussion in describing the differences between government and nongovernment economic analysis: (1) financing government versus nongovernment projects, (2) the typical length of government versus nongovernment project lives, (3) quantifying and valuing benefits and disbenefits, and (4) the general effects of politics on economic analysis.

Project Financing

Governmental organizations and market-driven firms differ in the way investments in equipment, facilities, and other projects are financed. In general, firms rely on monies from investors (through stock and bond issuance), private lenders, and retained earnings from operations. These sources serve as the pool from which investment dollars for projects come. Management's job is to match financial resources with projects in a way that keeps the firm growing, results in an efficient and productive environment, and continues to attract investors and future lenders of capital.

On the other hand, the government sector often uses taxation and municipal bond issuance as the source of investment capital. In government, taxation and revenue from operations are adequate to finance only modest projects. However, public projects tend to be large in scale (roadways, bridges, etc.), which means that for many public projects 100% of the investment costs must be borrowed. To prevent excessive public borrowing and to assure timely debt repayment, the U.S. government, through constitutional and legislative channels, has restricted government debt. These restrictions include:

1. Local government bodies are limited in their borrowing to a specified percentage of the assessed property value in their taxation district.

2. For new construction, borrowed funds attained through the sale of bonds require the approval of local voters (sometimes by a two-thirds majority). For example, a $20 million bond proposition for a new municipal jail might increase property taxes in the city's tax district by $1.50 for every $1000 of assessed property value. These added tax revenues would then be used to retire the debt on the bonds.

3. Repayment of public debt must be made following a specific plan over a pre-set period of time. For monies borrowed by issuing bonds, interest payments and maturity dates are set at the time of issuance.

Limitations on the use and sources of borrowed monies make funding public sector projects much different from in the private sector. Private sector firms are seldom able to borrow 100% of required funds for projects, as can be done in the public sector, but at the same time, private entities do not face restrictions on debt retirement or the uncertainty of voter approval. Passing the bond proposition is the public *go-ahead*.

When projects are financed by bonds, it may become very difficult to shift funds between projects or to add funds for a project. Thus public agencies sometimes use a **design to cost** methodology, rather than minimizing life-cycle costs (the approach that has been used throughout the text). The goal is to design the best possible school, road, bridge, and so on that can be built without exceeding the fixed budgeted cost.

Project Duration

Government projects often have longer lives than those in the private sector. In the private sector, projects most often have a projected or intended life ranging between 5 and 15 years. Some markets and technologies change more rapidly and some more slowly, but a majority of projects fall in this interval. Complex advanced manufacturing technologies, like computer-aided manufacturing or flexible automated manufacturing cells, typically have project lives at the longer end of this range.

Government projects typically have lives in the range of 20 to 50 years (or longer). Typical projects include federal highways, city water/sewer infrastructure, county dumps, and state libraries and museums. These projects, by nature, have a longer useful life than a typical private-sector project. There are exceptions to this rule because private firms invest in facilities and other long-range projects, and government entities also invest in projects with shorter-term lives. But, in general, investment duration in the government sector is longer.

Government projects, because they tend to be long range and large scale, usually require substantial funding in the early stages. Highway, water/sewer, and dam projects can require millions of dollars in design, surveying, and construction costs. Therefore, it is in the best interest of decision makers who are advocates of such projects to spread that first cost over as many years as possible to reduce the annual cost of capital recovery. Using longer project lives to downplay the effects of a large first cost increases the desirability of the project, as measured by the B/C ratio. Another aspect closely associated with managing the size of the capital recovery cost in a B/C ratio analysis is the interest rate used for discounting. Lower interest rates reduce the capital recovery cost of having money tied up in a project. Example 16–6 illustrates the effects that project life and interest rate can have on the analysis and acceptability of a project.

EXAMPLE 16–6

Consider a project that has been approved by local voters to build a new junior high school, needed because of increased (and projected) population growth. Analyze the project with interest rates of 3, 10, and 15% and with horizons of 15, 30, and 60 years.

Building first costs (design, planning, and construction)	$10,000,000
Initial cost for roadway and parking facilities	5,500,000
First cost to equip and furnish facility	500,000
Annual operating and maintenance costs	350,000
Annual savings from rented space	400,000
Annual benefits to community	1,600,000

SOLUTION

With this project we examine the effect that varying project lives and interest rates have on the economic value of a public project. In each case, the formula is

$$\text{Conventional B/C ratio} = \frac{1,600,000 + 400,000}{(10,000 + 5,500,000 + 500,000)(A/P, i, n) + 350,000}$$

The B/C ratio for each combination of project life and interest rate is tabulated as follows:

Conventional Benefit–Cost Ratio for Various Combinations of Project Life and Interest Rate

Project Life (years)	Interest 3%	Interest 10%	Interest 15%
15	1.24	0.86	0.69
30	1.79	1.03	0.76
60	2.24	1.08	0.77

Build ←   → Do not build

From these numbers one can see the effect of project life and interest on the analysis and recommendation. At the lower interest rate, the project has B/C ratios above 1.0 in all cases of project life, while at the higher rate the ratios are all less than 1.0. At an interest rate of 10% the recommendation to invest changes from *no* at a life of 15 years to *yes* at 30 and 60 years. A higher interest rate discounts the benefits in later years more heavily, so that they may not matter. In this case, the benefits from Years 31 to 60 add only 0.01 to the B/C ratio at 15% and 0.05 at 10%. At 3% those benefits add 0.45 to the ratio. By manipulating these two parameters (project life and interest rate), it is possible to reach entirely different conclusions regarding the desirability of the project.

EXAMPLE 16–7 · (Example 16–6 Revisited)

Looking at the data in Example 16–6, develop a spreadsheet table and graph that illustrates changes to the B/C ratio as interest rate changes between 1% and 15% and project life varies from 5 to 60 years.

SOLUTION

Notice in Figure 16–3 the rightmost curve where values to the right and below represent a B/C ratio less than 1 and those to the left and above greater than 1. This line, called the efficient frontier, can be used to show decision makers the combinations of interest rate and project life that lead to a barely acceptable B/C ratio of 1. This figure provides much more information than the simple results table in Example 16–6. For instance, regardless of project life, the junior high school is not recommended if interest is above ∼10.5%. Also, if the life of the school is less than 10 years, the B/C ratio is not favorable.

	A	B	C	D	E	F	G	H	I	J	K	L	M
1	$10,000	building first cost											
2	5,500	initial roadway cost											
3	500	first cost to equip		=-PV($A9,B$8,($A$6+$A$5-$A$4))/($A$3+$A$2+$A$1)									
4	350	annual costs		all others copied from this reference cell									
5	400	annual savings											
6	1,600	annual benefit				**Conventional B/C Ratio**							
7						Project life (years)							
8	Interest	5	10	15	20	25	30	35	40	45	50	55	60
9	1%	0.50	0.98	1.43	1.86	2.27	2.66	3.03	3.39	3.72	4.04	4.35	4.64
10	2%	0.49	0.93	1.33	1.69	2.01	2.31	2.58	2.82	3.04	3.24	3.42	3.58
11	3%	0.47	0.88	1.23	1.53	1.80	2.02	2.22	2.38	2.53	2.65	2.76	2.85
12	4%	0.46	0.84	1.15	1.40	1.61	1.78	1.92	2.04	2.14	2.22	2.28	2.33
13	5%	0.45	0.80	1.07	1.29	1.45	1.59	1.69	1.77	1.83	1.88	1.92	1.95
14	6%	0.43	0.76	1.00	1.18	1.32	1.42	1.50	1.55	1.59	1.63	1.65	1.67
15	7%	0.42	0.72	0.94	1.09	1.20	1.28	1.34	1.37	1.40	1.42	1.44	1.45
16	8%	0.41	0.69	0.88	1.01	1.10	1.16	1.20	1.23	1.25	1.26	1.27	1.28
17	9%	0.40	0.66	0.83	0.94	1.01	1.06	1.09	1.11	1.12	1.13	1.14	1.14
18	10%	0.39	0.63	0.78	0.88	0.94	0.97	0.99	1.01	1.02	1.02	1.03	1.03
19	11%	0.38	0.61	0.74	0.82	0.87	0.90	0.91	0.92	0.93	0.93	0.93	0.94
20	12%	0.37	0.58	0.70	0.77	0.81	0.83	0.84	0.85	0.85	0.86	0.86	0.86
21	13%	0.36	0.56	0.67	0.72	0.76	0.77	0.78	0.79	0.79	0.79	0.79	0.79
22	14%	0.35	0.54	0.63	0.68	0.71	0.72	0.73	0.73	0.73	0.74	0.74	0.74

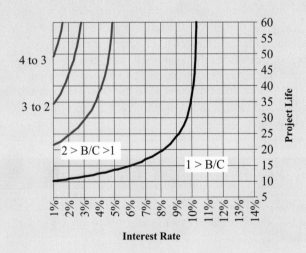

FIGURE 16–3

B/C ratios graph for Exp. 16–6

Examples 16–6 and 16–7 demonstrate that we must ensure that a long life and a low interest rate for a public project are truly appropriate and not chosen solely to make a marginal project look better.

Quantifying and Valuing Benefits and Disbenefits

The junior high school in Example 16–6 included annual benefits to the community of $1.6 million. If you were evaluating the school, how would you estimate this? Many public

TABLE 16–1 Example Benefits and Disbenefits for Public Investments

Public Project	Primary Benefits	Primary Disbenefits
New airport outside city	More flights, new businesses	Increased travel time to airport, more traffic on outer belt
Interstate bypass around town	Quicker commute times, reduced congestion on surface roads	Lost sales to businesses on surface roads, lost agricultural lands
New metro subway system	Faster commute times, less pollution	Lost jobs due to bus line closing, less access to service (fewer stops)
Creation of a city waste disposal facility versus sending waste out of state	Less costly; faster and more responsive to customers	Objectionable sight and smells, lost market value to homeowners, lost forestland
Construction of a nuclear power plant	Cheaper energy costs, new industry in area	Environmental risks

sector projects like the junior high school and the examples in Table 16–1 have consequences that are difficult to state in monetary terms. These include both project benefits and **disbenefits** (negative aspects of the project).

First the number of people affected by the project have to be counted—now and through the project's horizon. Then a dollar value for each person is required. For the junior high school it may be easy to estimate the number of students. But how much better will the educational outcomes be with the new school, and how valuable is that improvement?

On the other hand, consider the levees around New Orleans that needed to be rebuilt in the aftermath of Katrina. An economic evaluation of the different alternatives required estimating the number of residents that would be protected by improving the levees in New Orleans, which is extremely difficult. This required estimating not only how the rebuilding of the city might progress but also storm surges from future hurricanes, whose frequency and intensity may be changing. Once the number of people and homes had been estimated over the next 30 to 100 years, it was necessary to put a value on property, on disrupted lives, and on human lives.

While many individuals find it difficult to put a dollar value on a human life, there are many public projects whose main intent is to reduce the number of deaths due to floods, cancer, auto accidents, and other causes. Those projects are often justified by the value of preventing deaths. Thus, valuing human lives is an inescapable part of public-sector engineering economy.

Because the benefits and disbenefits of public projects are often difficult to quantify and value, the estimated values will have more uncertainty than is typical for private-sector projects. Thus those who favor a project and those who oppose it will often push to have values used that support their position.

Project Politics

To some degree political influences are felt in nearly every decision made in any organization. Predictably, some individual or group will support particular interests over competing views. This actuality exists in both firms and government organizations. In government the effects of politics are continuously felt at all levels because of the large-scale and multipurpose nature of projects, because government decision making involves the use of the citizenry's common pool of money, and because individuals and groups have different values and views. For example, how important is economic development relative to environmental protection?

The guideline for public decision making, as set forward in the Preamble to the Constitution, is to *promote the general welfare* of citizens. However, it is impossible to please everyone all the time. The term "general welfare" implies that the architects of this document understood that the political process would produce opposing viewpoints, but at the same time they empowered decision makers to act in a representative way.

Since government projects tend to be large in scale, the time required to plan, design, fund, and construct them is usually several years. However, the political process tends to produce government leaders who support short-term decision making (because many government terms of office, either elected or appointed, are relatively short). Therein lies another difference between firm and government decision making—short-term decision making, long-term projects.

Because government decision makers are in the public eye more than those in the private sector, governmental decisions are generally more affected by "politics." Thus, the decisions that public officials make may not always be the best from an *overall* perspective. If a particular situation exposes a public official to ridicule, he may choose an expedient action to eliminate negative exposure (whereas a more careful analysis might have been better). Or, such a decision maker may placate a small, but vocal, political group over the interest of the majority of citizens by committing funds to a favored project (at the expense of other better projects). Or, a public decision maker may avoid controversy by declining to make a decision on an important, but politically charged, issue (whereas it would be in the overall interest of the citizenry if action were taken). Indeed, the role of politics in government decision making is more complex and far ranging than in the private sector.

EXAMPLE 16–8   (Example 16–4 Revisited)

Consider again Example 16–4, where we evaluated designs for two coal-burning power plants. What social, political, environmental, regulatory, or business considerations might come into play as the designed plants move toward construction and operation?

Solution

Remember that government projects are often opposed and supported by different groups in the populace. Thus, decision makers become very aware of potential political aspects when they are considering such projects.

For the electric power plant decision, several political considerations may affect any evaluation of funding this project.

- The governor has been a strong advocate of workers' rights and has received abundant campaign support from organized labor (which is especially important in an industrialized state). By championing this project, the governor should be seen as pro-labor, thereby benefiting his bid for reelection, even if the project is not funded.
- The regulated electric utility providers in the state are strongly against this project, claiming that it would directly compete with them and take away some of their biggest customers. The providers have a strong lobby and key contacts with the state's utilities commission. A senior state senator has already protested that this project is the first step toward "rampant socialism in this great state."
- Business leaders in the municipalities where the two facilities would be constructed favor the project because it would create more jobs and increase the tax base. These leaders promote the project as a win–win opportunity for government and industry, where the state can benefit by reducing costs, and the electric utilities can improve their service by focusing more effectively on residential customers and their needs.
- The lieutenant governor is promoting this project, proclaiming that it is an excellent example of "initiating proactive and creative solutions to the problems that this state faces."
- Federal and state regulatory agencies are closely watching this project with respect to the Clean Air Act. Speculation is that the state plans to use a high-sulfur grade of local coal exclusively. Thus "stack scrubbers" would be required, or the high-sulfur coal would have to be mixed with lower-sulfur coal imported from other states to bring the overall air emissions in line with federal standards. The governor is using this opportunity to make the point that "the people of this state don't need regulators to tell us if we can use our own coal!"
- The state's coal operators and mining unions strongly support this project. They see the increased demands for coal and the governor's pro-labor advocacy as very positive. They plan to lobby the legislature strongly in favor of the project.
- Land preservation and environmental groups are strongly opposing the proposed project. They have studied the potential negative impacts of this project on the land and on water and air quality, as well as on the ecosystem and wildlife, in the areas where the two facilities would be constructed. Environmentalists have started a public awareness campaign urging the governor to act as the "chief steward" of the state's natural beauty and resources.

Will the project be funded? We can only guess. Clearly, however, we can see the competing influences that can be, and often are, part of decision making in the public sector.

SUMMARY

Economic analysis and decision making in government is notably different from these processes in the private sector because the basic objectives of the public and private sectors are fundamentally different. Government investments in projects seek to maximize benefits to the *greatest number of citizens,* while minimizing the *disbenefits to citizens* and *costs to the government.* Private firms, on the other hand, are focused primarily on maximizing stockholder wealth.

Several factors, not affecting private firms, enter into the decision-making process in government. The source of capital for public projects is limited primarily to taxes and bonds. Government bonds issued for project construction are subject to legislative restrictions on debt not required for private firms. Also, raising tax and bond monies involves sometimes long and politically charged processes not present in the private sector. In addition, government projects tend to be larger in scale than the projects of competitive firms and the government projects affect many more people and groups in the population. The benefits and disbenefits to the many people affected are difficult to quantify and value, which is unlike the private sector, where products and services are sold and the revenue to the firm is clearly defined. All these factors slow down the process and make investment decision analysis more difficult for government decision makers than for those in the private sector. Another difference between the public and private sectors lies in how the interest rate (MARR) is set for economic studies. In the private sector, considerations for setting the rate include the cost of capital and opportunity costs. In government, establishing the interest rate is complicated by uncertainty in specifying the cost of capital and the issue of assigning opportunity costs to taxpayers or to the government.

The benefit–cost ratio is widely used to evaluate and justify government-funded projects. This measure of merit is the ratio of the equivalent worth of benefits to the equivalent worth of costs. This ratio can be calculated by PW, AW, or FW methods. A B/C ratio *greater than 1.0* indicates that a project should be invested in if funding sources are available. For considering *mutually exclusive alternatives,* an incremental analysis is required. This method results in the recommendation of the project with the highest investment cost that can be incrementally justified. Two versions of the B/C ratio, the *conventional* and *modified* B/C ratios, produce identical recommendations. The conventional B/C ratio treats annual operating and maintenance costs as a cost in the denominator, while the modified B/C ratio subtracts those costs from the benefits in the numerator.

PROBLEMS

Objective and Viewpoint

16-1 Public-sector economic analysis and decision making is often called "a multi-actor or multi-stakeholder decision problem." Explain.

16-2 Compare the general underlying objective of public decision making versus private decision making. List two ethical issues unique to each.

16-3 The text recommends a viewpoint that is appropriate in public decision making. What is suggested? What example is given to highlight the dilemma of viewpoint in public decision making? Provide another example.

16-4 In government projects, what is meant by the phrase "most of the benefits are local"? What conflict does this create for the federal government in funding the projects from public monies?

16-5 List the potential costs, benefits, and disbenefits that should be considered in evaluating a potential nuclear power plant. What stakeholder viewpoints will need to be considered?

16-6 A municipal landfill and incineration facility is planned.
(*a*) Name at least three benefits, three disbenefits, and three costs. What stakeholder viewpoints will need to be considered?
(*b*) Burning landfill materials produces environmental and health issues. List the primary issues in each case.

16-7 An interstate bypass will completely circle a city.
(*a*) Name at least three benefits, three disbenefits, and three costs. What stakeholder viewpoints will need to be considered?
(*b*) Routes for the bypass use private land seized under *eminent domain*. Defend your position on the ethics of eminent domain.

16-8 A light rail system will connect the airport, the city center, and a cluster of high-density housing on the other side of the river. Name at least three benefits, three disbenefits, and three costs. What stakeholder viewpoints will need to be considered?

16-9 Improvements at a congested intersection require that the government acquire the properties on the four corners: two gas stations, a church, and a bank. Construction will take a year, and the costs will be shared 70% by the state and 30% by the city. Traffic through the intersection is mainly commuters, local residents, and deliveries to and by local businesses. There is some through traffic from other parts of the metropolitan area.

Identify the benefits, disbenefits, and costs that must be considered in evaluating this project. From the city's viewpoint, which must be included? From the state's viewpoint? What viewpoint should be used to evaluate this project?

16-10 The state may eliminate a railroad grade crossing
(E) by building an overpass. The new structure, together with the needed land, would cost $4 million. The analysis period is assumed to be 30 years because either the railroad or the highway above it will be relocated by then. Salvage value of the bridge (actually, the net value of the land on either side of the railroad tracks) 30 years hence is estimated to be $750,000. A 5% interest rate is to be used.

About 1000 vehicles per day are delayed by trains at the grade crossing. Trucks represent 30%, and 70% are other vehicles. Time for truck drivers is valued at $25 per hour and for other drivers at $7.50 per hour. Average time saving per vehicle will be 2 minutes if the overpass is built. No time saving occurs for the railroad.

The railroad spends $75,000 annually for crossing guards. During the preceding 10-year period, the railroad has paid out $1.0 million in settling lawsuits and accident cases related to the grade crossing. The proposed project will entirely eliminate both these expenses. The state estimates that the new overpass will save it about $8000 per year in expenses directly due to the accidents. The overpass, if built, will belong to the state.

Should the overpass be built? If the overpass is built, how much should the railroad be asked to contribute to the state as its share of the $4 million construction cost?

Selecting Rate

16-11 Discuss the alternative concepts that can be employed when setting the discounting rate for economic analysis in the public sector. What is the authors' final recommendation for setting this rate?

16-12 Is the 7% interest rate specified in OMB A94 a
(A) real or a nominal interest rate? Should it be used with costs expressed as constant-value dollars or with costs inflated using carefully selected inflation rates?

16-13 The city's landfill department has a capital budget of $600,000.

(a) What is the government's opportunity cost of capital if it has the following independent projects to consider? What does this indicate about which projects should be recommended?

(b) If B/C ratios at 7% are the basis, what projects are recommended?

(c) Which recommendation is better?

Project	First Cost	Rate of Return (%)	B/C Ratio at 7%
A	$100,000	23	1.30
B	200,000	22	1.40
C	300,000	17	1.50
D	200,000	19	1.35
E	100,000	18	1.56

16-14 The state's fish and game department has a capi-
(A) tal budget of $9 million. What is the government's opportunity cost of capital if it has the following projects to consider? What does this indicate about which projects should be recommended?

Project	First Cost	Rate of Return (%)	B/C Ratio at 7%
A	$2,000,000	9	1.23
B	1,000,000	14	1.42
C	2,000,000	10	1.17
D	3,000,000	16	1.45
E	2,000,000	13	1.56
F	3,000,000	15	1.35
G	3,000,000	12	1.32
H	1,000,000	11	1.26

16-15 A municipal bond has a face value of $1000. Interest of $35 is paid every 6 months. The bond has a life of 10 years. What is the effective rate of interest on this bond? Is this rate adjusted for inflation? What is the municipal government's cost of capital for this bond? Estimate the rate, considering all levels of government.

16-16 A proposed bridge would cost $4 million to build
(A) and $180,000 per year in maintenance. The bridge
(E) should last 40 years. Benefits to the driving public are estimated to be $900,000 per year. Damages (not paid) to adjacent property owners due to noise are estimated to be $250,000 per year. The interest rate that should be used to evaluate this project is unclear.

(a) Calculate the breakeven annual interest rate that results in a B/C ratio of 1.

(b) What is the breakeven interest rate if the noise disbenefits are included? Should they be included?

Contributed by D. P. Loucks, Cornell University

B/C and Modified Ratio

16-17 What is the essential difference between the *conventional* and *modified* versions of the benefit–cost ratio? Is it possible for these two measures to provide conflicting recommendations regarding invest/do-not-invest decisions?

16-18 Consider the following investment opportunity:
(A)

Initial cost	$100,000
Additional cost at end of Year 1	150,000
Benefit at end of Year 1	0
Annual benefit per year at end of Years 2–10	20,000

With interest at 5%, what is the benefit–cost ratio for this project?

16-19 Calculate the conventional and modified benefit–cost ratios for the following project.

Required first costs	$1,500,000
Annual benefits to users	$500,000
Annual disbenefits to users	$25,000
Annual cost to government	$180,000
Project life	40 years
Interest rate	4%

16-20 A government agency has estimated that a flood con-
(A) trol project has costs and benefits that are parabolic,

according to the equation

$$(\text{Present worth of benefits})^2$$
$$- 22(\text{Present worth of cost}) + 44 = 0$$

where both benefits and costs are stated in millions of dollars. What is the present worth of cost for the optimal size project?

16-21 For the data given in Problem 16-19, for handling benefits and costs, demonstrate that the calculated B/C ratio is the same using each of the following methods: present worth, annual worth, and future worth.

16-22 Chungyang Dam is being constructed across the
(A) Hungshui River in southern China. The dam will pro-
(G) duce electricity to serve over 500,000 people in the region. The initial cost is 3.7 billion yuan and annual operating costs are 39.2 million yuan. A major overhaul of the electric generation facilities estimated to cost 650 million yuan will occur at the end of Year 25. The dam and generating plant have no salvage value, but must be torn down and removed at the end of Year 50 for 175 million yuan. Ishan Electric has a MARR of 10%.

(a) What annual benefit in yuan is needed for a B/C ratio of 1?

(b) Use the Internet and current exchange rates to find the annual benefit in U.S. dollars.

(c) What are the top 3 sources for electricity generated from renewable power in the U.S.? List three benefits and three disbenefits for each source.

Contributed by Paul R. McCright, University of South Florida

Incremental Analysis

16-23 The city engineer has prepared two plans for roads in the city park. Both plans meet anticipated requirements for the next 40 years. The city's minimum attractive rate of return is 7%.

Plan *A* is a three-stage development program: $300,000 is to be spent now followed by $250,000 at the end of 15 years and $300,000 at the end of 30 years. Annual maintenance will be $75,000 for the first 15 years, $125,000 for the next 15 years, and $250,000 for the final 10 years.

Plan *B* is a two-stage program: $450,000 is required now, followed by $50,000 at the end of 15 years. Annual maintenance will be $100,000 for the first 15 years and $125,000 for the subsequent years. At the end of 40 years, this plan has a salvage value of $150,000.

Use a conventional benefit–cost ratio analysis to determine which plan should be chosen.

16-24 The Highridge Water District needs an additional
Ⓐ supply of water from Steep Creek. The engineer has selected two plans for comparison:

Gravity plan: Divert water at a point 10 miles up Steep Creek and carry it through a pipeline by gravity.

Pumping plan: Divert water at a point near the district and pump it through 2 miles of pipeline. The pumping plant can be built in two stages: half now and half 10 years later.

	Gravity	Pumping
Initial investment	$2,800,000	$1,400,000
Investment in 10th year	0	200,000
Operation, maintenance, replacements, per year	10,000	25,000
Average annual power cost		
First 10 years	0	50,000
Next 25 years	0	100,000

Use a 35-year analysis period and 5% interest. Salvage values can be ignored. Use the conventional benefit–cost ratio method to select the more economical plan.

16-25 The Arkansas Department of Transportation may build a new highway between Texarkana and Fort Smith, currently a distance of 181 miles. Design 1 is a four-lane highway built entirely on the existing route. Design 2 includes a significant rerouting through a mountainous region that would reduce the mileage to 166 miles. Design 3 is a fully access-controlled interstate-quality highway with more rerouting, which would reduce the total mileage to 148 miles.

Benefits for this project depend on mileage saved times the number of vehicles, plus the estimated value for the larger number of trips that will occur with the shorter and faster routes. The estimated benefits and costs of the three potential designs are shown in the table. Doing nothing yields no costs and no benefits. Using incremental analysis for the B/C ratio, a planning horizon of 75 years, and a MARR of 6%, which design would you recommend?

	Initial Cost	Annual Maintenance Cost	Annual Benefit
Design 1	$ 456M	$17M	$107M
Design 2	810M	28M	198M
Design 3	1552M	58M	287M

Contributed by Paul R. McCright, University of South Florida

16-26 Two different routes are being considered for a
Ⓐ mountain highway. The **high road** would require building several bridges and would navigate around the highest mountain points, thus requiring more roadway. The **low road** would construct several tunnels for a more direct route through the mountains. Projected travel volume for this new section of road is 2500 cars per day. Use the *modified* B/C ratio to determine which alternative should be recommended. Assume that project life is 45 years and $i = 6\%$.

	High Road	Low Road
Construction cost/mile	$200,000	$450,000
Number of miles required	35	10
Annual benefit/car-mile	$0.015	$0.045
Annual O&M costs/mile	$2,000	$10,000

16-27 A 50-meter tunnel must be constructed for a new city aqueduct. One alternative is to build a full-capacity tunnel now for $500,000. The other alternative is to build a half-capacity tunnel now for $300,000 and then to build a second parallel half-capacity tunnel 20 years hence for $300,000. The cost to repair the tunnel lining every 10 years is $20,000 for the full-capacity tunnel and $16,000 for each half-capacity tunnel.

Determine whether the full-capacity tunnel or the half-capacity tunnel should be constructed now. Solve the problem by the conventional benefit–cost ratio analysis, using a 4% interest rate and a 50-year analysis period. There will be no tunnel lining repair at the end of the 50 years.

16-28 A two-lane highway between two cities, 10 miles
Ⓐ apart, is to be converted to a four-lane divided freeway. The average daily traffic (ADT) on the new freeway is forecast to average 20,000 vehicles per day over the next 20 years. Trucks represent 5% of the total traffic. Annual maintenance on the existing

highway is $1500 per lane-mile. The existing accident rate is 4.58 per million vehicle-miles (MVM). Three alternate plans are under consideration.

Plan A: Improve along the existing development by adding two lanes adjacent to the existing lanes at a cost of $450,000 per mile. This will reduce auto travel time by 2 minutes and truck travel time by 1 minute. The estimated accident rate is 2.50 per MVM. Annual maintenance is $1250 per lane-mile.

Plan B: Improve the highway along the existing alignment with grade improvements at a cost of $650,000 per mile. Plan *B* adds two lanes and would reduce auto and truck travel time by 3 minutes each. The accident rate on this improved road is estimated to be 2.40 per MVM. Annual maintenance is $1000 per lane-mile.

Plan C: Construct a new freeway on a new alignment at a cost of $800,000 per mile. This plan would reduce auto travel time by 5 minutes and truck travel time by 4 minutes. Plan *C* is 0.3 mile longer than *A* or *B*. The estimated accident rate for *C* is 2.30 per MVM. Annual maintenance is $1000 per lane-mile. Plan *C* includes abandoning the existing highway with no salvage value.

Incremental operating cost

Autos	6¢ per mile
Trucks	18¢ per mile

Time saving

Autos	3¢ per minute
Trucks	15¢ per minute

Average accident cost $1200

If a 5% interest rate is used, which of the three proposed plans should be adopted?

16-29 Evaluate these mutually exclusive alternatives with a horizon of 20 years and a MARR of 15%.

	A	B	C
Initial investment	$9500	$18,500	$22,000
Annual savings	3200	5,000	9,800
Annual costs	1000	2,750	6,400
Salvage value	6000	4,200	14,000

Use each of these approaches:

(*a*) *Conventional* B/C ratio

(*b*) *Modified* B/C ratio

(*c*) Present worth analysis

(*d*) Internal rate of return analysis

(*e*) Payback period

16-30 The Fishery and Wildlife Agency of Ireland is considering four mutually exclusive design alternatives for a major salmon hatchery. This agency of the Irish government uses the following B/C ratio for decision making (EW = EUAW):

$$\text{B/C ratio} = \frac{\text{EW(Net benefits)}}{\text{EW(Capital recovery cost)} + \text{EW(O\&M cost)}}$$

Using an interest rate of 8% and a project life of 30 years, recommend which design is best.

(Values in 1000s)

	Irish Fishery Design Alternatives			
	A	B	C	D
First cost	$9500	$12,500	$14,000	$15,750
Annual benefits	2200	1,500	1,000	2,500
Annual O&M costs	550	175	325	145
Annual disbenefits	350	150	75	700
Salvage value	1000	6,000	3,500	7,500

16-31 Six mutually exclusive investments have been identified for evaluation by means of the benefit–cost ratio method. Assume a MARR of 10% and an equal project life of 25 years for all alternatives.

	Mutually Exclusive Alternatives					
Annualized	1	2	3	4	5	6
Net costs to sponsor ($M)	15.5	13.7	16.8	10.2	17.0	23.3
Net benefits to users ($M)	20.0	16.0	15.0	13.7	22.0	25.0

(*a*) Use annual worth and the B/C ratio to identify the best alternative.

(*b*) Is there an easier way to select the best alternative?

(*c*) If this were a set of *independent* alternatives, how would you conduct a comparison?

16-32 A section of state highway needs repair. At present, the traffic volume is low and few motorists would benefit. However, traffic is expected to increase, with resulting increased motorist benefits. The repair work will produce benefits for 10 years after it is completed.

Should the road be repaired and, if so, when should the work be done? Use a 15% MARR.

(Costs in 1000s)

Year	Repair Now	Repair in 2 Years	Repair in 4 Years	Repair in 5 Years
0	−$150			
1	5			
2	10	−$150		
3	20	20		
4	30	30	−$150	
5	40	40	40	−$150
6	50	50	50	50
7	50	50	50	50
8	50	50	50	50
9	50	50	50	50
10	50	50	50	50
11	0	50	50	50
12	0	50	50	50
13	0	0	50	50
14	0	0	50	50
15	0	0	0	50

16-33 The local highway department is analyzing reconstruction of a mountain road. The vehicle traffic increases each year, hence the benefits to the motoring public also increase. Based on a traffic count, the benefits are projected as follows:

Year	End-of-Year Benefit
2017	$10,000
2018	12,000
2019	14,000
2020	16,000
2021	18,000
2022	20,000
	and so on, increasing $2000 per year

The reconstructed pavement will cost $275,000 when it is installed and will have a 15-year useful life. The construction period is short, hence a beginning-of-year reconstruction will result in the listed end-of-year benefits. Assume a 6% interest rate. The reconstruction, if done at all, must be done not later than 2023. When is the first year that this project is justified?

Challenges for Public Sector

16-34 Describe how a decision maker can use each of the following to "skew" the results of a B/C ratio analysis in favor of his or her own position on funding projects.

(*a*) Conventional versus modified ratios

(*b*) Interest rates

(*c*) Project duration

(*d*) Benefits, costs, and disbenefits

(*e*) When presenting your results, is it ethical to pick a method that matches your preferred choice?

16-35 Briefly describe your sources and methods for estimating the value of

(*a*) a saved hour of commuting time.

(*b*) converting 15 miles of unused railroad tracks near a community of 300,000 into a new bike path.

(*c*) reducing annual flood risks from the Mississippi River for St. Louis by 5%.

(*d*) a new engineering or engineering technology major at your university.

16-36 If you favored Plan *B* in Problem 16-23, what value of MARR would you use in the computations? Explain.

16-37 Discuss potential data sources and methods for estimating each of the costs, benefits, and disbenefits identified in Problem 16-5 for the nuclear power plant.

16-38 For the municipal landfill and incineration facility in Problem 16–6, discuss potential data sources and methods for estimating each of the benefits, disbenefits, and costs.

16-39 For the interstate bypass in Problem 16-7, discuss potential data sources and methods for estimating each of the benefits, disbenefits, and costs.

16-40 For the light rail system in Problem 16-8, discuss potential data sources and methods for estimating each of the benefits, disbenefits, and costs.

16-41 Discuss potential data sources and methods for estimating each of the costs, benefits, and disbenefits identified in Problem 16-9 for the congested intersection.

16-42 Think about a major government construction project under way in your state, city, or region. Are the decision makers who originally analyzed and initiated the project currently in office? How can politicians use "political posturing" with respect to government projects?

16-43 Big City Carl, a local politician, is pushing a new dock and pier system at the river to attract commerce. A committee appointed by the mayor (an opponent of Carl's) has developed the following estimates.

Cost to remove current facilities	$ 750,000
New construction costs	2,750,000
Annual O & M costs	185,000
Annual benefits from new commerce	550,000
Annual disbenefits to sportsmen	35,000
Project life	20 years
Interest rate	8%

(a) Using the *conventional* B/C ratio, determine whether the project should be funded.

(b) After studying the numbers given by the committee, Big City Carl argued that the project life should be *at least* 25 years and more likely closer to 30 years. How did he arrive at this estimate, and why is he making this statement?

(c) If Carl suggests that each number is 5% "better," then what B/C ratio would he support?

16-44 The federal government proposes to construct a multipurpose water project to provide water for irrigation and municipal use. In addition, there are flood control and recreation benefits. The benefits are given below. The annual benefits are one-tenth of the decade benefits. The operation and maintenance cost is $15,000 per year. Assume a 50-year analysis period with no net project salvage value.

(a) If an interest rate of 5% is used, and a benefit–cost ratio of unity, what capital expenditure can be justified to build the water project now?

(b) If the interest rate is changed to 8%, how does this change the justified capital expenditure?

(Benefits in 1000s)

| Purpose | Decades | | | | |
	First	Second	Third	Fourth	Fifth
Municipal	$ 40	$ 50	$ 60	$ 70	$110
Irrigation	350	370	370	360	350
Flood control	150	150	150	150	150
Recreation	60	70	80	80	90
Totals	$600	$640	$660	$660	$700

Minicases

16-45 Research and report on how an agency of your state evaluates public project proposals. What interest rate and what economic measures are used? Are factors that include life-cycle costs or environmental effects included?

16-46 Research and report on how an agency of your municipality evaluates public project proposals. What interest rate and what economic measures are used?

Data for 16-47 to 16-49 adds detail to the chapter-opening vignette. This material was contributed by *William R. Truran, U.S. Department of Defense, and Peter A. Cerenzio, Cerenzio & Panaro Consulting Engineers*

Daily and intermediate cover material is 20% of the landfill's usable space. Density of solid waste is 1500 pounds per cubic yard. LFG recovery rates:

3000 cubic feet per ton for municipal solid waste (MSW)

1500 cubic feet per ton for construction and demolition waste (C&D)

Assume waste composition is 80% MSW and 20% C&D. Methane content in the LFG is 50% for MSW and 20% for C&D.

Heating value

For methane: 1,030,000 BTU/1000 cubic feet
For fuel oil: 138,800 BTU/gallon

Assumed furnace efficiency is 88% for methane and 82% for fuel oil.

Cost of fuel oil is $2.50/gallon.

Heating load for residential dwelling is
100 million BTU per year.

1.17×10^4 BTU per kilowatt-hours

Methane BTUs valued at equivalence to $0.05/kWh

16-47 An economic analysis is needed for a new municipal solid waste landfill. This includes determining the potential economic benefit from the gas that is generated when solid waste decomposes. The landfill is proposed at 14 acres, with a design capacity of 1 million cubic yards of capacity. The final capping system will require a 3-foot layer over the 14 acres, and the waste flow rate is 120,000 tons per year. Calculate:

(a) The life of the landfill.
(b) The average annual methane production (assume all methane production has ceased by 15 years after landfill closure).
(c) The dollar value of annual methane converted to electricity (neglect collection and energy production costs).

16-48 A developer has proposed a 650-unit residential
G development adjacent to the landfill. The proposal includes using the gas generated from the landfill to heat the homes (and mitigate odors). To determine the economic feasibility of the proposal, the value of the gas for heating purposes must be determined. Determine whether the quantity is sufficient for heating the development. Is this economically more attractive than using fuel oil? Is it operationally feasible? What impact might the project have on the environment? Are there any disbenefits from the project to the homeowners?

16-49 A 4.6-acre landfill must be evaluated for economic
G viability. As part of the cost–benefit analysis, the cost to extract and treat the landfill gas must be determined. Using the following, design a landfill gas extraction system and estimate the cost to implement such a system:

Landfill dimensions are 1000 feet by 200 feet.
Area of influence of a landfill gas extraction well is a 50-foot radius.
Cost of well construction is $3000 each.
Cost of wellhead (necessary for each well) is $2500 each.
Cost of collection header piping 8″ diameter HDPE is $35/linear foot.

Cost of condensate knockouts (located at low points) is $5000 each.
Cost of blower/flare station is $500,000.

CASES

The following cases from *Cases in Engineering Economy* (www.oup.com/us/newnan) are suggested as matched with this chapter.

CASE 16 **Great White Hall**
Proposal comparison using B/C analysis for RFP with unclear specifications.

CASE 41 **Metropolitan Highway**
Realistic variety of benefits and costs. Requires that assumptions be made.

CASE 40 **Olympic Bid Perspectives**
Public sector with data and questions from three perspectives.

CASE 42 **Protecting the Public**
P(injury) and fraction of contact wearers must be "guesstimated." Considers opportunity losses to other venues in visitor-days.

CASE 43 **Bigstate Bridging the Gap**
Difficulty ranges from comparing three alternative bridges with different lives to considering growth in traffic and uncertain construction costs.

CASE 44 **Sunnyside—Up or Not?**
Uncertain growth rates over 30 years and setting utility rates.

CASE 45 **Transmission Intertie**
Electric power project with primary and secondary benefits.

CASE 47 **Bigstate Highway Department**
Capital budgeting including mutually exclusive alternatives. Includes uncertain value of a life.

CASE 48 **Dot Puff Project Selection**
Public-sector capital budgeting with added constraint on man-years required.

17

ACCOUNTING AND ENGINEERING ECONOMY

A Tale of Three Engineers

Accounting is more than bookkeeping. Accounting analyzes and reports on the amount and availability of an organization's or an individual's financial resources. Understanding basic accounting is important for your professional development as an engineer as well as the management of your personal finances. Three examples demonstrate how engineers use accounting in both their careers and personal lives.

Moving Up the Career Ladder. Our first engineer excelled technically and was quickly asked to accept project management and then line management duties. The engineer had the responsibility for both justifying engineering decisions and project budgets. Quantifying and accounting for financial performance and efficiency was an essential component of the position. Whether you work for a large, medium, or small firm, enter academia, or form your own company, knowledge of accounting will help you obtain a promotion, as it did for our first engineer.

Investing for the Future. Our second engineer used accounting for personal finances. Not only are monthly and annual expenses tracked, but there is a budget and a balance sheet. The fundamental accounting equation "Assets = Liabilities + Equity" still applies. This engineer had a student loan (liability) immediately after graduation, and now has a mortgage (liability) on a house (asset). This engineer's savings plan includes the employer's defined contribution plan. This engineer planned finances immediately after graduation and in 5-year increments. Personal expenses and income are regularly checked against a short- and long-term strategy. Savings and investments included a down payment for a house, an emergency fund, and now focuses on a long-term retirement nest egg. Knowledge of accounting is being applied to manage, control, and secure a financial future. (Appendix 9A introduces investing and defined benefit and defined contribution plans. Appendix 10A discusses how diversification reduces risk. Appendix 12A introduces budgeting.)

Engineering Entrepreneurs. Our third engineer is actually a group of engineering students from the University of California, Davis. After the unexpected death of their research advisor, the students formed a small corporation and applied for funding from a government agency. They believed they had an excellent idea and submitted a proposal and budget justification. Two months later, the student business received an award letter for over $1 million. The key to their successful bid was an innovative idea, a sound technical plan and budget, and the accounting skills to manage the finances and required reporting.

What happened to those students? They finished their engineering degrees. One of them became a faculty member and teaches engineering economics. She shares with students that engineers are inherently technical, but when armed with financial knowledge and management skills, engineering students can become great entrepreneurs. ■ ■ ■

Contributed by Jani Macari Pallis, University of Bridgeport and Cislunar Aerospace, Inc.

QUESTIONS TO CONSIDER

1. How can you develop accounting or financial analysis skills before you enter the workforce?
2. Accounting and ethics both played large roles in the last fiscal crisis and in many smaller scandals. How can knowledge of accounting help you guard against ethical lapses?
3. Create annual budgets for the next 5 years. Use these budgets to connect your current and planned assets, liabilities, and equity.

After Completing This Chapter...

The student should be able to:

- Describe the links between engineering economy and accounting.
- Describe the objectives of general accounting, explain what financial transactions are, and show how they are important.
- Use a firm's balance sheet and associated financial ratios to evaluate the firm's health.
- Use a firm's income statement and associated financial ratios to evaluate the firm's performance.
- Use traditional absorption costing to calculate product costs.
- Understand the greater accuracy in product costs available with activity-based costing (ABC).

Key Words

absorption costing	equity	profit and
accounting data	fundamental	loss statement
assets	accounting equation	profit margin
balance sheet	general accounting	quick assets
business transaction	income statement	quick ratio
cost accounting	indirect cost	working capital
current ratio	interest coverage	
direct cost	liabilities	

Engineering economy focuses on the financial aspects of projects, while accounting focuses on the financial aspects of firms. Thus the application of engineering economy is much easier if one has some understanding of accounting principles. In fact, one important accounting topic, depreciation, was the subject of an earlier chapter.

THE ROLE OF ACCOUNTING

Accounting data are used to value capital equipment, to decide whether to make or buy a part, to determine costs and set prices, to set indirect cost rates, and to make product mix decisions. Accounting is used in private-sector firms and public-sector agencies, but for simplicity this chapter uses "the firm" to designate both. Accountants track the costs of projects, products, and services, which are the basis for estimating future costs and revenues.

The engineering economy, accounting, and managerial functions are interdependent. As shown in Figure 17–1, data and communications flow between them. Whether carried out by a single person in a small firm or by distinct divisions in a large firm, all are needed.

- Engineering economy analyzes the economic impact of design alternatives and projects over their life cycles.
- Accounting determines the dollar impact of past decisions, reports on the economic viability of a unit or firm, and evaluates potential funding sources.
- Management allocates available investment funds to projects, evaluates unit and firm performance, allocates resources, and selects and directs personnel.

Accounting	**Management**	**Engineering Economy**
About past	About past and future	About future
Analyzing	Capital budgeting	Feasibility of alternatives
Summarizing	Decision making	Collect & analyze data
Reporting	Setting goals	Estimate
Financial indicators	Assessing impacts	Evaluate projects
Economic trends	Analyzing risk	Recommend
Cost acquisitions	Planning	Audit
	Controlling	Identify needs
	Record keeping	Trade-offs & constraints

← Data and Communication →   ← Data and Communication →
← Budgeting   Data and Communication   Estimating →

FIGURE 17–1 The accounting, managerial, and engineering economy functions.

Accounting for Business Transactions

A **business transaction** involves two parties and the exchange of dollars (or the promise of dollars) for a product or service. Each day, millions of transactions occur between firms and their customers, suppliers, vendors, and employees. Transactions are the lifeblood of the business world and are most often stated in monetary terms. The accounting function records, analyzes, and reports these exchanges.

Transactions can be as simple as payment for a water bill, or as complex as the international transfer of millions of dollars of buildings, land, equipment, inventory, and other assets. Also, with transactions, one business event may lead to another—all of which need to be accounted for. Consider, for example, the process of selling a robot or a bulldozer. This simple act involves several related transactions: (1) releasing equipment from inventory, (2) shipping equipment to the purchaser, (3) invoicing the purchaser, and finally (4) collecting from the purchaser.

Transaction accounting involves more than just reporting: it includes finding, synthesizing, summarizing, and analyzing data. For the engineering economist, historical data housed in the accounting function are the foundation for estimates of future costs and revenues.

Most accounting is done in nominal or *stable* dollars. Higher market values and costs due to inflation are less objective than cost data, and with a going concern, accountants have decided that objectivity should be maintained. Similarly, most assets are valued at their acquisition cost adjusted for depreciation and improvements. To be conservative, when market value is lower than this adjusted cost, the lower value is used. If a firm must be liquidated, then current market value must be estimated.

The accounting function provides data for **general accounting** and **cost accounting.** This chapter's presentation begins with the balance sheet and income statement, which are the two key summaries of financial transactions for general accounting. This discussion includes some of the basic financial ratios used for short- and long-term evaluations. The chapter concludes with a key topic in cost accounting—allocating indirect expenses.

THE BALANCE SHEET

The primary accounting statements are the balance sheet and the income statement. The balance sheet describes the firm's financial condition at a specific time, while the income statement describes the firm's performance over a period of time—usually a year.

The balance sheet lists the firm's assets, liabilities, and equity on a specified date. This is a picture of the organization's financial health or a snapshot in time. Usually, balance sheets are taken at the end of the quarter and fiscal year. The balance sheet is based on the **fundamental accounting equation:**

$$\text{Assets} = \text{Liabilities} + \text{Equity} \qquad (17\text{-}1)$$

Figure 17–2 illustrates the basic format of the balance sheet. Notice in the balance sheet, as in Equation 17–1, that assets are listed on the left-hand side and liabilities and equity are on the right-hand side. The fact that the firm's resources are *balanced* by the sources of funds is the basis for the name of the balance sheet.

Balance Sheet for Engineered Industries, December 31, 2016 (all amounts in $1000s)

Assets		Liabilities	
Current assets		Current liabilities	
Cash	$ 1940	Accounts payable	$ 1150
Accounts receivable	950	Notes payable	80
Securities	4100		
Inventories	1860	Accrued expense	950
(*minus*) Bad debt provision	−80	Total current liabilities	2180
Total current assets	8770		
		Long-term liabilities	1200
		Total liabilities	**3380**
Fixed assets			
Land	335		
Plant and equipment	6500		
(*minus*) Accumulated depr.	−2350		
Total fixed assets	4485	**Equity**	
		Preferred stock	110
Other assets		Common stock	650
Prepays/deferred charges	140	Capital surplus	930
Intangibles	420	Retained earnings	8,745
Total other assets	560	Total equity	10,435
Total assets	**13,815**	**Total liabilities and equity**	**13,815**

FIGURE 17–2 Sample balance sheet.

Assets

In Equation 17–1 and Figure 17–2, **assets** are items owned by the firm and have monetary value. Liabilities are what the firm owes. **Equity** represents funding from the firm and its owners (the shareholders). In Equation 17–1, assets are always balanced by the sum of the liabilities and the equity. The value for retained earnings is set so that equity equals assets minus liabilities.

On a balance sheet, assets are listed in order of decreasing liquidity, that is, according to how quickly each one can be converted to cash. Thus, *current assets* are listed first, and within that category in order of decreasing liquidity are listed cash, receivables, securities, and inventories. *Fixed assets,* or *property, plant, and equipment,* are used to produce and deliver goods and/or services, and they are not intended for sale. Items such as prepayments and intangibles such as patents are listed last.

The term "receivables" comes from the manner of handling billing and payment for most business sales. Rather than requesting immediate payment for every transaction by check or credit card, most businesses record each transaction and then bill for all transactions. The total that has been billed less payments already received is called accounts receivable, or receivables.

Liabilities

On the balance sheet, **liabilities** are divided into two major classifications—short term and long term. The *short-term* or *current liabilities* are expenses, notes, and other payable accounts that are due within one year from the balance sheet date. *Long-term liabilities* include mortgages, bonds, and loans with later due dates. For Engineered Industries in Figure 17–2, total current and long-term liabilities are $2,180,000 and $1,200,000, respectively. Often in performing engineering economic analyses, the **working capital** for a project must be estimated. The total amount of working capital available may be calculated with Equation 17-2 as the difference between current assets and current liabilities.

$$\text{Working capital} = \text{Current assets} - \text{Current liabilities} \tag{17-2}$$

For Engineered Industries, there would be $8,770,000 - $2,180,000 = $6,590,000 available in working capital.

Equity

Equity is also called *owner's equity* or *net worth.* It includes the par value of the owners' stockholdings and the capital surplus, which are the excess dollars brought in over par value when the stock was issued. The capital surplus can also be called *additional paid-in capital*, or APIC. Retained earnings are dollars a firm chooses to retain rather than paying out as dividends to stockholders.

Retained earnings within the equity component is the dollar quantity that always brings the balance sheet, and thus the fundamental accounting equation, into balance. For Engineered Industries, *total equity* value is listed at $10,435,000. From Equation 17–1 and the assets, liabilities, and equity values in Figure 17–2, we can write the balance as follows:

$$\text{Assets} = \text{Liabilities} + \text{Equity}$$

$$\text{Assets (current, fixed, other)} = \text{Liabilities (current and long-term)} + \text{Equity}$$

$$8,770,000 + 4,485,000 + 560,000 = 2,180,000 + 1,200,000 + 10,435,000$$

$$\$13,815,000 = \$13,815,000$$

An example of owner's equity is ownership of a home. Most homes are purchased by means of a mortgage loan that is paid off at a certain interest rate over 15 to 30 years. At any point in time, the difference between what is owed to the bank (the remaining balance on the mortgage) and what the house is worth (its appraised market value) is the *owner's equity*. In this case, the loan balance is the *liability,* and the home's value is the *asset*—with *equity* being the difference. Over time, as the house loan is paid off, the owner's equity increases.

The balance sheet is a very useful tool that shows one view of the firm's financial condition at a particular point in time.

Financial Ratios Derived from Balance Sheet Data

One common way to evaluate the firm's health is through ratios of quantities on the balance sheet. Firms in a particular industry will typically have similar values, and exceptions will often indicate firms with better or worse performance. Two common ratios used to analyze the firm's current position are the current ratio and the quick (or acid-test) ratio.

A firm's **current ratio** is the ratio of current assets to current liabilities, as in Equation 17-3.

$$\text{Current ratio} = \text{Current assets/Current liabilities} \tag{17-3}$$

This ratio provides insight into the firm's solvency over the short term by indicating its ability to cover current liabilities. Historically, firms aim to be at or above a ratio of 2.0; however, this depends heavily on the industry as well as the individual firm's management practices and philosophies. The current ratio for Engineered Industries in Figure 17–2 is above 2 (8,770,000/2,180,000 = 4.02).

Both working capital and the current ratio indicate the firm's ability to meet currently maturing obligations. However, neither describes the type of assets owned. The **acid-test ratio** or **quick ratio** becomes important when one wishes to consider the firm's ability to pay debt "instantly." The acid-test ratio is computed by dividing a firm's **quick assets** (cash, receivables, and market securities) by total current liabilities, as in Equation 17-4.

$$\text{Quick ratio} = \text{Quick assets/Total current liabilities} \qquad (17\text{-}4)$$

Current inventories are excluded from quick assets because of the time required to sell these inventories, collect the receivables, and subsequently have the cash on hand to reduce debt. For Engineered Industries in Figure 17–2, the calculated acid-test ratio is well above 1 [(1,940,000 + 950,000 + 4,100,000)/2,180,000 = 3.21].

Working capital, current ratio, and quick ratio are all indications of the firm's financial health (status). A thorough financial evaluation would consider all three, including comparisons with values from previous periods and with broad-based industry standards. When trends extend over multiple periods, the trends may be more important than the current values.

THE INCOME STATEMENT

The **income statement** or **profit and loss statement** summarizes the firm's revenues and expenses over a month, quarter, or year. Rather than being a snapshot like the balance sheet, the income statement encompasses a *period* of business activity. The income statement is used to evaluate revenue and expenses that occur in the interval *between* consecutive balance sheet statements. The income statement reports the firm's *net income (profit)* or *loss* by subtracting expenses from revenues. If revenues minus expenses is positive in Equation 17-5, there has been a profit, if negative a loss has occurred.

$$\text{Revenues} - \text{Expenses} = \text{Net profit (Loss)} \qquad (17\text{-}5)$$

To aid in analyzing performance, the income statement in Figure 17–3 separates operating and nonoperating activities and shows revenues and expenses for each. Operating revenues are made up of sales revenues (minus returns and allowances), while nonoperating revenues come from rents and interest receipts.

Operating expenses produce the products and services that generate the firm's revenue stream of cash flows. Typical operating expenses include cost of goods sold, selling and promotion costs, depreciation, general and administrative costs, and lease payments. *Cost of goods sold (COGS)* includes the labor, materials, and indirect costs of production.

Engineers design production systems, specify materials, and analyze make/buy decisions. All these items affect a firm's cost of goods sold. Good engineering design focuses not only on technical functionality but also on cost-effectiveness as the design *integrates*

**Income Statement for Engineered Industries for End of Year 2017
(all amounts in $1000)**

Operating revenues and expenses	
Operating revenues	
Sales	$ 18,900
(*minus*) Returns and allowances	−870
Total operating revenues	**18,030**
Operating expenses	
Cost of goods and services sold	
Labor	6140
Materials	4640
Indirect cost	2,280
Selling and promotion	930
Depreciation	450
General and administrative	2,160
Lease payments	510
Total operating expense	**17,110**
Total operating income	**920**
Nonoperating revenues and expenses	
Rents	20
Interest receipts	300
Interest payments	−120
Total nonoperating income	**200**
Net income before taxes	**1120**
Income taxes	−390
Net profit (loss) for 2017	**730**

FIGURE 17–3 Sample income statement.

the entire production system. Also of interest to the engineering economist is *depreciation* (see Chapter 11)—which is the systematic "writing off" of a capital expense over a period of years. This noncash expense is important because it represents a decrease in value of the firm's capital assets.

The operating revenues and expenses are shown first, so that the firm's operating income from its products and services can be calculated. Also shown on the income statement are nonoperating expenses such as interest payments on debt.

From the data in Figure 17–3, Engineered Industries has total expenses (operating = $17,110,000 and nonoperating = $120,000) of $17,230,000. Total revenues are $18,350,000 (= $18,030,000 + $20,000 + $300,000). The net after-tax profit for year 2016 shown in Figure 17–3 as $730,000, but it can also be calculated using Equation 17-5 as

$$\text{Net profits (Loss)} = \text{Revenues} - \text{Expenses [before taxes]}$$

$$\$1,120,000 = 18,350,000 - 17,230,000 \text{ [before taxes] and with}$$

$$\$390,000 \text{ taxes paid}$$

thus

$$\$730,000 = 1,120,000 - 390,000 \text{ [after taxes]}$$

Financial Ratios Derived from Income Statement Data

The **profit margin** (Equation 17-6) equals net profits divided by net sales revenue. Net sales revenue equals sales minus returns and allowances.

$$\text{Profit margin} = \text{Net profit/Net sales revenue} \qquad (17\text{-}6)$$

This ratio provides insight into the cost efficiency of operations as well as a firm's ability to convert sales into profits. For Engineered Industries in Figure 17–3, the profit margin is $730{,}000/18{,}030{,}000 = 0.040 = 4.0\%$. Like other financial measures, the profit margin is best evaluated by comparisons with other time periods and industry benchmarks, and trends may be more significant than individual values.

Interest coverage, as given in Equation 17-7, is calculated as the ratio of total income to interest payments—where *total income* is total revenues minus all expenses except interest payments.

$$\text{Interest coverage} = \text{Total income/Interest payments} \qquad (17\text{-}7)$$

The interest coverage ratio (which for industrial firms should be at least 3.0) indicates how much revenue must drop to affect the firm's ability to finance its debt. With an interest coverage ratio of 3.0, a firm's revenue would have to decrease by two-thirds (unlikely) before it became impossible to pay the interest on the debt. The larger the interest coverage ratio the better. Engineered Industries in Figure 17–3 has an interest coverage ratio of

$$(18{,}350{,}000 - 17{,}110{,}000)/120{,}000 = 10.3$$

Linking the Balance Sheet, Income Statement, and Capital Transactions

The balance sheet and the income statement are separate but linked documents. Understanding how the two are linked together helps clarify each. Accounting describes such links as the *articulation* between these reports.

The balance sheet shows a firm's assets, liabilities, and equity at a particular point in time, whereas the income statement summarizes revenues and expenses over a time interval. These tabulations can be visualized as a snapshot at the period's beginning (a balance sheet), a video summary over the period (the income statement), and a snapshot at the period's end (another balance sheet). The income statement and changes in the balance sheets summarize the business transactions that have occurred during that period.

There are many links between these statements and the cash flows that make up business transactions, but for engineering economic analysis the following are the most important.

1. Overall profit or loss (income statement) and the starting and ending equity (balance sheets).
2. Acquisition of capital assets.
3. Depreciation of capital assets.

The overall profit or loss during the year (shown on the income statement) is reflected in the change in retained earnings between the balance sheets at the beginning and end of the year. To find the change in retained earnings (RE), one must also subtract any dividends distributed to the owners and add the value of any new capital stock sold:

$$RE_{beg} + \text{Net income/Loss} + \text{New stock} - \text{Dividends} = RE_{end}$$

When capital equipment is purchased, the balance sheet changes, but the income statement does not. If cash is paid, then the cash asset account decrease equals the increase in the capital equipment account—there is no change in total assets. If a loan is used, then the capital equipment account increases, and so does the liability item for loans. In both cases the equity accounts and the income statement are unchanged.

The depreciation of capital equipment is shown as a line on the income statement. The depreciation for that year equals the change in accumulated depreciation between the beginning and the end of the year—after subtraction of the accumulated depreciation for any asset that is sold or disposed of during that year.

Example 17–1 applies these relationships to the data in Figures 17–1 and 17–2.

EXAMPLE 17–1

For simplicity, assume that Engineered Industries will not pay dividends in 2017 and did not sell any capital equipment. It did purchase $400,000 in capital equipment. What can be said about the values on the balance sheet at the end of 2017, using the linkages just described?

SOLUTION

First, the net profit of $730,000 will be added to the retained earnings from the end of 2016 to find the new retained earnings at the end of 2017:

$$RE_{12/31/2017} = \$730,000 + \$8,745,000 = \$9,475,000$$

Second, the plant and equipment assets shown at the end of 2017 would increase from $6,500,000 to $6,900,000.

Third, the accumulated depreciation would increase by the $450,000 in depreciation shown in the 2017 income statement from the $2,350,000 posted in 2016. The new accumulated depreciation on the 2017 balance sheet would be $2,800,000. Combined with the change in the amount of capital equipment, the new fixed asset total for 2017 would equal:

$$\$335,000 + \$6,900,000 - \$2,800,000 = \$4,435,000$$

TRADITIONAL COST ACCOUNTING

A firm's *cost-accounting system* collects, analyzes, and reports operational performance data (costs, utilization rates, etc.). Cost-accounting data are used to develop product costs, to determine the mix of labor, materials, and other costs in a production setting, and to evaluate outsourcing and subcontracting possibilities.

Direct and Indirect Costs

Costs incurred to produce a product or service are traditionally classified as either **direct** or **indirect (overhead)**. Direct costs come from activities directly associated with the final product or service produced. Examples include material costs and labor costs for engineering design, component assembly, painting, and drilling.

Some organizational activities are difficult to link to specific projects, products, or services. For example, the receiving and shipping areas of a manufacturing plant are used by all incoming materials and all outgoing products. Materials and products differ in their weight, size, fragility, value, number of units, packaging, and so on, and the receiving and shipping costs depend on all these factors. Also, different materials arrive together and different products are shipped together, so these costs are intermingled and often cannot be tied directly to each product or material.

Other costs, such as the organization's management, sales, and administrative expenses, are difficult to link directly to individual products or services. These indirect or overhead expenses also include machine depreciation, engineering and technical support, and customer warranties.

Indirect Cost Allocation

To allocate indirect costs to different departments, products, and services, accountants use quantities such as direct-labor hours, direct-labor costs, material costs, and total direct cost. One of these is chosen to be the burden vehicle. The total of all indirect or overhead costs is divided by the total for the burden vehicle. For example, if direct-labor hours is the burden vehicle, then overhead will be allocated based on overhead dollar per direct-labor hour. Then each product, project, or department will *absorb* (or be allocated) overhead costs, based on the number of direct-labor hours each has.

This is the basis for calling traditional costing systems **absorption costing**. For decision making, the problem is that the absorbed costs represent average, not incremental, performance.

Four common ways of allocating overhead are direct-labor hours, direct-labor cost, direct-materials cost, and total direct cost. The first two differ significantly only if the cost per hour of labor differs for different products. Example 17–2 uses direct-labor and direct-materials cost to illustrate the different choices of burden vehicle.

EXAMPLE 17–2

Industrial Robots does not manufacture its own motors or computer chips. Its premium product differs from its standard product in having heavier-duty motors and more computer chips for greater flexibility.

As a result, Industrial Robots manufactures a higher fraction of the standard product's value itself, and it purchases a higher fraction of the premium product's value. Use the following data to allocate $850,000 in overhead on the basis of labor cost and materials cost.

	Standard	Premium
Number of units per year	750	400
Labor cost (each)	$400	$500
Materials cost (each)	50	900

SOLUTION

First, the labor and material costs for the standard product, the premium product, and in total are calculated.

	Standard	Premium	Total
Number of units per year	750	400	
Labor cost (each)	$ 400	$ 500	
Materials cost (each)	550	900	
Labor cost	300,000	200,000	$500,000
Materials cost	412,500	360,000	772,500

Then the allocated cost per labor dollar, $1.70, is found by dividing the $850,000 in overhead by the $500,000 in total labor cost. The allocated cost per material dollar, $1.100324, is found by dividing the $850,000 in overhead by the $772,500 in materials cost. Now, the $850,000 in allocated overhead is split between the two products using labor costs and material costs.

	Standard	Premium	Total
Labor cost	$300,000	$200,000	$500,000
Overhead/labor	1.70	1.70	
Allocation by labor	510,000	340,000	850,000
Material cost	412,500	360,000	0
Overhead/material	1.1003	1.1003	
Allocation by material	453,884	396,116	850,000

If labor cost is the burden vehicle, then 60% of the $850,000 in overhead is allocated to the standard product. If material cost is the burden vehicle, then 53.4% is allocated to the standard product. In both cases, the $850,000 has been split between the two products. Using total direct costs would produce another overhead allocation between these two values. However, for decision making about product mix and product prices, incremental overhead costs must be analyzed. All the allocation or burden vehicles are based on an average cost of overhead per unit of burden vehicle.

Problems with Traditional Cost Accounting

Allocation of indirect costs can distort product costs and the decisions based on those costs. To be accurate, the analyst must determine which indirect or overhead expenses will be changed because of an engineering project. In other words, what are the incremental cash flows? For example, vacation and sick leave accrual may be part of overhead, but will they change if the labor content is changed? The changes in costs incurred must be estimated. Loadings, or allocations, of overhead expenses cannot be used.

This issue has become very important because in some firms, automation has reduced direct-labor content to less than 5% of the product's cost. Yet in some of these firms, the basis for allocating overhead is still direct-labor hours or cost.

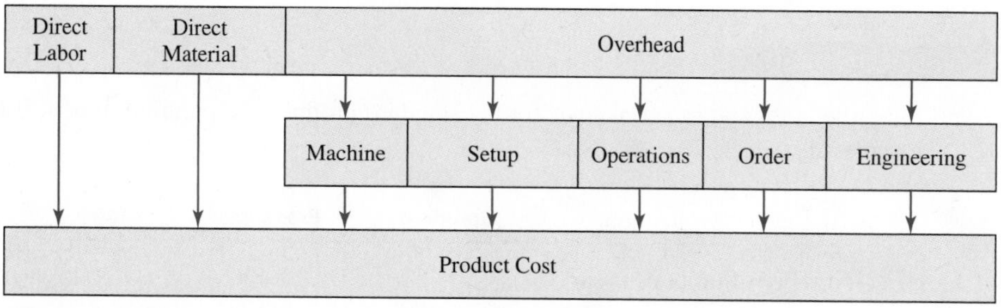

FIGURE 17–4 Activity-based costing versus traditional overhead allocation. (Based on an example by Kim LaScola Needy.)

Other firms are shifting to activity-based costing (ABC), where each activity is linked to specific cost drivers, and the number of dollars allocated as overhead is minimized (see Liggett, Trevino, and Lavelle, 1992). Figure 17–4 illustrates the difference between activity-based costing and traditional overhead allocations (see Tippet and Hoekstra, 1993).

Other Problems to Watch For

Project managers have often accused centralized accounting systems of being too slow or being "untimely." Because engineering economy is not concerned with the problem of daily project control, this is a less critical issue. However, if an organization establishes multiple files and systems so that project managers (and others) have the timely data they need, then the level of accuracy in one or all systems may be low. As a result, analysts making cost estimates will have to consider other internal data sources.

There are several cases in which data on equipment or inventory values may be questionable. When inventory is valued on a "last in, first out" basis, the remaining inventory may be valued too low. Similarly, land valued at its acquisition cost is likely to be significantly undervalued. Finally, capital equipment may be valued at either a low or a high value, depending on allowable depreciation techniques and company policy.

PROBLEMS

Accounting

17-1 Why is it important for engineers and managers to understand accounting principles? Name a few ways that they can do so.

17-2 Explain the accounting function within a firm. What does this function do, and why is it important? What types of data does it provide?

17-3 Manipulation of financial data by the Enron Corporation was revealed in October 2001. Firm executives
(E) were sentenced to prison. Arthur Andersen, which had been one of the "Big 5" accounting firms and Enron's auditor, surrendered its licenses to practice as certified public accountants in August 2002. This and other scandals led to the passage of the federal Sarbanes–Oxley (SarbOx) legislation in July 2002. What are the key components of this law?

17-4 How would the information and activities in Figure 17–1 relate to potential contributions by external professional services that firms will often hire

(a) consulting engineers?

(b) management consulting firms?

(c) accounting firms for auditing?

17-5 Consider Figure 17–1 and external professional services that firms will often hire. What ethical ques-
(E) tions seem likely to arise when a firms hires

(a) consulting engineers for design?

(b) management consulting firms?

(c) accounting firms for auditing?

17-6 Insolvency or cash flow problems in the U.S. banking industry started the financial crisis of 2007. Have significant changes in accounting standards and practices been forced through legislation? If so, what are these changes?

17-7 Accounting and finance are required topics for most management degrees, and many engineers do become managers during their careers. After graduation with a B.S. in engineering, what courses are available to you from your or another nearby university?

17-8 Using the table of contents, compare the coverage of this engineering economy text with a text for the introductory course in corporate finance (typically a junior course in BBA programs). What topics do the texts have in common? Identify the topics that seem to be covered in only one—note which topics for which course.

Balance Sheet

17-9 Develop short definitions for the following terms: balance sheet, income statement, and fundamental accounting equation.

17-10 Explain the difference between short-term and long-term liabilities.

17-11 Calculate the equity of the Gravel Construction Company if it has $10 million worth of assets. Gravel has $1.3 million in current liabilities and $2.5 million in long-term liabilities.

17-12 Matbach Industries has $930,000 in current assets and $470,000 in fixed assets less $180,000 in accumulated depreciation. The firm's current liabilities total $370,000, and the long-term liabilities $115,000.

    (*a*) What is the firm's equity?

    (*b*) If the firm's stock and capital surplus total $305,000, what is the value for retained earnings?

17-13 CalcTech has $1.3M in current assets and $550,000 in fixed assets less $200,000 in accumulated depreciation. The firm's current liabilities total $180,000, and the long-term liabilities $205,000.

    (*a*) What is the firm's equity?

    (*b*) If the firm's stock and capital surplus total $202,000, what is the value for retained earnings?

17-14 First Step Baby Monitor Company has current assets of $5.5 million and current liabilities of $2.2 million. Give the company's working capital and current ratio.

17-15 From the following data, taken from the balance sheet of Petra's Widget Factory, determine the working capital, current ratio, and quick ratio.

Cash	$110,000
Net accounts and notes receivable	325,000
Retailers' inventories	210,000
Prepaid expenses	6,000
Accounts and notes payable (short term)	300,000
Accrued expenses	187,000

17-16 For Gee-Whiz Devices, calculate the following: working capital, current ratio, and quick ratio.

Gee-Whiz Devices Balance Sheet Data

Cash	$100,000
Market securities	45,000
Net accounts and notes receivable	150,000
Retailers' inventories	200,000
Prepaid expenses	8,000
Accounts and notes payable (short term)	315,000
Accrued expenses	90,000

17-17 Turbo Start has current assets totaling $1.5 million (this includes $500,000 in current inventory) and current liabilities totaling $400,000. Find the current ratio and quick ratio. Are the ratios at desirable levels? Explain.

17-18 (*a*) For Evergreen Environmental Engineering (EEE), determine the working capital, current ratio, and quick ratio. Evaluate the company's economic situation with respect to its ability to pay off debt.

EEE Balance Sheet Data ($1000s)

Cash	$120,000
Securities	40,000
Accounts receivable	110,000
Inventories	300,000
Prepaid expenses	3,000
Accounts payable	351,000
Accrued expenses	89,000

(b) The entries to complete EEE's balance sheet include:

More EEE Balance Sheet Data ($1000s)

Long-term liabilities	$ 220,000
Land	25,000
Plant and equipment	510,000
Accumulated depreciation	210,000
Stock	81,000
Capital surplus	15,000
Retained earnings	Value not given

Construct EEE's balance sheet.

(c) What are EEE's values for total assets, total liabilities, and retained earnings?

17-19 (a) For J&W Graphics Supply, compute the current ratio. Is this a financially healthy company? Explain.

J&W Graphics Supply Balance Sheet Data ($1000s)

Assets	
Cash	$1740
Accounts receivable	2500
Inventories	900
Bad debt provision	−75
Liabilities	
Accounts payable	1050
Notes payable	500
Accrued expenses	125

(b) The entries of complete J&W's balance sheet include:

More J&W Balance Sheet Data ($1000s)

Long-term liabilities	$950
Land	475
Plant and equipment	3100
Accumulated depreciation	1060
Stock	680
Capital surplus	45
Retained earnings	Value not given

Construct J&W's balance sheet.

(c) What are J&W's values for total assets, total liabilities, and total earnings?

17-20 For Sutton Manufacturing, determine the current ratio and the quick ratio. Are these values acceptable? Why or why not?

Sutton Manufacturing Balance Sheet Data ($1000s)

Assets		Liabilities	
Current assets		Current liabilities	
Cash	$ 870	Notes payable	$ 500
Accounts receivable	450	Accounts payable	600
Inventory	1200	Accruals	200
Prepaid expenses	60	Taxes payable	30
		Current portion long-term debt	100
Total current assets	2580		
Net fixed assets		Total current liabilities	1430
Land	1200	Long-term debt	2000
Plant and equipment	3800	Officer debt (subordinated)	200
(less accumulated depreciation)	−1000	Total liabilities	3630
Other assets		Equity	
Notes receivable	200	Common stock	1670
Intangibles	120	Capital surplus	400
		Retained earnings	1200
		Total equity	3270
Total assets	6900	Total liabilities and equity	6900

17-21 If a firm has a current ratio less than 2.0 and a quick ratio less than 1.0, will the company eventually go bankrupt and out of business? Explain your answer.

17-22 What is the advantage of comparing financial statements across periods or against industry benchmarks over looking at statements associated with a single date or period?

Income Statement

17-23 Laila's Surveying Inc. had revenues of $1.5 million in 2016. Expenses totaled $900,000. What was her net profit (or loss) if the tax rate is 38%?

17-24 Bohr Paint Company has annual sales of $12 million per year. If there is a profit of $5000 per day with 7 days per week operation, what is the total yearly business expense? All calculations are on a before-tax basis.

17-25 For Magdalen Industries, compute the net income before taxes and net profit (or loss). Taxes for the year were $7.5 million.

(a) Calculate net profit for the year.

(b) Construct the income statement.

(c) Calculate the interest coverage and net profit ratio. Is the interest coverage acceptable? Explain why or why not.

Magdalen Industries Income Statement Data ($M)

Revenues	
Total operating revenue	$124
(including sales of $48 million)	
Total nonoperating revenue	36
Expenses	
Total operating expenses	70
Total nonoperating expenses	35
(interest payments)	

17-26 Find the net income of Turbo Start (Problem 17-17) given the following data from the balance sheet and income statement.

Turbo Start Data ($1000s)

Accounts payable	$ 200
Selling expense	500
Sales revenue	5000
Owner's equity	2400
Income taxes	600
Cost of goods sold	3000
Accounts receivable	500

17-27 The general ledger of the Fly-Buy-Nite (FBN) Engineering Company contained the following account balances. Construct an income statement. What is the net income before taxes and the net profit (or loss) after taxes? FBN has a tax rate of 35%.

	Amount ($1000s)
Administrative expenses	$ 2,750
Subcontracted services	15,000
Development expenses	900
Interest expense	200
Sales revenue	35,000
Selling expenses	4,500

17-28 For Andrew's Electronic Instruments, calculate the interest coverage and net profit ratio. Is Andrew's business healthy?

Income Statement for Andrew's Electronics for End of Year 2010 ($1000s)

Revenues	
Operating revenues	
Sales	$395
(*minus*) Returns	−15
Total operating revenues	380
Nonoperating revenues	
Interest receipts	50
Stock revenues	25
Total nonoperating revenues	75
Total revenues, R	455
Expenses	
Operating expenses	
Cost of goods and services sold	
Labor	200
Materials	34
Indirect cost	68
Selling and promotion	20
Depreciation	30
General and administrative	10
Lease payments	10
Total operating expenses	372
Nonoperating expenses	
Interest payments	22
Total nonoperating expenses	22
Total expenses, E	394
Net income before taxes, R − E	61
Incomes taxes	30
Net profit (loss) for the year 2010	31

Linking Balance Sheet and Income Statement

17-29 Use your own experiences or a search engine to provide specific examples where the balance sheet and income statement are linked.

17-30 Lithium batteries are popular and widely used. Lithium is typically mined from salt flats where water is scarce, yet lithium mining requires large amounts of water. Are the environmental costs of lithium captured in the financial statements of companies that produce lithium batteries?

17-31 Sutton Manufacturing (balance sheet at the end of last year in Problem 17-20) had the following entries in this year's income statement.

Depreciation $420,000

Profit   480,000

      In addition, we also know the firm purchased $800,000 of equipment with cash. The firm paid $200,000 in dividends this year.

      What are the entries in the balance sheet at the end of this year for

(a) plant and equipment?

(b) accumulated depreciation?

(c) retained earnings?

17-32 Magdalen Industries (Problem 17-25) had the following entries in its balance sheet at the end of last year.

Plant and equipment $15 million

(less accumulated depreciation)   8 million

Retained earnings   60 million

      In addition to the income statement data for this year in Problem 17-25, we also know that the firm purchased $3 million of equipment with cash and that depreciation expenses were $2 million of the $70 million in operating expenses listed in Problem 17-26. The firm paid no dividends this year.

What are the entries in the balance sheet at the end of this year for

(a) plant and equipment?

(b) accumulated depreciation?

(c) retained earnings?

Allocating Costs

17-33 Categorize each of the following costs as direct or indirect. Assume that a traditional costing system is in place.

Machine run costs	Cost to market the product
Machine depreciation	Cost of storage
Material handling costs	Insurance costs
Cost of materials	Cost of product sales force
Overtime expenses	Engineering drawings
Machine operator wages	Machine labor
Utility costs	Cost of tooling and fixtures
Support (administrative)	
staff salaries	

17-34 RLW-II Enterprises estimated that indirect manufacturing costs for the year would be $75 million and that 15,000 machine-hours would be used.

(a) Compute the predetermined indirect cost application rate using machine hours as the burden vehicle.

(b) Determine the total cost of production for a product with direct material costs of $2 million, direct-labor costs of $2.5 million, and 200 machine-hours.

17-35 Philippe Francois Inc. produces concrete sundials—5000 were produced in a recent production run. The run required 1500 machine hours, three "set-ups" of the mixing equipment, and 100 hours of final inspection time. Costs are estimated as follows: $50 per machine hour, $3000 per "set-up," and $20 per inspection hour. Direct materials and direct labor total $95 per sundial. Indirect expenses total 45% of the direct material and labor costs.

(a) Determine the "cost to produce" each unit.

(b) At a 15% profit level what is the unit sales price?

(c) With six runs per year what is the total annual profit for this product?

17-36 LeGaroutte Industries makes industrial pipe manufacturing equipment. Use direct-labor hours as the burden vehicle, and compute the total cost per unit for each model given in the table. Total manufacturing indirect costs are $15,892,000, and there are 100,000 units manufactured per year for Model S, 50,000 for Model M, and 82,250 for Model G.

Item	Model S	Model M	Model G
Direct-material costs	$3,800,000	$1,530,000	$2,105,000
Direct-labor costs	600,000	380,000	420,000
Direct-labor hours	64,000	20,000	32,000

17-37 Par Golf Equipment Company produces two types of golf bag: the standard and deluxe models. The total indirect cost to be allocated to the two bags is $35,000. Determine the net revenue that Par Golf can expect from the sale of each bag.

(a) Use direct-labor cost to allocate indirect costs.

(b) Use direct-materials cost to allocate indirect costs.

Data Item	Standard	Deluxe
Direct-labor cost	$60,000	$70,000
Direct-material cost	40,000	47,500
Selling price	70	105
Units produced	1800	1400

Minicase

17-38 Find a real world example where a firm has included
Ⓖ environmental remediation costs on its balance sheet
or in a note to the balance sheet. Briefly describe
what was done. What percentage of the firm's assets
and net profit are these costs?

CASES

The following cases from *Cases in Engineering Economy* (www.oup.com/us/newnan) are suggested as matched with this chapter.

CASE **4** **Balder-Dash Inc.**
Standard cost and allocated costs versus true
marginal cost (IEs).

CASE **52** **Aunt Allee's Jams and Jellies**
Product costing needs activity based costs.

INTRODUCTION TO SPREADSHEETS

C omputerized spreadsheets are available nearly everywhere, and they can be easily applied to economic analysis. In fact, spreadsheets were originally developed to analyze financial data, and they are often credited with initiating the explosive growth in demand for desktop computing.

A spreadsheet is a two-dimensional table, whose cells can contain numerical values, labels, or formulas. The software automatically updates the table when an entry is changed, and there are powerful tools for copying formulas, creating graphs, and formatting results.

THE ELEMENTS OF A SPREADSHEET

A spreadsheet is a two-dimensional table that labels the columns in alphabetical order A to Z, AA to AZ, BA to BZ, etc. The rows are numbered from 1 to 65,536 or higher. Thus a *cell* of the spreadsheet is specified by its column letter and row number. For example, A3 is the third row in column A and AA6 is the sixth row in the twenty-seventh column. Each cell can contain a label, a numerical value, or a formula.

A *label* is any cell where the contents should be treated as text. Arithmetic cannot be performed on labels. Labels are used for variable names, row and column headings, and explanatory notes. In Excel any cell that contains more than a simple number, such as 3.14159, is treated as a label, unless it begins with an =, which is the signal for a formula. Thus 2*3 and B1+B2 are labels. Meaningful labels can be wider than a normal column. One solution is to allow those cells to "wrap" text, which is one of the "alignment" options. The table heading row (row 8) in Example A–1 has turned this on by selecting row 8, right-clicking on the row, and selecting wrap text under the alignment tab.

A *numerical value* is any number. Acceptable formats for entry or display include percentages, currency, accounting, scientific, fractions, date, and time. In addition the number of decimal digits, the display of $ symbols, and commas for "thousands" separators can be adjusted. The format for cells can be changed by selecting a cell, a block of cells, a row, a column, or the entire spreadsheet. Then right-click on the selected area, and a menu that includes "format cells" will appear. Then number formats, alignment, borders, fonts, and patterns can be selected.

Formulas must begin with an =, such as =3*4^2 or =B1+B2. They can include many functions—financial, statistical, trigonometric, etc. (and others can be defined by the user). The formula for the "current" cell is displayed in the formula bar at the top of the spreadsheet. The value resulting from the formula is displayed in the cell in the spreadsheet.

Often the printed-out spreadsheet will be part of a report or a homework assignment and the formulas must be explained. Here is an easy way to place a copy of the formula in an adjacent or nearby cell. (1) Convert the cell with the formula to a label by inserting a space before the = sign. (2) Copy that label to an adjacent cell by using cut and paste. Do not drag the cell to copy it, as any formula ending with a number (even an address like B4) will have the number automatically incremented. (3) Convert the original formula back into a formula by deleting the space.

DEFINING VARIABLES IN A DATA BLOCK

The cell A1, top left corner, is the HOME cell for a spreadsheet. Thus, the top left area is where the data block should be placed. This data block should have every variable in the spreadsheet with an adjacent label for each. This data block supports a basic principle of good spreadsheet modeling, which is to use variables in your models.

The data block in Example A–1 contains *entered data*—the loan amount (A1), the number of payments (A2), and the interest rate (A3), and *computed data*—the payment (A4). Then instead of using the loan amount of $5000 in a formula, the cell reference A1 is used. Even if a value is referenced only once, it is better to include it in the data block. By using one location to define each variable, you can change any value at one place in the spreadsheet and have the entire spreadsheet instantly recomputed.

Even for simple homework problems you should use a data block.

1. You may be able to use it for another problem.
2. Solutions to simple problems may grow into solutions for complex problems.
3. Good habits, like using data blocks, are easy to maintain once they are established.
4. It makes the assumptions clear if you've estimated a value or for grading.

In the real world, data blocks are even more important. Most problems are solved more than once, as more and more accurate values are estimated. Often the spreadsheet is revised to add other variables, time periods, locations, etc. Without data blocks, it is hard to change a spreadsheet and the likelihood of missing a required change skyrockets.

If you want your formulas to be easier to read, you can name your variables. *Note*: In Excel, the cell's location or name is displayed at the left of the formula bar. Variable names can be entered here. They will then automatically be applied if cell addresses are entered by the point and click method. If cell addresses are entered as A2, then A2 is what is displayed. To change a displayed A2 to the name of the cell (LoanAmount), the process is to click on Insert, click on Name, click on Apply, and then select the names to be applied.

COPY COMMAND

The copy command and relative/absolute addressing make spreadsheet models easy to build. If the range of cells to be copied contains only labels, numbers, and functions, then the copy command is easy to use and understand. For example, the formula =EXP(1.9) would be copied unchanged to a new location. However, cell addresses are usually part of the range being copied, and their absolute and relative addresses are treated differently.

An *absolute address* is denoted by adding $ signs before the column and/or row. For example in Figure A–1a, $A$4 is the absolute address for the interest rate. When an absolute address is copied, the column and/or row that is fixed is copied unchanged. Thus $A$4 is completely fixed, $A4 fixes the column, and A$4 fixes the row. One common use for absolute addresses is any data block entry, such as the interest rate. When entering or editing a formula, changing between A4, $A$4, A$4, $A4, and A4 is most easily done using the F4 key, which scrolls an address through the choices.

In contrast, a *relative address* is best interpreted as directions from one cell to another. For example in Figure A–1a, the balance due in year t equals the balance due in year $t - 1$ minus the principal payment in Year t. Specifically for the balance due in Year 1, D10 contains =D9−C10. From cell D10, cell D9 is one row up and C10 is one column to the left, so the formula is really (contents of 1 up) minus (contents of 1 to the left). When a cell containing a relative address is copied to a new location, it is these directions that are copied to determine any new relative addresses. So if cell D10 is copied to cell F14, the formula is =F13−E14.

Thus to calculate a loan repayment schedule, as in Figure A–1, the row of formulas is created and then copied for the remaining years.

EXAMPLE A–1

Four repayment schedules for a loan of $5000 to be repaid over 5 years at an interest rate of 8% were shown in Table 3–1. Use a spreadsheet to calculate the amortization schedule for the constant principal payment option.

SOLUTION

The first step is to enter the loan amount, number of periods, and interest rate into a data block in the top left part of the spreadsheet. The next step is to calculate the constant principal payment amount, which was given as $1252.28 in Table 3–1. The factor approach to finding this value is given in Chapter 3 and the spreadsheet function is explained in Chapter 4.

The next step is to identify the columns for the amortization schedule. These are the year, interest owed, principal payment, and balance due. Because some of these labels are wider than a normal column, the cells are formatted so that the text wraps (row height increases automatically). The initial balance is shown in the Year-0 row.

Next, the formulas for the first year are written, as shown in Figure A–1a. The interest owed (cell B10) equals the interest rate ($A$4) times the balance due for Year 0 (D9). The principal payment (cell C10) equals the annual payment ($A$6) minus the interest owed and paid (B10). Finally, the balance due (cell D10) equals the balance due for the previous year (D9) minus the principal payment (C10). The results are shown in Figure A–1a.

Now cells A10 to D10 are selected for Year 1. By dragging down on the right corner of D10, the entire row can be copied for Years 2 through 5. Note that if you use cut and paste, then you must complete the year column separately (dragging increments the year, but cutting and pasting does not). The results are shown in Figure A–1b.

	A	B	C	D	E
1	Entered Data				
2	5000	Loan Amount			
3	5	Number of Payments			
4	8%	Interest Rate			
5	Computed Data				
6	$1,252.28	Loan Payment			
7					
8	Year	Interest Owed	Principal Payment	Balance Due	
9	0			5000.00	
10	1	400.00	852.28	4147.72	=D9−C10
11					
12		=$A$4*D9		=$A$6−B10	

(a)

	A	B	C	D	E
8	Year	Interest Owed	Principal Payment	Balance Due	
9	0			5000.00	
10	1	400.00	852.28	4147.72	
11	2	331.82	920.46	3227.25	
12	3	258.18	994.10	2233.15	
13	4	178.65	1073.63	1159.52	
14	5	92.76	1159.52	0.00	=D13−C14
15					
16		=$A$4*D13		=$A$6−B14	

(b)

FIGURE A-1 *(a)* Year 1 amortization schedule. *(b)* Completed amortization schedule.

This appendix has introduced the basics of spreadsheets. Chapter 2 uses spreadsheets and simple bar charts to draw cash flow diagrams. Chapters 4 to 15 each have spreadsheet sections. These are designed to develop spreadsheet modeling skills and to reinforce your understanding of engineering economy. As spreadsheet packages are built around using the computer mouse to click on cells and items in charts, there is usually an intuitive connection between what you would like to do and how to do it. The best way to learn how to use the spreadsheet package is to simply play around with it. In addition, as you look at the menu choices, you will find new commands that you hadn't thought of but will find useful.

TIME VALUE OF MONEY CALCULATIONS USING SPREADSHEETS AND CALCULATORS

Spreadsheets and financial calculators are perfect tools for solving economic problems with financial functions that are very similar to the engineering economy factors—but more powerful. Now there are even smart phone apps with the same capabilities. Some types of programmable calculators can be used in the FE exam, making interest tables unnecessary.

This appendix focuses on the use of spreadsheets and calculators to solve simple time value of money problems. Included are the following topics:

- Basics of a 5-Button Spreadsheet Calculator
- Spreadsheets, Calculators, and the FE Exam
- Additional Spreadsheet and Calculator Capabilities
- More Complex Examples
- Possible Errors and Their Solutions

BASICS OF A 5-BUTTON SPREADSHEET CALCULATOR

Figure B-1 shows a variety of cash flow diagrams and a five button calculator template that can be used to solve economic problems quickly and easily.

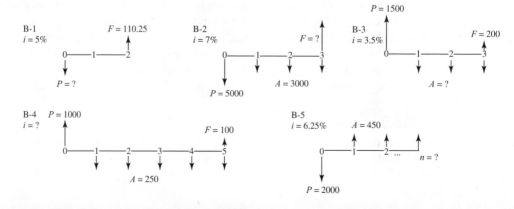

For example B-1, the input variables from the cash flow diagram are placed in cells B2 through F2, with the interest rate in cell B2, the number of periods in cell C2, etc. Cell H2 calculates the PV using the appropriate cell addresses for the input variables. The equations used in column H are shown in column I. Spreadsheets such as Excel and Google docs use this type of format; others such as Open Office and Quattro Pro use a slightly different format for equivalent functions. Any of the spreadsheet functions may be copied to a new row and then new data can be inserted to solve similar problems. Thus, the five *annuity functions* in Figure B-1 are collectively a 5-button financial calculator. The same inputs may be keyed into a financial or programmable calculator to obtain identical results. When using a calculator, it is important to enter all variables if you do not know what value was entered last. You could have an unknown value residing in the calculator's memory. Additional examples of using the 5-button spreadsheet calculator can be found in Chapters 3 and 4.

EXAMPLE B1–B5

5-BUTTON SOLUTION

	A	B	C	D	E	F	G	H	I
1	Example	i	N	PMT	PV	FV	Solve for	Answer	Formula
2	B-1	5.0%	2	0		110.25	PV	-$100.00	=PV(B2,C2,D2,F2)
3	B-2	7.0%	3	-3000	-5000		FV	$15,770	=FV(B3,C3,D3,E3)
4	B-3	3.5%	3		1500	200	PMT	-$599.79	=PMT(B4,C4,E4,F4)
5	B-4		5	-250	1000	100	RATE	5.15%	=RATE(C5,D5,E5,F5)
6	B-5	6.25%		450	-2000	0	N	5.37	=NPER(B6,D6,E6,F6)

FIGURE B-1 Cash flow diagrams and the 5-button spreadsheet calculator.

EXAMPLE B-1 FACTOR SOLUTION

The present value of a cash flow of $110.25, two years in the future, at an interest rate of 5% is −$100.00. Note that this follows the convention of positive numbers for inflows and negative numbers for outflows. Use of the spreadsheet or calculator avoids looking up factors, which is a primary source of errors.

Using interest factors,

B-1 $\quad P = 110.25(P/F, 5\%, 2) = 110.25(0.9070) = \$100.00$

EXAMPLE B-2 FACTOR SOLUTION

The FV function can determine the future value of both a P and an A in a single calculation. Note that negative values of PV and PMT return a positive value for FV of $15,770.

Solving B-2 using factors and the interest tables requires two lookups, two multiplications, and one addition, because the tabulated factors can only convert from one type of cash flow to another.

B-2 $\quad F = 3000(F/A, 7\%, 3) + 5000(F/P, 7\%, 3) = 3000(3.215) + 5000(1.225) = \$15,770$

EXAMPLE B-3 FACTOR SOLUTION

A uniform cash flow can be determined using the PMT function, with P and F being entered in a single calculation.

Using factors requires several steps, as shown here. If the 3.5% table were not available, this would also require interpolation between the 3% and the 4% tables. Due to the round off of the interest factors to 4 significant digits, the factor answer is slightly less precise.

B-3 $\quad A = 1500(A/P, 3.5\%, 3) + 200(A/F, 3.5\%, 3) = 1500(0.3569) + 200(0.3219)$

$\qquad = \$599.73$

EXAMPLE B-4 FACTOR SOLUTION

An unknown interest rate can be solved in a single computation, even though there are three types of cash flows: a P, an F, and an A. The RATE function requires at least one positive cash flow (inflow) and at least one negative cash flow (outflow). Using factors and interest tables would require solving the following equation (or a similar one):

B-4 $\quad 0 = 1000 - 250(P/A, i, 5) + 100(P/F, i, 5)$

This requires either an iterative solution or using interpolation.

$$\text{at } 5\%, 0 = 1000 - 250(4.329) + 100(0.7835) = -3.90$$
$$\text{at } 6\%, 0 = 1000 - 250(4.212) + 100(0.7473) = 21.7$$

Interpolating, $i = 5.15\%$

Spreadsheets and calculators are much more efficient than factors when solving for unknowns that require interpolation, such as the interest rate.

EXAMPLE B-5 FACTOR SOLUTION

The NPER function will display the answer in fractional periods; in this case, 5.37 periods. Using factors to solve for the number of periods again requires the use of interpolation or iterative solutions.

B-5 $\quad 0 = -2000 + 450(P/A, 6.25\%, n)$

$\qquad (P/A, 6.25\%, n) = 2000/450 = 4.444$

This will require interpolation to find two variables, one interpolation to find a set of values for n at 6.25% (because there is no 6.25% table available), and another for determining the answer. These will not be performed here, but yield an answer of 5.37 periods.

There are many advantages to using spreadsheets and financial calculators instead of, or in addition to, using the interest tables. These include:

- Time is saved because problems are entered directly without having to look up factors in tables, then copying those factors. Problems are solved with the use of a few key strokes.
- There is less chance for error because there are fewer steps. Many errors are made in looking up factors due to looking down the wrong column or row. Transcribing errors, or finding a factor but writing it down wrong, are minimized.

- There is no need to interpolate when determining an interest rate or number of periods. This greatly speeds this type of calculation.
- Students who use a spreadsheet will leave an electronic, student accessible paper trail. Once a spreadsheet is successfully built, such as the one in Figure B-1, students can access these saved files for studying, performing homework, and perhaps during a test. This can aid in students' learning and comprehension by focusing time and attention on setting up problems rather than spending time doing financial arithmetic.

SPREADSHEETS, CALCULATORS, AND THE FE EXAM

The naming of variables is slightly different between engineering economy, spreadsheets, and calculators. Table B-1 shows the typical naming conventions.

Spreadsheets use Equation B-1 to solve for the unknown variable. Equation B-2 rewrites this in standard factor notation.

$$PMT \left[\frac{1 - (1 + i)^{-n}}{i} \right] + FV(1 + i)^{-n} + PV = 0 \tag{B-1}$$

$$A(P/A, i, n) + F(P/F, i, n) + P = 0 \tag{B-2}$$

Financial calculators use Equation B-3 to solve for the unknown variable. Note that Equation B-3 is written so that i is entered as a percentage, so 8 is entered for 8%, which is how the interest rate is entered for financial calculators. How the interest rate is entered is the single difference between how spreadsheets and calculators solve time value of money problems.

$$PMT \left[\frac{1 - (1 + i/100)^{-n}}{i/100} \right] + FV(1 + i/100)^{-n} + PV = 0 \tag{B-3}$$

Since the factors are all positive, one of *PMT*, *FV*, or *PV* (or *A*, *F*, and *P*) must be different in sign from the other two. If two positive cash flows are entered to solve for the third cash flow, then that third value must be negative to solve the equation. These equations are the foundation for the sign convention that is used in financial calculators and the spreadsheet annuity functions PV, FV, PMT, RATE, and NPER.

There are three classes of calculators that are of interest. First, there are many *financial* or *business* calculators that have buttons labeled i, N, *PMT*, *PV*, and *FV* or equivalents. The

TABLE B–1 Naming Conventions in Engineering Economy, Spreadsheets, and Calculators

Variable	Engineering Economy	Spreadsheets	Typical Calculator Button
Present value	P	PV	PV
Future value	F	FV	FV
Uniform series	A	PMT	PMT
Interest rate	i	RATE	I/Y
Number of periods	n	NPER	N

button for the uniform series cash flow, *A*, is labeled *PMT*, because one of the most important uses of these calculators is for calculating loan payments (*PMT*). Examples include the Texas Instrument BAII Plus and BAII Plus Professional, the Sharp EL-738C, and the Hewlett Packard 10bII. We no longer recommend the HP 12C for new purchases since when solving for *n*, the 12C rounds fractional values of *n* up to an integer value (see Example B–8). These calculators also typically have the capability to find the NPV of complex cash flow patterns, but we suggest using spreadsheets for more complex patterns.

There are also *graphing* calculators that have the time value of money calculations in a menu. There are menu entries for *i*, *N*, *PMT*, *PV*, and *FV* or equivalents. Example calculators include the Texas Instrument 83 and 84 Plus.

Finally, there are programmable scientific calculators where Equation B-3 can be entered to create the menu that is built into the graphing calculators. For example, Chapter 17 of the manuals for the Hewlett Packard 33s and 35s details this. These chapters use a non-standard notation of *B* or balance for *PV* and *P* or payment for *PMT*. We suggest using standard notation and substituting *P* for *B* and *A* for *P*. Note that on the HP 35s this changes the checksum from 382E to 8DD6. The HP 33s and HP 35s calculators are currently allowed for use on the Fundamentals of Engineering (FE) exam, while the financial and graphing calculators are not.

There are too many variations of labeling and physical layout of the keys and too many financial calculators for a text to provide details. However, the website www.TVMCalcs.com provides good tutorials on financial calculators, and calculator manuals are available for download from the manufacturers.

Some calculators are shipped from the factory with a setting of 12 months per year, so that *n* will be entered as years for loans with monthly payments. We recommend that this be changed to 1 payment per "year or period" and left on this setting. Then you know to enter *n* as the number of periods or payments, and there is no confusion for problems with different length periods.

Phone apps are now widely available that perform as financial calculators, making it possible to carry the capability anywhere.

ADDITIONAL SPREADSHEET AND CALCULATOR CAPABILITIES

When initially building a spreadsheet, equations must be entered into a cell to solve the problem. In Figure B-1, the following equation is entered into cell H2 to solve for the present value:

$$= \text{PV}(\text{rate}, \text{nper}, \text{pmt}, [\text{fv}], [\text{type}])$$

The [ ] denotes that these are optional values where the default is zero. If nothing is entered for the future value, it will be assumed to have a zero value. If no value is entered for the type, periodic cash flows (*A*) will be assumed to occur at the end of the time periods.

[type] = 0 (default)	End of period cash flows
[type] > 0	Beginning period cash flows

Most financial calculators can be set to either end of period (default) or beginning of period cash flows.

The RATE function requires a combination of positive and negative inputs in order to calculate an interest rate.

$$= \text{RATE}(\text{nper}, \text{pmt}, \text{pv}, [\text{fv}], [\text{type}], [\text{guess}])$$

Note that the RATE function contains an optional input *guess*, where you can guess what the rate will be. If you do not enter a value, the default is 10%. If the RATE function does not provide an answer, change the guess. An entry is usually not required. For more see the last section of this Appendix.

MORE COMPLEX EXAMPLES

The following examples demonstrate problems that are more complex, but still make use of the basic spreadsheet annuity functions.

- Example B–6 requires conversion of annual rates and periods to monthly.
- Example B–7 is a PV problem including an initial payment and a salvage value.
- Example B–8 solves a discounted payback problem.
- Example B–9 includes cash flows at the beginning of the period.
- Example B–10 is an EAC problem with multiple cash flows.
- Example B–11 involves the doubling, tripling, and quadrupling of money.

EXAMPLE B–6

Kris is buying a used car and needs to find the monthly payment. The loan is for $14,000, and the dealer is willing to offer financing for 5 years at 6.5% interest. What is the monthly payment?

SOLUTION

First, we need to convert the period and interest rate to months. There are 60 ($= 5 \times 12$) monthly payments at a monthly interest rate of $6.5\%/12 = 0.5417\%$. The required monthly payment is $273.93.

	A	B	C	D	E	F	G	H	I
1	Problem	*i*	*n*	*PMT*	*PV*	*FV*	Solve for	Answer	Formula
2	B-6	0.5417%	60		-14000	0	PMT	$273.93	=PMT(B2,C2,E2,F2)
3									PMT(rate, nper, pv, [fv], [type])

This problem can also be solved using a financial or programmable calculator. The keys on most calculators are the same as the spreadsheet variables, so $PMT(i, n, P, F) = PMT(0.5417\%, 60, -14000, 0) = \$273.93$. Note that the calculator requires the same sign convention as spreadsheets.

EXAMPLE B–7

Automating a process will cost $8000 now, but it will save $2200 annually for 5 years. The machinery will have a salvage value of $1500. What is the present worth of the machinery if the interest rate is 10%?

SOLUTION

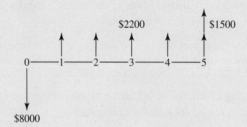

Let us first solve this with the tabulated factors.

$$PW = -8000 + 2200(P/A, 10\%, 5) + 1500(P/F, 10\%, 5)$$

$$= -8000 + 2200(3.791) + 1500(0.6209)$$

$$= -8000 + 8340 + 931 = \$1271$$

The spreadsheet can find the combined present value of the $2200 annual savings and the salvage value of $1500 because the interest rate and number of periods are the same.

	A	B	C	D	E	F	G	H	I
1	Problem	i	n	PMT	PV	FV	Solve for	Answer	Formula
2	B-7	10.0%	5	2200		1500	PV	-9271.11	=PV(B2,C2,D2,F2)
3					-8000		PV	1271.11	=-G2+E3

Notice that the present value of the future cash flows is negative when the A and F values are entered as positive numbers. This comes from solving Equation B-1. Since in this case we want the *positive* PV of the returns to the project, we can change the sign of the answer in step 2, subtracting the $8000 initial cash flow to find the PV of $1271.

Using a calculator, $PV(i, n, A, F) = PV(10, 5, 2200, 1500) = -9271.11$. The calculator finds the combined present worth of the $2200 annual savings and the salvage value of $1500. The initial cash flow of $8000 is subtracted to find the PV of $1271.11. Like the spreadsheet, positive values for A and F produce a negative value for P.

EXAMPLE B–8

A firm might purchase a computerized quality control system. The proposed system will cost $76,000, but is expected to save $22,000 each year in reduced overtime. The firm requires that all cost reduction projects have a discounted payback of no more than 4 years with a 10% interest rate. Should the firm invest in the new system?

SOLUTION

	A	B	C	D	E	F	G	H
1	Problem	i	n	PMT	PV	FV	Solve for	Answer
2	B-8	10.0%		22000	-76000	0	NPER	4.45

The time for the system to pay for itself can be found using the NPER function; the discounted payback period is 4.45 years. Because a proposed project can have a discounted payback of no more than 4 years, the project would not be approved.

The problem can also be solved with a calculator using $N(i, A, P, F) = N(10\%, 22000, -76000, 0) = 4.45$ years. Note that the HP 12C calculator will report 5 years. In fact, the HP 12C will report $N = 5$ years when the payback period is as small as 4.005 periods.

EXAMPLE B–9

Some equipment is needed for a 4-year project. It can be leased for $15,000 annually, or it can be purchased for $60,000 at the beginning and sold for $35,000 at the end. What is the rate of return for owning the equipment rather than leasing it?

SOLUTION

Because the lease payments occur at the beginning of each period, [type] is set to any number above 0 (such as 1) to automatically shift the cash flows into *beginning* rather than *end* of period. This affects only the A and not the F.

	A	B	C	D	E	F	G	H
1	Problem	i	n	PMT	PV	FV	Solve for	Answer
2	B-9		4	15000	-60000	35000	RATE	23.49%

A financial calculator must also be shifted into *begin* rather than *end* mode, then enter $i(n, A, P, F) = i(4, 15000, -60000, 35000) = 23.49\%$.

Without knowledge that the rate is nearly 25%, using the tabulated factors would require multiple trial solutions to solve the following 3-factor equation.

$$0 = -60,000 + 15,000(F/P, i, 1)(P/A, i, 4) + 35,000(P/F, i, 4)$$

EXAMPLE B–10

What is the EAC to keep a piece of new equipment operating? The warranty covers repair costs for the first 2 years, but expected repair costs are $1500 per year for the rest of the 7-year life. Normal maintenance is expected to cost $1800 annually. There is an overhaul costing $8500 at the end of year 4. The firm's interest rate is 8%.

SOLUTION

Whether solved with tables, spreadsheets, or a financial calculator, the first step is to draw the cash flow diagram.

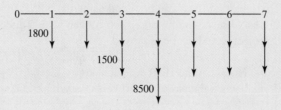

Let us first solve this with the tabulated factors.

$$EAC = 1800 + 1500(F/A, 8\%, 5)(A/F, 8\%, 7) + 8500(P/F, 8\%, 4)(A/P, 8\%, 7)$$

$$= 1800 + 1500(5.867)(0.1121) + 8500(0.7350)(0.1921)$$

$$= 1800 + 986.5 + 1200 = \$3986.5$$

With a spreadsheet, it is easiest to separately calculate the EUAC values for the repair costs and the overhaul.

	A	B	C	D	E	F	G	H
		i	n	*PMT*	*PV*	*FV*	Solve for	Answer
1	Problem							
2	Repair	8%	5	1500	0		FV	-8799.90
3		8%	7		0	-8799.90	PMT	986.23
4								
5	Overhaul	8%	4	0		8500	PV	-6247.75
6		8%	7		-6247.75	0	PMT	1200.02
7								
8	Maint.			1800			EAC	$3,986.25

This is very similar with a financial calculator. For the repair costs, $FV(i, n, A, P) = FV(8, 5, 1500, 0) = -8,799.90$. Since the i and P values are unchanged, they do not need to be re-entered. The only entry needed is $N = 7$, since the F was just calculated. $PMT(i, n, P, F) = PMT(8\%, 7, 0, -8799.90) = \$986.23$.

For the overhaul costs the i value need not be re-entered, $PV(i, n, A, F) = PV(8\%, 4, 0, 8500) = -6,247.75$. Now N is changed to 7 and F to 0, then solving for $PMT(i, n, P, F) = PMT(8\%, 7, -6247.75, 0) = \$1200.02$.

Maintenance is $1800 per year.

Adding the three values is the same as before, and the EAC is $3986.25.

EXAMPLE B–11

How long does it take for your money to double if you can earn 7% on your investment? Triple? Quadruple?

SOLUTION

	A	B	C	D	E	F	G	H
1	Problem	i	n	PMT	PV	FV	Solve for	Answer
2	B-11	7.0%		0	-1	2	NPER	10.24
3		7.0%		0	-1	3	NPER	16.24
4		7.0%		0	-1	4	NPER	20.49

Using a financial calculator, the setup is similar; for doubling, $n(i, A, P, F) = n(7, 0, -1, 2) = 10.24$.

The problem can also be solved using functions, but requires interpolation, or can be solved mathematically.

$$F = P(1 + i)^n$$

$$2 = 1(1.07)^n$$

$$\ln(2) = n \ln(1.07)$$

$$n = \ln(2)/\ln(1.07) = 0.6931/0.06766 = 10.24$$

POSSIBLE ERRORS AND THEIR SOLUTIONS

RATE and IRR

When using RATE or IRR, the #NUM! error may occur. Both of these functions are iterative; they home in on an answer using multiple attempts or iterations. The default starting point (guess) is 10%, and if a specific answer cannot be found within 20 iterations, the error value appears. This usually happens for interest rates that are very high or very negative, but even moderate answers can cause difficulty. The problem is usually solved by trying values for [guess] that approximate the answer.

EXAMPLE B–12

An account having $800,000 will be used to make a series of 20 annual payments of $150,000 each. What interest rate is needed to fund the payments?

SOLUTION

A simple rate problem can unexpectedly result in an error message, but are usually corrected by inserting a [guess]. Notice that the guess does not need to be very accurate.

	A	B	C	D	E	F	G	H	I
1	Problem	i	n	PMT	PV	FV	Solve for	Answer	Formula
2	B-12		20	150,000	-800,000	0	RATE	#NUM!	=RATE(C2,D2,E2,F2)
3			20	150,000	-800,000	0	RATE	18.1%	=RATE(C3,D3,E3,F3,0,30%)

NPER

The function NPER is also iterative—like RATE and IRR. Unfortunately, NPER does not have a [guess] option to help. NPER sometimes returns an error if the value is very large. The answer can sometimes be found by using GOAL SEEK, as shown in Example B–13.

EXAMPLE B–13

An account having $600,000 will be used to make a series of annual payments of $45,000 each. The account returns 7.5% per year. How many years will payments be made?

SOLUTION

Using NPER returns an error showing that it could not solve the problem. NPER is increasing very rapidly as the interest increases by a very small amount. This can be solved by finding a value of *n* where the PV or FV = 0. GOAL SEEK returns a value of 617 plus years. SOLVER can also be used to provide the same answer.

	A	B	C	D	E	F	G	H	I
1	Problem	*i*	*n*	*PMT*	*PV*	*FV*	Solve for	Answer	Formula
2	B-13	7.5%		45,000	-600,000	0	NPER	#NUM!	=NPER(B2,D2,E2)
3		7.5%	617.3	45,000	-600,000	0	FV	0	=FV(B5,C5,D5,E5)

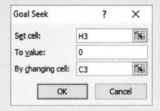

Goal Seek   ? ×

Set cell: H3

To value: 0

By changing cell: C3

OK   Cancel

PROBLEMS

Present Worth

B-1   A process redesign will cost $70,000 now, but it will save $18,000 annually for 6 years. The new machinery will have a salvage value of $12,500. What is the present worth of the machinery, if the interest rate is 12%?

B-2 Ⓐ   Moving to a new facility will save $400,000 annually for 20 years. The cost of building the facility and moving is $2.5 million now. The facility will have a salvage value of $100,000. What is the facility's PW, if the interest rate is 12%?

B-3   A lottery pays the winner $1 million in 20 equal annual payments of $50,000. The first payment will be made at the end of the second year. What is the present worth if the winner's annual interest rate is 5.25%?

Equivalent Annual Worth

B-4 Ⓐ   You need to save $20,000 to buy a car in 3 years. At a 2% nominal annual interest rate, how much do you need to save each month?

B-5   You can receive lottery winnings of either $800,000 now or $100,000 per year for the next 10 years. If your interest rate is 5% per year, which do you prefer?

B-6 Ⓐ   A new road will cost $45 million to build, and $2 million annually to maintain and operate over its 50-year life. The roadbed and right-of-way are estimated to have a salvage value of $15 million. If the state highway department uses an interest rate of 5%, what is the EUAC for the road?

B-7   A new car will cost $24,000 to buy and $5500 annually to operate. If it is sold for $9300 after 6 years, what is the EUAC? Assume that the owner's interest rate is 3% for the time value of money.

Price of a Bond

B-8 A $1000 bond has 15 more years to maturity, an interest rate of 6%, and it pays interest semi-annually. If the current market interest rate is 5%, what is the price of the bond?
(A)

B-9 A $1000 bond has 12 more years to maturity, an interest rate of 6%, and it pays interest semi-annually. If the current market interest rate is 8%, what is the price of the bond?

B-10 An investor purchased a 10-year, $20,000 corporate bond just over 12 months ago. The bond pays 6.5% interest, due semi-annually. The current market interest rate for similar bonds has increased to 7.25%. What price will this bond sell for today?
(A)

Loan Payment

B-11 What is the monthly payment for a 5-year car loan at a nominal interest rate of 8.5%? The loan's initial balance is $16,885.

B-12 A couple is buying a house, but they need a mortgage for $150,000. A 15-year loan can be obtained for 3.25% annual interest. What is the monthly payment?
(A)

Number of Periods

B-13 A student owing $17,565 on a credit card has decided to use only a debit card in the future. The nominal annual rate on the credit card is 13.8%. If the student makes monthly payments of $250, how long is it until the credit card is paid off?

B-14 A homeowner may install solar panels on the roof of her house. The total installed cost is $15,000 after federal and state credits. This solar system will meet the monthly electricity need, which costs $120
(A)
(G)

per month. The homeowner's interest rate is 3% annually.

(a) If electricity prices remain constant, how long will it be until the system pays for itself?

(b) If electricity prices increase in the future, how will this affect the installation paying for itself?

Internal Rate of Return

B-15 An automated storage and retrieval system will cost $135,000, but it will save $33,000 annually in labor costs. The system's salvage value is expected to be $20,000 when it is renovated in 10 years. What is the rate of return on the project? Do people need to be laid off to justify the installation of the new system?
(E)

B-16 Moving to a new facility will save $400,000 annually for 20 years. The cost of building the facility and moving is $2.5 million now. The facility will have a salvage value of $100,000. What is the rate of return on this project?
(A)

B-17 A new road will cost $45 million to build, and $2 million annually to maintain and operate over its 50-year life. The roadbed and right-of-way are estimated to have a salvage value of $15 million. If user benefits are estimated at $5.5 million annually, what is the rate of return on this road?

B-18 Some equipment is needed for a construction project. It can be leased for $150,000 annually, or it can be purchased for $900,000 at the beginning and sold for $225,000 at the end of 3 years. What is the rate of return for owning the equipment rather than leasing it?
(A)

COMPOUND INTEREST TABLES

Values of Interest Factors When *n* Equals Infinity

Single Payment:

$(F/P, i, \infty) = \infty$

$(P/F, i, \infty) = 0$

Arithmetic Gradient Series:

$(A/G, i, \infty) = 1/i$

$(P/G, i, \infty) = 1/i^2$

Uniform Payment Series:

$(A/F, i, \infty) = 0$

$(A/P, i, \infty) = i$

$(F/A, i, \infty) = \infty$

$(P/A, i, \infty) = 1/i$

1/4%

Compound Interest Factors

1/4%

	Single Payment		Uniform Payment Series				Arithmetic Gradient		
	Compound Amount Factor	Present Worth Factor	Sinking Fund Factor	Capital Recovery Factor	Compound Amount Factor	Present Worth Factor	Gradient Uniform Series	Gradient Present Worth	
	Find F Given P	Find P Given F	Find A Given F	Find A Given P	Find F Given A	Find P Given A	Find A Given G	Find P Given G	
n	F/P	P/F	A/F	A/P	F/A	P/A	A/G	P/G	n
1	1.003	.9975	1.0000	1.0025	1.000	0.998	0.000	0.000	1
2	1.005	.9950	.4994	.5019	2.003	1.993	0.499	0.995	2
3	1.008	.9925	.3325	.3350	3.008	2.985	0.998	2.980	3
4	1.010	.9901	.2491	.2516	4.015	3.975	1.497	5.950	4
5	1.013	.9876	.1990	.2015	5.025	4.963	1.995	9.901	5
6	1.015	.9851	.1656	.1681	6.038	5.948	2.493	14.826	6
7	1.018	.9827	.1418	.1443	7.053	6.931	2.990	20.722	7
8	1.020	.9802	.1239	.1264	8.070	7.911	3.487	27.584	8
9	1.023	.9778	.1100	.1125	9.091	8.889	3.983	35.406	9
10	1.025	.9753	.0989	.1014	10.113	9.864	4.479	44.184	10
11	1.028	.9729	.0898	.0923	11.139	10.837	4.975	53.913	11
12	1.030	.9705	.0822	.0847	12.167	11.807	5.470	64.589	12
13	1.033	.9681	.0758	.0783	13.197	12.775	5.965	76.205	13
14	1.036	.9656	.0703	.0728	14.230	13.741	6.459	88.759	14
15	1.038	.9632	.0655	.0680	15.266	14.704	6.953	102.244	15
16	1.041	.9608	.0613	.0638	16.304	15.665	7.447	116.657	16
17	1.043	.9584	.0577	.0602	17.344	16.624	7.944	131.992	17
18	1.046	.9561	.0544	.0569	18.388	17.580	8.433	148.245	18
19	1.049	.9537	.0515	.0540	19.434	18.533	8.925	165.411	19
20	1.051	.9513	.0488	.0513	20.482	19.485	9.417	183.485	20
21	1.054	.9489	.0464	.0489	21.534	20.434	9.908	202.463	21
22	1.056	.9465	.0443	.0468	22.587	21.380	10.400	222.341	22
23	1.059	.9442	.0423	.0448	23.644	22.324	10.890	243.113	23
24	1.062	.9418	.0405	.0430	24.703	23.266	11.380	264.775	24
25	1.064	.9395	.0388	.0413	25.765	24.206	11.870	287.323	25
26	1.067	.9371	.0373	.0398	26.829	25.143	12.360	310.752	26
27	1.070	.9348	.0358	.0383	27.896	26.078	12.849	335.057	27
28	1.072	.9325	.0345	.0370	28.966	27.010	13.337	360.233	28
29	1.075	.9301	.0333	.0358	30.038	27.940	13.825	386.278	29
30	1.078	.9278	.0321	.0346	31.114	28.868	14.313	413.185	30
36	1.094	.9140	.0266	.0291	37.621	34.387	17.231	592.499	36
40	1.105	.9049	.0238	.0263	42.014	38.020	19.167	728.740	40
48	1.127	.8871	.0196	.0221	50.932	45.179	23.021	1 040.055	48
50	1.133	.8826	.0188	.0213	53.189	46.947	23.980	1 125.777	50
52	1.139	.8782	.0180	.0205	55.458	48.705	24.938	1 214.588	52
60	1.162	.8609	.0155	.0180	64.647	55.653	28.751	1 600.085	60
70	1.191	.8396	.0131	.0156	76.395	64.144	33.481	2 147.611	70
72	1.197	.8355	.0127	.0152	78.780	65.817	34.422	2 265.557	72
80	1.221	.8189	.0113	.0138	88.440	72.427	38.169	2 764.457	80
84	1.233	.8108	.0107	.0132	93.343	75.682	40.033	3 029.759	84
90	1.252	.7987	.00992	.0124	100.789	80.504	42.816	3 446.870	90
96	1.271	.7869	.00923	.0117	108.349	85.255	45.584	3 886.283	96
100	1.284	.7790	.00881	.0113	113.451	88.383	47.422	4 191.242	100
104	1.297	.7713	.00843	.0109	118.605	91.480	49.252	4 505.557	104
120	1.349	.7411	.00716	.00966	139.743	103.563	56.508	5 852.112	120
240	1.821	.5492	.00305	.00555	328.306	180.312	107.586	19 398.985	240
360	2.457	.4070	.00172	.00422	582.745	237.191	152.890	36 263.930	360
480	3.315	.3016	.00108	.00358	926.074	279.343	192.670	53 820.752	480

1/2%

Compound Interest Factors

1/2%

	Single Payment		Uniform Payment Series				Arithmetic Gradient		
	Compound Amount Factor	Present Worth Factor	Sinking Fund Factor	Capital Recovery Factor	Compound Amount Factor	Present Worth Factor	Gradient Uniform Series	Gradient Present Worth	
	Find F Given P	Find P Given F	Find A Given F	Find A Given P	Find F Given A	Find P Given A	Find A Given G	Find P Given G	
n	F/P	P/F	A/F	A/P	F/A	P/A	A/G	P/G	n
1	1.005	.9950	1.0000	1.0050	1.000	0.995	0	0	1
2	1.010	.9901	.4988	.5038	2.005	1.985	0.499	0.991	2
3	1.015	.9851	.3317	.3367	3.015	2.970	0.996	2.959	3
4	1.020	.9802	.2481	.2531	4.030	3.951	1.494	5.903	4
5	1.025	.9754	.1980	.2030	5.050	4.926	1.990	9.803	5
6	1.030	.9705	.1646	.1696	6.076	5.896	2.486	14.660	6
7	1.036	.9657	.1407	.1457	7.106	6.862	2.980	20.448	7
8	1.041	.9609	.1228	.1278	8.141	7.823	3.474	27.178	8
9	1.046	.9561	.1089	.1139	9.182	8.779	3.967	34.825	9
10	1.051	.9513	.0978	.1028	10.228	9.730	4.459	43.389	10
11	1.056	.9466	.0887	.0937	11.279	10.677	4.950	52.855	11
12	1.062	.9419	.0811	.0861	12.336	11.619	5.441	63.218	12
13	1.067	.9372	.0746	.0796	13.397	12.556	5.931	74.465	13
14	1.072	.9326	.0691	.0741	14.464	13.489	6.419	86.590	14
15	1.078	.9279	.0644	.0694	15.537	14.417	6.907	99.574	15
16	1.083	.9233	.0602	.0652	16.614	15.340	7.394	113.427	16
17	1.088	.9187	.0565	.0615	17.697	16.259	7.880	128.125	17
18	1.094	.9141	.0532	.0582	18.786	17.173	8.366	143.668	18
19	1.099	.9096	.0503	.0553	19.880	18.082	8.850	160.037	19
20	1.105	.9051	.0477	.0527	20.979	18.987	9.334	177.237	20
21	1.110	.9006	.0453	.0503	22.084	19.888	9.817	195.245	21
22	1.116	.8961	.0431	.0481	23.194	20.784	10.300	214.070	22
23	1.122	.8916	.0411	.0461	24.310	21.676	10.781	233.680	23
24	1.127	.8872	.0393	.0443	25.432	22.563	11.261	254.088	24
25	1.133	.8828	.0377	.0427	26.559	23.446	11.741	275.273	25
26	1.138	.8784	.0361	.0411	27.692	24.324	12.220	297.233	26
27	1.144	.8740	.0347	.0397	28.830	25.198	12.698	319.955	27
28	1.150	.8697	.0334	.0384	29.975	26.068	13.175	343.439	28
29	1.156	.8653	.0321	.0371	31.124	26.933	13.651	367.672	29
30	1.161	.8610	.0310	.0360	32.280	27.794	14.127	392.640	30
36	1.197	.8356	.0254	.0304	39.336	32.871	16.962	557.564	36
40	1.221	.8191	.0226	.0276	44.159	36.172	18.836	681.341	40
48	1.270	.7871	.0185	.0235	54.098	42.580	22.544	959.928	48
50	1.283	.7793	.0177	.0227	56.645	44.143	23.463	1 035.70	50
52	1.296	.7716	.0169	.0219	59.218	45.690	24.378	1 113.82	52
60	1.349	.7414	.0143	.0193	69.770	51.726	28.007	1 448.65	60
70	1.418	.7053	.0120	.0170	83.566	58.939	32.468	1 913.65	70
72	1.432	.6983	.0116	.0166	86.409	60.340	33.351	2 012.35	72
80	1.490	.6710	.0102	.0152	98.068	65.802	36.848	2 424.65	80
84	1.520	.6577	.00961	.0146	104.074	68.453	38.576	2 640.67	84
90	1.567	.6383	.00883	.0138	113.311	72.331	41.145	2 976.08	90
96	1.614	.6195	.00814	.0131	122.829	76.095	43.685	3 324.19	96
100	1.647	.6073	.00773	.0127	129.334	78.543	45.361	3 562.80	100
104	1.680	.5953	.00735	.0124	135.970	80.942	47.025	3 806.29	104
120	1.819	.5496	.00610	.0111	163.880	90.074	53.551	4 823.52	120
240	3.310	.3021	.00216	.00716	462.041	139.581	96.113	13 415.56	240
360	6.023	.1660	.00100	.00600	1 004.5	166.792	128.324	21 403.32	360
480	10.957	.0913	.00050	.00550	1 991.5	181.748	151.795	27 588.37	480

3/4%

Compound Interest Factors

3/4%

	Single Payment		Uniform Payment Series				Arithmetic Gradient		
	Compound Amount Factor	Present Worth Factor	Sinking Fund Factor	Capital Recovery Factor	Compound Amount Factor	Present Worth Factor	Gradient Uniform Series	Gradient Present Worth	
	Find F Given P F/P	Find P Given F P/F	Find A Given F A/F	Find A Given P A/P	Find F Given A F/A	Find P Given A P/A	Find A Given G A/G	Find P Given G P/G	
n									n
1	1.008	.9926	1.0000	1.0075	1.000	0.993	0	0	1
2	1.015	.9852	.4981	.5056	2.008	1.978	0.499	0.987	2
3	1.023	.9778	.3308	.3383	3.023	2.956	0.996	2.943	3
4	1.030	.9706	.2472	.2547	4.045	3.926	1.492	5.857	4
5	1.038	.9633	.1970	.2045	5.076	4.889	1.986	9.712	5
6	1.046	.9562	.1636	.1711	6.114	5.846	2.479	14.494	6
7	1.054	.9490	.1397	.1472	7.160	6.795	2.971	20.187	7
8	1.062	.9420	.1218	.1293	8.213	7.737	3.462	26.785	8
9	1.070	.9350	.1078	.1153	9.275	8.672	3.951	34.265	9
10	1.078	.9280	.0967	.1042	10.344	9.600	4.440	42.619	10
11	1.086	.9211	.0876	.0951	11.422	10.521	4.927	51.831	11
12	1.094	.9142	.0800	.0875	12.508	11.435	5.412	61.889	12
13	1.102	.9074	.0735	.0810	13.602	12.342	5.897	72.779	13
14	1.110	.9007	.0680	.0755	14.704	13.243	6.380	84.491	14
15	1.119	.8940	.0632	.0707	15.814	14.137	6.862	97.005	15
16	1.127	.8873	.0591	.0666	16.932	15.024	7.343	110.318	16
17	1.135	.8807	.0554	.0629	18.059	15.905	7.822	124.410	17
18	1.144	.8742	.0521	.0596	19.195	16.779	8.300	139.273	18
19	1.153	.8676	.0492	.0567	20.339	17.647	8.777	154.891	19
20	1.161	.8612	.0465	.0540	21.491	18.508	9.253	171.254	20
21	1.170	.8548	.0441	.0516	22.653	19.363	9.727	188.352	21
22	1.179	.8484	.0420	.0495	23.823	20.211	10.201	206.170	22
23	1.188	.8421	.0400	.0475	25.001	21.053	10.673	224.695	23
24	1.196	.8358	.0382	.0457	26.189	21.889	11.143	243.924	24
25	1.205	.8296	.0365	.0440	27.385	22.719	11.613	263.834	25
26	1.214	.8234	.0350	.0425	28.591	23.542	12.081	284.421	26
27	1.224	.8173	.0336	.0411	29.805	24.360	12.548	305.672	27
28	1.233	.8112	.0322	.0397	31.029	25.171	13.014	327.576	28
29	1.242	.8052	.0310	.0385	32.261	25.976	13.479	350.122	29
30	1.251	.7992	.0298	.0373	33.503	26.775	13.942	373.302	30
36	1.309	.7641	.0243	.0318	41.153	31.447	16.696	525.038	36
40	1.348	.7416	.0215	.0290	46.447	34.447	18.507	637.519	40
48	1.431	.6986	.0174	.0249	57.521	40.185	22.070	886.899	48
50	1.453	.6882	.0166	.0241	60.395	41.567	22.949	953.911	50
52	1.475	.6780	.0158	.0233	63.312	42.928	23.822	1 022.64	52
60	1.566	.6387	.0133	.0208	75.425	48.174	27.268	1 313.59	60
70	1.687	.5927	.0109	.0184	91.621	54.305	31.465	1 708.68	70
72	1.713	.5839	.0105	.0180	95.008	55.477	32.289	1 791.33	72
80	1.818	.5500	.00917	.0167	109.074	59.995	35.540	2 132.23	80
84	1.873	.5338	.00859	.0161	116.428	62.154	37.137	2 308.22	84
90	1.959	.5104	.00782	.0153	127.881	65.275	39.496	2 578.09	90
96	2.049	.4881	.00715	.0147	139.858	68.259	41.812	2 854.04	96
100	2.111	.4737	.00675	.0143	148.147	70.175	43.332	3 040.85	100
104	2.175	.4597	.00638	.0139	156.687	72.035	44.834	3 229.60	104
120	2.451	.4079	.00517	.0127	193.517	78.942	50.653	3 998.68	120
240	6.009	.1664	.00150	.00900	667.901	111.145	85.422	9 494.26	240
360	14.731	.0679	.00055	.00805	1 830.8	124.282	107.115	13 312.50	360
480	36.111	.0277	.00021	.00771	4 681.5	129.641	119.662	15 513.16	480

1%   Compound Interest Factors   1%

	Single Payment		Uniform Payment Series				Arithmetic Gradient		
	Compound Amount Factor	Present Worth Factor	Sinking Fund Factor	Capital Recovery Factor	Compound Amount Factor	Present Worth Factor	Gradient Uniform Series	Gradient Present Worth	
	Find F Given P	Find P Given F	Find A Given F	Find A Given P	Find F Given A	Find P Given A	Find A Given G	Find P Given G	
n	F/P	P/F	A/F	A/P	F/A	P/A	A/G	P/G	n
1	1.010	.9901	1.0000	1.0100	1.000	0.990	0	0	1
2	1.020	.9803	.4975	.5075	2.010	1.970	0.498	0.980	2
3	1.030	.9706	.3300	.3400	3.030	2.941	0.993	2.921	3
4	1.041	.9610	.2463	.2563	4.060	3.902	1.488	5.804	4
5	1.051	.9515	.1960	.2060	5.101	4.853	1.980	9.610	5
6	1.062	.9420	.1625	.1725	6.152	5.795	2.471	14.320	6
7	1.072	.9327	.1386	.1486	7.214	6.728	2.960	19.917	7
8	1.083	.9235	.1207	.1307	8.286	7.652	3.448	26.381	8
9	1.094	.9143	.1067	.1167	9.369	8.566	3.934	33.695	9
10	1.105	.9053	.0956	.1056	10.462	9.471	4.418	41.843	10
11	1.116	.8963	.0865	.0965	11.567	10.368	4.900	50.806	11
12	1.127	.8874	.0788	.0888	12.682	11.255	5.381	60.568	12
13	1.138	.8787	.0724	.0824	13.809	12.134	5.861	71.112	13
14	1.149	.8700	.0669	.0769	14.947	13.004	6.338	82.422	14
15	1.161	.8613	.0621	.0721	16.097	13.865	6.814	94.481	15
16	1.173	.8528	.0579	.0679	17.258	14.718	7.289	107.273	16
17	1.184	.8444	.0543	.0643	18.430	15.562	7.761	120.783	17
18	1.196	.8360	.0510	.0610	19.615	16.398	8.232	134.995	18
19	1.208	.8277	.0481	.0581	20.811	17.226	8.702	149.895	19
20	1.220	.8195	.0454	.0554	22.019	18.046	9.169	165.465	20
21	1.232	.8114	.0430	.0530	23.239	18.857	9.635	181.694	21
22	1.245	.8034	.0409	.0509	24.472	19.660	10.100	198.565	22
23	1.257	.7954	.0389	.0489	25.716	20.456	10.563	216.065	23
24	1.270	.7876	.0371	.0471	26.973	21.243	11.024	234.179	24
25	1.282	.7798	.0354	.0454	28.243	22.023	11.483	252.892	25
26	1.295	.7720	.0339	.0439	29.526	22.795	11.941	272.195	26
27	1.308	.7644	.0324	.0424	30.821	23.560	12.397	292.069	27
28	1.321	.7568	.0311	.0411	32.129	24.316	12.852	312.504	28
29	1.335	.7493	.0299	.0399	33.450	25.066	13.304	333.486	29
30	1.348	.7419	.0287	.0387	34.785	25.808	13.756	355.001	30
36	1.431	.6989	.0232	.0332	43.077	30.107	16.428	494.620	36
40	1.489	.6717	.0205	.0305	48.886	32.835	18.178	596.854	40
48	1.612	.6203	.0163	.0263	61.223	37.974	21.598	820.144	48
50	1.645	.6080	.0155	.0255	64.463	39.196	22.436	879.417	50
52	1.678	.5961	.0148	.0248	67.769	40.394	23.269	939.916	52
60	1.817	.5504	.0122	.0222	81.670	44.955	26.533	1 192.80	60
70	2.007	.4983	.00993	.0199	100.676	50.168	30.470	1 528.64	70
72	2.047	.4885	.00955	.0196	104.710	51.150	31.239	1 597.86	72
80	2.217	.4511	.00822	.0182	121.671	54.888	34.249	1 879.87	80
84	2.307	.4335	.00765	.0177	130.672	56.648	35.717	2 023.31	84
90	2.449	.4084	.00690	.0169	144.863	59.161	37.872	2 240.56	90
96	2.599	.3847	.00625	.0163	159.927	61.528	39.973	2 459.42	96
100	2.705	.3697	.00587	.0159	170.481	63.029	41.343	2 605.77	100
104	2.815	.3553	.00551	.0155	181.464	64.471	42.688	2 752.17	104
120	3.300	.3030	.00435	.0143	230.039	69.701	47.835	3 334.11	120
240	10.893	.0918	.00101	.0110	989.254	90.819	75.739	6 878.59	240
360	35.950	.0278	.00029	.0103	3 495.0	97.218	89.699	8 720.43	360
480	118.648	.00843	.00008	.0101	11 764.8	99.157	95.920	9 511.15	480

1¼%

Compound Interest Factors

1¼%

	Single Payment		Uniform Payment Series				Arithmetic Gradient		
	Compound Amount Factor	Present Worth Factor	Sinking Fund Factor	Capital Recovery Factor	Compound Amount Factor	Present Worth Factor	Gradient Uniform Series	Gradient Present Worth	
	Find F Given P	Find P Given F	Find A Given F	Find A Given P	Find F Given A	Find P Given A	Find A Given G	Find P Given G	
n	F/P	P/F	A/F	A/P	F/A	P/A	A/G	P/G	n
1	1.013	.9877	1.0000	1.0125	1.000	0.988	0	0	1
2	1.025	.9755	.4969	.5094	2.013	1.963	0.497	0.976	2
3	1.038	.9634	.3292	.3417	3.038	2.927	0.992	2.904	3
4	1.051	.9515	.2454	.2579	4.076	3.878	1.485	5.759	4
5	1.064	.9398	.1951	.2076	5.127	4.818	1.976	9.518	5
6	1.077	.9282	.1615	.1740	6.191	5.746	2.464	14.160	6
7	1.091	.9167	.1376	.1501	7.268	6.663	2.951	19.660	7
8	1.104	.9054	.1196	.1321	8.359	7.568	3.435	25.998	8
9	1.118	.8942	.1057	.1182	9.463	8.462	3.918	33.152	9
10	1.132	.8832	.0945	.1070	10.582	9.346	4.398	41.101	10
11	1.146	.8723	.0854	.0979	11.714	10.218	4.876	49.825	11
12	1.161	.8615	.0778	.0903	12.860	11.079	5.352	59.302	12
13	1.175	.8509	.0713	.0838	14.021	11.930	5.827	69.513	13
14	1.190	.8404	.0658	.0783	15.196	12.771	6.299	80.438	14
15	1.205	.8300	.0610	.0735	16.386	13.601	6.769	92.058	15
16	1.220	.8197	.0568	.0693	17.591	14.420	7.237	104.355	16
17	1.235	.8096	.0532	.0657	18.811	15.230	7.702	117.309	17
18	1.251	.7996	.0499	.0624	20.046	16.030	8.166	130.903	18
19	1.266	.7898	.0470	.0595	21.297	16.849	8.628	145.119	19
20	1.282	.7800	.0443	.0568	22.563	17.599	9.088	159.940	20
21	1.298	.7704	.0419	.0544	23.845	18.370	9.545	175.348	21
22	1.314	.7609	.0398	.0523	25.143	19.131	10.001	191.327	22
23	1.331	.7515	.0378	.0503	26.458	19.882	10.455	207.859	23
24	1.347	.7422	.0360	.0485	27.788	20.624	10.906	224.930	24
25	1.364	.7330	.0343	.0468	29.136	21.357	11.355	242.523	25
26	1.381	.7240	.0328	.0453	30.500	22.081	11.803	260.623	26
27	1.399	.7150	.0314	.0439	31.881	22.796	12.248	279.215	27
28	1.416	.7062	.0300	.0425	32.280	23.503	12.691	298.284	28
29	1.434	.6975	.0288	.0413	34.696	24.200	13.133	317.814	29
30	1.452	.6889	.0277	.0402	36.129	24.889	13.572	337.792	30
36	1.564	.6394	.0222	.0347	45.116	28.847	16.164	466.297	36
40	1.644	.6084	.0194	.0319	51.490	31.327	17.852	559.247	40
48	1.845	.5509	.0153	.0278	65.229	35.932	21.130	759.248	48
50	1.861	.5373	.0145	.0270	68.882	37.013	21.930	811.692	50
52	1.908	.5242	.0138	.0263	72.628	38.068	22.722	864.960	52
60	2.107	.4746	.0113	.0238	88.575	42.035	25.809	1 084.86	60
70	2.386	.4191	.00902	.0215	110.873	46.470	29.492	1 370.47	70
72	2.446	.4088	.00864	.0211	115.675	47.293	30.205	1 428.48	72
80	2.701	.3702	.00735	.0198	136.120	50.387	32.983	1 661.89	80
84	2.839	.3522	.00680	.0193	147.130	51.822	34.326	1 778.86	84
90	3.059	.3269	.00607	.0186	164.706	53.846	36.286	1 953.85	90
96	3.296	.3034	.00545	.0179	183.643	55.725	38.180	2 127.55	96
100	3.463	.2887	.00507	.0176	197.074	56.901	39.406	2 242.26	100
104	3.640	.2747	.00474	.0172	211.190	58.021	40.604	2 355.90	104
120	4.440	.2252	.00363	.0161	275.220	61.983	45.119	2 796.59	120
240	19.716	.0507	.00067	.0132	1 497.3	75.942	67.177	5 101.55	240
360	87.543	.0114	.00014	.0126	6 923.4	79.086	75.840	5 997.91	360
480	388.713	.00257	.00003	.0125	31 017.1	79.794	78.762	6 284.74	480

1¹/₂ %　　Compound Interest Factors　　1¹/₂ %

	Single Payment		Uniform Payment Series				Arithmetic Gradient		
	Compound Amount Factor	Present Worth Factor	Sinking Fund Factor	Capital Recovery Factor	Compound Amount Factor	Present Worth Factor	Gradient Uniform Series	Gradient Present Worth	
	Find F Given P	Find P Given F	Find A Given F	Find A Given P	Find F Given A	Find P Given A	Find A Given G	Find P Given G	
n	F/P	P/F	A/F	A/P	F/A	P/A	A/G	P/G	n
1	1.015	.9852	1.0000	1.0150	1.000	0.985	0	0	1
2	1.030	.9707	.4963	.5113	2.015	1.956	0.496	0.970	2
3	1.046	.9563	.3284	.3434	3.045	2.912	0.990	2.883	3
4	1.061	.9422	.2444	.2594	4.091	3.854	1.481	5.709	4
5	1.077	.9283	.1941	.2091	5.152	4.783	1.970	9.422	5
6	1.093	.9145	.1605	.1755	6.230	5.697	2.456	13.994	6
7	1.110	.9010	.1366	.1516	7.323	6.598	2.940	19.400	7
8	1.126	.8877	.1186	.1336	8.433	7.486	3.422	25.614	8
9	1.143	.8746	.1046	.1196	9.559	8.360	3.901	32.610	9
10	1.161	.8617	.0934	.1084	10.703	9.222	4.377	40.365	10
11	1.178	.8489	.0843	.0993	11.863	10.071	4.851	48.855	11
12	1.196	.8364	.0767	.0917	13.041	10.907	5.322	58.054	12
13	1.214	.8240	.0702	.0852	14.237	11.731	5.791	67.943	13
14	1.232	.8118	.0647	.0797	15.450	12.543	6.258	78.496	14
15	1.250	.7999	.0599	.0749	16.682	13.343	6.722	89.694	15
16	1.269	.7880	.0558	.0708	17.932	14.131	7.184	101.514	16
17	1.288	.7764	.0521	.0671	19.201	14.908	7.643	113.937	17
18	1.307	.7649	.0488	.0638	20.489	15.673	8.100	126.940	18
19	1.327	.7536	.0459	.0609	21.797	16.426	8.554	140.505	19
20	1.347	.7425	.0432	.0582	23.124	17.169	9.005	154.611	20
21	1.367	.7315	.0409	.0559	24.470	17.900	9.455	169.241	21
22	1.388	.7207	.0387	.0537	25.837	18.621	9.902	184.375	22
23	1.408	.7100	.0367	.0517	27.225	19.331	10.346	199.996	23
24	1.430	.6995	.0349	.0499	28.633	20.030	10.788	216.085	24
25	1.451	.6892	.0333	.0483	30.063	20.720	11.227	232.626	25
26	1.473	.6790	.0317	.0467	31.514	21.399	11.664	249.601	26
27	1.495	.6690	.0303	.0453	32.987	22.068	12.099	266.995	27
28	1.517	.6591	.0290	.0440	34.481	22.727	12.531	284.790	28
29	1.540	.6494	.0278	.0428	35.999	23.376	12.961	302.972	29
30	1.563	.6398	.0266	.0416	37.539	24.016	13.388	321.525	30
36	1.709	.5851	.0212	.0362	47.276	27.661	15.901	439.823	36
40	1.814	.5513	.0184	.0334	54.268	29.916	17.528	524.349	40
48	2.043	.4894	.0144	.0294	69.565	34.042	20.666	703.537	48
50	2.105	.4750	.0136	.0286	73.682	35.000	21.428	749.955	50
52	2.169	.4611	.0128	.0278	77.925	35.929	22.179	796.868	52
60	2.443	.4093	.0104	.0254	96.214	39.380	25.093	988.157	60
70	2.835	.3527	.00817	.0232	122.363	43.155	28.529	1 231.15	70
72	2.921	.3423	.00781	.0228	128.076	43.845	29.189	1 279.78	72
80	3.291	.3039	.00655	.0215	152.710	46.407	31.742	1 473.06	80
84	3.493	.2863	.00602	.0210	166.172	47.579	32.967	1 568.50	84
90	3.819	.2619	.00532	.0203	187.929	49.210	34.740	1 709.53	90
96	4.176	.2395	.00472	.0197	211.719	50.702	36.438	1 847.46	96
100	4.432	.2256	.00437	.0194	228.802	51.625	37.529	1 937.43	100
104	4.704	.2126	.00405	.0190	246.932	52.494	38.589	2 025.69	104
120	5.969	.1675	.00302	.0180	331.286	55.498	42.518	2 359.69	120
240	35.632	.0281	.00043	.0154	2 308.8	64.796	59.737	3 870.68	240
360	212.700	.00470	.00007	.0151	14 113.3	66.353	64.966	4 310.71	360
480	1 269.7	.00079	.00001	.0150	84 577.8	66.614	66.288	4 415.74	480

1³/4%

Compound Interest Factors

1³/4%

	Single Payment		Uniform Payment Series				Arithmetic Gradient		
	Compound Amount Factor	Present Worth Factor	Sinking Fund Factor	Capital Recovery Factor	Compound Amount Factor	Present Worth Factor	Gradient Uniform Series	Gradient Present Worth	
	Find F Given P	Find P Given F	Find A Given F	Find A Given P	Find F Given A	Find P Given A	Find A Given G	Find P Given G	
n	F/P	P/F	A/F	A/P	F/A	P/A	A/G	P/G	n
1	1.018	.9828	1.0000	1.0175	1.000	0.983	0	0	1
2	1.035	.9659	.4957	.5132	2.018	1.949	0.496	0.966	2
3	1.053	.9493	.3276	.3451	3.053	2.898	0.989	2.865	3
4	1.072	.9330	.2435	.2610	4.106	3.831	1.478	5.664	4
5	1.091	.9169	.1931	.2106	5.178	4.748	1.965	9.332	5
6	1.110	.9011	.1595	.1770	6.269	5.649	2.450	13.837	6
7	1.129	.8856	.1355	.1530	7.378	6.535	2.931	19.152	7
8	1.149	.8704	.1175	.1350	8.508	7.405	3.409	25.245	8
9	1.169	.8554	.1036	.1211	9.656	8.261	3.885	32.088	9
10	1.189	.8407	.0924	.1099	10.825	9.101	4.357	39.655	10
11	1.210	.8263	.0832	.1007	12.015	9.928	4.827	47.918	11
12	1.231	.8121	.0756	.0931	13.225	10.740	5.294	56.851	12
13	1.253	.7981	.0692	.0867	14.457	11.538	5.758	66.428	13
14	1.275	.7844	.0637	.0812	15.710	12.322	6.219	76.625	14
15	1.297	.7709	.0589	.0764	16.985	13.093	6.677	87.417	15
16	1.320	.7576	.0547	.0722	18.282	13.851	7.132	98.782	16
17	1.343	.7446	.0510	.0685	19.602	14.595	7.584	110.695	17
18	1.367	.7318	.0477	.0652	20.945	15.327	8.034	123.136	18
19	1.390	.7192	.0448	.0623	22.311	16.046	8.481	136.081	19
20	1.415	.7068	.0422	.0597	23.702	16.753	8.924	149.511	20
21	1.440	.6947	.0398	.0573	25.116	17.448	9.365	163.405	21
22	1.465	.6827	.0377	.0552	26.556	18.130	9.804	177.742	22
23	1.490	.6710	.0357	.0532	28.021	18.801	10.239	192.503	23
24	1.516	.6594	.0339	.0514	29.511	19.461	10.671	207.671	24
25	1.543	.6481	.0322	.0497	31.028	20.109	11.101	223.225	25
26	1.570	.6369	.0307	.0482	32.571	20.746	11.528	239.149	26
27	1.597	.6260	.0293	.0468	34.141	21.372	11.952	255.425	27
28	1.625	.6152	.0280	.0455	35.738	21.987	12.373	272.036	28
29	1.654	.6046	.0268	.0443	37.363	22.592	12.791	288.967	29
30	1.683	.5942	.0256	.0431	39.017	23.186	13.206	306.200	30
36	1.867	.5355	.0202	.0377	49.566	26.543	15.640	415.130	36
40	2.002	.4996	.0175	.0350	57.234	28.594	17.207	492.017	40
48	2.300	.4349	.0135	.0310	74.263	32.294	20.209	652.612	48
50	2.381	.4200	.0127	.0302	78.903	33.141	20.932	693.708	50
52	2.465	.4057	.0119	.0294	83.706	33.960	21.644	735.039	52
60	2.832	.3531	.00955	.0271	104.676	36.964	24.389	901.503	60
70	3.368	.2969	.00739	.0249	135.331	40.178	27.586	1 108.34	70
72	3.487	.2868	.00704	.0245	142.127	40.757	28.195	1 149.12	72
80	4.006	.2496	.00582	.0233	171.795	42.880	30.533	1 309.25	80
84	4.294	.2329	.00531	.0228	188.246	43.836	31.644	1 387.16	84
90	4.765	.2098	.00465	.0221	215.166	45.152	33.241	1 500.88	90
96	5.288	.1891	.00408	.0216	245.039	46.337	34.756	1 610.48	96
100	5.668	.1764	.00375	.0212	266.753	47.062	35.721	1 681.09	100
104	6.075	.1646	.00345	.0209	290.028	47.737	36.652	1 749.68	104
120	8.019	.1247	.00249	.0200	401.099	50.017	40.047	2 003.03	120
240	64.308	.0156	.00028	.0178	3 617.6	56.254	53.352	3 001.27	240
360	515.702	.00194	.00003	.0175	29 411.5	57.032	56.443	3 219.08	360
480	4 135.5	.00024		.0175	236 259.0	57.129	57.027	3 257.88	480

2%   Compound Interest Factors   2%

	Single Payment		Uniform Payment Series				Arithmetic Gradient		
	Compound Amount Factor	Present Worth Factor	Sinking Fund Factor	Capital Recovery Factor	Compound Amount Factor	Present Worth Factor	Gradient Uniform Series	Gradient Present Worth	
	Find F Given P	Find P Given F	Find A Given F	Find A Given P	Find F Given A	Find P Given A	Find A Given G	Find P Given G	
n	F/P	P/F	A/F	A/P	F/A	P/A	A/G	P/G	n
1	1.020	.9804	1.0000	1.0200	1.000	0.980	0	0	1
2	1.040	.9612	.4951	.5151	2.020	1.942	0.495	0.961	2
3	1.061	.9423	.3268	.3468	3.060	2.884	0.987	2.846	3
4	1.082	.9238	.2426	.2626	4.122	3.808	1.475	5.617	4
5	1.104	.9057	.1922	.2122	5.204	4.713	1.960	9.240	5
6	1.126	.8880	.1585	.1785	6.308	5.601	2.442	13.679	6
7	1.149	.8706	.1345	.1545	7.434	6.472	2.921	18.903	7
8	1.172	.8535	.1165	.1365	8.583	7.325	3.396	24.877	8
9	1.195	.8368	.1025	.1225	9.755	8.162	3.868	31.571	9
10	1.219	.8203	.0913	.1113	10.950	8.983	4.337	38.954	10
11	1.243	.8043	.0822	.1022	12.169	9.787	4.802	46.996	11
12	1.268	.7885	.0746	.0946	13.412	10.575	5.264	55.669	12
13	1.294	.7730	.0681	.0881	14.680	11.348	5.723	64.946	13
14	1.319	.7579	.0626	.0826	15.974	12.106	6.178	74.798	14
15	1.346	.7430	.0578	.0778	17.293	12.849	6.631	85.200	15
16	1.373	.7284	.0537	.0737	18.639	13.578	7.080	96.127	16
17	1.400	.7142	.0500	.0700	20.012	14.292	7.526	107.553	17
18	1.428	.7002	.0467	.0667	21.412	14.992	7.968	119.456	18
19	1.457	.6864	.0438	.0638	22.840	15.678	8.407	131.812	19
20	1.486	.6730	.0412	.0612	24.297	16.351	8.843	144.598	20
21	1.516	.6598	.0388	.0588	25.783	17.011	9.276	157.793	21
22	1.546	.6468	.0366	.0566	27.299	17.658	9.705	171.377	22
23	1.577	.6342	.0347	.0547	28.845	18.292	10.132	185.328	23
24	1.608	.6217	.0329	.0529	30.422	18.914	10.555	199.628	24
25	1.641	.6095	.0312	.0512	32.030	19.523	10.974	214.256	25
26	1.673	.5976	.0297	.0497	33.671	20.121	11.391	229.196	26
27	1.707	.5859	.0283	.0483	35.344	20.707	11.804	244.428	27
28	1.741	.5744	.0270	.0470	37.051	21.281	12.214	259.936	28
29	1.776	.5631	.0258	.0458	38.792	21.844	12.621	275.703	29
30	1.811	.5521	.0247	.0447	40.568	22.396	13.025	291.713	30
36	2.040	.4902	.0192	.0392	51.994	25.489	15.381	392.036	36
40	2.208	.4529	.0166	.0366	60.402	27.355	16.888	461.989	40
48	2.587	.3865	.0126	.0326	79.353	30.673	19.755	605.961	48
50	2.692	.3715	.0118	.0318	84.579	31.424	20.442	642.355	50
52	2.800	.3571	.0111	.0311	90.016	32.145	21.116	678.779	52
60	3.281	.3048	.00877	.0288	114.051	34.761	23.696	823.692	60
70	4.000	.2500	.00667	.0267	149.977	37.499	26.663	999.829	70
72	4.161	.2403	.00633	.0263	158.056	37.984	27.223	1 034.050	72
80	4.875	.2051	.00516	.0252	193.771	39.744	29.357	1 166.781	80
84	5.277	.1895	.00468	.0247	213.865	40.525	30.361	1 230.413	84
90	5.943	.1683	.00405	.0240	247.155	41.587	31.793	1 322.164	90
96	6.693	.1494	.00351	.0235	284.645	42.529	33.137	1 409.291	96
100	7.245	.1380	.00320	.0232	312.230	43.098	33.986	1 464.747	100
104	7.842	.1275	.00292	.0229	342.090	43.624	34.799	1 518.082	104
120	10.765	.0929	.00205	.0220	488.255	45.355	37.711	1 710.411	120
240	115.887	.00863	.00017	.0202	5 744.4	49.569	47.911	2 374.878	240
360	1 247.5	.00080	.00002	.0200	62 326.8	49.960	49.711	2 483.567	360
480	13 429.8	.00007		.0200	671 442.0	49.996	49.964	2 498.027	480

2^1/$_2$%

Compound Interest Factors

2^1/$_2$%

	Single Payment		Uniform Payment Series				Arithmetic Gradient		
	Compound Amount Factor	Present Worth Factor	Sinking Fund Factor	Capital Recovery Factor	Compound Amount Factor	Present Worth Factor	Gradient Uniform Series	Gradient Present Worth	
	Find F Given P	Find P Given F	Find A Given F	Find A Given P	Find F Given A	Find P Given A	Find A Given G	Find P Given G	
n	F/P	P/F	A/F	A/P	F/A	P/A	A/G	P/G	n
1	1.025	.9756	1.0000	1.0250	1.000	0.976	0	0	1
2	1.051	.9518	.4938	.5188	2.025	1.927	0.494	0.952	2
3	1.077	.9286	.3251	.3501	3.076	2.856	0.984	2.809	3
4	1.104	.9060	.2408	.2658	4.153	3.762	1.469	5.527	4
5	1.131	.8839	.1902	.2152	5.256	4.646	1.951	9.062	5
6	1.160	.8623	.1566	.1816	6.388	5.508	2.428	13.374	6
7	1.189	.8413	.1325	.1575	7.547	6.349	2.901	18.421	7
8	1.218	.8207	.1145	.1395	8.736	7.170	3.370	24.166	8
9	1.249	.8007	.1005	.1255	9.955	7.971	3.835	30.572	9
10	1.280	.7812	.0893	.1143	11.203	8.752	4.296	37.603	10
11	1.312	.7621	.0801	.1051	12.483	9.514	4.753	45.224	11
12	1.345	.7436	.0725	.0975	13.796	10.258	5.206	53.403	12
13	1.379	.7254	.0660	.0910	15.140	10.983	5.655	62.108	13
14	1.413	.7077	.0605	.0855	16.519	11.691	6.100	71.309	14
15	1.448	.6905	.0558	.0808	17.932	12.381	6.540	80.975	15
16	1.485	.6736	.0516	.0766	19.380	13.055	6.977	91.080	16
17	1.522	.6572	.0479	.0729	20.865	13.712	7.409	101.595	17
18	1.560	.6412	.0447	.0697	22.386	14.353	7.838	112.495	18
19	1.599	.6255	.0418	.0668	23.946	14.979	8.262	123.754	19
20	1.639	.6103	.0391	.0641	25.545	15.589	8.682	135.349	20
21	1.680	.5954	.0368	.0618	27.183	16.185	9.099	147.257	21
22	1.722	.5809	.0346	.0596	28.863	16.765	9.511	159.455	22
23	1.765	.5667	.0327	.0577	30.584	17.332	9.919	171.922	23
24	1.809	.5529	.0309	.0559	32.349	17.885	10.324	184.638	24
25	1.854	.5394	.0293	.0543	34.158	18.424	10.724	197.584	25
26	1.900	.5262	.0278	.0528	36.012	18.951	11.120	210.740	26
27	1.948	.5134	.0264	.0514	37.912	19.464	11.513	224.088	27
28	1.996	.5009	.0251	.0501	39.860	19.965	11.901	237.612	28
29	2.046	.4887	.0239	.0489	41.856	20.454	12.286	251.294	29
30	2.098	.4767	.0228	.0478	43.903	20.930	12.667	265.120	30
31	2.150	.4651	.0217	.0467	46.000	21.395	13.044	279.073	31
32	2.204	.4538	.0208	.0458	48.150	24.849	13.417	293.140	32
33	2.259	.4427	.0199	.0449	50.354	22.292	13.786	307.306	33
34	2.315	.4319	.0190	.0440	52.613	22.724	14.151	321.559	34
35	2.373	.4214	.0182	.0432	54.928	23.145	14.512	335.886	35
40	2.685	.3724	.0148	.0398	67.402	25.103	16.262	408.221	40
45	3.038	.3292	.0123	.0373	81.516	26.833	17.918	480.806	45
50	3.437	.2909	.0103	.0353	97.484	28.362	19.484	552.607	50
55	3.889	.2572	.00865	.0337	115.551	29.714	20.961	622.827	55
60	4.400	.2273	.00735	.0324	135.991	30.909	22.352	690.865	60
65	4.978	.2009	.00628	.0313	159.118	31.965	23.660	756.280	65
70	5.632	.1776	.00540	.0304	185.284	32.898	24.888	818.763	70
75	6.372	.1569	.00465	.0297	214.888	33.723	26.039	878.114	75
80	7.210	.1387	.00403	.0290	248.382	34.452	27.117	934.217	80
85	8.157	.1226	.00349	.0285	286.278	35.096	28.123	987.026	85
90	9.229	.1084	.00304	.0280	329.154	35.666	29.063	1 036.54	90
95	10.442	.0958	.00265	.0276	377.663	36.169	29.938	1 082.83	95
100	11.814	.0846	.00231	.0273	432.548	36.614	30.752	1 125.97	100

3% Compound Interest Factors 3%

	Single Payment		Uniform Payment Series				Arithmetic Gradient		
	Compound Amount Factor	Present Worth Factor	Sinking Fund Factor	Capital Recovery Factor	Compound Amount Factor	Present Worth Factor	Gradient Uniform Series	Gradient Present Worth	
n	Find F Given P F/P	Find P Given F P/F	Find A Given F A/F	Find A Given P A/P	Find F Given A F/A	Find P Given A P/A	Find A Given G A/G	Find P Given G P/G	n
1	1.030	.9709	1.0000	1.0300	1.000	0.971	0	0	1
2	1.061	.9426	.4926	.5226	2.030	1.913	0.493	0.943	2
3	1.093	.9151	.3235	.3535	3.091	2.829	0.980	2.773	3
4	1.126	.8885	.2390	.2690	4.184	3.717	1.463	5.438	4
5	1.159	.8626	.1884	.2184	5.309	4.580	1.941	8.889	5
6	1.194	.8375	.1546	.1846	6.468	5.417	2.414	13.076	6
7	1.230	.8131	.1305	.1605	7.662	6.230	2.882	17.955	7
8	1.267	.7894	.1125	.1425	8.892	7.020	3.345	23.481	8
9	1.305	.7664	.0984	.1284	10.159	7.786	3.803	29.612	9
10	1.344	.7441	.0872	.1172	11.464	8.530	4.256	36.309	10
11	1.384	.7224	.0781	.1081	12.808	9.253	4.705	43.533	11
12	1.426	.7014	.0705	.1005	14.192	9.954	5.148	51.248	12
13	1.469	.6810	.0640	.0940	15.618	10.635	5.587	59.419	13
14	1.513	.6611	.0585	.0885	17.086	11.296	6.021	68.014	14
15	1.558	.6419	.0538	.0838	18.599	11.938	6.450	77.000	15
16	1.605	.6232	.0496	.0796	20.157	12.561	6.874	86.348	16
17	1.653	.6050	.0460	.0760	21.762	13.166	7.294	96.028	17
18	1.702	.5874	.0427	.0727	23.414	13.754	7.708	106.014	18
19	1.754	.5703	.0398	.0698	25.117	14.324	8.118	116.279	19
20	1.806	.5537	.0372	.0672	26.870	14.877	8.523	126.799	20
21	1.860	.5375	.0349	.0649	28.676	15.415	8.923	137.549	21
22	1.916	.5219	.0327	.0627	30.537	15.937	9.319	148.509	22
23	1.974	.5067	.0308	.0608	32.453	16.444	9.709	159.656	23
24	2.033	.4919	.0290	.0590	34.426	16.936	10.095	170.971	24
25	2.094	.4776	.0274	.0574	36.459	17.413	10.477	182.433	25
26	2.157	.4637	.0259	.0559	38.553	17.877	10.853	194.026	26
27	2.221	.4502	.0246	.0546	40.710	18.327	11.226	205.731	27
28	2.288	.4371	.0233	.0533	42.931	18.764	11.593	217.532	28
29	2.357	.4243	.0221	.0521	45.219	19.188	11.956	229.413	29
30	2.427	.4120	.0210	.0510	47.575	19.600	12.314	241.361	30
31	2.500	.4000	.0200	.0500	50.003	20.000	12.668	253.361	31
32	2.575	.3883	.0190	.0490	52.503	20.389	13.017	265.399	32
33	2.652	.3770	.0182	.0482	55.078	20.766	13.362	277.464	33
34	2.732	.3660	.0173	.0473	57.730	21.132	13.702	289.544	34
35	2.814	.3554	.0165	.0465	60.462	21.487	14.037	301.627	35
40	3.262	.3066	.0133	.0433	75.401	23.115	15.650	361.750	40
45	3.782	.2644	.0108	.0408	92.720	24.519	17.156	420.632	45
50	4.384	.2281	.00887	.0389	112.797	25.730	18.558	477.480	50
55	5.082	.1968	.00735	.0373	136.072	26.774	19.860	531.741	55
60	5.892	.1697	.00613	.0361	163.053	27.676	21.067	583.052	60
65	6.830	.1464	.00515	.0351	194.333	28.453	22.184	631.201	65
70	7.918	.1263	.00434	.0343	230.594	29.123	23.215	676.087	70
75	9.179	.1089	.00367	.0337	272.631	29.702	24.163	717.698	75
80	10.641	.0940	.00311	.0331	321.363	30.201	25.035	756.086	80
85	12.336	.0811	.00265	.0326	377.857	30.631	25.835	791.353	85
90	14.300	.0699	.00226	.0323	443.349	31.002	26.567	823.630	90
95	16.578	.0603	.00193	.0319	519.272	31.323	27.235	853.074	95
100	19.219	.0520	.00165	.0316	607.287	31.599	27.844	879.854	100

$3^1/_2\%$

Compound Interest Factors

$3^1/_2\%$

	Single Payment		Uniform Payment Series				Arithmetic Gradient		
	Compound Amount Factor	Present Worth Factor	Sinking Fund Factor	Capital Recovery Factor	Compound Amount Factor	Present Worth Factor	Gradient Uniform Series	Gradient Present Worth	
	Find F Given P	Find P Given F	Find A Given F	Find A Given P	Find F Given A	Find P Given A	Find A Given G	Find P Given G	
n	F/P	P/F	A/F	A/P	F/A	P/A	A/G	P/G	n
1	1.035	.9662	1.0000	1.0350	1.000	0.966	0	0	1
2	1.071	.9335	.4914	.5264	2.035	1.900	0.491	0.933	2
3	1.109	.9019	.3219	.3569	3.106	2.802	0.977	2.737	3
4	1.148	.8714	.2373	.2723	4.215	3.673	1.457	5.352	4
5	1.188	.8420	.1865	.2215	5.362	4.515	1.931	8.719	5
6	1.229	.8135	.1527	.1877	6.550	5.329	2.400	12.787	6
7	1.272	.7860	.1285	.1635	7.779	6.115	2.862	17.503	7
8	1.317	.7594	.1105	.1455	9.052	6.874	3.320	22.819	8
9	1.363	.7337	.0964	.1314	10.368	7.608	3.771	28.688	9
10	1.411	.7089	.0852	.1202	11.731	8.317	4.217	35.069	10
11	1.460	.6849	.0761	.1111	13.142	9.002	4.657	41.918	11
12	1.511	.6618	.0685	.1035	14.602	9.663	5.091	49.198	12
13	1.564	.6394	.0621	.0971	16.113	10.303	5.520	56.871	13
14	1.619	.6178	.0566	.0916	17.677	10.921	5.943	64.902	14
15	1.675	.5969	.0518	.0868	19.296	11.517	6.361	73.258	15
16	1.734	.5767	.0477	.0827	20.971	12.094	6.773	81.909	16
17	1.795	.5572	.0440	.0790	22.705	12.651	7.179	90.824	17
18	1.857	.5384	.0408	.0758	24.500	13.190	7.580	99.976	18
19	1.922	.5202	.0379	.0729	26.357	13.710	7.975	109.339	19
20	1.990	.5026	.0354	.0704	28.280	14.212	8.365	118.888	20
21	2.059	.4856	.0330	.0680	30.269	14.698	8.749	128.599	21
22	2.132	.4692	.0309	.0659	32.329	15.167	9.128	138.451	22
23	2.206	.4533	.0290	.0640	34.460	15.620	9.502	148.423	23
24	2.283	.4380	.0273	.0623	36.666	16.058	9.870	158.496	24
25	2.363	.4231	.0257	.0607	38.950	16.482	10.233	168.652	25
26	2.446	.4088	.0242	.0592	41.313	16.890	10.590	178.873	26
27	2.532	.3950	.0229	.0579	43.759	17.285	10.942	189.143	27
28	2.620	.3817	.0216	.0566	46.291	17.667	11.289	199.448	28
29	2.712	.3687	.0204	.0554	48.911	18.036	11.631	209.773	29
30	2.807	.3563	.0194	.0544	51.623	18.392	11.967	220.105	30
31	2.905	.3442	.0184	.0534	54.429	18.736	12.299	230.432	31
32	3.007	.3326	.0174	.0524	57.334	19.069	12.625	240.742	32
33	3.112	.3213	.0166	.0516	60.341	19.390	12.946	251.025	33
34	3.221	.3105	.0158	.0508	63.453	19.701	13.262	261.271	34
35	3.334	.3000	.0150	.0500	66.674	20.001	13.573	271.470	35
40	3.959	.2526	.0118	.0468	84.550	21.355	15.055	321.490	40
45	4.702	.2127	.00945	.0445	105.781	22.495	16.417	369.307	45
50	5.585	.1791	.00763	.0426	130.998	23.456	17.666	414.369	50
55	6.633	.1508	.00621	.0412	160.946	24.264	18.808	456.352	55
60	7.878	.1269	.00509	.0401	196.516	24.945	19.848	495.104	60
65	9.357	.1069	.00419	.0392	238.762	25.518	20.793	530.598	65
70	11.113	.0900	.00346	.0385	288.937	26.000	21.650	562.895	70
75	13.199	.0758	.00287	.0379	348.529	26.407	22.423	592.121	75
80	15.676	.0638	.00238	.0374	419.305	26.749	23.120	618.438	80
85	18.618	.0537	.00199	.0370	503.365	27.037	23.747	642.036	85
90	22.112	.0452	.00166	.0367	603.202	27.279	24.308	663.118	90
95	26.262	.0381	.00139	.0364	721.778	27.483	24.811	681.890	95
100	31.191	.0321	.00116	.0362	862.608	27.655	25.259	698.554	100

4%

Compound Interest Factors

4%

	Single Payment		Uniform Payment Series				Arithmetic Gradient		
	Compound Amount Factor	Present Worth Factor	Sinking Fund Factor	Capital Recovery Factor	Compound Amount Factor	Present Worth Factor	Gradient Uniform Series	Gradient Present Worth	
	Find F Given P	Find P Given F	Find A Given F	Find A Given P	Find F Given A	Find P Given A	Find A Given G	Find P Given G	
n	F/P	P/F	A/F	A/P	F/A	P/A	A/G	P/G	n
1	1.040	.9615	1.0000	1.0400	1.000	0.962	0	0	1
2	1.082	.9246	.4902	.5302	2.040	1.886	0.490	0.925	2
3	1.125	.8890	.3203	.3603	3.122	2.775	0.974	2.702	3
4	1.170	.8548	.2355	.2755	4.246	3.630	1.451	5.267	4
5	1.217	.8219	.1846	.2246	5.416	4.452	1.922	8.555	5
6	1.265	.7903	.1508	.1908	6.633	5.242	2.386	12.506	6
7	1.316	.7599	.1266	.1666	7.898	6.002	2.843	17.066	7
8	1.369	.7307	.1085	.1485	9.214	6.733	3.294	22.180	8
9	1.423	.7026	.0945	.1345	10.583	7.435	3.739	27.801	9
10	1.480	.6756	.0833	.1233	12.006	8.111	4.177	33.881	10
11	1.539	.6496	.0741	.1141	13.486	8.760	4.609	40.377	11
12	1.601	.6246	.0666	.1066	15.026	9.385	5.034	47.248	12
13	1.665	.6006	.0601	.1001	16.627	9.986	5.453	54.454	13
14	1.732	.5775	.0547	.0947	18.292	10.563	5.866	61.962	14
15	1.801	.5553	.0499	.0899	20.024	11.118	6.272	69.735	15
16	1.873	.5339	.0458	.0858	21.825	11.652	6.672	77.744	16
17	1.948	.5134	.0422	.0822	23.697	12.166	7.066	85.958	17
18	2.026	.4936	.0390	.0790	25.645	12.659	7.453	94.350	18
19	2.107	.4746	.0361	.0761	27.671	13.134	7.834	102.893	19
20	2.191	.4564	.0336	.0736	29.778	13.590	8.209	111.564	20
21	2.279	.4388	.0313	.0713	31.969	14.029	8.578	120.341	21
22	2.370	.4220	.0292	.0692	34.248	14.451	8.941	129.202	22
23	2.465	.4057	.0273	.0673	36.618	14.857	9.297	138.128	23
24	2.563	.3901	.0256	.0656	39.083	15.247	9.648	147.101	24
25	2.666	.3751	.0240	.0640	41.646	15.622	9.993	156.104	25
26	2.772	.3607	.0226	.0626	44.312	15.983	10.331	165.121	26
27	2.883	.3468	.0212	.0612	47.084	16.330	10.664	174.138	27
28	2.999	.3335	.0200	.0600	49.968	16.663	10.991	183.142	28
29	3.119	.3207	.0189	.0589	52.966	16.984	11.312	192.120	29
30	3.243	.3083	.0178	.0578	56.085	17.292	11.627	201.062	30
31	3.373	.2965	.0169	.0569	59.328	17.588	11.937	209.955	31
32	3.508	.2851	.0159	.0559	62.701	17.874	12.241	218.792	32
33	3.648	.2741	.0151	.0551	66.209	18.148	12.540	227.563	33
34	3.794	.2636	.0143	.0543	69.858	18.411	12.832	236.260	34
35	3.946	.2534	.0136	.0536	73.652	18.665	13.120	244.876	35
40	4.801	.2083	.0105	.0505	95.025	19.793	14.476	286.530	40
45	5.841	.1712	.00826	.0483	121.029	20.720	15.705	325.402	45
50	7.107	.1407	.00655	.0466	152.667	21.482	16.812	361.163	50
55	8.646	.1157	.00523	.0452	191.159	22.109	17.807	393.689	55
60	10.520	.0951	.00420	.0442	237.990	22.623	18.697	422.996	60
65	12.799	.0781	.00339	.0434	294.968	23.047	19.491	449.201	65
70	15.572	.0642	.00275	.0427	364.290	23.395	20.196	472.479	70
75	18.945	.0528	.00223	.0422	448.630	23.680	20.821	493.041	75
80	23.050	.0434	.00181	.0418	551.244	23.915	21.372	511.116	80
85	28.044	.0357	.00148	.0415	676.089	24.109	21.857	526.938	85
90	34.119	.0293	.00121	.0412	827.981	24.267	22.283	540.737	90
95	41.511	.0241	.00099	.0410	1012.8	24.398	22.655	552.730	95
100	50.505	.0198	.00081	.0408	1237.6	24.505	22.980	563.125	100

4¹/₂ %

Compound Interest Factors

4¹/₂ %

	Single Payment		Uniform Payment Series				Arithmetic Gradient		
	Compound Amount Factor	Present Worth Factor	Sinking Fund Factor	Capital Recovery Factor	Compound Amount Factor	Present Worth Factor	Gradient Uniform Series	Gradient Present Worth	
	Find F Given P	Find P Given F	Find A Given F	Find A Given P	Find F Given A	Find P Given A	Find A Given G	Find P Given G	
n	F/P	P/F	A/F	A/P	F/A	P/A	A/G	P/G	n
1	1.045	.9569	1.0000	1.0450	1.000	0.957	0	0	1
2	1.092	.9157	.4890	.5340	2.045	1.873	0.489	0.916	2
3	1.141	.8763	.3188	.3638	3.137	2.749	0.971	2.668	3
4	1.193	.8386	.2337	.2787	4.278	3.588	1.445	5.184	4
5	1.246	.8025	.1828	.2278	5.471	4.390	1.912	8.394	5
6	1.302	.7679	.1489	.1939	6.717	5.158	2.372	12.233	6
7	1.361	.7348	.1247	.1697	8.019	5.893	2.824	16.642	7
8	1.422	.7032	.1066	.1516	9.380	6.596	3.269	21.564	8
9	1.486	.6729	.0926	.1376	10.802	7.269	3.707	26.948	9
10	1.553	.6439	.0814	.1264	12.288	7.913	4.138	32.743	10
11	1.623	.6162	.0722	.1172	13.841	8.529	4.562	38.905	11
12	1.696	.5897	.0647	.1097	15.464	9.119	4.978	45.391	12
13	1.772	.5643	.0583	.1033	17.160	9.683	5.387	52.163	13
14	1.852	.5400	.0528	.0978	18.932	10.223	5.789	59.182	14
15	1.935	.5167	.0481	.0931	20.784	10.740	6.184	66.416	15
16	2.022	.4945	.0440	.0890	22.719	11.234	6.572	73.833	16
17	2.113	.4732	.0404	.0854	24.742	11.707	6.953	81.404	17
18	2.208	.4528	.0372	.0822	26.855	12.160	7.327	89.102	18
19	2.308	.4333	.0344	.0794	29.064	12.593	7.695	96.901	19
20	2.412	.4146	.0319	.0769	31.371	13.008	8.055	104.779	20
21	2.520	.3968	.0296	.0746	33.783	13.405	8.409	112.715	21
22	2.634	.3797	.0275	.0725	36.303	13.784	8.755	120.689	22
23	2.752	.3634	.0257	.0707	38.937	14.148	9.096	128.682	23
24	2.876	.3477	.0240	.0690	41.689	14.495	9.429	136.680	24
25	3.005	.3327	.0224	.0674	44.565	14.828	9.756	144.665	25
26	3.141	.3184	.0210	.0660	47.571	15.147	10.077	152.625	26
27	3.282	.3047	.0197	.0647	50.711	15.451	10.391	160.547	27
28	3.430	.2916	.0185	.0635	53.993	15.743	10.698	168.420	28
29	3.584	.2790	.0174	.0624	57.423	16.022	10.999	176.232	29
30	3.745	.2670	.0164	.0614	61.007	16.289	11.295	183.975	30
31	3.914	.2555	.0154	.0604	64.752	16.544	11.583	191.640	31
32	4.090	.2445	.0146	.0596	68.666	16.789	11.866	199.220	32
33	4.274	.2340	.0137	.0587	72.756	17.023	12.143	206.707	33
34	4.466	.2239	.0130	.0580	77.030	17.247	12.414	214.095	34
35	4.667	.2143	.0123	.0573	81.497	17.461	12.679	221.380	35
40	5.816	.1719	.00934	.0543	107.030	18.402	13.917	256.098	40
45	7.248	.1380	.00720	.0522	138.850	19.156	15.020	287.732	45
50	9.033	.1107	.00560	.0506	178.503	19.762	15.998	316.145	50
55	11.256	.0888	.00439	.0494	227.918	20.248	16.860	341.375	55
60	14.027	.0713	.00345	.0485	289.497	20.638	17.617	363.571	60
65	17.481	.0572	.00273	.0477	366.237	20.951	18.278	382.946	65
70	21.784	.0459	.00217	.0472	461.869	21.202	18.854	399.750	70
75	27.147	.0368	.00172	.0467	581.043	21.404	19.354	414.242	75
80	33.830	.0296	.00137	.0464	729.556	21.565	19.785	426.680	80
85	42.158	.0237	.00109	.0461	914.630	21.695	20.157	437.309	85
90	52.537	.0190	.00087	.0459	1 145.3	21.799	20.476	446.359	90
95	65.471	.0153	.00070	.0457	1 432.7	21.883	20.749	454.039	95
100	81.588	.0123	.00056	.0456	1 790.9	21.950	20.981	460.537	100

5%   Compound Interest Factors   5%

	Single Payment		Uniform Payment Series				Arithmetic Gradient		
	Compound Amount Factor	Present Worth Factor	Sinking Fund Factor	Capital Recovery Factor	Compound Amount Factor	Present Worth Factor	Gradient Uniform Series	Gradient Present Worth	
	Find F Given P	Find P Given F	Find A Given F	Find A Given P	Find F Given A	Find P Given A	Find A Given G	Find P Given G	
n	F/P	P/F	A/F	A/P	F/A	P/A	A/G	P/G	n
1	1.050	.9524	1.0000	1.0500	1.000	0.952	0	0	1
2	1.102	.9070	.4878	.5378	2.050	1.859	0.488	0.907	2
3	1.158	.8638	.3172	.3672	3.152	2.723	0.967	2.635	3
4	1.216	.8227	.2320	.2820	4.310	3.546	1.439	5.103	4
5	1.276	.7835	.1810	.2310	5.526	4.329	1.902	8.237	5
6	1.340	.7462	.1470	.1970	6.802	5.076	2.358	11.968	6
7	1.407	.7107	.1228	.1728	8.142	5.786	2.805	16.232	7
8	1.477	.6768	.1047	.1547	9.549	6.463	3.244	20.970	8
9	1.551	.6446	.0907	.1407	11.027	7.108	3.676	26.127	9
10	1.629	.6139	.0795	.1295	12.578	7.722	4.099	31.652	10
11	1.710	.5847	.0704	.1204	14.207	8.306	4.514	37.499	11
12	1.796	.5568	.0628	.1128	15.917	8.863	4.922	43.624	12
13	1.886	.5303	.0565	.1065	17.713	9.394	5.321	49.988	13
14	1.980	.5051	.0510	.1010	19.599	9.899	5.713	56.553	14
15	2.079	.4810	.0463	.0963	21.579	10.380	6.097	63.288	15
16	2.183	.4581	.0423	.0923	23.657	10.838	6.474	70.159	16
17	2.292	.4363	.0387	.0887	25.840	11.274	6.842	77.140	17
18	2.407	.4155	.0355	.0855	28.132	11.690	7.203	84.204	18
19	2.527	.3957	.0327	.0827	30.539	12.085	7.557	91.327	19
20	2.653	.3769	.0302	.0802	33.066	12.462	7.903	98.488	20
21	2.786	.3589	.0280	.0780	35.719	12.821	8.242	105.667	21
22	2.925	.3419	.0260	.0760	38.505	13.163	8.573	112.846	22
23	3.072	.3256	.0241	.0741	41.430	13.489	8.897	120.008	23
24	3.225	.3101	.0225	.0725	44.502	13.799	9.214	127.140	24
25	3.386	.2953	.0210	.0710	47.727	14.094	9.524	134.227	25
26	3.556	.2812	.0196	.0696	51.113	14.375	9.827	141.258	26
27	3.733	.2678	.0183	.0683	54.669	14.643	10.122	148.222	27
28	3.920	.2551	.0171	.0671	58.402	14.898	10.411	155.110	28
29	4.116	.2429	.0160	.0660	62.323	15.141	10.694	161.912	29
30	4.322	.2314	.0151	.0651	66.439	15.372	10.969	168.622	30
31	4.538	.2204	.0141	.0641	70.761	15.593	11.238	175.233	31
32	4.765	.2099	.0133	.0633	75.299	15.803	11.501	181.739	32
33	5.003	.1999	.0125	.0625	80.063	16.003	11.757	188.135	33
34	5.253	.1904	.0118	.0618	85.067	16.193	12.006	194.416	34
35	5.516	.1813	.0111	.0611	90.320	16.374	12.250	200.580	35
40	7.040	.1420	.00828	.0583	120.799	17.159	13.377	229.545	40
45	8.985	.1113	.00626	.0563	159.699	17.774	14.364	255.314	45
50	11.467	.0872	.00478	.0548	209.347	18.256	15.223	277.914	50
55	14.636	.0683	.00367	.0537	272.711	18.633	15.966	297.510	55
60	18.679	.0535	.00283	.0528	353.582	18.929	16.606	314.343	60
65	23.840	.0419	.00219	.0522	456.795	19.161	17.154	328.691	65
70	30.426	.0329	.00170	.0517	588.525	19.343	17.621	340.841	70
75	38.832	.0258	.00132	.0513	756.649	19.485	18.018	351.072	75
80	49.561	.0202	.00103	.0510	971.222	19.596	18.353	359.646	80
85	63.254	.0158	.00080	.0508	1 245.1	19.684	18.635	366.800	85
90	80.730	.0124	.00063	.0506	1 594.6	19.752	18.871	372.749	90
95	103.034	.00971	.00049	.0505	2 040.7	19.806	19.069	377.677	95
100	131.500	.00760	.00038	.0504	2 610.0	19.848	19.234	381.749	100

6%

Compound Interest Factors

6%

	Single Payment		Uniform Payment Series				Arithmetic Gradient		
	Compound Amount Factor	Present Worth Factor	Sinking Fund Factor	Capital Recovery Factor	Compound Amount Factor	Present Worth Factor	Gradient Uniform Series	Gradient Present Worth	
	Find F Given P	Find P Given F	Find A Given F	Find A Given P	Find F Given A	Find P Given A	Find A Given G	Find P Given G	
n	F/P	P/F	A/F	A/P	F/A	P/A	A/G	P/G	n
1	1.060	.9434	1.0000	1.0600	1.000	0.943	0	0	1
2	1.124	.8900	.4854	.5454	2.060	1.833	0.485	0.890	2
3	1.191	.8396	.3141	.3741	3.184	2.673	0.961	2.569	3
4	1.262	.7921	.2286	.2886	4.375	3.465	1.427	4.945	4
5	1.338	.7473	.1774	.2374	5.637	4.212	1.884	7.934	5
6	1.419	.7050	.1434	.2034	6.975	4.917	2.330	11.459	6
7	1.504	.6651	.1191	.1791	8.394	5.582	2.768	15.450	7
8	1.594	.6274	.1010	.1610	9.897	6.210	3.195	19.841	8
9	1.689	.5919	.0870	.1470	11.491	6.802	3.613	24.577	9
10	1.791	.5584	.0759	.1359	13.181	7.360	4.022	29.602	10
11	1.898	.5268	.0668	.1268	14.972	7.887	4.421	34.870	11
12	2.012	.4970	.0593	.1193	16.870	8.384	4.811	40.337	12
13	2.133	.4688	.0530	.1130	18.882	8.853	5.192	45.963	13
14	2.261	.4423	.0476	.1076	21.015	9.295	5.564	51.713	14
15	2.397	.4173	.0430	.1030	23.276	9.712	5.926	57.554	15
16	2.540	.3936	.0390	.0990	25.672	10.106	6.279	63.459	16
17	2.693	.3714	.0354	.0954	28.213	10.477	6.624	69.401	17
18	2.854	.3503	.0324	.0924	30.906	10.828	6.960	75.357	18
19	3.026	.3305	.0296	.0896	33.760	11.158	7.287	81.306	19
20	3.207	.3118	.0272	.0872	36.786	11.470	7.605	87.230	20
21	3.400	.2942	.0250	.0850	39.993	11.764	7.915	93.113	21
22	3.604	.2775	.0230	.0830	43.392	12.042	8.217	98.941	22
23	3.820	.2618	.0213	.0813	46.996	12.303	8.510	104.700	23
24	4.049	.2470	.0197	.0797	50.815	12.550	8.795	110.381	24
25	4.292	.2330	.0182	.0782	54.864	12.783	9.072	115.973	25
26	4.549	.2198	.0169	.0769	59.156	13.003	9.341	121.468	26
27	4.822	.2074	.0157	.0757	63.706	13.211	9.603	126.860	27
28	5.112	.1956	.0146	.0746	68.528	13.406	9.857	132.142	28
29	5.418	.1846	.0136	.0736	73.640	13.591	10.103	137.309	29
30	5.743	.1741	.0126	.0726	79.058	13.765	10.342	142.359	30
31	6.088	.1643	.0118	.0718	84.801	13.929	10.574	147.286	31
32	6.453	.1550	.0110	.0710	90.890	14.084	10.799	152.090	32
33	6.841	.1462	.0103	.0703	97.343	14.230	11.017	156.768	33
34	7.251	.1379	.00960	.0696	104.184	14.368	11.228	161.319	34
35	7.686	.1301	.00897	.0690	111.435	14.498	11.432	165.743	35
40	10.286	.0972	.00646	.0665	154.762	15.046	12.359	185.957	40
45	13.765	.0727	.00470	.0647	212.743	15.456	13.141	203.109	45
50	18.420	.0543	.00344	.0634	290.335	15.762	13.796	217.457	50
55	24.650	.0406	.00254	.0625	394.171	15.991	14.341	229.322	55
60	32.988	.0303	.00188	.0619	533.126	16.161	14.791	239.043	60
65	44.145	.0227	.00139	.0614	719.080	16.289	15.160	246.945	65
70	59.076	.0169	.00103	.0610	967.928	16.385	15.461	253.327	70
75	79.057	.0126	.00077	.0608	1 300.9	16.456	15.706	258.453	75
80	105.796	.00945	.00057	.0606	1 746.6	16.509	15.903	262.549	80
85	141.578	.00706	.00043	.0604	2 343.0	16.549	16.062	265.810	85
90	189.464	.00528	.00032	.0603	3 141.1	16.579	16.189	268.395	90
95	253.545	.00394	.00024	.0602	4 209.1	16.601	16.290	270.437	95
100	339.300	.00295	.00018	.0602	5 638.3	16.618	16.371	272.047	100

7% Compound Interest Factors 7%

	Single Payment		Uniform Payment Series				Arithmetic Gradient		
	Compound Amount Factor	Present Worth Factor	Sinking Fund Factor	Capital Recovery Factor	Compound Amount Factor	Present Worth Factor	Gradient Uniform Series	Gradient Present Worth	
	Find F Given P	Find P Given F	Find A Given F	Find A Given P	Find F Given A	Find P Given A	Find A Given G	Find P Given G	
n	F/P	P/F	A/F	A/P	F/A	P/A	A/G	P/G	n
1	1.070	.9346	1.0000	1.0700	1.000	0.935	0	0	1
2	1.145	.8734	.4831	.5531	2.070	1.808	0.483	0.873	2
3	1.225	.8163	.3111	.3811	3.215	2.624	0.955	2.506	3
4	1.311	.7629	.2252	.2952	4.440	3.387	1.416	4.795	4
5	1.403	.7130	.1739	.2439	5.751	4.100	1.865	7.647	5
6	1.501	.6663	.1398	.2098	7.153	4.767	2.303	10.978	6
7	1.606	.6227	.1156	.1856	8.654	5.389	2.730	14.715	7
8	1.718	.5820	.0975	.1675	10.260	5.971	3.147	18.789	8
9	1.838	.5439	.0835	.1535	11.978	6.515	3.552	23.140	9
10	1.967	.5083	.0724	.1424	13.816	7.024	3.946	27.716	10
11	2.105	.4751	.0634	.1334	15.784	7.499	4.330	32.467	11
12	2.252	.4440	.0559	.1259	17.888	7.943	4.703	37.351	12
13	2.410	.4150	.0497	.1197	20.141	8.358	5.065	42.330	13
14	2.579	.3878	.0443	.1143	22.551	8.745	5.417	47.372	14
15	2.759	.3624	.0398	.1098	25.129	9.108	5.758	52.446	15
16	2.952	.3387	.0359	.1059	27.888	9.447	6.090	57.527	16
17	3.159	.3166	.0324	.1024	30.840	9.763	6.411	62.592	17
18	3.380	.2959	.0294	.0994	33.999	10.059	6.722	67.622	18
19	3.617	.2765	.0268	.0968	37.379	10.336	7.024	72.599	19
20	3.870	.2584	.0244	.0944	40.996	10.594	7.316	77.509	20
21	4.141	.2415	.0223	.0923	44.865	10.836	7.599	82.339	21
22	4.430	.2257	.0204	.0904	49.006	11.061	7.872	87.079	22
23	4.741	.2109	.0187	.0887	53.436	11.272	8.137	91.720	23
24	5.072	.1971	.0172	.0872	58.177	11.469	8.392	96.255	24
25	5.427	.1842	.0158	.0858	63.249	11.654	8.639	100.677	25
26	5.807	.1722	.0146	.0846	68.677	11.826	8.877	104.981	26
27	6.214	.1609	.0134	.0834	74.484	11.987	9.107	109.166	27
28	6.649	.1504	.0124	.0824	80.698	12.137	9.329	113.227	28
29	7.114	.1406	.0114	.0814	87.347	12.278	9.543	117.162	29
30	7.612	.1314	.0106	.0806	94.461	12.409	9.749	120.972	30
31	8.145	.1228	.00980	.0798	102.073	12.532	9.947	124.655	31
32	8.715	.1147	.00907	.0791	110.218	12.647	10.138	128.212	32
33	9.325	.1072	.00841	.0784	118.934	12.754	10.322	131.644	33
34	9.978	.1002	.00780	.0778	128.259	12.854	10.499	134.951	34
35	10.677	.0937	.00723	.0772	138.237	12.948	10.669	138.135	35
40	14.974	.0668	.00501	.0750	199.636	13.332	11.423	152.293	40
45	21.002	.0476	.00350	.0735	285.750	13.606	12.036	163.756	45
50	29.457	.0339	.00246	.0725	406.530	13.801	12.529	172.905	50
55	41.315	.0242	.00174	.0717	575.930	13.940	12.921	180.124	55
60	57.947	.0173	.00123	.0712	813.523	14.039	13.232	185.768	60
65	81.273	.0123	.00087	.0709	1 146.8	14.110	13.476	190.145	65
70	113.990	.00877	.00062	.0706	1 614.1	14.160	13.666	193.519	70
75	159.877	.00625	.00044	.0704	2 269.7	14.196	13.814	196.104	75
80	224.235	.00446	.00031	.0703	3 189.1	14.222	13.927	198.075	80
85	314.502	.00318	.00022	.0702	4 478.6	14.240	14.015	199.572	85
90	441.105	.00227	.00016	.0702	6 287.2	14.253	14.081	200.704	90
95	618.673	.00162	.00011	.0701	8 823.9	14.263	14.132	201.558	95
100	867.720	.00115	.00008	.0701	12 381.7	14.269	14.170	202.200	100

8%

Compound Interest Factors

8%

	Single Payment		Uniform Payment Series				Arithmetic Gradient		
	Compound Amount Factor	Present Worth Factor	Sinking Fund Factor	Capital Recovery Factor	Compound Amount Factor	Present Worth Factor	Gradient Uniform Series	Gradient Present Worth	
	Find F Given P	Find P Given F	Find A Given F	Find A Given P	Find F Given A	Find P Given A	Find A Given G	Find P Given G	
n	F/P	P/F	A/F	A/P	F/A	P/A	A/G	P/G	n
1	1.080	.9259	1.0000	1.0800	1.000	0.926	0	0	1
2	1.166	.8573	.4808	.5608	2.080	1.783	0.481	0.857	2
3	1.260	.7938	.3080	.3880	3.246	2.577	0.949	2.445	3
4	1.360	.7350	.2219	.3019	4.506	3.312	1.404	4.650	4
5	1.469	.6806	.1705	.2505	5.867	3.993	1.846	7.372	5
6	1.587	.6302	.1363	.2163	7.336	4.623	2.276	10.523	6
7	1.714	.5835	.1121	.1921	8.923	5.206	2.694	14.024	7
8	1.851	.5403	.0940	.1740	10.637	5.747	3.099	17.806	8
9	1.999	.5002	.0801	.1601	12.488	6.247	3.491	21.808	9
10	2.159	.4632	.0690	.1490	14.487	6.710	3.871	25.977	10
11	2.332	.4289	.0601	.1401	16.645	7.139	4.240	30.266	11
12	2.518	.3971	.0527	.1327	18.977	7.536	4.596	34.634	12
13	2.720	.3677	.0465	.1265	21.495	7.904	4.940	39.046	13
14	2.937	.3405	.0413	.1213	24.215	8.244	5.273	43.472	14
15	3.172	.3152	.0368	.1168	27.152	8.559	5.594	47.886	15
16	3.426	.2919	.0330	.1130	30.324	8.851	5.905	52.264	16
17	3.700	.2703	.0296	.1096	33.750	9.122	6.204	56.588	17
18	3.996	.2502	.0267	.1067	37.450	9.372	6.492	60.843	18
19	4.316	.2317	.0241	.1041	41.446	9.604	6.770	65.013	19
20	4.661	.2145	.0219	.1019	45.762	9.818	7.037	69.090	20
21	5.034	.1987	.0198	.0998	50.423	10.017	7.294	73.063	21
22	5.437	.1839	.0180	.0980	55.457	10.201	7.541	76.926	22
23	5.871	.1703	.0164	.0964	60.893	10.371	7.779	80.673	23
24	6.341	.1577	.0150	.0950	66.765	10.529	8.007	84.300	24
25	6.848	.1460	.0137	.0937	73.106	10.675	8.225	87.804	25
26	7.396	.1352	.0125	.0925	79.954	10.810	8.435	91.184	26
27	7.988	.1252	.0114	.0914	87.351	10.935	8.636	94.439	27
28	8.627	.1159	.0105	.0905	95.339	11.051	8.829	97.569	28
29	9.317	.1073	.00962	.0896	103.966	11.158	9.013	100.574	29
30	10.063	.0994	.00883	.0888	113.283	11.258	9.190	103.456	30
31	10.868	.0920	.00811	.0881	123.346	11.350	9.358	106.216	31
32	11.737	.0852	.00745	.0875	134.214	11.435	9.520	108.858	32
33	12.676	.0789	.00685	.0869	145.951	11.514	9.674	111.382	33
34	13.690	.0730	.00630	.0863	158.627	11.587	9.821	113.792	34
35	14.785	.0676	.00580	.0858	172.317	11.655	9.961	116.092	35
40	21.725	.0460	.00386	.0839	259.057	11.925	10.570	126.042	40
45	31.920	.0313	.00259	.0826	386.506	12.108	11.045	133.733	45
50	46.902	.0213	.00174	.0817	573.771	12.233	11.411	139.593	50
55	68.914	.0145	.00118	.0812	848.925	12.319	11.690	144.006	55
60	101.257	.00988	.00080	.0808	1 253.2	12.377	11.902	147.300	60
65	148.780	.00672	.00054	.0805	1 847.3	12.416	12.060	149.739	65
70	218.607	.00457	.00037	.0804	2 720.1	12.443	12.178	151.533	70
75	321.205	.00311	.00025	.0802	4 002.6	12.461	12.266	152.845	75
80	471.956	.00212	.00017	.0802	5 887.0	12.474	12.330	153.800	80
85	693.458	.00144	.00012	.0801	8 655.7	12.482	12.377	154.492	85
90	1 018.9	.00098	.00008	.0801	12 724.0	12.488	12.412	154.993	90
95	1 497.1	.00067	.00005	.0801	18 701.6	12.492	12.437	155.352	95
100	2 199.8	.00045	.00004	.0800	27 484.6	12.494	12.455	155.611	100

9%   Compound Interest Factors   9%

	Single Payment		Uniform Payment Series				Arithmetic Gradient		
	Compound Amount Factor	Present Worth Factor	Sinking Fund Factor	Capital Recovery Factor	Compound Amount Factor	Present Worth Factor	Gradient Uniform Series	Gradient Present Worth	
	Find F Given P	Find P Given F	Find A Given F	Find A Given P	Find F Given A	Find P Given A	Find A Given G	Find P Given G	
n	F/P	P/F	A/F	A/P	F/A	P/A	A/G	P/G	n
1	1.090	.9174	1.0000	1.0900	1.000	0.917	0	0	1
2	1.188	.8417	.4785	.5685	2.090	1.759	0.478	0.842	2
3	1.295	.7722	.3051	.3951	3.278	2.531	0.943	2.386	3
4	1.412	.7084	.2187	.3087	4.573	3.240	1.393	4.511	4
5	1.539	.6499	.1671	.2571	5.985	3.890	1.828	7.111	5
6	1.677	.5963	.1329	.2229	7.523	4.486	2.250	10.092	6
7	1.828	.5470	.1087	.1987	9.200	5.033	2.657	13.375	7
8	1.993	.5019	.0907	.1807	11.028	5.535	3.051	16.888	8
9	2.172	.4604	.0768	.1668	13.021	5.995	3.431	20.571	9
10	2.367	.4224	.0658	.1558	15.193	6.418	3.798	24.373	10
11	2.580	.3875	.0569	.1469	17.560	6.805	4.151	28.248	11
12	2.813	.3555	.0497	.1397	20.141	7.161	4.491	32.159	12
13	3.066	.3262	.0436	.1336	22.953	7.487	4.818	36.073	13
14	3.342	.2992	.0384	.1284	26.019	7.786	5.133	39.963	14
15	3.642	.2745	.0341	.1241	29.361	8.061	5.435	43.807	15
16	3.970	.2519	.0303	.1203	33.003	8.313	5.724	47.585	16
17	4.328	.2311	.0270	.1170	36.974	8.544	6.002	51.282	17
18	4.717	.2120	.0242	.1142	41.301	8.756	6.269	54.886	18
19	5.142	.1945	.0217	.1117	46.019	8.950	6.524	58.387	19
20	5.604	.1784	.0195	.1095	51.160	9.129	6.767	61.777	20
21	6.109	.1637	.0176	.1076	56.765	9.292	7.001	65.051	21
22	6.659	.1502	.0159	.1059	62.873	9.442	7.223	68.205	22
23	7.258	.1378	.0144	.1044	69.532	9.580	7.436	71.236	23
24	7.911	.1264	.0130	.1030	76.790	9.707	7.638	74.143	24
25	8.623	.1160	.0118	.1018	84.701	9.823	7.832	76.927	25
26	9.399	.1064	.0107	.1007	93.324	9.929	8.016	79.586	26
27	10.245	.0976	.00973	.0997	102.723	10.027	8.191	82.124	27
28	11.167	.0895	.00885	.0989	112.968	10.116	8.357	84.542	28
29	12.172	.0822	.00806	.0981	124.136	10.198	8.515	86.842	29
30	13.268	.0754	.00734	.0973	136.308	10.274	8.666	89.028	30
31	14.462	.0691	.00669	.0967	149.575	10.343	8.808	91.102	31
32	15.763	.0634	.00610	.0961	164.037	10.406	8.944	93.069	32
33	17.182	.0582	.00556	.0956	179.801	10.464	9.072	94.931	33
34	18.728	.0534	.00508	.0951	196.983	10.518	9.193	96.693	34
35	20.414	.0490	.00464	.0946	215.711	10.567	9.308	98.359	35
40	31.409	.0318	.00296	.0930	337.883	10.757	9.796	105.376	40
45	48.327	.0207	.00190	.0919	525.860	10.881	10.160	110.556	45
50	74.358	.0134	.00123	.0912	815.085	10.962	10.430	114.325	50
55	114.409	.00874	.00079	.0908	1 260.1	11.014	10.626	117.036	55
60	176.032	.00568	.00051	.0905	1 944.8	11.048	10.768	118.968	60
65	270.847	.00369	.00033	.0903	2 998.3	11.070	10.870	120.334	65
70	416.731	.00240	.00022	.0902	4 619.2	11.084	10.943	121.294	70
75	641.193	.00156	.00014	.0901	7 113.3	11.094	10.994	121.965	75
80	986.555	.00101	.00009	.0901	10 950.6	11.100	11.030	122.431	80
85	1 517.9	.00066	.00006	.0901	16 854.9	11.104	11.055	122.753	85
90	2 335.5	.00043	.00004	.0900	25 939.3	11.106	11.073	122.976	90
95	3 593.5	.00028	.00003	.0900	39 916.8	11.108	11.085	123.129	95
100	5 529.1	.00018	.00002	.0900	61 422.9	11.109	11.093	123.233	100

10%   Compound Interest Factors   10%

	Single Payment		Uniform Payment Series				Arithmetic Gradient		
	Compound Amount Factor	Present Worth Factor	Sinking Fund Factor	Capital Recovery Factor	Compound Amount Factor	Present Worth Factor	Gradient Uniform Series	Gradient Present Worth	
	Find F Given P	Find P Given F	Find A Given F	Find A Given P	Find F Given A	Find P Given A	Find A Given G	Find P Given G	
n	F/P	P/F	A/F	A/P	F/A	P/A	A/G	P/G	n
1	1.100	.9091	1.0000	1.1000	1.000	0.909	0	0	1
2	1.210	.8264	.4762	.5762	2.100	1.736	0.476	0.826	2
3	1.331	.7513	.3021	.4021	3.310	2.487	0.937	2.329	3
4	1.464	.6830	.2155	.3155	4.641	3.170	1.381	4.378	4
5	1.611	.6209	.1638	.2638	6.105	3.791	1.810	6.862	5
6	1.772	.5645	.1296	.2296	7.716	4.355	2.224	9.684	6
7	1.949	.5132	.1054	.2054	9.487	4.868	2.622	12.763	7
8	2.144	.4665	.0874	.1874	11.436	5.335	3.004	16.029	8
9	2.358	.4241	.0736	.1736	13.579	5.759	3.372	19.421	9
10	2.594	.3855	.0627	.1627	15.937	6.145	3.725	22.891	10
11	2.853	.3505	.0540	.1540	18.531	6.495	4.064	26.396	11
12	3.138	.3186	.0468	.1468	21.384	6.814	4.388	29.901	12
13	3.452	.2897	.0408	.1408	24.523	7.103	4.699	33.377	13
14	3.797	.2633	.0357	.1357	27.975	7.367	4.996	36.801	14
15	4.177	.2394	.0315	.1315	31.772	7.606	5.279	40.152	15
16	4.595	.2176	.0278	.1278	35.950	7.824	5.549	43.416	16
17	5.054	.1978	.0247	.1247	40.545	8.022	5.807	46.582	17
18	5.560	.1799	.0219	.1219	45.599	8.201	6.053	49.640	18
19	6.116	.1635	.0195	.1195	51.159	8.365	6.286	52.583	19
20	6.728	.1486	.0175	.1175	57.275	8.514	6.508	55.407	20
21	7.400	.1351	.0156	.1156	64.003	8.649	6.719	58.110	21
22	8.140	.1228	.0140	.1140	71.403	8.772	6.919	60.689	22
23	8.954	.1117	.0126	.1126	79.543	8.883	7.108	63.146	23
24	9.850	.1015	.0113	.1113	88.497	8.985	7.288	65.481	24
25	10.835	.0923	.0102	.1102	98.347	9.077	7.458	67.696	25
26	11.918	.0839	.00916	.1092	109.182	9.161	7.619	69.794	26
27	13.110	.0763	.00826	.1083	121.100	9.237	7.770	71.777	27
28	14.421	.0693	.00745	.1075	134.210	9.307	7.914	73.650	28
29	15.863	.0630	.00673	.1067	148.631	9.370	8.049	75.415	29
30	17.449	.0573	.00608	.1061	164.494	9.427	8.176	77.077	30
31	19.194	.0521	.00550	.1055	181.944	9.479	8.296	78.640	31
32	21.114	.0474	.00497	.1050	201.138	9.526	8.409	80.108	32
33	23.225	.0431	.00450	.1045	222.252	9.569	8.515	81.486	33
34	25.548	.0391	.00407	.1041	245.477	9.609	8.615	82.777	34
35	28.102	.0356	.00369	.1037	271.025	9.644	8.709	83.987	35
40	45.259	.0221	.00226	.1023	442.593	9.779	9.096	88.953	40
45	72.891	.0137	.00139	.1014	718.905	9.863	9.374	92.454	45
50	117.391	.00852	.00086	.1009	1 163.9	9.915	9.570	94.889	50
55	189.059	.00529	.00053	.1005	1 880.6	9.947	9.708	96.562	55
60	304.482	.00328	.00033	.1003	3 034.8	9.967	9.802	97.701	60
65	490.371	.00204	.00020	.1002	4 893.7	9.980	9.867	98.471	65
70	789.748	.00127	.00013	.1001	7 887.5	9.987	9.911	98.987	70
75	1 271.9	.00079	.00008	.1001	12 709.0	9.992	9.941	99.332	75
80	2 048.4	.00049	.00005	.1000	20 474.0	9.995	9.961	99.561	80
85	3 299.0	.00030	.00003	.1000	32 979.7	9.997	9.974	99.712	85
90	5 313.0	.00019	.00002	.1000	53 120.3	9.998	9.983	99.812	90
95	8 556.7	.00012	.00001	.1000	85 556.9	9.999	9.989	99.877	95
100	13 780.6	.00007	.00001	.1000	137 796.3	9.999	9.993	99.920	100

12% Compound Interest Factors 12%

	Single Payment		Uniform Payment Series				Arithmetic Gradient		
	Compound Amount Factor	Present Worth Factor	Sinking Fund Factor	Capital Recovery Factor	Compound Amount Factor	Present Worth Factor	Gradient Uniform Series	Gradient Present Worth	
	Find F Given P	Find P Given F	Find A Given F	Find A Given P	Find F Given A	Find P Given A	Find A Given G	Find P Given G	
n	F/P	P/F	A/F	A/P	F/A	P/A	A/G	P/G	n
1	1.120	.8929	1.0000	1.1200	1.000	0.893	0	0	1
2	1.254	.7972	.4717	.5917	2.120	1.690	0.472	0.797	2
3	1.405	.7118	.2963	.4163	3.374	2.402	0.925	2.221	3
4	1.574	.6355	.2092	.3292	4.779	3.037	1.359	4.127	4
5	1.762	.5674	.1574	.2774	6.353	3.605	1.775	6.397	5
6	1.974	.5066	.1232	.2432	8.115	4.111	2.172	8.930	6
7	2.211	.4523	.0991	.2191	10.089	4.564	2.551	11.644	7
8	2.476	.4039	.0813	.2013	12.300	4.968	2.913	14.471	8
9	2.773	.3606	.0677	.1877	14.776	5.328	3.257	17.356	9
10	3.106	.3220	.0570	.1770	17.549	5.650	3.585	20.254	10
11	3.479	.2875	.0484	.1684	20.655	5.938	3.895	23.129	11
12	3.896	.2567	.0414	.1614	24.133	6.194	4.190	25.952	12
13	4.363	.2292	.0357	.1557	28.029	6.424	4.468	28.702	13
14	4.887	.2046	.0309	.1509	32.393	6.628	4.732	31.362	14
15	5.474	.1827	.0268	.1468	37.280	6.811	4.980	33.920	15
16	6.130	.1631	.0234	.1434	42.753	6.974	5.215	36.367	16
17	6.866	.1456	.0205	.1405	48.884	7.120	5.435	38.697	17
18	7.690	.1300	.0179	.1379	55.750	7.250	5.643	40.908	18
19	8.613	.1161	.0158	.1358	63.440	7.366	5.838	42.998	19
20	9.646	.1037	.0139	.1339	72.052	7.469	6.020	44.968	20
21	10.804	.0926	.0122	.1322	81.699	7.562	6.191	46.819	21
22	12.100	.0826	.0108	.1308	92.503	7.645	6.351	48.554	22
23	13.552	.0738	.00956	.1296	104.603	7.718	6.501	50.178	23
24	15.179	.0659	.00846	.1285	118.155	7.784	6.641	51.693	24
25	17.000	.0588	.00750	.1275	133.334	7.843	6.771	53.105	25
26	19.040	.0525	.00665	.1267	150.334	7.896	6.892	54.418	26
27	21.325	.0469	.00590	.1259	169.374	7.943	7.005	55.637	27
28	23.884	.0419	.00524	.1252	190.699	7.984	7.110	56.767	28
29	26.750	.0374	.00466	.1247	214.583	8.022	7.207	57.814	29
30	29.960	.0334	.00414	.1241	241.333	8.055	7.297	58.782	30
31	33.555	.0298	.00369	.1237	271.293	8.085	7.381	59.676	31
32	37.582	.0266	.00328	.1233	304.848	8.112	7.459	60.501	32
33	42.092	.0238	.00292	.1229	342.429	8.135	7.530	61.261	33
34	47.143	.0212	.00260	.1226	384.521	8.157	7.596	61.961	34
35	52.800	.0189	.00232	.1223	431.663	8.176	7.658	62.605	35
40	93.051	.0107	.00130	.1213	767.091	8.244	7.899	65.116	40
45	163.988	.00610	.00074	.1207	1 358.2	8.283	8.057	66.734	45
50	289.002	.00346	.00042	.1204	2 400.0	8.304	8.160	67.762	50
55	509.321	.00196	.00024	.1202	4 236.0	8.317	8.225	68.408	55
60	897.597	.00111	.00013	.1201	7 471.6	8.324	8.266	68.810	60
65	1 581.9	.00063	.00008	.1201	13 173.9	8.328	8.292	69.058	65
70	2 787.8	.00036	.00004	.1200	23 223.3	8.330	8.308	69.210	70
75	4 913.1	.00020	.00002	.1200	40 933.8	8.332	8.318	69.303	75
80	8 658.5	.00012	.00001	.1200	72 145.7	8.332	8.324	69.359	80
85	15 259.2	.00007	.00001	.1200	127 151.7	8.333	8.328	69.393	85
90	26 891.9	.00004		.1200	224 091.1	8.333	8.330	69.414	90
95	47 392.8	.00002		.1200	394 931.4	8.333	8.331	69.426	95
100	83 522.3	.00001		.1200	696 010.5	8.333	8.332	69.434	100

15%

Compound Interest Factors

15%

	Single Payment		Uniform Payment Series				Arithmetic Gradient		
	Compound Amount Factor	Present Worth Factor	Sinking Fund Factor	Capital Recovery Factor	Compound Amount Factor	Present Worth Factor	Gradient Uniform Series	Gradient Present Worth	
	Find F Given P	Find P Given F	Find A Given F	Find A Given P	Find F Given A	Find P Given A	Find A Given G	Find P Given G	
n	F/P	P/F	A/F	A/P	F/A	P/A	A/G	P/G	n
1	1.150	.8696	1.0000	1.1500	1.000	0.870	0	0	1
2	1.322	.7561	.4651	.6151	2.150	1.626	0.465	0.756	2
3	1.521	.6575	.2880	.4380	3.472	2.283	0.907	2.071	3
4	1.749	.5718	.2003	.3503	4.993	2.855	1.326	3.786	4
5	2.011	.4972	.1483	.2983	6.742	3.352	1.723	5.775	5
6	2.313	.4323	.1142	.2642	8.754	3.784	2.097	7.937	6
7	2.660	.3759	.0904	.2404	11.067	4.160	2.450	10.192	7
8	3.059	.3269	.0729	.2229	13.727	4.487	2.781	12.481	8
9	3.518	.2843	.0596	.2096	16.786	4.772	3.092	14.755	9
10	4.046	.2472	.0493	.1993	20.304	5.019	3.383	16.979	10
11	4.652	.2149	.0411	.1911	24.349	5.234	3.655	19.129	11
12	5.350	.1869	.0345	.1845	29.002	5.421	3.908	21.185	12
13	6.153	.1625	.0291	.1791	34.352	5.583	4.144	23.135	13
14	7.076	.1413	.0247	.1747	40.505	5.724	4.362	24.972	14
15	8.137	.1229	.0210	.1710	47.580	5.847	4.565	26.693	15
16	9.358	.1069	.0179	.1679	55.717	5.954	4.752	28.296	16
17	10.761	.0929	.0154	.1654	65.075	6.047	4.925	29.783	17
18	12.375	.0808	.0132	.1632	75.836	6.128	5.084	31.156	18
19	14.232	.0703	.0113	.1613	88.212	6.198	5.231	32.421	19
20	16.367	.0611	.00976	.1598	102.444	6.259	5.365	33.582	20
21	18.822	.0531	.00842	.1584	118.810	6.312	5.488	34.645	21
22	21.645	.0462	.00727	.1573	137.632	6.359	5.601	35.615	22
23	24.891	.0402	.00628	.1563	159.276	6.399	5.704	36.499	23
24	28.625	.0349	.00543	.1554	184.168	6.434	5.798	37.302	24
25	32.919	.0304	.00470	.1547	212.793	6.464	5.883	38.031	25
26	37.857	.0264	.00407	.1541	245.712	6.491	5.961	38.692	26
27	43.535	.0230	.00353	.1535	283.569	6.514	6.032	39.289	27
28	50.066	.0200	.00306	.1531	327.104	6.534	6.096	39.828	28
29	57.575	.0174	.00265	.1527	377.170	6.551	6.154	40.315	29
30	66.212	.0151	.00230	.1523	434.745	6.566	6.207	40.753	30
31	76.144	.0131	.00200	.1520	500.957	6.579	6.254	41.147	31
32	87.565	.0114	.00173	.1517	577.100	6.591	6.297	41.501	32
33	100.700	.00993	.00150	.1515	664.666	6.600	6.336	41.818	33
34	115.805	.00864	.00131	.1513	765.365	6.609	6.371	42.103	34
35	133.176	.00751	.00113	.1511	881.170	6.617	6.402	42.359	35
40	267.864	.00373	.00056	.1506	1 779.1	6.642	6.517	43.283	40
45	538.769	.00186	.00028	.1503	3 585.1	6.654	6.583	43.805	45
50	1 083.7	.00092	.00014	.1501	7 217.7	6.661	6.620	44.096	50
55	2 179.6	.00046	.00007	.1501	14 524.1	6.664	6.641	44.256	55
60	4 384.0	.00023	.00003	.1500	29 220.0	6.665	6.653	44.343	60
65	8 817.8	.00011	.00002	.1500	58 778.6	6.666	6.659	44.390	65
70	17 735.7	.00006	.00001	.1500	118 231.5	6.666	6.663	44.416	70
75	35 672.9	.00003		.1500	237 812.5	6.666	6.665	44.429	75
80	71 750.9	.00001		.1500	478 332.6	6.667	6.666	44.436	80
85	144 316.7	.00001		.1500	962 104.4	6.667	6.666	44.440	85

18%

Compound Interest Factors

18%

	Single Payment		Uniform Payment Series				Arithmetic Gradient		
	Compound Amount Factor	Present Worth Factor	Sinking Fund Factor	Capital Recovery Factor	Compound Amount Factor	Present Worth Factor	Gradient Uniform Series	Gradient Present Worth	
	Find F Given P	Find P Given F	Find A Given F	Find A Given P	Find F Given A	Find P Given A	Find A Given G	Find P Given G	
n	F/P	P/F	A/F	A/P	F/A	P/A	A/G	P/G	n
1	1.180	.8475	1.0000	1.1800	1.000	0.847	0	0	1
2	1.392	.7182	.4587	.6387	2.180	1.566	0.459	0.718	2
3	1.643	.6086	.2799	.4599	3.572	2.174	0.890	1.935	3
4	1.939	.5158	.1917	.3717	5.215	2.690	1.295	3.483	4
5	2.288	.4371	.1398	.3198	7.154	3.127	1.673	5.231	5
6	2.700	.3704	.1059	.2859	9.442	3.498	2.025	7.083	6
7	3.185	.3139	.0824	.2624	12.142	3.812	2.353	8.967	7
8	3.759	.2660	.0652	.2452	15.327	4.078	2.656	10.829	8
9	4.435	.2255	.0524	.2324	19.086	4.303	2.936	12.633	9
10	5.234	.1911	.0425	.2225	23.521	4.494	3.194	14.352	10
11	6.176	.1619	.0348	.2148	28.755	4.656	3.430	15.972	11
12	7.288	.1372	.0286	.2086	34.931	4.793	3.647	17.481	12
13	8.599	.1163	.0237	.2037	42.219	4.910	3.845	18.877	13
14	10.147	.0985	.0197	.1997	50.818	5.008	4.025	20.158	14
15	11.974	.0835	.0164	.1964	60.965	5.092	4.189	21.327	15
16	14.129	.0708	.0137	.1937	72.939	5.162	4.337	22.389	16
17	16.672	.0600	.0115	.1915	87.068	5.222	4.471	23.348	17
18	19.673	.0508	.00964	.1896	103.740	5.273	4.592	24.212	18
19	23.214	.0431	.00810	.1881	123.413	5.316	4.700	24.988	19
20	27.393	.0365	.00682	.1868	146.628	5.353	4.798	25.681	20
21	32.324	.0309	.00575	.1857	174.021	5.384	4.885	26.300	21
22	38.142	.0262	.00485	.1848	206.345	5.410	4.963	26.851	22
23	45.008	.0222	.00409	.1841	244.487	5.432	5.033	27.339	23
24	53.109	.0188	.00345	.1835	289.494	5.451	5.095	27.772	24
25	62.669	.0160	.00292	.1829	342.603	5.467	5.150	28.155	25
26	73.949	.0135	.00247	.1825	405.272	5.480	5.199	28.494	26
27	87.260	.0115	.00209	.1821	479.221	5.492	5.243	28.791	27
28	102.966	.00971	.00177	.1818	566.480	5.502	5.281	29.054	28
29	121.500	.00823	.00149	.1815	669.447	5.510	5.315	29.284	29
30	143.370	.00697	.00126	.1813	790.947	5.517	5.345	29.486	30
31	169.177	.00591	.00107	.1811	934.317	5.523	5.371	29.664	31
32	199.629	.00501	.00091	.1809	1 103.5	5.528	5.394	29.819	32
33	235.562	.00425	.00077	.1808	1 303.1	5.532	5.415	29.955	33
34	277.963	.00360	.00065	.1806	1 538.7	5.536	5.433	30.074	34
35	327.997	.00305	.00055	.1806	1 816.6	5.539	5.449	30.177	35
40	750.377	.00133	.00024	.1802	4 163.2	5.548	5.502	30.527	40
45	1 716.7	.00058	.00010	.1801	9 531.6	5.552	5.529	30.701	45
50	3 927.3	.00025	.00005	.1800	21 813.0	5.554	5.543	30.786	50
55	8 984.8	.00011	.00002	.1800	49 910.1	5.555	5.549	30.827	55
60	20 555.1	.00005	.00001	.1800	114 189.4	5.555	5.553	30.846	60
65	47 025.1	.00002		.1800	261 244.7	5.555	5.554	30.856	65
70	107 581.9	.00001		.1800	597 671.7	5.556	5.555	30.860	70

20%

Compound Interest Factors

20%

	Single Payment		Uniform Payment Series				Arithmetic Gradient		
	Compound Amount Factor	Present Worth Factor	Sinking Fund Factor	Capital Recovery Factor	Compound Amount Factor	Present Worth Factor	Gradient Uniform Series	Gradient Present Worth	
	Find F Given P	Find P Given F	Find A Given F	Find A Given P	Find F Given A	Find P Given A	Find A Given G	Find P Given G	
n	F/P	P/F	A/F	A/P	F/A	P/A	A/G	P/G	n
1	1.200	.8333	1.0000	1.2000	1.000	0.833	0	0	1
2	1.440	.6944	.4545	.6545	2.200	1.528	0.455	0.694	2
3	1.728	.5787	.2747	.4747	3.640	2.106	0.879	1.852	3
4	2.074	.4823	.1863	.3863	5.368	2.589	1.274	3.299	4
5	2.488	.4019	.1344	.3344	7.442	2.991	1.641	4.906	5
6	2.986	.3349	.1007	.3007	9.930	3.326	1.979	6.581	6
7	3.583	.2791	.0774	.2774	12.916	3.605	2.290	8.255	7
8	4.300	.2326	.0606	.2606	16.499	3.837	2.576	9.883	8
9	5.160	.1938	.0481	.2481	20.799	4.031	2.836	11.434	9
10	6.192	.1615	.0385	.2385	25.959	4.192	3.074	12.887	10
11	7.430	.1346	.0311	.2311	32.150	4.327	3.289	14.233	11
12	8.916	.1122	.0253	.2253	39.581	4.439	3.484	15.467	12
13	10.699	.0935	.0206	.2206	48.497	4.533	3.660	16.588	13
14	12.839	.0779	.0169	.2169	59.196	4.611	3.817	17.601	14
15	15.407	.0649	.0139	.2139	72.035	4.675	3.959	18.509	15
16	18.488	.0541	.0114	.2114	87.442	4.730	4.085	19.321	16
17	22.186	.0451	.00944	.2094	105.931	4.775	4.198	20.042	17
18	26.623	.0376	.00781	.2078	128.117	4.812	4.298	20.680	18
19	31.948	.0313	.00646	.2065	154.740	4.843	4.386	21.244	19
20	38.338	.0261	.00536	.2054	186.688	4.870	4.464	21.739	20
21	46.005	.0217	.00444	.2044	225.026	4.891	4.533	22.174	21
22	55.206	.0181	.00369	.2037	271.031	4.909	4.594	22.555	22
23	66.247	.0151	.00307	.2031	326.237	4.925	4.647	22.887	23
24	79.497	.0126	.00255	.2025	392.484	4.937	4.694	23.176	24
25	95.396	.0105	.00212	.2021	471.981	4.948	4.735	23.428	25
26	114.475	.00874	.00176	.2018	567.377	4.956	4.771	23.646	26
27	137.371	.00728	.00147	.2015	681.853	4.964	4.802	23.835	27
28	164.845	.00607	.00122	.2012	819.223	4.970	4.829	23.999	28
29	197.814	.00506	.00102	.2010	984.068	4.975	4.853	24.141	29
30	237.376	.00421	.00085	.2008	1 181.9	4.979	4.873	24.263	30
31	284.852	.00351	.00070	.2007	1 419.3	4.982	4.891	24.368	31
32	341.822	.00293	.00059	.2006	1 704.1	4.985	4.906	24.459	32
33	410.186	.00244	.00049	.2005	2 045.9	4.988	4.919	24.537	33
34	492.224	.00203	.00041	.2004	2 456.1	4.990	4.931	24.604	34
35	590.668	.00169	.00034	.2003	2 948.3	4.992	4.941	24.661	35
40	1 469.8	.00068	.00014	.2001	7 343.9	4.997	4.973	24.847	40
45	3 657.3	.00027	.00005	.2001	18 281.3	4.999	4.988	24.932	45
50	9 100.4	.00011	.00002	.2000	45 497.2	4.999	4.995	24.970	50
55	22 644.8	.00004	.00001	.2000	113 219.0	5.000	4.998	24.987	55
60	56 347.5	.00002		.2000	281 732.6	5.000	4.999	24.994	60

25%

Compound Interest Factors

25%

	Single Payment		Uniform Payment Series				Arithmetic Gradient		
	Compound Amount Factor	Present Worth Factor	Sinking Fund Factor	Capital Recovery Factor	Compound Amount Factor	Present Worth Factor	Gradient Uniform Series	Gradient Present Worth	
	Find F Given P	Find P Given F	Find A Given F	Find A Given P	Find F Given A	Find P Given A	Find A Given G	Find P Given G	
n	F/P	P/F	A/F	A/P	F/A	P/A	A/G	P/G	n
1	1.250	.8000	1.0000	1.2500	1.000	0.800	0	0	1
2	1.563	.6400	.4444	.6944	2.250	1.440	0.444	0.640	2
3	1.953	.5120	.2623	.5123	3.813	1.952	0.852	1.664	3
4	2.441	.4096	.1734	.4234	5.766	2.362	1.225	2.893	4
5	3.052	.3277	.1218	.3718	8.207	2.689	1.563	4.204	5
6	3.815	.2621	.0888	.3388	11.259	2.951	1.868	5.514	6
7	4.768	.2097	.0663	.3163	15.073	3.161	2.142	6.773	7
8	5.960	.1678	.0504	.3004	19.842	3.329	2.387	7.947	8
9	7.451	.1342	.0388	.2888	25.802	3.463	2.605	9.021	9
10	9.313	.1074	.0301	.2801	33.253	3.571	2.797	9.987	10
11	11.642	.0859	.0235	.2735	42.566	3.656	2.966	10.846	11
12	14.552	.0687	.0184	.2684	54.208	3.725	3.115	11.602	12
13	18.190	.0550	.0145	.2645	68.760	3.780	3.244	12.262	13
14	22.737	.0440	.0115	.2615	86.949	3.824	3.356	12.833	14
15	28.422	.0352	.00912	.2591	109.687	3.859	3.453	13.326	15
16	35.527	.0281	.00724	.2572	138.109	3.887	3.537	13.748	16
17	44.409	.0225	.00576	.2558	173.636	3.910	3.608	14.108	17
18	55.511	.0180	.00459	.2546	218.045	3.928	3.670	14.415	18
19	69.389	.0144	.00366	.2537	273.556	3.942	3.722	14.674	19
20	86.736	.0115	.00292	.2529	342.945	3.954	3.767	14.893	20
21	108.420	.00922	.00233	.2523	429.681	3.963	3.805	15.078	21
22	135.525	.00738	.00186	.2519	538.101	3.970	3.836	15.233	22
23	169.407	.00590	.00148	.2515	673.626	3.976	3.863	15.362	23
24	211.758	.00472	.00119	.2512	843.033	3.981	3.886	15.471	24
25	264.698	.00378	.00095	.2509	1 054.8	3.985	3.905	15.562	25
26	330.872	.00302	.00076	.2508	1 319.5	3.988	3.921	15.637	26
27	413.590	.00242	.00061	.2506	1 650.4	3.990	3.935	15.700	27
28	516.988	.00193	.00048	.2505	2 064.0	3.992	3.946	15.752	28
29	646.235	.00155	.00039	.2504	2 580.9	3.994	3.955	15.796	29
30	807.794	.00124	.00031	.2503	3 227.2	3.995	3.963	15.832	30
31	1 009.7	.00099	.00025	.2502	4 035.0	3.996	3.969	15.861	31
32	1 262.2	.00079	.00020	.2502	5 044.7	3.997	3.975	15.886	32
33	1 577.7	.00063	.00016	.2502	6 306.9	3.997	3.979	15.906	33
34	1 972.2	.00051	.00013	.2501	7 884.6	3.998	3.983	15.923	34
35	2 465.2	.00041	.00010	.2501	9 856.8	3.998	3.986	15.937	35
40	7 523.2	.00013	.00003	.2500	30 088.7	3.999	3.995	15.977	40
45	22 958.9	.00004	.00001	.2500	91 831.5	4.000	3.998	15.991	45
50	70 064.9	.00001		.2500	280 255.7	4.000	3.999	15.997	50
55	213 821.2			.2500	855 280.7	4.000	4.000	15.999	55

30%

Compound Interest Factors

30%

	Single Payment		Uniform Payment Series				Arithmetic Gradient		
	Compound Amount Factor	Present Worth Factor	Sinking Fund Factor	Capital Recovery Factor	Compound Amount Factor	Present Worth Factor	Gradient Uniform Series	Gradient Present Worth	
	Find F Given P	Find P Given F	Find A Given F	Find A Given P	Find F Given A	Find P Given A	Find A Given G	Find P Given G	
n	F/P	P/F	A/F	A/P	F/A	P/A	A/G	P/G	n
1	1.300	.7692	1.0000	1.3000	1.000	0.769	0	0	1
2	1.690	.5917	.4348	.7348	2.300	1.361	0.435	0.592	2
3	2.197	.4552	.2506	.5506	3.990	1.816	0.827	1.502	3
4	2.856	.3501	.1616	.4616	6.187	2.166	1.178	2.552	4
5	3.713	.2693	.1106	.4106	9.043	2.436	1.490	3.630	5
6	4.827	.2072	.0784	.3784	12.756	2.643	1.765	4.666	6
7	6.275	.1594	.0569	.3569	17.583	2.802	2.006	5.622	7
8	8.157	.1226	.0419	.3419	23.858	2.925	2.216	6.480	8
9	10.604	.0943	.0312	.3312	32.015	3.019	2.396	7.234	9
10	13.786	.0725	.0235	.3235	42.619	3.092	2.551	7.887	10
11	17.922	.0558	.0177	.3177	56.405	3.147	2.683	8.445	11
12	23.298	.0429	.0135	.3135	74.327	3.190	2.795	8.917	12
13	30.287	.0330	.0102	.3102	97.625	3.223	2.889	9.314	13
14	39.374	.0254	.00782	.3078	127.912	3.249	2.969	9.644	14
15	51.186	.0195	.00598	.3060	167.286	3.268	3.034	9.917	15
16	66.542	.0150	.00458	.3046	218.472	3.283	3.089	10.143	16
17	86.504	.0116	.00351	.3035	285.014	3.295	3.135	10.328	17
18	112.455	.00889	.00269	.3027	371.518	3.304	3.172	10.479	18
19	146.192	.00684	.00207	.3021	483.973	3.311	3.202	10.602	19
20	190.049	.00526	.00159	.3016	630.165	3.316	3.228	10.702	20
21	247.064	.00405	.00122	.3012	820.214	3.320	3.248	10.783	21
22	321.184	.00311	.00094	.3009	1 067.3	3.323	3.265	10.848	22
23	417.539	.00239	.00072	.3007	1 388.5	3.325	3.278	10.901	23
24	542.800	.00184	.00055	.3006	1 806.0	3.327	3.289	10.943	24
25	705.640	.00142	.00043	.3004	2 348.8	3.329	3.298	10.977	25
26	917.332	.00109	.00033	.3003	3 054.4	3.330	3.305	11.005	26
27	1 192.5	.00084	.00025	.3003	3 971.8	3.331	3.311	11.026	27
28	1 550.3	.00065	.00019	.3002	5 164.3	3.331	3.315	11.044	28
29	2 015.4	.00050	.00015	.3001	6 714.6	3.332	3.319	11.058	29
30	2 620.0	.00038	.00011	.3001	8 730.0	3.332	3.322	11.069	30
31	3 406.0	.00029	.00009	.3001	11 350.0	3.332	3.324	11.078	31
32	4 427.8	.00023	.00007	.3001	14 756.0	3.333	3.326	11.085	32
33	5 756.1	.00017	.00005	.3001	19 183.7	3.333	3.328	11.090	33
34	7 483.0	.00013	.00004	.3000	24 939.9	3.333	3.329	11.094	34
35	9 727.8	.00010	.00003	.3000	32 422.8	3.333	3.330	11.098	35
40	36 118.8	.00003	.00001	.3000	120 392.6	3.333	3.332	11.107	40
45	134 106.5	.00001		.3000	447 018.3	3.333	3.333	11.110	45

35%

Compound Interest Factors

35%

	Single Payment		Uniform Payment Series				Arithmetic Gradient		
	Compound Amount Factor	Present Worth Factor	Sinking Fund Factor	Capital Recovery Factor	Compound Amount Factor	Present Worth Factor	Gradient Uniform Series	Gradient Present Worth	
	Find F Given P	Find P Given F	Find A Given F	Find A Given P	Find F Given A	Find P Given A	Find A Given G	Find P Given G	
n	F/P	P/F	A/F	A/P	F/A	P/A	A/G	P/G	n
1	1.350	.7407	1.0000	1.3500	1.000	0.741	0	0	1
2	1.822	.5487	.4255	.7755	2.350	1.289	0.426	0.549	2
3	2.460	.4064	.2397	.5897	4.173	1.696	0.803	1.362	3
4	3.322	.3011	.1508	.5008	6.633	1.997	1.134	2.265	4
5	4.484	.2230	.1005	.4505	9.954	2.220	1.422	3.157	5
6	6.053	.1652	.0693	.4193	14.438	2.385	1.670	3.983	6
7	8.172	.1224	.0488	.3988	20.492	2.508	1.881	4.717	7
8	11.032	.0906	.0349	.3849	28.664	2.598	2.060	5.352	8
9	14.894	.0671	.0252	.3752	39.696	2.665	2.209	5.889	9
10	20.107	.0497	.0183	.3683	54.590	2.715	2.334	6.336	10
11	27.144	.0368	.0134	.3634	74.697	2.752	2.436	6.705	11
12	36.644	.0273	.00982	.3598	101.841	2.779	2.520	7.005	12
13	49.470	.0202	.00722	.3572	138.485	2.799	2.589	7.247	13
14	66.784	.0150	.00532	.3553	187.954	2.814	2.644	7.442	14
15	90.158	.0111	.00393	.3539	254.739	2.825	2.689	7.597	15
16	121.714	.00822	.00290	.3529	344.897	2.834	2.725	7.721	16
17	164.314	.00609	.00214	.3521	466.611	2.840	2.753	7.818	17
18	221.824	.00451	.00158	.3516	630.925	2.844	2.776	7.895	18
19	299.462	.00334	.00117	.3512	852.748	2.848	2.793	7.955	19
20	404.274	.00247	.00087	.3509	1 152.2	2.850	2.808	8.002	20
21	545.769	.00183	.00064	.3506	1 556.5	2.852	2.819	8.038	21
22	736.789	.00136	.00048	.3505	2 102.3	2.853	2.827	8.067	22
23	994.665	.00101	.00035	.3504	2 839.0	2.854	2.834	8.089	23
24	1 342.8	.00074	.00026	.3503	3 833.7	2.855	2.839	8.106	24
25	1 812.8	.00055	.00019	.3502	5 176.5	2.856	2.843	8.119	25
26	2 447.2	.00041	.00014	.3501	6 989.3	2.856	2.847	8.130	26
27	3 303.8	.00030	.00011	.3501	9 436.5	2.856	2.849	8.137	27
28	4 460.1	.00022	.00008	.3501	12 740.3	2.857	2.851	8.143	28
29	6 021.1	.00017	.00006	.3501	17 200.4	2.857	2.852	8.148	29
30	8 128.5	.00012	.00004	.3500	23 221.6	2.857	2.853	8.152	30
31	10 973.5	.00009	.00003	.3500	31 350.1	2.857	2.854	8.154	31
32	14 814.3	.00007	.00002	.3500	42 323.7	2.857	2.855	8.157	32
33	19 999.3	.00005	.00002	.3500	57 137.9	2.857	2.855	8.158	33
34	26 999.0	.00004	.00001	.3500	77 137.2	2.857	2.856	8.159	34
35	36 448.7	.00003	.00001	.3500	104 136.3	2.857	2.856	8.160	35

40%

Compound Interest Factors

40%

	Single Payment		Uniform Payment Series				Arithmetic Gradient		
	Compound Amount Factor	Present Worth Factor	Sinking Fund Factor	Capital Recovery Factor	Compound Amount Factor	Present Worth Factor	Gradient Uniform Series	Gradient Present Worth	
	Find F Given P	Find P Given F	Find A Given F	Find A Given P	Find F Given A	Find P Given A	Find A Given G	Find P Given G	
n	F/P	P/F	A/F	A/P	F/A	P/A	A/G	P/G	n
1	1.400	.7143	1.0000	1.4000	1.000	0.714	0	0	1
2	1.960	.5102	.4167	.8167	2.400	1.224	0.417	0.510	2
3	2.744	.3644	.2294	.6294	4.360	1.589	0.780	1.239	3
4	3.842	.2603	.1408	.5408	7.104	1.849	1.092	2.020	4
5	5.378	.1859	.0914	.4914	10.946	2.035	1.358	2.764	5
6	7.530	.1328	.0613	.4613	16.324	2.168	1.581	3.428	6
7	10.541	.0949	.0419	.4419	23.853	2.263	1.766	3.997	7
8	14.758	.0678	.0291	.4291	34.395	2.331	1.919	4.471	8
9	20.661	.0484	.0203	.4203	49.153	2.379	2.042	4.858	9
10	28.925	.0346	.0143	.4143	69.814	2.414	2.142	5.170	10
11	40.496	.0247	.0101	.4101	98.739	2.438	2.221	5.417	11
12	56.694	.0176	.00718	.4072	139.235	2.456	2.285	5.611	12
13	79.371	.0126	.00510	.4051	195.929	2.469	2.334	5.762	13
14	111.120	.00900	.00363	.4036	275.300	2.478	2.373	5.879	14
15	155.568	.00643	.00259	.4026	386.420	2.484	2.403	5.969	15
16	217.795	.00459	.00185	.4018	541.988	2.489	2.426	6.038	16
17	304.913	.00328	.00132	.4013	759.783	2.492	2.444	6.090	17
18	426.879	.00234	.00094	.4009	1 064.7	2.494	2.458	6.130	18
19	597.630	.00167	.00067	.4007	1 419.6	2.496	2.468	6.160	19
20	836.682	.00120	.00048	.4005	2 089.2	2.497	2.476	6.183	20
21	1 171.4	.00085	.00034	.4003	2 925.9	2.498	2.482	6.200	21
22	1 639.9	.00061	.00024	.4002	4 097.2	2.498	2.487	6.213	22
23	2 295.9	.00044	.00017	.4002	5 737.1	2.499	2.490	6.222	23
24	3 214.2	.00031	.00012	.4001	8 033.0	2.499	2.493	6.229	24
25	4 499.9	.00022	.00009	.4001	11 247.2	2.499	2.494	6.235	25
26	6 299.8	.00016	.00006	.4001	15 747.1	2.500	2.496	6.239	26
27	8 819.8	.00011	.00005	.4000	22 046.9	2.500	2.497	6.242	27
28	12 347.7	.00008	.00003	.4000	30 866.7	2.500	2.498	6.244	28
29	17 286.7	.00006	.00002	.4000	43 214.3	2.500	2.498	6.245	29
30	24 201.4	.00004	.00002	.4000	60 501.0	2.500	2.499	6.247	30
31	33 882.0	.00003	.00001	.4000	84 702.5	2.500	2.499	6.248	31
32	47 434.8	.00002	.00001	.4000	118 584.4	2.500	2.499	6.248	32
33	66 408.7	.00002	.00001	.4000	166 019.2	2.500	2.500	6.249	33
34	92 972.1	.00001		.4000	232 427.9	2.500	2.500	6.249	34
35	130 161.0	.00001		.4000	325 400.0	2.500	2.500	6.249	35

45%

Compound Interest Factors

45%

	Single Payment		Uniform Payment Series				Arithmetic Gradient		
	Compound Amount Factor	Present Worth Factor	Sinking Fund Factor	Capital Recovery Factor	Compound Amount Factor	Present Worth Factor	Gradient Uniform Series	Gradient Present Worth	
	Find F Given P	Find P Given F	Find A Given F	Find A Given P	Find F Given A	Find P Given A	Find A Given G	Find P Given G	
n	F/P	P/F	A/F	A/P	F/A	P/A	A/G	P/G	n
1	1.450	.6897	1.0000	1.4500	1.000	0.690	0	0	1
2	2.103	.4756	.4082	.8582	2.450	1.165	0.408	0.476	2
3	3.049	.3280	.2197	.6697	4.553	1.493	0.758	1.132	3
4	4.421	.2262	.1316	.5816	7.601	1.720	1.053	1.810	4
5	6.410	.1560	.0832	.5332	12.022	1.876	1.298	2.434	5
6	9.294	.1076	.0543	.5043	18.431	1.983	1.499	2.972	6
7	13.476	.0742	.0361	.4861	27.725	2.057	1.661	3.418	7
8	19.541	.0512	.0243	.4743	41.202	2.109	1.791	3.776	8
9	28.334	.0353	.0165	.4665	60.743	2.144	1.893	4.058	9
10	41.085	.0243	.0112	.4612	89.077	2.168	1.973	4.277	10
11	59.573	.0168	.00768	.4577	130.162	2.185	2.034	4.445	11
12	86.381	.0116	.00527	.4553	189.735	2.196	2.082	4.572	12
13	125.252	.00798	.00362	.4536	276.115	2.204	2.118	4.668	13
14	181.615	.00551	.00249	.4525	401.367	2.210	2.145	4.740	14
15	263.342	.00380	.00172	.4517	582.982	2.214	2.165	4.793	15
16	381.846	.00262	.00118	.4512	846.325	2.216	2.180	4.832	16
17	553.677	.00181	.00081	.4508	1 228.2	2.218	2.191	4.861	17
18	802.831	.00125	.00056	.4506	1 781.8	2.219	2.200	4.882	18
19	1 164.1	.00086	.00039	.4504	2 584.7	2.220	2.206	4.898	19
20	1 688.0	.00059	.00027	.4503	3 748.8	2.221	2.210	4.909	20
21	2 447.5	.00041	.00018	.4502	5 436.7	2.221	2.214	4.917	21
22	3 548.9	.00028	.00013	.4501	7 884.3	2.222	2.216	4.923	22
23	5 145.9	.00019	.00009	.4501	11 433.2	2.222	2.218	4.927	23
24	7 461.6	.00013	.00006	.4501	16 579.1	2.222	2.219	4.930	24
25	10 819.3	.00009	.00004	.4500	24 040.7	2.222	2.220	4.933	25
26	15 688.0	.00006	.00003	.4500	34 860.1	2.222	2.221	4.934	26
27	22 747.7	.00004	.00002	.4500	50 548.1	2.222	2.221	4.935	27
28	32 984.1	.00003	.00001	.4500	73 295.8	2.222	2.221	4.936	28
29	47 826.9	.00002	.00001	.4500	106 279.9	2.222	2.222	4.937	29
30	69 349.1	.00001	.00001	.4500	154 106.8	2.222	2.222	4.937	30
31	100 556.1	.00001		.4500	223 455.9	2.222	2.222	4.938	31
32	145 806.4	.00001		.4500	324 012.0	2.222	2.222	4.938	32
33	211 419.3			.4500	469 818.5	2.222	2.222	4.938	33
34	306 558.0			.4500	681 237.8	2.222	2.222	4.938	34
35	444 509.2			.4500	987 795.9	2.222	2.222	4.938	35

50%

Compound Interest Factors

50%

	Single Payment		Uniform Payment Series				Arithmetic Gradient		
	Compound Amount Factor	Present Worth Factor	Sinking Fund Factor	Capital Recovery Factor	Compound Amount Factor	Present Worth Factor	Gradient Uniform Series	Gradient Present Worth	
	Find F Given P	Find P Given F	Find A Given F	Find A Given P	Find F Given A	Find P Given A	Find A Given G	Find P Given G	
n	F/P	P/F	A/F	A/P	F/A	P/A	A/G	P/G	n
1	1.500	.6667	1.0000	1.5000	1.000	0.667	0	0	1
2	2.250	.4444	.4000	.9000	2.500	1.111	0.400	0.444	2
3	3.375	.2963	.2105	.7105	4.750	1.407	0.737	1.037	3
4	5.063	.1975	.1231	.6231	8.125	1.605	1.015	1.630	4
5	7.594	.1317	.0758	.5758	13.188	1.737	1.242	2.156	5
6	11.391	.0878	.0481	.5481	20.781	1.824	1.423	2.595	6
7	17.086	.0585	.0311	.5311	32.172	1.883	1.565	2.947	7
8	25.629	.0390	.0203	.5203	49.258	1.922	1.675	3.220	8
9	38.443	.0260	.0134	.5134	74.887	1.948	1.760	3.428	9
10	57.665	.0173	.00882	.5088	113.330	1.965	1.824	3.584	10
11	86.498	.0116	.00585	.5058	170.995	1.977	1.871	3.699	11
12	129.746	.00771	.00388	.5039	257.493	1.985	1.907	3.784	12
13	194.620	.00514	.00258	.5026	387.239	1.990	1.933	3.846	13
14	291.929	.00343	.00172	.5017	581.859	1.993	1.952	3.890	14
15	437.894	.00228	.00114	.5011	873.788	1.995	1.966	3.922	15
16	656.814	.00152	.00076	.5008	1 311.7	1.997	1.976	3.945	16
17	985.261	.00101	.00051	.5005	1 968.5	1.998	1.983	3.961	17
18	1 477.9	.00068	.00034	.5003	2 953.8	1.999	1.988	3.973	18
19	2 216.8	.00045	.00023	.5002	4 431.7	1.999	1.991	3.981	19
20	3 325.3	.00030	.00015	.5002	6 648.5	1.999	1.994	3.987	20
21	4 987.9	.00020	.00010	.5001	9 973.8	2.000	1.996	3.991	21
22	7 481.8	.00013	.00007	.5001	14 961.7	2.000	1.997	3.994	22
23	11 222.7	.00009	.00004	.5000	22 443.5	2.000	1.998	3.996	23
24	16 834.1	.00006	.00003	.5000	33 666.2	2.000	1.999	3.997	24
25	25 251.2	.00004	.00002	.5000	50 500.3	2.000	1.999	3.998	25
26	37 876.8	.00003	.00001	.5000	75 751.5	2.000	1.999	3.999	26
27	56 815.1	.00002	.00001	.5000	113 628.3	2.000	2.000	3.999	27
28	85 222.7	.00001	.00001	.5000	170 443.4	2.000	2.000	3.999	28
29	127 834.0	.00001		.5000	255 666.1	2.000	2.000	4.000	29
30	191 751.1	.00001		.5000	383 500.1	2.000	2.000	4.000	30
31	287 626.6			.5000	575 251.2	2.000	2.000	4.000	31
32	431 439.9			.5000	862 877.8	2.000	2.000	4.000	32

60% Compound Interest Factors 60%

	Single Payment		Uniform Payment Series				Arithmetic Gradient		
	Compound Amount Factor	Present Worth Factor	Sinking Fund Factor	Capital Recovery Factor	Compound Amount Factor	Present Worth Factor	Gradient Uniform Series	Gradient Present Worth	
n	Find F Given P F/P	Find P Given F P/F	Find A Given F A/F	Find A Given P A/P	Find F Given A F/A	Find P Given A P/A	Find A Given G A/G	Find P Given G P/G	n
1	1.600	.6250	1.0000	1.6000	1.000	0.625	0	0	1
2	2.560	.3906	.3846	.9846	2.600	1.016	0.385	0.391	2
3	4.096	.2441	.1938	.7938	5.160	1.260	0.698	0.879	3
4	6.554	.1526	.1080	.7080	9.256	1.412	0.946	1.337	4
5	10.486	.0954	.0633	.6633	15.810	1.508	1.140	1.718	5
6	16.777	.0596	.0380	.6380	26.295	1.567	1.286	2.016	6
7	26.844	.0373	.0232	.6232	43.073	1.605	1.396	2.240	7
8	42.950	.0233	.0143	.6143	69.916	1.628	1.476	2.403	8
9	68.719	.0146	.00886	.6089	112.866	1.642	1.534	2.519	9
10	109.951	.00909	.00551	.6055	181.585	1.652	1.575	2.601	10
11	175.922	.00568	.00343	.6034	291.536	1.657	1.604	2.658	11
12	281.475	.00355	.00214	.6021	467.458	1.661	1.624	2.697	12
13	450.360	.00222	.00134	.6013	748.933	1.663	1.638	2.724	13
14	720.576	.00139	.00083	.6008	1 199.3	1.664	1.647	2.742	14
15	1 152.9	.00087	.00052	.6005	1 919.9	1.665	1.654	2.754	15
16	1 844.7	.00054	.00033	.6003	3 072.8	1.666	1.658	2.762	16
17	2 951.5	.00034	.00020	.6002	4 917.5	1.666	1.661	2.767	17
18	4 722.4	.00021	.00013	.6001	7 868.9	1.666	1.663	2.771	18
19	7 555.8	.00013	.00008	.6011	12 591.3	1.666	1.664	2.773	19
20	12 089.3	.00008	.00005	.6000	20 147.1	1.667	1.665	2.775	20
21	19 342.8	.00005	.00003	.6000	32 236.3	1.667	1.666	2.776	21
22	30 948.5	.00003	.00002	.6000	51 579.2	1.667	1.666	2.777	22
23	49 517.6	.00002	.00001	.6000	82 527.6	1.667	1.666	2.777	23
24	79 228.1	.00001	.00001	.6000	132 045.2	1.667	1.666	2.777	24
25	126 765.0	.00001		.6000	211 273.4	1.667	1.666	2.777	25
26	202 824.0			.6000	338 038.4	1.667	1.667	2.778	26
27	324 518.4			.6000	540 862.4	1.667	1.667	2.778	27
28	519 229.5			.6000	865 380.9	1.667	1.667	2.778	28

FUNDAMENTALS OF ENGINEERING (FE) EXAM PRACTICE PROBLEMS

From the NCEES website (ncees.org)

The National Council of Examiners for Engineering and Surveying (NCEES) is a national nonprofit organization dedicated to advancing professional licensure for engineers and surveyors. It develops, administers, and scores the examinations used for engineering and surveying licensure in the United States.

The Fundamentals of Engineering (FE) exam is the first step toward professional licensure. Effective January 2014, the exam transitioned to a computer-based testing (CBT) platform, administered in testing centers during four time periods each year. For more details on the exam, study materials, and examination content areas, please reference the NCEES website.

The set of homework problems in this Appendix has been developed in the multiple-choice style of the FE exam. With the new exam format, *Engineering Economy* and *Engineering Ethics* topics are found in all FE exams, although the number of questions and extent of coverage vary from discipline to discipline. Following are sections from the *NCEES Exam Specifications* for each discipline—highlighting the sections and number of questions on engineering economy and engineering ethics.

Chemical Engineering Exam
13. Process Design and Economics (8–12 questions)

A. Process flow diagrams and piping and instrumentation diagrams
B. Equipment selection (e.g., sizing and scale-up)
C. Cost estimation
D. Comparison of economic alternatives (e.g., net present value, discounted cash flow, rate of return, expected value and risk)
E. Process design and optimization (e.g., sustainability, efficiency, green engineering, inherently safer design, evaluation of specifications)

16. Ethics and Professional Practice (2–3 questions)

A. Codes of ethics (professional and technical societies)
B. Agreements and contracts
C. Ethical and legal considerations
D. Professional liability
E. Public protection issues (e.g., licensing boards)

Civil Engineering
4. Ethics and Professional Practice (4–6 questions)

A. Codes of ethics (professional and technical societies)
B. Professional liability
C. Licensure
D. Sustainability and sustainable design
E. Professional skills (e.g., public policy, management, and business)
F. Contracts and contract law

5. Engineering Economics (4–6 questions)

A. Discounted cash flow (e.g., equivalence, PW, equivalent annual worth, FW, rate of return)
B. Cost (e.g., incremental, average, sunk, estimating)
C. Analyses (e.g., breakeven, benefit-cost, life cycle)
D. Uncertainty (e.g., expected value and risk)

Electrical and Computer Engineering
3. Ethics and Professional Practice (3–5 questions)

A. Codes of ethics (professional and technical societies)
B. NCEES Model Law and Model Rules
C. Intellectual property (e.g., copyright, trade secrets, patents)

4. Engineering Economics (3–5 questions)

A. Time value of money (e.g., present value, future value, annuities)
B. Cost estimation
C. Risk identification
D. Analysis (e.g., cost-benefit, trade-off, breakeven)

Environmental Engineering
3. Ethics and Professional Practice (5–8 questions)

A. Codes of ethics (professional and technical societies)
B. Agreements and contracts
C. Ethical and legal considerations
D. Professional liability
E. Public protection issues (e.g., licensing boards)

F. Regulations (e.g., water, wastewater, air, solid/hazardous waste, groundwater/soils)

4. Engineering Economics (4–6 questions)

A. Discounted cash flow (e.g., life cycle, equivalence, PW, equivalent annual worth, FW, rate of return)

B. Cost (e.g., incremental, average, sunk, estimating)

C. Analyses (e.g., breakeven, benefit-cost)

D. Uncertainty (expected value and risk)

Industrial and Systems Engineering
3. Ethics and Professional Practice (5–8 questions)

A. Codes of ethics and licensure

B. Agreements and contracts

C. Professional, ethical, and legal responsibility

D. Public protection and regulatory issues

4. Engineering Economics (10–15 questions)

A. Discounted cash flows (PW, EAC, FW, IRR, amortization)

B. Types and breakdown of costs (e.g., fixed, variable, direct and indirect labor)

C. Cost analyses (e.g., benefit-cost, breakeven, minimum cost, overhead)

D. Accounting (financial statements and overhead cost allocation)

E. Cost estimation

F. Depreciation and taxes

G. Capital budgeting

Mechanical Engineering
4. Ethics and Professional Practice (3–5 questions)

A. Codes of ethics

B. Agreements and contracts

C. Ethical and legal considerations

D. Professional liability

E. Public health, safety, and welfare

5. Engineering Economics (3–5 questions)

A. Time value of money

B. Cost, including incremental, average, sunk, and estimating

C. Economic analyses

D. Depreciation

Other Disciplines
5. Ethics and Professional Practice (3–5 questions)

A. Codes of ethics

B. NCEES Model Law

C. Public protection issues (e.g., licensing boards)

7. Engineering Economics (7–11 questions)

A. Time value of money (e.g., present worth, annual worth, future worth, rate of return)

B. Cost (e.g., incremental, average, sunk, estimating)

C. Economic analyses (e.g., breakeven, benefit-cost, optimal economic life)

D. Uncertainty (e.g., expected value and risk)

E. Project selection (e.g., comparison of unequal life projects, lease/buy/make, depreciation, discounted cash flow)

PROBLEMS

Chapter 1

Decision Making

D-1 Engineering economic analysis provides useful input in all of the following situations except which one?

   (*a*) Determining how much we should pay for a machine that will provide a savings.

   (*b*) Determining the priority of investing our company's retained earnings.

   (*c*) Illustrating the economic advantages of one alternative over other feasible choices.

   (*d*) Convincing management that one person should be hired over another.

D-2 Engineering economic analysis can be described by the following statement:

   (*a*) Involves a systematic analysis of relevant costs and benefits.

   (*b*) Involves a comparison of competing alternatives.

   (*c*) Supports a rational economic decision-making objective.

   (*d*) All of the above.

D-3 To which of the following questions does an engineering economy analysis provide useful input?

   (*a*) Has the mechanical or electrical engineer chosen the most economical motor size given functional requirements?

   (*b*) Has the civil or mechanical engineer chosen the best thickness for insulating a building?

   (*c*) Has the biomedical engineer chosen the best materials for the company's artificial knee product?

   (*d*) All of the above

D-4 Which of the following job functions potentially conducts and utilizes engineering economic analysis in decision making?

   (*a*) Senior technical design engineer.

   (*b*) Midlevel manager of business and finance.

   (*c*) Senior management for new product development.

   (*d*) All of the above.

D-5 Engineering economic analysis can be described by the following statement:

   (*a*) Involves a systematic analysis of relevant costs and benefits.

   (*b*) Involves a comparison of competing alternatives.

   (*c*) Supports a rational economic decision-making objective.

   (*d*) All of the above.

Engineering Ethics

D-6 Engineers are acting in the most ethical way in which of these situations?

   (*a*) They are making sure the company's interests are protected.

   (*b*) They feel good about the decisions that they've made.

   (*c*) They act to protect the interests of society in general.

   (*d*) They ensure that their own best interest is protected.

D-7 To act as an ethical engineer, you should accept fees for engineering work in which situation?

   (*a*) If you need the money to keep your business open and thriving.

(b) If you are competent to complete all aspects of the job.

(c) If the contract is a cost plus contract.

(d) If there were no other engineers who bid on the job.

D-8 A registered professional engineer (PE) has as a primary obligation to protect which of the following entities?

(a) The government

(b) The PE's company

(c) The PE's country

(d) The general public

D-9 Engineers should act in ethical ways for what reason?

(a) It creates a feel-good situation for them.

(b) The engineering code of conduct requires it.

(c) They may be considered for a raise because of it.

(d) They may be violating the law if they don't.

D-10 A design engineer is responsible for an important subelement of a large project at a firm. The project has fallen behind schedule, and the important client is very angry and threatening to sue. The boss is expecting the engineer's design review to go well so that the project can be shipped by the end of the week. The engineer notices during final design review that an element of the design is wrong and will create a major safety issue for the entire system. To rework that portion of the design will take several months. The engineer should do which of the following?

(a) Sign off on the drawings because of the threatened lawsuit and because the project is so far behind.

(b) Do not sign off on the drawings, and let the boss know what is found at the final design review.

(c) Tell the boss that to sign off on the design now, the engineer must have an immediate raise.

(d) Sign off and keep an eye on how construction goes. Maybe the engineer can correct the safety issues before the project is fully operational.

D-11 As team leader for your unit, you function as both engineer and manager. One of your roles is to approve major purchases, and you have been contacted by a new supplier in your area. The new company has invited you to expensive dinners, has offered trips to vacation spots to attend "product shows," and has recently been sending to your private address personal items such as collectible art, coins, sports tickets, and golf club memberships. You are unsure how to handle this situation. You should do which of the following?

(a) Accept all of the gifts because you know that everyone else is doing it and that this is your chance to get a share of the action.

(b) Decline the gifts and other offers that would be considered outside the scope of ordinary business or professional contact.

(c) Knowing that the gifts will not influence your purchasing decisions, accept the items with no guilt.

(d) Accept the gifts but make sure that your boss and others on your team share in the bounty.

Cost Problems

D-12 A company produces several product lines. One of those lines generates the following annual cost and production data:

Manufacturing/Materials costs	$200,000
General/Administrative costs	50,000
Direct-labor costs	170,000
Other overhead costs	60,000
Annual production demand	10,000 units

The company adds 40% to its production cost in selling to the retailer. The retailer in turn adds a 50% profit margin when selling to its customers. How much would it cost a retailer to buy 100 units of the product?

(a) $4800

(b) $6720

(c) $7200

(d) $1008

D-13 An agribusiness is deciding what crop to plant in this area for the next growing season. The local agricultural extension office has provided the following data (in $100):

Crop	Cost per acre	Income/acre at 100% Yield	Estimated Yield(%)
A	$30	$45	80
B	45	75	65
C	15	25	90

Using a 200-acre plot as an example (subtract 10% for unusable areas of the field), which crop should be planted this year, and what is the total profit?

(a) A; $108,000

(b) C; $135,000

(c) C; $150,000

(d) B; $540,000

Use the data below for Problems D-14 to D-17.

A textbook publisher produces a textbook for $25 per book and sells a lot of 160 to the Campus Bookstore for $50 per unit. The bookstore sells the textbook new for $75 and used for $60. This edition of the book is used for 2 years (4 semesters). The bookstore sells all textbooks that it has at the beginning each semester, and it repurchases 50% of those at the end of each semester for $30.

D-14 What is the total cost to the textbook publisher?

(a) $25

(b) $50

(c) $4000

(d) $8000

D-15 What is the total cost to all parties (publisher, bookstore, students) over the life of this 160-unit lot of textbooks?

(a) $12,000

(b) $20,400

(c) $32,400

(d) $36,900

D-16 What is the net profit for the bookstore (sales − costs) over the life of this 160-unit lot of textbooks?

(a) $7900

(b) $11,600

(c) $11,900

(d) $14,000

D-17 What is the net profit for the publisher (sales − costs) over the life of this 160-unit lot of textbooks?

(a) $4000

(b) $8000

(c) $11,900

(d) $12,000

Chapter 2

Cost Concepts

D-18 A firm bought a used machine 2 years ago for $1500. When new, the machine cost $8000. Today it could be sold for $500. Which of the following statements is true?

(a) The fixed cost for operating the machine can be ignored in any analysis.

(b) The $8000 purchase price is not included in the analysis.

(c) The $1500 paid 2 years ago is included in the analysis.

(d) The variable cost of ownership is the difference between what was paid and what the machine is now worth ($1500 − $500 = $1000).

D-19 When considering two alternatives that are described only in terms of the cost of ownership, the breakeven point cannot be described by which of the following statements?

(a) The difference in initial cost between the two alternatives.

(b) The level of production (or activity) of each alternative under consideration is equivalent.

(c) Fixed plus variable costs of each alternative are equivalent.

(d) A rational decision maker should be indifferent between the two alternatives.

D-20 If JMJ Industries realizes a profit of $4.00 per unit sold, what is the fixed-cost portion of their production costs? Their variable costs are $1.50 per unit, and they sell 1000 units per year at a price of $6.00 per unit.

(a) There are no fixed costs in this type of problem.

(b) $250

(c) $500

(d) $2000

D-21 Consider the following production data for Alternatives *A* and *B* in a firm that uses a 10% interest rate.

	Alt. A	Alt. B
Annual fixed cost per unit	$2 million	$3.5 million
Annual variable cost per unit	850	250

If the company is going to produce 4000 units annually, which alternative should be chosen?

(a) Neither alternative should be chosen because the negative cash flows are greater than the positive cash flows for both alternatives.

(b) This problem cannot be solved because there is not enough data given.

(c) Alt. *A*

(d) Alt. *B*

D-22 A company has annual fixed costs of $2,500,000 and variable costs of 0.15¢ per unit produced. For the firm to break even if they charge $1.85 for their product, the level of annual production is nearest to what value?

(a) 375,000 units

(b) 1,315,789 units

(c) 1,351,351 units

(d) 1,562,500 units

Estimating in Engineering Economy

D-23 Which statement is not true with respect to estimating the economic impacts of proposed engineering projects?

(a) Order-of-magnitude estimates are used for high-level planning.

(b) Order-of-magnitude estimates are the most accurate type at about –3 to 5%.

(c) Increasing the accuracy of estimates requires added time and resources.

(d) Estimators tend to underestimate the magnitude of costs and to overestimate benefits.

D-24 The Department of Transportation is accepting bids for materials to provide "signage and safety" for a new 25-mile section of a 4-lane highway. DOT estimates are:

Lane paint	$500 per mile per lane
Reflective lane markers	$6 per 25 feet
Mile markers	$18 per unit
Flexible roadside delineators	$32 per unit; spaced at 1000 feet
Emergency boxes	$500 per unit spaced at 2.5 miles
Signage (various messages)	$1000 per mile; based on historical costs

The bids that DOT receives should be in what range of costs?

(a) Less than $100,000

(b) Between $100,000 and $200,000

(c) Between $200,000 and $250,000

(d) Greater than $250,000

D-25 A 250-gallon reactor cost $780,000 when it was constructed 20 years ago. What would a 750-gallon model cost today if the power-sizing exponent is 0.56 and the construction cost index for such facilities has increased from 141 to 556 over the last 20 years? Choose the closest value.

(a) $0.37 million

(b) $1.66 million

(c) $1.44 million

(d) $5.69 million

D-26 A half-million-square-foot warehouse facility is being considered by your company for a location in Kansas City, KS. You have a bid for a similar type 25,000-ft^2 facility in Washington, DC, at $4,375,000. If the warehouse construction cost index value for KS is 0.75 and for DC it is 1.34, what range should the KS bid fall into if you assume that construction costs are linear across size?

(a) Less than $30 million

(b) Between $30 million and $60 million

(c) Between $60 million and $100 million

(d) Greater than $100 million

D-27 A company has major clients in all 50 states. Fifteen (15) of the states have 4 clients, 10 have 3 clients, 20 have 6 clients and 5 have 10. The total number of clients the firm has is closest to what number?

(a) 23

(b) 50

(c) 250

(d) 260

Chapter 3

Simple Interest

D-28 Which of the following statements is not true?

(a) Simple interest is to be used only in simple decision situations.

(b) Compounded interest involves computing interest on top of interest.

(c) Simple interest is rare in practical situations of borrowing and loaning.

(d) If the interest is not stated as being simple or compounded, we assume the latter.

D-29 If you borrow $1000 from the bank at 5% simple interest per month due back in 2 years, what is the size of your monthly payments?

(a) $25

(b) $50

(c) $500

(d) $1200

D-30 Your quarterly payments on a loan are $500 and the interest that you are paying is 1% per quarter simple interest. The size of the principal that you have borrowed is closest to which value?

(a) $5000

(b) $12,500

(c) $20,000

(d) $50,000

D-31 The principal that you borrowed for a recent purchase was $15,000. You will pay the purchase off through a simple interest loan at 8% per year due in 3 years. The amount that is due at the end of 3 years is closest to what value?

(a) $1200

(b) $3600

(c) $16,200

(d) $18,900

D-32 If $10,000 is borrowed today at 5% simple interest, how much is due at the end of 10 years?

(a) $5000

(b) $10,000

(c) $15,000

(d) $16,289

Compound Interest, Single Cash Flows

D-33 A savings account's value today is $150 and it earns interest at 1% per month. How much will be in the account one year from today? Which of the following is correct to solve for the unknown value?

(a) $P = 150(1 + 0.01)^{12}$

(b) $150 = F(F/P, 12\%, 1)$

(c) $F = 150(1.12)^{-12}$

(d) $F = 150(F/P, 1\%, 12)$

D-34 To calculate how many years (n) an investment (P) must be kept in an account that earns interest at $i\%$, in order to triple in amount, which of the following expressions should be used?

(a) $n = -P + F(P/F, i\%, n)$

(b) $n = [\log(F/P)]/[\log(1 + i\%)]$

(c) $n = [\ln(-P + F)]/[\ln i\%]$

(d) $n = -F(1 + i\%)^P$

D-35 If you invest $40,000 in a stock whose value grows at 2% per year, your investment is nearest what value after 5 years?

(a) $40,800

(b) $43,296

(c) $44,164

(d) $64,420

D-36 An account pays interest at 1.5% per month. If you deposit $5000 at the beginning of this year, how much could you withdraw at the end of next year?

(a) $5151

(b) $7148

(c) $49,249

(d) $265,545

D-37 A machine will need to be replaced 15 years from today for $10,000. How much must be deposited now into an account that earns 5% per year to cover the replacement cost?

(a) $1486

(b) $4810

(c) $6139

(d) $10,000

D-38 You deposit $5000 into an account that will grow to $14,930 in 6 years. Your rate of return on this investment is closest to what value?

(a) 18%

(b) 20%

(c) 22%

(d) 25%

D-39 An investment company owns land now worth $500,000 that is increasing in value each year. If the land value doubles in 7 years, what is the yearly rate of return nearest to?

(a) 0.0%

(b) 2.0%

(c) 7.0%

(d) 10.5%

D-40 Your friend withdrew $630,315 from an account into which she had invested $350,000. If the account paid interest at 4% per year, she kept her money in the account for how many years?

(a) 1.8 years

(b) 6.5 years

(c) 12.5 years

(d) 15 years

D-41 Annual revenues in our company are $1.5 million this year. If they are expected to grow at a compounded rate of 20% per year, what will they be 10 years from now?

(a) $3.89 million

(b) $9.29 million

(c) $10.9 million

(d) $57.51 million

D-42 A student inherits $50,000 and invests it in government bonds that will average 3% annual interest. What is the value of the investment after 50 years?

(*a*) $67,195

(*b*) $219,195

(*c*) $355,335

(*d*) $5,869,545

D-43 A zero-coupon bond has a face (par) value of $10,000. The bond is sold at a discount for $6500 and held for 3 years, at which time it is sold. If an annual rate of 10% is earned over that 3-year period, how much was the bond sold for?

(*a*) $8652

(*b*) $13,310

(*c*) $16,859

(*d*) $25,937

Nominal and Effective Interest Rates

D-44 A rate of 2% per quarter compounded quarterly is closest to what annual compounded interest rate?

(*a*) 2.00%

(*b*) 8.00%

(*c*) 8.24%

(*d*) 24.00%

D-45 A nominal interest rate of "12% per year compounded yearly" is closest to:

(*a*) 1% per month effective rate

(*b*) 3% per quarter effective rate

(*c*) 6% per 6 months effective rate

(*d*) 12% per year effective rate

D-46 An interest rate expressed as "1.5% per month" is exactly the same as:

(*a*) 4.5% per quarter effective interest

(*b*) 18% effective interest per year

(*c*) 18% per year compounded monthly

(*d*) None of the above

D-47 A deposit of $50,000 is made into an account that pays 10% compounded semiannually. How much would be in the account after 10 years?

(*a*) $81,445

(*b*) $129,685

(*c*) $132,665

(*d*) $336,375

D-48 A mining firm must deposit funds in a "reclamation" account each quarter. The account must have $25 million on deposit when a project reaches its horizon in 10 years. The account pays interest at a rate of 2% per quarter. How much is the quarterly deposit?

(*a*) $41,500

(*b*) $172,575

(*c*) $228,325

(*d*) $414,000

D-49 A deposit of $1000 compounds to $2500 in 5 years. The interest on this account compounds quarterly. What is the closest nominal annual rate of return?

(*a*) 2.50%

(*b*) 4.70%

(*c*) 18.75%

(*d*) 20.11%

D-50 If your local bank indicates that it pays interest on passbook savings accounts at a rate of 2.25%, the nominal and effective interest rates are nearest which the following?

(*a*) 2.25%, 2.25%

(*b*) 2.25%, 2.28%

(*c*) 2.25%, 27%

(*d*) 27%, 2.25%

Continuous Compounding Interest Rate

D-51 How long will it take money to triple in an account if it pays interest expressed as 8% nominal annual compounded continuously?

(*a*) 13.73 years

(*b*) 14.27 years

(*c*) Your money will never triple.

(*d*) None of the above

D-52 A deposit of $500 per year (beginning of year) is made for a period of 2 years in an account that earns 6% nominal interest compounded continuously. How much is in the account after 2 years?

(*a*) $917

(*b*) $1062

(*c*) $1092

(*d*) $1095

D-53 A firm offers to sell a zero-coupon bond (no semiannual payments) to you today. When it matures in 5 years you will receive the par value of $10,000. If the firm pays interest at 15% compounded continuously on the bond how much would you pay for it today?

(*a*) $4724

(*b*) $4972

(c) $10,000

(d) $21,170

D-54 If $1000 is invested annually at 6% continuous compounding for each of 10 years, how much is in the account after the last deposit?

(a) $1822

(b) $10,000

(c) $13,181

(d) $13,295

Chapter 4

Uniform Cash Flow Series

D-55 You place $100 per month into an account that earns 1% per month. Which of the following expressions can be used to calculate the account's value after 3 years?

(a) $P = 100(P/A, 1\%, 3)$

(b) $F = 100(P/A, 1\%, 36)(F/P, 1\%, 36)$

(c) $F = 100[(1 + 0.01)^n - 1]/0.01$

(d) $F = 100(F/A, 12.68\%, 3)$

D-56 A machine must be replaced in 7 years at a cost of $7500. How much must be deposited at the end of each year into an account that earns 5% in order to have accumulated enough to pay for the replacement?

(a) $471

(b) $596

(c) $791

(d) $921

D-57 Winners of the PowerState Lottery can take $30 million now or payments of $2.5 million per year for the next 15 years. These are equivalent at what annual interest rate? The answer is closest to what value?

(a) 1%

(b) 2%

(c) 3%

(d) 5%

D-58 You deposit $1000 in a retirement investment account today that earns 1% per month. In addition, you deposit $50 at the end of every month starting this month and continue to do so for 30 years. The amount that has accumulated in this account at the end of 30 years is nearest to

(a) $35,949

(b) $42,027

(c) $174,748

(d) $210,698

D-59 Suppose $10,000 is deposited into an account that earns 10% per year for 5 years. At that point in time, uniform end-of-year withdrawals are made such that the account is emptied after the 15th withdrawal. The size of these annual withdrawals is closest to what value?

(a) $2118

(b) $2621

(c) $3410

(d) $16,105

D-60 A manufacturer borrows $85,000 for machinery. The loan is for 10 years at 12% per year. What is the annual payment on the machinery?

(a) $4843

(b) $8500

(c) $13,834

(d) $15,045

D-61 How many years would you have to put $100 per year into an account that earns 15% annually to accumulate $6508?

(a) 17 years

(b) 21 years

(c) 30 years

(d) 65 years

D-62 A $10,000 face value municipal bond pays $1000 interest at the end of every year. If there are 12 more years of payments, at what price today would the bond yield 18% over the next 12 years?

(a) $1372

(b) $4793

(c) $6165

(d) $10,000

Gradient Cash Flows

D-63 Today $5000 is deposited in an account that earns 2.5% per quarter. Additional deposits are made at the end of every quarter for the next 20 years. The deposits start at $100 and increase by $50 each year thereafter. The amount that has accumulated in this account at the end of 20 years can be expressed as follows.

(a) $= 5000(P/F, 2.5\%, 20) + 100(F/A, 2.5\%, 20\%) + 50$
    $(P/G, 2.5\%, 20 - 1)$

(b) $= 5000(F/P, 10\%, 80) + 150(P/G, 10\%, 80)$
    $(F/P = 10\%, 80)$

$(c) = 5000(P/F, 10.38\%, 20) + 100(P/A, 10.38\%, 20)$

$\quad\quad + 50(P/G, 10.38\%, 20)(P/F,\ 10.38\%, 1)$

$(d) = 5000(F/P, 10.38\%, 80) + 100(F/A,\ 2.5\%, 80)$

$\quad\quad + 50(P/G, 2.5\%, 80)(F/P,\ 2.5\%, 80)$

D-64 An investment returns the following end-of-year cash flows: Year 1, $0; Year 2, $1500; Year 3, $3000; Year 4, $4500; and Year 5, $6000. Given a 10% interest rate, what is the present worth?

(a) $5970

(b) $6597

(c) $9357

(d) $10,293

D-65 A project's annual revenues will be $50,000 the first year and will decrease by $1500 per year over its 20-year life. If the firm's interest rate is 12%, what is the project's present worth?

(a) $305,998

(b) $373,450

(c) $384,654

(d) $440,902

D-66 A cash flow series is described by the following: $10,000 + $250(t)$, where t is the number of compounding periods. The present worth of this series at the end of five periods, where interest is 2% per t, is nearest what value?

(a) $11,250

(b) $50,620

(c) $56,432

(d) $60,620

D-67 Revenue from sales of a training video for the first year are estimated to be $350,000. In addition, revenue is expected to decrease by $25,000 per year over the life of the video (which is 10 years). If interest is 10%, the present worth of the revenue over the life of the video is nearest what value?

(a) $100,000

(b) $125,000

(c) $1,578,475

(d) $2,723,025

Chapter 5
Present Worth Analysis

D-68 A project is being considered that has a first cost of $12,500, creates $5000 in annual cost savings, requires $3000 in annual operating costs, and has a salvage value of $2000 after a project life of 3 years.

If interest is 10% per year, which formula calculates the project's present worth?

(a) PW $= 12,500(P/F, 10\%, 1) + (-5000 + 3000)$
$(P/A, 10\%, 3) - 2000(F/P, 10\%, 3)$

(b) PW $= -12,500 + (5000 - 3000)(P/A, 10\%, 3)$
$-2000(P/F, 10\%, 3)$

(c) PW $= 12,500(F/P, 10\%, 3) + (5000 - 3000)$
$(F/A, 10\%, 3) + 2000$

(d) PW $= -12,500 + 5000(P/A, 10\%, 3) - 3000$
$(P/A, 10\%, 3) + 2000(P/F, 10\%, 3)$

D-69 A new packing machine will cost $57,000. The existing machine can be sold for $5000 now and the new machine for $7500 after its 10-year useful life. If the new machine reduces annual expenses by $5000, what is the present worth at 25% of this investment?

(a) −$18,388

(b) −$33,340

(c) −$34,145

(d) −$38,340

D-70 A vendor is offering an extended repair contract on a machine. The firm's experience is that this will cover repair costs over the next 4 years of $200, $200, $400, and $500. At 6%, what is the extended repair contract worth now?

(a) $1040

(b) $1089

(c) $1099

(d) $1300

D-71 Annual disbursements for maintenance of critical heavily used equipment will be $25,000 for the next 10 years, and then $35,000 into infinity. What is the present worth of the maintenance cost cash flow stream if interest is 15%?

(a) $166,667

(b) $183,147

(c) $192,367

(d) $233,334

D-72 Manufacturing costs from a scrapped poor-quality product are $6000 per year. An investment in an employee training program can reduce this cost. Program A reduces the cost by 75% and requires an investment of $12,000. Program B reduces the cost by 95% and will cost $20,000. Based on low turnover at the plant, either program should be effective for the next 5 years. If interest is 20%, the present worth of the two programs is nearest what values? (Consider cost reduction a positive cash flow.)

(a) A: −$25,460; B: −$37,049

(b) A: $1460; B: −$2951

(c) A: $5060; B: $1609

(d) A: $13,460; B: $17,049

Chapter 6
Annual Worth Analysis

D-73 New product tracking equipment costs $120,000 and will have a $10,000 salvage value when disposed of in 10 years. Annual repair costs begin at $5000 in the fifth year and increase by $500 per year thereafter until disposed of. If interest is 10%, what is the closest equivalent annual cost of ownership?

(a) $21,505

(b) $21,766

(c) $21,844

(d) $23,109

D-74 Your company is considering two alternatives:

	Alt. I	Alt. II
First cost	$42,500	$70,000
Annual maintenance	6,000	4,000
Annual savings if implemented	18,500	20,000
Salvage value	12,000	25,000
Useful life of alternative	3 years	6 years

What is the annual dollar advantage of Alt. II over Alt. I at an interest rate of 15%?

(a) Alt. II has no annual advantage over Alt I.

(b) $3020

(c) $3500

(d) $7436

D-75 Specialized bits (costing $50,000) used in the mining industry have a useful life of 5000 hours of operation and can be traded in when a new bit is purchased for 10% of first cost. The drilling machine that uses the bit is used 1000 hours per year. What is the equivalent uniform annual cost of these bits at 2.5%?

(a) $8559

(b) $9510

(c) $9828

(d) $10,920

D-76 A new chemical remediation tank is needed. Current technology tanks, which cost $150,000, must be drained and treated every 2 years at a cost of $30,000; the tanks will last 10 years, and each will have a salvage value of 5% of first cost. A tank with new technology has just come on the market. There are no periodic maintenance costs, and a tank will last 20 years. If the new tanks cost $325,000, what minimum salvage value, as a percentage of first cost, would be required for this technology to be a better option? Use a 12% interest rate.

(a) 10%

(b) 36%

(c) 57%

(d) 72%

D-77 A beautiful bridge is being built over the river that runs through a major city in your state. The cost of the bridge is estimated at $600 million. Annual costs of the bridge will be $200,000, and the bridge is estimated to last a very long time. If accountants in city hall use 3% as the interest rate for analysis, what is the annualized cost of the bridge project?

(a) $18 million

(b) $18.2 million

(c) $20,000 million

(d) $219,500 million

Chapter 7
Rate of Return Analysis

D-78 Which of the following equations can be used to find the internal rate of return (i) for a project that has initial investment of P, net annual cash flows of A, and salvage value of S after n years?

(a) $0 = -P + A(P/A, i\%, n) + S(P/A, i\%, n)$

(b) $(P - A)(P/A, i\%, n) = S(P/F, i\%, n)$

(c) $-A = -P(A/P, i\%, n) - S(A/F, i\%, n - 1)$

(d) $0 = -P(F/P, i\%, n) + A(F/A, i\%, n) + S$

D-79 The rate of return on an investment of $1500 that doubles in value over a 4-year period, and produces a $300 annual cash flow, is nearest to which value?

(a) 15%

(b) 20%

(c) 25%

(d) 30%

D-80 The interest rate that makes Alternatives A and B equivalent is in what range?

Year	Alt. A	Alt. B
0	−$1000	−$3000
1	100	500
2	100	550
3	100	600
4	200	650
5	200	700

(a) Less than 2%

(b) Between 2 and 5%

(c) Greater than 5%

(d) There is no interest rate that equates these two cash flow series.

D-81 A corporate bond with a face (par) value of $10,000 will mature 7 years from today (it was issued 3 years ago). The bond just after the 6th interest payment is being sold for $6950. The bond's interest rate is 4% nominal annual, payable semiannually. The yield of the bond if held to maturity is in what range?

(a) Less than 4%

(b) Between 4 and 6%

(c) Between 6 and 10%

(d) Greater than 10%

D-82 A firm borrowed $50,000 from a mortgage bank. The terms of the loan specify quarterly payments for a 10-year period. If payments to the bank are $3750 per quarter, what effective annual interest rate is the firm paying?

(a) Less than 1%

(b) 7%

(c) 28%

(d) 31%

Chapter 8
Incremental Cash Flows and Analysis

D-83 You are given the cash flow series for two projects, Alt. A and Alt. B.

Year	0	1	2	3	4	5	6
Alt. A ($)	−I1	X	X	X	X	X	X+S1
Alt. B ($)	−I2	Y	Y	Y	Y	Y	Y+S2

Assume $I2 > I1$ and X, Y, S1, and S2 are positive; the incremental rate of return (i) on the additional investment in Alt. B can be calculated with the following expression.

(a) $0 = -I2 + Y(P/A, i\%, 6) + S2(P/F, i\%, 6)$

(b) $0 = -(I2 - I1) + (Y - X)(P/A, i\%, 5)$
$+ (S2 - S1)(P/F, i\%, 6)$

(c) $0 = -(I2 - I1)(F/P, i\%, 6)$
$+ (Y - X)(F/A, i\%, 6) + (S2 - S1)$

(d) $0 = -(I2 - I1) + (Y - X) + [(Y + S2) - (X + S1)]$

D-84 A firm is considering two mutually exclusive alternatives ($i = 8\%$):

	Alt. Alpha	Alt. Omega
First cost	$10,000	$30,000
Annual maintenance	2,800	2,000
Annual savings if implemented	5,000	6,500
Salvage value	2,000	5,000
Useful life of alternative	4 years	8 years

If Alt. Alpha will be replaced with a "like alternative" at the end of 4 years, what is the present worth of the incremental cash flows associated with going from an investment in Alpha to an investment in Omega?

(a) −$6201

(b) −$5942

(c) −$5028

(d) $852

D-85 Project 1 requires an initial investment on $50,000 and has an internal rate of return (IRR) of 18%. A mutually exclusive alternative, Project 2, requires an investment of $70,000 and has an IRR of 23%. Which of the following statements is true concerning the rate of return on the incremental $20,000 investment?

(a) It is less than 18%.

(b) It is between 18 and 23%.

(c) It is greater than 23%.

(d) It cannot be determined from the data given.

D-86 Alternative Uno has a first cost of $10,000 and annual expenses of $3000, whereas Alternative Dos has a first cost of $35,000, annual expenses of $2000, and a recurring cost of $5000 every 10 years. If both alternatives have an infinite life, which of the following equations can be used to solve for the rate of return on the incremental investment?

(a) $0 = -\$25{,}000 + \$1000/i - \$5000(A/F, i, 10)$

(b) $0 = -\$25{,}000 + \$1000/i + \$5000(A/P, i, 10)/i$

(c) $0 = -\$25{,}000 + \$1000/i - \$5000(A/F, i, 10)/i$

(d) $0 = +\$25{,}000 - \$1000/i + \$5000(A/F, i, 10)$

D-87 Compare two competing, mutually exclusive new machines that have only cost data given and tell which of the following statements is true regarding the present worth of the incremental investment at your investment interest rate.

(a) If it is greater than zero, we chose the alternative with the largest initial investment expense.

(b) The internal rate of return will always be equal to the investment rate of return.

(c) Neither machine is chosen if there is only cost data and the present worth is less than zero.

(d) If it is less then zero, we chose the alternative with the smallest initial investment expense.

Chapter 9

Future Worth Analysis

D-88 The future worth of a project with initial cost P, positive annual cash flows of A, salvage value S, and interest rate of i over a life of n years can be calculated using which statement?

(a) $FW = -P(F/P, i\%, n) + A(F/A, i\%, n)$
$+ S(F/P, i\%, n)$

(b) $FW = P(F/P, i\%, n) + A(F/A, i\%, n) + S$

(c) $FW = -P(P/F, i\%, n) + A(F/A, i\%, n)$
$- A[(P/A, i\%, n) + S$

(d) $FW = -P(F/P, i\%, n) + A(F/A, i\%, n) + S$

D-89 A firm has been investing retained earnings to establish a building fund. The firm has retained $1.2 million, $1.0 million, and $950,000, respectively, 3, 2, and 1 year ago. This year the firm has $1.8 million to invest. If the firm earns 18% on invested funds, what is the value of the project that can be undertaken using the funds as a 25% down payment?

(a) $6.28 million

(b) $7.42 million

(c) $25.1 million

(d) $29.7 million

D-90 A firm is considering the purchase of a software analysis package that costs $450,000. Annual licensing fees are $25,000 (payable at the beginning of each year, starting in Year 1). The firm is bidding

on a large 4-year government project where the new software will be used. If the firm uses an interest rate of 20%, what value for software costs should be put in the bid?

(a) $514,725

(b) $527,650

(c) $1,067,540

(d) $1,094,346

D-91 Which of the following is a true statement regarding the future worth of a single investment alternative?

(a) It will be equal to both the present worth and the annual worth if the same discounting interest rate is used.

(b) Choose to invest if the calculated amount is less than zero at the investment rate of return.

(c) It will yield a recommendation consistent with the present worth and annual worth methods if the same discounting interest rate is used.

(d) It cannot be used to evaluate single investment alternatives.

D-92 Using the data for Uno and Dos from Problem 86, where the lives of both alternatives is 10 years, give the future worth on the incremental investment if the interest rate used is 10%.

(a) −$20,000

(b) −$20,783

(c) −$43,910

(d) −$53,910

Benefit–Cost Ratio Analysis

D-93 The annual benefits associated with construction of an outer belt highway are estimated at $10.5 million by a local planning commission. The initial construction costs will be $400 million, and the project's useful life is 50 years. Annual maintenance costs are $500,000 with periodic rebuilding costs of $10 million every 10 years. If interest is 2%, the benefit–cost ratio is closest to what value?

(a) 0.25

(b) 0.75

(c) 1.11

(d) 1.35

D-94 A city needs a new pedestrian bridge over a local stream. The city uses an interest rate of 5%, and the project life is 30 years. The following data (in millions of dollars) summarizes the bids that were received.

	Bid A	Bid B
Construction materials costs	$4.20	$6.20
Construction labor costs	0.60	0.70
Construction overhead costs	0.35	0.03
Initial administrative and legal costs	0.60	0.01
Annual operating costs	0.05	0.075
Annual revenue from operation	Unknown	0.40
Other annual benefits to the city	0.22	0.25

What would the annual revenue of Bid *A* have to be for the two projects to be equivalent? Choose the closest value.

(*a*) 0.10

(*b*) 0.20

(*c*) 0.30

(*d*) 0.40

D-95 When using the benefit–cost method of analyzing a project, which of the following is a true statement?

(*a*) It will always produce a recommendation consistent with the simple payback period method.

(*b*) It will always produce a recommendation consistent with present worth, future worth, and annual worth methods.

(*c*) It can be used only to evaluate projects from the public sector (such as bridges and roadways).

(*d*) None of the above.

D-96 Project *A* has a first cost of $950,000 and will produce a $50,000 net annual benefit over its 50-year life. Project *B* costs $1,250,000 and produces an $85,000 net annual benefit. If interest is 3% per year, the benefit–cost ratios of Projects *A* and *B* are nearest what values?

(*a*) 0.52, 0.67

(*b*) 0.74, 0.57

(*c*) 1.35, 1.75

(*d*) 2.63, 3.40

D-97 Using the data for Projects *A* and *B* in Problem 96, the benefit-cost ratio on the incremental investment is nearest what value?

(*a*) 0.17

(*b*) 0.33

(*c*) 3.00

(*d*) 5.83

Sensitivity and Breakeven Analysis

D-98 BVM manufactured and sold 25,000 small statues this past year. At that volume, the firm was exactly in a breakeven situation in terms of profitability. BVM's unit costs are expected to increase by 30% next year. What additional information is needed to determine how much the production volume/sales would have to increase next year to just break even in terms of profitability?

(*a*) Costs per unit

(*b*) Sales price per unit and costs per unit

(*c*) Total fixed costs, sales price per unit, and costs per unit

(*d*) No data is needed, the volume increase is $25,000 + 25,000(0.30) = 32,500$ units.

D-99 Process *A* has fixed costs of $10,000 and unit costs of $4.50 each, and Process *B* has fixed costs of $25,000 and unit costs of $1.50 each. At what level of annual production would the two processes have the same cost?

(*a*) 50 units

(*b*) 500 units

(*c*) 5000 units

(*d*) 50,000 units

D-100 A seasonal bus tour firm has 5 buses with a capacity of 60 people each. Each customer pays $25 for a one-day tour. Records show $360,000 in fixed costs per season, incremental costs of $5 per customer, and an average daily occupancy of 80%. The number of days of operation necessary each season to break even is closest to which value?

(*a*) 50 days

(*b*) 75 days

(*c*) 100 days

(*d*) 120 days

D-101 Alternative I has a first cost of $50,000, will produce an $18,000 net annual benefit over its 10-year life and be salvaged for $5,000. Alternative II costs $150,000 and has a salvage value of $50,000 after its 10-year useful life. If interest is 15%, what is the minimum amount of annual benefit that Alternative II must produce to make it the preferred choice?

(*a*) This value can not be determined from the data given.

(*b*) $23,500

(*c*) $31,450

(*d*) $35,708

D-102 Use the table to determine which project is best if it is known for sure that annual sales will be $7 million. All values are in millions (PW and Sales) of dollers.

Ann.	Estimated PW ($M)		
Sales	Proj. 1	Proj. 2	Proj. 3
$0	−75	−10	0
5M	125	15	150
10M	325	40	300

(a) Project 1

(b) Project 2

(c) Project 3

(d) None, since Projects 1 and 2 have negative values.

Chapter 10
Uncertainty and Probability

D-103 An interest rate of 15% is used to evaluate a new system that has a first cost of $212,400, annual operating and maintenance costs of $41,200, annual savings of $94,600, a life of 6 years, and a salvage value of $32,500. After initial evaluation, the firm receives word from the vendor that the first cost is 5% higher than originally quoted. The percentage error in the system's present worth from this is closest to what value?

(a) 5%

(b) 15%

(c) 100%

(d) 300%

D-104 A machine has a first cost of $10,000 and annual costs of $3500. There is no salvage value, and interest is 10%. If the project's useful life is described by the following data, what is the annual worth of costs?

	Useful Life (years)			
	4	5	6	7
Prob. of life (%)	5	22	41	32

(a) $3500

(b) $5127

(c) $5554

(d) $5796

D-105 Three estimators have estimated a project with a 10-year life.

	Estimate 1	Estimate 2	Estimate 3
First cost	$10,000	$17,500	$15,000
Net annual cash flows	7,500	8,000	6,000
Salvage value	3,500	0	10,000

Use an interest rate of 20% to determine the project's expected present worth. The value is closest to which of the following?

(a) $16,066

(b) $16,612

(c) $31,660

(d) $31,607

D-106 The first cost (FC), life (n), and annual benefits (A) for a prospective project are uncertain. Optimistic (OP), most likely (ML), and pessimistic (PS) estimates are given. If the interest rate is 25%, what is the present worth difference between a total worst-case scenario and a total best-case scenario?

	Estimate		
Parameter	Pessimistic	Most Likely	Optimistic
First cost	$150,000	$100,000	$80,000
Annual cash flows	25,000	45,000	50,000
Project life, in years	5	7	10

(a) $15.8

(b) $42.2

(c) $181.3

(d) $282.5

D-107 Which of the following statements, related to the use of decision tree analyses to model a problem and recommend a solution, is not true?

(a) In modeling the decision, the sequence flows from left to right, with later outcomes and decisions shown to the right of earlier decision and outcomes.

(b) Branches at a decision point are "pruned off" if they maximize the benefit to the decision maker relative to other choices.

(c) In analyzing the best path, sequence flows from right to left as inferior branches are pruned at decision points.

(d) Expected value at outcome points is calculated by multiplying the effect of each branch by the probability of that branch event.

Chapter 11
Depreciation of Capital Assets

D-108 The correct percentage to use to calculate the depreciation allowance for a MACRS 3-year property for Year 2 is which of the following?

(a) 14.4%

(b) 32.0%

(c) 33.3%

(d) 44.5%

D-109 With reference to the straight-line depreciation method, which statement is false?

(a) An equal amount of depreciation is allocated in each year.

(b) The book value of the asset decrements by a fixed amount each year.

(c) The depreciation life (n) is set based on the MACRS property classes.

(d) The asset is depreciated down to a book value equal to the salvage value.

D-110 A 7-year MACRS property has a cost basis for depreciation of $50,000. The estimated salvage (market) value is $10,000 after its 10-year useful life. The depreciation charge for the 4th year and book value of the asset after the 4th year of depreciation are closest to what values?

(a) $5760; $44,240

(b) $6245; $12,496

(c) $6245; $15,620

(d) $6245; $43,755

D-111 A $100,000 asset has a $20,000 salvage value after its 10-year useful life. The depreciation allowance using straight-line depreciation is closest to what value?

(a) $2000

(b) $8000

(c) $10,000

(d) $12,000

D-112 The book value of an asset that is listed as a 10-year MACRS property is $49,500 after the first year. If the asset's estimated salvage (market) value is $5000

after its 15-year useful life, what was the asset's original cost basis?

(a) $50,000

(b) $52,105

(c) $55,000

(d) $61,875

Chapter 12
Calculating Income Taxes

D-113 Which of the following is true?

(a) Tax rates are based on two flat-rate schedules, one for individuals and one for businesses.

(b) When businesses subtract expenses, they always include capital costs.

(c) For businesses, taxable income is total income less depreciation and ordinary expenses.

(d) When quantifying depreciation allowance, one must always divide first cost by MACRS 3-year life.

D-114 If a corporation has taxable income of $60,000, which of the following expressions is used to calculate the federal tax owed?

(a) Flat 15% of taxable income

(b) Flat 25% of taxable income

(c) $7,500 + 25% over $50,000

(d) $13,750 + 34% over $75,000

D-115 This past year CLL Industries had income from operations of $8.2 million and expenses of $1.8 million. Allowances for depreciated capital expenses were $400,000. What is CLL's taxable income and federal taxes owed for operations last year? Choose the closest values.

(a) $6.0 million; $1.93 million

(b) $6.0 million; $2.04 million

(c) $6.4 million; $1.93 million

(d) $6.4 million; $2.04 million

D-116 Annual data for a firm for this tax year are:

Revenues	$45 million
Operating and maintenance costs	7 million
Labor/Salary costs	15 million
Overhead and administrative costs	3 million
Depreciation allowance	8 million

Next year the firm can increase revenue by 20% and costs will increase by 2%. If the depreciation allowance stays the same, what will be the change in firm's after-tax net profit? Choose the closest answer.

(a) $5.2 million

(b) $5.4 million

(c) $7.9 million

(d) $13.3 million

D-117 Widget Industries erected a facility costing $1.56 million on land bought for $1 million. The firm used straight-line depreciation over a 39-year period; it installed $2.5 million worth of plant and office equipment (all classified as MACRS 7-year property). Gross income from the first year of operations (excluding capital expenditures) was $8.2 million, and $5.8 million was spent on labor and materials. How much did Widget pay in federal income taxes for the first year of operation?

(a) $680,935

(b) $1,002,750

(c) $1,321,815

(d) $2,788,000

Chapter 13
Replacement Analysis

D-118 Which of the following is a valid reason to consider replacing an existing asset?

(a) It has become obsolete and does not perform its intended function.

(b) There is a newer asset available that is technologically superior.

(c) It has become very costly to keep the asset in working order.

(d) All of the above.

D-119 The replacement of a typical existing asset (defender) that is beyond its minimum economic life with a new asset (challenger) should be done when?

(a) It should never be done because the existing asset is working.

(b) The defender's marginal cost is greater than the challenger's equivalent annual cost.

(c) The challenger's average future cost becomes less than the existing asset.

(d) The defender's equivalent annual cost equals the challenger's equivalent annual cost.

D-120 A factory asset has a first cost of $100,000, annual costs of $15,000 the first year and increasing by 7.5% per year, and a salvage (market) value that decreases

by 20% per year over its 5-year life. The minimum cost economic life of the asset is what value? Interest is 10%.

(a) 2 years

(b) 3 years

(c) 4 years

(d) 5 years

D-121 The minimum cost life of a new replacement machine is 6 years with a minimum equivalent annual cost of $6000. Given the existing machine's marginal cost data for the next 4 years, when should the existing machine be replaced?

Year	Total Marginal Cost
1	$5400
2	5800
3	6200
4	8000

(a) After Year 1

(b) After Year 2

(c) After Year 3

(d) After 6 years

D-122 A material handling system was purchased 3 years ago for $120,000. Two years ago it required substantial upgrading at a cost of $15,000. It once again is requiring an upgrading cost of $25,000. Alternately, a new system can be purchased today at a cost of $200,000. The existing machine could be sold today for $50,000. In an economic analysis, what first cost should be assigned to the existing system?

(a) $50,000

(b) $65,000

(c) $75,000

(d) $80,000

Chapter 14
Inflation Effects

D-123 If the real growth of money interest rate for the past year has been 4% and the general inflation has been 2.5%, the combined (market) interest rate is closest to what value?

(a) 1.5%

(b) 6.5%

(c) 6.6%

(d) 10.0%

D-124 To convert actual (inflated) dollars to constant purchasing power dollars (where n = difference in time between today and purchasing base) that occur at the same point in time, one must:

(a) Multiply by (P/F, inflation rate %, n).

(b) Multiply by (F/P, inflation rate %, n).

(c) Multiply by (P/F, combined rate %, n).

(d) Divide by (P/F, real interest rate %, n).

D-125 The cost of a material was $2.00 per ounce 5 years ago. If prices have increased (inflated) at an average rate of 4% per year, what is the cost per ounce now?

(a) $2.04

(b) $2.08

(c) $2.43

(d) $8.00

D-126 A deposit of $1000 is made into an account that promises a minimum of 2% per year increase in purchasing power. If general price inflation is 3, 1, and 5%, respectively, over the next 3 years, the minimum value that will be in the account at Year 3 is closest to what amount?

(a) $1061

(b) $1092

(c) $1157

(d) $1177

D-127 The cost of materials was $1000 per lot when the cost index is 145. Today the cost index is 210. What is the cost per lot?

(a) $690

(b) $1000

(c) $1448

(d) Cannot be determined with the data given.

Chapter 15
Capital Budgeting

D-128 Which of these statements can be said of projects considered to be mutually exclusive?

(a) The projects are equivalent or mutual in terms of their cash flows.

(b) Neither alternative should be chosen.

(c) All projects can be chosen as long as they meet minimum economic criterion.

(d) The selection of one in the set eliminates selection of others in that same set.

Use the following data for Problems 129–131
A firm has identified four projects for possible funding in the next budget cycle.

Project	First Cost ($M)	PW($M)
A	12	225
B	18	250
C	24	320
D	30	400

D-129 The projects are independent. What is the total number of possible investment strategies (combinations) that the firm can use?

(a) 4

(b) 6

(c) 10

(d) 15

D-130 If the projects have the following contingencies, what is the total number of possible investment strategies (combinations) that the firm can use? Project A can be invested in only by itself; Project D is chosen only if Project B is chosen.

(a) 0

(b) 2

(c) 4

(d) 6

D-131 If the projects are independent, and the project budget (capital limit) is $32 million, what investment combination maximizes PW?

(a) A and C

(b) D alone

(c) Invest in all projects to maximize PW.

(d) A and B

D-132 A firm is considering the "make vs. buy" question for a subcomponent. If the part is made in-house, the production data would be: first cost = $350,000; annual costs for operation = $45,000; salvage value = $15,000; project life = 5 years; interest = 10%; and material cost per unit = $8.50. If annual production is 10,000 units, the maximum amount that the firm should be willing to pay to an outside vendor for the subcomponent is nearest?

(a) $10 per unit

(b) $16 per unit

(c) $22 per unit

(d) $28 per unit

D-133 A firm is considering whether to buy specialized equipment that would cost $200,000 and have annual costs of $15,000. After 5 years of operation, the equipment would have no salvage value. The same equipment can be leased for $50,000 per year (annual costs included in the lease), payable at the beginning of each year. If the firm uses an interest rate of 5% per year, the annual cost advantage of leasing over purchasing is nearest what value?

(a) $2494

(b) $8694

(c) $11,200

(d) $12,758

Chapter 17
Accounting

D-134 Which of the following summarizes a firm's revenues and expenses over a month, quarter, or year?

(a) Balance sheet

(b) Statement of assets and liabilities

(c) Income statement

(d) None of the above

D-135 Which of the following financial ratios provides insight into a firm's solvency over the short term by indicating its ability to cover current liabilities?

(a) Acid-test ratio

(b) Quick ratio

(c) Current ratio

(d) Net profit ratio

D-136 Which statement is most accurate related to an activity-based costing (ABC) system?

(a) ABC spreads the firm's indirect costs based on volume-based activities.

(b) ABC seeks to assign costs to the activities that drive those costs.

(c) ABC gives an inaccurate view of a firm's costs and should never be used.

(d) ABC is called ABC because it is an easy method to use.

D-137 A firm's balance sheet has the following data:

Cash on hand	$450,000
Market securities	25,000
Net accounts and notes receivable	125,000
Retailers' inventories	560,000
Prepaid expenses	48,500
Accounts and notes payable (short term)	700,000
Accumulated liabilities	120,000

The firm's current ratio and acid-test ratio are closest to what values?

(a) 1.42; 0.73

(b) 1.42; 0.79

(c) 1.47; 0.73

(d) 1.47; 0.79

D-138 A firm's income statement has the following data (in $10,000):

Total operating revenues	$1200
Total nonoperating revenues	500
Total operating expenses	925
Total nonoperating expenses	125

If the firm's incomes taxes were $60,000 what was the net profit (loss)?

(a) All necessary data is not given, one cannot calculate net profit (loss).

(b) −$53,500

(c) $1,150,000

(d) $6,440,000

D-139 Annual manufacturing cost data (1000s) for four product lines are as follows:

Data	Line 1	Line 2	Line 3	Line 4
Annual production	4000	3500	9800	675
Cost of direct materials	$800	$650	$1200	$2500
Cost of direct labor	$3500	$3750	$600	$320

Rank the product lines from lowest to highest in terms of manufacturing cost per unit. Total indirect costs of $10.8 million are allocated based on total direct cost.

(a) 1-2-3-4

(b) 3-1-2-4

(c) 3-2-1-4

(d) 3-4-1-2

SELECTED ANSWERS TO END-OF-CHAPTER PROBLEMS

Chapter 1

1-2 yes on a, b, & e

1-18 a & c Max output – input; b & d Minimize input

1-54 $85.80 individual; $76 team

1-56 8' in length & diameter

1-58 itemized $9314; Std. mileage $10,170; breakeven 14,463 miles

1-60 a) $212.50 at 60 mph; $203.57 at 70 mph
b) $183.33 at 60 mph; $185.71 at 70 mph
c) $179.17 at 60 mph; $189.29 at 70 mph

1-62 775 units/yr

1-68 a) $1^{st} = 7.5$; $2^{nd} = 7.0$; $3^{rd} = 6.5$
b) $1^{st} = 6.3$; $2^{nd} = 7.5$; $3^{rd} = 8.3$

Chapter 2

2-2 average $13, $12.67, $12.40, & $12.14; marginal $13, $12, $12, $11 for a/b/c/d

2-4 1 shift $483/unit; 2 shifts $497.72

2-6 33,333

2-8 avg./marginal a) $60/0, b) $50/$50, c) $50/$78.95, d) $55.79/$78.95

2-10 a) $66.67, b) $80, c) $40

2-12 a) 277,778, b) $20,000 in yr 1

2-18 A for 0 to 4000, C for 4000 to 20,000, B for 20,000 to 30,000

2-20 4167 units

2-34 a) $7000, b) $4000, c) stainless $1500 less

2-38 a) $160,000, b) Current process = $200,000/yr
New process = $40,000/yr

2-44 a) $766,250, b) $1,034,438

2-46 a) $125/ft^2, b) i. $500,000

2-48 $0.97 million

2-50 $1007

2-52 a) $11,012

2-54 $227,950

2-56 $10,458

2-58 65% rate

2-60 $T_1 = 676$ person hours; $T_{20} = 195$

2-62 $N = 17.87$ so by 18th unit

2-64 costs reduced by 55%

2-74 $17M

Chapter 3

3-2 a) 6200

3-4 $110,000

3-6 a) 33.3 yrs, b) 23.45 yrs.

3-12 $7052.1

3-14 $1470

3-16 $112,095

3-18 a) $5105, c) $10,613, e) $526,005

3-20 a) 9.54%, c) 3.71%

3-22 $4909

3-24 a) $7010, b) $7011, c) $7010

3-26 12.4%

3-28 $3 more interest

3-30 17.5 years

3-32 8.623Q

3-34 $424,925

3-36 $9438

3-38 12.7%

3-40 9.31%

3-42 9%/yr

3-44	15.0%
3-46	nominal 4.08%, eff. 4.12%
3-48	134.4%
3-52	nominal 21%, eff. 23.37%
3-56	$36,000

Chapter 4

4-2	a) $116,550, b) $1,723,170
4-4	$60,833
4-6	$696.60
4-8	$3,803,000
4-10	35
4-12	50 months
4-14	1.5%/month
4-16	B = $634, C = $51.05, V = $228.13
4-18	$42,563
4-20	$7778.35
4-22	a) $1827, b) $409.43
4-24	$14,763
4-26	$4411
4-28	$488.78
4-30	$792.73
4-32	loan $118,596, house $148,245, save $33,149
4-48	10%
4-50	$792.28
4-52	a) $2189, b) $543, c) $1887
4-54	$8609
4-56	$292,870
4-58	18th bday $23,625; AW = $2658
4-62	$589.50
4-64	$494
4-66	$1496.91
4-74	$721,824
4-76	a) $186,154, b) $201,405
4-78	a) $168,682, b) $181,818
4-80	1st choice at $291,750; not $277,070
4-88	a) $294, b) 18%, c) 19.56%
4-90	16.67%
4-92	a) 1.25%, pmt = $343.10, i_a = 16.1%
4-100	payment $1542.55, last yr interest $1415.19
4-102	payment $77.46, last interest $0.39
4-110	$169,054
4-112	$217,750
4-118	$39,123

Chapter 5

5-2	$219,012
5-4	automate for a PW of $55,800
5-6	max. $18,102

5-8	$4985
5-10	$8156
5-12	$3247
5-14	19.4%
5-16	$18,356
5-24	a) Buy B, PW = $23,359
	b) Select A, PW = $10,193
5-26	$10,653 for $Cost_B$ is < $13,000 for A
5-28	$98,040 for B is higher
5-30	Quicksilver $80,867 < Foxhill $83,112 < Almaden $87,229
5-32	2-stage $20,098,000 < 1-stage $23,962,000
5-36	A for $683.10 >B for $503.50
5-38	B saves $5.06M
5-40	3-yr cost $116 < PW cost $121.33 of 3 1-yr
5-42	SuperBlower $143,243 < Sno-Mover $155,728
5-48	$1,200,000
5-50	$1265.60
5-52	$87,938
5-54	Cap. $Cost_A$ $957,920 < Cap. $Cost_B$ $1,008,830
5-58	$5,474,000
5-60	$389,150
5-64	B $14,404 >C $11,409 >A $4987
5-66	A $7.74, B $7.76, C $6.86, D $3.70, E $32.43,
	F $12.43; pick E
5-68	C »A & B, E »D; E $686.4 >C $638
5-70	12 yrs, C $93,497 >B $55,846 >A $53,255
5-74	Spartan $.14/$ invested
5-76	$19,438
5-80	$165,178
5-84	$216,286
5-86	$27,368
5-88	$272.43
5-90	$3602.75
5-92	$1985
5-94	$802
5-96	$4010

Chapter 6

6-2	$3572
6-4	$9287
6-6	$6.92
6-10	$606,300
6-12	$5231
6-14	$89.86
6-16	$4321
6-18	$n = 36$
6-20	$195.03
6-22	$756.49
6-24	$27,113

6-26 $1745

6-28 $42,817 quarterly ($42,799 by spreadsheet)

6-30   EUAC = $1828 >0

6-32   Buy for $4954 < lease for $5500

6-34   new payment of $29,832 is less

6-36 $EUAW_L$ = $6264; $EUAW_S$ = −$6392

6-38   existing EUAC is $3019 < new of $4170

6-40   Hydro-clean's 150,000 < $183,870

6-42 $EUAW_A$ −$752 better than $EUAW_B$ −$1336

6-44   Hybrid $6345 < Midsize $7190

6-46 $EUAC_Y$ $1106 < $EUAC_x$ $1252

6-48   Rt 105 = $4619, Rt 205 = $1029, Rt 305 = $149

6-50   C at $538 >B at $234 >A at $200

6-52   36-month at $28.13

6-54   A at $504 >B at $421

6-60   pay $513 more than receives for car

6-62   pmt = $579.98; $19,065 is owed after 2 yrs.

6-64   a) car price = $7527, b) save for 5.52 months, c) 0.25% per month or 3% APR

6-66   a) $550.36, b) beginning month so all principal, c) $545.45

6-68   a) payment $764.39, b) 168.8 months or 14.07 yrs, c) 84 months

Chapter 7

7-2   8.003%

7-4   a) 38%

7-6   nominal 15%, effective 16.08%

7-8   12%

7-10   a) 9.4%, b) 4.8%

7-12   5.4%

7-14   50%

7-16   11.65% → don't purchase

7-18   13.23%

7-20   a) 8.18%

7-22   7.82%

7-24   a) 7.70%, b) 8.00%

7-28   nominal 6.60%, effective 6.71%

7-30   7.5%

7-32   a) 7.65%, b) 7.82%

7-34   8 years

7-36   a) 6.0%, b) 6.17%

7-38   14.4%

7-40   effective 15.6%

7-42   a) 15.97%, b) 9.65%, c) 6.94%

7-44   15.72%

7-46   12.9%

7-48   13.7%

7-52   16.0%

7-54   If 71 years used, then 6.83%

7-56   32.5%

7-62   do A since incremental B – A is 6.1%

7-64   do A since incremental B – A is 4.3%

7-66   do A since incremental A – B is 23.15%

7-68   incremental RFID – bar code has negative incremental rate so do bar codes

7-70   27.9%

7-72   a) buy generator since 10.71% >8%, b) 11.42%

7-74   B – A 27.3% → B, C – B 15.1% → B

7-76   8.3% → B

7-78   11.65% Do A

7-80 $126,348

7-82   11.29% for $1,000,133

Appendix 7A

7A-2   unique root at 21.2%

7A-4   roots 0% & 218.8%, MIRR of 11.3%

7A-6   unique root at 11.5%

7A-8   unique root at 10.0%

7A-10   unique root at 13.5%

7A-12   −28.2%

7A-14   a) roots 10% & 30%, MIRR of 8.8%, b) Not ethical

7A-16   ignore negative root of −71.8%; IRR = 39.0%; MIRR = 16.46%

7A-18   ignore negative root of −63.7%, use root of 11.6%

7A-20   unique root of 8.4%

Chapter 8

8-2   X for 0 −7.7%, Y for 7.7 – 12.1%, then do nothing

8-4   b) neutralization with 26% incremental ROR

8-6   b) A with 10.8% incremental ROR

8-8   b) Slate with 27.0% incremental ROR

8-10   b) used since 12.6% incremental ROR for new

8-12   b) B with 9.2% incremental ROR

8-14   b) A with 9.6% incremental ROR

8-16   b) A with 7.6% incremental ROR

8-18   b) company B

8-20   b) Sort-Of with 18% incremental ROR

8-22   a) C for 0 – 5%, B for 5 – 7.5%, then do nothing
b) C for 0 – 5%, B for 5.0 – 10.0%, A for 10 – 100%

8-24   5-yr for 0 – 11.79%, 3-yr for 11.79 – 12.79%, 2-yr for 12.79 – 16%, then 1-yr for all higher rates

8-26   b) C $539 >B $234 >A $200

8-28   b) B with 7.2% incremental ROR

8-30   D for 0 – 50%, A for higher rates

8-34   Dallas

8-36   assuming a 20-yr life expectancy & 9% pick C

8-38 C since incremental over A is 15.8% and incremental of D over C is 7.8%

8-40 Option B

Chapter 9

9-2 $3431

9-4 $144,373

9-6 $357,526

9-8 $78,381

9-10 £1.3 × 10^{35}

9-12 $1830

9-14 $407,768

9-16 $6108

9-18 $2094

9-20 $12,458

9-22 $122,758

9-24 a) $454,722, b) $437,547

9-26 $112,309

9-28 $83.76/share

9-30 B

9-32 10-story

9-34 A incremental B/C over B is 1.053

9-36 A incremental over D is 1.07

9-38 $2,021,010

9-40 off-campus, incremental B/C is 0.195

9-44 2.6 year payback period >2 year

9-46 9.18 years

9-48 a) Payback$_A$ = 5 yrs, Payback$_B$ = 7 yrs, b) $216.3M, c) pick B with EUAW = $7.08M vs. A with $4.81M

9-50 a) B with $4.31 FW, b) B – C with 1.02 & D – B with 0.94, c) C with 3.3 is shortest payback

9-52 a) 0.79, b) Z has shortest payback of 2.4 yrs, c) X has 0% return, Z dominates Y

9-54 a) C has shortest payback of 2 yrs, b) B – C incremental ratio is 1.34 & A – B has $175 in costs and $100 in benefits

9-60 6.9 years

9-62 13.9 years

9-64 15 days

9-66 70%

9-68 $10,005

9-70 $7.922M

9-72 10.2 years

9-74 $175 min Fiasco resale

9-76 68.9 cars

9-78 13.5 years

9-80 $102,241

9-82 4 weeks

9-84 a) 6 years, b) 7.8 years, c) both correct, but breakeven considers more

Appendix 9A

9A-2 a) $36,411, b) $29,672, c) employer match $2720, employee $33,691

9A-4 a) 7.7%, b) 4.7%

9A-6 a) 35.5%, b) 31.5%

9A-8 41.7 yrs.

9A-10 a) $75,805, b) $37,903, c) $1,505,521

9A-12 $746,648

9A-14 7.66%

Chapter 10

10-2 EUAC$_{10,000}$ $4727, up to $6671, down to $2784

10-4 13.0%

10-6 a) $4.28M, $3.39M, $2.25M, b) $3.35M

10-8 P(250) = .1, P(300) = .4, P(350) = .4

10-10

Sales	Prob.	Unit Profit	Prob.
5000	0.2	$24.00	0.3
7000	0.7	$32.00	0.5
10,000	0.1	$38.00	0.2

10-12 Bill is 0.75s above mean >Al is 0.25s

10-14 Annual Profit, Prob.; $120,000, 6%; $160,000, 10%; $190,000, 4%; $168,000, 21%; $224,000, 35%; $266,000, 14%; $240,000, 3%; $320,000, 5%; $380,000, 2%

10-16 a) P($18,000, 12 yr) = 0.033 & 8 other combinations, b) 25.4%, 7.47%, −4.55%

10-18 315 days

10-20 $1100

10-22 A 1.90 >1.85 B

10-24 −$0.20

10-26 a) $3.11M, b) 10 yrs, c) $3.02M, d) don't match since PW nonlinear in time

10-28 $1322

10-30 3m for $48,120 E(EUAC)

10-32 $355,000

10-34 $212,520 both ways

10-36 a) $45,900, b) first cost $440,000, net $86,000, PW $45,900, c) same

10-38 a) 12.18%, b) 8.28%, c) interest formulas nonlinear so don't match

10-40 B $6600 >$6400 A

10-42 repair $26,000 < as-is $28,000 < replace $30,000

10-44 repair $22,920 < repair & spillway $36,050 < add upstream dam $72,500

10-46 a) E(loss) $420, b) logical to avoid catastrophic loss or to satisfy mortgage lender

10-48 E(PW) $8212, σ$_{PW}$ $1158

10-50 σ$_{PW}$ $1550

10-52 $381,993

10-54 a) $5951, b) 0.3, $65,686

10-56 PW(loss) 0.45, range −$204.5K to $265K, std. dev. $127.9K

10-58 F, 1, 2, & 6 form efficient frontier

10-64 answers may vary, mean $3126, std. dev. $6232

Appendix 10A

10A-2 a) $12,134, b) $8137, c) employer match $2720, employee $9414

10A-4 a) 4.0%, b) 2.0%

10A-6 60.5 yrs.

10A-8 a) $50,477, b) $455,402

10A-10 64 years; enjoy retirement

Chapter 11

11-2 $300,000, $480,000, $288,000, $172,800, $172,800, $86,400

11-4 1 $91,485, 2-39 $146,148, 40 $54,891

11-6 D_2 $33,337.5, BV_2 $16,665

11-8 a) D_2 $11,200, b) D_2 $7333

11-10 D_2 a) $15,200, b) $20,267, c) $18,240

11-12 D_2 a) $5000, b) $8000, c) $8182, d) $12,245

11-14 D_2 $7500, BV_2 $96,000

11-16 D_2 a) $12,245, b) $5000, c) $7778, d) $9375

11-18 A SOYD, B Units of prod., C MACRS, D DDB, E 150% DB

11-20 D_2 a) $105,000, b) $100,000

11-22 D_2 $5600, BV_2 $53,800

11-24 D_2 SL $15,000, SOYD $21,429, MACRS $24,490

11-26 PW SL $720, SOYD $766

11-28 PW SOYD $87,885, UOP $88,231

11-32 PW a) $138,991, b) $149,123, c) $174,920, d) $149,453

11-34 D_2 $15,384

11-36 $19,102

11-40 D_2 $20,512, D_4 $9416

11-42 D_2 a) $2500, b) $3000, c) $4445

11-44 a) D_9 SL $1667, SOYD $333, MACRS $0, b) PW SL $11,846, SOYD $12,615, MACRS $12,634

11-46 D_3 a) $8550, b) $4250, c) $8100

11-48 D_4, BV_8 a) $18,375, $0, b) $13,500, $42,000, c) $17,182, $22,364, d) $15,360, $25,166

11-50 1) 5-yr, $3264, $0; 2) 7-yr, $5247, $4017; 3) 15-yr, $1111.5, $16,827

11-54 Sale at $25,000: GDS $5800 recapture, SL $10,000 loss

11-56 a) $8000 recapture, b) $2000 loss, c) $3762.5 recapture w/50% final depr.

11-58 a) $11,521 recapture w/50% final depr., b) $37,667 loss, c) $19,000 loss, d) $25,438 loss

11-60 a) $52,479 loss w/50% final depr., b) $63,333 loss, c) $41,429 loss, d) $44,297 loss

11-62 $1462.5

11-64 % is $1800, cost is $3000

11-66 a) yr 2 $135,000, b) rate $1448/MBF

11-68 a) $200,000, b) $118,750

11-70 $102,500

Chapter 12

12-4 marginal rates 15%, 25%, 39%, 34%, 35%; average rates 15%, 17.9%, 33.0%, 34.0%, 35%

12-6 $3.75M

12-8 $106,604, 44.86%

12-10 10.9%

12-12 a) 2.27 yrs, b) 3.02 yrs, 23.95%

12-14 a) 15.4%, b) 9.8%, c) 7.9%

12-16 −$212,700

12-20 10.94%

12-22 11.6%

12-24 EUAC = $2675, EUAB = $2500

12-26 $7594

12-28 12.88%

12-30 a) 34.9%, b) 25.2%

12-32 9.9%

12-34 a) 2.24 yrs, b) 29%

12-36 3.24 yrs

12-38 $3275

12-40 $46,197

12-44 a) yr 1 & 5 $5925, yrs 2 − 4 $6181, b) $15K capital gain & $30,393 recaptured, c) $30,871

12-46 8.40%

12-48 −$11,028

12-50 $23,100

12-52 $283,372

12-54 $50,007

12-56 218 days

12-68 Alt. 1 a) $10,338, b) $1215, c) incremental for 2 is 9.2%, d) $69,552, e) incremental for 2 is 0.91

12-70 PW A is $234,000 higher

12-72 incremental A − C 36.2%, incremental B − A 16.7%, pick A

12-74 Alt. III has lowest PW cost at $6801

Appendix 12A

12A-2 $28,712.50; $28,937.50

12A-4 $1800

12A-6 $18,143.75

12A-8 a) $315, b) 299

12A-10 2.42%

12A-14 $3000

12A-18 $2500

12A-20 $417 tax savings

12A-24 $6000

12A-28 $52,931

12A-30 a) 70.9%, b) 17.1%, c) 12.0%, d) $1080, e) $82,353

12A-34 a) yes, 26.6% < 43%, b) $333,162

Chapter 13

13-8 All except initial cost

13-10 4 yrs, $17,240

13-12 1 yr, $5450

13-14 yr 3, $6615

13-16 2 yrs, $26,264

13-18 $14,539, 3 yrs

13-20 6 yrs, $2717

13-22 10 yr $3519

13-24 $5767, yr 9

13-26 12 yrs, $4437

13-30 1) $6650, 2) $6450, 3) $6350, 4) $6450

13-32 a) 1 yr, $3000, b) 4 yrs, $3100, c) 2 yrs more defender

13-34 challenger 5 yr $6989. Keep defender 3 more yrs

13-36 upgrade defender $5659 < challenger $7840

13-38 $6500, yr 3

13-40 keep indefinitely

13-42 a) 4 yrs, $2650, b) 3 yrs, $2406, c) 3 yrs

13-43 $684,939

13-44 high efficiency cheaper by $22

13-46 own cab $3496 < $6750 lease cab

13-48 defender $1.93M < challenger $3.544M

13-50 defender $5111 < $6109 challenger

13-52 old $240 < new $738.5

13-54 a) −$27,975, b) −$25,500, c) −$21,375 defender, −$76,500 challenger

13-56 C for −$3626

13-58 5 more yrs, from beginning 8 yrs $5642

Chapter 14

14-4 constant $ & real rate or inflated $ & market rate

14-6 $155.80

14-8 5.14%

14-10 $29,670

14-12 14.11 km/liter

14-14 11.7%

14-16 3.19%

14-18 $1.34, 2.97%

14-20 PW benefits at 11.3% equiv. $i = $14,540 < cost

14-22 a) 0.4266%, b) 5.24%, c) $50,000

14-24 3.17, or 3 to 4 pads

14-26 $1411

14-28 1.59%

14-30 $1184

14-32 $2029, $1940

14-34 yr 4 $66,915, 0.96%

14-36 $6.20M, $5.52M

14-38 $2,272,010, $717,500

14-40 35.1 months

14-44 a) $59,436, b) $70,200

14-46 $388,832

14-48 $85,449, $70,156, $59,717, $52,235

14-50 B at $571,154 < A at $599,258

14-52 $12,109 at 10.21%

14-54 A $199,650 >C $193,958 >B $167,323

14-56 C at $5688 PW, A at $6000

14-60 a) 150%, b) 66.8%, c) 31.9%, d) 27.3%, e) 6.8%

14-62 a) 7.9%, b) 4.6%, c) 3.1%, d) 2.2%

14-64 a) 2.41%, b) 267.0

14-68 Z $31,562

14-70 a) $18,116, b) $16,105

14-72 a) $89,250; $93,266; $97,463, b) $19,632, c) $391,843, d) $553,367

14-74 1.01%

14-76 a) 3.5%, b) 2.52%, c) 0.51%

14-78 a) $74,292, b) $41,134, c) $132,022

14-80 a) $38.61, b) $97.08

14-82 6.84%

14-84 $2698 EUAC

Chapter 15

15-4 a) 10.52%, b) 9.36%

15-6 a) 9.5%, b) 8.0%

15-8 10.2%, 9.1%

15-10 7.79%, 5.67%

15-18 243.29

15-20 26.0%

15-22 13.4%

15-24 12.4% > 8%; yes

15-26 19.46% < 20%; No

15-28 a) all but 4, b) do 5, 9, 10, 2, 1, and 7, c) First 6, d) 14.0%

15-30 a) 1B, 2B, 3A, & 4, b) $95K 3A, 2A, & 1B, c) last accepted is 13.7% & first foregone 11.2%, d) $95K 3A, 2A, & 1B same as b)

Chapter 16

16-12 real, use with constant value $s

16-14 B, D, E, & F ranked as top 4 by B/C & IRR for $9M, G at 12% is opportunity cost

16-16 a) 18.0%, b) 11.6%

16-18 0.56

16-20 7.5 optimal PW

16-22 a) 418.8M Yuan, b) $62.4M

16-24 $2,963,740 gravity

16-26 1.07 for low road, 0.61 for high − low

16-28 A vs. existing 1.71, B vs. A 2.17, C vs. B 2.42, so pick C
16-30 B vs. nothing 1.10, A vs. B 3.28, D vs. A −0.54, pick A
16-32 now −$4,571, 2 yrs $40,172, 4 yrs $52,751, 5 yrs $50,192
16-44 a) $879,628, b) $578,621

Chapter 17
17-12 a) $735,000, b) $430,000
17-14 a) $3,300,000, b) 2.5
17-16 a) $98,000, b) 1.22, c) 0.73
17-18 a) $130,000, b) 1.295, c) 0.614
17-20 a) 1.80, b) 0.92
17-24 $10,175,000
17-26 $900
17-28 a) 3.77, b) 0.08

17-32 a) $18M, b) $10M, c) $69M
17-34 a) $5000/hour, b) $15,500,000
17-36 S $132, M $93, G $84

Appendix B
B-2 $498,144
B-4 $539.52
B-6 $4.39M
B-8 $1104.65
B-10 $19,021
B-12 $1054.00
B-14 150.1 months = 12.5 years
B-16 15.07%
B-18 −14.98%, so lease

REFERENCES

American Telephone and Telegraph Co. *Engineering Economy*, 3rd. McGraw-Hill, 1977.

Baasel, W. D. *Preliminary Chemical Engineering Plant Design*, 2nd. Van Nostrand Reinhold, 1990.

Bernhard, R. H. "A Comprehensive Comparison and Critique of Discounting Indices Proposed for Capital Investment Evaluation," *The Engineering Economist*. Vol. 16, No. 3, 1971, pp. 157–186.

Block, S. "Are There Differences in Capital Budgeting Procedures between Industries? An Empirical Study," *The Engineering Economist*. Vol. 50, No. 1, 2005, pp. 55–67.

Dunn, E., and M. Norton, *Happy Money: The Science of Happier Spending*. Simon and Schuster, 2013.

Elizandro, D. W., and J. O. Matson. "Taking a Moment to Teach Engineering Economy," *The Engineering Economist*. Vol. 52, No. 2, 2007, pp. 97–116.

The Engineering Economist. A quarterly journal of the Engineering Economy Divisions of ASEE and IIE.

Eschenbach, T. G. "Multiple Roots and the Subscription/Membership Problem," *The Engineering Economist*. Vol. 29, No. 3, Spring 1984, pp. 216–223.

Eschenbach, T. *Engineering Economy: Applying Theory to Practice*, 3rd. Oxford University Press, 2011.

Eschenbach, T. G., E. R. Baker, and J. D. Whittaker. "Characterizing the Real Roots for P, A, and F with Applications to Environmental Remediation and Home Buying Problems," *The Engineering Economist*. Volume 52, No. 1, 2007, pp. 41–65.

Eschenbach, T. G., and J. P. Lavelle. "Technical Note: MACRS Depreciation with a Spreadsheet Function: A Teaching and Practice Note," *The Engineering Economist*. Vol. 46, No. 2, 2001, pp. 153–161.

Eschenbach, T. G., and J. P. Lavelle. "How Risk and Uncertainty Are/Could/Should Be Presented in Engineering Economy," *Proceedings of the 11th Industrial Engineering Research Conference*. IIE, Orlando, May 2002, CD.

Eschenbach, T. G., N. A. Lewis, J. C. Hartman, and L. E. Bussey (with chapter author H. L. Nachtmann). *The Economic Analysis of Industrial Projects* 3rd, Oxford University Press, 2015.

Eschenbach, T. G., N. A. Lewis, and J. C. Hartman. "Technical Note: Waiting Cost Models for Real Options." *The Engineering Economist*, Vol. 54, No. 1, 2009, pp. 1–21.

Fish, J. C. L. *Engineering Economics*. McGraw-Hill, 1915.

Green, D. W., and H. P. Perry. *Perry's Chemical Engineers' Handbook*, 8th, McGraw-Hill, 2007.

Lavelle, J. P., H. R. Liggett, and H. R. Parsaei, editors. *Economic Evaluation of Advanced Technologies: Techniques and Case Studies*. Taylor & Francis, 2002.

Lewis, N. A., T. G. Eschenbach, and J. C. Hartman, "Can We Capture the Value of Option Volatility?" *The Engineering Economist*, Vol. 53, No. 3, July – September 2008, pp. 230–258, winner of Grant Award for best article in volume 53.

Liggett, H. R., J. Trevino, and J. P. Lavelle. "Activity-Based Cost Management Systems in an Advanced Manufacturing Environment," *Economic and Financial Justification of Advanced Manufacturing Technologies*. H. R. Parsaei, W. G. Sullivan, and T. R. Hanley, editors. Elsevier Science, 1992.

Lorie, J. H., and L. J. Savage. "Three Problems in Rationing Capital," *The Journal of Business*. Vol. 28, No. 4, 1955, pp. 229–239.

Newnan, D. G. "Determining Rate of Return by Means of Payback Period and Useful Life," *The Engineering Economist*. Vol. 15, No. 1, 1969, pp. 29–39.

Sundaram, M., editor. *Engineering Economy Exam File*. Oxford University Press, 2014.

Tippet, D. D., and P. Hoekstra. "Activity-Based Costing: A Manufacturing Management Decision-Making Aid," *Engineering Management Journal*. Vol. 5, No. 2, June 1993, American Society for Engineering Management, pp. 37–42.

Wellington, A. M. *The Economic Theory of Railway Location*. JohnWiley & Sons, 1887.

Economic Criteria

Method of Analysis	Fixed Input	Fixed Output	Neither Input Nor Output Fixed
PRESENT WORTH	Maximize PW of Benefits	Minimize PW of Costs	Maximize (PW of Benefits − PW of Costs), or Maximize Net Present Worth
ANNUAL CASH FLOW	Maximize Equivalent Uniform Annual Benefits (EUAB)	Minimize Equivalent Uniform Annual Cost (EUAC)	Maximize (EUAB − EUAC)
FUTURE WORTH	Maximize FW of Benefits	Minimize FW of Costs	Maximize (FW of Benefits − FW of Costs), or Maximize Net Future Worth
BENEFIT–COST RATIO	Maximize Benefit–Cost Ratio	Maximize Benefit–Cost Ratio	*Two Alternatives:* Compute the incremental Benefit–Cost ratio ($\Delta B/\Delta C$) on the increment of *investment* between the alternatives. If $\Delta B/\Delta C \geq 1$, choose higher-cost alternative; if not, choose lower-cost alternative. *Three or more Alternatives:* Incremental analysis is required (see Ch. 9).
RATE OF RETURN	*Two Alternatives:* Compute the incremental rate of return (ΔROR) on the increment of *investment* between the alternatives. If $\Delta ROR \geq$ minimum attractive rate of return, choose the higher-cost alternative; if not, choose lower-cost alternative. *Three or more Alternatives:* Incremental analysis is required (see Ch. 8).		